Neuroimaging of Pain

Luca Saba
Editor

Neuroimaging of Pain

Springer

Editor
Luca Saba
University of Cagliari
Cagliari
Italy

ISBN 978-3-319-48044-2 ISBN 978-3-319-48046-6 (eBook)
DOI 10.1007/978-3-319-48046-6

Library of Congress Control Number: 2017935826

Printed on acid-free paper

This Springer imprint is published by Springer Nature
The registered company is Springer International Publishing AG
The registered company address is: Gewerbestrasse 11, 6330 Cham, Switzerland

Ed elli a me: «Ritorna a tua scïenza,
che vuol, quanto la cosa è più perfetta,
più senta il bene, e così la doglienza.

And he to me: "the more a thing is perfect,
the more keenly it will feel
both pleasure and pain"

Dante Alighieri, Italian poet (1265–1321)
Divine Comedy
Canto VI, lines 106–108.

The greatest enemy of knowledge is not
ignorance, it is the illusion of knowledge.

Stephen Hawking,
Theoretical Physicist and
Cosmologist (1942–)

This book is dedicated to my students, residents and fellows as a way of thanking them for what they teach to me every day.

Preface

Pain is a complex, multifactorial subjective and conscious experience; an interpretation of the nociceptive input influenced by memories, emotional, pathological, genetic, and cognitive factors. Resultant pain is therefore not always related linearly to the nociceptive drive or input, neither is it solely for vital protective function. By its very nature, pain is therefore difficult to assess, investigate, manage, and treat.

Until the advent of modern noninvasive human brain imaging methodologies about 20 years ago, our understanding of the role of the brain in pain processing was limited. In the last two decades, advances in brain imaging techniques have had a profound influence on our understanding of pain processing. In the early 1990s, human whole-brain functional imaging studies first showed multiple brain areas involved in pain processing, whereas other studies have revealed the involvement of forebrain neurotransmitters in pain modulation.

Recently, new advances in human brain imaging techniques allowed a better understand of the functional connectivity in pain pathways, as well as the functional and anatomical alterations that occur in chronic pain patients. Modern imaging techniques have permitted rapid progress in the understanding of networks in the brain related to pain processing and those related to different types of pain modulation.

The future is bright for what brain imaging can contribute to our understanding of pain. Especially in combination with cellular, genetic, and molecular approaches, imaging techniques might have a major impact in the diagnosis and differentiation of chronic pain problems and the evaluation of the effectiveness of therapeutic interventions.

The purpose of this book is to cover all the imaging techniques and new exciting methods like new tracers, biomarker, metabolomic and gene-array profiling, together with cellular, genetic, and molecular approaches for the analysis of the pain with the most world renowned scientists in these fields.

Cagliari, Italy
November 2016

Luca Saba

Acknowledgements

It is not possible to overstate my gratitude to the many individuals who helped to produce this book. I would like to thank Gabriele Finco for the interesting discussions about the pain circuits and Paolo Siotto for the continuous support and many suggestions. Special thanks to my wife, *Tiziana*; without your continuous encouragement this book would not have been completed, I have to say that this is also your book! Thank you for your great work.

I also wish to thank all the authors who contributed with their effort to this book. I am very proud that most of them are young researchers who devoted their time to this project with enthusiasm. Without them nothing of this would have been possible.

I have received considerable support and cooperation from individual at Springer, particularly Gregory Baer. Thank you Greg, you have really helped to minimize the obstacles I have encountered during this project!

Cagliari, Italy
November 2016

Luca Saba

Contents

The Economical Impact of Pain

1

Darrell J. Gaskin, Patrick Richard and Joseph Walburn

Abstract

Pain afflicts about 20% of adults globally. The source of pain ranges from a variety of health conditions, including arthritis, osteoarthritis, rheumatoid arthritis, migraine, fibromyalgia, back pain, cancer and chronic pain. Pain is very costly to society and individuals. It typically requires costly ongoing medical treatment and increases the costs of treating other health conditions. In addition to medical care costs, pain has indirect costs because it reduces labor market productivity through presenteeism and absenteeism. Workers with pain are sometimes unable to function at their full capabilities. Also, pain can cause workers to missed days or reduce their participation in the labor market from full-time- to part time or even drop out of the labor market altogether. Gaskin and Richard (J Pain 13(8):715–724, 2012, [1]) estimated the annual cost of chronic and persistent pain in the United States ranged from $560 to $635 billion in 2010. Gustavsson et al. (Eur J Pain, 2012, [2]) estimated that on average adults in Sweden incurred an annual costs of medical care due to chronic pain conditions of €2650 and loss annually €6429 in labor market productivity. This study reviews estimates of the costs of pain across a variety of countries and for several conditions that cause pain. The bottom line is pain regardless of the country or condition is costly.

Keywords

Costs of pain · Labor market costs · Direct medical care costs · Individual costs · Societal costs

D.J. Gaskin (✉)
Department of Health Policy and Management, Center for Health Disparities Solutions, Johns Hopkins Bloomberg School of Public Health, Baltimore, USA
e-mail: dgaskin1@jhu.edu

P. Richard
Department of Preventive Medicine and Biostatistics (PMB), School of Medicine (SOM), Uniformed Services University of the Health Sciences, Bethesda, USA

J. Walburn
Johns Hopkins Bloomberg School of Public Health, Baltimore, USA

L. Saba (ed.), *Neuroimaging of Pain*, DOI 10.1007/978-3-319-48046-6_1

1 Introduction

The number of individuals with a history of pain including chronic pain globally is high and is growing. Most recent estimates show that about 20% of adults worldwide suffer from pain associated with various conditions such as arthritis, osteoarthritis, rheumatoid arthritis, migraine, and fibromyalgia [3]. In terms of global incidence, about 10% of adults are diagnosed with pain every year [3]. The rising cost of pain globally has already posed significant challenges to policymakers, public payers, private insurers, employers, patients and their families. Thus, estimating the economic burden of pain is important to allocate resources accordingly, expand insurance programs, and understand its impact on patients and families' financial circumstances, and employer's productivity. While pain imposes a substantial financial burden on individuals, families and society at large around the world, because of significant treatment costs and productivity loss associated with the different conditions that cause pain there is a paucity of research on the global economic burden of pain. Globally, there has been one study that has estimated the global burden of low back pain in specific settings using estimates from the global burden of disease 2010 study [1]. However, most studies have examined the economic burden of pain, either direct medical care costs or indirect costs, for specific countries and for specific conditions.

Specifically, this chapter presents a critical review of the literature on the cost of pain. This review contains a particular assessment of the conceptual and methodological strengths and limitations of the literature on the economics of pain to inform future research on this topic. However, this review does not include studies of cost-benefit analyses (CBA), cost–effectiveness analyses (CEA) or cost utility analyses (CUA) of specific interventions—pharmacologic or non-pharmacologic—to prevent and treat pain or conditions associated with pain. This critical review of the literature on the global costs of pain is organized into four sections. Following the introduction, the next section presents a conceptual approach in estimating the economic burden of pain. The third section presents the methodology used to complete the literature on this topic. The fourth section presents the results of the search and assesses the strengths of the evidence on the global costs of pain. The fourth section discusses the implications of current findings in informing policymaking, clinical decision-making and programs to prevent and control pain globally. Finally, this section identifies the gaps in the literature on this topic for future research and concludes.

2 A Cost of Illness Approach

In general, studies use a cost-of-illness (COI) framework to estimate the economic burden of medical care. The COI consists of estimating the annual "incremental" medical care costs for individuals with the condition compared to those without the condition, which results in an "incremental" medical care cost per person due to the condition. Conceptually, the COI approach reflects the estimation of the maximum amount of medical care costs that could potentially be saved if the condition (pain in this case) could be prevented or eliminated. The estimation of the economic burden of health conditions such as pain encompasses three major categories: (1) direct medical care costs, (2) direct non-medical costs and (3) indirect costs such as productivity loss.

Direct medical care costs result from ambulatory care received (physician care, hospital outpatient, office based from non-physician providers, laboratory and radiological tests, etc.), hospital services (inpatient including surgery and emergency services), and pharmaceutical drugs. Direct medical care costs may include out-of-pocket costs incurred by individuals as well as payments to healthcare providers made on their behalf by insurers, but they do not usually include health insurance premiums. In other words, total medical care costs include payments made for hospital-based services such as inpatient, emergency room, outpatient (hospital, clinic, and office-based visits), prescription drugs, and other services (e.g. home health services, vision care services, ambulance services, dental care, and medical equipment). Direct non-medical costs consist of measuring the opportunity cost of time to receive treatment associated with the condition such as transportation costs and time spent in waiting rooms. Indirect costs consist of (1) productivity loss associated with pain due to inability to perform work at full capacity even when present (presenteeism) or days missed from work because of the illness (absenteeism); and (2) mortality costs due to premature deaths associated with pain or pain-related conditions.

Studies of economic burden use either a prevalence-based or an incidence-based approach. A prevalence-based approach provides an estimate of the economic burden of the condition for a specific time period, often one year, regardless of the onset of the condition. For instance, the study by Gaskin and Richard [1] used a prevalence-based approach to estimate the direct and indirect costs of pain in the United States. Although these studies provide useful information about the annual medical care costs associated with some of the conditions that cause pain they are limited by their inability to identify the specific sources of increased medical care costs associated with the condition. In contrast, incidence-based economic burden represents the total medical care costs within a set time, most often one year, for a condition.

3 Methodology

We performed a comprehensive search of articles and abstracts listed in PubMed, MEDLINE, National Bureau of Economic Research (NBER) and other relevant international journals, among others. These databases were searched for English language articles published between 2005 and 2015 that collected data from study participants living in the United States or abroad. We excluded studies that were outside of the time period. We used the following search terms: "cost", "economic cost", "economic burden", "direct costs", and "indirect costs" with specific names of diseases associated with pain. Diseases searched for include arthritis, osteoarthritis, rheumatoid arthritis, migraine, and fibromyalgia. We also used terms such as "chronic pain", "oncogenic pain", and "breakthrough pain" (see table below for more details). Some of these key words were also cross-searched. The online search was supplemented by a manual search from the reference list of retrieved articles to identify additional papers. Articles retrieved from the search were classified into the appropriate sections depending on the conditions they referenced.

4 Results of Literature Review

The titles and abstracts of the articles were screened to determine their relevance to the search criteria. Studies were eligible for inclusion in this review if they were research articles published in peer-reviewed journals of English language about the costs of pain in the United States and abroad between 2005 and 2015. After elimination of duplicates, we identified 24 non-duplicate articles based on the inclusion and exclusion criteria noted above. One of the articles evaluated the costs of two separate conditions. Each article was critically analyzed to compare study characteristics, methods, and findings. We also sorted the findings by conditions and by year (see Table 1). We included all studies that met the inclusion criteria regardless of the sample size, study design, and types of data such as observational or RCTs. We evaluated the studies by considering the types of costs such as direct, direct non-medical, indirect costs, their study design, data collection and analytic methods, and adjustment for confounding factors, such as comorbidity, and other clinical characteristics. We also considered the different perspectives such as payer, employer, or societal.

In total, 24 non-duplicate studies were used to examine the literature on the economic burden of pain in the US and internationally. Results are presented by different types of conditions associated with pain such as migraine, fibromyalgia, osteoarthritis, rheumatoid arthritis, back pain and

Table 1 Studies on the Cost of Pain for Selected Conditions, published between 2005 and 2015

Author	Title	Year	Categories	Country
Hazard E, Munakata J, Bigal ME, Rupnow M, Lipton R	The burden of migraine in the United States: current and emerging perspectives on disease management and economic analysis. Value in health	2009	Migraine	United States
Munakata J, Hazard E, Serrano D, Klingman D, et al.	Economic burden of transformed migraine: Results from the American migraine prevalence and prevention (AMPP) study	2004–2006	Migraine	United States
Stokes M, Becker WJ, Lipton RB, et al.	Cost of health care among patients with chronic and episodic migraine in Canada and the USA: Results from the International burden of migraine study (IBMS)	2009	Migraine	United States, Canada
White LA, Birnbaum HG, Kaltenboeck A, et al.	Employees with fibromyalgia: Medical comorbidity, healthcare costs, and work loss	2005	Fibromyalgia	United States
Berger A, Dukes E, Martin S, Edelsberg J, Oster G	Characteristics and healthcare costs of patients with fibromyalgia syndrome	2002–2005	Fibromyalgia	United States
Kleinman N, Harnett J, Melkonian A, et al.	Burden of fibromyalgia and comparisons with osteoarthritis in the workforce	2001–2008	Fibromyalgia, Osteoarthritis	United States
Wier LM, Andrews RM	The national hospital bill: the most expensive conditions by payer, 2008, HCUP Statistical Brief #107. Agency for Healthcare Research and Quality	2008	Osteoarthritis	United States
Berger A, Hartrick C, Edelsberg J, et al.	Direct and indirect economic costs among private-sector employees with osteoarthritis	2007	Osteoarthritis	United States
Kotlarz H, Gunnarsson CL, Fang H, et al.	Osteoarthritis and absenteeism costs: evidence from US National Survey Data	1996–2005	Osteoarthritis	United States
Hermans J, Koopmanschap MA, Bierma-Zeinstra SM, et al.	Productivity costs and medical costs among working patients with knee osteoarthritis	2009–2010	Osteoarthritis	Netherlands
Kavanaugh A	Economic consequences of established rheumatoid arthritis and its treatment	2006	Rheumatoid arthritis	United States
De Azevedo AV, Ferraz MB, Ciconelli RM	Indirect costs of rheumatoid arthritis in Brazil	2005	Rheumatoid arthritis	Brazil
Kobelt G, Woronoff AS, Richard B, et al.	Disease status, costs, and quality of life of patients with rheumatoid arthritis in France: The ECO-PR study	2005	Rheumatoid arthritis	France
Neovius M, Simard JF, Askling J	How large are the productivity losses in contemporary patients with RA, and how soon in relation to diagnosis do they develop?	2007	Rheumatoid arthritis	Sweden
Lundkvist J, Kastäng F, Kobelt G	The burden of rheumatoid arthritis and access to treatment: health burden and costs	2006	Rheumatoid arthritis	Europe
Mehra M, Hill K, Nicholl D, et al.	The burden of chronic low back pain with and without a neuropathic component: a healthcare resource use and cost analysis	2008	Back-pain	United States

(continued)

Table 1 (continued)

Author	Title	Year	Categories	Country
Crow WT, Willis DR	Estimating cost of care for patients with acute low back pain: A retrospective review of patient records	2002–2005	Back-pain	United States
Martin BI, Deyo RA, Mirza SK, et al.	Expenditures and health status among adults with back and neck problems (1997–2005)	1997–2005	Back-pain	United States
Ricci JA, Stewart WF, Chee E, et al.	Back pain exacerbations and lost productive time costs in United States workers	2003–2004	Back-pain	United States
Schmidt CO, Schweikert B, Wenig CM, et al.	Modeling the prevalence and cost of back pain with neuropathic components in the general population	2004–2005	Back-pain	Germany
Abernethy AP, Wheeler JL, Fortner BV	A health economic model of breakthrough pain	2008	Breakthrough pain	United States
Kronborg C, Handberg G, Axelsen F	Health care costs, work productivity and activity impairment in non-malignant chronic pain patients	2005–2006	Chronic pain	Denmark
Lalonde L, Choinière M, Martin E, et al.	Costs of moderate to severe chronic pain in primary care patients—a study of the ACCORD program.	2009–2010	Chronic pain	Canada
Gustavsson A1, Bjorkman J, Ljungcrantz C, et al.	Socio-economic burden of patients with a diagnosis related to chronic pain—Register data of 840,000 Swedish patients	2004–2009	Chronic pain	Sweden

breakthrough pain. We reviewed studies that estimated the direct medical care costs, direct non-medical and indirect costs of pain and conditions associated with pain. We also present estimates by different types of expenditures such as hospital, out-of-pocket costs, and prescriptions drugs expenditures. This review shows that there is a fairly large body of work documenting the cost of pain for specific conditions such as migraine, arthritis, osteoarthritis, rheumatoid arthritis, migraine, and fibromyalgia. No studies within the period of interest were found that evaluated the costs associated with oncogenic pain, and only one study was found on breakthrough pain.

4.1 Migraine

Migraine is often mischaracterized as having a low burden, mainly because of its episodic nature and lack of information of effects on life expectancy and long-term disability [5]. However, despite of this common perception, migraine is disabling disorder that places a significant burden on patients, their families and the healthcare system as a whole because of its high prevalence [5]. Migraine affects 18% of women, and 6% of men, and is most prevalent in people aged 25–55. As most studies in the epidemiologic burden of migraine have been conducted in the US, it has been found that about one in four households in the US has at least one member that experiences migraine headaches. This is significant, as an estimated 35 million of US citizens have suffered from this condition and symptoms commonly occur about four times each month [5]. The high prevalence of this condition and the fact people typically experience this condition in the peak of their productive years are associated a high economic burden [4, 5].

We found three studies that have estimated the economic burden of migraine [6–8]. Two of them focused exclusively on the economic burden of migraine in the United Sates and the other one analyzed data from the United States and Canada [6–8]. The first two studies used data from the American Migraine Prevalence and

Prevention (AMPP) Study to estimate the economic burden of migraine including both direct medical care and indirect costs. The AMPP survey is a 5-year longitudinal, national study of headache in the US that consists of three phases [7]. The first, an initial screening to identify individuals in the population with self-reported severe headache took place in 2004. The second phase, a survey identifying headache symptoms and their impact on individuals suffering from this condition took place a year later, in 2005. The third phase consisted of a follow-up phase, which included 3 additional annual questionnaires surveying changes in symptoms, impairment, and resource utilization. The data collected in the 2006 follow-up phase of the AMPP survey was used to estimate the economic burden of migraine in the US by both studies. Summarizing the results from both studies, the authors found that the average annual total costs, including both direct and indirect costs of migraine were about $7750 for patients that experienced chronic migraine compared to about $1758 for those that suffered from episodic migraine. Costs due to lost productivity accounted for the majority of total costs, amounting to about $5392 per person per year among people with chronic migraine, and about $978 among those with episodic migraine [6, 7].

Similarly, Stokes et al. [8] used an international, cross-sectional, web-based study conducted between February and April 2009 to determine the economic costs of migraine pain in Canada and the US populations. Study participants were selected based on questions that accessed ICHD-II diagnostic criteria for migraine. Costs were compared across participants classified into either chronic or episodic migraine categories. Unit costs of medical care were calculated based on publicly available data. Medication costs were estimated using average wholesale prices and daily dosage assumptions based on expert medical opinions. Medicare fees were used to estimate Medicare and private health plan costs for physician visits, diagnostic testing, and other services [8]. The authors found that, in the US, total mean headache costs for chronic migraine patients were $1036, and costs for episodic migraine patients were $383. Canadian prices were lower, with mean costs for chronic migraine patients totaling $471, and costs for episodic migraine totaling $172 [8].

4.2 Fibromyalgia

There are three studies that have estimated the economic burden of fibromyalgia and they were all conducted in the United States [9–11]. Although it only affects about 2% of the population, fibromyalgia is another condition that is characterized by widespread pain [9]. This condition, typically experienced by women, is associated with significant burden in terms of costs and quality of life because it impacts patients during their working years. Additionally, there are high medical care costs associated with this condition because treatment of fibromyalgia often requires treating the many comorbid conditions that are associated with fibromyalgia.

A study by White et al. [9] examined an administrative claims database of 31 self-insured companies to determine the economic burden of this illness. The study focused on 16 of the companies that included disability insurance information, representing about 850,000 employees. Direct medical care costs were calculated using total costs for treatment as reported by insurers. Prescription payments were added to medical costs in order to determine total direct costs of health care in 2005. Indirect costs were calculated based on disability claims, medical claims, and wages. Employees with fibromyalgia had average annual medical and drug costs of $7286 compared to $3915 in controls. The largest portion of this number was due to medical costs, which totaled $5656 per person with fibromyalgia. Employees with fibromyalgia missed 29.8 days on average in the year 2005, which accounts for about 15% of total working days for the year. This number was about three times the average work loss of controls. Average annual indirect costs were $2913 for those with fibromyalgia compared to $1359 for controls.

A second study data on 33,176 patients with fibromyalgia with from the PharMetrics Patient

Centric Database, which includes data on facility, professional service, and retail pharmacy claims for 85 US health plans to examine the economic burden of fibromyalgia [10]. Data analysis from this study showed that patients with fibromyalgia were more likely to experience comorbidities as a result of their condition, and were more likely to utilize medications, both for pain and for other problems. Patients with fibromyalgia also had four times more doctor visits, and twice as many outpatient visits than controls. On average, patients with fibromyalgia had total healthcare costs three times higher than comparison patients, spending a mean of $9573 in 12 months. Median costs were also higher among patients with fibromyalgia totaled $4247 compared to $822 in the controls.

Likewise, Kleinman et al. [11] used data from the Human Capital Management Services Research Reference (HCMSRR) Database, which contains information for nearly 700,000 employees from self-insured employers, to estimate the economic burden of fibromyalgia. The database contains data from 2001 to 2008. Costs were determined based on claims data and were adjusted for inflation to 2008 dollars. The authors used several regression models to estimate direct medical care costs including prescription drug costs and indirect costs based on sick leave, number of days absent because of the condition, long-term disability and worker's compensation [11]. Overall, the authors found that the total average costs among employees with fibromyalgia were $8452 compared to $4013 among those without fibromyalgia. Breaking costs down by different categories, they also found that adjusted mean of direct medical care costs were $5419 among employees with fibromyalgia compared to $2261 among employees without fibromyalgia Prescription drug costs were $1452 among patients with fibromyalgia compared to $560 among employees without fibromyalgia. Cost of sick leave, short-term disability, long-term disability, and worker's compensation totaled to $1739 in employees with fibromyalgia compared to $838 among employees without fibromyalgia.

4.3 Osteoarthritis

Osteoarthritis is the most common form of arthritis, characterized primarily by pain [12]. For instance, Osteoarthritis, highly prevalent among older adults, affects about 27 million adults in the US and 8.5 million in the UK. The public health impact of osteoarthritis is significant because the disease can last for decades and there is a lack of effective treatment to address the pain associated with this condition. According to a Statistical Brief published by researchers from the Healthcare Costs and Utilization (HCUP) project, total hospital expenditures for osteoarthritis in the US were about $40 billion in 2008, one of the most expensive conditions [13]. This statistical brief found that osteoarthritis makes up about 3.5% of the nation's hospital bills, and accounts for 70% of all arthritis related inpatient hospitalizations. Although hospital charges are typically higher than costs to the hospital for stays, and are generally high estimates due to negotiated discounts for hospital services.

In addition to this statistical brief, we found 5 studies that have examined the economic burden of osteoarthritis including four in the US and one in the Netherlands during the time period under examination. Kleinman et al. [11] in their estimation of the economic burden of fibromyalgia also computed the costs of osteoarthritis using data from the HCMSRR database. The authors found that the adjusted mean of total costs for osteoarthritis were $11,253. Adjusted means of healthcare and prescription drug costs of osteoarthritis were $8201 and $1018 respectively. Combined costs of sick leave, short-term disability, long-term disability and workers' compensation were $2089.

White et al. [9] also examined the costs of osteoarthritis in comparison to a group of patients with fibromyalgia. Average costs for employees with osteoarthritis totaled to $10,861, which were nearly double the costs for the control sample, which totaled $5274. Average total direct costs, which were the combined medical and drug costs, were $8325. Average medical costs were $6984, while prescription drug costs were $1341. Average indirect costs for osteoarthritis totaled to $1359, with the majority of the indirect costs due to disability claims instead of medically related absence.

Another study conducted in the United States used data from the HPM Database to estimate the economic burden of osteoarthritis for adult employees 18 year or older who suffered from osteoarthritis during the year of 2007 [14]. Employees who had two or more outpatient encounters on different days or one or more inpatient stays with mention of osteoarthritis were considered as having osteoarthritis. Those in the database without osteoarthritis were matched on age and gender to those with the disease. Healthcare costs were determined based on osteoarthritis—related pharmacotherapy, all other pharmacotherapy, outpatient visits, hospitalizations, and all other healthcare. Indirect costs were determined based on payments for worker's compensation claims, and short-term disability, in addition to the estimated cost of absenteeism. The authors multiplied the number of missed work hours by the mean hourly wage of US full-time civilian employees, $21 per hour in 2007, to estimate the costs of absenteeism. The authors found that mean total healthcare costs were $17751 among patients with osteoarthritis compared to $5057 among those in the comparison group. Health care received where osteoarthritis was the specific diagnosis under consideration accounted for $6024 or 34%, of total healthcare costs among patients with osteoarthritis. Mean total indirect costs were $5002, nearly two times higher for osteoarthritis patients compared $2120 for the control group.

A study by Kotlarz et al. [15] used pooled data from 1996 to 2005 from the US Medical Expenditure Panel Survey (MEPS) to estimate the costs of absenteeism associated with of osteoarthritis. The authors used logistic regressions to predict the probability of missing any work and subsequently predicted the number of days of work missed because of osteoarthritis using multivariable regression models conditional on those who missed at least 1 day from work because of illness. Aggregate costs were calculated using prevalence rates of arthritis from the *Morbidity and Mortality Weekly Report*. Aggregate costs as determined by this study were $5.5 billion per year for women and $4.8 billion per year for men, totaling to $10.3 billion overall.

A cross-sectional study conducted in the Netherlands examined data on subjects enrolled in a previous randomized control trial that investigated the cost-effectiveness of intraarticular hyaluronic acid on treating knee OA [16]. The authors used baseline before the trial intervention occurred combined with survey data from the Productivity and Disease Questionnaire to measure productivity costs of knee osteoarthritis. 40% of the patients in the study reported productivity loss due to knee symptoms compared to regular workdays. There was an average productivity loss of 14% while being present at work, and 20% reported absences from work in the last three months due to knee symptoms. productivity and medical costs amounted to €871 per patient per month. The authors found that the average total medical and productivity loss burden amounted to €871 per patient per month. The average medical costs were €149 and the average knee-related productivity costs were €722 per person per month, making up the majority of total costs. The majority of productivity lost costs were due to lost productivity while attending work, which totaled €448 per person per month. Costs of household work loss were €77, amounting to 6.2 lost hours of work.

4.4 Rheumatoid Arthritis

The economic burden of rheumatoid arthritis, a chronic progressive autoimmune disease characterized by joint destruction and functional disability, is widely studied in the US and in Europe [17–21]. However, this review does not include any studies on the economic burden of rheumatoid arthritis in the United States because most US-based studies were conducted prior to 2005. Advances in treatment of this disease have only made it more expensive to manage. In particular, the development of a new biologic drug has contributed to a rise in prices, as it costs up to $16,000 per patient to use this treatment in 2006, compared to older treatments, which cost about $500 for a year of treatment. This review comprises five studies on the economic burden of rheumatoid arthritis including one study in Brazil and four studies in Europe.

Although rheumatoid arthritis only affects about 1% of the population in Brazil but this condition has an important burden on patients and society in terms of indirect costs [18]. For instance, a 2008 study that examined the indirect costs of associated with rheumatoid arthritis at an outpatient clinic for of the Division of Rheumatology at the Federal University of São Paulo found a total of USD $466,107 for the population, or about $2428 per patient per year [18]. Neovius et al. [20] used data from the 2007 National Patient Register in Sweden to estimate the indirect costs of patients with rheumatoid arthritis. Using a human capital approach, the authors estimated the cost of sick leave, disability pension and unemployment associated rheumatoid arthritis. Mean sex-specific wage for men and women was used to value each day. The estimated costs of sick leave and disability pension in this population were 4.2 billion SEK (€391 million; 4.0 million days), compared to the 1.9 billion SEK in costs for the matched general population cohort.

Similarly, a study conducted in France on the economic burden of rheumatoid arthritis used an anonymous survey to collect additional data from patients registered with one of the two national patient associations [19]. Patients were asked general demographic information and information about their consumption of healthcare services, out-of-pocket expenses, informal care received, and work loss. The mean cost per patient was €11,658 while the cost to society totaled €21,690. Mean indirect costs totaled €3200. More broadly, a 2007 review contained imputed costs for European countries by using economic data on the cost of rheumatoid arthritis, estimated from published studies in various countries with costs inflated to 2006 prices [21]. International statistics were also used to determine comparable data that could be extrapolated to countries without data, in addition to epidemiological data on the prevalence of disease in each country. Calculations were based on adjusted mean annual costs per patient for different resource types, extrapolated to the populations of patients in each country. Results from this analysis showed the estimated average annual total costs due to RA in Europe were €13,463. In Eastern Europe estimated total costs were €4889 compared to €17,153 in Western Europe. Indirect costs in Europe were €4300. Indirect costs in Eastern Europe were €108, and €5872 in Western Europe.

4.5 Back Pain

The economic burden of lower back pain is the most studied among all the other conditions included in this review, partly because of its prevalence. There are five studies that have estimated the economic burden of lower back pain including four of them in the United and one of them in Germany [22–26]. Diagnosis of chronic lower back pain is usually based on length of time that low back pain has persisted, the location of the pain, and the existence of other symptoms [22]. For many patients with lower back pain, the pain does not persist for longer than a few days to a few weeks, but for others the pain persists for months. The prevalence of the number of people with chronic lower back pain who have a neuropathic component ranges from 17 to 54%. This percentage is likely to be underestimated given the uncertainty surrounding the diagnosis of this condition. In patients with chronic lower back pain, pain is typically much more severe for those with a neuropathic component compared to those without neuropathic component, leading to higher economic burden [22].

For instance, a 2012 study of chronic lower back pain with a neuropathic component examined the condition's costs in the general population [22]. This study examined medical and pharmaceutical claims data from the PharMetrics IMS LifeLink Health Plan Claims Database during the year of 2008. This database is the largest non-payer-owned integrated claims database of commercial insurers in the US, and contains medical and pharmacy claims for over 55 million members across 90 different health plans nationwide. Generalized Linear Models were used to compare resource utilizations across patient groups. Over the 12-month follow-up period during this study, total medical costs of chronic lower back pain were approximately $96 million. Chronic lower back pain patients with a neuropathic component accounted for 96% of the total costs. The authors found that the average annual cost per patient was $2426 for all patients with lower back pain. However, patients with a neuropathic component had higher mean annual costs, which were $2577 compared to $1007 in those without a neuropathic component.

A second study was conducted on patients with lower back pain existing for less than 6 months by examining medical records from 2002 to 2005 for patients at the Family Practice Residency Clinic at Florida Hospital East Orlando [23]. The study used electronic medical records on 1810 patient to compare patients who received standard medical care to those who received a combination of standard medical care and osteopathic manipulative treatment (OMT). Average healthcare utilization rates were calculated within 90-day episodes of care, and average costs were calculated per patient. Costs of care were found using the 2006 Medicare fee schedule for Orlando, Florida. Drug costs were based on December 2006 prices provided by Walgreen Co, a national drug store, at the average wholesale price, using generic prices where available. Physical therapy costs were based on the cost of four weeks of therapy in the Florida Hospital Physical Therapy Clinic. Those in the OMT group generated overall lower costs per episode of care, with costs of about $227 compared to $265 in the non-OMT group. Prescription costs in the OMT group were about $54 compared to $73 in controls.

Martin et al. [24] examined data from respondents to the MEPS Household Component Survey from 1997 to 2005 in order to determine healthcare expenditures for people treated for back and neck pain. Total expenditures were calculated by multiplying mean expenditures by population size. All expenditures were adjusted for 2005 inflation. In 2005, the mean age and sex adjusted medical expenditures among respondents with spine problems was $6096 while it was only $3516 for those without spine problems. The mean incremental increase in expenditures attributed to spine problems was $2580 per person. When multiplying this number by the estimated number of people in 2005 with spine problems, a total of $85.9 billion in additional healthcare expenditures are attributed to those with spine problems. The majority of the incremental expenditures between patients with spine problems compared to those without any spine problems was due to outpatient services (36%) and inpatient services (28%), as well as prescription drugs (23%).

Back pain cannot be accounted for by just direct costs, as the indirect costs associated with work loss are also significant. Ricci et al. [25] studied the indirect costs associated with back pain survey data, by asking respondents about their absenteeism and presenteeism in the previous two weeks. The average number of days with back pain during this period was 8.6, with a mean severity of 3.5 on a 10-point scale. 16.8% of US workers aged 40–65 years reported lost productive time (LPT) due to back pain, with presenteeism accounting for the majority (79.6%) of the LPT. Employees with exacerbations were more likely to report LPT due to back pain. The estimated total LPT amounted to $7.40 billion to employers per year, explained mostly by presenteeism (85.4%). Workers with back pain were also more likely to report health absenteeism and presenteeism due to other health problems, leading to costs of $23.51 billion, which were $16.88 billion more than costs due to LPT in controls with neither back pain nor arthritis. A fifth study compared participants across three grades of pain, based on intensity and functional ability was conducted in Germany [26]. German participants were asked to recall their healthcare utilization due to back pain. This was then used to calculate cost differences between pain grades. The higher the grade the severe was the pain. Average total costs due to back pain amounted to

€2456, with each pain grade having differing costs. Patients with pain grade 3 accounted for over half of the total costs, while 36% of the patients in pain grade 1 accounted for 12% of total costs. People with pain classified as neuropathic had costs 37% higher than overall mean costs of back pain. Pain grades 1, 2, and 3 had total costs of €702, €2456, and €4092, respectively.

4.6 Breakthrough Pain

Breakthrough pain is one form of cancer related pain. Unlike general ongoing background pain related with cancer, breakthrough pain is characterized by sudden intense flare-ups of pain that lasts for about 30 min [27]. For diagnostic purposes, breakthrough pain is defined as occurring for four or fewer episodes within a 24-hour period. Among cancer patients, estimated prevalence of this condition varies from 24 to 95%. The costs of breakthrough pain are relatively unexplored. In one study of cancer outpatients attending scheduled oncology visits, breakthrough pain was found to be a predictor of higher costs in cancer patients. Patients with breakthrough pain reported overall direct medical costs averaging $825 per month and $1080 in pain related costs, compared to $750 for pain related costs in patients with non-breakthrough pain.

The largest study designed to capture the costs of breakthrough pain measured costs using a computerized telephone survey of 1000 community-based cancer patients [27]. Patients with breakthrough pain in this study had more hospitalizations and longer hospital stays for each event. Total costs per year reported for pain related hospitalizations for patients with breakthrough pain were $1.7 million compared to $192,000 for patients with breakthrough pain. Breakthrough pain patients also had higher average yearly emergency room visit costs of $84,000, greater physician visit costs of $103,000, and higher total pain-related costs per year of $12,000.

4.7 Chronic Pain

For overall chronic pain, there is a lack of data available for the US. Most of the data available reflects individual pain-associated conditions rather than chronic pain as a whole. In the US, Gaskin and Richard [1] use the 2008 Medical Expenditure Panel Survey to estimate (1) the total US healthcare costs attributable to pain and (2) the annual costs of pain associated with lower worker productivity [2]. They defined persons with chronic pain as: (1) persons who reported that they experienced pain that limited their ability to work, (2) persons who were diagnosed with joint pain or arthritis, or (3) persons who had a disability that limited their ability to work. The SF-12 pain question in the MEPS asked respondents whether, during the past 4 weeks, pain interfered with normal work outside the home and housework. The joint pain question inquired whether respondents had experienced pain, swelling, or stiffness around a joint in the last 12 months. This includes pain caused by bursitis, gout, strains, and other injuries. The question for arthritis determined whether the person had ever been diagnosed with arthritis, and if so was it osteoarthritis or rheumatoid. The question about functional disability inquired whether respondents had any work or housework limitation. The authors found that the total costs ranged from $560 to $635 billion in 2010 dollars. The direct medical care costs due to pain ranged from $261 to $300 billion. The indirect costs (i.e. value of lost productivity) due to pain ranged from $299 to $335 billion.

A cross-sectional study of the costs of chronic pain was conducted from December 2005 to January 2006 sampled patients on the waiting list for treatment at the Multidisciplinary Pain Clinic at Odense University Hospital in Denmark [28]. Telephone interviews were conducted by personnel at the multidisciplinary pain clinic at Odense University Hospital. Data on health service use was obtained form The National Health Insurance Service Registry, which holds data on healthcare service use covered by Danish national health insurance. All records, including service type, cost, and delivery date, were extracted from the register. The Danish prescription register was used to obtain prescription drug use information. Patients were also asked to describe the amounts they spend on alternative treatments, such as reflexology, acupuncture, or hypnosis.

Unit costs of healthcare services were based on prevailing national insurance rates. 200 patients were examined in total, and regression analysis was used to determine costs based on different characteristics. Costs were about DKK 17,500 higher per person in the year after reported pain onset than costs 2–9 years before the onset of pain. After suffering from pain for over a year, the annual healthcare costs were about DKK 8000 higher per person than in years prior to the onset of pain. Prescription drug costs were DKK 2466 in patients during the years after pain onset. 79% of participants reported that they received alternative treatment outside of general health services. Annual expenditures by users of alternative treatments ranged widely, between DKK 300 and DKK 3000. As such, average expenditure on alternative treatment was DKK 2978 per participant per year.

The costs of chronic noncancer pain were reported for patients in Canada [29]. The study, conducted in the Réseau universitaire intégré de santé de l'Université de Montréal, encompassed six areas in the province of Quebec. This represents over 40% of the population in this province. Among these patients, a random sample, stratified by region and weighted by the number of pharmacies within each region, was conducted between May 2009 and January 2010. Eligible patients were those 18 years or older, who suffered from noncancer pain for at least 6 months and 2 days per week or greater, who rated their pain as greater than or equal to 4 on a 0–10 scale, who had active analgesic prescriptions from primary care providers, and who spoke and read French or English. Patients reporting their primary source of pain as migraine were not eligible to participate in the study.

All direct healthcare costs were estimated based on healthcare resources used in the year preceding recruitment. Outpatient physician visits, tests, and interventions were documented from the RAMQ database, in which a service code is assigned to each of the components, in addition to the specialty of the service provider, dates, locations, and amount reimbursed by the RAMQ. A pain specialist was used to identify tests and interventions provided to patients that related to chronic noncancer pain. Nonpharmalogical healthcare services were documented using telephone interviews intended to capture how frequently participants used one of 23 different therapies in the past 6 months. Telephone interviews were also used to assess use of over-the-counter medicines. Productivity costs were assessed for those who were currently employed at time of interview, and for those who were on temporary or permanent disability. Absenteeism and presenteeism both were assessed using telephone interviews. Mean direct healthcare costs averaged CAD $7374 per patient, CAD $10,524, and CAD $9546 for patients with mild, moderate, and severe pain disability, respectively, once adjusted for age, sex, pain duration, and Charlson comorbidity scores. Productivity costs were CAD $3005, CAD $5083, and CAD $5385 for each pain severity group. Total costs were CAD $12,913, CAD $17,970, and CAD $17,292 for each group.

Finally, a Swedish study extracted retrospective data on 837,896 patients from three Swedish Administrative registries: the Vega register from Western Sweden capturing 1.56 million inhabitants, the national prescriptions register from the National Board of Health and Welfare, and the national social insurance register held by the Swedish Insurance Agency [2]. Sweden's 10-digit personal code numbers allowed the data from the three registers to be linked, and patients with a diagnosis in the Vega register were used in the study. Monthly costs of care were calculated by multiplying the number of resources used by their unit costs, as collected form regional price lists. Annual mean direct costs due to chronic pain conditions were €2650, while total indirect costs were €6429. The condition with the highest direct costs was Cancer, with direct costs totaling €5988. The condition with the highest indirect costs associated with chronic pain was intervertebral disc disorder, with costs totaling €15,724.

5 Discussions

As noted above, we found three studies that examined the economic burden conducted in the US and in Canada [5–7]. While the first two studies conducted in the US provide valuable information about the economic burden of migraine, they have some limitations that are worth mentioning. The authors use survey data to estimate the medical care costs of migraine. Self-reported response biases may limit the internal validity of these studies. It is also important to note that the military or institutionalized populations were not included in these surveys. Migraine is highly prevalent in the military [30]. Also, the study did not account for comorbid conditions that may have influenced resource use. While the study by Stokes et al. [8] attempted to address some of these limitations by using an international, cross-sectional and web-based data study but the authors computed costs by using publicly available US and Canadian sources that describe the costs of specific aspects of healthcare. As such, the results are affected by variation in unit cost estimates [8]. The second limitation relates to the method used to collect data. Resource use data was collected using an online survey, which required participants to have access to an email account. This would limit generalization of the study results, as the study represents those who have access to Internet. There is also the potential for bias in the control group due to confounders that were not accounted for. Further, the form used to survey participants did not account for all possible types of physician and health professional visits, which could have led to inaccurate cost estimates. The study questionnaire did not collect geographic data on participants' healthcare utilization, and since costs likely vary by region, it is possible that if participants were concentrated in a particular area and the study wouldn't be representative of the whole population. The questionnaire also had more choices available for medication options for US respondents, which could have inflated US costs in relation to Canadian.

In terms of fibromyalgia, a limitation of all three studies in this area is that classification of fibromyalgia may be inconsistent, as the ICD-9 code used to measure fibromyalgia was broad enough that it might capture other conditions, potentially reducing validity of the study findings. Validation of the coding on medical claims was not possible for these studies, which limited the investigators' ability to determine disease severity. White et al. [9] did not determine if higher costs associated with fibromyalgia are due to polypharmacy and the use of drugs from multiple classes, both of which are common with the disease in the study that examined the characteristics and healthcare costs of patients with fibromyalgia syndrome, it was difficult to quantify over-the-counter medications or medications received for other pain conditions since pharmacy records were used. Pharmacy records are also limited in that they do not determine which pain medications were actually prescribed for fibromyalgia related pain, as there are a number of pain medications prescribed for non-pain related illnesses such as depression and seizure disorders. While the study by Kleinman et al. [11] improve on the study design methodology limitations noted above it did not account for out-of-pocket costs. Further, the composition of the study population differed from the US population.

A common limitation of studies on the indirect costs of osteoarthritis and rheumatoid arthritis is that they were limited to measures in the study databases [12–21]—namely absenteeism, short-term disability, and worker's compensation. These studies did not account for presenteeism and reduction in wages as a result of the condition, significant sources of lost productivity. For example, the methodology used to assess indirect costs measured only losses resulting form absenteeism, underestimating overall costs by not assessing presenteesim. This is especially true because of evidence suggesting that poor performance while at work is the main factor in productivity loss in relation to arthritis. Underestimation may have also occurred because the study was unable to account for lost productivity in unpaid labor, and rheumatoid arthritis as a barrier to finding a job for unemployed individuals. Another limitation is in data collection and accuracy. Productivity losses information was based on self-reported absenteeism data for up to a month to a year before interview. Some of these studies also only include individuals in the labor force, who may differ greatly from those who are not. The healthcare claims database also does not include information on hourly wages, and the average wage used may not accurately depict the actual wages earned by employees. Finally, information on direct costs was limited to what was included in healthcare claims, which also does not account for self-care, which mainly includes the use of over-the-counter drugs. The study by Hermans et al. [16] was further limited by the sample size of 117 patients. Also, the study examined only a conservatively treated group of working people with mild to moderate knee pain, limiting the study's generalizability.

Although the study by Neovius et al. [20] used a larger data sets but Sweden's generous welfare system, which would benefit rheumatoid arthritis patients, limits this study's generalizability to other countries whose welfare systems may not be as comprehensive. Also, this study underestimates overall cost of sick leave because it did not account for sick leave episodes less than 14 days. However, it is likely that this underestimation will be small, as other studies have shown that sick leave episodes under 10 days in rheumatoid arthritis patients account for .2% of total days. Also, as most other studies in this literature that have examined the indirect costs of conditions associated with pain they do not provide a complete picture of lost productive time and indirect costs, as it does not include premature death or presenteeism as a source of productivity loss. Also, other studies such as Kobelt et al. [19] have questionnaire to assess resource consumption. Use of a questionnaire to assess resource consumption may have introduced some bias into the study. Questionnaires were given through a patient association with more than 3000 members. These patients could have been different form the general population in that they may have been more likely to be more ill or older, or they could have better access to information. Also, those who chose to complete the questionnaires may have had more education or on different treatments than those who did not complete the questionnaire.

A common limitation of studies on back pain is their inability to capture non-prescription or over-the-counter pain treatments and providers' characteristics [22–26]. Also, the greater probability that patients will seek treatment may have led to this study underrepresenting patients with no neuropathic component. Some of these studies attempted to account chronic lower back pain patients with a neuropathic component by including sets of comorbidity measures. However, the use of commercial insurance databases would lead to exclusion of many subjects who are unemployed, have lost employment, or have reached retirement age. Others are also limited by the use of data from a specific region of Florida [22], or underestimated observed prevalence of spine problems or to non-institutionalized and civilian populations [23]. The study by Ricci et al. [25] by the 2-week study period, which provides a snapshot of individuals with recent back pain. This study design, although useful in capturing recent back pain episodes, does not provide a picture of the total population with back pain. There is also the possibility that this study design has an overrepresented sample of individuals with more frequent or longer episodes of back pain. The study design is further limiting since the sample group was restricted to workers 40–65 years of age, and excluded workers who did not meet the investigators' criteria for clinically important back pain, eliminating a number of workers who initially screened positive for back pain symptoms. The investigators expressed three final concerns with their study design. First, investigators describe their sample size as "modest," expressing concern that there are uncertainties in statistical confidence around the prevalence and cost estimates for back pain as a result. Also, their LPT estimates were not intended to capture other effects associated with workplace costs described as "ripple effect" costs, such as impact on coworkers' productivity. The final concern was that they did not gather information on causes of back pain, eliminating the possibility that they could determine specific contributions of different conditions to back pain. Finally, a limitation of the study by Schmidt et al. [26] is that although it indicates that neuropathic pain accounts for the majority of costs associated with back pain, neuropathic pain was not directly assessed in the general population for this analysis. This necessitates further population-based studies to directly assess neuropathic pain and its costs in the general population. Also, this study had an overrepresentation of individuals undergoing specific pain therapies, although the investigators attempted to account for this in their modeling. Furthermore, using a postal survey introduces the potential for recall bias, although other studies have shown that it is appropriate to assess healthcare utilization within a 3-month period in order to minimize recall failures.

We found three studies that have estimated the economic burden of chronic pain [28, 29]. A common limitation of these studies is that over the counter drugs and some prescription drugs are included in the estimates and they tend to rely on self-reported survey data, which is subject to recall bias [27]. There are some limitations with the study by Lalonde et al. [29] that must be taken into consideration. First, only patients who had active analgesic prescriptions from primary care physicians were recruited for the study, so costs cannot be extrapolated to those who have no active analgesic prescriptions, or prescriptions from specialists. Further, assumptions were made to estimate annual costs, especially productivity costs, annual over-the-counter medication use, and use of complementary healthcare services. This could have led to overestimation or underestimation of resource utilization and costs. Frequency of use was also not precisely documented. The final limitation is that although the study took into account direct costs associated with different self-management techniques, it did not consider the impact on caregiver time. Furthermore, the study by Gustavsson et al., likely includes a sample of patients that do not actually have pain symptoms, although they have diagnoses commonly associated with chronic pain [2]. This limitation exists because the ICD-10 system does not differentiate between patients that have no chronic pain, and those who do. Also, this study could not differentiate between healthcare utilization due to pain and utilization due to other reasons, and had no way to capture costs due to over the counter medication purchased for pain management. Understand the economic burden of pain, including healthcare expenditures, productivity loss, morbidity, and premature deaths, and relationships between costs and outcomes is an important issue for policymakers and researchers alike. The treatment of conditions associated with pain results in the loss of economic resources and opportunities for patients, families, employers, and the society overall. Further research should focus on estimating the global burden of pain.

In summary despite their shortcomings, the cost of illness studies reviewed in this chapter demonstrate the magnitude of the economic costs of pain to individuals, family, payers and society. These costs are not only healthcare related but include substantial costs due to lost productivity. Non-monetary costs associated with impact of pain on the leisure time, household chores, and overall quality of life are not quantified. We recognized these costs could be substantial. Hence, the overall societal costs are considerable and worth societal effort to develop ways to reduce and better managed pain by providing resources for research, training, and education to improve pain prevention and care.

Disclaimer The views expressed are those of the authors and do not necessarily reflect the official views of the Uniformed Services University of the Health Sciences or the U.S. Department of Defense.

References

1. Gaskin DJ, Richard P. The economic costs of pain in the United States. J Pain. 2012;13(8):715–24. doi:10.1016/j.jpain.2012.03.009.
2. Gustavsson A1, Bjorkman J, Ljungcrantz C, et al. (2012) Socio-economic burden of patients with a diagnosis related to chronic pain—register data of 840,000 Swedish patients. Eur J Pain. doi:10.1016/j.ejpain.2011.07.006.
3. International Association for the Study of Pain: unrelieved pain is a major global healthcare problem. http://www.iasp-pain.org/PublicationsNews/PAIN.aspx?navItemNumber=570. Accessed 14 Oct 2015.

4. Driscoll T, Jacklyn G, Orchard J, et al. The global burden of occupationally related low back pain: estimates from the Global burden of disease 2010 study. Ann Rheum Dis. 2014;73:975–81.
5. Bigal M, Lipton R. The epidemiology, burden, and comorbidities of migraine. Neurol Clin. 2009; 27:321–34.
6. Hazard E, Munakata J, Bigal ME, Rupnow M, Lipton R. The burden of migraine in the United States: current and emerging perspectives on disease management and economic analysis. Value Health. 2009;12:55–64.
7. Munakata J, Hazard E, Serrano D, Klingman D, et al. Economic burden of transformed migraine: results from the American migraine prevalence and prevention (AMPP) study. Headache. 2009. doi:10.1111/j.1526-4610.2009.01369.x.
8. Stokes M, Becker WJ, Lipton RB, et al. Cost of health care among patients with chronic and episodic migraine in Canada and the USA: results From the International Burden of Migraine Study (IBMS). Headache. 2011. doi:10.1111/j.1526-4610.2011.01945.x.
9. White LA, Birnbaum HG, Kaltenboeck A, et al. Employees with fibromyalgia: medical comorbidity, healthcare costs, and work loss. J Occup Environ Med. 2008. doi:10.1097/JOM.0b013e31815cff4b.
10. Berger A, Dukes E, Martin S, Edelsberg J, Oster G. Characteristics and healthcare costs of patients with fibromyalgia syndrome. Int J Clin Pract. 2007. doi:10.1111/j.1742-1241.2007.01480.x.
11. Kleinman N, Harnett J, Melkonian A, et al. Burden of fibromyalgia and comparisons with osteoarthritis in the workforce. J Occup Environ Med. 2009. doi:10.1097/JOM.0b013e3181bb808b.
12. Neogi T. The epidemiology and impact of pain in osteoarthritis. Osteoarthritis Cartilage. 2013. doi:10.1016/j.joca.2013.03.018.
13. Wier LM, Andrews RM (2011) The national hospital bill: the most expensive conditions by payer, 2008, HCUP Statistical Brief #107. Agency for Healthcare Research and Quality.
14. Berger A, Hartrick C, Edelsberg J, et al. Direct and indirect economic costs among private-sector employees with osteoarthritis. J Occup Environ Med. 2011. doi:10.1097/JOM.0b013e3182337620.
15. Kotlarz H, Gunnarsson CL, Fang H, et al. Osteoarthritis and absenteeism costs: evidence from US National survey data. J Occup Environ Med. 2010. doi:10.1097/JOM.0b013e3181cf00aa.
16. Hermans J, Koopmanschap MA, Bierma-Zeinstra SM, et al. Productivity costs and medical costs among working patients with knee osteoarthritis. Arthritis Care Res. 2012. doi:10.1002/acr.21617.
17. Kavanaugh A. Economic consequences of established rheumatoid arthritis and its treatment. Best Pract Res Clin Rheumatol. 2007. doi:10.1016/j.berh.2007.05.005.
18. De Azevedo AV, Ferraz MB, Ciconelli RM. Indirect costs of rheumatoid arthritis in Brazil. Value Health. 2008. doi:10.1111/j.1524-4733.2008.00332.x.
19. Kobelt G, Woronoff AS, Richard B, et al. Disease status, costs, and quality of life of patients with rheumatoid arthritis in France: the ECO-PR study. Joint Bone Spine. 2008. doi:10.1016/j.jbspin.2007.07.015.
20. Neovius M, Simard JF, Askling J. How large are the productivity losses in contemporary patients with RA, and how soon in relation to diagnosis do they develop? Ann Rheum Dis. 2011. doi:10.1136/ard.2010.136812.
21. Lundkvist J, Kastäng F, Kobelt G. The burden of rheumatoid arthritis and access to treatment: health burden and costs. Eur J Health Econ. 2008. doi:10.1007/sl0198-007-0088-8.
22. Mehra M, Hill K, Nicholl D, et al. The burden of chronic low back pain with and without a neuropathic component: a healthcare resource use and cost analysis. J Med Econ. 2012. doi:10.3111/13696998.2011.642090.
23. Crow WT, Willis DR. Estimating cost of care for patients with acute low back pain: a retrospective review of patient records. J Am Osteopath Assoc. 2009;109:229–33.
24. Martin BI, Deyo RA, Mirza SK, et al. Expenditures and health status among adults with back and neck problems (1997-2005). JAMA. 2008;. doi:10.1001/jama.299.6.656.
25. Ricci JA, Stewart WF, Chee E, et al. Back pain exacerbations and lost productive time costs in United States workers. Spine. 2006;31:3052–60.
26. Schmidt CO, Schweikert B, Wenig CM, et al. Modeling the prevalence and cost of back pain with neuropathic components in the general population. Eur J Pain. 2009. doi:10.1016/j.ejpain.2008.12.003.
27. Abernethy AP, Wheeler JL, Fortner BV. A health economic model of breakthrough pain. Am J Manag Care. 2008;14:S129–40.
28. Kronborg C, Handberg G, Axelsen F. Health care costs, work productivity and activity impairment in non-malignant chronic pain patients. Eur J Health Econ. 2009. doi:10.1007/s10198-008-0096-3.
29. Lalonde L, Choinière M, Martin E, et al. Costs of moderate to severe chronic pain in primary care patients—a study of the ACCORD Program. J Pain Res. 2014. doi:10.2147/JPR.S55388.
30. Neely ET, Midgette LA, Scher AI. Clinical review and epidemiology of headache disorders in US service members: with emphasis on post-traumatic headache. Headache. 2009;49(7):1089–96. doi:10.1111/j.1526-4610.2009.001460.x.

Author Biographies

Darrell J. Gaskin, Ph.D. is the William C. and Nancy F. Richardson Professor in Health Policy and Director of the Center for Health Disparities Solutions at the Johns Hopkins Bloomberg School of Public Health in the Department of Health Policy and Management. He is an internationally known expert in healthcare disparities, access to care for vulnerable populations and safety net hospitals. He seeks to identify and understand barriers to care for vulnerable populations; to develop and promote policies and practices that improve access to care for the poor, minorities and other vulnerable populations; and to eliminate race, ethnic, socioeconomic and geographic disparities in health and health care

Patrick Richard, Ph.D. is Associate Professor of Health Economics in the Department of Preventive Medicine & Biostatistics (PMB) within the School of Medicine (SOM) at the Uniformed Services University of the Health Sciences (USU). Dr. Richard is the Principle Investigator (PI) and Associate Investigator (AI) of several research projects with a primary focus on financial risks of health insurance, the economics of racial/ethnic disparities in health and healthcare, labor markets outcomes, and economic burden of chronic conditions such as depression, cancer, diabetes, chronic pain, traumatic brain injury (TBI) and Alzheimer's disease for Service members and their dependents/caregivers and Veterans. He has an established national and international reputation through teaching, publications and scholarship.

Joseph Walburn is a graduate of the Johns Hopkins Bloomberg School of Public Health, obtaining his MSPH in Health Policy. Through work with the Hopkins Center for Health Disparities Solutions, and the Maryland Department of Health and Mental Hygiene, he has further developed his understanding of the barriers to health care for vulnerable populations. Upon receiving his degree, he hopes to apply his skills to developing health policy that addresses these barriers to receiving health care, and that promotes access among underserved populations.

Metabolomics and Pain

2

Luigi Barberini, Luca Saba, Antonio Noto,
Claudia Fattuoni and Gabriele Finco

Abstract

Metabolomics is a new way for the Systems Biology application to the Medicine, it is supported by the recent advancements in technology for the analytical description of the molecules mixtures in biological fluids, and it is becoming the revolutionary approach to the modern "personalized medicine" for therapies and treatments. Important "insights" come from the metabolomics application to the pain condition description and we will discuss about several classes of molecules and metabolites and several canonical pathways involved in the pain physiology revealed by the metabolomics approach: Sphingolipids, Glycerophospholipids, Steroid hormones. It is important to remark some pitfalls of metabolomics approach, not only for the pain description and treatments, but also for all the medical applications; especially the lack of a generalized application in all the laboratories of the Standard operative procedures (SOP) for the samples preparation and models realization. Nevertheless, Metabolomic can give us an exciting way to progress towards understanding the basic mechanisms of pain in humans and it also can represent a robust approach to some important aspects of this problem as the appropriateness of pharmacological treatments for all the pain condition, stable or progressive in acute or chronic conditions; this allows us to be confident about the paradigm of the metabolomics approach. A final remarkable point will regard the next-generation approaches of Big Data and metabolomics: integrating genomic, proteomic and metabolomic measurements, we will have the possibility to better understand at holistic level the biochemical process of the pain and to identify robust biological markers for pain-related diseases, diagnosis and treatments, efficacy monitoring: this will lead us to the "therapeutic omics approach" with the connection

L. Barberini (✉) · L. Saba · A. Noto · C. Fattuoni ·
G. Finco
Cittadella Universitaria di Monserrato, University
of Cagliari, Sestu (Ca) 09042, Cagliari, Italy
e-mail: barberini@unica.it

L. Saba (ed.), *Neuroimaging of Pain*, DOI 10.1007/978-3-319-48046-6_2

between the Genotype and the Phenotype, about "what could happen" and "what is happened". These items should give the readers an overview of the situation about metabolomics and pain studies and to stimulate for a deeper approach by means of the bibliography reported.

Keywords
Metabolomics · 1H-NMR spectroscopy · GC-MS spectroscopy · Sphingolipids · Glycerophospholipids · Steroid hormones

1 Introduction

Pain is a subjective condition that cannot be objectively measured. It is important to introduce and develop methods for analytical description of the alterations induced by pain, in both physiological and non-physiological conditions. Recently, metabolomics application to the problem of pain description is fast increasing and producing new data in animal models of pain as in humans' models. The success for this process has been the ability of the modern methods in metabolomics to extract information from noisy but highly informative biofluids. Information can be extracted from noisy but highly informative biofluids such urines and plasma with a withdrawal process that has low or null impact and invasivity for the patients. In this chapter, we will present the innovation of some technological platforms used for the metabolomics and the application in the study of pain. These innovations are leading researchers towards important scientific discoveries: it is evident, for example that nociceptive and neuropathic pains have different underlying pathophysiologic mechanisms and, therefore, they should respond to diverse drugs. Furthermore, it is worth mentioning that a definite diagnosis of pain is difficult and complicated at the moment. It can be reached only with a combination of clinical examination and appropriate laboratory tests. But the possibility to produce low costs test and highly reproducible clinical examinations are recently increasing with the Modern Metabolomics and this leads to more efficient screening programs for study of pain condition in humans.

2 Modern Methods of Metabolomics

Modern Metabolomics (M.M.) represents a solid environment for the fruitful application of the Systems Biology in Medicine, the new way of metabolomics; new technologies specifically developed for metabolomics, like high sensitivity mass spectrometry and magnetic resonance, have the capability to reveal important information about the physiological process in the living being. The holistic approach is the most interesting application in M.M. as it can help our understanding of the multifactorial etiology through the simultaneous analysis of thousands of metabolites and the definition of specific "metabo-types". The identification of metabolites can be detected through the use of databases that classify them according to the biochemical characteristics, such as the Human Metabolome Database (HMDB) and METLIN. The analysis of the data generated in metabolomics studies of holistic is exceptionally complex and requires the use of specific software, such MetaboAnalyst3.0 or XCMS (some of the software named in this chapter of the book and used in the papers presented to discuss about metabolomics and Pain) to detect variations of biological interest.

2.1 Informatics Support to the Metabolomics Analysis

There is no doubt about the fact that the last important implementation for successful application of metabolomics approach in Medicine has been the development of the algorithms analysis and the diffusion of the informatics tools by the Web. Web availability of the data analysis has represented a sort of standardization method and procedure for the data comparison. An example of this is MetobaAnalyst3.0 Web site with the great variety of tools for the metabolomics data analysis. MetaboAnalyst3.0 (Metabolomics Pathway Analysis) is a user-friendly, web-based tool dedicated to the analysis and visualization of metabolomic data within the biological context of metabolic pathways. MetaboAnalyst3.0 combines several advanced pathway analysis procedures along with the analysis of pathway topological characteristics to help in identifying the most relevant canonical metabolic pathways involved in a given metabolomic study. The network visualization is presented in graphical style, easy to understand that supports intuitive and interactive data exploration. Additional features include the implementation of various univariate statistical procedures that can be accessed when users click on any metabolite node on a pathway map.

With this Web tool, the authors mean to provide a user-friendly analytical pipeline for high-throughput metabolomics studies. In particular, MetaboAnalyst3.0 aims to offer a variety of commonly used procedures for metabolomic data pre-processing, normalization, univariate and multivariate statistical analysis. The current implementation, the 3.0 releases of software and procedures, focuses on exploratory statistical analysis, functional interpretation and advanced statistics for exploration and pilot studies. Particular attention has been put in the treatment of several data formats and data types, originated by the most diffused current technological platforms as NMR, GC-MS and LC-MS spectra. Data are then processed, depending on their type, with particular attention to normalization; this is an important step to highly the part of interest of the data. The web service currently supports pathway analysis (including pathway enrichment analysis and pathway topology analysis) and the possibility to explores pathways for several model organisms, including Human, Mouse, Rat, Cow, Zebrafish, Drosophila, Malaria, Budding yeast, *E. coli*, etc., with a total of 1600 pathways. Animal models have been important for exploration of many diseases mechanisms, as in the neurodegenerative pathologies: Parkinson's disease has an important Drosophila models for the description of this degenerative syndrome in the brain.

2.2 Technological Platforms

2.2.1 Samples Preparation

Samples preparation is an important task for the metabolomics study. Usually performed by human operator, and so "operator dependent" in a certain way, this procedure has several steps depending on the samples matrices and platform applied for the analysis (Fig. 1).

Recently, robotics applied to the samples preparation allows for a greater level of reliability, reducing the operator-related errors and variability in the experiments (Figs. 2 and 3). Technology is still expensive but it will ensure in the next future an unprecedented repeatability in the measurements.

Automation and process control procedures start from the sample acquisition and also with

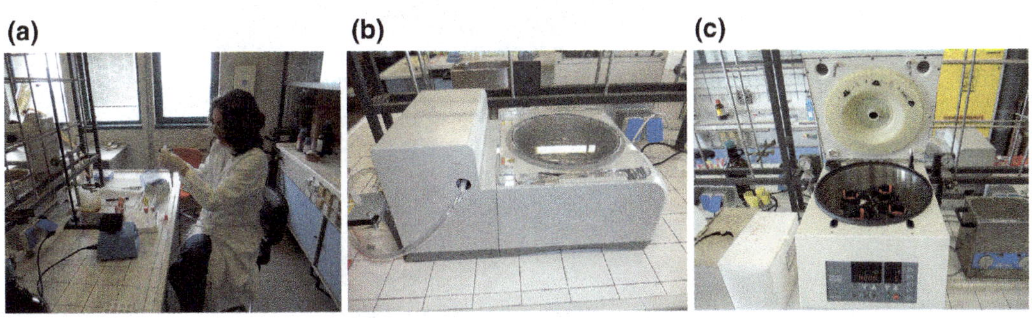

Fig. 1 **a–c** Typical preparative laboratory for metabolomics and instrumentations

Fig. 2 Robotization of samples preparation

the compounds for preparative procedures acquisition (Fig. 4). In the fully automatized process, labelling of single sample before analysis is performed in order to track all the different phases that could result in an alienation of the sample from the stock (outlier).

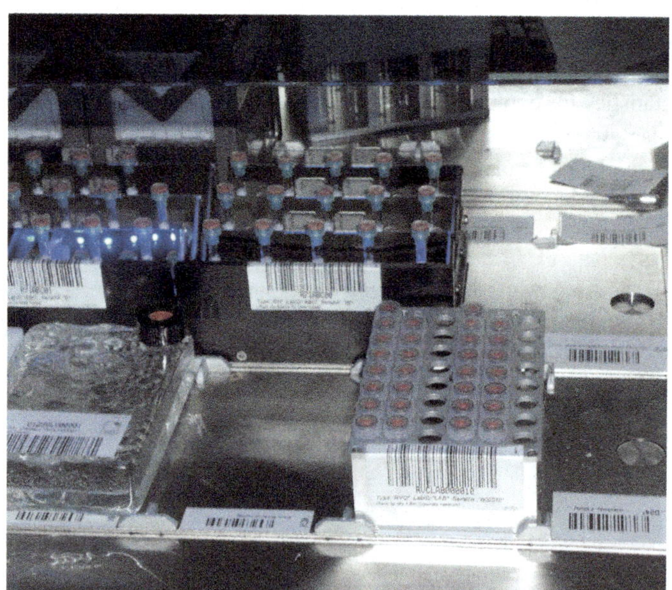

Fig. 3 Automation in samples preparation and submission to the analysis procedures

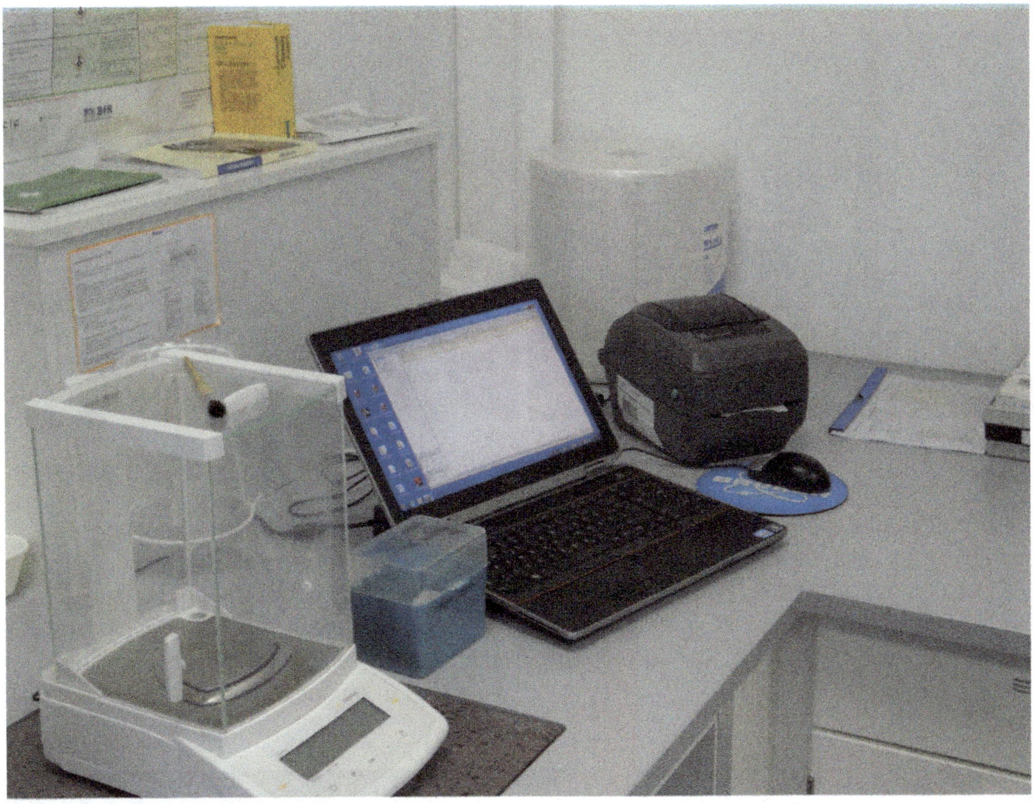

Fig. 4 Labelling of samples from the entering in the analytic laboratory. LIMS, Laboratory Information Management System

2.2.2 Nuclear Magnetic Resonance Technology (NMR)

Nuclear magnetic resonance (NMR) is a physical phenomenon in which nuclei in a magnetic field absorb and re-emit electromagnetic radiation. This energy is at a specific resonance frequency which depends on the strength of the magnetic field and the magnetic properties of the isotope of the atoms and on the structure of molecule. In this way, it can be used to identify and quantify molecules and their concentrations in mixtures.

NMR has substantial advantages for mixture analysis: first of all it is highly reproducible and it is fully quantitative with one calibration standard; usually, it needs a little sample preparation and it makes available the structural information. Further, it has a high dynamic range and this leads to a multimarket approach for samples characterization. It can be used for untargeted and targeted analysis in one experiment with a profile of low cost per sample. It is important to notify that high throughput is possible with a complete standardization under push button automation. This allows a retrospective use of older data in new statistical models or quantification allowing multiple solutions on one standardized platform.

Modern NMR technology allows the installation of powerful systems into very little laboratories; the active shielding technology allows for a little confinement space to require for the big magnetic fields produced. Usually, automation in all the management operations allows for unattended laboratory with the use of robotic also for the long-time charging of samples in the machine; temperature controls preserve the stack of samples before the analysis (Fig. 5).

Automation of analysis procedures, based on powerful software for the automatic recognition

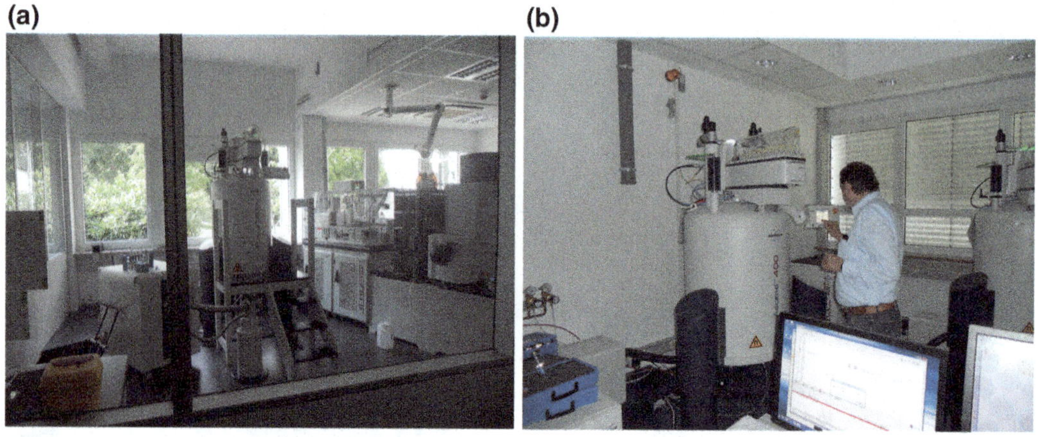

Fig. 5 **a**, **b** Typical laboratories of metabolomics

of metabolites, allows for a prompt and complete reporting of the analysis (Fig. 6).

NMR is a robust technology for the metabolomics approach with potential space of development extremely important for the Medical Metabolomics.

2.2.3 Gas Chromatography–Mass Spectrometry (GC-MS)

Along with the NMR technology there is also the Mass Spectrometry technology coupled to several platform of chromatography separation of molecules in mixtures (Fig. 7).

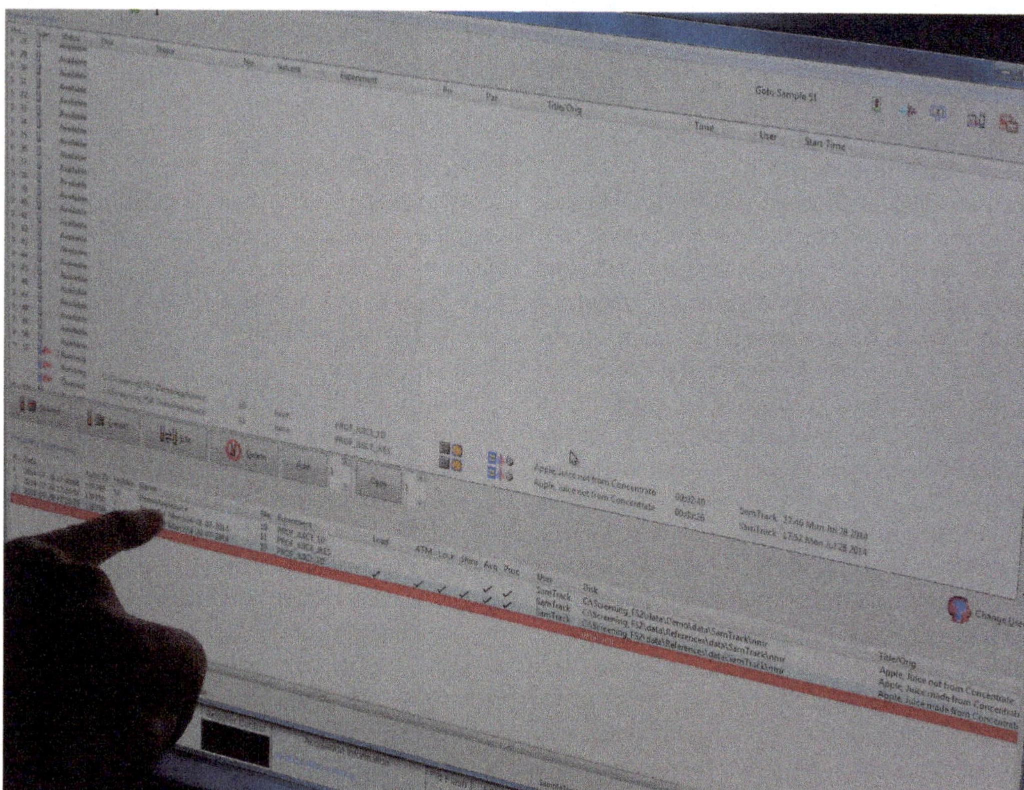

Fig. 6 Reporting of data analysis

Fig. 7 A GC-MS system in a metabolomics laboratory

Gas chromatography–mass spectrometry (GC-MS) combines gas chromatography and mass spectrometry to identify different substance. There is a wide range of applications for the GC-MS including drugs detection, environmental analysis and identification of unknown samples. GC-MS is also used in airport security to detect illegal substances in luggage or on human beings. GC-MS has been widely diffused in the metabolomics due to the relatively low cost and high sensitivity in the substance identification. It is widely used in the applications for the environmental monitoring and clean-up and into sports anti-doping analysis.

Applications for medicine includes the study of several congenital metabolic diseases also known as inborn errors in metabolism are now detectable by newborns screening tests using gas chromatography–mass spectrometry. Due to the sensitivity of the technology GC-MS can determine compounds in urine even in low concentration.

The GC-MS is composed of two blocks: the gas chromatograph and the mass spectrometer. The gas chromatograph utilizes capillary columns, with different properties (length, diameter, film thickness, etc.), for the sample separation. The difference in the chemical properties between different molecules in a mixture will promote separation of the molecules as the sample travels the length of the column. The molecules are retained by the column and released at different characteristic times, and this allows the system to capture, ionize, accelerate, deflect and detect the ionized molecules separately. The mass spectrometer breaks each molecule into ionized fragments and detects these fragments using their mass-to-charge ratio (m/z) that is characteristic parameter of identification.

2.2.4 Liquid Chromatography–Mass Spectrometry (LC-MS)

The mass spectrometer can be coupled to a liquid chromatograph; Liquid chromatography–mass spectrometry (LC-MS, or alternatively HPLC-MS) is an analytical chemistry technique that combines the physical separation capabilities of liquid chromatography (HPLC) with the mass analysis capabilities of mass spectrometry (MS). LC-MS is a powerful technique that has very high sensitivity, making it useful in many metabolomics applications. Its application is oriented towards the separation, general detection and potential identification of chemicals of particular molecules with high polarity and mass in complex mixtures. Particular systems of samples preparation in LC-MS can be used for rapid mass-directed purification of specific substances in such mixtures that are important in pharmaceutical, food and other industries.

2.2.5 Hybrid Mass Spectrometry

In order to achieve much more sensitivity in the mass spectrometry, the detection systems can be organized in different sections: coupling two single detectors in a single system a "Tandem" mass spectrometry is realized. Tandem mass spectrometry, also named as MS/MS or MS^2, involves multiple steps of mass spectrometry selection, with inner fragmentation sections located between these stages. In a tandem mass spectrometer, ions are created in the ion source and separated for their m/z ratio in the first stage of mass spectrometry; then, ions of a particular m/z (precursor ions) are selected and further fragmented (product ions). The resulting ions are then separated again and detected in the second stage of mass spectrometry. Hybrid technology can be used in this modern instruments and notation like QqQ (Triple quadrupole mass spectrometer) or QTOF, Quadrupole time-of-flight mass spectrometer (also QqTOF) are used to indicate different analysers used in these systems. Widely diffused in metabolomics, the triple quadrupole mass spectrometer, for example, is a tandem mass spectrometer consisting of two quadrupole mass analysers in series, with an intermeddle quadrupole (non-mass-resolving) between them as dissociation section.

But all of these configurations can be used for particular purposes in metabolomics. Modern systems reveal the analysts at low concentration, even in the presence of highly concentrated metabolites providing high sensitivity over a wide dynamic range.

3 Pain: General Definition and "Function"

Pain is the functional action used in the superior organism for the occurring tissues damaging signalling; in this condition we can talk of "physiologic" pain. Otherwise, when pain become self-consistence and it lose the function of tissue damaging alert it becomes "pathological" and becoming in turn a real disease (pain syndrome). Pain has had a fundamental role in the human being survival as message of the need to undertake a reaction to maintain physical integrity. For these reasons pain's receptors are widely diffused in all tissues and they are able to identify different kinds of potentially dangerous stimuli.

Epidemiological studies have revealed that there are different kinds of pain: the primary classification of pain regards the temporal scale of the pain sensation evolution, chronic and acute pain, but it is also frequently used as a definition that considers the anatomical location of pain. Usually, acute pain begins suddenly and is usually sharp in quality. It serves as a warning of disease or a threat to the body. Acute pain might be caused by many events or circumstances. Acute pain might be mild and last for just a moment, or it might be severe and last for weeks or months. In most cases, acute pain does not last longer than months, and it disappears when the underlying cause of pain has been treated or has healed. Unrelieved acute pain, however, might lead to chronic pain.

As the pain persists despite the fact that the injury has healed it is defined as chronic. Pain signals can remain active in the nervous system for a long time, sometimes months, or years. Physical effects include tense muscles, limited mobility, a lack of energy and changes in appetite. Emotional effects include depression, anger, anxiety and fear of re-injury. Such a fear might hinder a person's ability to return to normal work or leisure activities.

Common chronic pain conditions are related to headache low back pain cancer pain, arthritis pain, neurogenic pain (pain resulting from damage to nerves), psychogenic pain (pain not due to past disease or injury or any visible sign of damage inside) [1].

Chronic pain might have originated with an initial trauma/injury or infection, or there might be an ongoing cause of pain. However, some people suffer chronic pain in the absence of any past injury or evidence of body damage. The experience of pain sensation is complex, related to sociocultural characteristics, such as gender, ethnicity and age.

Chronic pain can be able to induce changing in all the systems of living creatures, modifications of functionalities and abilities. These modification can be defined as negative reactions, that can be consider as tentative to reduce the pain uncomfortable state, or positive reactions with the final effects to maintain and strengthen the pain sensation. This last situation is dramatic and terrible for patients, and sometime it may lead to suicide. Modification occurs in our mental schemes and in our behavioural mechanisms. Modification occurs in our phenotype. Modification occurs in all our vital reactions and homeostasis conditions. All these changes related to neurological, endocrinological and immunological systems could be monitored by means of several diagnostic techniques, such as metabolomics and magnetic resonance imaging.

4 Pain Assessment and Treatment

Everyone reacts uniquely to a given painful stimulus, on the basis of past experience and what is called his own "pain threshold", and each person is able to assess, according to its parameter, how strong her pain is and then it should be able to objectify through a measurement. Each individual learns the meaning of pain through the own experiences related to injury during the first years of life. Being an unpleasant experience, the somatic component of pain is also accompanied by an emotional stress. Therefore, the pain is always subjective and it is very important that the patient learns to measure his pain and record it in a daily diary for the cases of chronic pain. Furthermore, depending on whether its intensity is mild, moderate or severe, drugs that are to be used should be different and administered at different doses.

5 Metabolomics and Pain

5.1 Preclinical Models

Important information about the pain perception (in animal models) arises from the application of metabolomics approach. As previously reported, there are several classifications of the pain. Pathogenetics of pain proposes the classification: idiopathic pain, nociceptive pain, and neuropathic pain. Several research groups have published interesting papers shedding lights on the biological deep mechanism of pain perception with a particular attention to new biomarkers of pain detection. Generally speaking, an important role seems to be played by the sphingosine class molecules. The research into biomarkers discovery has recently received again a lot of attention. This is not only reflected by the increasing number of working groups in this topic, but also the sheer number of publications (more than 500,000) tells its own story. At least in part, the overwhelming interest might be rested on the current revitalization of an old concept in medicine—personalized healthcare, but undeniably also because of recent advances in diagnostic technology. Particularly, in the metabolomics approach for the Systems Biology!

Researchers, mainly two groups, the one led by Gary Siuzdak at The Scripps Research Institute, and the other one led by Marianne Manchester at the University of California, San Diego (UCSD) found multiple changes in the proinflammatory sphingomyelin/ceramide pathway in the spinal cord of rats with nerve injury-induced continuous pain; in fact, the team of scientists using the metabolomics approach has revealed the importance of the "*N,N*-dimethylsphingosine" (DMS), a breakdown of small molecules in the cellular membranes of the nervous system and not previously associated with pain; this important discovery could lead to an innovative treatment for the pain after the important and pioneer work of Rasmussen in the 2004 [2]. In their paper published by Patti et al., they revealed that endogenous metabolite *N,N*-dimethylsphingosine induces mechanical hypersensitivity in vivo. When administered to control animals, DMS caused painful hypersensitivity that mimicked the effects of nerve injury. The study, published online in January on Nature Chemical Biology, identifies DMS as a new pain mediator, and should increases interest in the sphingomyelin/ceramide pathway as candidate targets for pain treatment. These results could be an important target for the pharmacological research in the field of pain perception cures.

"We think this is a big step forward in the understanding and treatment of neuropathic pain, and is also a solid demonstration of the power of metabolomics", has declared Gary J. Patti. In the same paper Patti and colleagues show evidences that the ceramide pathway is also involved in neuropathic pain.

The modern metabolomics, projected towards the personalized medicine by means of the Systems Biology, can represent an important tool for the Medicine. Metabolomics aims to survey a great amount of the molecules in a given tissue (sugars, amino acids, hormones, lipids, organic acids) especially by means of integration of technological platforms. Thanks to the new level of sensitivity achieved with new NMR systems or in the hybrid mass spectrometers, thousands of chemical components can be identified in biofluids and in tissue extracts. In some cases some of these compounds remain unknown, or it is difficult to identify their biological functions. The use of metabolomics to discover the pain biology, and the discovery of a novel pain mediator by this method makes for a "compelling story", said Daniela Salvemini of Saint Louis University School of Medicine in Missouri. Salvemini told that the study "confirms and extends the importance of the ceramide pathway in pain". Salvemini and others have shown previously that ceramide, and its metabolite sphingosine-1-phosphate (S1P), mediate inflammatory pain in rodent models. Ceramide and S1P function as second messengers that sensitize nociceptive neurons in response to nerve growth factor (NGF) and the inflammatory cytokine tumour necrosis factor-α (TNF-α) [3, 4]. Blocking S1P or its receptor can relieve nociceptor hyperexcitability and pain [5]. In the spine, ceramide is upregulated in astrocytes and microglia by chronic morphine treatment. There is evidence that ceramide and S1P contribute to opioid-induced hyperalgesia and tolerance, and inhibiting production of the metabolites blocks the ill effects of long-term opioid treatment. On the other hand, some experiments have demonstrated an antinociceptive role for S1P in the spine.

In the past, scientists, who want to understand what makes the difference diseased cells from healthy cells, have often tried differences in genomics and proteomics of the subjects. Metabolomics, however, concerns the differences in the levels of metabolites, small molecules, such as sugars, vitamins and amino acids, which serve as the basic building blocks of cellular processes. "These are the molecules that are actually processed during cellular activity and monitoring provides them with more direct information about what is happening at the biochemical level", continue Patti [5]. Metabolomics is increasingly used to find biochemical markers of disease. The modern metabolomics can represent an important tool for the Medicine, so we could talk of Metabolomics Medicine. "…the search for biomarkers in pain is, like in many other fields, now increasingly concerned with 'omics' research".

Now we are on our way to properly organize the information about the metabolites networks alteration induced by Chronic Pain. Mechanisms of Systems Biology underlining the Medicine of Pain are not yet well understood, but the preclinical models are important for a more insight of the problem.

5.2 Humans Models

As previously reported, pain nature diagnosis is still a complex task and a wrong or untimely identification can lead to an uncomfortable state for the patient, to inappropriate treatment and possible pharmaceutical adverse reactions. Also, we must consider the increasing costs for the community just related to inappropriate treatments.

It is important to project the preclinical models towards the bedside of patients in order to produce results for the Community. Previous and early studies had shown that DMS, revealed in the papers of Gatti, is produced in some cancer cell lines and human brain tissue, but its roles were not understood and it was not easily connected to the pain perception [6, 7].

By means of the metabolomics it is possible to replicate test and validate Biological Systems model in human being also. It is easy to get information about the human metabolisms with sampling biofluids easy to withdraw. Several hypotheses can be tested exploring the metabolomics connections between canonical pathways. So, many groups around the world are applied into these tasks and time reduction in the medical investigation can be attended in many areas. Crucial for the interpretation of the models will be the ability of the researchers to separate the compartments of the contributions to the metabolites changes. In this way, we will be able to get models of diseases for a better understanding of the mechanism of induction of the pathology and about the ability to select a proper drug treatments. It is really interesting the result is obtained and presented in a recent paper from Finco et al. [1] that sheds light on the possibility to discriminate by means of metabolomics approach between nociceptive (NC) and neuropathic pain (NP), for example. This is important for the selection of the proper drug to submit to patients. Urinary samples were analysed with 1H NMR spectroscopy technological platform and compared with a control population (C). The application of multivariate discriminant analysis on the urine spectral profiles allowed the authors to successfully classify nociceptive and neuropathic pain with high sensibility and specificity. From this study it is possible to conclude that urine is good biofluid to study metabolic alterations induced by a chronic pain state; due to the fact that urine collects informations at the end of the catabolic process chain this biofluid is often "noisy" in terms of overlapping contributions due several pathological and physiological condition. The goodness of the discrimination model depends on the intensity of perturbation and on the sample "population" size. But it is important to put in evidence that we have a powerful tool of pain diagnosis to apply the proper treatment. Metabolites for this preliminary study in charge for the NC-NP-C are choline, phosphocholine, alanine and taurine. Some of these metabolites are involved in the neural membranes characterization but, generally speaking, they are hubs connected and related to several generic canonical pathways. But metabolomics approach is able to give to researchers a method to increase samples number and explorative capability in human models.

5.2.1 Some Particular Aspects: Appropriateness of Pharmacological Treatments and the Paradigm of the Metabolomics Approach

From the paper of Su et al. [8] metabolomics approach was applied to the study of effects of herbal medicine (namely Shaofu Zhuyu formula concentrated-granule, SFZYFG) treatment to Primary dysmenorrhea (PD), a pain condition characterized by painful menstrual cramps without any organic pathology.

Using tandem mass spectrometry (MS/MS) platform the authors analysed changes of metabolic profiling in plasma and urine samples in a population of PD patients and healthy controls before and after a 3-month SFZYFG treatment. In this study, thirty-five metabolites were identified and quantified for the contribution to PD progress. These promising identified biomarkers underpinning the metabolic pathway including sphingolipids metabolism, steroid hormone biosynthesis, and glycerophospholipid metabolism are altered in PD patients. Starting from the metabolites quantification the canonical pathways mainly involved in the PD evolution were identified by using the web tool "Pathway Analysis" within the MetaboAnalyst3.0 platform [9–11].

This is what we mean as "explorative" ability from the analysis to the speculative application of hypothesis test.

5.2.2 Metabolites and Pathways Analysis for Pain

Resuming the most important conclusions from the papers examined we can discuss about several classes of molecules and metabolites and several canonical pathways involved in the pain physiology and revealed by the metabolomics approach. We can start with the Sphingolipids (also named as glycosylceramides); they are a class of lipids with a backbone of sphingoid bases discovered in brain extracts in the 1870s. These compounds play important roles in signal transmission and cell recognition [12]. Disorders of sphingolipids metabolism have particular impact on neural tissue functionality. Another important class of molecules are the Glycerophospholipids (also named as phosphoglycerides). These molecules are phospholipids with an alcoholic molecule of glycerol, the alcohol to which two fatty acids and a phosphoric acid are attached as esters. This basic structure is a phosphatide, an important intermediate in the synthesis of many phosphoglycerides. The glycerophospholipid composition of neural membranes greatly alters their functionality. Again an important clue related to the membrane properties alteration of neuronal cells. Marked alterations in neural membrane glycerophospholipid composition have been reported to occur in neurological disorders. These alterations result in changes in membrane fluidity and permeability and these processes, along with the accumulation of lipid peroxides and compromised energy metabolism, may lead to the neuro-degeneration revealed in neurological disorders like Parkinsonism and Alzheimer disease. So we get important informations but we must test the specificity of the informations obtained.

An important class of molecules that could increase the specificity power of metabolomics analysis are "Steroid hormones" ; this class of molecules can be grouped into 2 classes:

sex steroids
corticosteroids

Within those 2 classes there are 5 types of molecules according to the receptors to which they bind: glucocorticoids and mineralocorticoids (corticosteroids) and androgens, estrogens and progestogens (sex steroids). We have to mention that we have Vitamin D derivatives that can be considered as a sixth class closely related to the hormone system with homologous receptors. They have some of the characteristics of true steroids as receptor ligands.

Steroid hormones operate in the control of metabolism, inflammation, immune functions, salt and water balance, development of sexual characteristics and the ability to withstand illness and injury. The term steroid describes both hormones produced by the body and artificially produced medications that duplicate the action for the naturally occurring steroids. Steroids are widely studied with the metabolomic approach in many of the paper proposed in this chapter, but some questions still remain open. Which are the best instruments to use all these informations and to get a deeply informative picture about the pain mechanism in humans being? Are we really ready for the new age of the Systems Biology in Medicine?

6 Conclusions. Strategies and Challenges for Next-Generation Metabolomic Analyses in Pain Studies

It is obvious that we are only collecting "preliminary informations" and we are learning about the modalities to operate data mining in Systems Biology methods applied to the Medicine by the prospective of the metabolomics. Before to generalize the data and the information we must achieve a higher level in the standards operative procedures adopted for the analysis; also we should standardize the pre-processing of the data and the Multivariate models to propose for the generalization and the discussion. What we see for the future?

6.1 Platforms Standardization for Data Comparison

Standardization is important in order to achieve an optimal environmental for data comparison in metabolomics: standard for samples preparation procedures, standard for analysis condition for all the technological platforms. This approach will ensure a better sensitivity and specificity for the analysis and a generalization of innovative methods for specific study. For example, for many years lipidomics has been considered as specific field of investigation for mass spectrometry; recently protocols based on particular sequences in NMR has discovered a pathway for these quantitative analysis really important in some application as the Pain Diagnosis. NMR-based Lipidomics can shed a light into the pathways and networks of cellular lipids in biological systems, giving a powerful tool for some important answer about the membrane behaviour in chronic inflammation conditions and Pain status [4, 7, 13, 14]. With a stable standardization platform many advances may arise in Human Metabolomics with major application in epidemiology, translational and clinical research. Some aspects of the metabolomics, as early disease recognition, disease staging, patient stratification and personalized treatment, all these aspects will be much more solid for the comparison all over the world with the aim to get a personalized long-term health modelling.

6.2 New Data Analysis Algorithms

Omic technologies are increasingly being applied to study complex biochemical and physiological states. Analysis of small molecules or metabolites, metabolomics, has been widely used to characterize organismal phenotypes including identification of biomarkers associated with autism, infant birth weight, metabolic syndrome and cancer. Next-generation approaches integrating genomic, proteomic and metabolomic measurements have shown promise to aid researchers to better understand otherwise recalcitrant biochemical process and identify robust biological markers for disease diagnosis and treatment efficacy monitoring.

Robust interpretation of experimental results measuring discreet biological domains remains a significant challenge in the face of complex biochemical regulation processes such as organismal versus tissue versus cellular metabolism, epigenetics, and protein post-translational modification. Integration of analyses carried out across multiple measurement or omic platforms is an emerging approach to help address these challenges. Key challenges remain for metabolomic researchers including large-scale studies data normalization, multivariate analysis, visualization and omics data integration.

Implementation of data normalization approaches including internal standard and quality control based methods maybe required to effectively remove analytical batch effects. Emerging methods incorporating replicated measurements to carry out LOESS or other nonlinear based smoothing models have shown promise to deal with complex analytical modes of variance.

Omic integration methods are required to combine and analyse biological measurements carried out across multiple platforms within a biological context. Leading approaches for omic integration include biochemical pathway, network-based and empirical correlation-based methods.

Given the aforementioned challenges, advanced data analysis tools are required to carry out effective omic and specifically metabolomic data interpretation. Modern data analysis tools are necessary to allow researchers to implement analysis pipelines incorporating data normalization, integration, multivariate analysis and ultimate interpretation with in a biochemical context. An emerging approach termed network mapping shows promise to effectively integrate statistical, multivariate and functional domain knowledge to calculate richly connected biochemical networks which can highlight metabolic perturbations specific to researchers' areas of interest.

The paradigm of systems biology emerged with the diffusion of system-level experiments: understanding complex biological systems requires understanding and modelling characteristics that are fundamentally determined by the organization of their constituent parts, emergent phenomena created by the interactions of those elements defined as hubs and spokes depending on the level of connection and interconnection. Especially for the metabolites defined as "hub nodes" there is an increased interest in medicine because they importance in the comprehension of the pathologies.

Network-driven approach is a powerful theoretical technique to analyse metabolome, to unveil the underlying hierarchical structure and to predict their behaviour under different conditions. Each metabolite gives contribution to several canonical pathways. In the metabolome some metabolites can exhibit a co-variation stronger than others. These correlations can have different influence on different metabolic pathways depending on the "position" of the metabolites. These co-variations can be described as different level of "connectivity" between metabolites. This connectivity is the expression of the metabolome dynamic that results in a pattern of statistic dependencies (functional connectivity) of some metabolites in order to realize a "functional connectome". Hub nodes are among the most intriguing structural features of metabolic networks. Hubs have attracted much attention in network science since they often correspond to nodes that have special integrative or control functions. It is likely that neuronal hubs have a privileged role in organizing network dynamics and exert strong influence on the state of more peripheral nodes. Due to their structural and functional connections, hub nodes integrate a highly diverse set of signals and are in a "position" to control the flow of information between relatively segregated parts of the metabolic network. So, we can have a modular structure in the metabolites "community" (secondary approach to the metabolites functional). Since much of the "between-modules" (modularity property) information flow travels through hubs, the rate at which they relay signals would have a large impact on system-wide communication. Criteria for hub identification vary across different studies. In some cases, hubs are identified as "highly connected nodes", that is, primarily on the basis of node degree or strength or on the clustering index. Because of their position on many of the network's shortest paths, any perturbation of the state of a hub node would be able to spread quickly across the network. As with any untargeted "omics" scheme, the metabolomics experiments presented in this chapter produced reams of data: besides DMS, 732 other compounds showed at least a twofold change in injured animals. "We need a prioritization scheme", Patti said. His hope is to profile the metabolome in a variety of pain models, as well as in human tissues, and compare the results. Towards that end, Patti [7, 13], and their colleagues have developed software to enable meta-analysis of metabolomics data. The way of prioritize the schemes is probably the network approach of the Systems Biology [15, 16].

References

1. Finco G, Locci E, Mura P, Massa R, Noto A, Musu M, et al. Can urine metabolomics be helpful in differentiating neuropathic and nociceptive pain? A proof-of concept study. PLoS ONE 2016;11(3): e0150476. doi:10.1371/journal.pone.0150476.
2. Rasmussen PV, Sindrup SH, Jensen TS, Bach FW. Symptoms and signs in patients with suspected neuropathic pain. Pain. 2004;110:461–9.
3. Zhang Yu, Meng Xianze, Li Aihui, Xin Jiajia, Berman Brian M, Lao Lixing, Tan Ming, Ren Ke, Zhanga Rui-Xin. Electroacupuncture alleviates affective pain in an inflammatory pain rat model. Eur J Pain. 2012;16(2):170–81.
4. Zhang YH, Vasko MR, Nicol GD. Ceramide, a putative second messenger for nerve growth factor, modulates the TTX-resistant Na+ current and delayed rectifier K+ current in rat sensory neurons. J Physiol. 2002;544:385–402.
5. Chi XX, Nicol GD. The sphingosine 1-phosphate receptor, S1PR1, plays a prominent but not exclusive role in enhancing the excitability of sensory neurons. J Neurophysiol. 2010;104:2741–8.

6. Chen YJ, Hill S, Huang H, Taraboletti A, Cho K, Gallo R, et al. Inflammation triggers production of dimethylsphingosine from oligodendrocytes. Neuroscience. 2014;279:113–21. doi:10.1016/j.neuroscience.2014.08.011.
7. Connor SC, Gray RA, Hodson MP, Clayton NM, Haselden JN, Chessell IP, et al. An NMR-based metabolic profiling study of inflammatory pain using the rat FCA model. Metabolomics. 2007;3:29–39.
8. Su S, Duan J, Wang P, Liu P, Guo J, Shang E, Qian D, Tang Y, Tang Z. Metabolomic study of biochemical changes in the plasma and urine of primary dysmenorrhea patients using UPLC-MS coupled with a pattern recognition approach. J Proteome Res. 2013 Feb 1;12(2):852–65. doi:10.1021/pr300935x (Epub 2013 Jan 18).
9. Xia J, Sinelnikov I, Han B, Wishart DS. MetaboAnalyst 3.0—making metabolomics more meaningful. Nucl Acids Res. 2015;43:W251–7.
10. Xia J, Mandal R, Sinelnikov I, Broadhurst D, Wishart DS. MetaboAnalyst 2.0—a comprehensive server for metabolomic data analysis. Nucl Acids Res. 2012;40:W127–33.
11. Xia J, Psychogios N, Young N, Wishart DS. MetaboAnalyst: a web server for metabolomic data analysis and interpretation. Nucl Acids Res. 2009;37:W652–60.
12. Coste J, Delecoeuilleire G, de Cohen LA, et al. Clinical course and prognostic factors in acute low back pain: an inception cohort study in primary care practice. BMJ 1994;308:577–80.
13. Patti GJ, Yanes O, Shriver LP, Courade JP, Tautenhahn R, Manchester M, Siuzdak G. Metabolomics implicates altered sphingolipids in chronic pain of neuropathic origin. Nat Chem Biol. 2012;8(3):232–4.
14. Joseph EK, Levine JD. Mitochondrial electron transport in models of neuropathic and inflammatory pain. Pain 2006;121(1–2):105–14.
15. Selected Abstracts of the 11th International Workshop on Neonatology • *From the womb to the adult* • Cagliari (Italy) • October 26th–31st 2015. ABS 42. Orthogonalization methods in metabolomics. Hypothesis and tests for validation 2022 L. Barberini, D. Grapov, A. Noto, A. Dessì, C. Fattuoni, F. Palmas, A. Casanova • Cagliari (Italy) and Ballwin (MO, USA).
16. Selected Abstracts of the 11th International Workshop on Neonatology 2022 *From the womb to the adult* • Cagliari (Italy) • October 26th–31st 2015. ABS 104. Strategies and challenges for next generation metabolomic analyses • D. Grapov • Ballwin (MO, USA).

For Further Insights

17. http://www.painresearchforum.org/news/13084-metabolomics-uncovers-new-driver-neuropathic-pain

Brain Neuroanatomy

3

Adam G. Thomas

Abstract
Unlike certain brain functions, the perception of pain involves multiple different areas of the brain, often working together in functional networks. As such, in order to understand how research, clinical conditions and treatments are related to brain structure and location, an overview of whole brain anatomy is essential. This chapter will commence with a summary of anatomical terms used in relation to the brain and in particular brain imaging. The basic anatomical organisation of different brain compartments will be described, followed by a description of the primary sensory pathways involved in pain perception. A more detailed look at different brainstem areas involved in pain perception and modulation will be followed by an in-depth description of the thalamic nuclei and their cortical connections. Multiple cortical and subcortical areas are involved in both the perception of pain and the response to it. The anatomical localisation of these regions will be described. Finally, the emerging concepts of different functional networks of brain regions relating to attention and emotional aspects of pain processing will be described.

Keywords
Thalamic nuclei · Functional anatomy · Brainstem · Angiography · Tractography · Norepinephric · Raphe magnus · Cortical regions

1 Introduction

Unlike certain brain functions, the perception of pain involves multiple different areas of the brain, often working together in functional networks. As such, in order to understand how research, clinical conditions and treatments are related to brain structure and location, an overview of whole brain anatomy is essential. This chapter will commence with a summary of anatomical terms used in relation to the brain and in particular brain imaging.

A.G. Thomas (✉)
Department of Neuroradiology, Nottingham University Hospitals NHS Trust, Queens Medical Centre Campus, Derby Road, Nottingham NG7 2UH, England
e-mail: adam.thomas@nuh.nhs.uk

L. Saba (ed.), *Neuroimaging of Pain*, DOI 10.1007/978-3-319-48046-6_3

The basic anatomical organisation of different brain compartments will be described, followed by a description of the primary sensory pathways involved in pain perception. A more detailed look at different brainstem areas involved in pain perception and modulation will be followed by an in depth description of the thalamic nuclei and their cortical connections. Multiple cortical and subcortical areas are involved in both the perception of pain and the response to it. The anatomical localisation of these regions will be described. Finally, the emerging concepts of different functional networks of brain regions relating to attention and emotional aspects of pain processing will be described.

2 Neuroanatomical Localisation and Terms

Anatomical figures in this chapter will be based around clinical neuroimaging studies. It is important to be familiar with the conventional way in which clinical studies and anatomical information is described. Most brain studies rely on cross sectional imaging of the brain, presented in 3 different anatomical planes: sagittal, axial and coronal (Fig. 1).

Neuroradiological convention is to present axial images as if viewing the patient from the feet, looking up to the top of the head. As such the patient/subject's right side is presented on the left side of the image. This is also the case when viewing coronal images. It is conventional to present sagittal images with the nose/front of the face to the left of the image and the back of the head at the right (Fig. 2).

As a word of caution however, it should be noted that frequently in psychology literature (particularly in the presentation of functional MRI data) this convention is not followed, with the right side of the patient being to the right of the image and the legends of any published image should be carefully inspected.

The naming of brain structures frequently uses anatomical descriptors such as anterior/posterior,

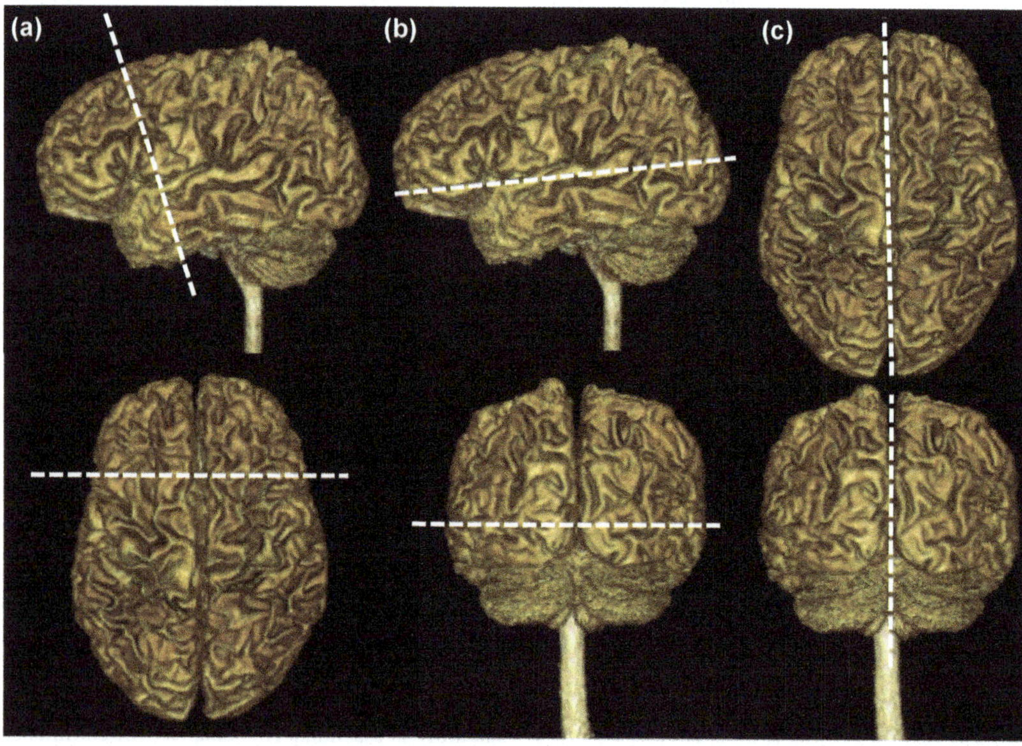

Fig. 1 3d surface shaded display showing the coronal plane (**a** *top row* viewed from side, *bottom row*, viewed from above), axial plane (**b** *top row* viewed from side, *bottom row* viewed from front) and sagittal plane (**c** *top row* viewed from above, *bottom row* viewed from front)

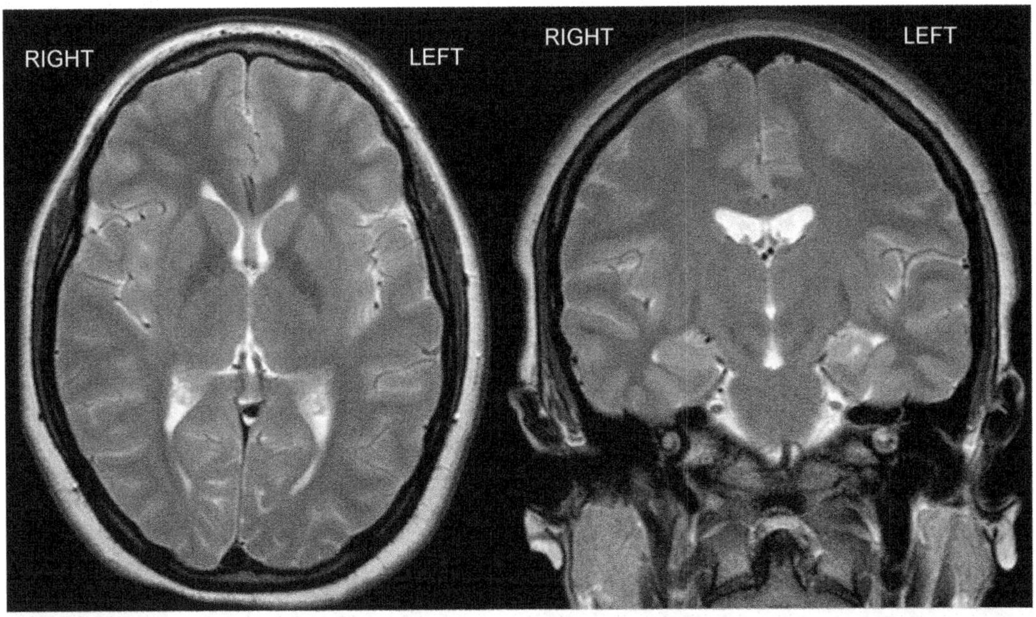

Fig. 2 Axial T2 (*left*) and coronal T2 (*right*) showing conventional radiological anatomical orientation

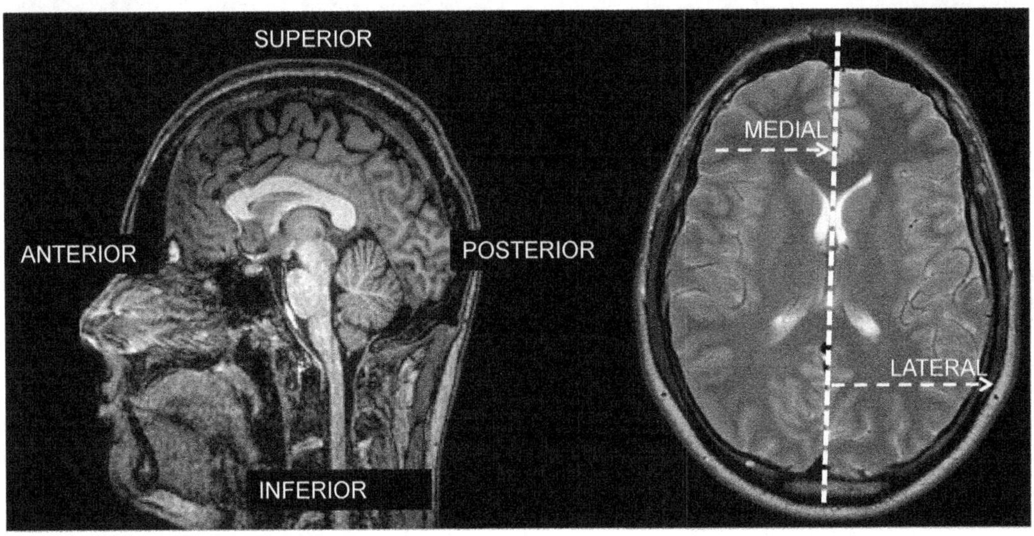

Fig. 3 Anatomical labelling. Sagittal T1 volume (MPRAGE) *top left*, axial T2, *bottom right*. The term 'medial' is used for structures close to the *midline*; lateral is used for structures further away from the *midline*

superior/inferior and right/left. These directions in relation to brain imaging are shown in Fig. 3.

The names of some structures relates to their embryological origin, reflecting the trilaminar disc created during gastrulation, before the complex folding that occurs to form the central nervous system [1]. The term dorsal and ventral are still used when describing anatomical localisation in the spine. The use of the term ventral and dorsal in the brain refers to the basal/inferior

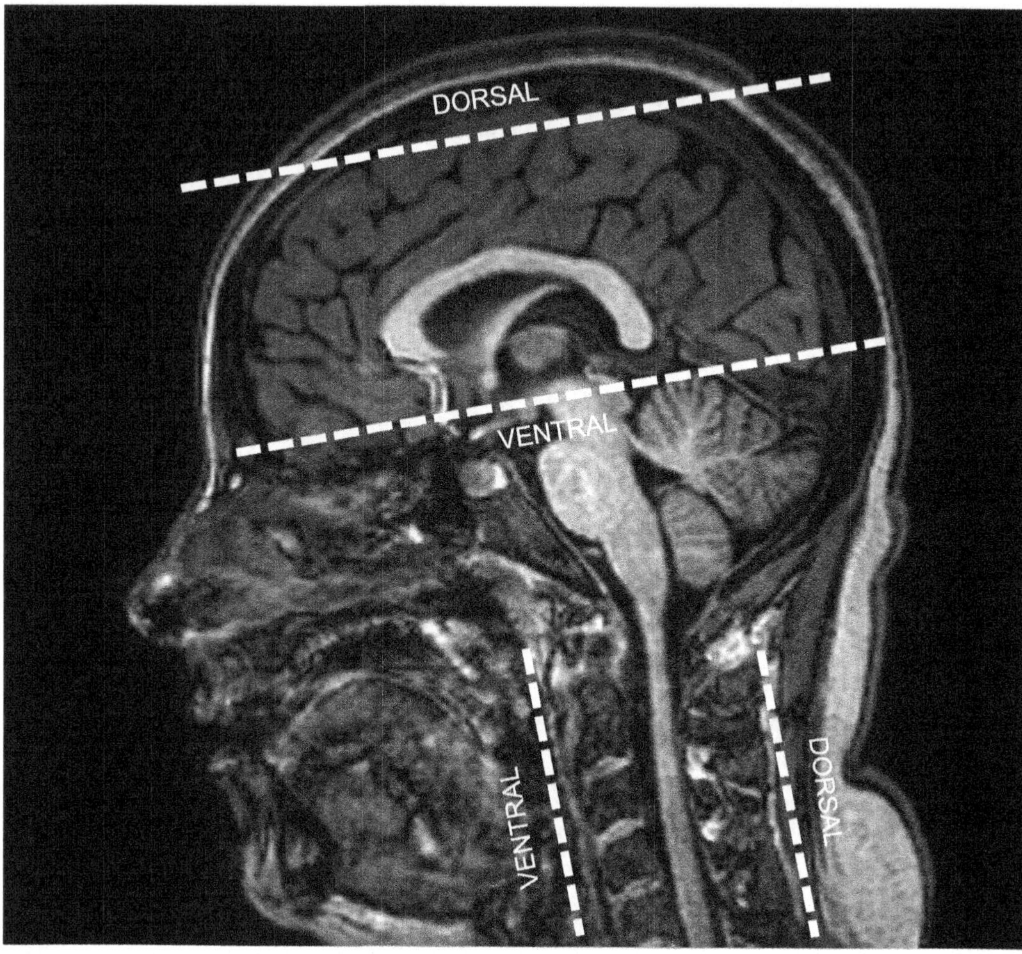

Fig. 4 Sagittal T1 volume showing different orientations of the terms 'dorsal' and 'ventral' in the spine and brain, reflecting folding of the common embryological precursor

surface and superior surfaces respectively whereas in the brainstem it refers to the anterior and posterior surfaces respectively (Fig. 4).

3 Basic Brain Structure

The brain is divided into two main compartments by the tentorium cerebelli, a fibrous dural reflection that separates the posterior fossa, beneath the tentorium, from the cerebrum, lying above it. This division reflects the embryological origin of different structures: all structures above the tentorium develop from the embryological diencephalon and prosencephalon, whereas infratentorial structures derive from the mesencephalon and metencephalon. In the mature brain the posterior fossa contains the brainstem and cerebellum, whereas the supratentorial compartment contains the cerebral cortex, deep grey matter and hypothalamus (Fig. 5) [1].

The mature adult brain is arranged with the heavily folded cortex around the outside or surface with white matter beneath it. There is further grey matter deep within the brain in the form of the basal ganglia, amgydala and thalami. The cerebellum is similarly arranged with the cerebellar cortex arranged around the outside or surface and deep grey matter nuclei situated centrally (Fig. 6) [2].

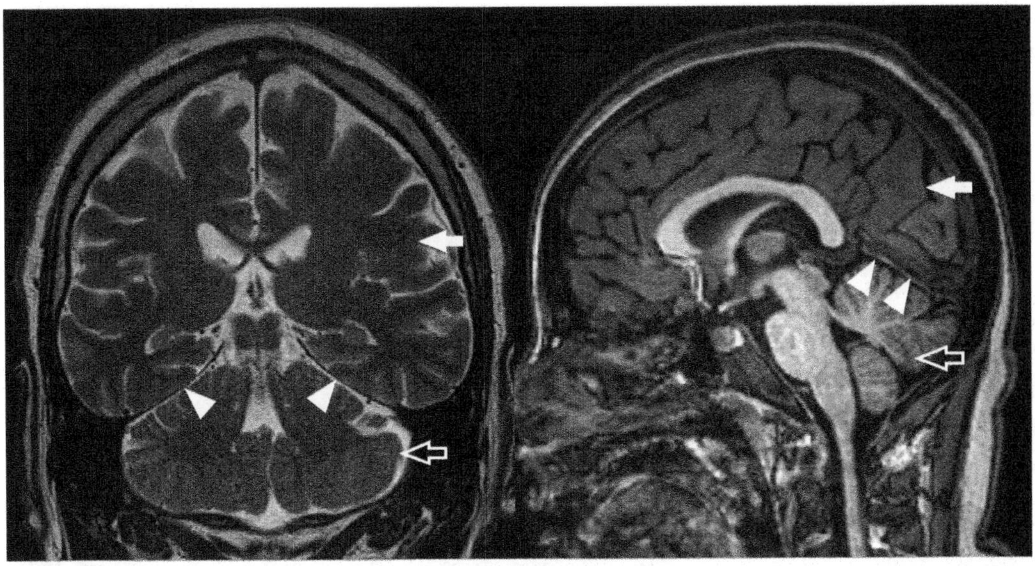

Fig. 5 Coronal T2 image (*left*), sagittal T1 volume (MPRAGE) (*right*). The supratentorial compartment (*solid white arrow*) is located above the tentorium cerebelli (*white arrowheads*); the infratentorial compartment is located beneath the tentorium (*open white arrow*)

The basal ganglia include the globus pallidus, caudate nucleus and putamen. The latter two are sometimes referred to as the corpus striatum, whereas the combination of the globus pallidus and putamen are referred to as the lentiform nucleus. The hockey-stick-shaped white matter between the caudate and lentiform nucleus and thalamus is the internal capsule (Fig. 7) [2].

4 Supratentorial Landmarks

The brain can be divided in lobes with broadly similar functions based on external landmarks formed by various prominent sulci. The frontal lobe is separated from the parietal lobe by the central sulcus, sometimes called the Rolandic fissure. In addition to being responsible for executive functions such as planning, motivation etc., the frontal lobe contains the primary and supplementary motor areas and Broca's area, responsible for speech production in the inferior frontal gyrus. The parietal lobe contains the primary somatosensory cortex, located in the postcentral gyrus but also performs many complex integrative functions including calculation, orientation in space and visual processing (Fig. 8) [3].

The largest 'sulcus' visible on the side of the brain is the Sylvian fissure. This landmark separates the temporal lobe, below, from the frontal and parietal lobes superiorly. The temporal lobe contains the primary auditory cortex and medially the limbic system structures involved in memory formation—primarily the hippocampus and associated gyri. The temporal lobes also contribute to visual processing (particularly visual form—e.g. face recognition) and are also important in understanding of speech—Wernicke's area is located at the junction of the superior temporal lobe and parietal lobe at the back of the Sylvian fissure (Fig. 9).

The occipital lobe is located in a paramedian location posteriorly and contains the primary visual cortex. The junction with the parietal lobe is marked on the medial surface by the parieto-occipital sulcus; the junction with the temporal and parietal lobe on the surface of the brain is less well defined (Fig. 10).

The limbic system is often referred to as a separate lobe of the brain although is made of multiple separate structures. It is one of the oldest

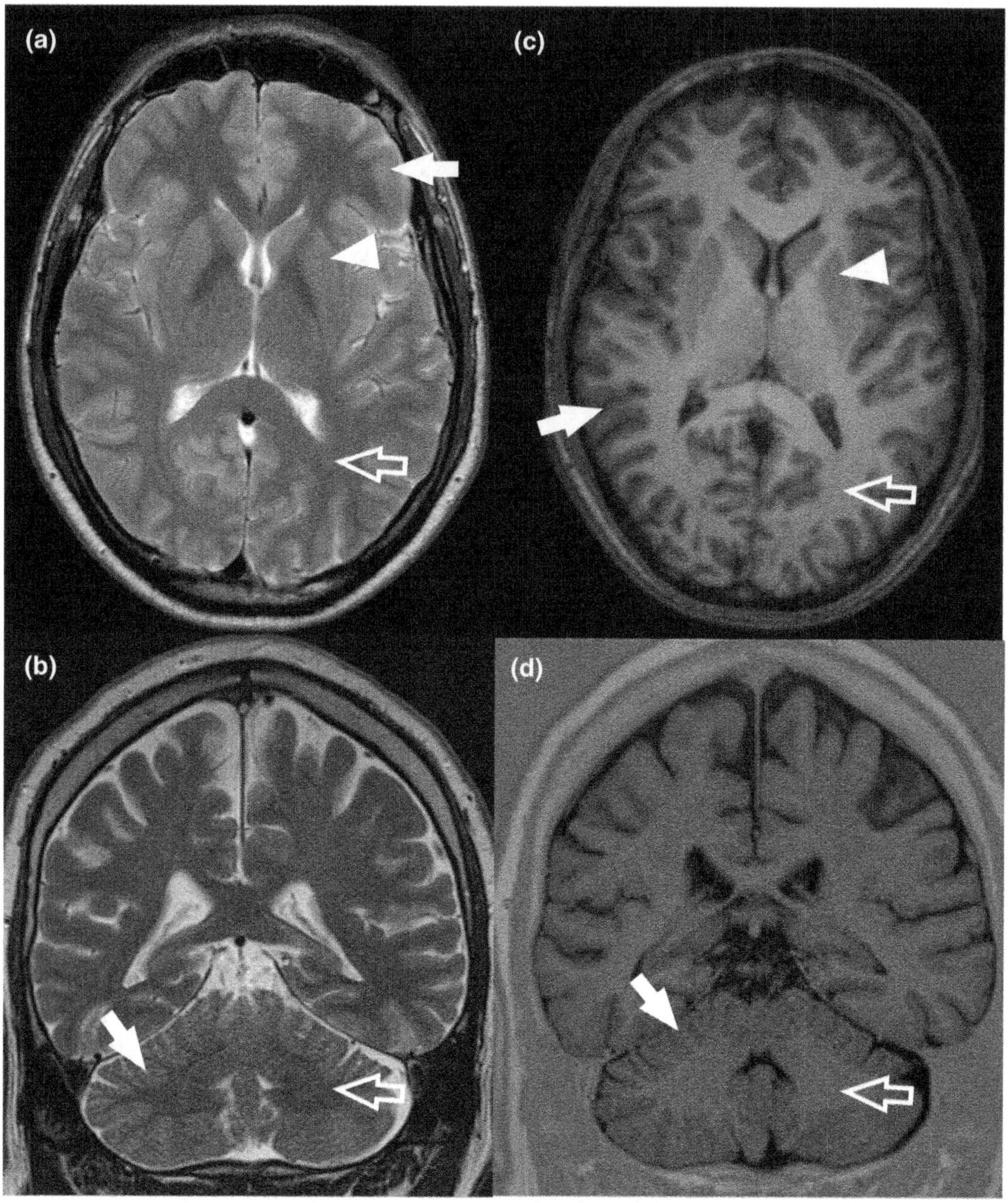

Fig. 6 Axial and coronal T2 (**a**, **b**), axial T1 volume (**c**) and coronal T1 inversion recovery (**d**). The cortical grey matter (*solid white arrows*) is arranged around the periphery of the cerebrum and cerebellum. The deep grey matter of the basal ganglia (*arrowhead*) is located centrally. The white matter (*open white arrows*) is situated beneath the cortical grey matter and surrounds the basal ganglia

parts of the brain (phylogenetically) and contains mesocortex (4 cortical layers as opposed to the 6 found in the neocortex). In addition to the amygdala and hippocampus located in the medial temporal lobe/temporal lobe uncus, the limbic system incorporates the fornix (the hippocampal outflow tract), the cingulate gyrus—the gyrus wrapping around the corpus callosum and also the insular cortex, deep within the Sylvian fissure (Fig. 11) [4]. Many of these structures are part of

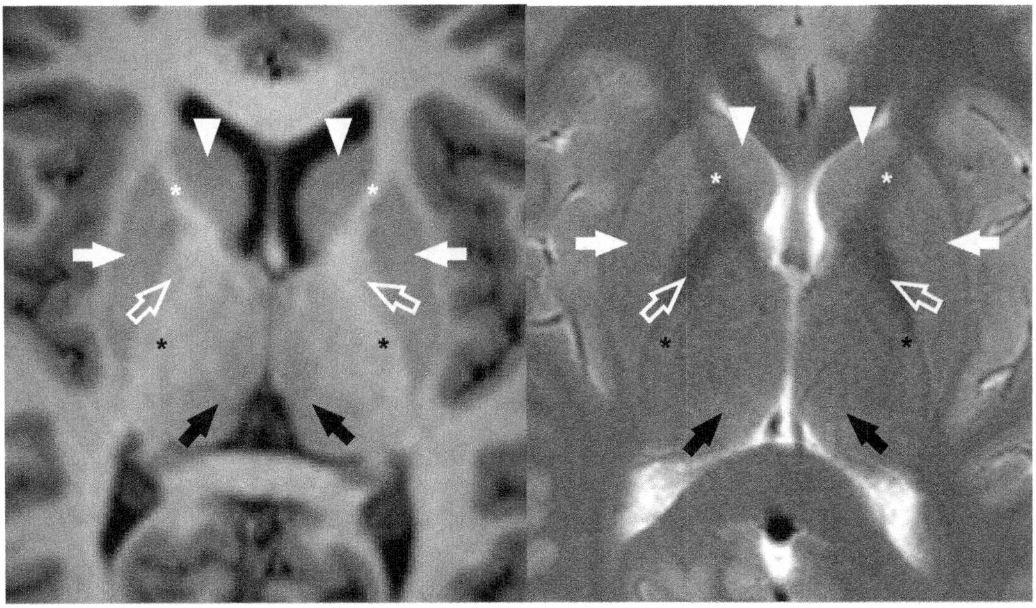

Fig. 7 Axial T1 volume (*left*) and T2 (*right*) showing the heads of the caudate nucleus (*white arrowheads*), the putamen (*white arrows*), the globus pallidus (*open white arrow*, best visualised on T2 due to hypointense mineralisation), the thalami (*black arrows*), the anterior limb of the internal capsule (*white asterisks*) running between the caudate and putamen and the posterior limb of the internal capsule (*black asterisks*) running between the putamen and thalamus

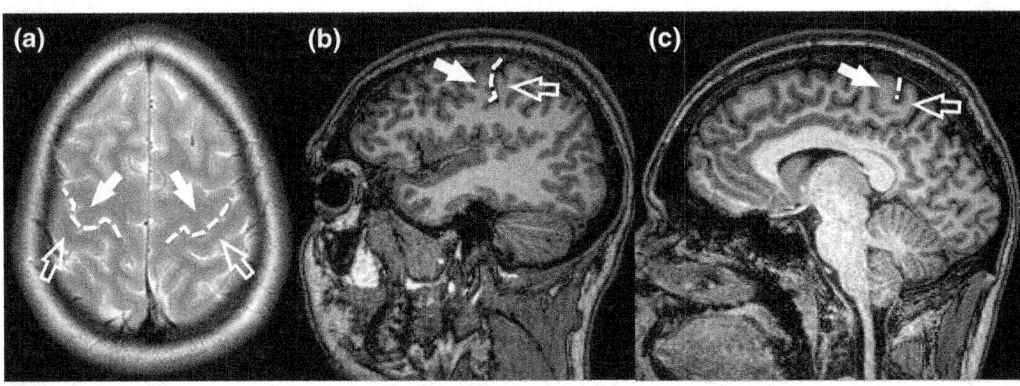

Fig. 8 Axial T2 (**a**), parasagittal MPRAGE T1 volume (**b**, **c**) showing position of the central sulcus (*dashed line*), separating the pre-central gyrus (*solid white arrow*) from postcentral gyrus (*open white arrow*). The frontal lobe is anterior to the central sulcus; the parietal is posterior to the central sulcus. Note how far posteriorly the frontal lobe extends on the sagittal imaging

the Papez circuit involved in the laying down of new memories [5]; the amygdala is also involved in fight-or-flight responses and emotional responses [6].

The fluid filled spaces within the centre of the brain are called the ventricular system and contain cerebrospinal fluid (CSF). The ventricles are continuous with each other via various foraminae: the lateral ventricles and third ventricles in the supratentorial compartment are connected via the foramen of Monro; the third and fourth ventricle are connected via the aqueduct of Sylvius which

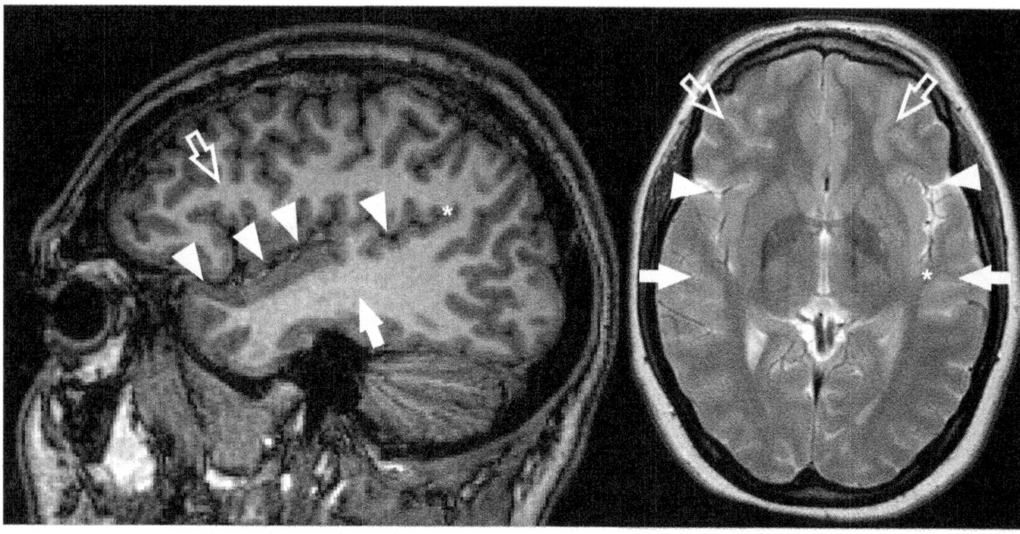

Fig. 9 Para-sagittal T1 volume (*left*) and axial T2 (*right*). The sylvian fissure (*white arrowheads*) separates the temporal lobe (*solid white arrow*) from the frontal lobe anteriorly (*open white arrow*) and parietal lobe posteriorly (*black arrow*). Wernicke's area is shown by the *asterisk*

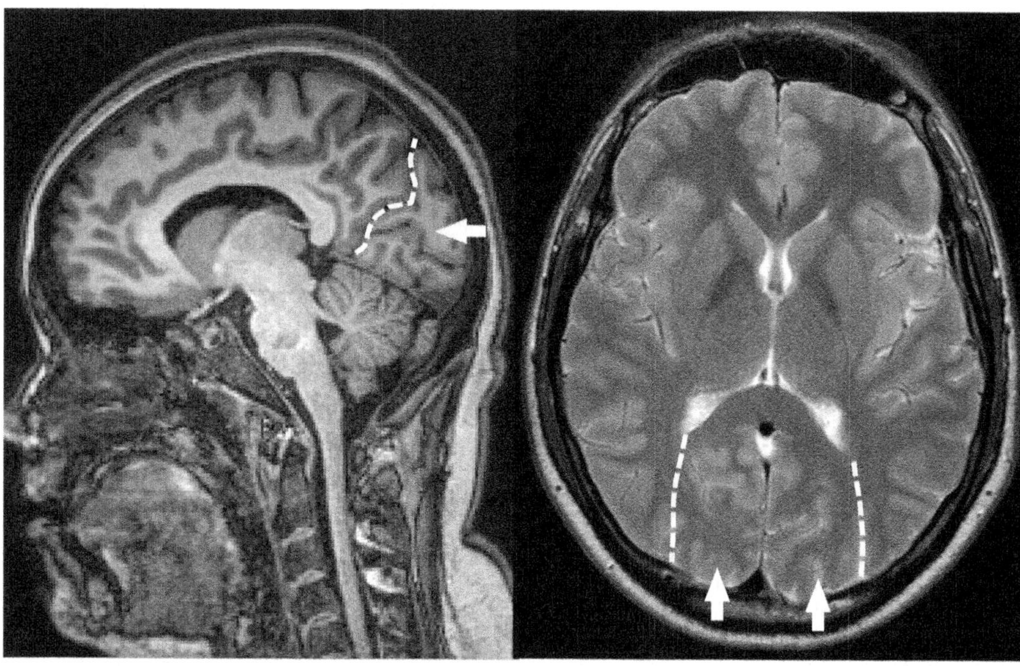

Fig. 10 Parasagittal T1 volume (*left*) and axial T2 (*right*) showing the occipital lobes (*white arrows*). The boundary between the parietal and occipital is clearly visible in the sagittal plane, marked by the parieto-occipital sulcus (*dashed line, left*). The boundary in the axial plane is less well defined and is approximated by the territory of supply of the posterior cerebral artery (approximate location shown by *dashed lines, right*)

runs through the midbrain. The fourth ventricle is connected to the central canal of the spinal cord inferiorly and also the subarachnoid space over the cerebral convexities via two exit foramina: the foramen of Magendie in the midline and foramina of Luschka, laterally (Fig. 12) [2].

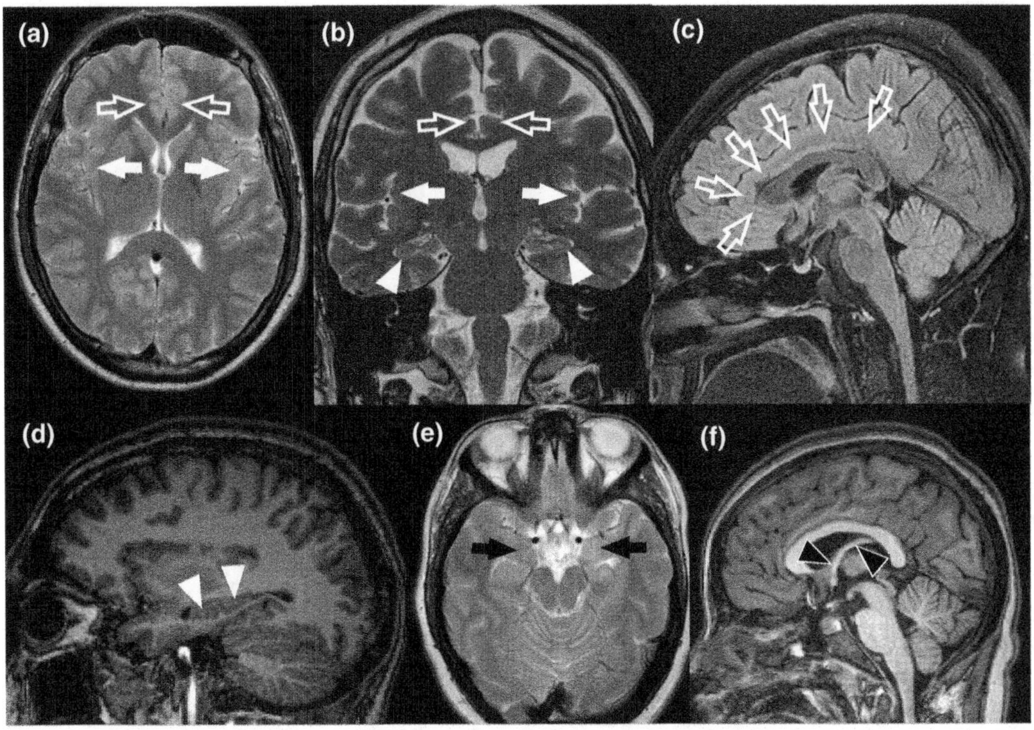

Fig. 11 Limbic system. Axial T2 (**a**, **e**), coronal T2 (**b**), sagittal FLAIR (**c**) and sagittal T1 volume (**d**, **f**). The insular cortex is shown by the *closed white arrows*; the cingulate cortex (*open white arrows*) wraps around the corpus callosum in the midline. The hippocampus is demonstrated in the medial temporal lobes (*white arrowheads*). The amygdala is the small grey matter nucleus in the medial temporal lobe immediately anterior to the hippcampus (*black arrows*). The outflow tract of the hippocampus, the fornix, is shown by the *black arrowheads* (**f**)

White matter

Although it appears fairly homogenous on standard structural imaging (e.g. T1 and T2 weighted MRI) the cerebral white matter is a highly ordered structure containing multiple different fibres and tracts running in different directions. These larger tracts are elegantly displayed by diffusion tensor imaging (DTI) with colour coding representing the main direction of the tracts displayed (directionally encoded colour—DEC) (Fig. 13) [7].

Commissures are tracts that connect the two different hemispheres of the brain, the larges of which is the corpus callosum. Other commissures include the anterior commissure, connecting the temporal lobes; the posterior commissure, at the posterior limit of the third ventricle, the hippocampal commissure (where the two fornices temporarily join) and the habenular commissure in the pineal region (Fig. 14).

Fasciculi are tracts connecting different regions of the brain in the same hemisphere. In the supratentorial brain these include the superior and inferior longitudinal fasciculi—running in an anteroposterior direction; and the uncinate fasciculus, connecting the ipsilateral frontal and temporal lobes via the external capsule (Fig. 15) [7].

Major tracts run in a superoinferior direction connecting the supratentorial brain with the cerebellum, brainstem and spinal cord. The largest of these are the corticospinal tracts—the motor output of the brain to the spinal cord (Fig. 16) [8]. Ascending sensory pathways through the brainstem and supratentorial brain will be considered later.

Although difficult to visualise, even with DTI, it is important to be aware of the functional concept of

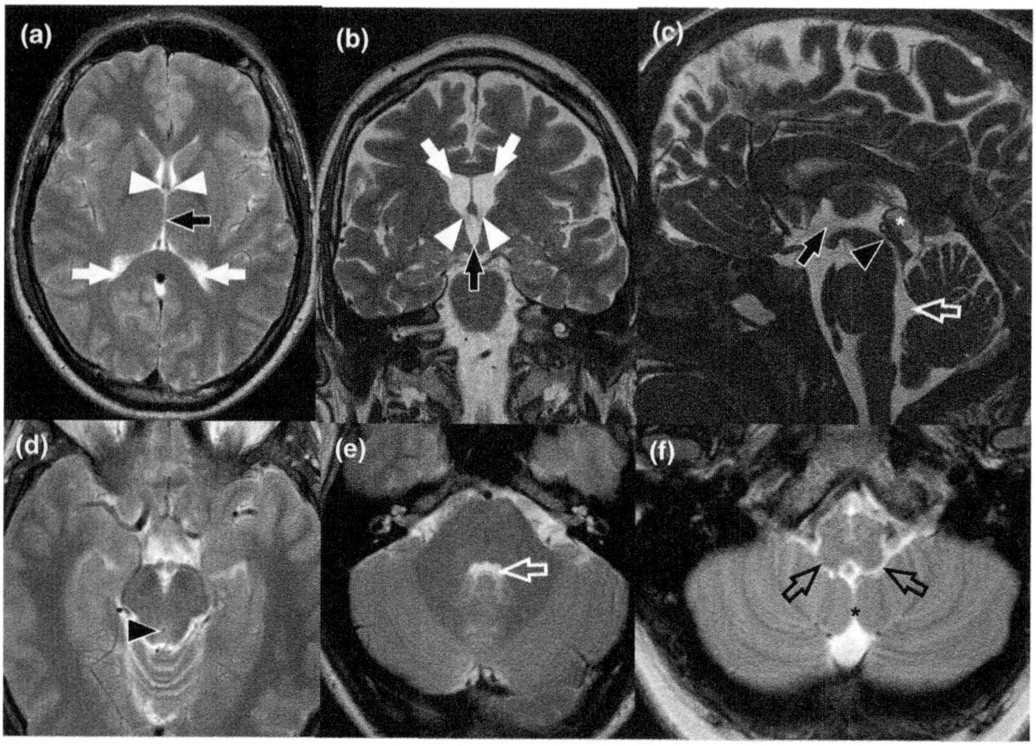

Fig. 12 Axial T2 (**a**, **d–f**), coronal T2 (**b**), sagittal 3d T2 (**c**). CSF is produced in the lateral ventricles (*solid white arrows*) and passes through the foramina of Monro (*white arrowheads*) into the third ventricle (*black solid arrow*). From there it passes through the cerebral aqueduct (of Sylvius) (*black arrowheads*) into the fourth ventricle (*white open arrow*). From the fourth ventricle it passes into the subarachnoid space around the surface of the brain and spinal cord via the foramina of Magendie (*black asterisk*) and Luschka (*open black arrows*). Incidental pineal cyst noted (*white asterisk*)

cortical–subcortical loops connecting the cortex and deep grey matter structures. These connections between the thalami and basal ganglia and the cortex are reciprocal and continually supply feedback to cortical areas, also integrating sensory information from the spinal cord in the thalami. For example fibres from the primary motor cortex terminate on the putamen which via either the direct or indirect pathways connect to the internal segment of the globus pallidus (integrating input from the subthalamic nucleus and substantia nigra); the GP interna output travels to the anterior/lateral nuclei of the thalami which in turn project back to the primary motor cortex. This system serves to modulate and fine tune movement [2, 3]. The thalami contain multiple internal nuclei with different reciprocal cortical connections; their functional importance with regard to pain perception will be considered later.

5 Infratentorial Landmarks

The brainstem can be divided into three segments, running cranially to caudally: the midbrain, pons and medulla. The cerebellum is located dorsal to the brainstem spanning the distance from the intercollicular sulcus of the midbrain to the obex (inferior limit) of the fourth ventricle behind the medulla (Figs. 17 and 18) [9].

The midbrain contains the IIIrd and IVth cranial nerve nuclei, the structures of the substantia nigra and red nuclei. The pons contains the reticular activating system and a diffuse network of other pontine nuclei along with the cranial nerve nuclei of the Vth to the VIIth nerves (at the pontomedullary junction). The medulla contains

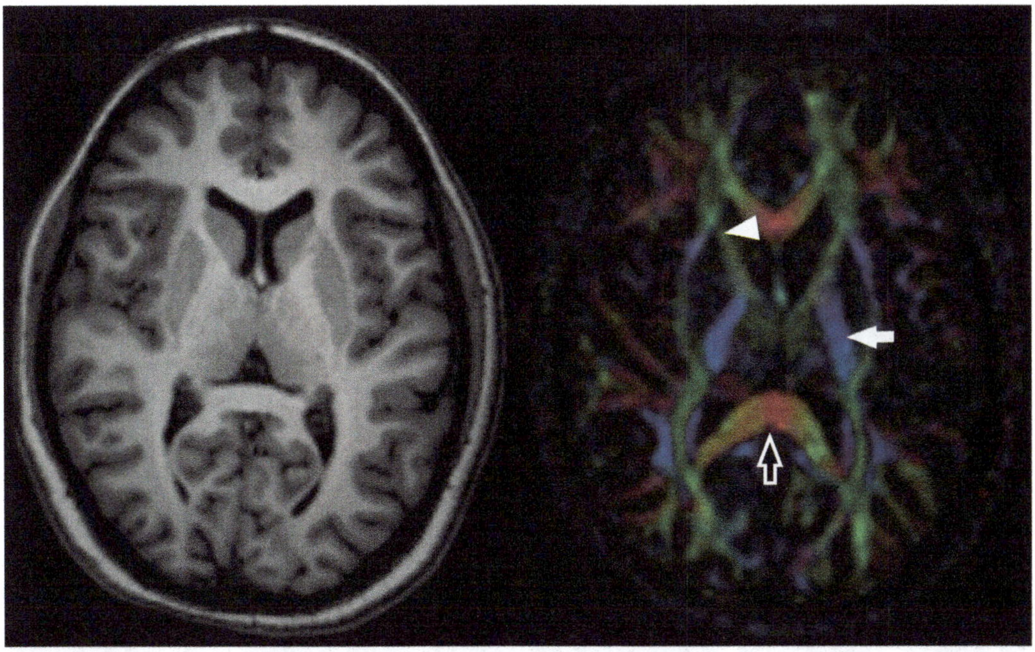

Fig. 13 *Left*, axial T1 volume, *right* DTI directionally encoded colour map. The white matter on the T1 image appears featureless but DTI allows identification of the orientation of fibre tracts: the corticospinal tracts (*white arrow*) are demonstrated in a superoinferior orientation coloured *blue*; the corpus callosum (*open white arrow*) is coloured *red* passing from left to right; the anterior limb of the internal capsule (*arrowhead*) is oriented in the anteroposterior direction and coloured *green*

the remaining VIIIth–XIIth cranial nerve nuclei along with important cardiorespiratory centres.

In addition to the corticospinal tracts running through the brainstem to the spinal cord, other longitudinal tracts can be identified. The medial longitudinal fasciculi connect the cranial nerve nuclei of the IIIrd, IVth and VIth nuclei, and along with the parapontine reticular formation are involved in coordinating eye movements. The sensory tracts running through the brainstem to the thalamus will be described in detail later.

6 Blood Supply of the Brain

The arterial supply to the brain can be split into vessels arising from the carotid, or 'anterior' circulation and those arising from the vertebrobasilar, or 'posterior' circulation. These two systems communicate via a vascular ring situated at the base of the brain called the circle of Willis. The major blood vessels supplying the supratentorial brain are the anterior, middle and posterior cerebral arteries. The largest branch is the middle cerebral artery (MCA) which supplies the majority of the frontal, temporal and parietal lobes along with the basal ganglia. The anterior cerebral arteries (ACAs) supply the medial part of the cerebral hemispheres anteriorly. The posterior cerebral arteries (PCAs) supply the medial surface of the brain posteriorly (the occipital lobes) and also the posterior aspect of the thalami (Fig. 19) [10].

The posterior fossa structures are supplied either directly, or via branches of the vertebral and basilar arteries. The three largest cerebellar arteries are the posterior–inferior cerebellar artery (PICA), the anterior–inferior cerebellar artery (AICA) and superior cerebellar artery (SCA). Much of the brainstem is supplied by small, perforating branches of the basilar artery directly (Fig. 20).

The venous drainage system is made up of the deep cerebral veins and dural venous sinuses. Superficial cortical veins also drain into the dural venous sinuses, which eventually join together at

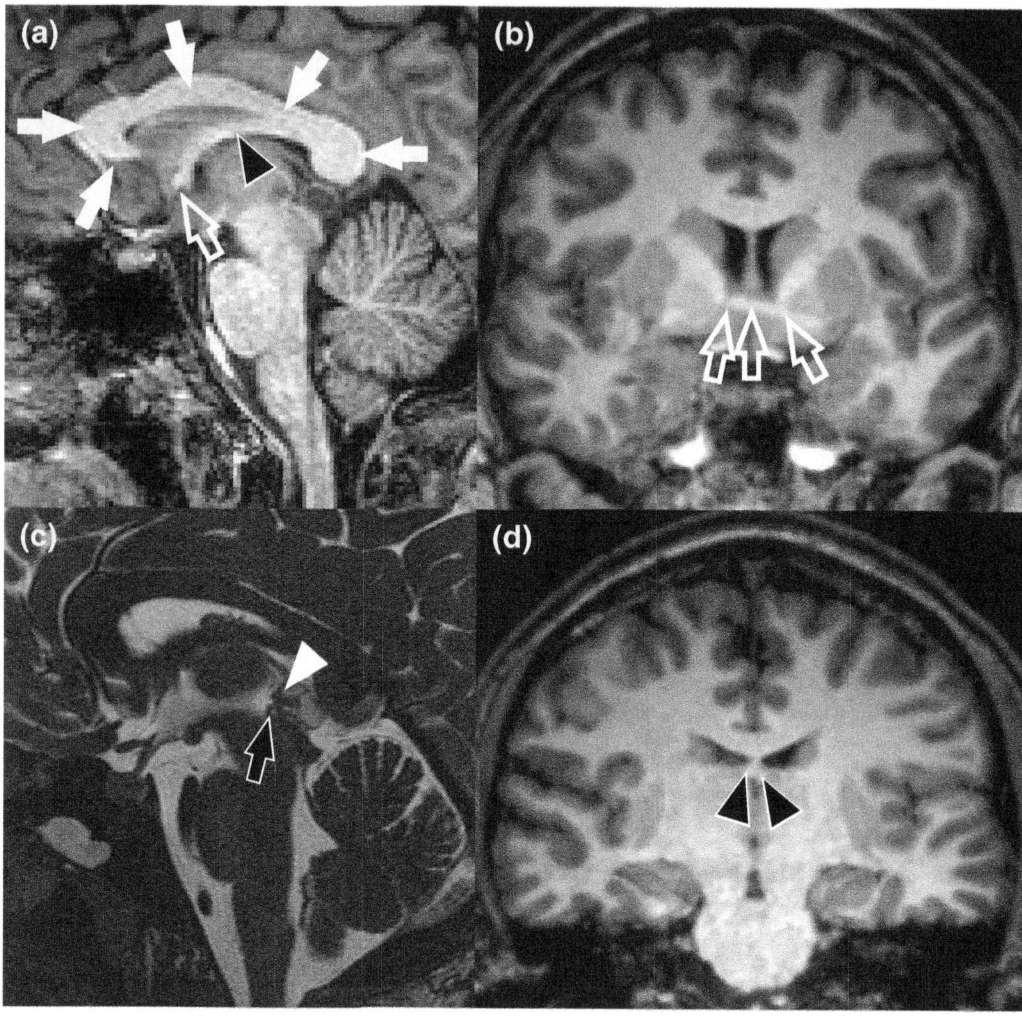

Fig. 14 Sagittal T1 (**a**), coronal T1 (**b**, **d**) and sagittal high-resolution 3d T2 (**c**). The components of the corpus callosum are named from anterior to posterior (*white arrows*) the rostrum, genu, body, isthmus and splenium. The anterior commissure (*open white arrows*, **a**, **b**) connects the two temporal lobes and can be seen immediately anterior to the fornix. The hippocampal commissure connects the fornices (*black arrowheads*, **a**, **d**). The posterior commissure (*black arrow*, **c**) and habenular commissure (*white arrowhead*, **c**) are found at the posterior margin of the third ventricle

the jugular foramen where venous blood enters the jugular veins (Fig. 21).

7 Sensory Pathways—Connecting the Spinal Cord to the Cerebral Cortex

As described elsewhere, sensory information from the periphery arrives at the medulla, via the spinal cord separated into two broad categories of modality: the spinothalamic system, carrying information regarding crude touch, pain and temperature, and the dorsal column system containing information regarding discriminatory touch, light touch/vibration and joint position sense. The spinothalamic system crosses at, or within one or two levels of entering the spinal cord so that in the medulla the right spinothalamic tract represents relevant sensory information from the contralateral, left side of the body. These are already second

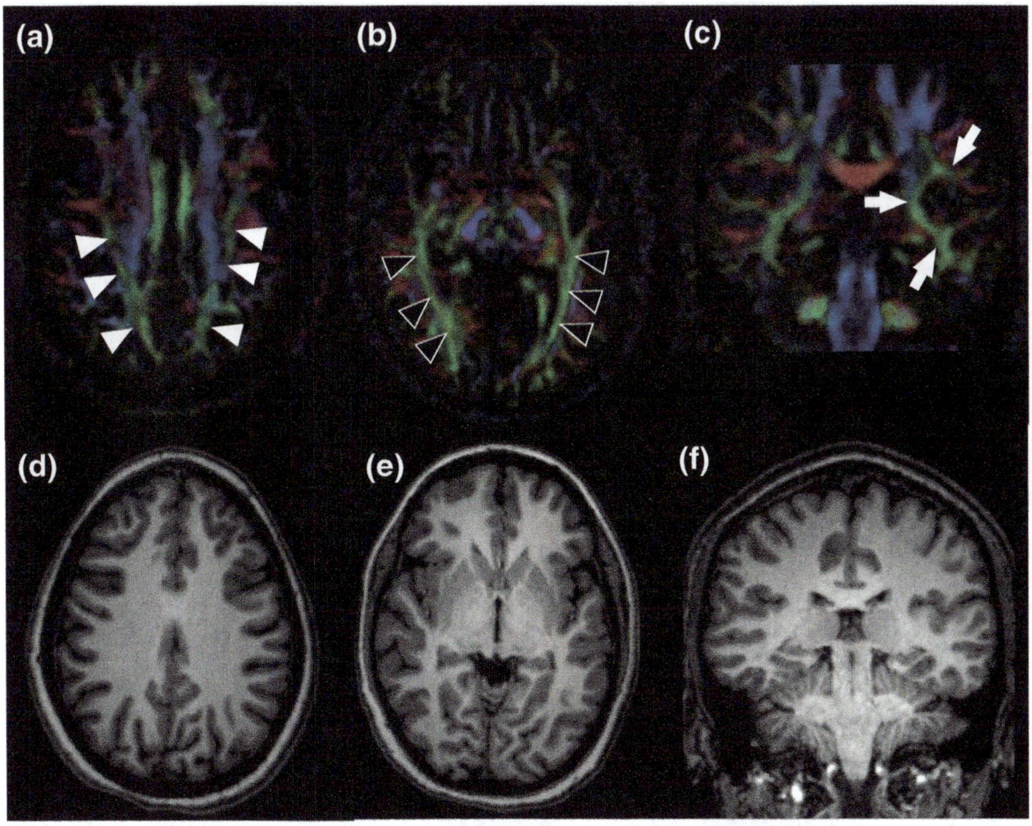

Fig. 15 Axial (**a**, **b**) and coronal (**c**) directionally encoded DTI colour maps; **d**, **e** (axial) and **f** (coronal) T1 volumes. Part of the superior longitudinal fasciculus is shown in *green* in a (*white arrowheads*) connecting the frontal and parietal lobes; the inferior longitudinal fasciculus connecting the occipital and temporal lobes is shown in *green* in **b** (*black arrowheads*). The arcuate fasciculus is shown in **c** in *green* (*white arrows*) connecting the temporal and parietal lobes

order neurones, having synapsed at, or within a few levels of entry. In distinction, the dorsal column system remains ipsilateral to the side of entry up until the level of the medulla where the first order neurones terminate on one of two nuclei—the nucleus gracilis, located medially (carrying dorsal column information from the lower half of the body) and the nucleus cuneatus, located laterally (carrying dorsal column information from the upper half of the body) (Fig. 22) [11].

The output of the nuclei gracilis and cuneatus is the medial lemniscus—these fibres now cross to the contralateral side of the brainstem in a paramedian location and are found in an antero-posterior configuration at the level of the obex of the fourth ventricle (Fig. 23).

Having crossed, the medial lemnisci ascend through the pontine tegmentum, located still near the midline, anterior to the fourth ventricle and pontine reticular formation. They are now joined by the spinothalamic tracts located laterally (Fig. 24).

As the sensory tracts ascend further into the midbrain they move more laterally, now being found near the lateral edges of the midbrain; the spinothalamic tracts rotate to take a more posterior position, still in close apposition with the medial lemnisci (Fig. 25) [2, 3, 12].

The destination of these second order neurones in the medial lemnisci and spinothalamic tracts, carrying sensory information from the body, is the ventroposterolateral (VPL) nucleus

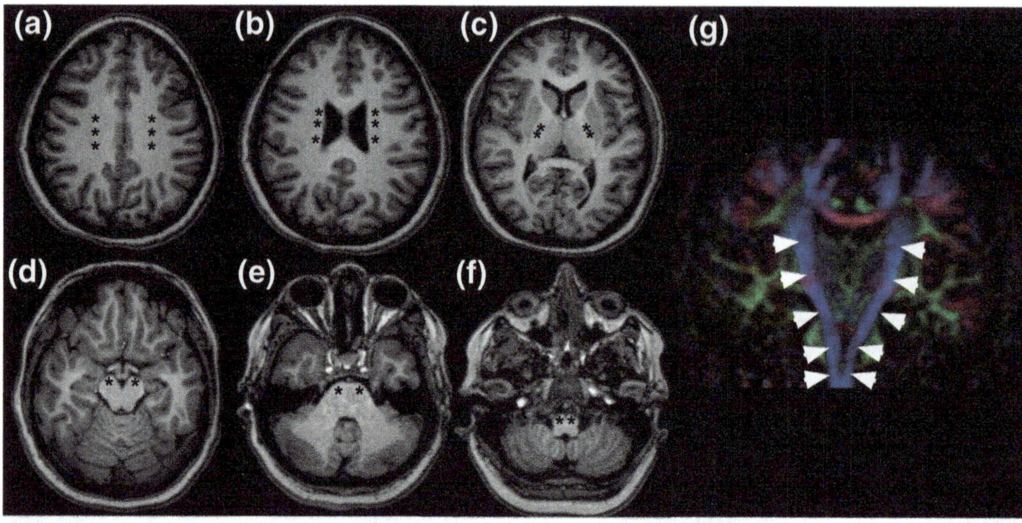

Fig. 16 Axial T1 volume showing location of the corticospinal tracts (*black asterisks*) at the level of the centrum semiovale (**a**), corona radiata (**b**), posterior limb of the internal capsule (**c**), cerebral peduncles (**d**), pons (**e**) and medullary pyramids (**f**). The tracts are outlined in *blue* (*white arrowheads*) in the coronal directionally encoded DTI map (**g**)

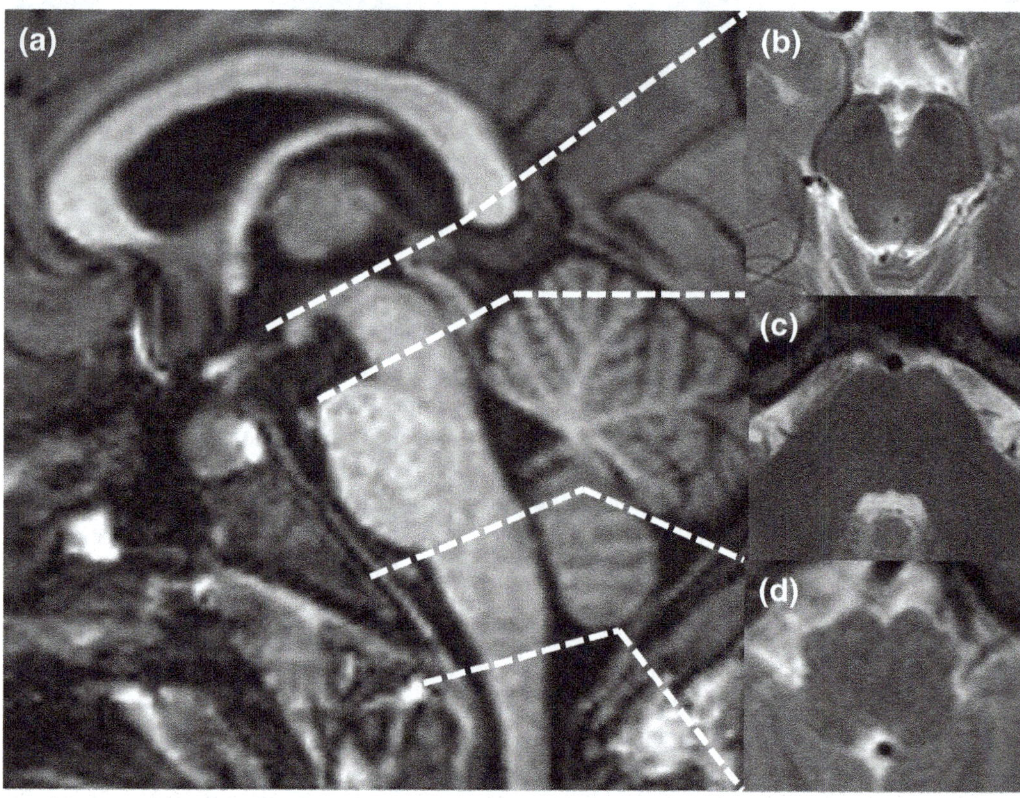

Fig. 17 Sagittal T1 volume (**a**) and axial T2 (**b–d**) showing components of the brainstem: the midbrain (**b**), the pons (**c**) and medulla (**d**)

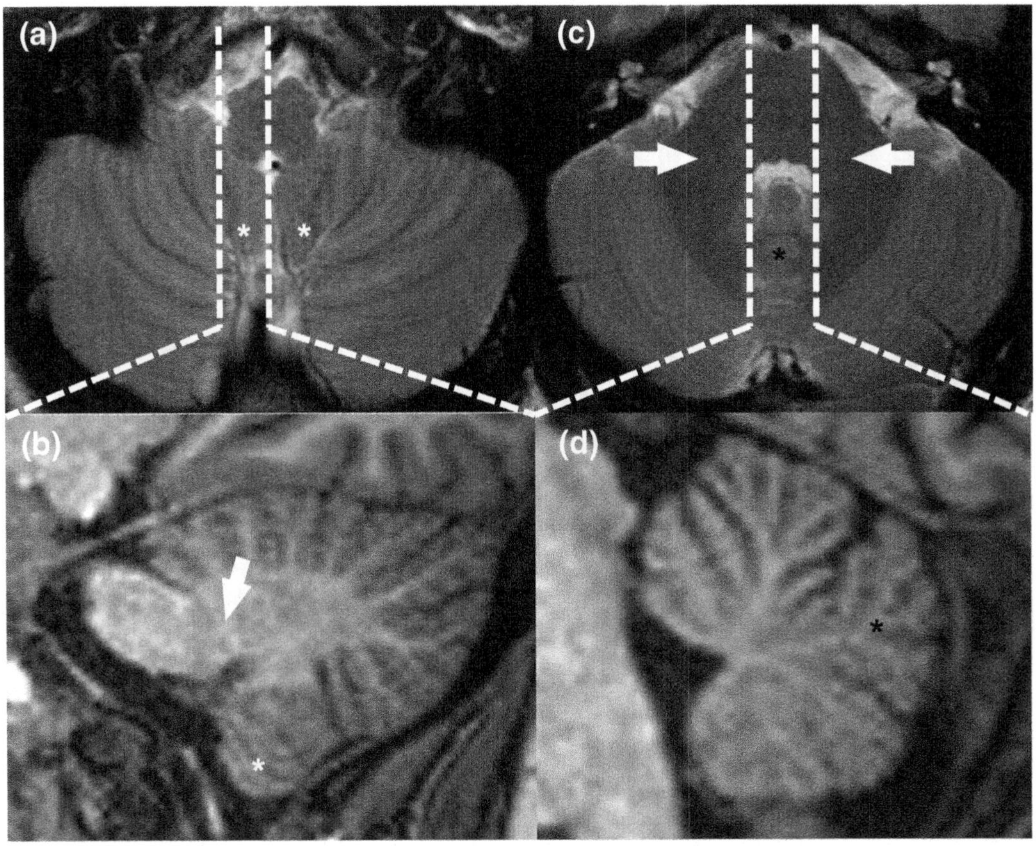

Fig. 18 Axial T2 (**a**, **c**) and parasagittal (**b**) and sagittal T1 volume (**d**) showing the cerebellar tonsils (*white asterisk*), cerebellar vermis in the midline between the two hemispheres (*black asterisk*) and the middle cerebellar peduncles (*white arrows*) connecting the cerebellum to the pons

of the thalamus (facial sensation will be described in detail subsequently) (Fig. 26).

Having synapsed in the VPL nucleus third order neurones project to the primary somatosensory cortex, located in the postcentral gyrus of the parietal lobe via the internal capsule. The posterior limb of the internal capsule, containing the somatosensory projection fibres runs between the thalamus medially and lentiform nucleus laterally (Fig. 27).

From the internal capsule, fibres project to the primary somatosensory cortex (S1), located along the postcentral gyrus in the parietal lobe. Some fibres also terminate in the secondary somatosensory cortex in the parietal operculum which will be described later, along with the somatosensory association cortex more posteriorly in the parietal lobes (Fig. 28).

The representation of the type and intensity of sensory stimuli is encoded in S1 with anatomical regions showing differing areas of activity according to the sensory homunculus. The foot is located medially, dipping down into the interhemispheric fissure, whereas the hand is located on the superior convexity; the region for facial sensation is located more inferiorly on the lateral surface (Fig. 29) [11].

8 Facial Sensation

Although the broad pattern of anatomical arrangement is the same for facial as body sensation some specific differences are highlighted, particularly given the prevalence of facial pain as a clinical problem.

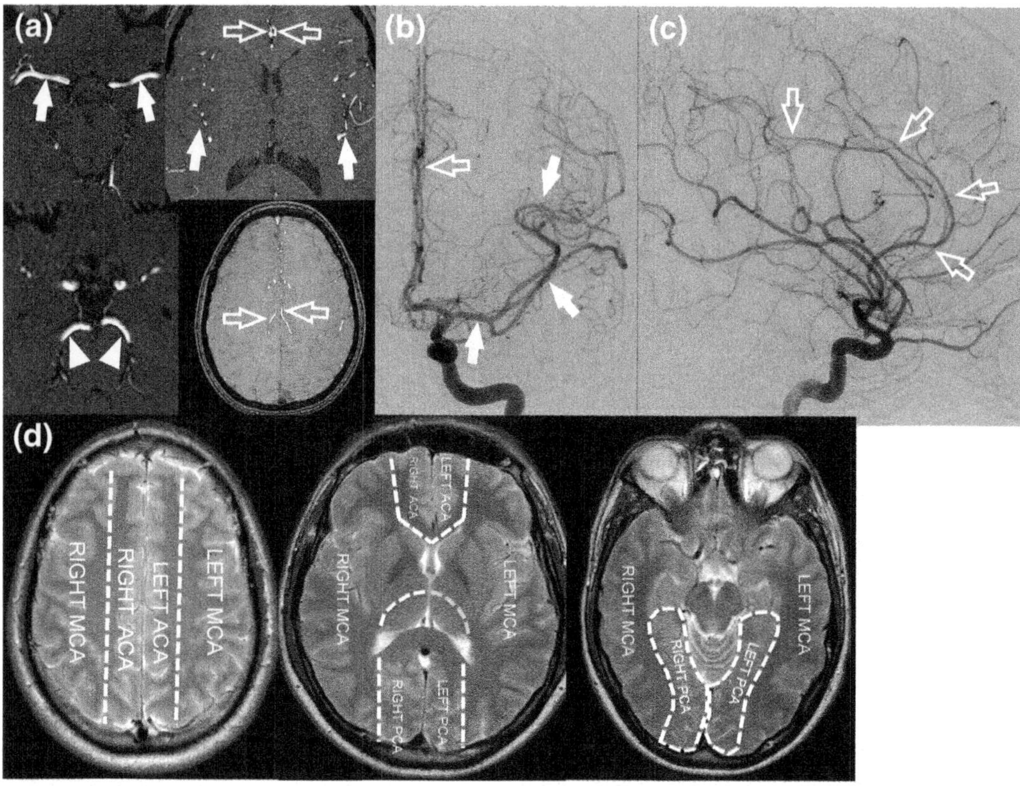

Fig. 19 Anterior circulation. Axial time-of-flight MR (**a**), anteroposterior view (**b**) and lateral view (**c**) cerebral angiogram (internal carotid artery injection) and axial T2 weighted images (**d**). The MCAs are shown by the *solid white arrows*, the ACAs by the *open white arrows* and the PCAs by the *white arrowheads*. The supratentorial vascular territories are outlined on the axial T2 weighted images

The trigeminal nerve is the main sensory nerve supplying sensation to the head/face via its three afferent branches—the ophthalmic division (V_i) covering the forehead/orbits; the maxillary division (V_{ii}) covering sensation from the mid-face and the mandibular division (V_{iii}). These join together to form the Gasserian, or trigeminal ganglion in Meckel's cave, a small continuation of the subarachnoid space from the pre-pontine cistern, medial to the temporal lobes (Fig. 30). This contains the sensory cell bodies of the sensory fibres with afferent branches extending proximally, along the cisternal portion of the trigeminal nerve to terminate on the trigeminal nucleus in the brainstem [3].

Several blood vessels are found in close proximity to the trigeminal nerve as it crosses the pre-pontine cistern/enters the pons. These are usually the superior cerebellar or anterior inferior cerebellar arteries, or sometimes a prominent petrosal vein branch. Contact, and particularly displacement, of the proximal, relatively unmyelinated portion of the trigeminal nerve (the so-called root entry zone) is associated with trigeminal neuralgia (Fig. 31) [13].

The trigeminal nucleus spans the length of the brainstem and is divided into three parts: the mesencephalic nucleus (in the midbrain), the main sensory (and adjacent motor) nucleus in the pons and the inferior extension into the medulla/upper cervical cord, called the spinal nucleus and associated tract (Fig. 32).

The mesencephalic nucleus plays a relatively small role, receiving muscular stretch information from muscles of mastication and is primarily involved in the jaw jerk reflex. The main sensory

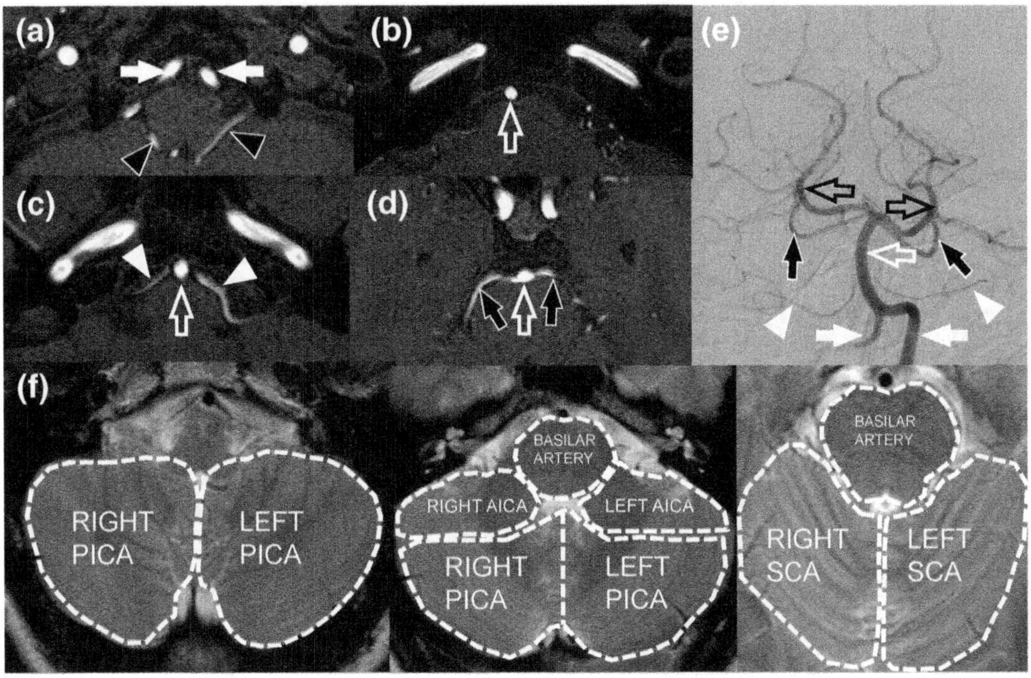

Fig. 20 Axial time-of-flight MRA at the level of the medulla (**a**), pons (**b**), pons/midbrain junction (**c**) and midbrain (**d**). AP oblique vertebrobasilar cerebral angiogram (**e**) and axial T2 weighted images with posterior fossa vascular territories (**f**). Solid white arrows show the distal vertebral arteries, *black arrowheads* show the PICAs, *open white arrow* shows the basilar artery, *white arrowheads* show the AICAs, *solid black arrows* show the SCAs, *open black arrows* show the PCAs, the terminal branches of the basilar artery

nucleus at the level of the pons is the cranial correlate of the nucleus cuneatus/gracilis in the medulla; it receives general somatosensory afferent input (i.e. light/discriminate touch, etc.) including jaw joint position sense from the ipsilateral side of the head. After synapsing efferents cross the midline as the trigeminal lemniscus and joint the medial lemniscus, ascending through the brainstem towards the thalamus. The destination nucleus in the thalamus for facial sensation is the ventroposteromedial (VPM) nucleus, located immediately adjacent to the VPL nucleus, dealing with body sensation. A small subsection of second order efferents do not cross and ascend to the ipsilateral VPM, carrying intraoral sensation, called the dorsal trigeminal tract (Fig. 33) [2, 3].

Afferent fibres containing pain and temperature information enter the pons and travel inferiorly, in the trigeminal tract to the spinal nucleus of the trigeminal nerve. Different subsections of the spinal nucleus deal with different sensory modalities but pain and temperature in particular are dealt with by the most inferior 'caudal' nucleus, which can extend down as far as the C3–4 segment. After synapsing, second order neurones cross to the contralateral ascending spinothalamic tract and then travel superiorly to the VPM nucleus of the contralateral thalamus (Fig. 34) [2].

9 Other Cranial Nerves with Sensory Components

In addition to specialist sensory information (e.g. taste), there is a contribution to general and pain sensation, in particular around the ear/external auditory canal from the facial (VIIth), glossopharyngeal (IXth) and vagus (Xth) nerves. Sensory afferent fibres from all these nerves synapse on the spinal trigeminal nucleus and tract and form second order neurone connections

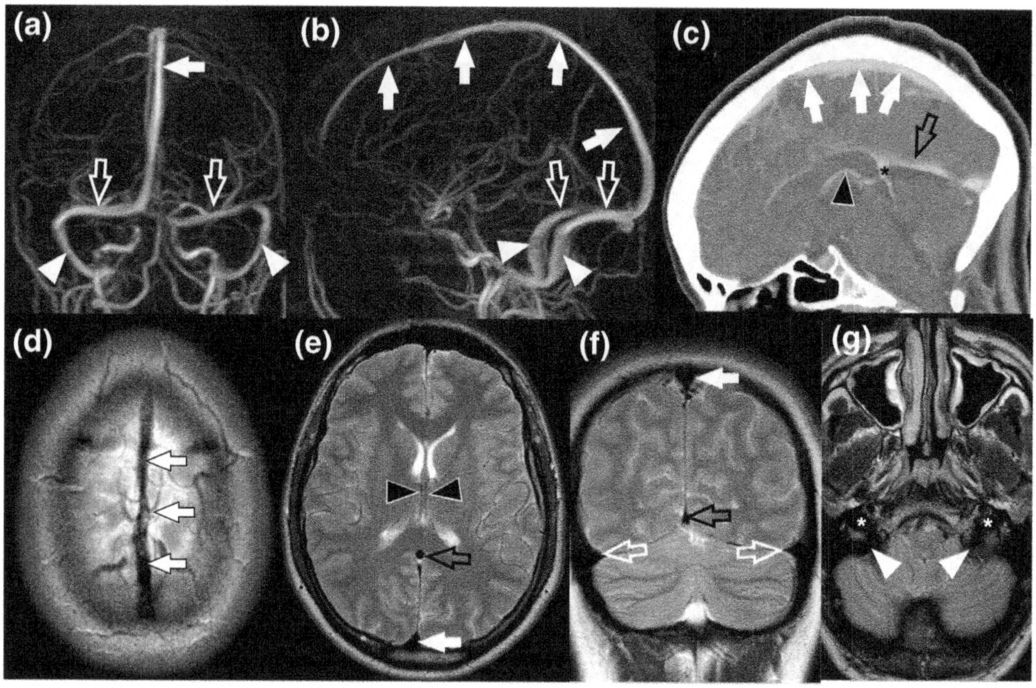

Fig. 21 Cerebral venous system. Phase contrast MR venogram (**a**), AP maximum intensity projection (**b**), lateral maximum intensity projection (**c**), sagittal CT venogram (**d**, **e**, **g**), axial T2 weighted images (**f**), coronal T2 weighted image. *Solid white arrow* Superior sagittal sinus; *open white arrow* transverse sinuses; *white arrowheads* sigmoid sinuses; *open black arrow* straight sinus; *black arrowheads* internal cerebral veins; *black asterisk* vein of Galen; *white asterisk* jugular foramen

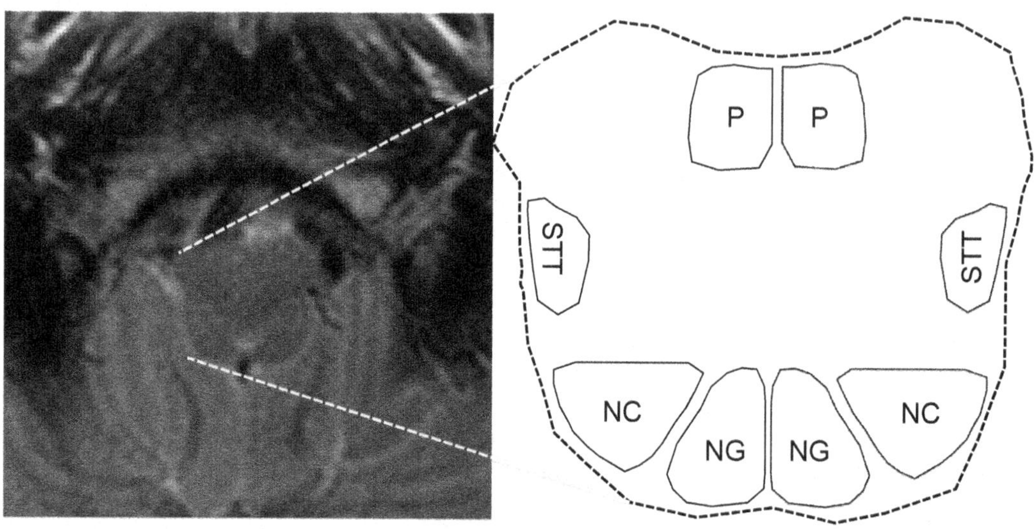

Fig. 22 Axial T2 weighted image with schematic drawing showing the relative positions of the dorsal column nuclei (*NG* Nucleus gracilis, *NC* nucleus cuneatus), the spinothalamic tracts (*STT*) and the corticospinal tracts in the medullary pyramids (*P*)

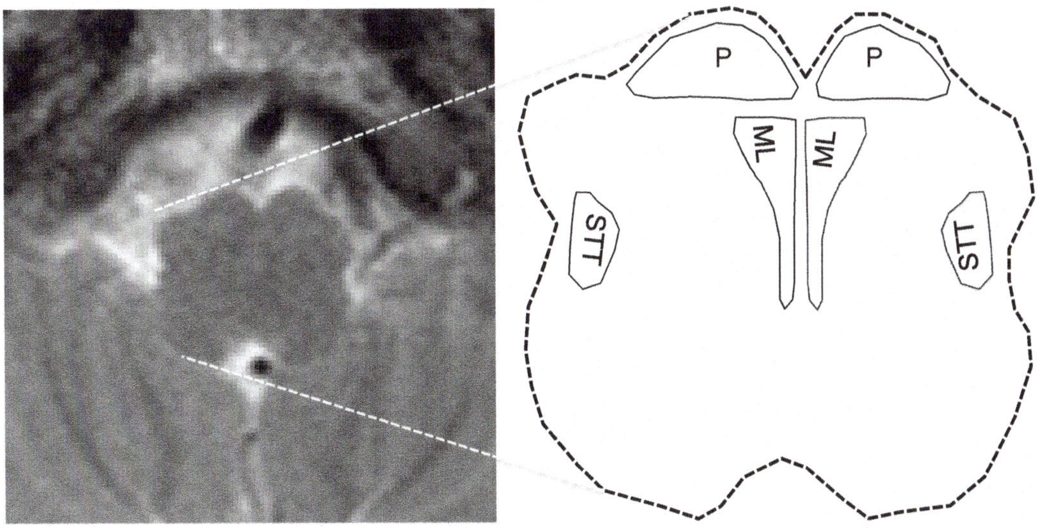

Fig. 23 The medullary pyramids (*P*) are easier to identify anteriorly. The spinothalamic tract (*STT*) remains located laterally. The medial lemnisci (*ML*) have formed in the midline as the outflow tract of the nucleus gracilis and cuneatus

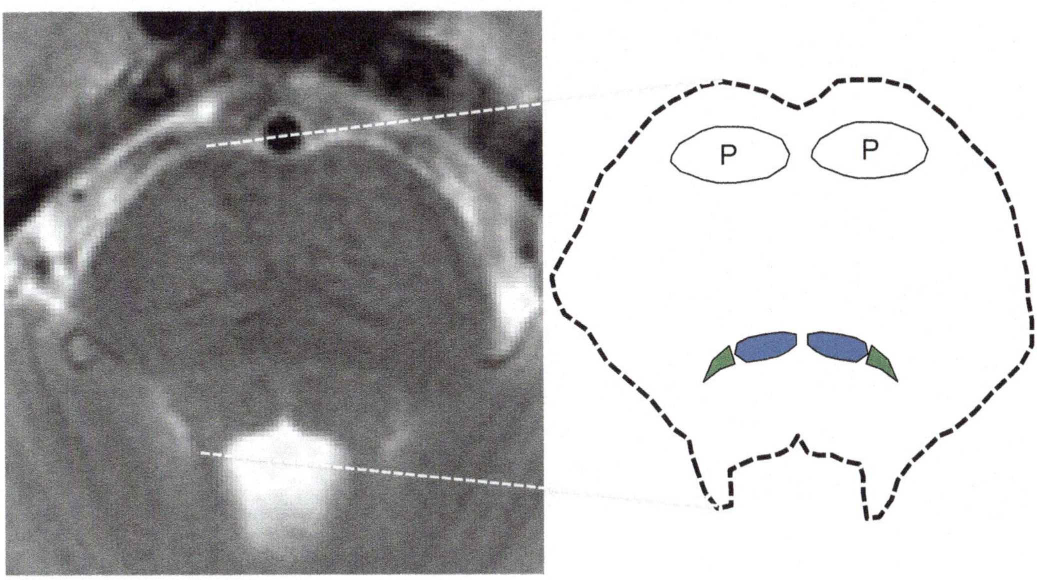

Fig. 24 The corticospinal tracts (*P*) are located in the anterior pons. The medial lemnisci (*blue*) have are now joined laterally by the spinothalamic tracts (*green*) in the pontine tegmentum

as described for the trigeminal nerve above. The more 'vague' visceral efferent fibres (e.g. sensation from mucous membranes in the gut, pharynx, larynx etc.) as well as taste information terminate in a different brainstem nucleus called the nucleus (and tract) of solitarius, located in the dorsomedial medulla at the level of the pyramids (Fig. 35) [14]:

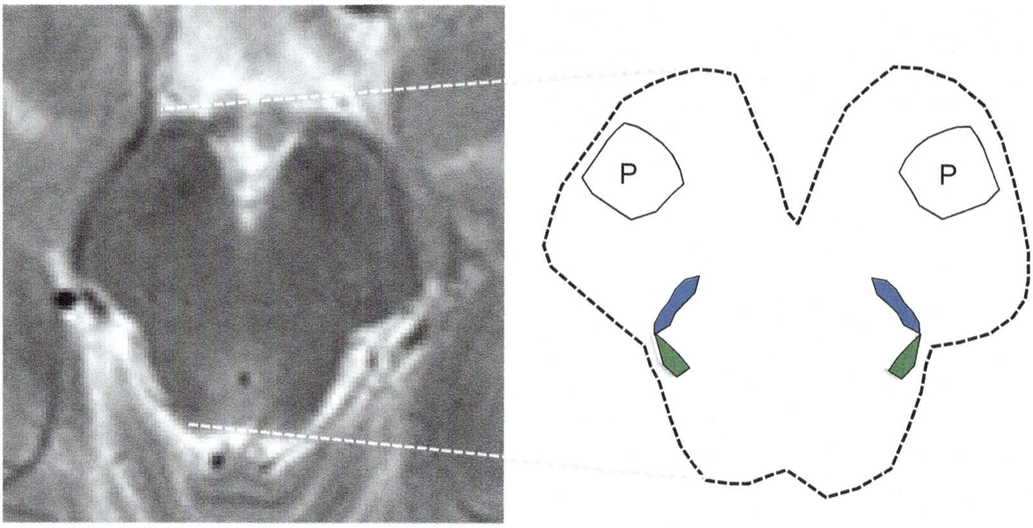

Fig. 25 The location of the corticospinal tracts (*P*) is shown in the cerebral peduncles. The medial lemnisci (*blue*) and spinothalamic tracts (*green*) are now positioned more dorsolaterally

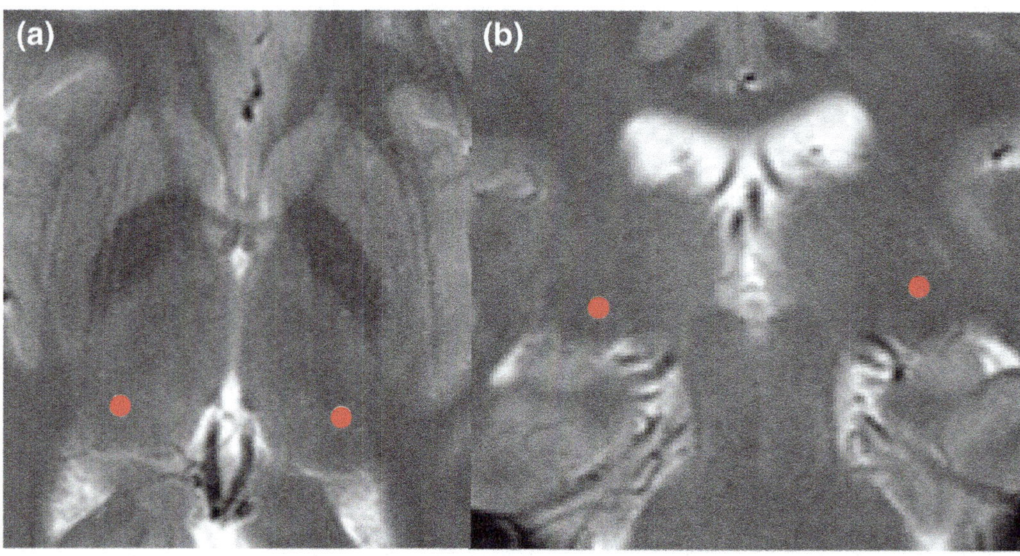

Fig. 26 Axial (**a**) and coronal (**b**) T2 weighted images showing the location of the ventral posterolateral (VPL) nucleus of the thalamus shown by the *red circles*

10 Brainstem Interactions of the Ascending Sensory System

The brainstem contains a multitude of nuclei, situated in the tegmentum—the tissue ventral to the fourth ventricle and cerebral aqueduct. Although difficult to identify on standard neuroimaging the approximate location is shown in Fig. 36.

The nuclei are interspersed with many ascending and descending pathways making multiple different connections and having many different functions. The group together are referred to as the reticular formation. The reticular formation receives direct input from the ascending, uncrossed

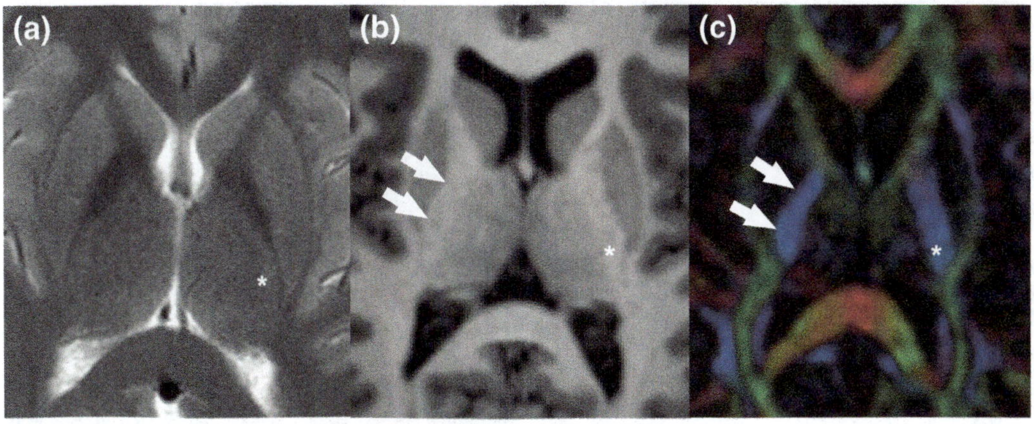

Fig. 27 Axial T2 (**a**), T1 volume (**b**) and directionally encoded colour FA map (**c**) showing the location of the somatosensory projection (*white asterisk*) in the posteriormost aspect of the posterior limb of the internal capsule (*white arrows*)

spinotectal tract—and have two output tracts—the medial (originating in the pons) and lateral (originating in the medulla) reticulospinal tracts. This output keeps spinal reflex arcs in a state of tonic inhibition but may facilitate reflexes such as withdrawal at a subconscious level in response to a painful stimulus [2, 3, 11].

The reticular formation alerts the brain to the presence of a potentially noxious stimulus via the reticulothalamic pathway—uncrossed neurones that pass from the reticular system to the intralaminar nuclei of the thalamus (see later). The reticular formation and intralaminar thalamic nuclei are together referred to as the reticular activating system (RAS).

Ascending spinothalamic tract fibres also synapse with a variety of other brainstem/diencephalic structures in a combination of three tracts that are phylogenetically older than the 'direct' anterolateral spinothalamic system and sometimes referred to as the paleospinothalamic pathway [11]. The spinotectal tract is crossed at the level of entry to the spinal cord and terminates in the region of the superior colliculus (tectal plate of the midbrain) and serves to turn the head and eyes in the direction of the stimulus (Fig. 37).

The spinohypothalamic pathway also crosses with the spinothalamic tract near/at the level of spinal cord entry and activates the autonomic reflex responses to noxious stimuli, such as elevating the heart rate and blood pressure (Fig. 38).

The third component of the paleospinothalamic pathway is the spinomesencephalic tract. This is also a crossed component of the ascending spinothalamic tract that terminates in the midbrain periaqueductal grey matter (PAG). A small component also terminates on the adjacent parabrachial nucleus of the midbrain which has outputs directly to the amygdala involved in the emotional response to pain (Fig. 39) [15].

The function of the PAG is thought to lie in the modulation of ascending pain information. The PAG is another structure that is difficult to identify on standard neuroimaging but forms a horseshoe of grey matter ventral and lateral to the cerebral aqueduct in the midbrain (Fig. 40).

In response to spinothalamic input the PAG can produce an inhibitory effect on transmission through the spinothalamic system at the level of the dorsal horn/entry level in the spinal cord. Fibres run inferiorly from the PAG to the nucleus raphe magnus, a serotonergic midline pontine raphe nucleus whose output runs inferiorly through the spinal cord and inhibits transmission of painful sensory input in the dorsolateral spinal cord (Fig. 41) [16].

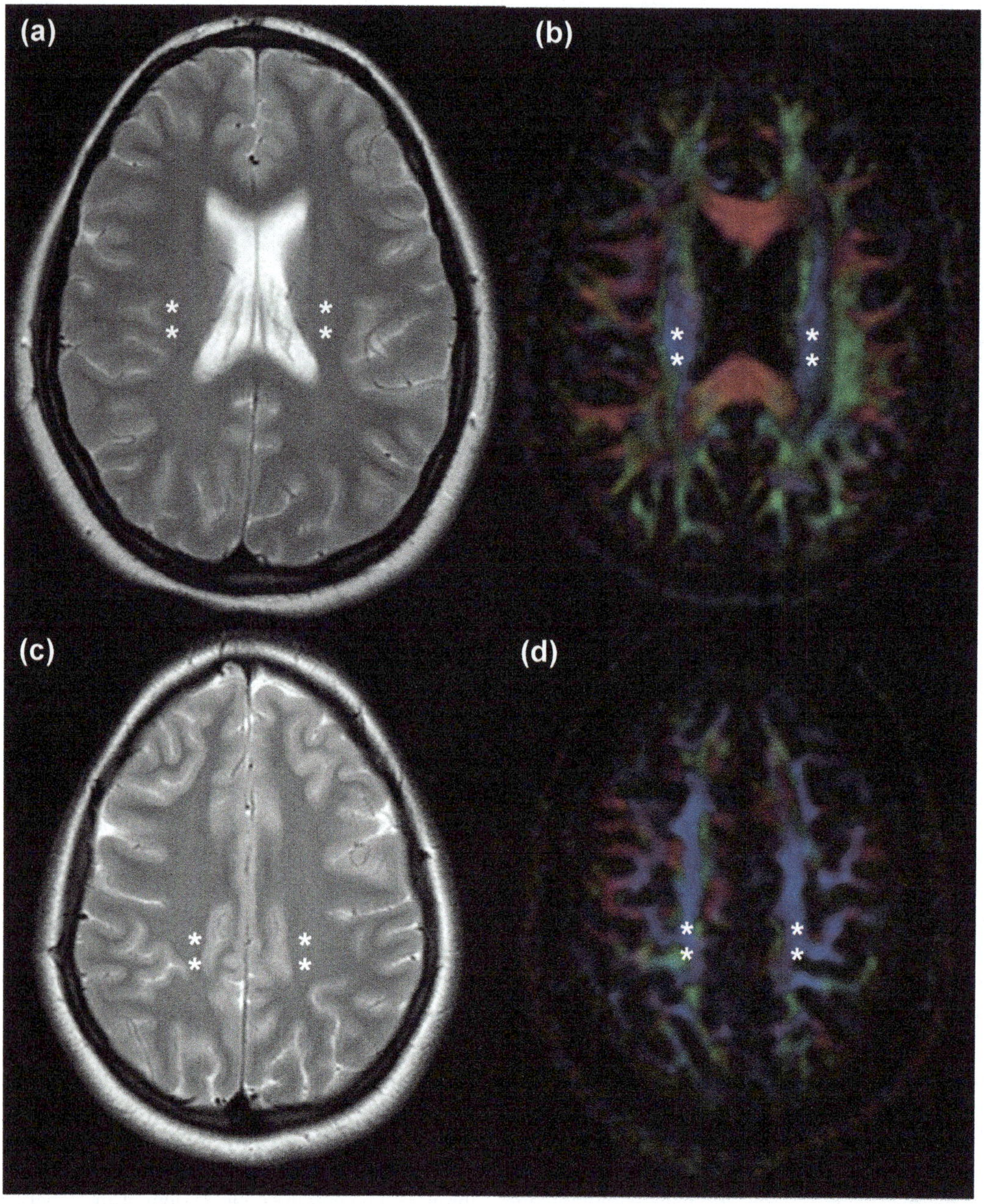

Fig. 28 Third order neurons (*asterisks*) project from the VPL nucleus of the thalamus to the primary somatosensory cortex via the corona radiata (axial T2, **a** and colour-coded FA map, **b**) and centrum semiovale (axial T2, **c** and colour-coded FA map, **d**)

The PAG is known to receive many ascending sensory inputs and also descending input from the supratentorial brain. Similarly there are many reciprocal outputs back to the reticular formation and also superiorly back to the supratentorial brain, amygdala and hypothalamus. It would appear the PAG has an integrative role in influencing via balance of outputs the

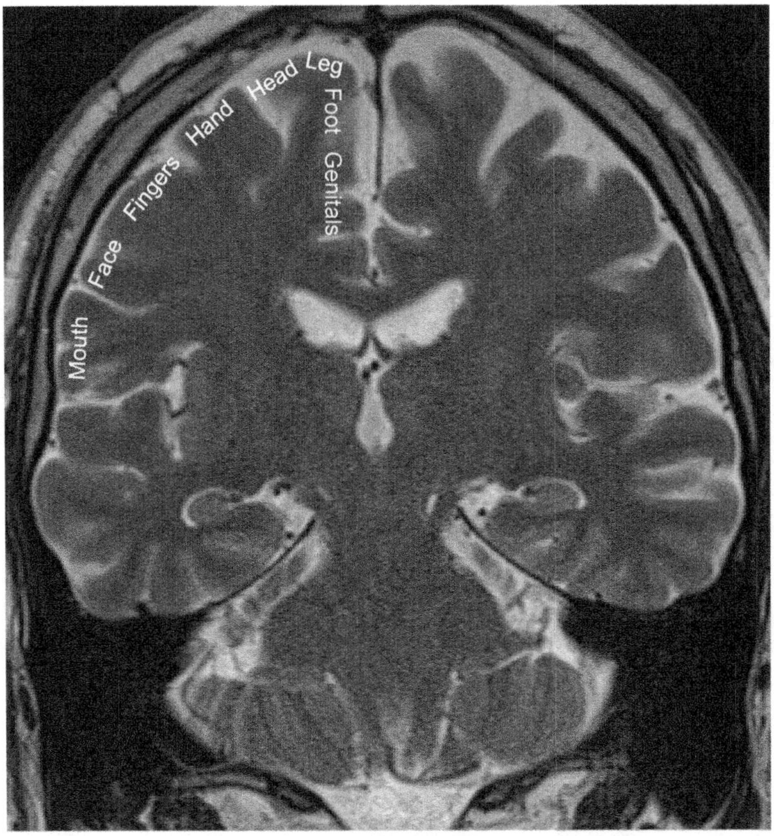

Fig. 29 Coronal T2 weighted images through the postcentral gyrus showing the primary somatosensory cortical homonculus—i.e. the approximate distribution of regions of cortical sensation by body part

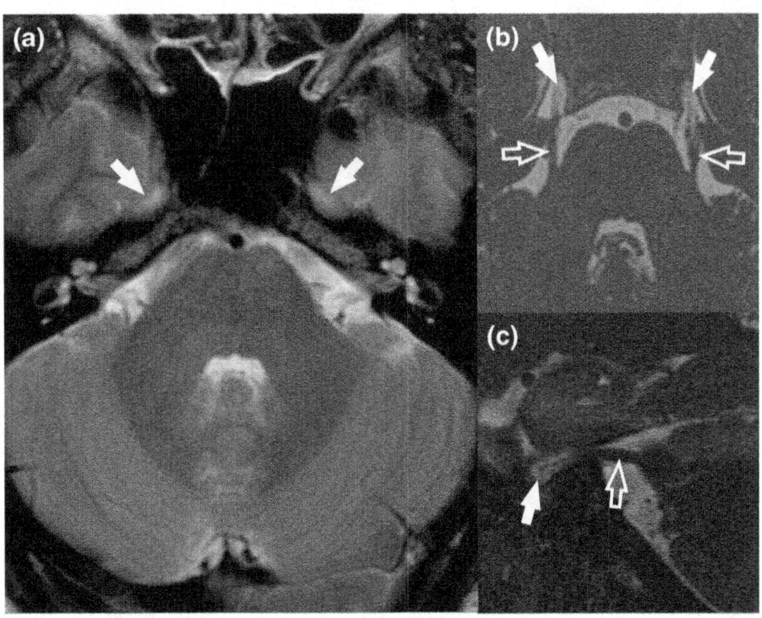

Fig. 30 Axial T2 (**a**), high-resolution 3d T2 axial (**b**) and sagittal (**c**) showing CSF in Meckel's cave (*white arrow*). The cisternal portions of the trigeminal nerve can be seen crossing the pre-pontine cistern from the pons (*open white arrow*)

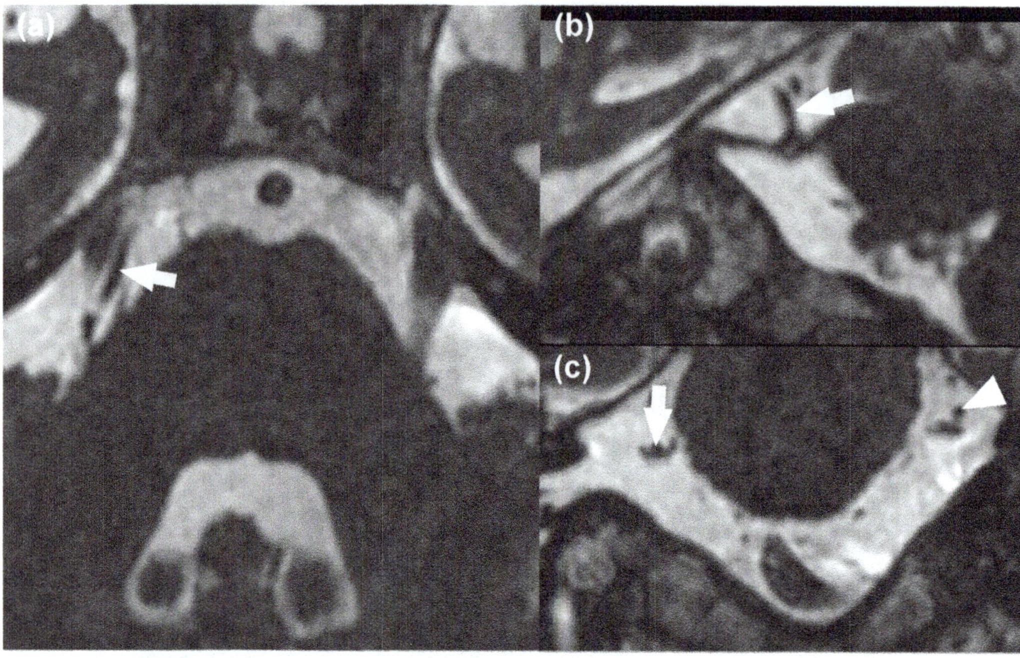

Fig. 31 Axial (**a**), sagittal (**b**) and coronal (**c**) 3d T2 weighted images in a patient with right-sided trigeminal neuralgia showing compression of the cisternal portion of the right trigeminal nerve by the right superior cerebellar artery (*white arrow*); compare to the normal appearance on the left side (*white arrowhead*)

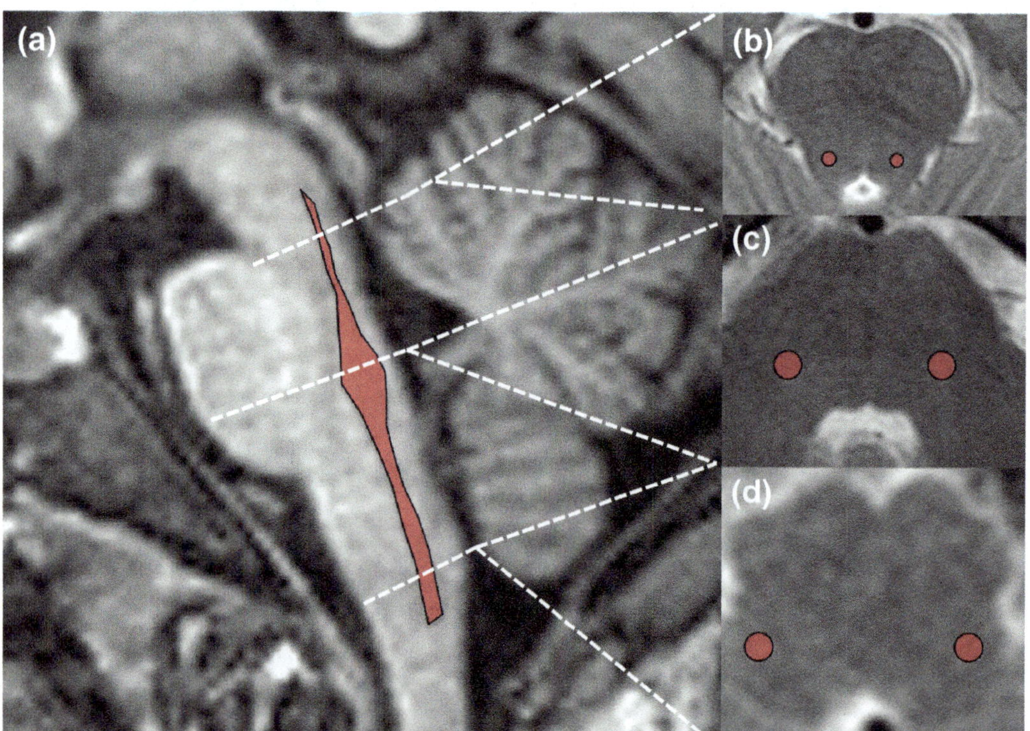

Fig. 32 Sagittal T1 volume (**a**) showing one of the spinal trigeminal nuclei (*red*) with corresponding axial T2 images at level of the mesencephalic nucleus (**b**), main sensory nucleus (**c**) and spinal nucleus (**d**) which are all in continuity

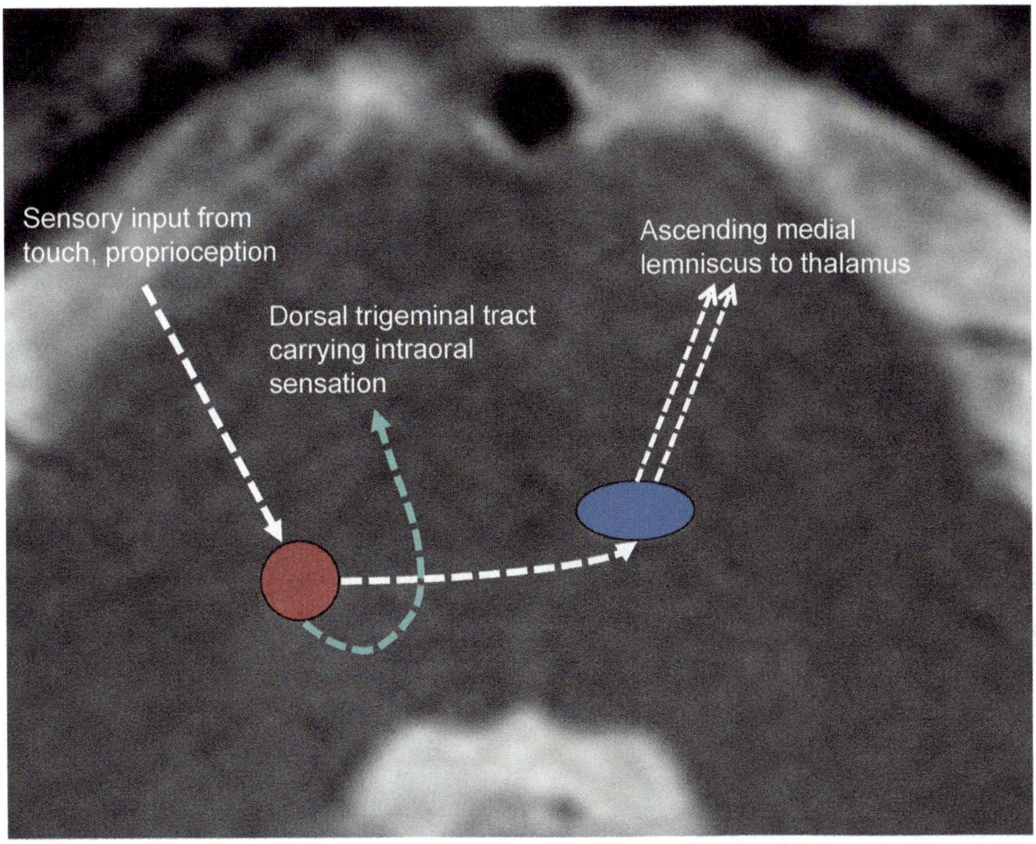

Fig. 33 Light touch and proprioception information synapses in the main sensory nucleus in the pons. 2nd order neurons mainly cross to join a subsection of the contralateral medial lemniscus called the trigeminal lemniscus to ascend to the thalamus. A small subgroup carrying intraoral sensation remain ipsilateral and ascend as the dorsal trigeminal tract heading for the ipsilateral VPM nucleus in the thalamus

fight-or-flight response to a threat. There are connections with the locus ceruleus in the upper pons which is one of the main norepinephric outputs back to the hypothalamus, thalamus and supratentorial brain; downward norepinephric modulation of spinal pain sensation transmission comes from the lateral reticular formation in the medulla (Fig. 42) [2].

Several nociceptive pathways also terminate in the cerebellum (the cuneocerebellar tract, the dorsal, ventral and rostral spinocerebellar tracts). As these pathways are primarily concerned with the subconscious maintenance of body posture they will not be considered further.

11 The Thalamus

The thalami are collections of deep grey matter nuclei situated deep within the brain, either side of the third ventricle. They act as a relay station for tracts both ascending from the spinal cord to the cortex, and vice versa from the cortex to the thalamus, back to the cord and other parts of the brain. The internal structure of the thalamus is not appreciable on routine neuroimaging but by understanding a schematic diagram this can be applied to anatomical imaging allowing at least approximate locations of individual nuclei to be extrapolated.

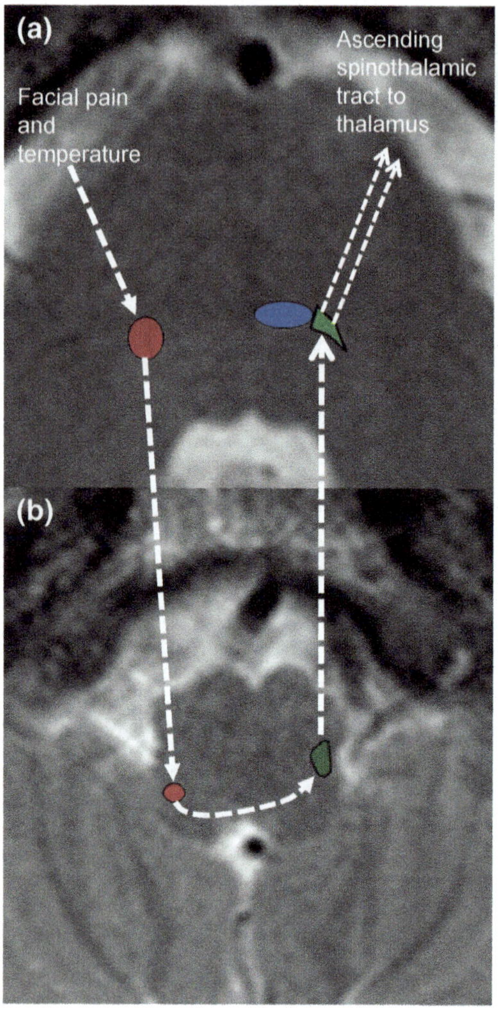

Fig. 34 Axial T2 images at the level of the pons (**a**) and medulla (**b**). Pain and temperature enters main sensory nucleus and descends to the ipsilateral spinal nucleus in the medulla where they synapse. 2nd order neurons cross to the contralateral spinothalamic tract (*green*) and ascend to the contralateral thalamus

The thalamus is made up of different groups of nuclei, separated by internal structures called 'laminae'. The internal laminae form a 'Y' shape with the anterior group of nuclei in between the fork of the Y and the medial and lateral groups situated either side of the stem (Fig. 43) [2, 11, 17, 18].

The different groups can also be divided into ventral and dorsal 'tiers' or layers. The primary nuclei involved in the sensory system are the ventral posteromedial and lateral nuclei (subtending sensation of the face and body respectively), located in the ventral tier of the lateral group of thalamic nuclei (Fig. 44).

Other thalamic nuclei have other specific cortical connections—for example the ventral anterior and lateral nuclei are part of the motor pathways connecting the motor cortex, basal ganglia and cerebellum; the medial and lateral geniculate nuclei form part of the auditory and visual pathways respectively. Other nuclei have a less specific output with wide-ranging connections to many different regions of the ipsilateral cerebral hemisphere. This is particularly the case with the intralaminar nuclei—nuclei found within the laminae that separate the main groups of thalamic nuclei. The largest, the centromedian nucleus and smaller parafascicular nucleus, are located medial to the VPL and VPM nuclei. These intralaminar nuclei receive input from the spinoreticular formation in the brainstem and can be considered the superior extension of reticular activating system itself within the thalamus. From here efferents are found to the corpus striatum (caudate and putamen), the primary and secondary somatosensory cortices and the insula/cingulate cortex (Fig. 45) [2].

The medial group of thalamic nuclei—the dorsomedial nucleus and laterodorsal nucleus in particular have extensive projections to the prefrontal cortex (particularly dorsolateral prefrontal cortex, orbitofrontal cortex) and parts of the limbic system, receiving afferent inputs from the amygdala (which itself is a target for output from the parabrachial nucleus in the midbrain, part of the spinomesencephalic pain pathway).

12 Cortical Regions Involved in the Pain Perception

The primary regions involved in sensation are the primary and somatosensory cortices (SI and SII). S1 is located in the posterior surface of the central sulcus, extending onto the cortical surface of the postcentral gyrus. This extends all the way down the postcentral gyrus from the medial

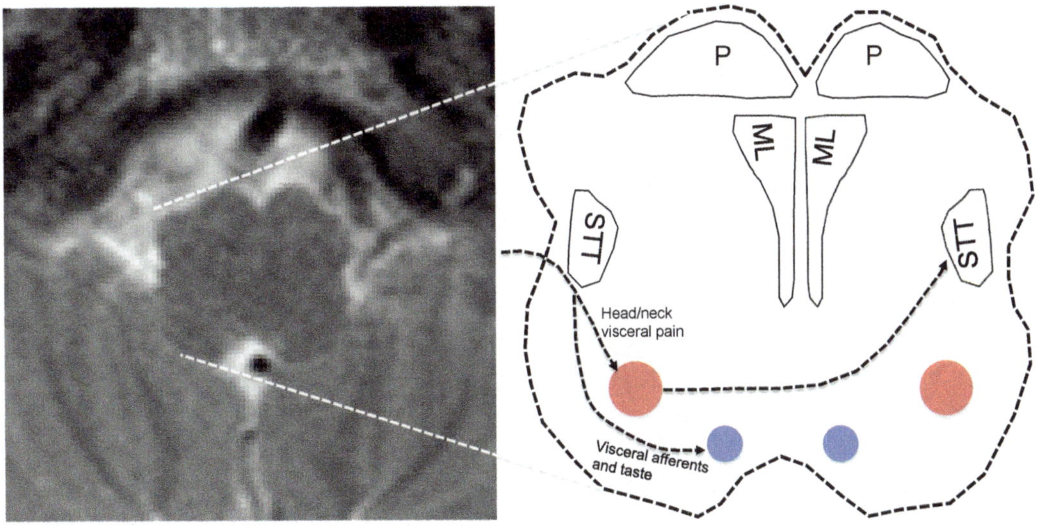

Fig. 35 Axial T2 and schematic of the medulla. Glossopharyngeal and vagus sensory afferents. Pain fibres synapse on the spinal trigeminal nucleus (*red circle*) from which fibres cross to the contralateral spinothalamic tract (STT) to ascend to the thalamus. Visceral afferents and taste information synapse on the nucleus/tract of solitarius (*blue circle*)

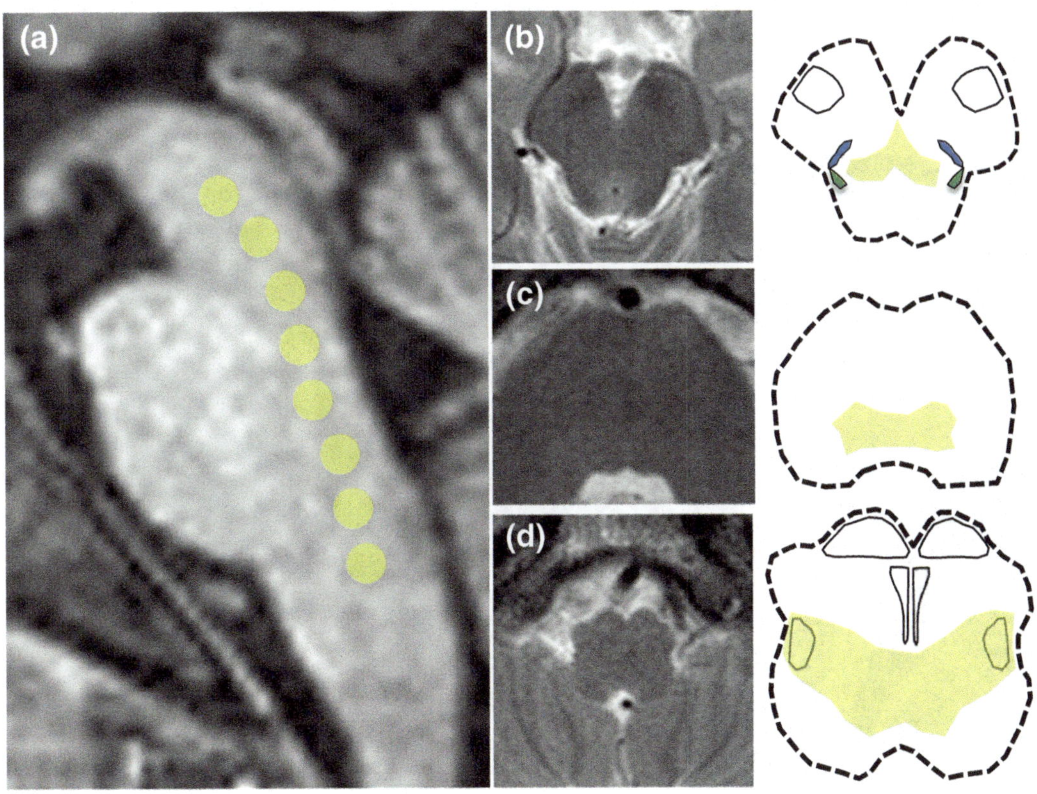

Fig. 36 Sagittal T1 volume (**a**), axial T2 at the level of the midbrain (**b**), pons (**c**) and medulla (**d**) with adjacent schematic showing the distribution of tegmental nuclei in *yellow*

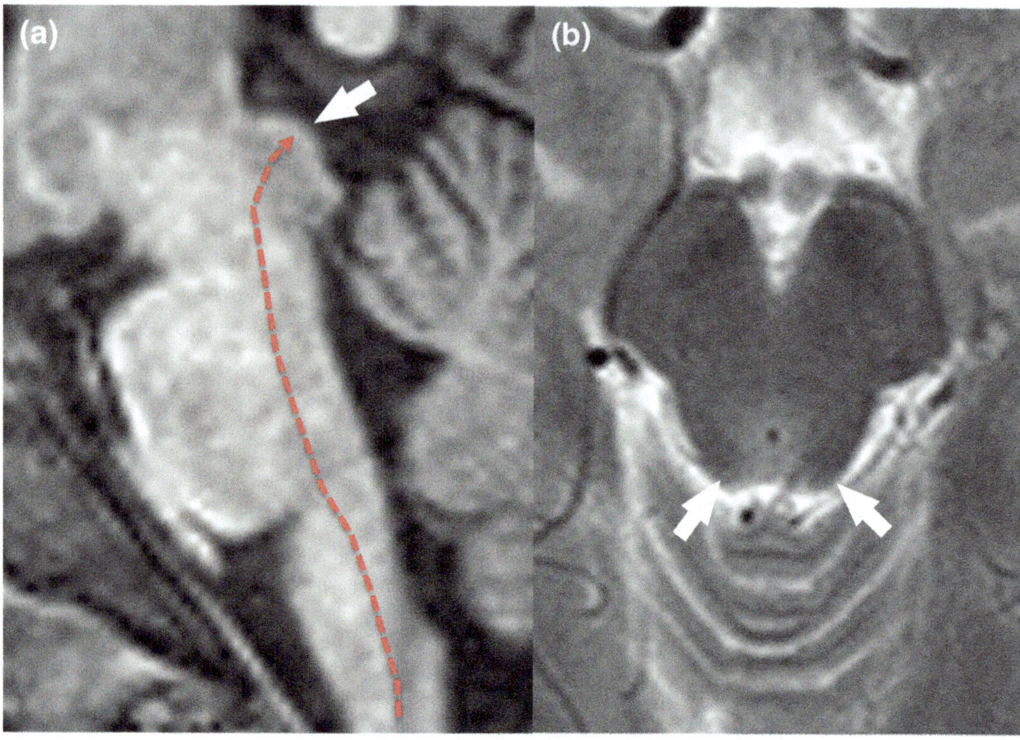

Fig. 37 Sagittal T1 volume (**a**) and axial T2 (**b**) showing position of the spinotectal tract terminating in the superior colliculus (*white arrows*). This serves to turn the head/gaze towards a painful stimulus

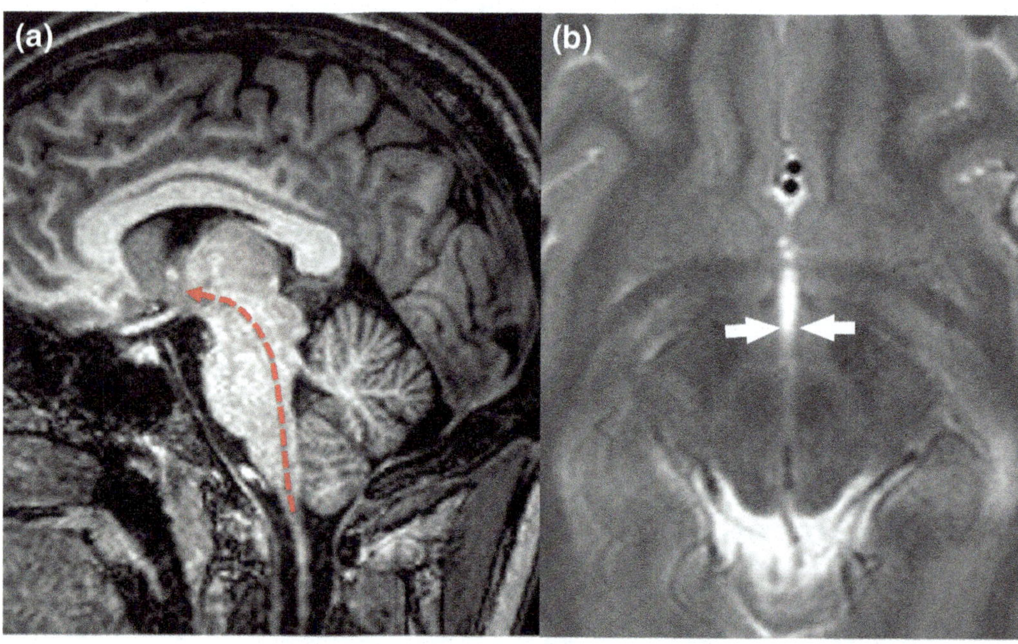

Fig. 38 Sagittal T1 volume (*left*, **a**) and axial T2 (*right*, **b**) showing the course of the spinohypothalamic tract (*dashed red line*). The hypothalamus is a collection of grey matter nuclei found lining the anterior inferior third ventricle (*white arrows*, **b**)

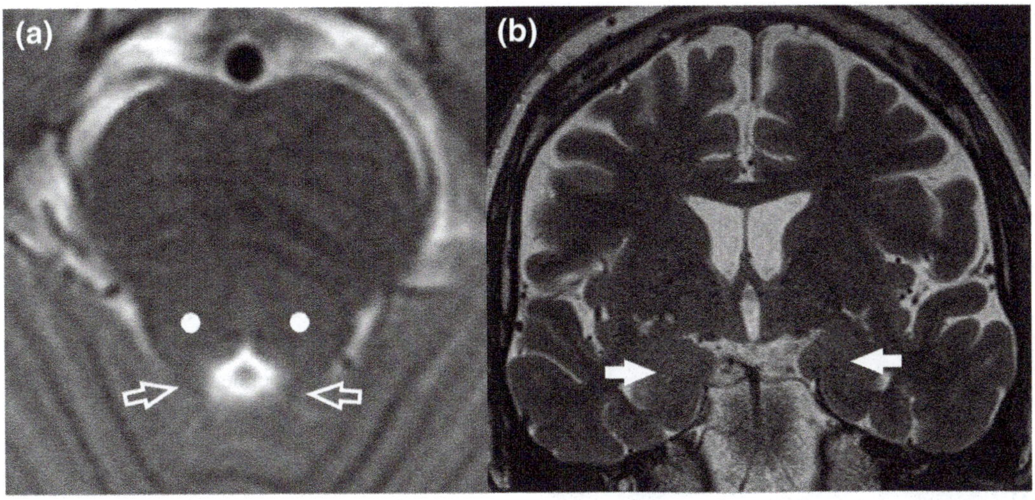

Fig. 39 Axial T2 (**a**, *left*) and coronal T2 (**b**, *right*) showing the position of the parabrachial nuclei in the brainstem (*white circles*) at the level of the superior cerebellar peduncles (*open white arrows*). These have bidirectional connections with the amygdalae (*white arrows*). Part of the spinomesencephalic tract terminates in the parabrachial nuclei and stimulate the amygdalae contributing to the emotional aspects of pain sensation. The parabrachial nuclei are also intimately connected with the respiratory nuclei in the pons

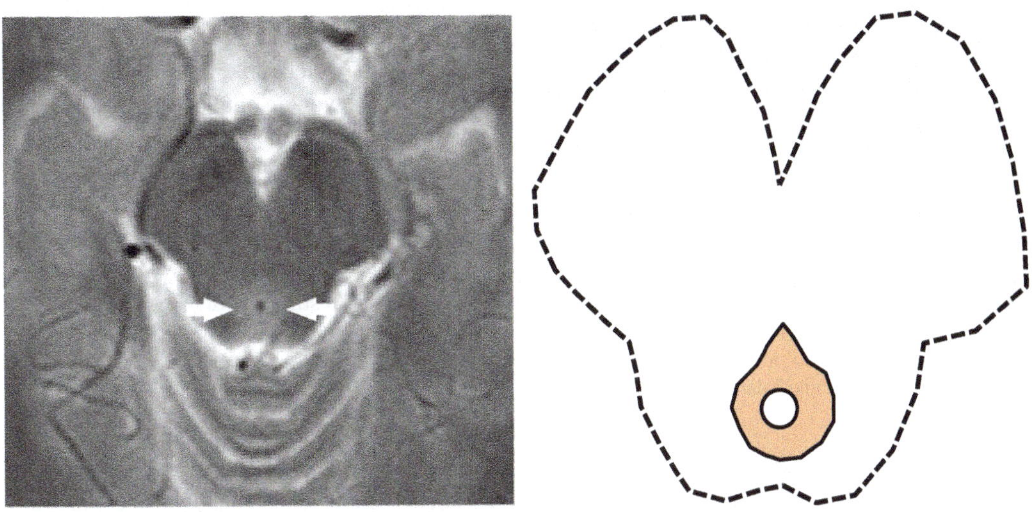

Fig. 40 Axial T2 image (*left*) and schematic (*right*) at the level of the midbrain showing position of the periaqueductal grey matter (*white arrows*)—shown in *orange* surrounding the cerebral aqueduct on the schematic image

interhemispheric fissure to the inferiormost aspect just above the lateral/posterior part of the Sylvian fissure (Fig. 46) [19].

SII is located at the inferiormost part of the postcentral gyrus, extending onto the superior surface of the superior temporal gyrus (Fig. 47).

SI is though to be involved in the detection of location and character of sensory stimulus with additional fine discriminatory functions such as the ability to identify an object via touch (stereognosis), whereas SII is thought to be more involved in memory aspects of sensory input.

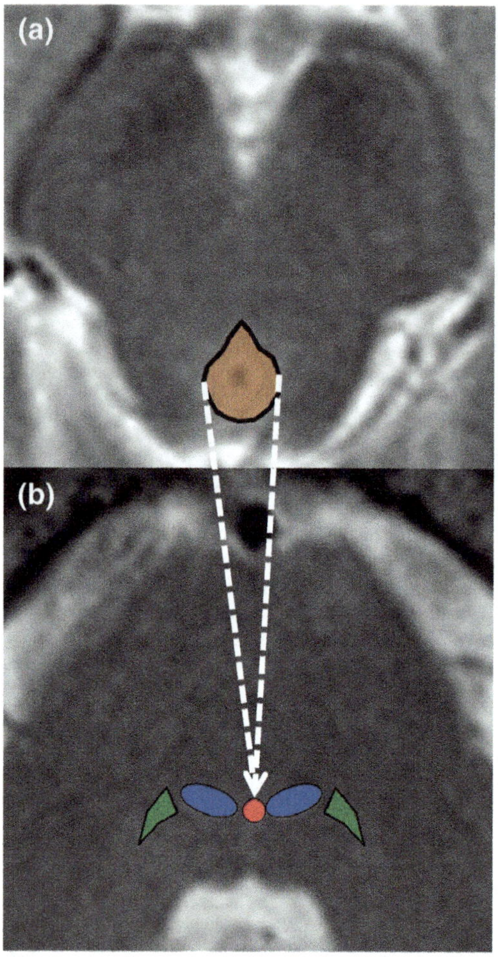

Fig. 41 Axial T2 weighted images at the level of the midbrain (**a**) and pons (**b**) showing the periaqueductal grey in *orange* (**a**) with output down on to the nucleus raphe magnus (*red circle*) in the pons. The medial lemniscus (*blue*) and spinothalamic tracts (*green*) are shown on either side

Perception of pain occurs both in SI where its character is appreciated but also/simultaneously in the anterior cingulum and insular cortices, via projection from the medial and intralaminar thalamic nuclei (Fig. 48) [20].

More recent, detailed fMRI work has shown that painful stimuli involve all areas of the insula and parietal operculum (SII) whereas other, non-noxious stimuli such as heat and cold discrimination may involve subregions of the SII and insular respectively [21].

13 Influence of Higher Order States on Pain Processing

It is well known that different emotional states can influence the experience of painful stimuli, with low mood being associated with enhanced pain perception, independent of the severity of the stimulus. Similarly, the degree to which one pays attention to painful stimuli can dramatically affect the degree to which pain is perceived; consider the professional ballet dancer who is able to 'ignore' painful sensations from the feet, or soldiers who are able to continue to fight in battle despite being severely injured. Recent functional MRI work has started to elucidate the neuroanatomical basis for these so-called 'attention/salience' networks and those involved in emotional states [22, 23].

The attention network appears to have two separate modes with a voluntary, or goal-directed mode operated by the cortex in the superior parietal lobes (Brodmann area 7) and frontal eye fields (Fig. 49). These functions are represented bilaterally—i.e. in both cerebral hemispheres [24].

A second, stimulus-driven network is right-lateralised and appears to reside in the inferior frontal lobe and inferior parietal lobule/superior temporal gyrus (Fig. 50).

Whilst these areas of the brain were primarily initially described in relation to visual attention, subsequent studies have shown they are also active in attending to other sensory inputs, including painful stimuli. It would appear that they affect the perception of pain through changing (either up or downregulating) activity in the ascending thalamocortical pain pathways. Recent work has shown positive correlation of activity between the superior parietal cortex

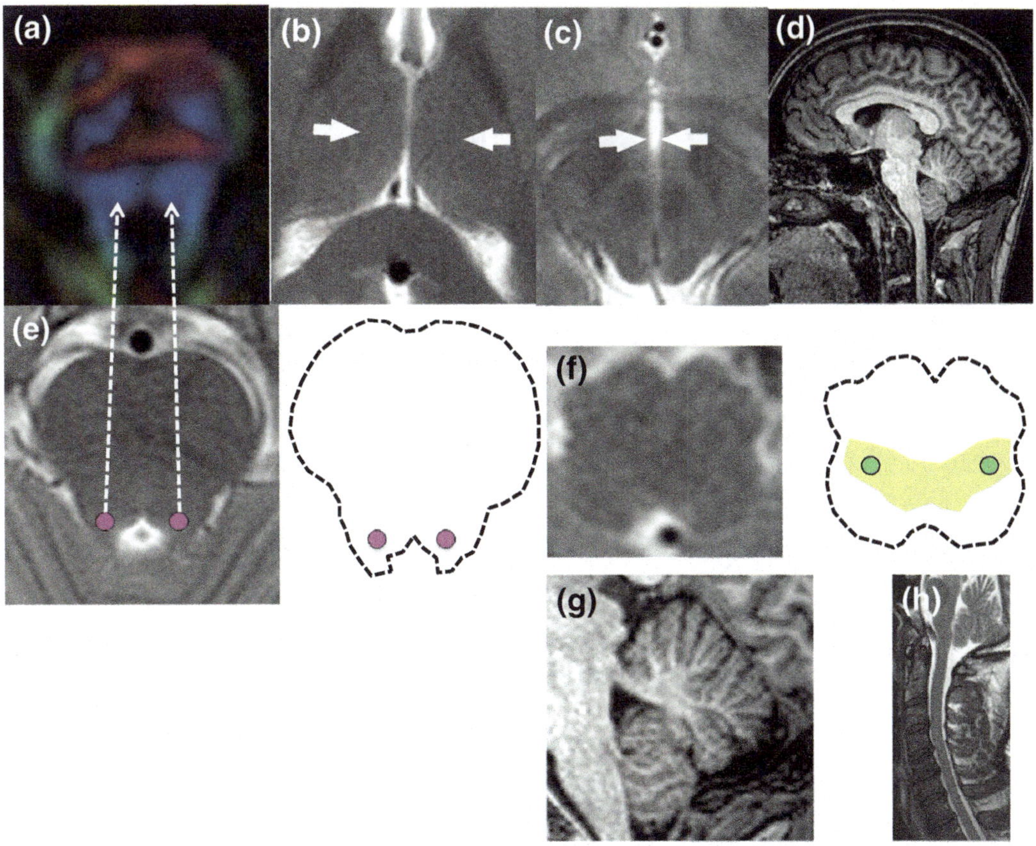

Fig. 42 Norepinephric pathways in the brain. The locus ceruleus (*pink circles* on axial T2 and graphic, **e**) is the origin of norepinephric neurons which pass superiorly up the central tegmental tracts (axial colour DTI map, *arrows*, **a**) to the intralaminar nuclei of the thalami (*arrows*, axial T2, **b**), the hypothalamus (*arrows*, axial T2, **c**) and the cerebral cortex generally (sagittal MPRAGE, **d**). The more inferior norepiephric centre in the dorsolateral medulla (*green circles*, graphic and axial T2, **f**) send neurons to the cerebellum (**g**, sagittal MPRAGE) and spinal cord (sagittal T2, **h**)

(BA 7) and the anterior insula cortex (involved in pain perception) when subjects changed the degree of attention they gave painful stimuli.

Mood-related changes in pain perception appear to act via a separate pathway. Activity in the lateral orbitofrontal cortex (BA 47) is associated with negative mood states, whereas activity in the medial orbitofrontal cortex (BA 45) is associated with positive mood states (Fig. 51) [24, 25].

Both these regions have extensive connections with other parts of the brain involved in sensory, and also pain processing; particularly the amygdala and periaqueductal grey matter (PAG). They are also connected to the anterior cingulum, a region of the brain associated with detecting the affective, or 'mood'-related component to a painful stimulus. It would appear that a negative mood (associated with increased activity in the lateral orbitofrontal cortex) may increase pain perception via facilitation of pain-related thalamocortical pathways, including the anterior cingulum, but also through reduction in inhibitory activity mediated at the dorsal horn level from the PAG and raphe nucleus.

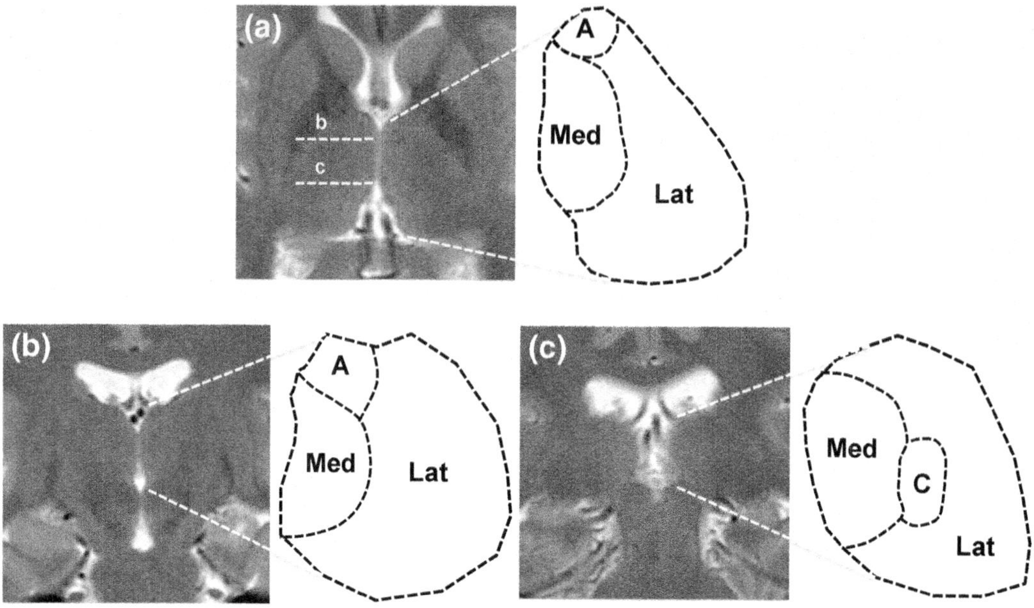

Fig. 43 The broad grouping of thalamic nuclei is shown on axial T2 image (**a**) with schematic; coronal sections at approximate locations (**b**) and **c** show the relative arrangement of the nuclear groups, including the intralaminar nuclei (**c**). *A* Anterior group, *Med* medial group, *lat* lateral group, *c* intralaminar nuclei

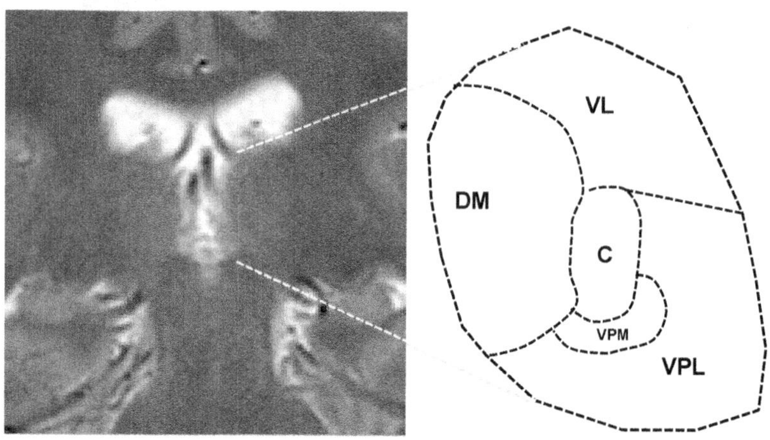

Fig. 44 Coronal T2 (*left*) and schematic showing the more detailed arrangement of the primary sensory nuclei within the ventrolateral thalamic group (the VPM and VPL). *DM* Dorsomedial, *VL* ventral lateral, *C* intralaminar, *VPM* ventroposteromedial, *VPL* ventroposterolateral

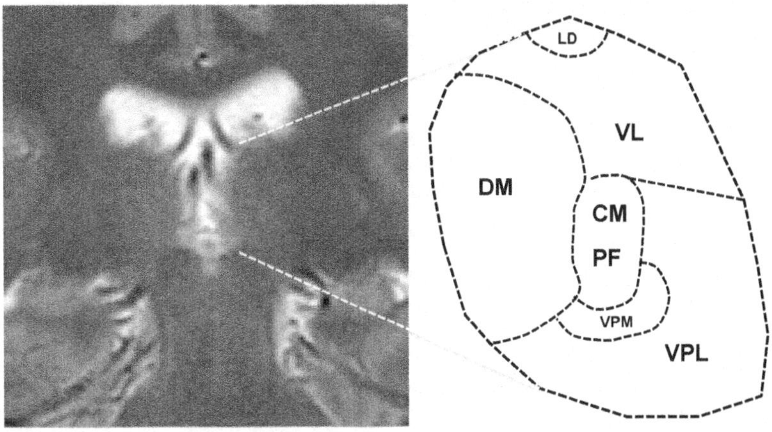

Fig. 45 Coronal T2 image and schematic showing subdivision of intralaminar nuclei—the centromedian and parafascicular nuclei. The laterodorsal nucleus is also shown on the superiormost surface of the thalamus. *LD* Laterodorsal, *DM* dorsomedial, *CM* centromedian, *PF* parafascicular, *VPM* ventroposteromedial, *VPL* ventroposterolateral

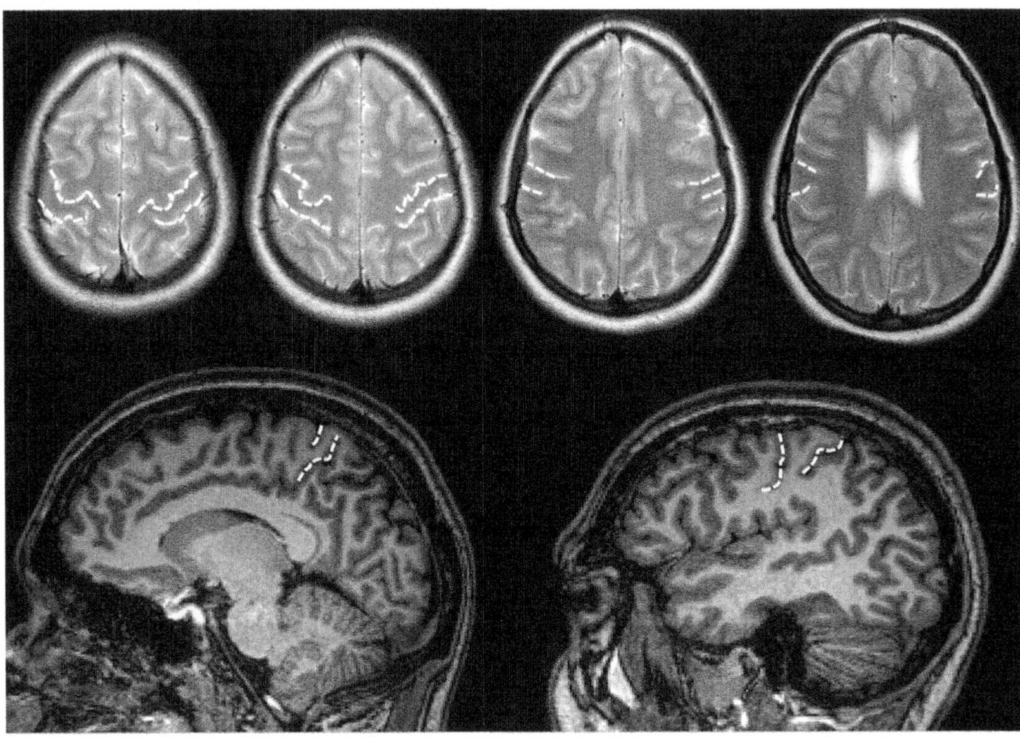

Fig. 46 Axial T2 (*top row*) and sagittal/parasagittal MPRAGE (*bottom row*) showing location of the primary sensory cortex (postcentral gyrus) between *dashed lines*

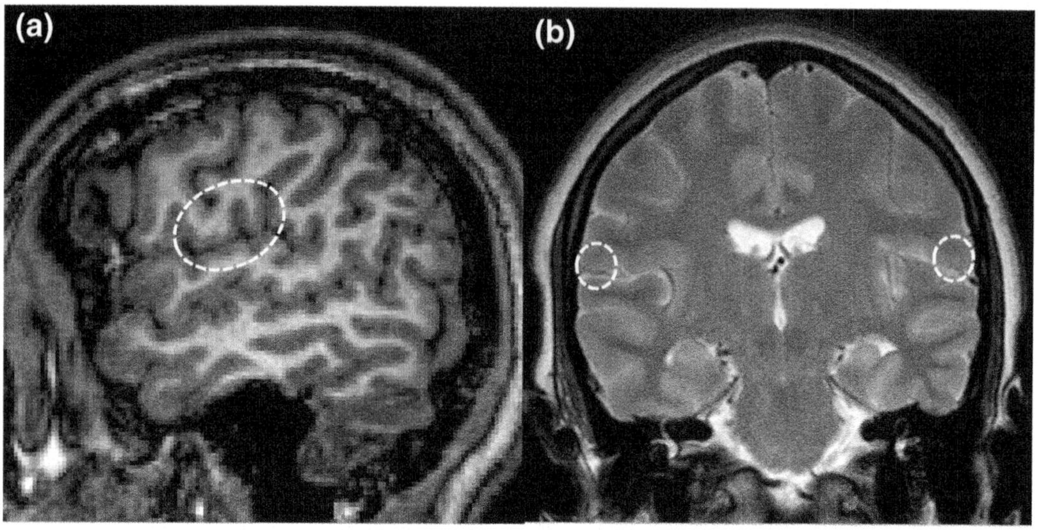

Fig. 47 Sagittal MPRAGE (**a**) and coronal T2 (**b**) showing the location of the secondary somatosensory cortex (*dashed circles*), spanning the sylvian fissure onto the superior temporal gyrus

14 Conclusion

Even the relatively basic neuroanatomy of sensory perception can appear complicated on first encounter with different brainstem tracts for different modalities of sensory perception, which move location at different levels. Add to this the extensive interconnectivity of the brainstem reticular system and the influence of widespread functional cortical networks and it can be easily seen that an appreciation of whole brain neuroanatomy is essential to understand this rapidly evolving field.

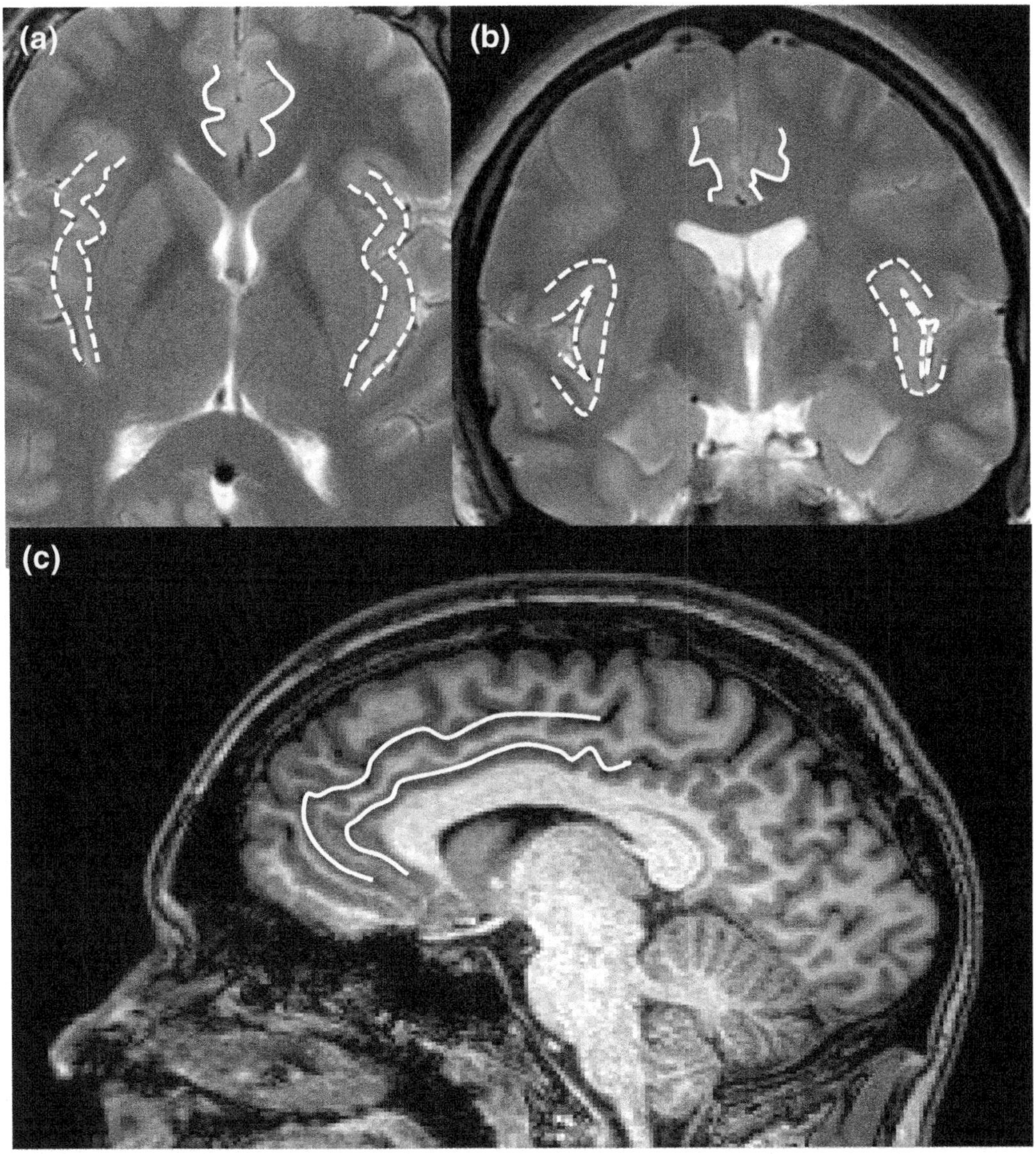

Fig. 48 Axial T2 (**a**), coronal T2 (**b**) and sagittal MPRAGE (**c**) showing the location of the insular cortex (between *dashed lines*) and anterior cingulate cortex (between *solid lines*)

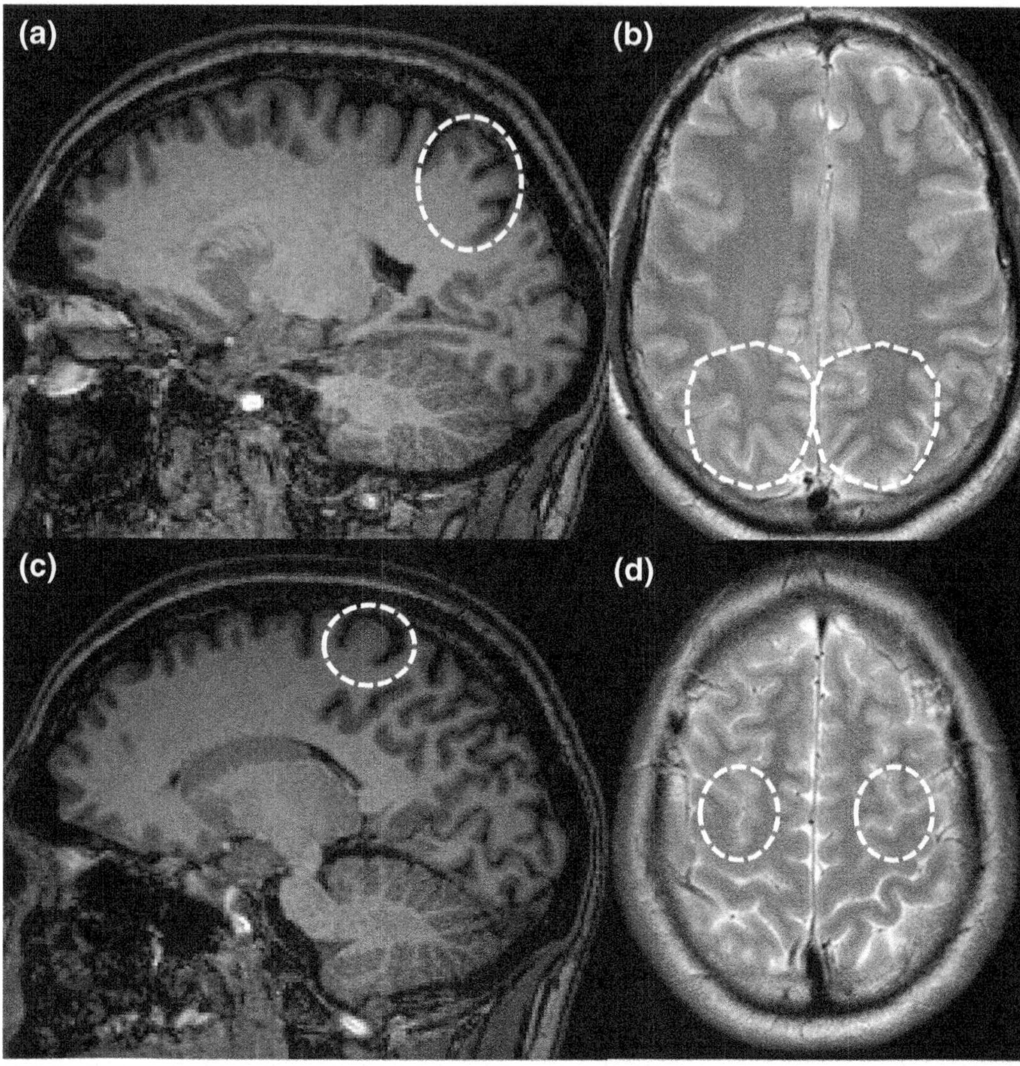

Fig. 49 Sagittal MPRAGE (**a**, **c**) and axial T2 (**b**, **d**) showing the location of Brodmann area 7 (**a**, **b**) and the frontal eye fields (**c**, **d**), *dashed circles*

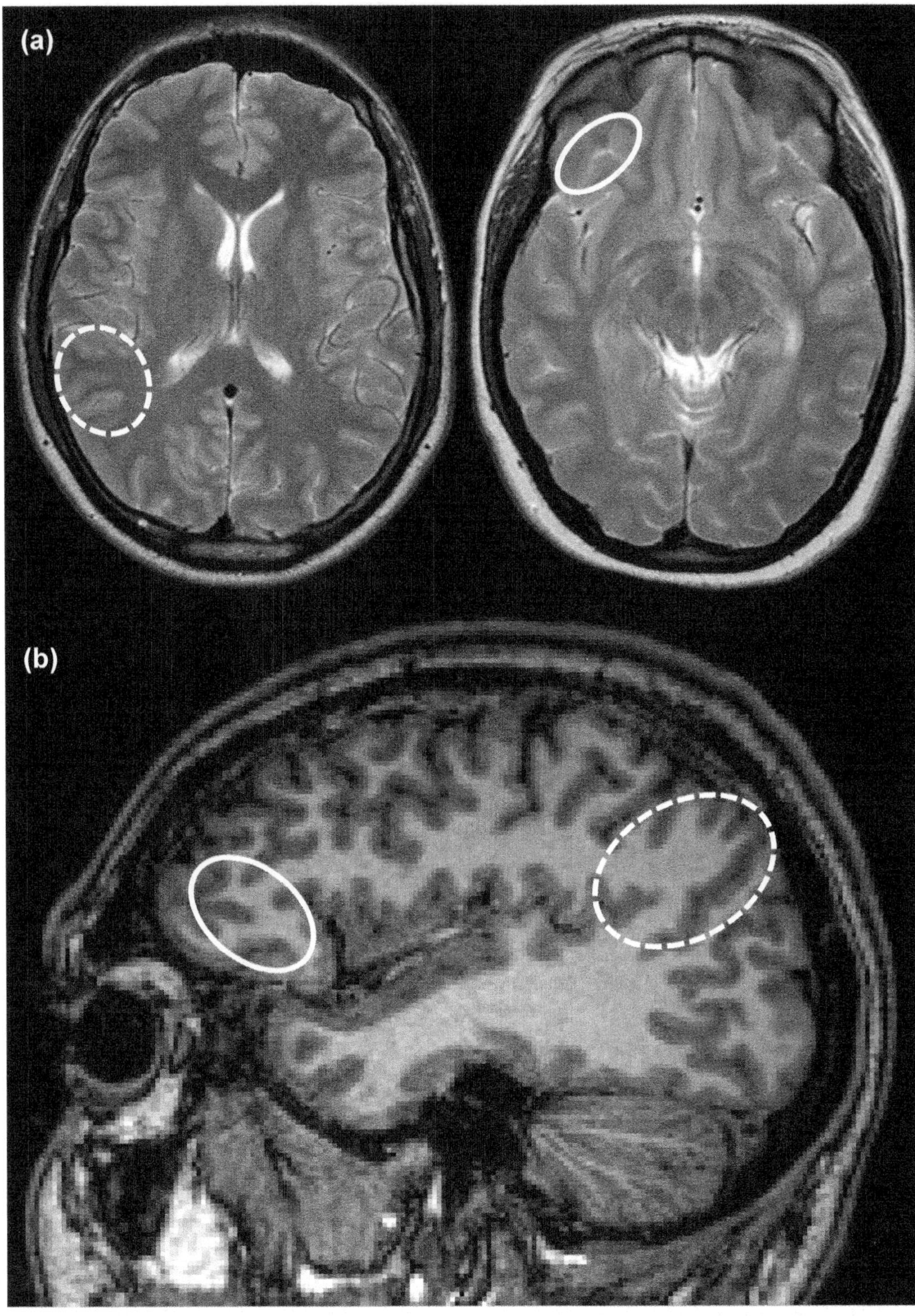

Fig. 50 Axial T2 (**a**) and parasagittal MPRAGE (**b**) showing the right-lateralised centres in the inferior parietal lobule (*dashed circle*) and inferior frontal gyrus (*solid circle*)

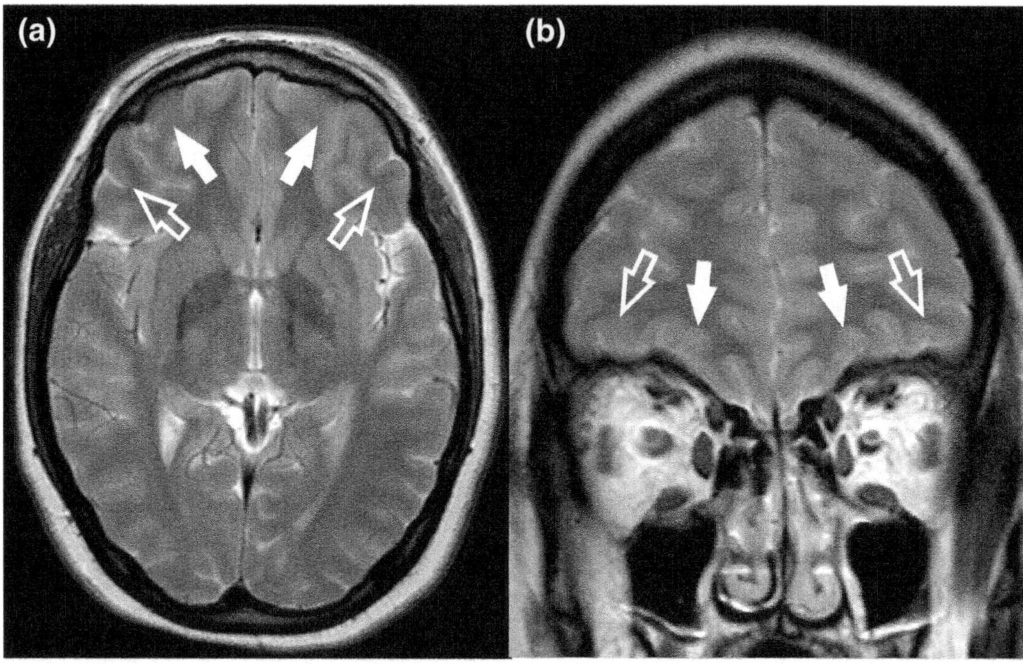

Fig. 51 Axial (**a**) and coronal (**b**) T2 images showing the location of the medial (*solid arrows*) and lateral (*open arrows*) orbitofrontal cortex

References

1. Schoewolf GC, Bleyl SB, Brauer PR, et al. Larsen's human embryology. 5th ed. London: Churchill Livingstone; 2014.
2. Nolte J. The human brain: an introduction to its functional anatomy. 6th ed. Missouri: Mosby; 2008.
3. Hathout GM. Clinical neuroradiology: a case-based approach. Cambridge: Cambridge University Press; 2008.
4. Thomas AG, Koumellis P, Dineen RA. The fornix in health and disease: an imaging review. Radiographics. 2011;31:1107–21.
5. Shah A, Jhawar SS, Goel A. Analysis of the anatomy of the Papez circuit and ajoining limbic system by fiber dissection techniques. J Clin Neurosci. 2012;19:289–98.
6. Whalen PJ, Phelps EJ, editors. The human amygdala. New York: Guilford Press; 2009.
7. Gerrish AC, Thomas AG, Dineen RA. Brain white matter tracts: functional anatomy and clinical relevance. Semin Ultrasound CT MR. 2014;35:432–44.
8. Hussain A, Utz MJ, Tian W, et al. Imaging and diseases of the ascending and descending pathways. Semin Ultrasound CT MR. 2014;35:474–86.
9. Barkovich AJ, Raybaud C. Pediatric neuroimaging. Philadelphia: Lippincott Williams and Wilkins; 2011.
10. Osborne AG. Diagnostic cerebral angiography. 2nd ed. Philadelphia: Lippincott Williams and Wilkins; 1999.
11. Patestas M, Gartner LP. A textbook of neuroanatomy. Hoboken: Wiley-Blackwell; 2006.
12. Kamali A, Kramer LA, Butler IJ, et al. Diffusion tensor tractography of the somatosensory system in the human brainstem: initial findings using high isotropic spatial resolution at 3.0T. Eur Radiol. 2009;19:1480–8.
13. Becker M, Kohler R, Vargas MI, et al. Pathology of the trigeminal nerve. Neuroimaging Clin N Am. 2008;18:283–307.
14. Sarrazin JL, Toulgoat F, Benoudiba F. The lower cranial nerves: IX, X, XI, XII. Diagn Interv Imaging. 2013;94:1051–62.
15. Neugebauer V. Amygdala pain mechanisms. Handb Exp Pharmacol. 2015;227:261–84.
16. Wilis WD, Westlund KN. Neuroanatomy of the pain system and of the pathways that modulate pain. J Clin Neurophysiol. 1997;14:2–31.
17. Bender B, Mänz C, Korn A, et al. Optimized 3D magnetization-prepared rapid acquisition of gradient echo: identification of thalamus substructures at 3T. AJNR Am J Neuroradiol. 2011;32:2110–5.
18. Tourdias T, Saranathan M, Levesque IR, et al. Visualization of intra-thalamic nuclei with optimized white-matter-nulled MPRAGE at 7T. Neuroimage. 2013;84:534–45.
19. Omori S, Isose S, Otsuru N, et al. Somatotopic representation of pain in the primary somatosensory cortex (s1) in humans. Clin Neurophysiol. 2013;124:1422–30.

20. Chen TL, Babiloni C, Ferretti A, et al. Human secondary somatosensory cortex is involved in the processing of somatosensory rare stimuli: an fMRI study. NeuroImage. 2008;40:1765–71.
21. Mazzola L, Faillenot I, Barral F-G, et al. Spatial segregation of somato-sensory and pain activations in the human operculo-insular cortex. NeuroImage. 2012;60:409–18.
22. Plega B, Villringer A. The human somatosensory system: from perception to decision making. Prog Neurobiol. 2013;103:76–97.
23. Hayes DJ, Northoff G. Common brain activations for painful and non-painful aversive stimuli. BMC Neurosci. 2012;13:60.
24. Villemure C, Schweinhardt P. Supraspinal pain processing: distinct roles of emotion and attention. Neuroscientist. 2010;16:276–84.
25. Kringelbach ML, Rolls ET. The functional neuroanatomy of the human orbitofrontal cortex: evidence from neuroimaging and neuropsychology. Prog Neurobiol. 2004;72:341–72.

Author Biography

Dr. Adam G. Thomas trained in medicine at University College London, graduating with honours. He obtained Membership of the Royal College of Physicians (UK) and Fellowship of the Royal College of Radiologists during postgraduate medical and radiology training and graduated with distinction from the Masters programme in Translational Neuroimaging at the University of Nottingham in 2012. He is a Consultant Neuroradiologist with expertise in paediatric neuroradiology at Nottingham University Hospitals, Queens Medical Centre, UK.

Physiopathology of Pain

4

Serge Marchand

Abstract

The different brain imaging techniques that have emerged in the last decades have raised major advancement in our understanding of the neurophysiological mechanisms implicated in pain in both healthy subjects and in patients suffering from different pain conditions. The new brain imaging protocols are developed based on the background of previous surgical, behavioral, psychophysical, and electrophysiological researches on nociception and pain in animal, healthy subject, and patients. Having a good background of normal and pathophysiological pain neurophysiology is essential for the design of research protocols that will take advantage of new brain imaging technologies to better investigate the complex phenomenon of pain. Pain is a dynamic phenomenon that is the end result of several factors. The association between nociceptive activity and pain perception depends on several intrinsic and extrinsic influences. For the same nociceptive stimulus, pain perception and related brain activity will greatly differ between subjects. Studies support that environment and genetic factors are both playing important roles and seem to be modality specific. The effect of environment on genetics, epigenetics (lasting changes in gene expression without alteration of DNA sequence), is essential to be taken into account in pain. Nerve injuries or even psychological factors could change the central nervous system by affecting DNA methylation and produce a "genomic" memory of pain in the adult cortex. Pain perception is then the result of inherited physiological and psychological factors that are influenced by and hopefully guide the development of new therapeutic approaches for the patients that are suffering.

S. Marchand (✉)
Department of Surgery/Division of Neurosurgery,
Université de Sherbrooke, Sherbrooke, Québec,
Canada
e-mail: Serge.Marchand@USherbrooke.ca

L. Saba (ed.), *Neuroimaging of Pain*, DOI 10.1007/978-3-319-48046-6_4

Keywords
Nociception · Pain pathways · Conditioned pain modulation · Diffuse noxious inhibitory control · Affective · Sensory · Central sensitization · Pain modulation

1 Introduction

The different brain imaging techniques that have emerged in the last decades have raised major advancement in our understanding of the neurophysiological mechanisms implicated in pain in both healthy subjects and in patients suffering from different pain conditions.

The new brain imaging protocols are developed based on the background of previous surgical, behavioral, psychophysical, and electrophysiological researches on nociception and pain in animal, healthy subject, and patients. Having a good background of normal and pathophysiological pain neurophysiology is essential for the design of research protocols that will take advantage of new brain imaging technologies to better investigate the complex phenomenon of pain.

Pain is a dynamic phenomenon that is the end result of several factors. The association between nociceptive activity and pain perception depends on several intrinsic and extrinsic influences. For the same nociceptive stimulus, pain perception and related brain activity will greatly differ between subjects. Using functional magnetic resonance imaging (fMRI) of the brain, Coghill and colleagues found that the more sensitive subjects exhibited more pain-induced activity in the primary somatosensory cortex, anterior cingulate cortex (ACC), and prefrontal cortex (PFC) than did less sensitive subjects [11]. Interestingly, they also found that the thalamus activity was not different between the two groups, supporting that the same nociceptive signal is transported to the thalamus. It is the sensory and affective pain-related brain structures that are encoding for these inter-individual differences in pain perception.

The importance of intrinsic factors in pain is supported by genetic predispositions to be less or more sensitive to pain [91]. In one study comparing 59 identical pair of twins with 39 fraternal twins, the authors conclude that 60% of the variance in cold pressor pain and 26% of the variance in heat pain was genetically mediated [60]. These results suggest that environment and genetic factors are both playing important roles and seem to be modality specific. The effect of environment on genetics, epigenetics (lasting changes in gene expression without alteration of DNA sequence), is also essential to be taken into account in pain [6]. Nerve injuries or even psychological factors could change the central nervous system by affecting DNA methylation and produce a "genomic" memory of pain in the adult cortex [21]. It could even explain the comorbidity between some psychiatric factors such as depression and pain [76]. These results support the importance of psychological factors such as mood, anxiety, catastrophizing, and personality in pain perception [84].

Pain perception is then the result of inherited physiological and psychological factors that are influenced by our environment. Together these factors are framing our reaction to different painful situations, but probably also our predisposition for pain chronification.

2 Theories of Pain Mechanisms

Researches are driven by theories. Most of the time we need a challenging new paradigm to emerge to stimulate new research protocols that will lead to new theories. The clinical approaches for the treatment of pain are based on these theories. It is then interesting to have a brief overview of the evolution of our understanding of pain mechanisms. It helps us realize that the evolution of pain treatments is highly related to motley of older and new pain theories to explain pain mechanisms.

2.1 Specificity Theory

The specificity theory was first introduced by Descartes during the seventeenth century [22] and refined with the modern physiology by Müller [59] and Frey [28] at the end of the nineteenth century. They proposed that the somatosensory system could be divided according to specific receptors for tactile, hot, cold, and pain receptors. With the specificity, we have a theoretical framework to explain how specific afferences from the periphery, Aδ and C fibers, are connecting to specific pathways, spinothalamic, and spinoreticular tracts from the spinal cord, that are sending their fibers to specific structures of the thalamus, ventrolateral, and ventromedian nuclei, to cortical structures that are related to sensory, primary and secondary somatosensory cortices, and affective, anterior cingulate and insular cortices, components of pain [9].

The specificity theory is still confirmed by the identification of specific receptors, fibers, pathways, and CNS structures that are responsible for our perception of these somatosensory modalities. Even if several studies are supporting that these pathways and higher center structures are definitely playing a role in pain perception, their anatomical identification is not sufficient to explain the complexity of pain. The mechanisms involved in different conditions, such as the increasing perception of pain following repetitive nociceptive stimuli (temporal summation) or of a larger surface (spatial summation) or some chronic pain conditions, clearly support that the specificity theory alone cannot explain the complexity of pain.

2.2 Pattern Theory

The pattern theory, introduced by Goldscheider [30], suggested that not only the type of fibers, the pathways, or the different anatomical structures but also the pattern of impulses in the nervous system would modulate pain perception. Based on this theory, it is easier to understand that a thermal stimulus can pass from a warm perception to burning hot if the stimulation persists at the same temperature (temporal summation) or is presented on a larger surface (spatial summation).

Changes in the activation patterns could help understand complex phenomenon such as allodynia, pain from a non-painful stimulus, or spontaneous pain in conditions where no apparent lesions are detectable. We understand that even small changes in the neuronal activity of spinal or supraspinal structures will be sufficient to produce what is now known as central sensitization. Central sensitization can be described as a plasticity of the central nervous system that will produce a reduction of the threshold to produce a painful sensation to the point that even a non-nociceptive stimulus will be perceived as painful (allodynia) or more painful than usual (hyperalgesia) and a receptive field expansion that will enable the non-injured tissue to produce pain (secondary hyperalgesia) [95].

2.2.1 Patterns and Brain Dynamics

Electroencephalographic (EEG) activity of pain perception revealed that synchronous gamma-band frequency (30–100 Hz) seems to play a major role in the cortical integration of multiple sensory modalities, including pain [50]. Because of their non-specific modality responses, it was suggested that gamma-band oscillation (GBO) is more related to salience or attention [37]. However, primary somatosensory cortex (S1) GBO is correlating to pain perception, even when salience is reduced by repetition [100]. These results on GBO and other recent pain imaging techniques are stressing out that activity patterns and not just anatomical locations are essential to render the complexity of pain perception.

2.3 Gate Control Theory

In 1965 the gate control theory by Melzack and Wall [56] came with another important part of the complex puzzle of pain: the fact that endogenous pain modulatory mechanisms could enhance or reduce pain perception. For instance, the gate control theory proposed that the stimulation of non-painful Aβ afferences could produce a localized analgesia by blocking the nociceptive afferences directly at their entry in the spinal cord. Moreover, even if the specific mechanisms were not explained in the gate control theory, Melzack and Wall already proposed that descending mechanisms from higher centers would influence this modulatory mechanism.

2.4 Diffuse Noxious Inhibitory Mechanisms

A few years after the gate control theory was proposed, Reynolds demonstrated that stimulation of the periaqueductal gray (PAG) in the brainstem produced a strong inhibition [69]. The role of the rostroventral medulla in the modulation of pain has since been well documented [25]. Regions such as the PAG and the nucleus raphe magnus (NRM) have been identified as important serotoninergic and noradrenergic descending inhibitory pathways. These inhibitory pathways then recruit enkephalinergic interneurons in the spinal cord to produce the analgesic response.

We had to wait until the end of the 70s before a model known as diffuse noxious inhibitory controls (DNIC) was proposed [48, 49]. This model is based on the observation that a localized nociceptive stimulation can produce a diffuse analgesic effect over the rest of the body, an analgesic approach known as counter-irritation. In the DNIC model, Le Bars et al. [48, 49] proposed that a nociceptive stimulus will send input to superior centers, but will also send afferences to the PAG and NRM of the brainstem, recruiting diffuse descending inhibitory output at all levels of the spinal cord.

Together, the gate control and DNIC have played a very important role in supporting that pain perception is not only the endpoint of nociceptive activations but will also be modulated by several endogenous mechanisms. Deficits of these mechanisms are probably responsible for several complex chronic pain conditions [98].

2.5 Pain as a Homeostatic Emotion

Another very interesting view of pain has been proposed by Bud Craig [14]. Rather than seeing pain as part of the exteroceptive sense of touch, he suggests that we have neuroanatomical and neurophysiological demonstrations that it is in fact a homeostatic signal. The human feeling of pain is then both a distinct sensation and a motivation at the same time. This model makes sense when we think that pain is described as a sensory, affective, and cognitive experience. Moreover, even the International Association for the Study of Pain (IASP) is describing pain as the result of an actual or potential lesion. All these descriptions are fitting homeostatic behavioral drives. Moreover, lesion of the somatosensory cortex is not affecting pain, while thalamic stimulations are producing analgesia [14]. The earliest brain activity following a nociceptive stimulus is in the posterior insula and mid-cingulate cortex [51], two regions that are playing a role in the affective reactions and in homeostasis.

3 From the Periphery to the Cortex

One approach to study the neurophysiology of pain is to follow the nociceptive signal from the periphery to the cortex. It is also important to appreciate the role of descending signal from the higher centers to the brainstem and periphery that will modulate the nociceptive signal at all the level of the central nervous system, changing our pain perception.

There is no direct relation between nociceptive activity and pain perception. The term "nociception" comes from Sherrington's observations regarding stimuli that are likely to affect the integrity of the organism [72]. It indicates potentially painful or algesic nerve information before it comes to consciousness or higher brain centers. Frequently a nociceptive stimulus will be translated in pain; however, several conditions can change this perception depending on the salience or significance of the information that reach consciousness at the same time [90]. In a neutral condition, pain is normally very salient. It is a protective mechanism. However, in an emergency situation or during an important distraction, pain salience may shift to a second order and will be felt as lower or even absent.

In Fig. 1, we can follow the nociceptive signal from the periphery to the cortex.

(1) The nociceptive signal (mechanical, chemical, or thermal) will recruit peripheral nociceptors that conduct the signal in the primary afferent neurons to the dorsal horn of the spinal cord.
(2) In the dorsal horn, the primary afferent neuron will make a synaptic contact with the secondary or projection neuron that will constitute the spinothalamic and spinoreticular tracts that immediately cross in the spinal cord and send contralateral afferent projections to higher centers.
(3) A large proportion of afferents will make a second synapse in the lateral and medial nuclei of the thalamus. **It is important to emphasize that the secondary neurons may also synapse with neurons in different nuclei of the brainstem including the PAG and the NRM, areas involved in descending endogenous pain modulation.
(4) From the thalamus and brainstem nuclei, the secondary neuron will project to tertiary neurons to the primary and secondary somatosensory cortices (SI, SII). The SI and SII are involved in the sensory quality of pain, which includes location, duration and intensity. Tertiary neurons also project to limbic structures, including the ACC and the insula, which are involved in the affective or emotional component of pain.

All synaptic contacts with excitatory and inhibitory neurons at all levels of the CNS are site of important integration regions that are the target of most pharmacological approaches.

3.1 The Role of Glial Cells

During the last century, all efforts to understand the CNS mechanisms implicated in pain were focused on neurons. We just start to realize that glial cells are not just there for support or protection, but are playing a major and active role in several CNS processes, including pain [20]. The role of astrocytes [40] and microglia [75] have been well documented in the development and persistency of pain, especially in models of neuropathic pain. In normal conditions, glial cells seem to play a limited role in pain, with no or few effects on pain threshold [75]. However, after an injury, microglia becomes reactive. Activation of microglia in the dorsal horn is concomitant with the development of neuropathic pain [83]. Microglia releases several factors that are pronociceptive and mediated through a complex signaling system involving cytokines [96]. Even the paradoxical opioid-induced hyperalgesia phenomenon could be related to an action of microglial cells in the spinal cord [24]. Interestingly, there are several microglia-targeting drugs that can be developed based on the results from animal researches [75, 87, 96].

Imaging techniques using PET-MRI radiolabelling of glia activation markers are able to demonstrate the importance of glial activation in structures such as the thalamus in some chronic pain conditions [53].

We can no longer focus on the neurophysiology of pain without taking into account the major role of the glial cells and their roles in the development and treatment of pain.

3.2 From the Periphery to the Spinal Cord

Even if imaging techniques are particularly aiming at the activity of the higher centers, it is important to remember what is happening from

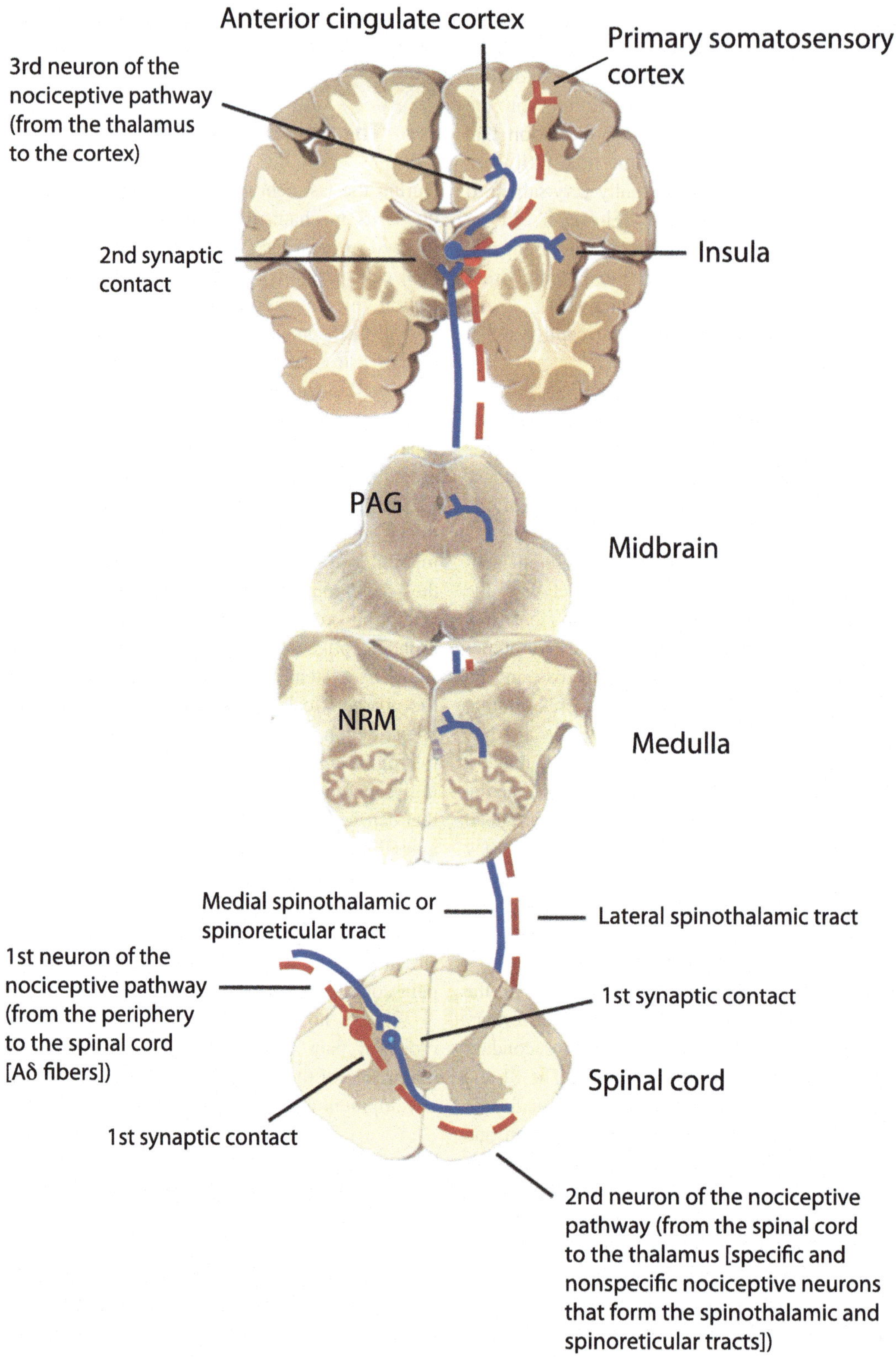
Anterior cingulate cortex
Primary somatosensory cortex
3rd neuron of the nociceptive pathway (from the thalamus to the cortex)
2nd synaptic contact
Insula
PAG
Midbrain
NRM
Medulla
Medial spinothalamic or spinoreticular tract
Lateral spinothalamic tract
1st neuron of the nociceptive pathway (from the periphery to the spinal cord [Aδ fibers])
1st synaptic contact
Spinal cord
1st synaptic contact
2nd neuron of the nociceptive pathway (from the spinal cord to the thalamus [specific and nonspecific nociceptive neurons that form the spinothalamic and spinoreticular tracts])

Fig. 1 Pain pathways: From the periphery to the cortex, we can follow the lateral spinothalamic (*broken red line*) and the spinoreticular (*full line*) from the periphery to the cortex. The lateral spinothalamic tract is projecting to the lateral thalamus nuclei and to the somatosensory cortex. The spinoreticular tract is projecting to the medial thalamus and different cortical structures associated to the affective component of pain including, but not restricted to, the insula and the cingulate cortex. These different pain pathways are activating the brain structures responsible for the complex pain-related perception

the periphery to make sense of what we found at the spinal and supraspinal level.

Afferent fibers originating in the periphery fall into three groups, namely Aβ, Aδ and C fibers.

3.2.1 Non-nociceptive Afferent Fibers

The Aβ fibers are large myelinated fibers that conduct at high speed (35–75 m/s) and usually transmit non-nociceptive signals. They do however also participate in pain modulation by recruiting inhibitory interneurons in the *substantia gelatinosa* of the dorsal horn of the spinal cord. This mechanism is one of the fundamental components of the gate control theory, whereby an innocuous stimulus will reduce the nociceptive input from the same region [56]. Besides playing a dynamic inhibitory role when recruited, the Aβ fibers seem also to play a tonic inhibitory role on the nociceptive input. Blocking the input from these large fibers will result in an increased response to nociceptive stimuli [67].

3.2.2 Nociceptive Fibers

Two other classes of fibers, the myelinated Aδ and the thin unmyelinated C fibers mainly transmit nociceptive messages. The Aδ fibers are myelinated and relatively large, conducting the signal relatively rapidly (5–30 m/s) from the periphery to the spinal cord. Because of this rapid conduction velocity, they are responsible for the sharp localization of pain and for the rapid spinal response, which can be measured in the laboratory as the nociceptive reflex. They represent the majority of the myelinated fibers. Two types of Aδ fibers exist depending on the specificity of their responses to different stimulation [8]: (1) the mechanonociceptors respond preferentially to intense and potentially harmful mechanical stimulation; and (2) the polymodal Aδ fibers respond to mechanical, thermal and chemical stimulations. Because of the rapid conduction velocity, the Aδ fibers are responsible for the first pain sensation, a rapid pinprick-like, sharp and transient sensation.

In contrast, the C fibers that have a slow conducting velocity (0.5–2 m/s) will mediate a second or dull aching pain. They represent three quarters of the sensory afferent input and are mostly recruited by nociceptive stimulation. Because of their slow conduction velocity, they are responsible for the second pain, a dull, diffuse and late sensation. However, they are also involved in non-nociceptive somatosensory information such as in the sensation of itch (pruritus) [74], and paradoxically, in the perception of pleasant touch, as documented in a patient with a rare disease linked to a deafferentation of the myelinated sensory fibers [62]!

3.2.3 First and Second Pain

The conduction velocity differences between the Aδ and C fibers can be appreciated when isolating the sensation of first and second pain (Fig. 2). Following a brief nociceptive stimulation, the Aδ fibers will rapidly transmit a brief and acute pinprick-like sensation perceived to be precisely located at the point of stimulation. Following this activity, C fibers will transmit their information, with a relatively long delay (100 ms to a second depending on the stimulus location). This second sensory input results in a more diffuse deep pain sensation.

It is possible to isolate first and second pain in the laboratory. Using a blood pressure cuff, we can temporarily block trophic factors present in the blood from reaching the nerves, resulting in a reduction of nerve conduction. The first fibers that will show reduced activity are those with largest diameter, including the Aδ fibers. This allows the activity of C fibers to be isolated and independently studied. Following this procedure, a nociceptive stimulation, independent of the nature of the stimulation, hot, cold or mechanical, will be perceived with a certain delay as a deeper pain sensation.

The application of capsaicin, the hot pepper extract, will produce a burning sensation due to the activation of the vanilloid receptors on C fibers. However, at higher doses, C fibers will be blocked as a result of a specific action on calcium ion channels, with resulting isolation of the Aδ fibers at the skin surface. This time, the same nociceptive sensation will be perceived as a sharp pinprick-like sensation without the second burning pain sensation.

Cortical representation of first and second pain has also been studied using magnetoencephalography (MEG). Among the regions activated, first pain was particularly related to activation of S1 whereas second pain was closely related to anterior cingulate cortex activation [66]. However, another study found no specific activations, but proposed that it's rather the recruitment of the same structures with different time windows.

3.3 The Spinal Cord: First and Important Step in the Central Nervous System

The first major distinction between nociceptive and non-nociceptive afferent fibers is that the latter ascend ipsilaterally (on the same side) to the brainstem before making synaptic contact with the second neuron and finally crossing to the opposite side before projecting to higher centers. For nociceptive fibers, the signal is transported to the dorsal horns of the spinal cord (or the brainstem for trigeminal afferent impulses) to make first synaptic contact with the secondary neurons (or projection neurons). The secondary neurons cross the spinal cord immediately under the central canal to form the spinothalamic contralateral projection tract.

The Aδ and C nociceptive fibers occupy the ventrolateral position in relation to the dorsal root. They make their way through Lissauer's tract, upward or downward, along one or more segments. Then, they ipsilaterally penetrate the dorsolateral portion of the dorsal horn.

3.3.1 Organization of the Spinal Cord

The gray matter of the spinal cord is divided into 10 cytoarchitectonic layers or laminae (known as Rexed laminae). The Aδ fibers mainly end in the first lamina and in the superficial portion of the second. Afferent fibers coming from the deep tissue and viscera, on the other hand, essentially end in laminae I and V [57]. C fibers mainly end in laminae I and II. As for the large myelinated Aβ fibers, they complete their journey in lamina III or deeper. In spite of their characteristic entry into the laminae, Aβ, Aδ, and C fibers establish connections among one another. The dorsal horn remains the preferred site for significant synaptic convergence. In fact, the same fiber from the dorsal horn of the spinal cord can receive cutaneous, muscular, and visceral afferent impulses [47]. The convergence of afferent impulses originating from different systems allows us to better understand the interaction that can exists between systems that seem independent at first. Therefore, muscular pain could be exacerbated by a new visceral pain, and vice versa.

The dorsal horns contain an important network of synaptic convergences, bringing together the collateral fibers and interneurons. Thus, passage through the sensory spine is an important step during which nociceptive information will be modulated. Its complex network of neurons, which includes endings of primary nociceptive neurons, secondary neurons, interneurons, and neurons of descending tracts, contains a multitude of neurotransmitters and a sizeable mosaic of receptors that will modulate the nociceptive afferent impulses before they are forwarded to the higher centers.

Three main categories of nerve cells in the CNS participate in nociception: nociceptive projection neurons, excitatory interneurons, and inhibitory interneurons.

3.3.2 Projection Neurons

Nociceptive projection neurons relay the message to the higher centers and are classified into two groups: specific nociceptive projection neurons and multireceptive projection neurons [32, 78]. Specific nociceptive neurons are neurons

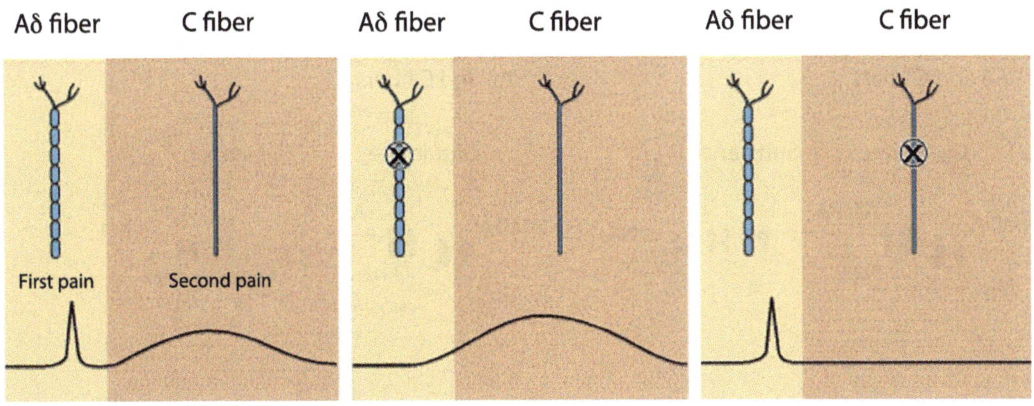

Fig. 2 Aβ, Aδ and C fibers

that receive their information only from primary afferent nociceptors. Therefore, they only respond to stimulations of mechanical or thermal origin of potentially painful intensity [2].

Multireceptive or wide-dynamic-range neurons gather information provided by the primary afferent nociceptors with mechanoreceptors. These are the neurons with small receptive fields that receive afferent impulses from Aδ and C fibers, and also from non-nociceptive Aβ fibers. Thus, these neurons of the dorsal horns of the spine respond in a graduated manner to stimulation of different intensity varying from non-nociceptive to nociceptive. Multireceptive neurons are dynamic and their receptive fields not only include excitatory area, but also inhibitory ones. Modification of these receptive fields plays an active role in certain types of chronic pain [47].

3.3.3 Pain Modulation in the Spinal Cord

An important challenge in pain imaging is to differentiate the signal from excitatory and inhibitory interneurons. The transmission of a nociceptive impulse is not summarized solely as the passage of nociceptive information between the first afferent neuron and the second projection neuron in the spinal cord. Excitatory and inhibitory interneurons actively participate in the modulation of nociceptive responses. As we saw a little earlier, glial cells also play a dominant role in nociceptive responses. The action of these excitatory and inhibitory neurons in spinal cord could lead to central sensitization, or hyperalgesia.

Hyperalgesia is defined as an exaggerated response to normally painful stimulation. In the 1950s, Hardy proposed that two kinds of hyperalgesia could affect the skin: primary hyperalgesia, occurring directly at the injury site, and secondary hyperalgesia, with its origins in the CNS [34]. Primary hyperalgesia can be explained by the release of different inflammatory factors in the periphery, which leads to the recruitment of nociceptors near the site of the injury (potassium, prostaglandins, bradykinin, histamine, substance P, and serotonin), which has the effect of recruiting nearby nociceptors and producing sensitization. The injury site as well as the neighboring tissues will thus have lower pain thresholds.

Secondary hyperalgesia, on the other hand, can be explained by a central phenomenon that is known by the general term "central sensitization" [95]. Repeated recruitment of C fibers after an injury can cause a series of events at the spinal level, which could have the effect of sensitizing the projection neurons in the dorsal horns of the spinal cord. High-frequency recruitment of C fibers will produce an increase in the action potential of the spinal neurons [23].

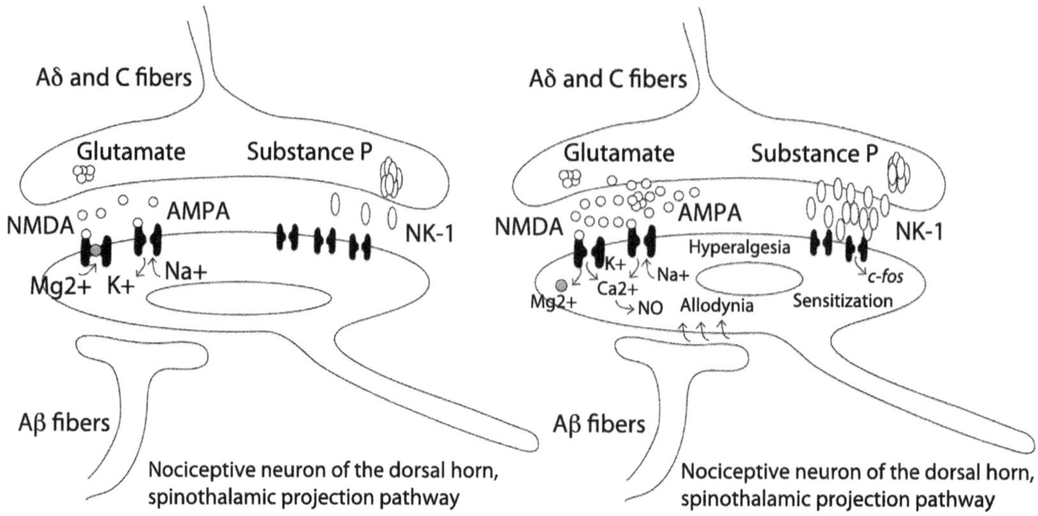

Fig. 3 Central sensitization: Following repetitive stimulation from a presynaptic neuron, high glutamate release will produce a cascade of postsynaptic activity that will produce long-lasting cellular sensitization resulting in central sensitization. Persisting central sensitization is proposed as being one the mechanisms implicated in chronic pain

Wind-up is a relatively short-lived transient phenomenon, but the repeated recruitment of C fibers can also lead to spinal sensitization, which may extend over several hours or even several days [86]. Thus, an intense, long-lasting stimulation will result in the recruitment of nociceptive fibers, including C fibers, which release excitatory amino acids (EAAs), glutamate, and peptides, such as substance P and CGRP. These neurotransmitters recruit postsynaptic glutamatergic receptors such as AMPA (α-amino-3--hydroxyl-5-methyl-4-isoxazole propionate) and NMDA in the case of EAAs, and neurokinin-1 receptors in the case of substance P. Prolonged stimulation of the NMDA receptors will produce long-lasting cellular sensitization through the activation of the gene transcription factors (c-fos and c-jun) (Fig. 3). These transcription factors induce the expression of some rapidly responding nuclear genes, in turn leading to nociceptor sensitization. This structural plasticity will have the effect of reducing the recruitment threshold of the nociceptors and thus producing hyperalgesia or allodynia, which could persist even after the injury has disappeared. The phenomenon of central sensitization allows us to better understand the importance of relieving pain as early as possible in order to avoid chronification.

3.4 Pain Pathways: From Spinal to Higher Centers

Generally, the nociceptive neurons of the dorsal horn follow a pathway through the anterolateral quadrant of the spinal cord. However, anterolateral cordotomy reveals that all fibers do not uniformly obey this rule. Indeed, this surgical intervention does not involve analgesia. Instead, it causes hypoalgesia, or a reduction in pain in response to a normally painful stimulus. In addition, in a few months, individuals who have undergone this intervention will partially recover their sensitivity [26]. Since regeneration is rather improbable in the CNS, the partial recovery of sensitivity suggests the contribution of more than one pathway. Other pathways are thus available to the nociceptive message, including the lateral spinothalamic tract, the dorsal column-medial lemniscus pathway, and the spinoreticulothalamic tract.

Two of these tracts are responsible for nociceptive afferents: the lateral spinothalamic tract and the spinoreticular tract (or medial spinothalamic tract). A third pathway, the dorsal column-medial lemniscus pathway, is mainly responsible for transporting non-nociceptive information originating from the Aβ fibers. The fibers of the dorsal column-medial lemniscus pathway are divided into gracile tracts (coming from the lower limbs) and cuneate tracts (coming from the upper limbs). These cells do not respond differently to nociceptive and non-nociceptive stimuli, and they project their afferent impulses into the ventrobasal complex [ventral posterolateral (VPL) and ventral posteromedial (VPM)] of the thalamus. They receive information about mild mechanical stimulations and joint movements. Nevertheless, anatomical and clinical studies have shown that the medial region of the dorsal column of the spinal cord play an important role in transporting visceral afferents, including nociceptive afferents coming from the viscera [93].

The lateral spinothalamic tract is, as its name indicates, in a lateral position, and it projects directly toward the lateral thalamic nuclei of the ventrobasal complex. The projections in the ventrobasal complex are also called the neospinothalamic tract. They generally have the characteristics of either specific or multireceptive nociceptors. The projection cells of the spinothalamic tract, mainly coming from laminae I and IV–VI [94], are projected toward the nuclei of the contralateral ventrobasal complex. Their receptive fields are generally contralateral and circumscribed. The fibers of the spinothalamic tract have rapid afferents with relatively precise receptive fields that project toward thalamic and then cortical regions with precise somatotopic representations. The spinothalamic tract, therefore, has the necessary qualities for localization and perception of the sensory-discriminative component of pain [92].

The spinoreticular tract is in a more medial position. Its projections in the medial complex of the thalamus are also called the paleospinothalamic tract. The majority of its afferents come from the deep laminae VII and VIII and are projected toward the medial nuclei of the thalamus and certain structures of the brainstem, including the PAG and NRM [92]. Unlike the spinothalamic tract, the spinoreticular tract has very large receptive fields that sometimes cover the whole body. The spinoreticular tract afferents mostly come from slow C fibers. Those projections lead toward the regions of the brainstem, thalamus, and cortex, which play major roles in memory and emotions. These qualities make it an ideal candidate for having a dominant role in the perception of the unpleasant or motivational–affective aspect of pain [92]. It is by the activation of the spinoreticular tract that we recruit descending analgesia (Diffuse noxious inhibitory control—DNIC) [19].

3.4.1 Visceral Pain: A Specific Pathway

The visceral system is a very sophisticated sensory system implicating the concomitant activity of two extrinsic innervations, vagal and spinal, as well as numerous intrinsic neurons [45]. For example, the intestine has a neuronal system that operates independently but also in relation with the rest of the CNS, known as the brain–gut axis. Several visceral pain syndromes, as the irritable bowel syndrome, present no clear lesion or dysregulation of the painful organ. The brain–gut axis seems to play an important role in these syndromes and may help to better understand the interaction between external events such as a stressful situation and an effect on the symptoms [42]. Emerging data are also stressing the importance of the microbiome, proposing the significance of a «microbiome-brain-gut axis» in some pathological conditions, including pain [4]. These results suggest that alteration in the gut microbial composition is associated with marked changes in behaviors such as mood, pain, and cognition, that are related to a bidirectional communication between the brain and the gut microbiota [79]. Understanding these interactions may lead to treatments acting on the microbiota that will affect brain functions.

3.4.2 Brain, Gut, and Emotions

As for somatic pain, chronic visceral pain is related to both peripheral and central sensitization. Excitatory and inhibitory descending pathways are also implicated in the visceral system, suggesting an important central influence of visceral sensitivity. Finally, the autonomic nervous system influence on visceral sensitivity may help explain the role of emotions on the modulation of visceral pain. Based on these observations, some chronic visceral pain presents the characteristics of neuropathic pain [45].

Interestingly, we all have experienced what we call a «gut-feeling». For example, a situation that feels uncomfortable without being able to clearly identify why.

William James, at the end of the nineteenth century [39], already proposed that body responses are fundamental to perceive emotions. More recent researches are supporting the importance of our body interoception informing us about states such as well-being or stress that is encoded in the insula and is playing an important role in general emotional states, including our analysis of a pain state [13, 17]. This close interaction between visceral afferences and the insular cortex (IC) may help understand why visceral pain has such important emotional effects.

3.5 Higher Center and Pain

We have known for a long time that pain is a complex sensory and emotional experience demanding the participation of the higher centers of the CNS. Nevertheless, the role of the cortex in pain perception has been demonstrated only recently, despite studies dating back to the beginning of the twentieth century from Head and Holmes proposing that it is only once the nociceptive information is sent to the cortex that we can really speak of pain, since pain is a perception [35, 36]. Because an animal cannot tell us its perception of pain, we must refer to its nociceptive behaviors, suspecting that these behaviors are generally responses to pain. The last few decades have been crucial in identifying the role of the different cortical regions in pain. Dividing the thalamic nuclei into groups that receive afferents from the sensory–discriminative tract and those that receive afferents from the motivational–affective tract can simplify the presentation of the cerebral structures implicated in pain perception.

3.5.1 Imaging Pain Response in Higher Centers

Imaging studies of pain are reporting activation in multiple brain regions including SI, SII, ACC/MCC, insula, PFC, cerebellum, and supplementary motor area (SMA) [18, 29]. In brain imaging, we study the experience of pain by trying to figure out the role of different structures by establishing a link between pain characteristics and the activation of some structures. The «pain matrix» proposed by Melzack [55] paved the way for the imaging studies that found different structures that are implicated in different components of the pain experience. Most of imaging studies are reporting activities in a number of brain sites including sensory (SI, SII), affective (ACC/MCC, insula, PFC), cognitive (ACC/MCC, PFC, SII), and motor (SMA, cerebellum) aspects of pain [18, 29]. However, it has been proposed that there is no specific pain matrix since activities in these regions can also be recorded by different stimulation modalities that are not painful and could then be more related to the salience of the stimuli rather than specific to pain [37].

3.5.2 Pain Matrices

Garcia-Larrea and Peyron [29] proposed that there are at least three pain matrices that are responsible for our complex pain experiences: (1) the nociceptive cortical matrix, (2) the perceptual matrix, and (3) the pain memory matrix.

(1) The nociceptive cortical matrix is projecting from the posterior thalamus nuclei to the posterior insula, medial parietal operculum, and mid-cingulate cortex. This first-order matrix is the earliest response to noxious stimuli.
(2) The perceptual matrix is composed of several cortical regions including the mid and anterior insula, anterior cingulate, PFC, and the posterior parietal area. This perceptual matrix

is different from the nociceptive matrix by the fact that it does not receive direct nociceptive inputs and it can be activated in context not involving pain. It is a context dependent matrix.

(3) The pain memory matrix is composed of several high-orders cortical structures such as the perigenual cingulate, the orbitofrontal cortex, the temporal lobe, and the anterolateral PFC. We know that important changes in pain perception can occur without any nociceptive stimuli and without changing the activities in the pain pathways from the thalamus to the somatosensory cortex for instance. A good example is the empathetic observation of someone in a painful situation that will make us grimacing and feeling as if we were experiencing their pain [38]. We recently found that we can even trigger our endogenous pain inhibitory mechanisms just by observing ourselves or someone else during experimental pain (cold pressor test) [31]. Manipulating the unpleasantness of a painful stimulus by giving it a different meaning, being the less intense or «more pleasant» versus being the most intense or «more unpleasant» in a series of stimuli will totally change our perceived intensity by changing the activity in higher order structures of this matrix [52].

3.6 Specificity of Brain Regions in Pain

Pain is a complex phenomenon constructed around several sensory, affective, and cognitive concepts that are interacting and adapting to our environment in relation with previous experiences. It is then artificial to present different brain structures as being responsible for a specific pain component since each of these structures is influencing how the other structures of the «pain matrix» will code the message. New imaging approaches that are not time-locked to painful stimuli, but rather are measuring more natural ongoing pain in patients using resting state imaging, are demonstrating how dynamic several brain regions are activated during this resting state [46, 88].

For the sake of understanding the neurophysiology and the pathophysiology of pain, it is still interesting to try to understand the neuroanatomical organization of the higher structures that are playing different roles in the experience of pain, keeping in mind that it is not a static but rather dynamic system.

Since the first studies of cerebral imagery of the regions that play a role in pain using positron emission tomography (PET) [77], several subsequent studies have confirmed the participation of the four principal cerebral centers (Fig. 4): the primary somatosensory cortex (SI), in the postcentral gyrus of the parietal lobe; the secondary somatosensory cortex (SII), in the parietal operculum; the anterior and medial cingulate cortex (ACC/MCC), in the cingulate gyrus; and the insula, in the lobe of the IC, which is found under the temporal and frontal lobes, in the Sylvian fissure [12]. Methods that involve making a lesion specific to structures or recording nerve cells in these same localized regions have only allowed us to have a fragmented view of the role of the cortex in pain. We have sufficient data to conclude that cortical structures such as SI contribute to the sensory–discriminative component of pain, whereas the frontal, cingulate, and insular cortical structures are involved in the motivational–affective component [12, 43, 77].

3.6.1 Sensory Discrimination: Primary Somatosensory Cortex (SI)

The spinothalamic tract, originating from the ventrobasal complex of the thalamus, projects toward the primary (SI) and secondary (SII) somatosensory cortices [92]. Injuries to these structures produce a loss of capacity to specify the location and intensity of nociceptive stimulation [9, 43], which confirms their role in the sensory–discriminative component of pain. However, it is important to note that injuries to the somatosensory cortex can sometimes produce the completely opposite effect, hyperalgesia [43]. This phenomenon can be explained by the destruction of the excitatory or inhibitory cortical

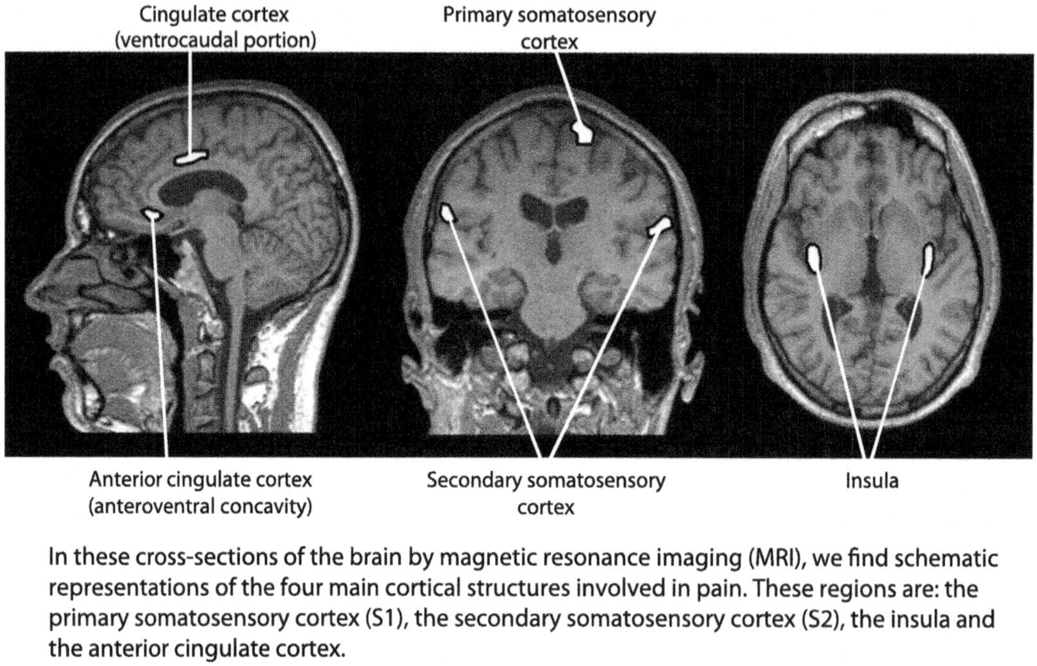

Fig. 4 Some of the cortical structures involved in pain: Schematic representations of the four main cortical structures involved in pain. These regions are the primary somatosensory cortex (S1), the secondary somatosensory cortex (S2), the insula, and the anterior cingulate cortex

regions, depending on the extent of the injury to the parietal cortex. Several studies have confirmed the role of the SI cortex in the sensory–discriminative component of pain because of its specific activity in certain cerebral imaging protocols aimed at isolating the nociceptive component of a stimulus [7].

3.6.2 Sensory Discrimination: Secondary Somatosensory Cortex (SII)

Although it appears to be less specifically involved than the SI cortex, the SII cortex also seems to play an important role in the sensory–discriminative component of the location and appreciation of the characteristics of nociceptive stimulation, despite receptive fields of variable and generally bilateral dimensions. In patients who have undergone a hemispherectomy because of chronic untreatable epilepsy attacks, the stimulation of the leg contralateral to the lesion causes the activation of the SI cortex on the same side as the stimulated leg, as opposed to the contralateral activation seen in healthy subjects [63]. This cortical reorganization brings to light the possible participation of networks between SI and SII in enabling a certain plasticity of the somatosensory cortex [44].

3.6.3 The Motivational–Affective Component of Pain

The ACC, MCC, and the insula are regions of the limbic system that play a dominant role in the motivational–affective component of pain. In addition, their wide receptive fields are covering large surfaces of the body, suggesting that these structures participate in general and interoceptive sensations [13, 16].

3.6.4 Motivational–Affective: Anterior Cingulate Cortex (ACC)

Several studies have highlighted the participation of the cingulate gyrus following painful stimulation [85, 101]. Clinical studies in patients who have had injuries to the ACC have revealed a reduction of both clinical [65] and experimental pain [77]. This region of the limbic system receives its afferents from the medial pathway and plays a dominant role in the motivational–affective component of pain. Visceral pain with a strong affective component, such as that associated with irritable bowel syndrome, preferentially activates this cerebral structure [10], highlighting its role in the affective component of pain. The ACC is highly related to the psychological construct we have about pain. The anticipation of pain is activating the ACC [90].

3.6.5 Motivational–Affective: Insular Cortex (IC)

The complex of the IC has several means of contact with the cortical structures that are classically associated with pain: SI, SII, and cingulate cortices. The insula has several contacts with the limbic structures such as the amygdala and perirhinal cortices, suggesting an important role in the affective component of pain. In some individuals, stimulation of the insular complex produces emotional sensations of fear, and injury to this same structure produces an absence of emotional responses to nociceptive stimulation [82]. The presence of thermoreceptive and nociceptive neurons in the IC has been clearly documented [16]. In one study on Thunberg's thermal grill illusion, which consists of a paradoxical perception of pain in contact with warm and cold juxtaposed bars that would only produce painless hot or cold sensations if they were touched individually, Craig and Bushnell [15] showed that the pain comes from a decrease in tonic inhibition of nociceptive neurons by the simultaneous presentation of hot and cold temperatures. This phenomenon is mainly produced in the insula and might occur with certain pains of central origin [15], which would explain the pain similar to a burn felt by patients with a thalamic syndrome. As we saw earlier, the insula is also the hub for homeostatic signal [14].

3.6.6 Cognitive Control: Prefrontal Cortex (PFC)

The PFC is directly connected to the limbic system and has been demonstrated to be responsible for regulating our emotions, including the motivational–affective aspect of pain [89]. Moreover, the PFC is in direct communication with descending pain modulation pathways, including the PAG. Roy and colleagues [70] demonstrated that this interconnection between the PFC and the PAG is playing a major role in learning and predicting errors, a circuit to learn how to avoid painful situations.

A study of Leknes and colleagues on the reappraisal of painful stimuli is a very nice demonstration of the importance of the PFC on pain [52]. In their study the same stimulus intensity was perceived as unpleasant or pleasant depending on the context where this stimulus was the worst or the least painful. In the worst painful situation, the nociceptive stimuli were presented alternatively with non-painful stimuli. In the least painful context, the same stimuli were presented alternatively with more intense stimuli. In the least painful condition, the subject's perception flipped from a negative to positive hedonics relative to the context. A complex circuitry triggered by the orbitofrontal and ventrobasal PFC was reducing the insula and ACC activity, but also activated the PAG pain modulation pathway.

3.7 Pain, a Multifaceted Perception Needing a Large Brain Network

In summary, our growing understanding of the role of the higher centers in pain allows us to realize the complex balance between the sensory and affective components. It is now easier than ever to accept the importance of the mutual influence between emotions and sensations in the pain experience. Certain higher centers (SI, SII) specialize in the sensory–discriminative component of pain to give precise information on the location, intensity, and all the other characteristics of the nociceptive stimulation. Other centers (ACC, IC) specialize in the emotional appreciation of pain. The affective component is not only associated with the intensity of the stimulation, but it also refers to other emotions, such as anticipation or fear [68]. For example, we may experience suffering when we attend to the pain of another person, especially when this person is dear to us. A study revealed that empathy for other people's suffering activates the same brain centers associated with the motivational–affective component of pain as if it were our own pain, but without the activity of the centers associated with the sensory–discriminative component [73]. Our perception of the pain of others is, therefore, quite real, in cerebral terms!

3.7.1 Resting State Activity: The Default Mode Network (DMN)

Brain imaging using PET or fMIR was based on a repetitive task, in our case a painful stimulus, subtracted from a control task where the subject was doing nothing. However, we know that doing nothing is not possible. Most of the time the subject will think either at the previous or coming task, or at some other things. A relatively recent tendency is to record what is happening during the baseline, or the resting state [33]. It was no surprise to realize that the brain is not at idle, but very active. The active regions included the medial PFC, medial temporal lobe, posterior cingulate cortex, and lateral parietal cortex [27]. Studies are supporting that this activity is related to connections between these structures for a co-activity or deactivation that may subserve salience, executive control, cognitive, and emotional functions [18].

Studies have shown that the DMN is abnormal in chronic pain patients. Abnormal DMN activity may help understand the focus on pain in chronic pain [1]. The default mode is associated with a "mind-wondering" that is contrasting to living the moment as proposed by the philosophy of contemplative meditation. Interestingly, the main nodes of the default mode network, medial prefrontal and posterior cingulate cortices, were relatively deactivated in experienced meditators [5]. Even relatively new meditators practicing mindfulness get the beneficial effects that are related to changes in the DFN including increased activation in the right dorsolateral PFC and in the left caudate/anterior insula and decreased activation in the rostral PFC and right parietal cortex [80].

A thorough understanding of the neuronal networks of the higher centers allows us to better grasp the nature of the physiology of pain and pathophysiology of certain types of chronic pain conditions. Based on these results, it is obvious that affective and cognitive components are playing major roles in several pain conditions, stressing the need to select an intervention that takes these aspects into account in the treatment of pain.

3.7.2 Interaction of Excitatory and Inhibitory Mechanisms

As we just described, excitatory mechanisms, such as central sensitization , can increase the nociceptive signal while inhibitory mechanisms will decrease the signal. Persistent pain can result from the recruitment of excitatory mechanisms such as central sensitization or the reduction of the efficacy of inhibitory mechanisms [54, 98]. Central sensitization is expressed as pain hypersensitivity, particularly dynamic tactile allodynia, secondary punctate hyperalgesia, aftersensations, and enhanced temporal summation. Quantitative sensory testing is generally used to characterize these abnormal sensations. On the other hand, efficacy of inhibitory mechanisms is tested using the response of conditioned pain modulation (CPM; also known as diffuse noxious inhibitory control—DNIC).

The recruitment of receptors implicated in the membrane depolarization (e.g., *N*-methyl-D-Aspartate—NMDA) will produce a neuronal hyperexcitability and the resulting pain will be related to endogenous pain excitatory mechanisms [23, 97]. On the other hand, a deficit of inhibitory mechanisms will be related to a reduced activity of descending serotonergic and noradrenergic pathways [58]. Even if two patients present apparently similar pain conditions, the implicated mechanisms may be different and will not respond to the same treatments. For instance, in the case of excitatory hyperactivity (central sensitization), anticonvulsant may be a good treatment choice. However, if a deficit of inhibitory mechanisms is implicated, better results may be obtained with antidepressant to trigger back serotonergic and noradrenergic endogenous inhibitory mechanisms (DNIC) [99].

Recent studies have highlighted the fact that relatively simple quantitative sensory testing is able to identify a deficit of excitatory (sensitization by temporal summation) versus a deficit of CPM that respond differently to different classes of drugs. For example, studies have shown that a deficit of CPM is a good predictor of the response to duloxetine, a noradrenergic, and serotonergic drug [99], while temporal summation was a good predictor of the response to pregabalin (blocker of neuronal hyperactivity in the class of anticonvulsant drugs) [61, 64]. Interestingly, the response is specific to the mechanisms; CPM efficacy was not a good predictor of pregabalin efficacy while temporal summation was not predicting the efficacy of duloxetine.

These results support that finding new approaches to detect the implicated mechanisms in chronic pain will help guiding the treatment. The different brain imaging techniques are part of the tools that will help identifying specific mechanisms and the specific effects of some treatments [18, 81].

3.8 Chronic Pain: A Central Sensitization Paradigm

Brain imaging studies are used to understand pain mechanisms in healthy subjects, but also to better characterize the mechanisms implicated in different chronic pain conditions. Central sensitization, which we can define as a pain that is maintained by the central nervous system, is probably one of the most accepted theories to understand how pain could persist for so long in patients that present no apparent injury. Understanding the mechanisms of central sensitization is important to help predict and reduce the occurrence of chronification, but also to offer treatments that are adapted to specific pathologies.

Memory and pain share common grounds. For instance, long-term potentiation (LTP), a lasting increase in synaptic strength that is necessary for learning and memory [3], is probably responsible for persisting lower pain threshold or spontaneous pain. It is comparable to central sensitization that is also a synaptic facilitation that is leading to reduced pain threshold and amplification of pain responses [41]. Interestingly, LTP can be induced in the pain pathways by high-frequency stimuli on Aδ or C fibers, but can also be activated by natural noxious stimuli, but only if descending, presumably inhibitory pathways are interrupted or weakened, suggesting an interaction between excitatory and inhibitory mechanisms [71].

4 Conclusion

Pain is a complex phenomenon. The neurophysiology of pain juxtaposes several different parameters. From the periphery to the higher centers, the nociceptive information goes through several steps: sensory conduction, transmission, modulation, and perception. It is then translated into pain behaviors that express suffering and help seeking.

To explain the course of the nerve impulse, we often use simplifications that follow a linear path. However, pain perception is much more than the mere expression of the nociceptive signal. The activity of modulation systems at all levels of the nervous system illustrates the difficulty in establishing a link between the activation of a nociceptor and the pain felt. The sensory aspect of pain is of importance, but the affective component is responsible for most of the pain modulation mechanisms.

Neurophysiological understanding of this modulation process allows us to put it to use for the treatment of pain. It helps maximize the efficacy of drug therapies and opens up a variety of nonpharmacological interventions for patients.

Brain imaging has revolutionized how we can study pain neurophysiology. We realize that pain pathways that were described using lesion methodologies or electrophysiology are confirmed and better understood. We also found new pathways or new regions that are linked to different behavioral processes linked to pain perception. Several regions of the higher centers playing different roles in the sensory–discriminative, motivational–affective, cognitive, homeostatic, and in the salience of the experience are interconnected and will constantly change our perception of pain. Better understanding this complexity is the only way to better understand the variability of pain responses between patients that seems to have comparable disease. They will also help understand that even if two patients present apparently similar pain conditions, they may not respond to the same treatments depending on the implicated mechanisms.

Brain imaging is not a fishing expedition. In most of the case, we are targeting specific regions in our analysis. However, in some conditions, brain activities could be recorded without being linked to a specific condition in order to understand what is happening during a more natural resting state condition.

The future is really bright for brain imaging. New techniques are emerging very rapidly and techniques such as MRI, PET, and electrophysiology are used in parallel to take advantage of their unique qualities.

References

1. Baliki MN, Mansour AR, Baria AT, Apkarian AV. Functional reorganization of the default mode network across chronic pain conditions. PLoS ONE. 2014;9(9):e106133. doi:10.1371/journal.pone.0106133.
2. Besson JM, Chaouch A. Peripheral and spinal mechanisms of nociception. Physiol Rev. 1987;67:67–186.
3. Bliss TV, Collingridge GL. A synaptic model of memory: long-term potentiation in the hippocampus. Nature. 1993;361(6407):31–9. doi:10.1038/361031a0.
4. Borre YE, Moloney RD, Clarke G, Dinan TG, Cryan JF. The impact of microbiota on brain and behavior: mechanisms & therapeutic potential. Adv Exp Med Biol. 2014;817:373–403. doi:10.1007/978-1-4939-0897-4_17.
5. Brewer JA, Worhunsky PD, Gray JR, Tang YY, Weber J, Kober H. Meditation experience is associated with differences in default mode network activity and connectivity. Proc Natl Acad Sci U S A. 2011;108(50):20254–9. doi:10.1073/pnas.1112029108.
6. Buchheit T, Van de Ven T, Shaw A. Epigenetics and the transition from acute to chronic pain. Pain Med. 2012;13(11):1474–90. doi:10.1111/j.1526-4637.2012.01488.x.
7. Bushnell MC, Duncan GH, Hofbauer RK, Ha B, Chen JI, Carrier B. Pain perception: is there a role for primary somatosensory cortex? Proc Natl Acad Sci U S A. 1999;96(14):7705–9.
8. Byers MR, Bonica JJ, Loeser JD. Peripheral pain mechanisms and nociceptor plasticity. Management of pain. New York: Lippincott Williams & Wilkins; 2001. p. 26–72.
9. Casey KL, Bushnell MC. The imaging of pain: background and rational. Pain imaging. Progress in pain research and management. Seattle: IASP Press; 2000. p. 1–29.
10. Chang L, Berman S, Mayer EA, Suyenobu B, Derbyshire S, Naliboff B, Vogt B, Fitzgerald L, Mandelkern MA. Brain responses to visceral and somatic stimuli in patients with irritable bowel syndrome with and without fibromyalgia. Am J Gastroenterol. 2003;98(6):1354–61.
11. Coghill RC, McHaffie JG, Yen YF. Neural correlates of interindividual differences in the subjective experience of pain. Proc Natl Acad Sci U S A. 2003;100(14):8538–42.
12. Coghill RC, Talbot JD, Evans AC, Meyer E, Gjedde A, Bushnell MC, Duncan GH. Distributed processing of pain and vibration by the human brain. J Neurosci. 1994;14(7):4095–108.
13. Craig AD. How do you feel? Interoception: the sense of the physiological condition of the body. Nat Rev Neurosci. 2002;3(8):655–66.
14. Craig AD. A new view of pain as a homeostatic emotion. Trends Neurosci. 2003;26(6):303–7.
15. Craig AD, Bushnell MC. The thermal grill illusion: unmasking the burn of cold pain. Science. 1994;265(5169): 252–55.
16. Craig AD, Chen K, Bandy D, Reiman EM. Thermosensory activation of insular cortex. Nat Neurosci. 2000;3(2):184–90.
17. Damasio A, Carvalho GB. The nature of feelings: evolutionary and neurobiological origins. Nat Rev Neurosci. 2013;14(2):143–52. doi:10.1038/nrn3403.
18. Davis KD, Moayedi M. Central mechanisms of pain revealed through functional and structural MRI. J Neuroimmune Pharmacol (The Official Journal of the Society on NeuroImmune Pharmacology). 2013;8(3):518–34. doi:10.1007/s11481-012-9386-8.
19. De Broucker T, Cesaro P, Willer JC, Le Bars D. Diffuse noxious inhibitory controls in man. Involvement of the spinoreticular tract. Brain. 1990;113(Pt 4):1223–34.
20. De Leo JA, Sorkin LS, Watkins LR, International Association for the Study of Pain. Immune and glial regulation of pain. Seattle: IASP Press; 2007.
21. Descalzi G, Ikegami D, Ushijima T, Nestler EJ, Zachariou V, Narita M. Epigenetic mechanisms of

chronic pain. Trends Neurosci. 2015;38(4):237–46. doi:10.1016/j.tins.2015.02.001.
22. Descartes R. Traité de l'homme. In: Descartes Oeuvres et lettres. Bibliothèque de la Pléiade, Gallimard; 1644. p. 803–873.
23. Eide PK. Wind-up and the NMDA receptor complex from a clinical perspective. Eur J Pain. 2000;4 (1):5–15.
24. Ferrini F, Trang T, Mattioli TA, Laffray S, Del'Guidice T, Lorenzo LE, Castonguay A, Doyon N, Zhang W, Godin AG, Mohr D, Beggs S, Vandal K, Beaulieu JM, Cahill CM, Salter MW, De Koninck Y. Morphine hyperalgesia gated through microglia-mediated disruption of neuronal Cl(-) homeostasis. Nat Neurosci. 2013;16 (2):183–92. doi:10.1038/nn.3295.
25. Fields HL, Basbaum A, Heinrich RL. Central nervous system mechanisms of pain modulation. In: McMahon SB, Koltzenburg M, editors. Wall and Melzack's Texbook of pain, vol. 5. Philadelphia: Elsevier Limited; 2006. p. 125–42.
26. Fields HL. Pain. New York: McGraw-Hill Book Company; 1987.
27. Fox MD, Snyder AZ, Vincent JL, Raichle ME. Intrinsic fluctuations within cortical systems account for intertrial variability in human behavior. Neuron. 2007;56(1):171–84. doi:10.1016/j.neuron. 2007.08.023.
28. Frey MV. Treatise on the sensory functions of the human skin. In: Handwerker HO, editor. Classical German contributions to pain research. Allemagne: Gesellschaft zum Studium des Schmerzes für Deutschland; 1897. p. 69–132.
29. Garcia-Larrea L, Peyron R. Pain matrices and neuropathic pain matrices: a review. Pain. 2013;. doi:10.1016/j.pain.2013.09.001.
30. Goldscheider A. The specific energy of the sensory nerves of the skin. In: Handwerker HO, editor. Classical German contributions to pain research. Allemagne: Gesellschaft zum Studium des Schmerzes für Deutschland; 1884. p. 47–69.
31. Gougeon V, Gaumond I, Goffaux P, Potvin S, Marchand S. Triggering descending pain inhibition by observing ourselves or a loved-one in pain. Clin J Pain. 2016;32(3):238–45. doi:10.1097/AJP. 0000000000000244.
32. Guilbaud G, Besson JM, Brasseur C. Physiologie du circuit de la douleur. In: Douleurs. Paris: Maloine; 1997.
33. Gusnard DA, Raichle ME, Raichle ME. Searching for a baseline: functional imaging and the resting human brain. Nat Rev Neurosci. 2001;2(10):685–94. doi:10.1038/35094500.
34. Hardy JD, Wolff GH, Goodell H. Pain sensation and reactions. Baltimore: Williams & Wilkins; 1952.
35. Head H, Holmes G. Sensory disturbances from sensory cerebral lesions. Brain. 1911;34:102–254.
36. Holmes G. Disorders of sensation produced by cortical lesions. Brain. 1927;50:413–27.
37. Iannetti GD, Mouraux A. From the neuromatrix to the pain matrix (and back). Experimental Brain Res. 2010;205(1):1–12. doi:10.1007/s00221-010-2340-1.
38. Jackson PL, Meltzoff AN, Decety J. How do we perceive the pain of others? A window into the neural processes involved in empathy. NeuroImage. 2005;24(3):771–9. doi:10.1016/j.neuroimage.2004. 09.006.
39. James W. Discussion: the physical basis of emotion. Psychol Rev. 1894;1:13. doi:10.1037/h0065078.
40. Ji RR, Kawasaki Y, Zhuang ZY, Wen YR, Decosterd I. Possible role of spinal astrocytes in maintaining chronic pain sensitization: review of current evidence with focus on bFGF/JNK pathway. Neuron Glia Biol. 2006;2(4):259–69. doi:10.1017/ S1740925X07000403.
41. Ji RR, Kohno T, Moore KA, Woolf CJ. Central sensitization and LTP: do pain and memory share similar mechanisms? Trends Neurosci. 2003;26 (12):696–705. doi:10.1016/j.tins.2003.09.017.
42. Jones MP, Dilley JB, Drossman D, Crowell MD. Brain-gut connections in functional GI disorders: anatomic and physiologic relationships. Neurogastroenterol Motil. 2006;18(2):91–103.
43. Kenshalo DR, Jr., Douglass DK, Bromm B, Desmedt JE. The role of the cerebral cortex in the experience of pain. In: Bromm B, Desmedt JE (eds) Pain and the brain: from nociception to cognition. Advances in pain research and therapy. New York: Raven Press; 1995. p. 21–34.
44. Khoshnejad M, Piche M, Saleh S, Duncan G, Rainville P. Serial processing in primary and secondary somatosensory cortex: a DCM analysis of human fMRI data in response to innocuous and noxious electrical stimulation. Neurosci Lett. 2014;577:83–8. doi:10.1016/j.neulet.2014.06.013.
45. Knowles CH, Aziz Q. Basic and clinical aspects of gastrointestinal pain. Pain 2009;141(3):191–209. doi:10.1016/j.pain.2008.12.011 S0304-3959(08) 00731-8 [pii].
46. Kucyi A, Salomons TV, Davis KD. Mind wandering away from pain dynamically engages antinociceptive and default mode brain networks. Proc Natl Acad Sci U S A. 2013;110(46):18692–7. doi:10. 1073/pnas.1312902110.
47. Le Bars D. The whole body receptive field of dorsal horn multireceptive neurones. Brain Res Brain Res Rev. 2002;40(1–3):29–44.
48. Le Bars D, Dickenson AH, Besson JM. Diffuse noxious inhibitory controls (DNIC). 1. Effects on dorsal horn convergent neurones in the rat. Pain. 1979;6(3):283–304.
49. Le Bars D, Dickenson AH, Besson JM. Diffuse noxious inhibitory controls (DNIC). II. Lack of effect on non-convergent neurones, supraspinal involvement and theoretical implications. Pain. 1979;6(3):305–27.
50. Lee MC, Mouraux A, Iannetti GD. Characterizing the cortical activity through which pain emerges

from nociception. J Neurosci. 2009;29(24):7909–16. doi:10.1523/JNEUROSCI.0014-09.2009.
51. Lee MC, Tracey I. Imaging pain: a potent means for investigating pain mechanisms in patients. Br J Anaesth. 2013;111(1):64–72. doi:10.1093/bja/aet174.
52. Leknes S, Berna C, Lee MC, Snyder GD, Biele G, Tracey I. The importance of context: when relative relief renders pain pleasant. Pain. 2013;154(3):402–10. doi:10.1016/j.pain.2012.11.018.
53. Loggia ML, Chonde DB, Akeju O, Arabasz G, Catana C, Edwards RR, Hill E, Hsu S, Izquierdo-Garcia D, Ji RR, Riley M, Wasan AD, Zurcher NR, Albrecht DS, Vangel MG, Rosen BR, Napadow V, Hooker JM. Evidence for brain glial activation in chronic pain patients. Brain. 2015;138 (Pt 3):604–15. doi:10.1093/brain/awu377.
54. Marchand S. Applied neurophysiology. In: Beaulieu P, Lussier D, Porreca F, Dickenson AH, editors. Pharmacology of pain. Seattle: IASP Press; 2010. p. 3–26.
55. Melzack R. Phantom limbs and the concept of a neuromatrix. Trends in Neuroscience. 1990;13:88–92.
56. Melzack R, Wall PD. Pain mechanisms: a new theory. Science. 1965;150:971–9.
57. Millan MJ. The induction of pain: an integrative review. ProgNeurobiol. 1999;57(1):1–164.
58. Millan MJ. Descending control of pain. Prog Neurobiol. 2002;66(6):355–474.
59. Müller J. On the sense of feeling. In: Handwerker HO, editor. Classical German contributions to pain research. Allemagne: Gesellschaft zum Studium des Schmerzes für Deutschland; 1837. p. 27–47.
60. Nielsen CS, Stubhaug A, Price DD, Vassend O, Czajkowski N, Harris JR. Individual differences in pain sensitivity: genetic and environmental contributions. Pain. 2008;136(1–2):21–9. doi:10.1016/j.pain.2007.06.008.
61. Niesters M, Proto PL, Aarts L, Sarton EY, Drewes AM, Dahan A. Tapentadol potentiates descending pain inhibition in chronic pain patients with diabetic polyneuropathy. Br J Anaesth. 2014;113(1):148–56. doi:10.1093/bja/aeu056.
62. Olausson H, Lamarre Y, Backlund H, Morin C, Wallin BG, Starck G, Ekholm S, Strigo I, Worsley K, Vallbo AB, Bushnell MC. Unmyelinated tactile afferents signal touch and project to insular cortex. Nat Neurosci. 2002;5(9):900–4.
63. Olausson H, Marchand S, Bittar RG, Bernier J, Ptito A, Bushnell MC. Central pain in a hemispherectomized patient. Eur J Pain. 2001;5(2):209–17. doi:10.1053/eujp.2001.0233.
64. Olesen SS, Graversen C, Bouwense SA, van Goor H, Wilder-Smith OH, Drewes AM. Quantitative sensory testing predicts pregabalin efficacy in painful chronic pancreatitis. PLoS ONE. 2013;8(3):e57963. doi:10.1371/journal.pone.0057963.
65. Pillay PK, Hassenbusch SJ. Bilateral MRI-guided stereotactic cingulotomy for intractable pain. Stereotact Funct Neurosurg. 1992;59:33–8.
66. Ploner M, Gross J, Timmermann L, Schnitzler A. Cortical representation of first and second pain sensation in humans. Proc Natl Acad Sci U S A. 2002;99(19):12444–8. doi:10.1073/pnas.182272899.
67. Price DD. Psychological and neural mechanics of pain. Seattle, WA: IASP Press; 1999.
68. Price DD. Psychological and neural mechanisms of the affective dimension of pain. Science. 2000;288 (5472):1769–72.
69. Reynolds DV. Surgery in the rat during electrical analgesia. Science. 1969;164(878):444–5.
70. Roy M, Shohamy D, Daw N, Jepma M, Wimmer GE, Wager TD. Representation of aversive prediction errors in the human periaqueductal gray. Nat Neurosci. 2014;17(11):1607–12. doi:10.1038/nn.3832.
71. Sandkuhler J. Understanding LTP in pain pathways. Mol Pain 2007;3:9. doi:10.1186/1744-8069-3-9 1744-8069-3-9 [pii].
72. Sherrington CS. The integrative action of the nervous system. Scribner's. 1906.
73. Singer T, Seymour B, O'Doherty J, Kaube H, Dolan RJ, Frith CD. Empathy for pain involves the affective but not sensory components of pain. Science. 2004;303(5661):1157–62.
74. Stander S, Steinhoff M, Schmelz M, Weisshaar E, Metze D, Luger T. Neurophysiology of pruritus: cutaneous elicitation of itch. Arch Dermatol. 2003;139(11):1463–70.
75. Suter MR, Wen YR, Decosterd I, Ji RR. Do glial cells control pain? Neuron Glia Biol. 2007;3 (3):255–68. doi:10.1017/S1740925X08000100.
76. Swiergiel AH, Juszczak GR, Stankiewicz AM. Genetic and epigenetic mechanisms linking pain and psychiatric disorders. Modern trends in pharmacopsychiatry. 2015;30:120–37. doi:10.1159/000435937.
77. Talbot JD, Marrett S, Evans AC, Meyer E, Bushnell MC, Duncan GH. Multiple representations of pain in human cerebral cortex. Science. 1991;251 (1999):1355–8.
78. Terman GW, Bonica JJ, Loeser JD. Spinal Mechanisms and their modulation. In: Loeser JD, Butler SH, Chapman CR, Turk DC, editors. Management of pain, vol. 3. Philadelphia: Lippincott Williams & Wilkins; 2001. p. 73–152.
79. Tillisch K. The effects of gut microbiota on CNS function in humans. Gut Microbes. 2014;5(3):404–10. doi:10.4161/gmic.29232.
80. Tomasino B, Fabbro F. Increases in the right dorsolateral prefrontal cortex and decreases the rostral prefrontal cortex activation after-8 weeks of focused attention based mindfulness meditation. Brain Cogn. 2016;102:46–54. doi:10.1016/j.bandc.2015.12.004.

81. Tracey I. "Seeing" how our drugs work brings translational added value. Anesthesiology. 2013;119(6):1247–8. doi:10.1097/ALN.0000000000000018.
82. Treede RD, Apkarian AV, Bromm B, Greenspan JD, Lenz FA. Cortical representation of pain: functional characterization of nociceptive areas near the lateral sulcus. Pain. 2000;87(2):113–9.
83. Tsuda M, Inoue K, Salter MW. Neuropathic pain and spinal microglia: a big problem from molecules in "small" glia. Trends Neurosci 2005;28(2):101–7. doi:10.1016/j.tins.2004.12.002 S0166-2236(04)00369-8 [pii].
84. Turk DC, Okifuji A. Psychological factors in chronic pain: evolution and revolution. J Consult Clin Psychol. 2002;70(3):678–90.
85. Vogt BA. Pain and emotion interactions in subregions of the cingulate gyrus. Nat Rev Neurosci. 2005;6(7):533–44. doi:10.1038/nrn1704.
86. von Hehn CA, Baron R, Woolf CJ. Deconstructing the neuropathic pain phenotype to reveal neural mechanisms. Neuron. 2012;73(4):638–52. doi:10.1016/j.neuron.2012.02.008.
87. Watkins LR, Maier SF. Glia: a novel drug discovery target for clinical pain. Nat Rev Drug Discov 2003;2(12):973–85. doi:10.1038/nrd1251 nrd1251 [pii].
88. Wiech K, Jbabdi S, Lin CS, Andersson J, Tracey I. Differential structural and resting state connectivity between insular subdivisions and other pain-related brain regions. Pain. 2014;155(10):2047–55. doi:10.1016/j.pain.2014.07.009.
89. Wiech K, Kalisch R, Weiskopf N, Pleger B, Stephan KE, Dolan RJ. Anterolateral prefrontal cortex mediates the analgesic effect of expected and perceived control over pain. J Neurosci. 2006;26(44):11501–9. doi:10.1523/JNEUROSCI.2568-06.2006.
90. Wiech K, Lin CS, Brodersen KH, Bingel U, Ploner M, Tracey I. Anterior insula integrates information about salience into perceptual decisions about pain. J Neurosci (The Official Journal of the Society for Neuroscience). 2010;30(48):16324–31. doi:10.1523/JNEUROSCI.2087-10.2010.
91. Williams FM, Scollen S, Cao D, Memari Y, Hyde CL, Zhang B, Sidders B, Ziemek D, Shi Y, Harris J, Harrow I, Dougherty B, Malarstig A, McEwen R, Stephens JC, Patel K, Menni C, Shin SY, Hodgkiss D, Surdulescu G, He W, Jin X, McMahon SB, Soranzo N, John S, Wang J, Spector TD. Genes contributing to pain sensitivity in the normal population: an exome sequencing study. PLoS Genet. 2012;8(12):e1003095. doi:10.1371/journal.pgen.1003095.
92. Willis WD. Nociceptive pathways: anatomy and physiology of nociceptive ascending pathways. Philos Trans R Soc Lond B Biol Sci. 1985;308(1136):253–70.
93. Willis WD, Al Chaer ED, Quast MJ, Westlund KN. A visceral pain pathway in the dorsal column of the spinal cord. Proc Natl Acad Sci USA. 1999;96(14):7675–9.
94. Willis WD, Kenshalo DR Jr, Leonard RB. The cells of origin of the primate spinothalamic tract. J Comp Neurol. 1979;188(4):543–74.
95. Woolf CJ. Central sensitization: implications for the diagnosis and treatment of pain. Pain. 2011;152(3 Suppl):S2–15. doi:10.1016/j.pain.2010.09.030.
96. Woolf CJ, Salter MW. Plasticity and pain: role of the dorsal horn. In: McMahon SB, Koltzenburg M, editors. Wall and Melzack's Texbook of pain, vol. 5. Philadelphia: Elsevier Limited; 2006. p. 91–105.
97. Woolf CJ, Thompson SW. The induction and maintenance of central sensitization is dependent on N-methyl-D-aspartic acid receptor activation; implications for the treatment of post-injury pain hypersensitivity states. Pain. 1991;44(3):293–9.
98. Yarnitsky D. Role of endogenous pain modulation in chronic pain mechanisms and treatment. Pain. 2015;156(Suppl 1):S24–31. doi:10.1097/01.j.pain.0000460343.46847.58.
99. Yarnitsky D, Granot M, Nahman-Averbuch H, Khamaisi M, Granovsky Y. Conditioned pain modulation predicts duloxetine efficacy in painful diabetic neuropathy. Pain. 2012;153(6):1193–8. doi:10.1016/j.pain.2012.02.021.
100. Zhang ZG, Hu L, Hung YS, Mouraux A, Iannetti GD. Gamma-band oscillations in the primary somatosensory cortex—a direct and obligatory correlate of subjective pain intensity. J Neurosci. 2012;32(22):7429–38. doi:10.1523/JNEUROSCI.5877-11.2012.
101. Zhuo M. Long-term potentiation in the anterior cingulate cortex and chronic pain. Philos Trans R Soc Lond B Biol Sci. 2014;369(1633):20130146. doi:10.1098/rstb.2013.0146.

Magnetic Resonance Imaging 5

Michele Anzidei, Fabrizio Boni, Vincenzo Noce,
Daniele Guerrieri, Beatrice Sacconi and Carlo Catalano

Abstract

Magnetic Resonance (MR) plays a leading role in pain imaging, offering optimal anatomic imaging and contributing functional and chemical studies of Central and Peripheral Nervous System. These tools have increased the comprehension of different chronic painful syndromes and the evaluation of treatment response to pharmacological or other therapeutic interventions. Furthermore, several neuro-MRI techniques, including functional magnetic resonance imaging (fMRI), MR spectroscopy (MRS) and diffusion-weighted imaging (DWI), have been demonstrated to depict nervous system pathologies associated with pain. Also, body MRI may be useful to depict several causes and manifestations of pain, local or diffuse, acute or chronic, covering the entire spectrum of disorders, supporting a multidisciplinary diagnosis process. In this section, after a brief discussion of MR basics, the main imaging procedures and their application in assessing the main painful syndromes will be deeply explored, with support of pictorial essays for each technique.

Keywords

Pain · MRI · Brain · Spinal cord · Nerves · Functional imaging

1 Introduction

Pain is defined by IASP Task Force as 'an unpleasant sensory and emotional experience associated with actual or potential tissue damage, or described in terms of such damage' [1]. Although many mechanisms of pain are not completely comprehended, the fast progress of imaging techniques in the last decades has provided remarkable advances in the understanding of painful conditions and their key mechanism [2]. Magnetic Resonance (MR) plays a leading role in pain imaging, offering optimal anatomic imaging and contributing functional and chemical studies of Central and Peripheral Nervous System. These tools have increased the comprehension of different chronic painful syndromes

M. Anzidei (✉) · F. Boni · V. Noce · D. Guerrieri · B. Sacconi · C. Catalano
Department of Radiological Sciences, University La Sapienza, Rome, Italy
e-mail: michele.anzidei@gmail.com

L. Saba (ed.), *Neuroimaging of Pain*, DOI 10.1007/978-3-319-48046-6_5

and the evaluation of treatment response to pharmacological or other therapeutic interventions. Furthermore, several neuro-MRI techniques, including functional magnetic resonance imaging (fMRI), MR spectroscopy (MRS) and diffusion-weighted imaging (DWI), have been demonstrated to depict nervous system pathologies associated with pain.

Neuroimaging studies identified many cortical regions involved in the perception of pain [3]; all the authors agree that pain is a complex neurological entity involving several networks and areas, comprehensively defined as 'pain neuromatrix,' including lateral components (primary and secondary somatosensory areas, S1\S2) and medial components (affective–cognitive–evaluative involving areas, like the anterior parts of insula, anterior cingulate cortex, and prefrontal cortex) [4]. These areas have been identified by using a study design defined as 'stimulate and see what you get'; the initial studies used PET to evaluate involved cortical regions by using painful stimuli. fMRI has been subsequently employed by using blood oxygen level-dependent (BOLD) method, reporting similar results to those of the first PET studies [5]. Other study designs have been developed, including those exploring other factors correlating with painful experience, such as anticipation, placebo effect, empathy, and differences between individuals in the pain perception; another extensive series of studies evaluated the effects of pharmacological and non-pharmacological therapies on pain, such as opioids, acupuncture, spinal cord stimulation, meditation, and hypnosis [6]. The list of pain conditions which can be evaluated by neuroimaging is large, including patients with chronic regional pain syndromes, chronic back pain, migraine, epilepsy, fibromyalgia, and chronic fatigue syndrome among others; the list of the available imaging techniques is also vast, including the already mentioned PET, SPECT, fMRI, MRS, DWI as well as diffusion tensor imaging (DTI) and pharmacological MRI (phMRI). The expectation for the next future would be to improve the capability of these imaging techniques in identifying the specific pain mechanisms in order to represent diagnostic tools for the individual patient evaluation, suggesting potential solutions for the single patient care.

Body MRI may be useful to depict several causes and manifestations of pain, local or diffuse, acute or chronic, covering the entire spectrum of disorders, supporting a multidisciplinary diagnosis process. For instance, wrist and hand MR imaging may play an important role in early diagnosis of rheumatoid arthritis and in developing the differential diagnosis for recent-onset arthritis [7].

In this section, after a brief discussion of MR basics, the main imaging procedures and their application in assessing the main painful syndromes will be deeply explored, with support of pictorial essays for each technique.

2 Magnetic Resonance Imaging

2.1 General Physical Principles

MRI takes advantage of the ubiquitous presence of hydrogen atoms or protons (H+) embedded in the water molecules and their magnetic dipoles in order to produce electrical signals by switching on and off external magnetic fields [8]. The entire process requires a magnet providing a constant magnetic field (B_0), a gradient system providing orthogonal fields for spatial localization of signals and a radiofrequency (RF) system with transmitter coils providing additional fields for spin excitation and receiver coils receiving MR signals over the imaging volume. When a volume is put inside a field B_0, all the H+ are aligned; by adding a set of RF and gradients, a selected slice of the volume is excited, the H+ are flipped out of their alignment and electric signals are transduced in the coils as the perturbed H + come back to their previous aligned state (relaxation) [9]. This is reached by using a specific MRI protocol including a pulse sequence, defining the characteristics of the RF pulse and a parameters set such as echo time (TE), repetition time (TR), matrix, field of view (FOV), and flip angles (FA). TE represents the time between RF excitation and first acquisition; TR is the time interval between subsequent RF excitation per slice. By varying TR and TE, several MRI sequences can be performed, with different contrasts as per longitudinal magnetization recovery (T1) and transverse magnetization decay (T2); more in detail, long TR and TE lead to T2-weighted imaging, whereas short TR and TE to T1-weighted imaging. Each MR signal decodes for intensity, spatial and phase information; the different signal information is collected in the so-called k-space, which can be subsequently converted into a readable image by using a mathematical process (two-dimensional Fourier transform).

There are a large number of different sequences, which basically are classified into two families, spin echo sequences (SE) and gradient echo sequences (GE). The word 'pulse sequence' refers to a determinate series of radiofrequency (RF) waves or electromagnetic gradients, administrated in order to create MR images.

A typical SE sequence requires a 90° along with a slice selective gradient, and 180° RF pulse, whereas GE sequences are characterized by lack of the refocusing 180° RF pulse and a FA equal or smaller than 90°. Among specific sequences, it is worth to mention Echo-planar imaging (EPI) sequences, representing the base for several advanced techniques that will be covered in this chapter, such as DTI and fMRI. A dedicated MRI study of the brain can include, beyond conventional sequences, advanced imaging sequences, such as fMRI, MRS, ASL, DTI, with special reference to the individual clinical case.

All these MRI techniques will be more specifically described in the present section, both in terms of technical principles as well as in the present and potential clinical applications in several pain conditions.

2.2 Conventional MR Sequences

All conventional MR pulse sequences are derived by classical Spin Echo (SE) and Gradient Echo (GE). The basis of conventional MR imaging in painful syndromes is to depict anatomy, to detect pathologic variants or abnormalities and to emphasize specific morphological changes in the tissue structure (inflammation, fatty degeneration, fibrosis, etc.).

2.2.1 Spin Echo Sequences

As aforementioned, SE sequences are composed of a fundamental sequence of 90°–180° RF pulses (Fig. 1). Modifying TR and TE, it is possible to highlight specific characteristics for each tissue, obtaining so-called T1-weighted (T1-w) and T2-weighted (T2-w) images.

More specifically, short TR and short TE (less than 700 and 30 ms, respectively) heighten differences in T1 properties between tissues (T1-weighting), while T2 weighting require long TE and TR (greater than 2000 and 80 ms, respectively); avoiding T1 and T2 dependence by using a long TR with a short TE it is possible to obtain a proton density sequence. Because of their long acquisition time, these classical sequences are actually less used; advances in MR technology have enabled a reduction in acquisition time with the use of fast SE sequences. In a fast SE (FSE) or turbo SE sequence (TSE), a single 90° pulse is applied to flip the net magnetization vector, after which multiple 180° rephasing pulses are applied (Fig. 2).

The 180° echo pulses generate the so-called echo train, and the total number of 180° RF pulses and echoes is referred to as the echo train length. By using FSE and TSE techniques a dramatic reduction of acquisition time is guaranteed. Main use of TSE is to acquire T2-w images, although it can be applied even to produce T1-w images. This is due to the significant reduction in scan time that can be achieved for long TR scans when modest echo train lengths are used. Echo train lengths less than 10 are typically used for brain and spine imaging, but very long echo trains (100 or more) can be used in abdominal imaging to acquire T2-weighted images in less than one second. These classic sequences (SE or TSE) with different weighting (T1, T2 or PD), allow to depict several causes and manifestations of pain, local or diffuse, acute or chronic, covering the entire spectrum of disorders, often allowing to get the diagnosis

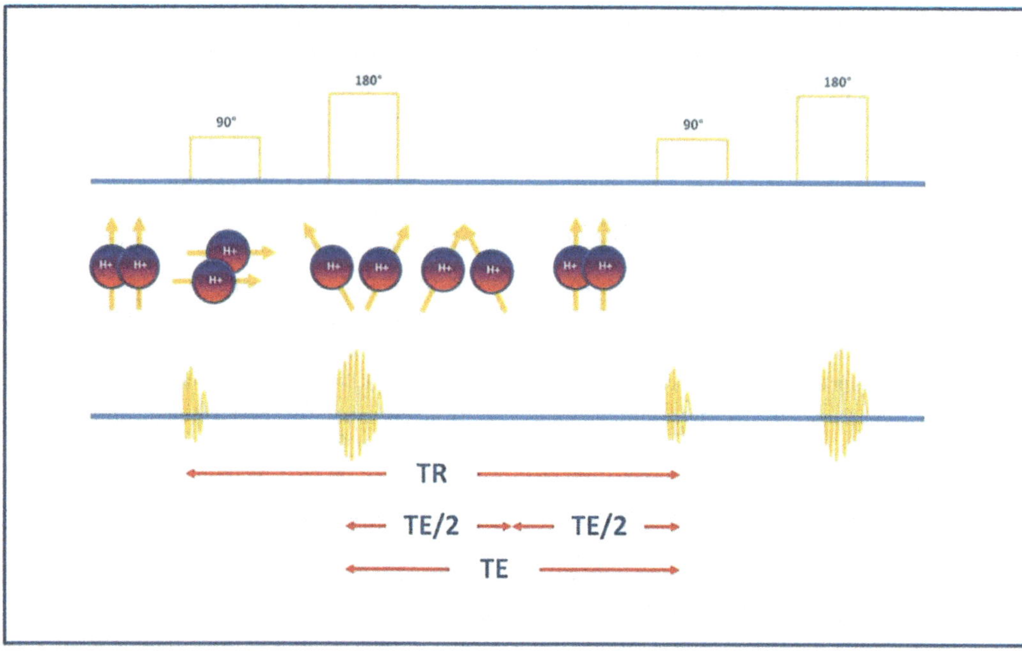

Fig. 1 Schematic representation of spin-echo sequences displaying radiofrequency pulses and consequent protons' excitement. *TR* Time of repetition. *TE* Time of echo

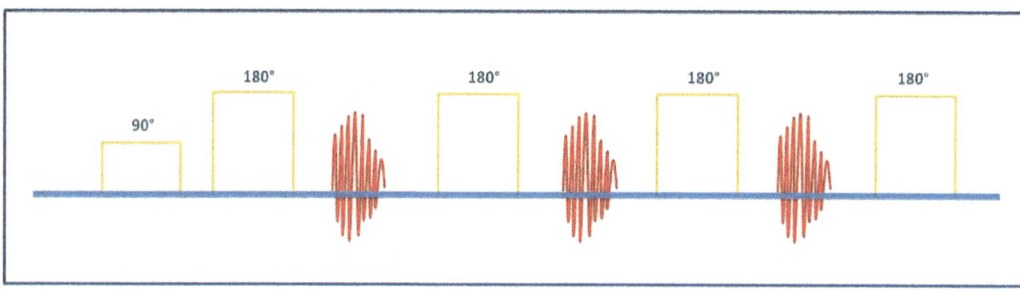

Fig. 2 Schematic representation of fast (or turbo) spin-echo sequences. The echo train is composed by a series of subsequent 180° pulses, permitting the acquisition of multiple lines of the volume for each TR, thus it allows an overall shortening of imaging time

without any other and more sophisticated processes (Figs. 3, 4, 5 and 6).

Use of paramagnetic contrast agent with T1-w sequences (TSE or GE) often offers more information, useful in diagnostic process (Figs. 7 and 8).

2.2.2 Gradient Echo Sequences

In order to reduce time of acquisition of standard SE, with a 180° pulse to refocus the protons, GE sequences (Fig. 9) employ gradient reversal pulses, in at least two directions, to generate echo signal.

The principal advantage of GRE technique is represented by its very short TR (that can be interpreted as faster time of acquisition). Table 1 resumes the main differences between classical SE and GRE sequences. GRE sequences may be classified coherent (refocused) or incoherent (spoiled) on the basis of the steady state phenomenon. This electromagnetic event is a

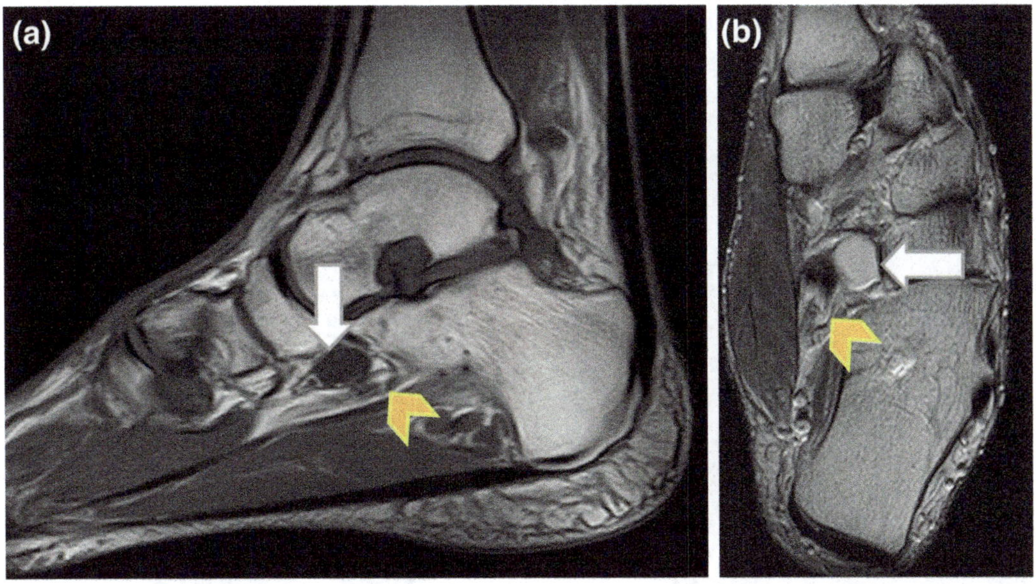

Fig. 3 Example of anatomic study by conventional T1-w and T2-w images. **a** Coronal T1-w of right ankle displaying fluid distension of synovial sheath, called ganglion cyst (*white arrow*), in tarsal tunnel. This abnormality causes tarsal tunnel syndrome, characterized by pain and paraesthesia in toes, sole, or heel, due to posterior neurovascular bundle (*yellow arrowhead*) compression. **b** Axial T2-w depicting a round fluid lesion (*white arrow*) in tarsal tunnel, conflicting with nervous–vascular structures

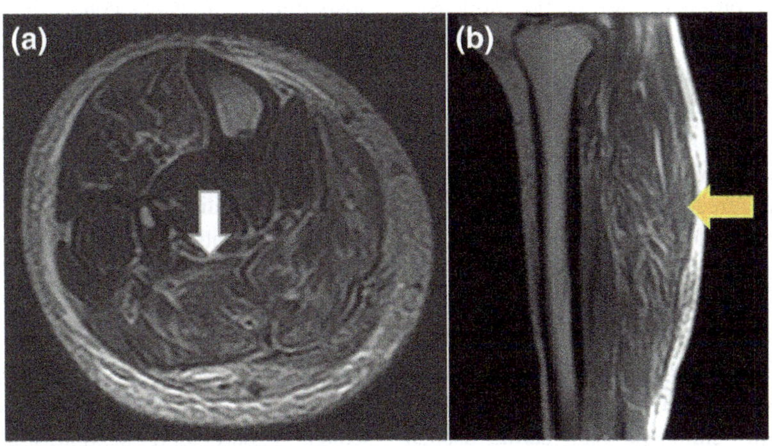

Fig. 4 Polymyositis. An autoimmune pathological process identifiable on MRI for a fatty infiltration around muscles, as well demonstrated on these T1-w (*white arrow*) and T2-w (*yellow arrow*)

consequence of extremely short TRs, usually shorter than T1 and T2 of the tissues imaged, causing a specific weighting called T2*, particularly sensitive and susceptible to magnetic field inhomogeneity. Coherent or partially refocused GRE sequences use a gradient to maintain the T2* and eventually produce T2-w images. The fundamental difference between partially refocused and fully refocused GRE sequences is that all the gradients in the latter are refocused. On the contrary, incoherent or spoiled GRE sequences utilize, after each echo, a specific RF

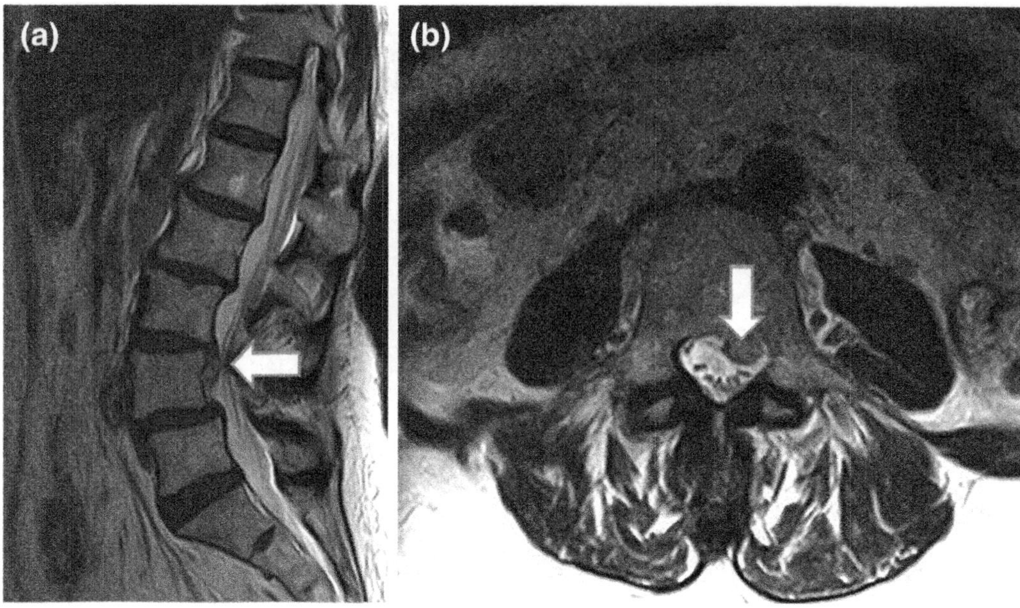

Fig. 5 Lumbar disc hernia. T2-w sagittal image (**a**) depicts a disc herniation (*white arrows*), with caudal migration, causing compression of left L4 nerve root at dural origin (**b**)

pulse or gradient, called spoiler, to null the T2* effect, thereby producing T1 or proton-density weighting. Partially refocused GRE images are mostly used in imaging of encephalic nerves and internal auditory canal structures (Fig. 10), while spoiled GRE sequences are widely employed in contrast-enhanced MRI.

Steady-state free precession (SSFP) represents a specific fully rephased GRE technique, with continuous repetition of short TRs (5 ms). SSFP guarantee the best temporal resolution among MR pulse sequences, high signal-to-noise ratio, but an extreme susceptibility to artifacts caused by magnetic field inhomogeneities. These sequences are employed in cardiovascular (Fig. 11) and gastroenteric imaging, due to high signal achieved from blood vessels and static fluids.

2.2.3 Fat and Fluid Signal Suppression

MRI permits the specific suppression or saturation of tissue signals; this tool is widely used to achieve a better characterization. Among the commonest structures that are suppressed in MRI there are fat tissue, cerebrospinal fluid, and silicon. Fat saturation is targeted to obtain diagnosis in case of lesions contacting adipocytes or intracytoplasmic fat, but it is advantageous even in order to enhance signal from other tissues, nulling the 'fat background' (e.g., edema imaging). At least three methods for fat nulling are routinely used in MRI. The commonest method for fat saturation is applying a frequency-selective saturation RF pulse or spoiler gradient to null fat signal, immediately subsequent to 90° initial pulse. This technology is fast, 'lipid specific' and because fat suppression is achieved by selective saturation pulse before normal acquisition, it can be used with any imaging sequences, particularly in contrast-enhanced MRI (Fig. 12); on the other hand, it may be hampered by artifacts due to magnetic field inhomogeneity and incomplete suppression.

The second method for fat signal nulling is Inversion Recovery (IR) imaging, obtained by application of an initial 180° to flip the net magnetization vector of fat tissue (STIR, Short Tau Inversion Recovery); the 90° pulse is the applied exactly at fat 'null interval' (140 ms) to suppress its signal (Fig. 13).

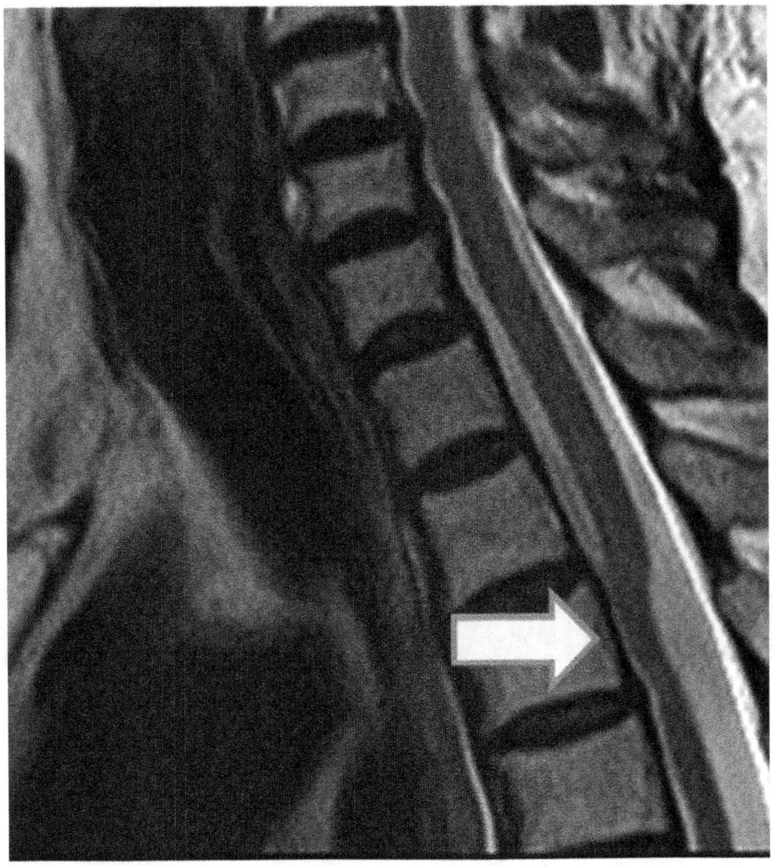

Fig. 6 Idiopathic spinal cord herniation. This condition is only encountered between T2 and T8 where the normal thoracic kyphosis leads to the thoracic cord being in close proximity to the ventral theca. The key feature is focal distortion and rotation of the cord with no CSF seen between it and the ventral theca (as well demonstrable on this T2-w image, *white arrow*)

STIR sequences are actually widely used in common MR protocols, from oncologic to musculoskeletal imaging, in order to detect solid lesions, as well as edema due to traumatic, functional or vascular injuries (Figs. 14 and 15). This technique is limited by long TR required, causing increase of acquisition duration.

The third method for fat suppression is the so-called 'Out-of-phase imaging'; using Spoiled GRE sequences this technique excites different precession rates between ^1H in fat (CH_2) and water (H_2O). When co-presence of CH_2 and H_2O happens in same volume, the signal is nulled. Note that this technique permits the suppression of 'microscopical fat' (steroids or triglycerides deposits), characteristically present in adenomatous cells and steatosis hepatocytes.

Suppression of signal applied to Cerebrospinal Fluid gains dramatic significance in neuroimaging; this technique is based on inversion-recovery SE sequences referred as FLuid-Attenunated Inversion Recovery (FLAIR). Nulling CSF permits detection of lesions otherwise not distinguishable, specifically if localized nearby sulci or ventricles, and edema in central nervous system, a signal of tissue damage from vascular cause or compression from a growing mass.

2.2.4 Proton Density Imaging

Setting a pulse sequence with long TR (2000–5000 ms) and short TE (10–20), differences between T1 and T2 relaxation times for different tissues will be minimized, while stronger signals

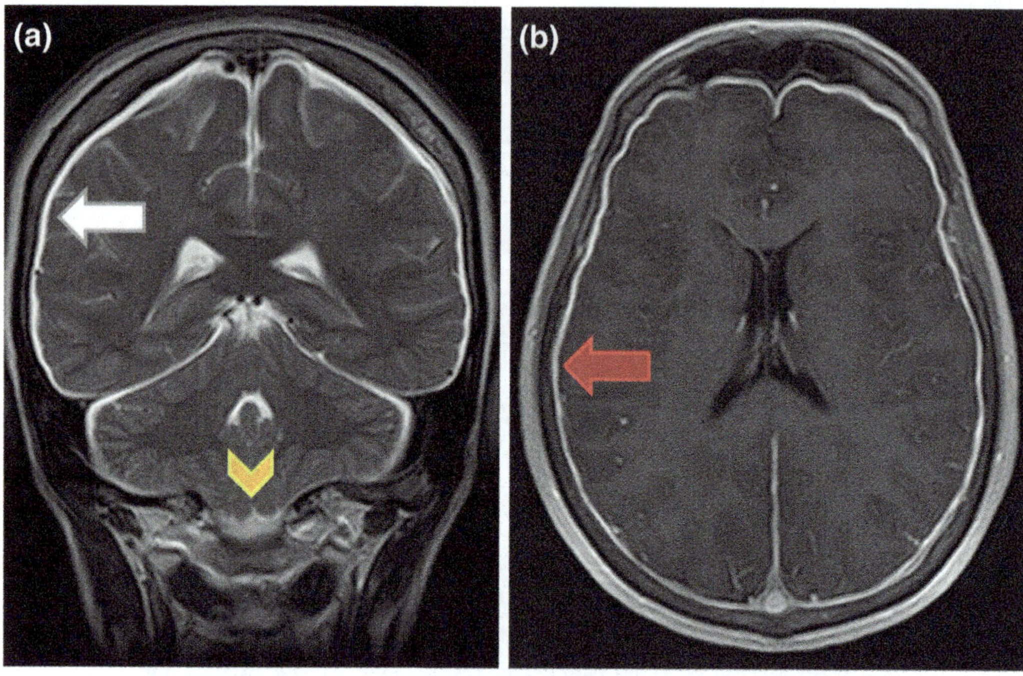

Fig. 7 Spontaneous intracranial hypotension. On the *left* (**a**, Flair) evidence of subdural effusion (*white arrow*) and cerebellar tonsillar ectopia (*yellow arrowhead*); on the *right* (**b**, T1-w post-Gd), diffuse pachymeningeal enhancement (*red arrow*)

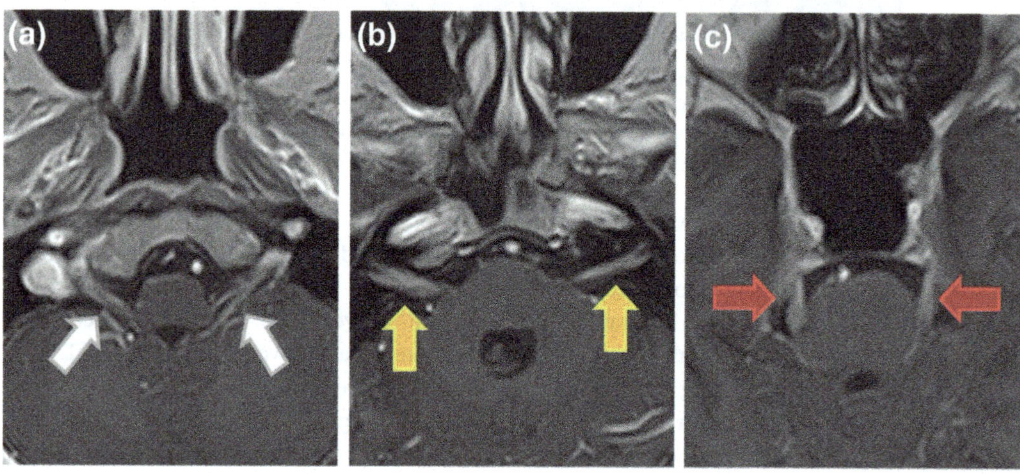

Fig. 8 Carcinomatous meningitis. On **a**, **b** and **c** images, T1-w post-Gd sequences showing significant enhancement of nerve sheaths in patient suffering from lung cancer (*white arrows* XI cranial nerves; *yellow arrows* VIII cranial nerves; *red arrows* V cranial nerves), indicating spread of malignant cells

are produced by structures with high density of hydrogen protons. This pulse sequence design is called proton density (PD) weighting. PD sequences are widely adopted in musculoskeletal imaging protocols, in order to accurately display and evaluate cartilage, and in neuroimaging, even though their usage might be superseded by more accurate FLAIR sequences.

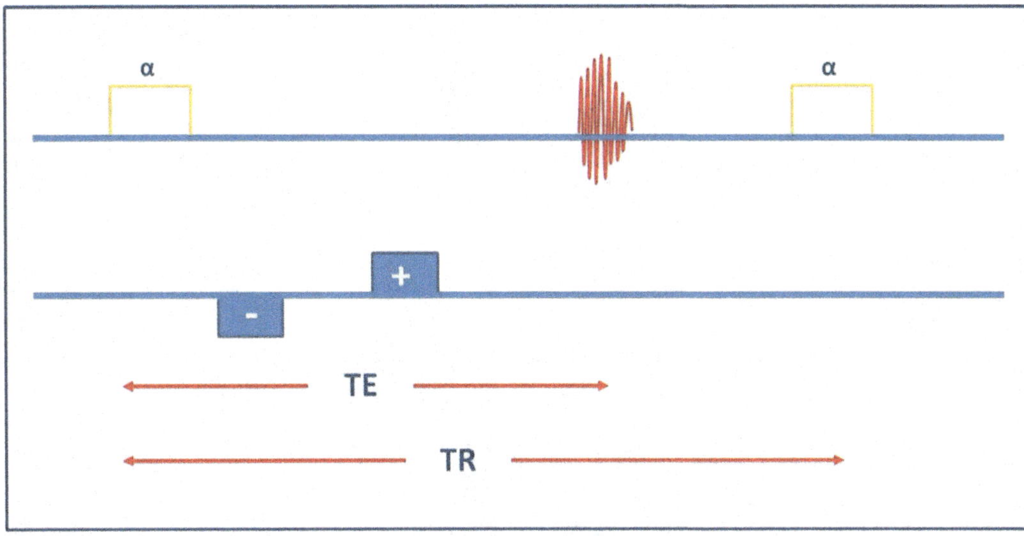

Fig. 9 Schematic representation of GE sequences. α slice-selective gradients are employed with limited flip angles (about 45°), dephasing and rephrasing of transverse magnetization are determined by alternated negative phase-encoding and positive frequency-encoding (readout) gradients

Table 1 Comparison between SE and GRE sequences

Comparison between SE and GRE sequences		
Parameter	SE	GRE
Rephasing system	RF pulse	Gradient variation
Flip angle	90° only	Variable
Efficiency at reducing magnetic inhomogeneity	Very efficient (true T2 effect)	Not very efficient (T2* weighted)
Acquisition time	Slow imaging	Fast imaging

2.2.5 Echo-Planar Imaging

This technique allows a significant shortening in acquisition time, since a single echo train is employed to acquire the entire volume. Echo-planar imaging (EPI) is available for both SE and GRE sequences; multiple lines of imaging data are acquired after a single RF excitation (single-shot EPI) or more RF pulses (multi-shot EPI). The latter guarantees higher spatial resolution and reduction of image distortion and signal loss due to susceptibility differences, T2 relaxation, and main field inhomogeneities [10]. Echo planar imaging is now a technique of choice for diffusion-weighted imaging (Fig. 16).

2.3 Dynamic Contrast-Enhanced MRI

Conventional static or phasic contrast-enhanced MRI is able to determinate vascularization of tissues, but gives no quantitative insight into hemodynamic processes, thus many dynamic contrast-enhanced (DCE) techniques have been designed, in order to provide further help in differential diagnoses of pathologic processes. The principal perfusional imaging technique is based on gadolinium bolus injection and subsequent acquisition of T1-weighted gradient echo sequence with short TE (<1.5 ms) and TR (<7 ms) and flip angle around 30° [11]. Due to high time resolution, these sequences permit to

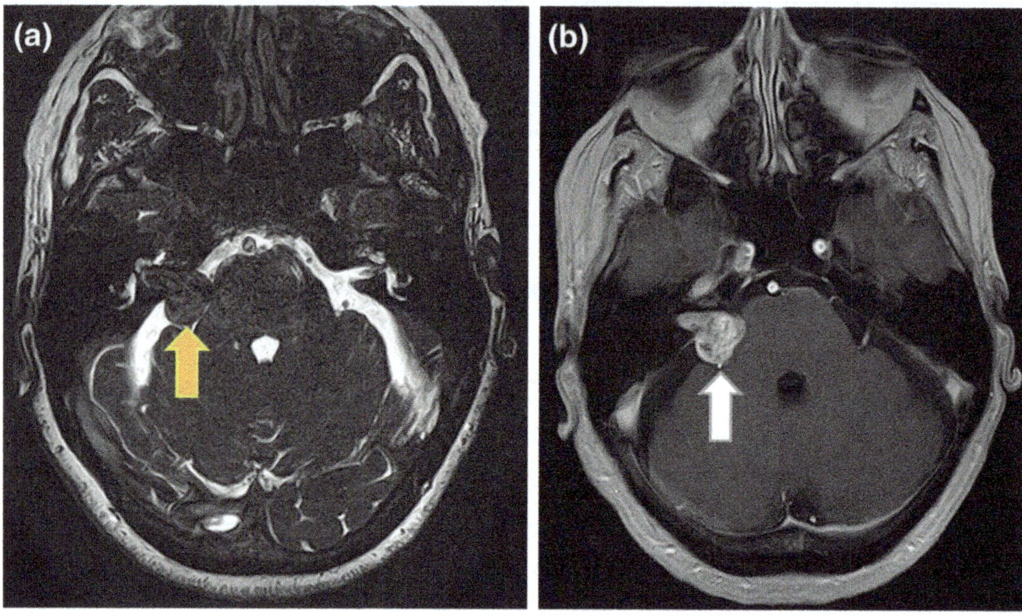

Fig. 10 Acoustic neuroma. This lesion classically present on imaging as a solid nodular mass with an intracanalicular component that often result in widening of the porus acusticus, as well demonstrated on CISS (constructive interference in steady state; *yellow arrow*) sequence (**a**) and confirmed on T1 post-Gd sequence (*white arrow*) (**b**)

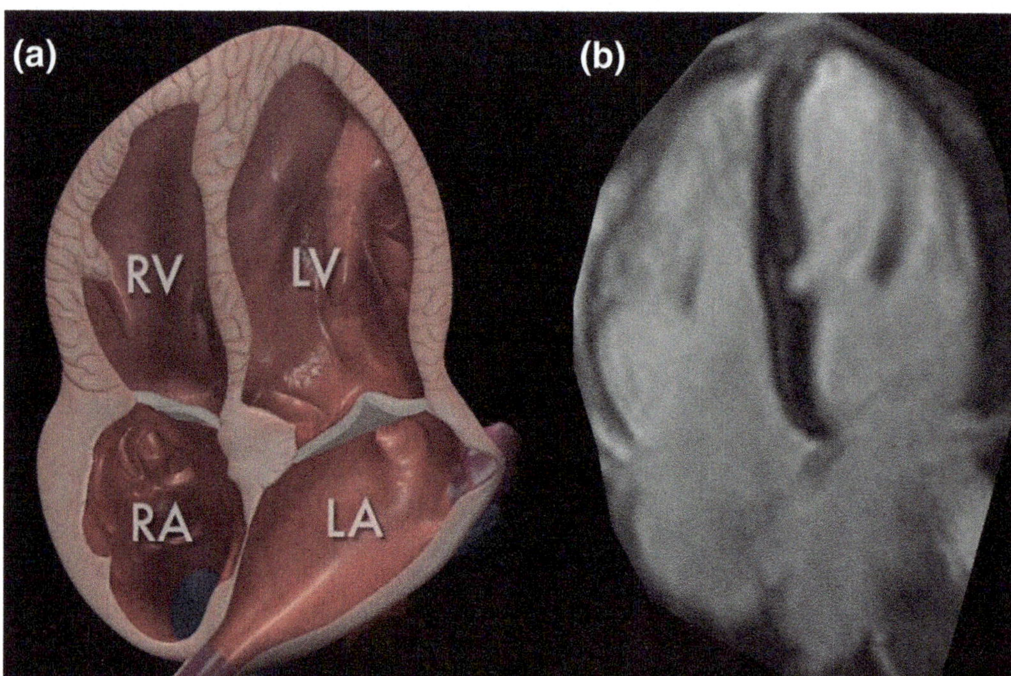

Fig. 11 Steady-state free precession (SSFP) imaging in cardiac MR. **a** 4 chambers plan of heart schematized. *RA* Right atrium. *LA* Left atrium. *RV* Right ventricle. *LV* Left ventricle. **b** 4 chambers SSFP imaging in patient with moderate right ventricular volume overload

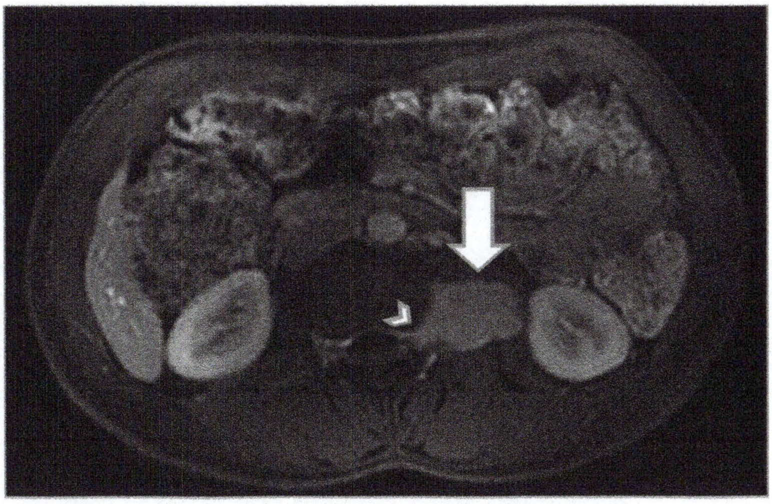

Fig. 12 T1-weighted fat-suppressed axial image, acquired after administration of Gadolinium, in a patient with left lumbar pain and paraesthesia. The study revealed a bulky left paravertebral mass with diffuse vascularization (*white arrow*) and 'dumbbell sign,' expression of relationship with the spinal cord. The final diagnosis was spinal schwannoma

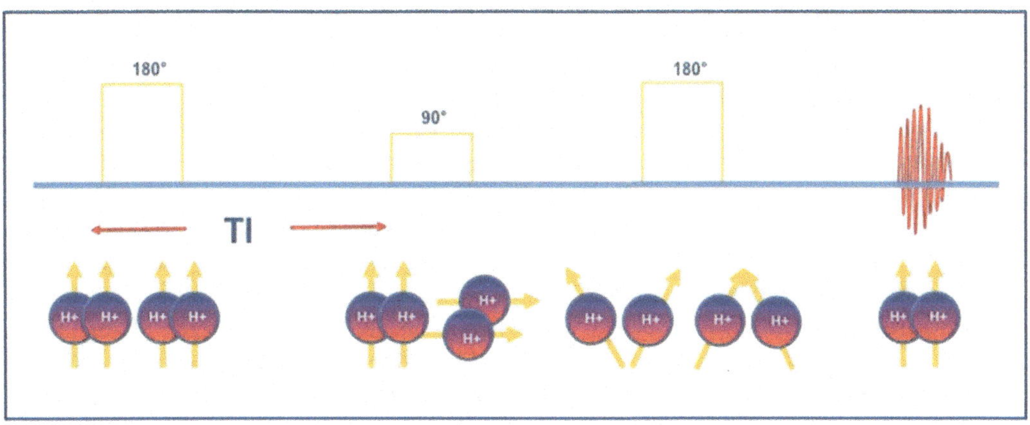

Fig. 13 Schematic representation of IR sequences. A 180° nulling RP is employed to suppress fat or water signal. After a determined inversion time (TI), a 90° pulse is applied to start the usual SE

acquire post-contrast consecutive images at several time points, in order to quantitatively evaluate contrast medium extraction. Through pharmacokinetic modeling of DCE data, a number of parameters can be determined, such as transfer constant (Ktrans) and fractional volume of the interstitial space (ve), deriving from these color map to display perfusion and permeability characteristics of tissues. The K-trans assessed in the first pass of contrast medium depicts cases where there is high permeability, while that measured in the steady state may better

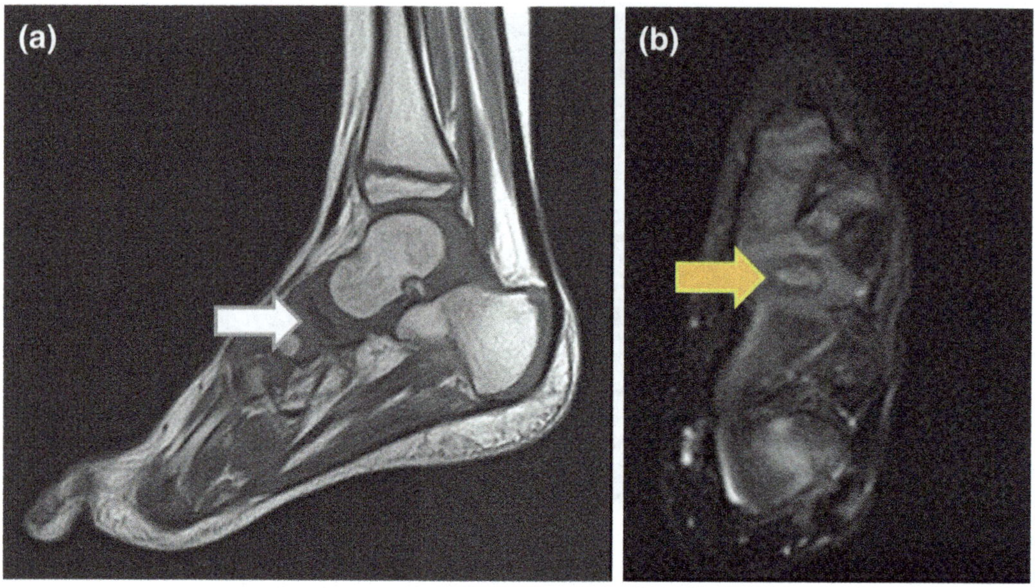

Fig. 14 Kohler's disease. Avascular necrosis of the navicular bone, detectable as an hypointensity on T1-w sequence (**a**, *white arrow*) and corresponding hyperintensity on STIR sequence (**b**, *yellow arrow*)

characterize situations of lower permeability, when Ktrans is dependent on surface area and is not flow limited (Fig. 17).

Many studies have addressed DCE-MRI to detection and characterization of neoplasms in many districts [12, 13], with great regard to CNS malignancy [14] (Fig. 18).

2.4 Magnetic Resonance Angiography (MRA)

Magnetic resonance angiography (MRA) has become an established imaging modality in management of vascular pathologies, from atherosclerotic condition (peripheral or carotid districts among the others) to vasculitis [15, 16]. MR Angiographic study can be performed with sequences with or without contrast medium. Both time-of-flight (TOF) and phase-contrast (PC) MRA are non-contrast techniques with intravascular blood detected by virtue of its movement compared with static surrounding tissues. Contrast-enhanced (CE) MRA relies on the T1 shortening effect of intravenously administered contrast media circulating in the blood. In TOF-MRA, vessel-to-background contrast is generated by the inflow of fresh, unsaturated blood in a saturated tissue slice [17]. Saturation of stationary background tissue is achieved by submitting it to radiofrequency pulses with a repetition time much shorter than tissue T1 values, thereby decreasing its longitudinal magnetization vector [18]. Because inflowing unsaturated blood still has a large longitudinal magnetization vector, it will be seen in the imaged slice as an area of high signal intensity. Intravascular protons are also subject to these saturation effects, which are proportional to the time protons reside in the imaging slice. Therefore short TR, slow flow, and course of the blood vessel in the imaging slice plane all unfavorably affect vessel-to-background contrast. TOF-MRA is possible by imaging successive, independent slices (2D TOF-MRA) or by imaging a volume that is later partitioned into separate slices (3D TOF-MRA). Although TOF-MRA is an attractive and entirely noninvasive method for imaging arteries, it is not widely applied because it suffers from a number of serious drawbacks. Currently, the main application of this sequence (particularly using 3D TOF MRA) regards the

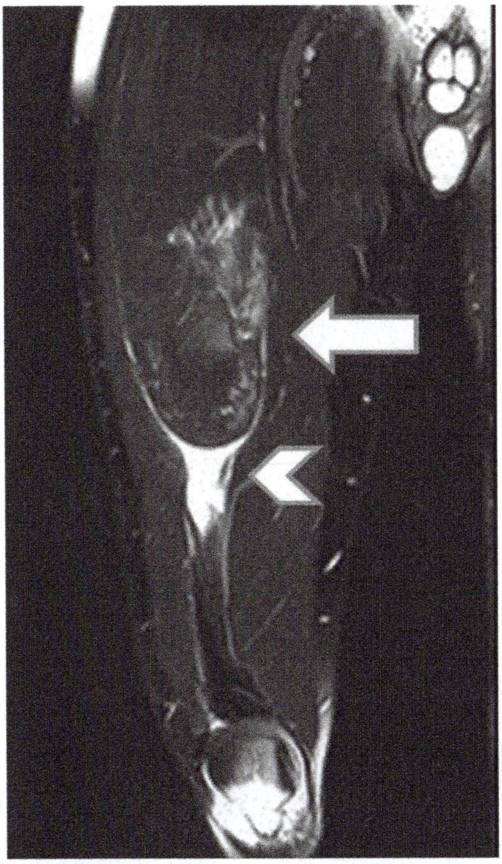

Fig. 15 Coronal T2-weighted STIR image in a young football player with severe right leg pain after accidental injury. Image shows diffuse oedema of the quadriceps (*white arrow*) and myotendinous junction complete tear with tendon retraction (*white arrowhead*)

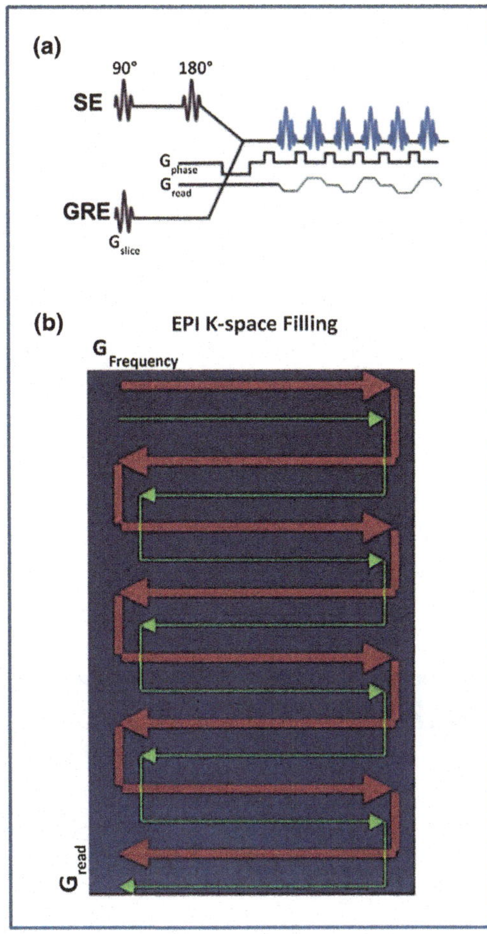

Fig. 16 **a** Schematic representation of echo-planar imaging for both SE and GRE techniques, with phase-encoding and frequency-encoding (readout) gradients rapidly turned on and off to shorten acquisition time. **b** K-space filling geometry in EPI

study of arterial circulation; in this context, TOF technique allows visualization of major intracranial arteries and peripheral branches in a relatively short time and generally does not require use of contrast agent, allowing a diagnostic depicting of pathological condition, sometimes associated with pain sensation, such as intracranial aneurysm (Fig. 19).

The clinical utility of TOF-MRA in other anatomical districts is limited by the long duration of image acquisition, a tendency for stenosis overestimation and different type of artifacts (motion artifacts, ghosting, flow void), but it can be useful as a backup modality in patients who cannot receive contrast medium.

Phase-contrast (PC)-MRA was developed as an alternative to TOF-MRA and uses an entirely different technique to generate vascular contrast. In PC-MRA, vessel-to-background contrast is generated by displaying the accumulated phase difference in transverse magnetization between moving protons in blood and stationary background tissues. PC MRA is based on the accumulated phase difference between mobile spins and stationary spins [19]. This characteristic renders PC acquisition more sensitive to slow flow, such as occurs in veins.

The introduction of MRI systems with higher gradients and fast breath-hold 3D sequences was

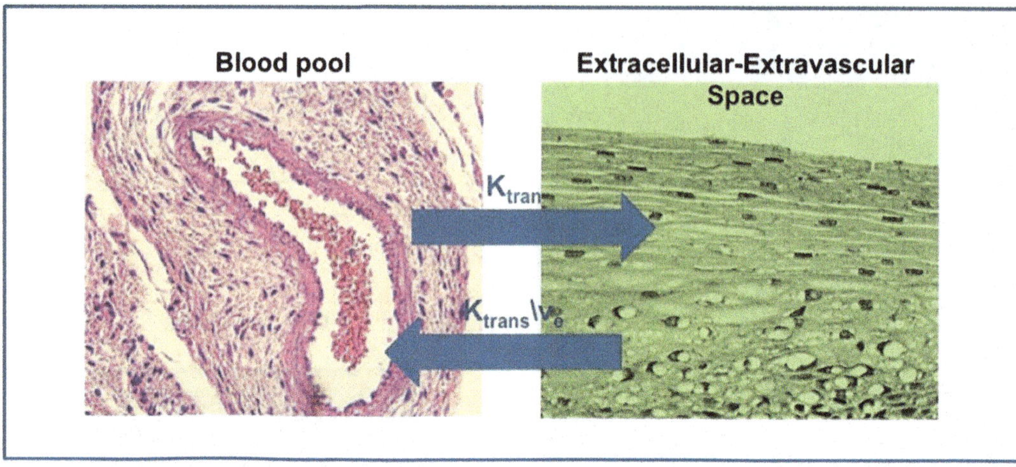

Fig. 17 Pharmacokinetic bicompartimental model on the basis of dynamic contrast-enhanced MRI. Ktrans is the tissue-specific constant that regulates contrast medium transfer to tumoral tissue. Ve expresses the percentage of tumoral tissue occupied by extracellular\extravascular space (interstitium)

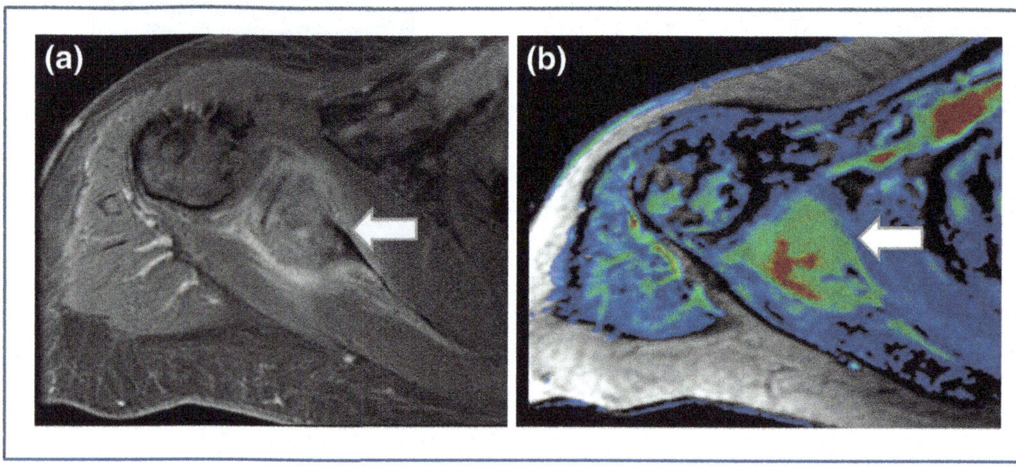

Fig. 18 **a** T1-weighted GRE fat-suppressed, dynamic contrast-enhanced, axial image of a painful metastatic lesion on right scapula. **b** Color map expressing Ktrans function for the same lesion, assessing a quantitative perfusional study

prerequisite for contrast-enhanced MR Angiography (CE-MRA). CE-MRA has improved spatial resolution compared with TOF, and PC-MRA and has helped to partially eliminate physiologic effects, such as turbulence leading to signal loss. This technique exploits the difference in the T1 relaxation times of blood and surrounding tissues when a rapid bolus infusion of a paramagnetic contrast agent is injected. These gadolinium-based agents exert a T1 shortening effect, generating a high intravascular SNR, which is largely unaffected by inflow. CE-MRA essentially needs a compromise between the desire for high spatial resolution and volumetric coverage (i.e., long acquisition duration), the desire to avoid disturbing venous enhancement (i.e., short acquisition duration), and high vessel-to-background contrast [20]. Rapid scan times are achievable by the use of fast-gradient sequences employing short TR. CE-MRA

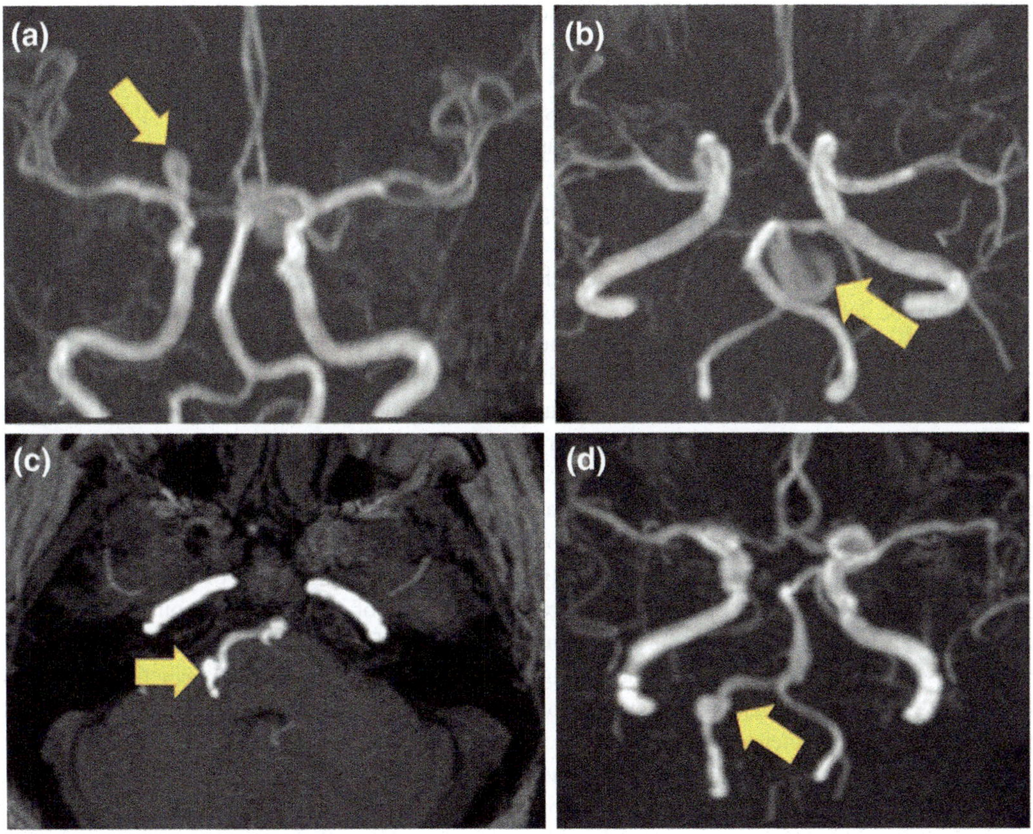

Fig. 19 Time-of-flight imaging of intracranial aneurysms in several localizations: right middle cerebral artery (**a**), basilar artery, ruptured (**b**) and right vertebral artery, at pontine level (**c**, **d**)

images are essentially a record of the vessel lumen, and timing of the scan is crucial to ensure high-quality images and avoid venous contamination. Because the time of peak arterial enhancement can vary substantially between patients, the CE-MRA examination must be tailored to the individual contrast arrival time. To determine the delay between the start of injection of contrast medium and the acquisition of central k-space profiles a 2D time-resolved test bolus technique can be used. However, more recently, bolus tracking techniques have been introduced to detect the arrival of contrast [21]; in contrast to injecting a small amount of contrast material in a separate test bolus scan, real-time bolus monitoring allows the operator to inject the total volume of contrast material, and to proceed with the 3D CE-MRA acquisition precisely when the desired signal enhancement in the arterial bed of interest has been detected by the scanner, or by visual feedback [22]. A following practical aspect of CE-MRA to consider is the vessel-to-background contrast. The T1 decrease into vessel, due to contrast medium injection, is not sufficient to selectively enhance arteries and to suppress background tissue; as a result, the signal of these tissues—and specifically, fat tissue—must be eliminated to present easily understandable images. The most commonly technique to do it is subtraction of non-enhanced 'mask' images, identical to the 3D CE-MRA volumes. Another important practical aspect to evaluate is venous enhancement and strategies to reduce this potential problem; this drawback is particularly prevalent in patients with cellulitis and AV malformations. There are different strategies to decrease the chance for disturbing venous enhancement, such as: increasing acquisition speed, separate acquisition for the lower leg station, specialized k-space filling algorithms, time-resolved acquisition strategy, infrasystolic venous compression. The most straightforward way of preventing venous enhancement is by shortening acquisition duration. This should be done, first of all, by lowering TR and TE to the shortest possible value. An important technical evolution was achieved with development of dedicated centric k-space filling algorithms; this is useful for CE-MRA because the time between arterial and venous opacification is usually shorter than the duration of a high spatial resolution 3D CE-MRA acquisition. The underlying principle is to collect central k-space profiles, which primarily determine image contrast, at peak contrast enhancement in the arteries of interest whereas veins are not or only minimally enhanced. When peripheral k-space profiles are read out, primarily information encoding details in the image is acquired. When centric k-space filling is combined with parallel imaging, the chances of venous enhancement decrease even further [23].

All these improvements make this technique an important diagnostic tool in management of patients with vasculature symptoms, with the aim to recognize or characterize pathological findings responsible for painful condition (Fig. 20).

More sophisticated techniques use repetitive centric k-space filling to obtain high spatial resolution MR angiograms with high temporal frame rate. Korosec et al. were first to describe this concept, which they named Time-Resolved Imaging of Contrast Kinetics (TRICKS). With TRICKS the contrast-sensitive central part of k-space is sampled more often than the peripheral resolution-sensitive views [24]. After the acquisition is finished, central k-space lines are combined with peripheral lines through a process of temporal interpolation such that a series of time-resolved 3D images of the vasculature are obtained.

2.5 Diffusion-Weighted Imaging

Molecular diffusion, or Brownian motion, refers to the notion that any type of molecule in a fluid (e.g., water) is randomly displaced as the molecule is agitated by thermal energy. Diffusion-Weighted imaging (DWI) is a relatively new imaging technique that probes differences in Brownian motion of water molecules between tissues, reflecting histological organization [25].

Diffusion weighting enables to distinguish between rapid diffusion of protons (unrestricted diffusion) and slow diffusion of protons (restricted diffusion). For diffusion-weighted imaging, either an echo-planar or a fast GRE sequence is used. DWI is inherently a low-resolution and low-SNR technique and its low quality issues are also exacerbated by its high sensitivity to physiological motion. To reduce motion sensitivity, single-shot echo-planar imaging (EPI) is commonly used. The simplest configuration of this pulse sequence uses a pair of large gradient pulses placed on both sides of the 180° refocusing pulse. The first gradient pulse dephases the magnetization across the sample (or voxel in imaging); and the second pulse rephases the magnetization. For stationary (non-diffusing) molecules, the phases induced by both gradient pulses will completely cancel, the magnetization will be maximally coherent, and there will be no signal attenuation from diffusion. In other words, if no net movement of spinning nuclei occurs between the applications of the gradient pulses, the first gradient dephases the spins and the second rephases them; therefore, high signal intensity is seen. If there is net movement, the protons are not affected by both gradients (they may undergo dephasing but not rephasing, or vice versa); therefore, the signal intensity is decreased. The amount of signal loss is directly proportional to the degree of water motion. Signal loss is proportional to the motion component in the same direction as the diffusion gradient, while no signal loss would occur if the motion was perpendicular to the gradient direction.

The gradient strength, or more often the diffusion weighting, may be expressed in terms of the *b*-value. The *b*-value is proportional to the product of the diffusion time interval and the square of the strength of the diffusion gradient. A larger *b*-value is achieved by increasing the gradient amplitude

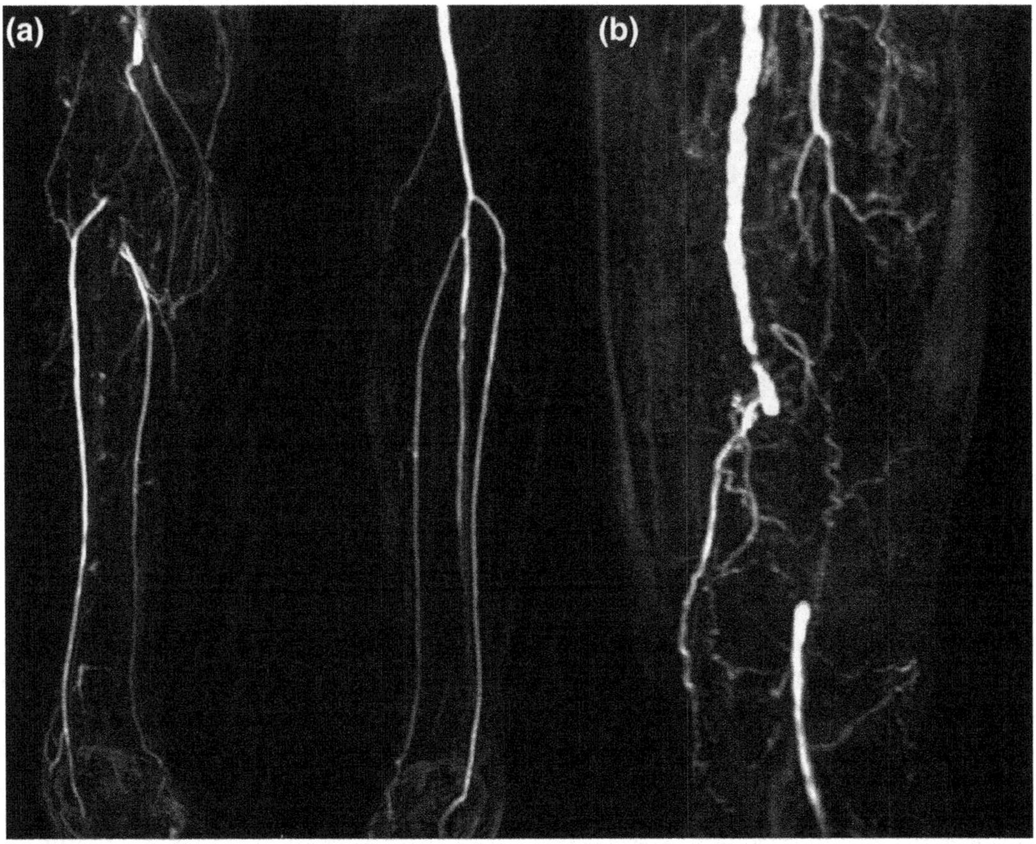

Fig. 20 Right popliteal artery occlusion in a patient with lower limb claudication. **a** Sagittal T1-weighted CE-MRA. **b** MIP reconstruction particular on right popliteal artery showing a chronic occlusion with collateral vessels development

and duration and by widening the interval between paired gradient pulses. To sense slow moving water molecules and smaller diffusion distances, *b*-value should be higher (e.g., $b = 550$ s/mm^2). All diffusion images should be compared with a reference image that is not diffusion weighted (a standard SE image), in which the strength of the diffusion gradient is zero.

The sensitivity of DWI to diffusion (characterized by its *b*-value) can be modified by altering the combination of gradient pulse amplitude, the time for which the gradients are applied and the time that elapses between their application. DWI has actually an important clinical application, however, imaging interpretation is not intuitive. To resolve this problem, let us assume that the diffusion has no restrictions and that its displacement distribution therefore can be described with a free diffusion physical model, which is a 3D isotropic Gaussian distribution. In this model, the physical diffusion coefficient D is replaced by the ADC, which is derived from the equation AD: $-b \ln(\mathrm{DWI}/b0)$, where DWI is the diffusion-weighted image intensity for a specific *b*-value and diffusion gradient direction, defined as in the previous section, and *b*0 is a reference image without diffusion weighting. In order to obtain an image of the ADC values, two acquisitions are necessary: one set obtained without application of a diffusion gradient (which have an appearance similar to that of T2-weighted images), and one obtained with a diffusion gradient. The ADC calculation is based on the negative logarithm of the ratio of those two image sets (images obtained with diffusion weighting compared with those obtained without

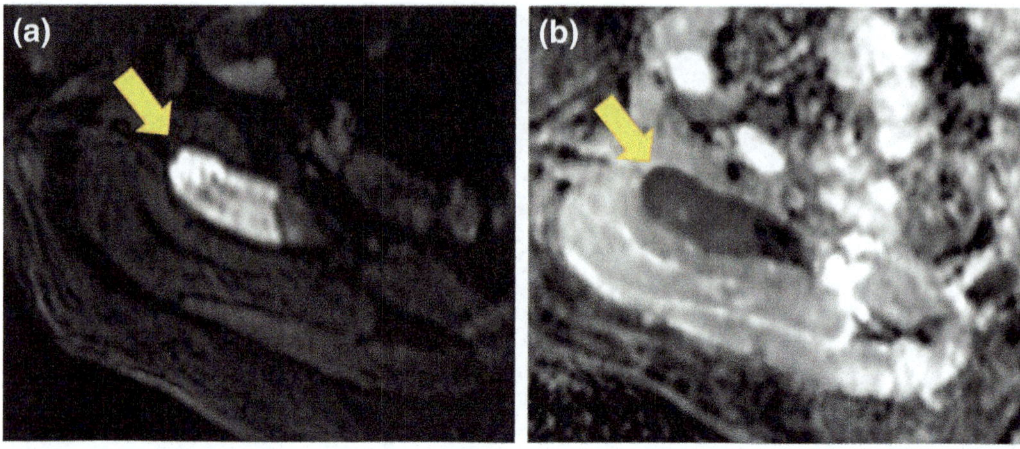

Fig. 21 **a** Painful right iliac bone metastasis with evidence of water diffusion restriction at DWI imaging (b = 1000 s \mm^2). **b** ADC map hypointensity confirms DWI detection

diffusion weighting). Even if stringent measures have been taken to avoid the effects of gross motion and flow, a diffusion-weighted image is still affected by MR properties other than that of diffusion, e.g., T2 weighting. To remove all effects other than that of diffusion, it is mandatory to use the apparent diffusion coefficient, as described. However, an ADC map created in this way by combining two images, with and without diffusion weighting or using two b-values, the lower of which is not large enough to remove the effects of perfusion, contains information about perfusion as well as diffusion components. To differentiate between perfusion and diffusion multiple b-values are needed.

The visualization of changes in the diffusion properties of tissue water with MR imaging has become a useful, multifaceted tool to characterize tissue structure and to identify and differentiate disease processes. DWI is routinely used in investigations of stroke in brain imaging [26]; it actually has a relatively new role in oncologic imaging, in which this technique has became an important tool used to characterize tumor cellularity. However, sometimes the appearance of high signal intensity on diffusion-weighted images also may be due to T2 effects, or so-called T2 shine-through. In clinical neurological practice, the absence of corresponding effects on ADC maps allows areas of restricted diffusion from recent stroke to appear dark and areas of unrestricted diffusion from older stroke to appear relatively bright. Therefore, DWI has always to be evaluated in comparison with ADC maps, in order to allow the determination of the age of a stroke (this concept should be always keep in mind during DWI evaluation in every clinical application). In oncologic management, DWI can use as a detection technique, able to identify some lesions (often causing unjustified pain, as in case of bone metastases), difficulty assessable with classic technique [27] (Fig. 21).

2.6 Diffusion Tensor Imaging

Diffusion tensor imaging (DTI) is an extension of DWI that allows data profiling based upon white matter tract orientation. In white matter (WM), diffusion follows the 'pathway of least resistance' along the WM tract; this direction of maximum diffusivity along the WM fibers is projected into the final image.

DTI is defined as a MRI technique that uses anisotropic diffusion to estimate the axonal—WM—organization of the brain, while fiber tractography (FT) is a 3D reconstruction technique to access neural tracts using data collected by DTI. In DTI, the anisotropy is utilized in order to estimate the axonal organization of the brain;

in fact, diffusion is anisotropic (directionally dependent) in WM fiber tracts, as axonal membranes and myelin sheaths present barriers to the motion of water molecules in directions not parallel to their own orientation. The direction of maximum diffusivity has been shown to coincide with the WM fiber tract orientation. This information is contained in the diffusion tensor, a mathematic model of diffusion in three-dimensional space [28]. The diffusion tensor describes the magnitude, the degree of anisotropy, and the orientation of diffusion anisotropy. By applying the appropriate magnetic field gradients, MR imaging may be sensitized to the random, thermally driven motion (diffusion) of water molecules in the direction of the field gradient. Since water tends to spread in the fibrous tissues (mainly in the WM) following the fiber orientation, diffusion tensor becomes an indicator of cognitive functional organization, allowing the identification of mutual connections between different functional centers and highlighting any possible alterations due to pathological situations [29]. Technically, tensor is a term used to describe a matrix of numbers derived from diffusion measurements in several different directions, from which it is possible to estimate the diffusivity in any arbitrary direction or determine the direction of maximum diffusivity [30]. The tensor matrix may be easily visualized as an ellipsoid whose diameter in any direction estimates the diffusivity in that direction and whose major principle axis is oriented in the direction of maximum diffusivity. The tensor model consists in a 3×3 matrix derived from the measurement of diffusivity in at least six noncollinear directions. At least six diffusion gradients and the corresponding ADC maps along six orthogonal directions (three orthogonal pure: x, y, z, and three combined: xy, xz, yz) in order to calculate the diffusion tensor. Using more than six encoding directions will improve the accuracy of the tensor measurement for any arbitrary orientation [31].

From a technical point of view, there are some considerations to make when assessing a diffusion tensor protocol. The protocol choice is moderately complicated by the wide spectrum of pulse sequence parameters that must be configured. The majority of DTI studies nowadays use b-values in the range of 700–1300 s/mm^2 (with a b-value of 1000 s/mm^2 being most common), leading to 30–50% signal reduction assuming the mean diffusivity of normal white matter is around $0.8–1.0 \times 10 - 3$ mm^2/s [32]. The determination of the optimum b-value is complicated by the involvement of many factors, including: SNR (the higher the SNR, the more accurately signal attenuation can be measured with higher b-values), echo time (the smaller the b-value, the shorter the achievable echo time), and other factors that are more difficult to assess such as eddy current and motion artifacts (in general, smaller b-values produce less artifacts). Measurements of diffusion anisotropy tend to be quite sensitive to image noise, which can also lead to biases in the anisotropy estimates [33]. The accuracy of DTI measures may be improved by either increasing the number of encoding directions or increasing the number of averages. Unfortunately, this increases the scan time for DTI data collection [34].

The information contained in the diffusion tensor can be “viewed” through creation of maps of appropriate diffusion indices derived from DTI data, of which the main are represent by mean diffusivity (MD) or in other term apparent diffusion coefficient (ADC), and fractional anisotropy (FA). Specifically, MD reflects the average magnitude of molecular displacement by diffusion (the more the MD value, the more the isotropic is the medium), while FA reflects the directionality of molecular displacement by diffusion and vary between 0 (isotropic diffusion, such as in CSF) and 1 (infinite anisotropic diffusion). In particular, the FA provides information regarding to the shape of rotation ellipsoid associated with the tensor: starting from the null value for FA is the shape is spherical (isotropic diffusion); higher values of FA correspond to shapes ellipsoid more elongated, up to reach the linear form for FA = 1 (maximum anisotropy).

Another important measure is the tensor orientation described by the major eigenvector direction. For diffusion tensors with high anisotropy, the major eigenvector direction is

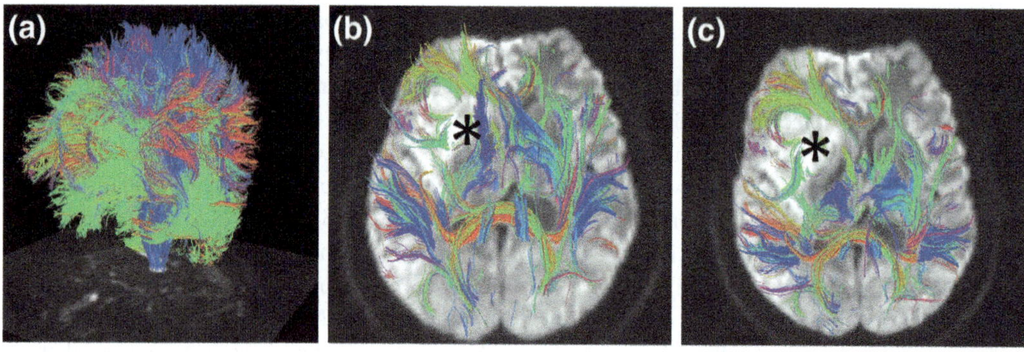

Fig. 22 **a** Full brain volume fiber tracking from DTI datasets in a patients with right frontal lobe glioma. Axial–oblique view (**b**) and axial view (**c**), with thin slice segmentation, demonstrating the fibers around the tumor (*asterisks*)

generally assumed to be parallel to the direction of white matter tract, which is often represented using an RGB (red-green-blue) color map to indicate the eigenvector orientations. The local eigenvector orientations can be used to identify and parcellate-specific WM tracts; thus DT-MRI has an excellent potential for applications that require high anatomical specificity. This technique, preparatory to the implementation of tractography, represents excellent ability of this technique in anatomic functional WM depiction make it an important and innovative tool in pre- and postoperative management of patient with brain lesions [35].

DTI has been reported in a broad spectrum of applications, such as in assessment of WM deformation determined by tumors, in pre-surgical planning (Fig. 22), in Alzheimer disease (to detect an early phase of disease), in schizophrenia, in focal cortical dysplasia, and in multiple sclerosis (for plaque assessment).

The primary reason is that water diffusion in tissues is highly sensitive to differences in the microstructural architecture of cellular membranes. Increases in the average spacing between membrane layers will increase the apparent diffusivity, whereas smaller spaces will lead to lower apparent diffusivities. This sensitivity makes DTI a powerful method for detecting microscopic differences in tissue properties. However, the interpretation of changes in the measured diffusion tensor is complex and should be performed with care; in particular, FA is highly sensitive to microstructural changes, but not very specific to the type of changes.

2.7 MR Spectroscopy

Magnetic resonance spectroscopy (MRS) is an imaging technique able to depict neurochemical function of a volume-of-interest (VOI) [36]. MRS technique has been initially developed to determinate chemical and biochemical properties of some compounds in solution; the introduction of gradient field technology in MRI permitted the evolution of in vivo spectroscopy. Biological and medical spectroscopy applications are mainly addressed to ^{1}H, ^{13}C, 19F, and ^{31}P isotopes. MRS technique requires high-intensity static magnetic fields (at least 1.5 T) in order to ensure better homogeneity and increased sensitivity. In vivo MR spectroscopy is analyzed on the basis of three different parameters: chemical shifts, signal intensities, and spin–spin (J) coupling [37]. While chemical shifts and J-couplings provide qualitative information on the chemical structure, signal intensities provide information on different concentrations of compounds. The chemical shift is defined as the ratio of the resonance frequency difference between the signal of interest (*n*) and a reference signal (*n*ref) relative to the operating frequency of the MR system (*n*0), and is expressed in units of ppm. Chemical shifts are used to characterize in vivo different compounds, employing the property of local chemical

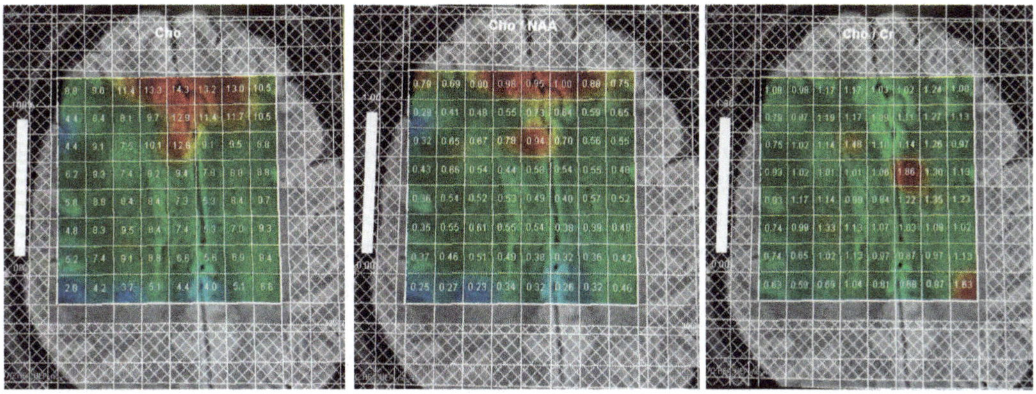

Fig. 23 Multiple voxel spectroscopy. *Panel* showing an assessment of different metabolites in patient with solid brain lesions

environment and molecular bonding to affect the distribution of electron density around a nucleus and to modify the resonance frequencies. Resonances of the methyl groups of N-acetylaspartate (NAA, $d = 2.01$ ppm) or creatine (Cr, $d = 3.02$ ppm) are typically used as internal chemical shift references for in vivo MRS applications. Resonance signals can be displayed as single peaks (singlets) or can be split into several signals (multiplets). MRS highlights the split resonances, caused by spin–spin coupling (J-coupling) between neighboring protons. The strength of coupling is defined by the J-coupling constant and can be extracted from the splitting pattern of the spectral multiplet. Intensities of resonance signals (measured as amplitudes or integrals of the areas under the signals) are proportional to the concentration and to the number of corresponding magnetically equivalent nuclei in a molecule.

2.7.1 MR Spectroscopic Imaging

MRS can be performed using different techniques. Opportune setting of TR and TE is crucial; in particular, some specific investigations require short echo times (TE = 20–35 ms) to detect metabolites with short relaxation times, such as glutamate, glutamine, myoinositol, glycine, GABA, and some amino acids. Cr is a relatively stable metabolite, requiring a long TE to be observed (100 ms or more); it is widely employed as internal control for deriving other metabolites concentrations. Single-voxel spectroscopy (SVS) is the most time-efficient modality, offering a spatial resolution in the order of 1–8 cm^3. It analyzes the signal from a given VOI of a tissue. Multiple voxel techniques, known as chemical shift imaging (CSI) or magnetic resonance spectroscopic imaging (MRSI), permit the derivation of metabolite maps (Fig. 23). The CSI is able to include larger volumes of tissue in the spectroscopy study with possible separate voxel-by-voxel analysis of the entire acquired volume. These voxels can be as small as 1 cm^3. MRS is hampered by several limitations, such as artifacts due to high differences in magnetic susceptibility (bone, air, large vessels, metals).

In the last years, several studies have expressed relationships between *N*-acetyl aspartate (NAA) and glucose metabolism in determinate brain areas (most importantly prefrontal cortex, anterior cingulate cortex and thalamus) and painful syndromes. Kupers and colleagues studied 13 healthy volunteers after painful heat stimulation to the right, revealing a GABA concentration increase of 15% in the rostral anterior cingulate cortex (rACC), by single-voxel ^{1}H-MRS at a 3T scanner with a short TE (20 ms) [38].

2.8 Susceptibility Weighted Imaging

This sequence is a high spatial resolution 3D gradient echo MR imaging technique with phase post-processing that accentuates the paramagnetic properties of blood products and is very sensitive in the detection of intravascular venous deoxygenated blood as well as extravascular blood products [39]. Because of its ability in blood products detection, SWI is becoming a new exciting tool used in studies of arterial venous malformations, occult venous disease, multiple sclerosis, trauma, tumors and functional brain imaging [40]. SWI exploits the loss of signal intensity created by disturbance of a homogeneous magnetic field; various paramagnetic, ferromagnetic, or diamagnetic substances such as air/tissue or air/bone interfaces can be responsible of these disturbance. Sensitivity to susceptibility effects increase as one progresses from fast spin-echo to routine spin-echo to gradient echo techniques, from T1- to T2- to T2*-weighting, from short-to-long echo times, and from lower to higher field strengths. After data acquisition, additional post-processing can accentuate the signal intensity loss caused by any susceptibility effects. The phase images are high-pass filtered and then transformed to a special mask that varies in amplitude between zero and unit; this mask is multiplied a few times into the original magnitude image, in order to generate enhanced contrast between paramagnetic substances and surrounding tissue. Actually, SWI has as a main application the identification of small amounts of hemorrhage/blood product or calcium, booth possibly unapparent on other MRI sequences [41].

2.9 BOLD Functional MRI

The rationale of blood oxygenation level-dependent (BOLD) magnetic resonance imaging is that blood flow would sensitively depict the tissue's activity [42]. The BOLD contrast method employs the electromagnetic property of deoxygenated hemoglobin (dHb) to suppress fMRI signal generated from neighboring water molecules. The increase in level of oxygenation of the blood required by tissue's activity is displayed by BOLD imaging as a decrease in dHb suppression, and thus as an increase in fMRI signal. Disadvantages of BOLD fMRI mainly consist in the unfeasibility to assess a baseline fMRI signal, since BOLD signal measures change between alternating states, and in the weakness of fMRI signal compared to PET imaging. Venous BOLD functional imaging employs T2*-weighted sequences, particularly sensitive to magnetic field inhomogeneities, thus able to read out changes in dHb\water interaction both in intravascular and extravascular spaces [43]. BOLD imaging can be performed with both SE and GE techniques; the latter are mostly employed because of their relatively high sensitivity at the cost of limited spatial resolution. Another fMRI technique is quantitative assessment of blood volume to a determinate tissue by administration of iron oxide contrast agents. Using T2*-w sequences, an increase in blood volume to a tissue induces an increase in the content of contrast agents, and consequently a decrease in MRI signal. The use of noninvasive neuroimaging methods including BOLD fMRI has been utilized to clarify the neural bases of many kinds of sensorimotor and mental processes in neuroscience. fMRI has an extensive role in neuroimaging. Initial pain neuroimaging experiences have studied location and pattern of neural activity evoked by a painful or non-painful stimulus. These studies revealed a common whole brain activation pattern (comprehending primary and secondary somatosensory cortex, cerebellum, anterior insular and cingulate cortices, basal ganglia, and both frontal regions and posterior parietal cortex) responding to mechanical, heat, cold, and electrical stimuli, and unique stimuli, referred to as a network or the 'pain matrix' (Fig. 24).

2.10 MR Thermometry in Focused Ultrasound Ablative Technique

The development of MRI techniques able to noninvasively measure temperature changes in tissue led physicians to look at these with high interest to enhance the guidance of thermal

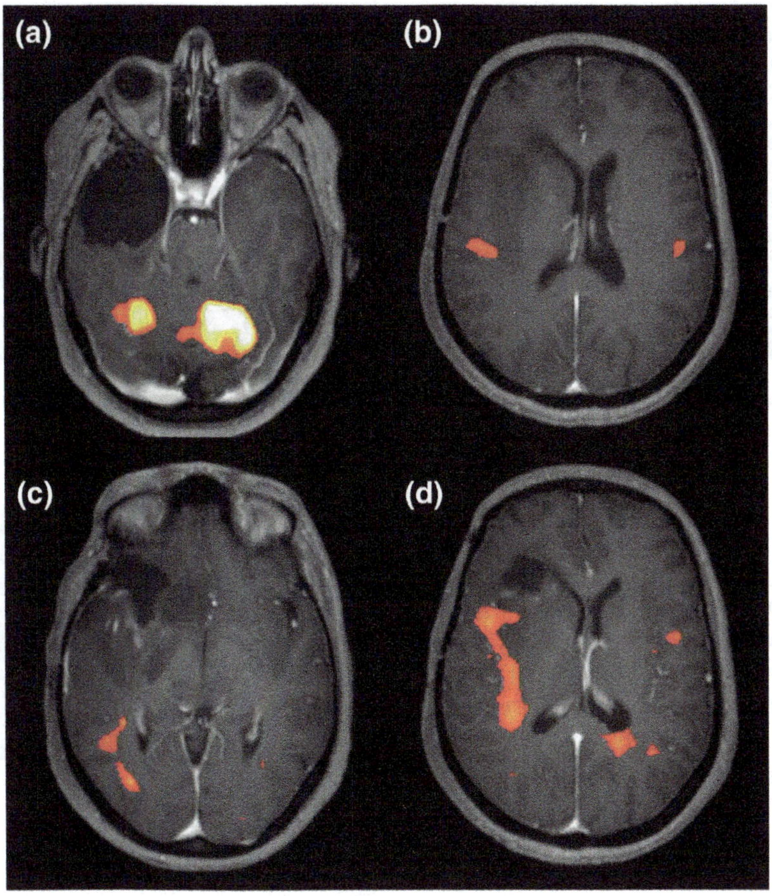

Fig. 24 Patient with recurrent cerebral tumor; pre-surgical evaluation with fMRI and BOLD sequences. Absence of close contiguity relations between the areas of activation (left hand **a**, **b** and left foot **c**, **d**) and recurrent lesion and the above described lesion known

therapy treatments (such as Magnetic Resonance guided Focused Ultrasound). A possible technique able to assess temperature tissue changes is diffusion weighted; molecular water mobility due to thermal Brownian motion is quantified by the molecular diffusion coefficient of water, which is, by definition, a temperature-dependent process and can be quantified using MRI via the apparent diffusion coefficient. By far, the most exploited and widely validated quantitative MRTI techniques are based on the temperature sensitivity of the water proton chemical shift [44]. The shift of the PRF is proportional to temperature over a large range of temperatures (0–100 °C), with a sensitivity of % 0.01 ppm/°C for bulk water. Similar to the previous method, the physical basis for the temperature-dependent PRF phenomenon is that a rise in temperature leads to a corresponding increase in molecular Brownian motion. The result of this is that, as temperature rises, hydrogen bonds between local water molecules bend, stretch, and break. Using MRI, this temperature-dependent PRF shift can be measured using chemical shift imaging (CSI) techniques to directly measure the frequency shift. However, the easiest method for fast, high-resolution estimation of temperature changes due to the PRF shift is based on indirect measurements via relating the difference in phase between subsequent images to the frequency shift. A shift in the PRF is linearly related to temperature and can be mapped rapidly with standard

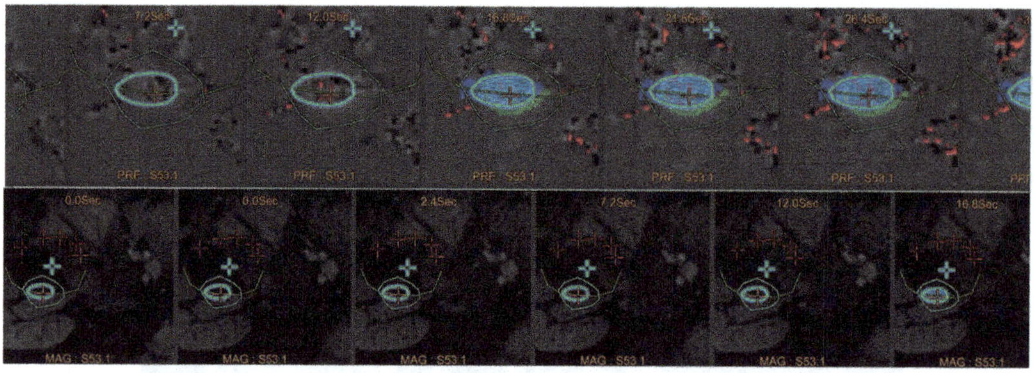

Fig. 25 MR thermography monitoring MRgFUS ablation by a real-time color map with blue area representing major heat

MR imaging sequences using phase differences. However, the conventional MR thermometry is insensitive to temperature changes in fat and is susceptible to motion artifacts including tissue swelling due to the need for image subtraction. The primary disadvantage of using standard CSI techniques is poor spatiotemporal resolution, limiting the ability to directly apply this technique for monitoring rapid heat delivery in a volume. In the clinical practice, an ablative non-invasive image-guided treatment (such as High-Intensity Focused Ultrasound, HIFU) may be improved by using of real-time thermal mapping with phase-difference fast-spoiled gradient-echo MR imaging, which is performed at the targeted region before, during, and immediately after sonication [45]. These images are automatically compared with a reference image obtained immediately before the sonications in order to generate a real-time thermal map (Fig. 25).

Thermal feedback is generated by real-time PRF while magnitude images highlight the temperature changes and the anatomy in the targeted area [46]. A temperature graph shows the temperature change on the temperature maps (Fig. 26).

The benefit of combining MR with the focused ultrasound treatment is real-time monitoring of the localization of the individual sonications, enabling the measurement of energy deposition and the temperature changes in the region being treated, and feedback on the effectiveness and safety of the sonications, allowing to obtain a quantitative real time image of an effect, with a comparative evaluation between temperature analysis, patient sensations, and sonication effects.

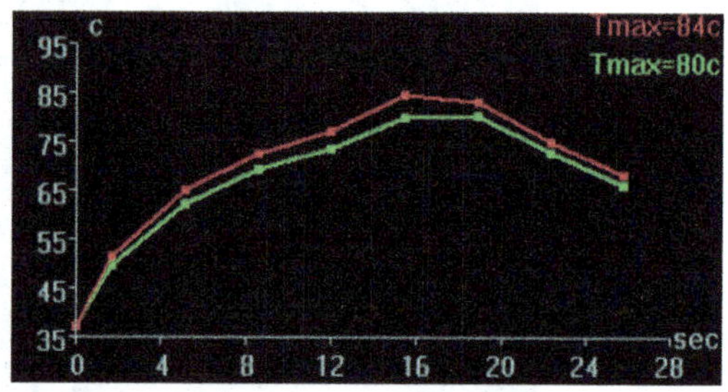

Fig. 26 Temperature versus time curve monitoring thermal energy administered by MrgFUS sonication

References

1. Merskey H, Bogduk N. Classification of chronic pain. IASP Task Force on Taxonomy. 2nd ed. Seattle: IASP Press; 1994.
2. Tracey I. Imaging pain. Br J Anaesth. 2008;101(1):32–9.
3. Albe-Fessard D, Berkley KJ, Kruger L, Ralston HJ, III, Willis WD, Jr. Diencephalic mechanisms of pain sensation. Brain Res. 1985;356:217–96. Apkarian AV, Bushnell MC, Treede RD, Zubieta JK. Human brain mechanisms of pain perception and regulation in health and disease. Eur J Pain. 2005;9:463–84.
4. Harris RE, Sundgren PC, Pang Y, et al. Dynamic levels of glutamate within the insula are associated with improvements in multiple pain domains in fibromyalgia. Arthritis Rheum. 2008;58(3):903–7.
5. Khan HS, Stroman PW. Inter-individual differences in pain processing investigated by functional

magnetic resonance imaging of the brainstem and spinal cord. Neuroscience. 2015;29(307):231–41.
6. Jongen JL, Hans G, Benzon HT, et al. Neuropathic pain and pharmacological treatment. Pain Pract. 2014;14(3):283–95.
7. Narváez JA, Narváez J, De Lama E et al. MR imaging of early rheumatoid arthritis. Radiographics. 2010;30(1):143–63; discussion 163-5.
8. Brown MA, Semelka RC. MR imaging abbreviations, definitions, and descriptions: a review. Radiology. 1999;213:647–62.
9. McRobbie DW, Moore EA, Graves MJ et al. MRI from picture to proton. 2nd ed. 2010. Cambridge: Cambridge University Press.
10. Mirowitz SA, Brown JJ, McKinstry RC, Li Tao. Principles and applications of echo-planar imaging: a review for the general radiologist. Mehdi Poustchi-Amin. RadioGraphics. 2001;21(3):767–79.
11. Choyke PL, Dwyer AJ, Knopp MV. Functional tumor imaging with dynamic contrast-enhanced magnetic resonance imaging. J Magn Reson Imaging. 2003;17(5):509–20.
12. Bali MA, Metens T, Denolin V, et al. Tumoral and nontumoral pancreas: correlation between quantitative dynamic contrast-enhanced MR imaging and histopathologic parameters. Radiology. 2011;261 (2):456–66.
13. Bhooshan N, Giger ML, Jansen SA, et al. Cancerous breast lesions on dynamic contrast-enhanced mr images: computerized characterization for image-based prognostic markers. Radiology. 2010;254(3):680–90.
14. Lacerda S, Law M. Magnetic resonance perfusion and permeability imaging in brain tumors. Neuroimaging Clin N Am. 2009;19(4):527–57.
15. Meaney JF, Ridgway JP, Chakraverty S, et al. Stepping-table gadolinium-enhanced digital subtraction MR angiography of the aorta and lower extremity arteries: preliminary experience. Radiology. 1999;211:59–67.
16. Ho KY, Leiner T, de Haan MW, et al. Peripheral MR angiography. Eur Radiol. 1999;9:1765–74.
17. Shigematsu Y, Korogi Y, Hirai T, et al. 3D TOF turbo MR angiography for intracranial arteries: phantom and clinical studies. J Magn Reson Imaging. 1999;10:939–44.
18. Wedeen VJ, Meuli RA, Edelman RR, et al. Projective imaging of pulsatile flow with magnetic resonance. Science. 1985;230:946–8.
19. Walker MF, Souza SP, Domoulin CL. Quantitative flow measurement in phase contrast MR angiography. JCAT. 1988;12:304–13.
20. Hentsch A, Aschauer MA, Balzer JO et al. Gadobutrolenhanced moving-table magnetic resonance angiography in patients with peripheral vascular disease: a prospective, multicentre blinded comparison with digital subtraction angiography. Eur Radiol. 2003;13:2103–14.
21. Luccichenti G, Cademartiri F, Ugolotti U, et al. Magnetic resonance angiography with elliptical ordering and fluoroscopic triggering of the renal arteries. Radiol Med (Torino). 2003;105:42–7.
22. Wang Y, Chen CZ, Chabra SG, et al. Bolus arterial-venous transit in the lower extremity and venous contamination in bolus chase threedimensional magnetic resonance angiography. Invest Radiol. 2002;37:458–63.
23. Willinek WA, Gieseke J, Conrad R, et al. Randomly segmented central k-space ordering in high-spatial-resolution contrast-enhanced MR angiography of the supraaortic arteries: initial experience. Radiology. 2002;225:583–8.
24. Korosec FR, Frayne R, Grist TM, et al. Time-resolved contrast-enhanced 3D MR angiography. Magn Reson Med. 1996;36:345–51.
25. Mori S, Barker PB. Diffusion magnetic resonance imaging: its principle and applications. Anat Rec. 1999;257:102–9.
26. Roberts TP, Rowley HA. Diffusion weighted magnetic resonance imaging in stroke. Eur J Radiol. 2003;45:185–94.
27. Charles-Edwards EM, de Souza NM. Diffusion-weighted magnetic resonance imaging and its application to cancer. Cancer Imaging. 2006;6:135–43.
28. Mori S, Zhang J. Principles of diffusion tensor primer imaging and its applications to basic neuroscience research. Neuron. 2006;51:527–39.
29. Hagmann BP, Jonasson L, Maeder P, et al. Understanding diffusion MR imaging techniques: from scalar diffusion-weighted imaging to diffusion tensor imaging. RadioGraphics. 2006;26:S205–23.
30. Alexander AL, Lee JE, Lazar M, Field AS. Diffusion tensor imaging of the brain. Neurotherapeutics. 2007;4(3):316–29.
31. Basser PJ, Matiello J, Le Bihan D. MR diffusion tensor spectroscopy and imaging. Biophys J. 1994;66:259–67.
32. Tournier JD, Mori S, Leemans A. Diffusion tensor imaging and beyond. Magn Reson Med. 2011;65 (6):1532–56.
33. Pierpaoli C, Basser PJ. Toward a quantitative assessment of diffusion anisotropy. Magn Reson Med. 1996;36:893–906.
34. Moseley ME, Cohen Y, Kucharczyk J, et al. Diffusion-weighted MR imaging of anisotropic water diffusion in cat central nervous system. Radiology. 1990;176:439–46.
35. Pajevic S, Pierpaoli C. Color schemes to represent the orientation of anisotropic tissues from diffusion tensor data: Application to white matter fiber tract mapping in the human brain. Magn Reson Med. 1999;42:526–40.
36. Tran T, Ross B, Lin A. Magnetic resonance spectroscopy in neuro-logical diagnosis. Neurol Clin. 2009;27(1):21–60.
37. Grachev ID, Fredrickson BE, Apkarian AV. Abnormal brain chemistry in chronic back pain: an in vivo proton magnetic resonance spectroscopy study. Pain. 2000;89:7–18.

38. Kupers R, Danielsen ER, Kehlet H, Christensen R, Thomsen C. Painful tonic heat stimulation induces GABA accumulation in the prefrontal cortex in man. Pain. 2009;142:89–93.
39. Haacke EM, Xu Y, Cheng YC, et al. Susceptibility weighted imaging (SWI). Magn Reson Med. 2004;52:612–8.
40. Sehgal V, Delproposto Z, Haacke EM, et al. Clinical applications of neuroimaging with susceptibility-weighted imaging. J Magn Reson Imaging. 2005;22:439–50.
41. Pauling L, Coryell CD. The magnetic properties and structure of hemoglobin, oxyhemoglobin and carbonmonoxyhemoglobin. Proc Natl Acad Sci U S A. 1936;22:210–6.
42. Vincent K, Moore J, Kennedy S, Tracey I. Blood oxygenation level dependent functional magnetic resonance imaging: current and potential uses in obstetrics and gynaecology. BJOG. 2008;116 (2):240–6.
43. Price RR, Allison J, Massoth RJ et al. Practical aspects of functional MRI (NMR Task Group #8). Med Phys. 2002;29(8).
44. McDannold N. Quantitative MRI-based temperature mapping based on the proton resonant frequency shift: review of validation studies. Int J Hyperther. 2005;21(6):533–46. 27.
45. Graham SJ, Chen L, Leitch M, et al. Quantifying tissue damage due to focused ultrasound heating observed by MRI. Magn Reson Med. 1999;41:321–8.
46. Kuroda K, Oshio K, Chung AH, Hynynen K, Jolesz FA. Temperature mapping using the water proton chemical shift: a chemical shift selective phase mapping method. Magn Reson Med. 1997;38 (5):845–51.

6 Anticipatory Brain Responses and Expectancy Effects on Pain: Theory, Research Findings and Functional Networks

Christopher Brown

Abstract

Pain is broadly recognised to have an important evolutionary function in preventing bodily harm, and hence neural learning mechanisms have evolved to prepare organisms to avoid physical danger. Learned pain expectancies trigger anticipatory neural responses that result in changes in perception, emotion and behaviour. Such changes are adaptive for avoiding acute injury but can be maladaptive in clinical conditions in which pain is chronic. This chapter will review the use of neuroimaging as a research method for understanding anticipation and expectancy effects on pain. These observations have inspired a body of work over the last two decades focussing on identifying the neural mechanisms by which cognitive expectancies influence pain perception. Brain responses to pain anticipation, and changes in subsequent nociceptive processing, have been investigated to identify possible neural mediators of expectancy effects on pain, and have proven to be relevant to understanding placebo analgesia and its opposite, the nocebo effect. The chapter will discuss the concept of uncertainty and its theoretically supported role in modulating pain anticipation and expectancy effects on pain. In particular, evidence suggests that certain expectations have the greatest impact on pain perception. Hence, identification of the neural mechanisms supporting (un)certainty in expectation would be of great interest in helping to develop novel therapeutic strategies for chronic pain. Towards this end, the chapter will explore the role of different functional networks in mediating the effects of expectation and certainty on pain.

Keywords

Expectancy · Anticipation · Placebo · Nocebo · Hyperalgesia · Analgesia

C. Brown (✉)
University of Liverpool, Department of Psychological Sciences, Eleanor Rathbone Building, Bedford Street South, Liverpool L69 7ZA, UK
e-mail: christopher.brown@liverpool.ac.uk

L. Saba (ed.), *Neuroimaging of Pain*, DOI 10.1007/978-3-319-48046-6_6

1 Introduction

1.1 Expectancy and Anticipation

Pain acts as a learning signal to the organism that certain objects or situations are dangerous and should be handled with care or avoided. This is well illustrated by clinical cases in which a rare genetic disorder can lead to complete insensitivity to pain; such individuals grow up with accumulated wounds, bruises, broken bones, and other health issues that lead to a reduced life expectancy [1]. Because of the importance of physical integrity to survival, emotional (fear, anxiety) and behavioural (avoidance) responses to threats such as pain can occur automatically [2]. This may be the evolutionary basis for loss aversion in economics [3], i.e. the tendency to prefer avoiding losses than making gains.

Indeed, from a neuroscience perspective, the experience of pain epitomises how the brain responds to threat. Until the mid-twentieth century, pain processing in the brain was thought to be a direct reflection of the afferent processes of nociception. However, with the introduction of the 'pain matrix' theory [4], the concept of pain changed to be understood as a result of complex information processing in the brain, which included sensory information such as the location, temporal characteristics and intensity of a stimulus, but also included important affective (i.e. emotional) and cognitive aspects [4, 5]. More recently, the concept of pain-specific cognitive and emotional networks has been replaced with a more nuanced view of shared brain functions involved with processing pain as well as other motivationally relevant stimuli [6]. These cognitive and affective processes are thought to be critical for threat avoidance.

Evolution has provided complex organisms with the ability to avoid physical danger through neural learning mechanisms that can associate past experiences and other sources of knowledge with environmental cues [7]. When such learned cues are processed by the brain, they trigger the cognitive and affective responses of pain expectancy and anticipation. The terms *expectancy* and *anticipation* are frequently regarded as synonyms, but for the purposes of this chapter expectancy is referred to as the encoding of a prediction of a forthcoming painful event. Anticipation, on the other hand, is regarded as the broader cascade of neurocognitive and other physiological events resulting from expectancy, which may result in changes in perception, emotion and motor behaviour.

1.2 Clinical Importance

Experimental research has shown that pain expectancies have a strong influence on the perception of acute experimental pain [8–10]. Expectancy can both enhance and diminish perceived pain. Positive, non-painful beliefs about an anticipated stimulus will result in reduced pain, whereas negative beliefs about a painful stimulus will exacerbate the painful sensation. The medical community became interested in these expectancy effects as clinical studies started to realise the power of placebos and researchers began to identify potential cognitive (e.g. expectancy) and corresponding biological mechanisms underlying the placebo response, including placebo analgesia [11]. At the same time, research measuring anticipatory responses to pain discovered that a range of pain-related brain processes were occurring predominantly prior to the experience of pain [12–14] and served as the basis for the modulation of pain perception by expectations [15, 16]. The precise anticipatory neurophysiological events that take place in response to pain expectancy, and that mediate the effects of expectancies on pain perception and behaviour, are the subject of much research that will be summarised in this chapter.

Pain expectancy and anticipatory responses to pain are thought to have a clinical impact; psychological models of chronic pain (experienced by around 20% of adults) have recognised the central importance of pain expectancies in determining outcomes in clinical populations [17]. Chronic pain is often unrelated to any identifiable tissue damage. Difficulty in treating chronic pain means that it is currently associated with more disability and related suffering worldwide than any other medical condition [18]. In the absence of a cure for many types of chronic pain, it has become a priority to better understand how patients can better live with their pain. Psychological models of chronic pain have recognised the central importance of pain expectancies in determining outcomes in clinical populations [17] such that the repeated experience of pain may cause severe limitations on activity and quality of life. As argued in this chapter, expectancies exert their influence on pain perception and behaviour via anticipatory responses. Hence, the identification of the brain mechanisms supporting the negative impact of pain anticipation is expected to improve the management of acute and chronic pain and hence relieve suffering associated with pain.

1.2.1 Focus Point 1: Learning Expectations

Learning from past experience is important for the prediction of future events and experiences. Pavlovian (or classical) and instrumental (or operant) conditioning tasks have been used to measure how organisms learn from experience and enable anticipation and response to motivationally significant events such as pain or rewarding stimuli. When an animal learns that certain environmental cues, such as auditory tones presented during an experiment, indicate that a painful shock is imminent, the animal is using the neural mechanisms of classical conditioning.

Historically, developments in the understanding of these mechanisms have emphasised that it is the information value of predictive cues (i.e. their ability to inform expectation), rather than stimulus–response contingencies, that determine behavioural outcomes. In the middle of the twentieth century, Woodworth [19] and Tolman [20] argued that classical conditioning requires that the animal develops an *expectation* about how the tones are related to the forthcoming stimulus. Bolles [21] further developed this idea with regard to instrumental conditioning by suggesting that behaviour depends on the *motivational value* of the expected outcome. On this basis, a 'reinforcement learning' (RL) model was formalised and popularised by Rescorla and Wagner [22], who postulated that learning occurs when organisms are 'surprised', i.e. when events violate expectations (a 'prediction error'). Expectancies are therefore thought to underlie many types of learning [23] in both animals and humans.

The influential Rescorla and Wagner model gave rise to more sophisticated, computational RL models. RL models all share in common the use of a scalar reinforcement signal (e.g. a rewarding or punishing signal) to direct learning. The most popular theorised RL signal, the temporal difference (TD) prediction error (Fig. 1), has been verified as an important learning signal as a result of neuroscience research in animals and humans. Temporal difference methods were introduced into the psychological and biological literature by Richard Sutton and Andrew Barto in the early 1980s [24]. TD learning is innovative for taking into account the timing of different events, allowing it to account for higher order conditioning. For example, in second-order conditioning, if stimulus B predicts an aversive outcome (e.g. electric shock) and stimulus A predicts stimulus B, then stimulus A also gains aversive predictive value. In TD learning, the goal of the learning system (the 'agent') is to estimate the values of different states or situations, in terms of all of the future rewards or punishments that they predict. This is a departure from Rescorla and Wagner's framework, in which predictions are only of the immediately forthcoming reward [25]. Multiple lines of evidence from electrophysiological recordings of behaving animals, lesion studies and pharmacological manipulation link the TD signal to the function of dopaminergic neurons in the midbrain [25]. In humans, neuroimaging has verified the existence of *subjective value* and *prediction error* computations, theorised in TD models to explain how the brain learns to predict harm, in regions innervated by dopaminergic afferents including the ventral striatum [7, 26].

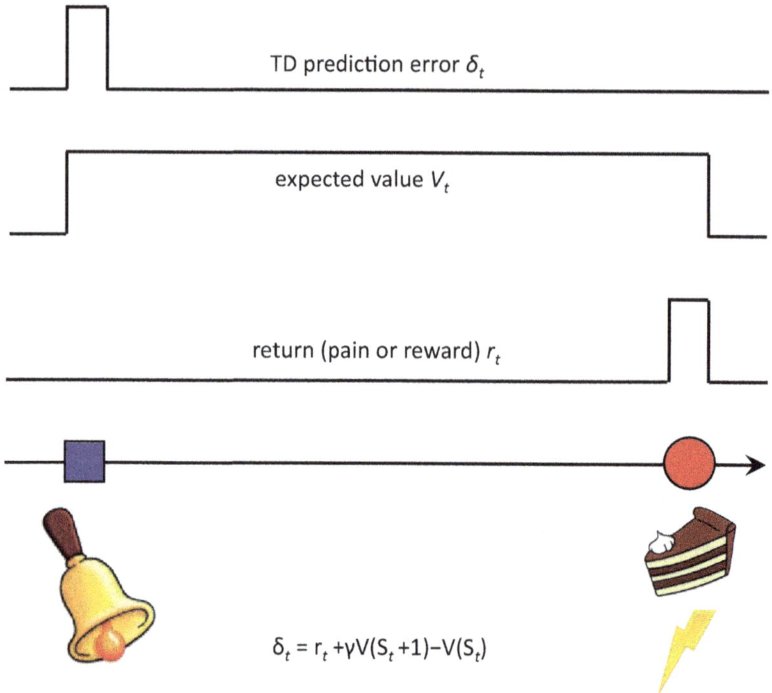

Fig. 1 Time course of different 'hidden variables' of interest in the temporal difference (TD) model of reinforcement learning. The bell predicts a rewarding cake or a punishing electric shock some time later. At the time of the cue, the phasic TD prediction error equals the magnitude of the predicted reward. The expected value signal also becomes positive at this time and stays elevated until the time of the expected reward. At the time of the reward, a phasic cue might signal the occurrence of the reward, but no prediction error occurs if the reward was predicted. Figure adapted with permission from [25]

1.2.2 Focus Point 2: Predictive Coding and Uncertainty

Contemporary theories of the brain regard it as an 'expectancy machine', in that expectations are required both to construct a perception of external events and also respond to those events [27]. This view accounts for how sensory perceptions are generated by the brain and how the generation of those perceptions can be biased by prior information (expectancies). This perspective on the brain began when Hermann von Helmholtz [28] proposed that the brain does not represent sensory images per se, but rather models the causes of those images. Because these causes cannot be perceived directly, they must be inferred from sensory impressions. However, as Friston discussed [27], the problem is that sensations can potentially have multiple causes that interact. Taking an example from vision, the retinal image size can be affected both by object size and distance from the observer. There is therefore inherent uncertainty in the causes of sensory impressions, which the brain must deal with to generate perceptions and guide actions. One solution to this problem is for the brain's model of the environment to contain prior expectations about how causes interact, for example the expectation that regardless of the distance from the observer, objects maintain a constant size. The brain cannot generate all of its prior beliefs de novo; instead it must estimate them from sensory data, by approximating the causes of sensory input as a prediction and then comparing this with the observed sensory data to generate a prediction error. This scheme is generally called predictive coding [29]. According to this scheme, during perception, predictions generated at higher levels of the processing hierarchy are fed back and 'subtracted' from incoming sensory signals, such that the neural information

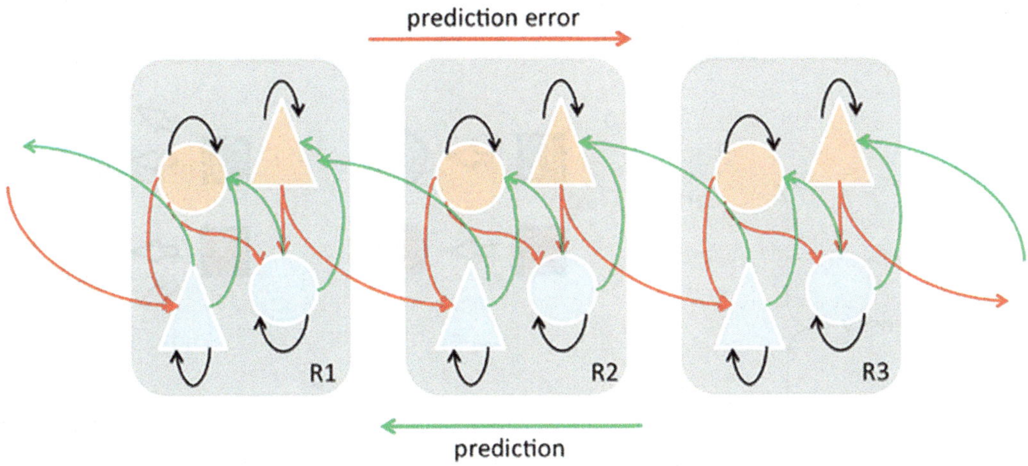

Fig. 2 A schematic of hierarchical predictive coding across three cortical regions; the 'lowest' (*R1*) on the *left* and the 'highest' (*R3*) on the *right*. *Light-blue cells* represent state units, *orange cells* represent error units. Note that predictions and prediction errors are sent and received from each level in the hierarchy. Feed-forward signals conveying prediction errors originate in superficial layers and terminate in deep (infragranular) layers of their targets. Conversely, feedback signals conveying predictions originate in deep layers and project to superficial layers. With permission from [160]

passed forward from stage to stage consists only of prediction error [30] (Fig. 2).

An implication of this framework is that optimal perception and behaviour depends on minimising prediction error. This can either be achieved by changing the brain's predictions to explain sensory input through the act of perception and learning, or by actively changing sensory input to fulfil the brain's predictions by acting on the world. Minimising prediction errors by changing sensory data must also increase the accuracy of predictions. In short, the agent will selectively sample the sensory inputs that it expects. This is known as *active inference*. As Friston explains [31], an intuitive example of this process would be feeling our way in darkness: we anticipate what we might touch next and then try to confirm those expectations. In the context of pain, expectations about the potential for bodily harm are therefore more likely to confirmed especially when the sensory data available are ambiguous, for example when changes in nociception from a treatment are subtle, thereby giving rise to placebo effects. While uncertainty in sensory inputs may bias perception towards expectations, a high degree of uncertainty in expectations is thought to weight perception towards the information provided by bottom-up sensory inputs [32]. There is therefore utility in investigating how neural processes are modulated by uncertainty in order to gain insight into how expectancy impacts on pain perception.

2 Measuring Brain Responses During Anticipation of Pain

The inherently subjective nature of expectations and pain renders their measurement problematic. The neural correlates of subjective states in humans have been investigated using non-invasive brain imaging techniques. This has been applied in experimental research studies that provoke the anticipation of pain through the induction of pain expectancy.

Pain expectancy can be elicited through verbal, auditory or visual cues in which participants predict pain based on the information contained in the cues [33] (Fig. 3). This basic approach has formed the backbone of numerous research studies that have investigated the brain mechanisms by which expectations modulate pain perception. Expectancies may be induced in a similar manner when studying placebo or nocebo

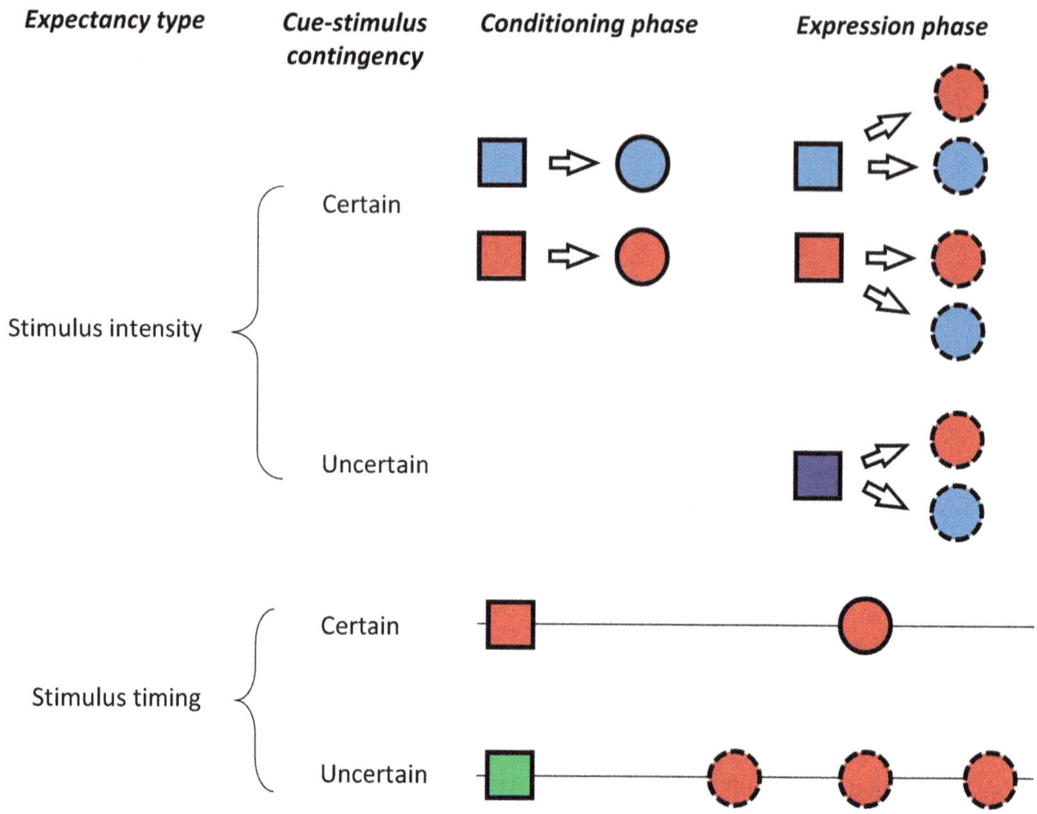

Fig. 3 Experimental paradigms for studying stimulus expectancies. Cues (*squares*) predicting high (*red*) or low (*blue*) intensity pain can be conditioned to corresponding painful or non-painful stimuli (*circles*). *Dashed outlines* When expectancies for stimulus intensity or timing are uncertain, the outcomes are explicitly probabilistic, while unexpected outcomes can also occur after certain expectations that are more likely to be biased by prior expectancies

responses in laboratory settings. For instance, the application of a placebo cream may act as a cue to induce expectations of pain relief [34]. Expectations can also be induced that are relatively certain or uncertain [35]. Certain expectations occur when cues have been associated with high probability to a particular outcome either by conditioning or verbal instruction, with conditioning procedures likely to be more powerful [36]. Experimental designs required to measure the effect of certain expectations on pain perception normally required the pain stimulus to be miscued outside of the awareness of the subject. Uncertain expectations can result when cues are explicitly associated with more than one possible outcome, such that pain stimuli are not required to be miscued as the cue is explicitly probabilistic.

There are a number of techniques available for imaging brain responses during anticipation and experience of pain. The relative advantages of the most popular methods will be discussed here: Positron emission tomography (PET), functional magnetic resonance imaging (fMRI), electroencephalography (EEG) and magnetoencephalography (MEG). The utility of PET and fMRI is their high spatial resolution (e.g. normally 1 mm for fMRI and 2–4 mm for PET). However, fMRI and $H_2^{15}O$ PET measure neuronal activity only indirectly via changes in blood oxygenation or blood flow, respectively. Dependency on haemodynamic changes in the brain limits the temporal resolution of fMRI to about 2–3 s. PET techniques are much slower. Unfortunately, this limitation means that it cannot be firmly established from fMRI and $H_2^{15}O$ PET studies whether neural processes that are modelled to take place post-stimulus (i.e. after nociceptor activation) are not in fact including neural activity occurring within the 2 s prior to nociception. The contamination of estimates of nociceptive responses with anticipatory neural activity can be mitigated if the stimulus presented is unpredictable (see Fig. 3); however, it is debatable whether imaging experiments in humans can achieve such purely unpredictable stimuli and it is therefore likely that estimates of nociceptive activity do contains anticipatory responses. As such, to study nociception responses with fMRI, anticipatory responses should be matched as best as possible between two experimental conditions being compared.

PET and fMRI may also not be measuring the immediate neuronal responses to nociception that are most closely related to afferent inputs. As an illustration, the nociceptive neuronal responses measured from laser-evoked potentials (LEPs) occurring in more posterior regions of the cingulate cortex than are recorded with fMRI [37]. As a consequence, the measurement of 'nociceptive' processing with fMRI is likely to be heavily biased towards other temporally correlated processes such as anticipation, or post-stimulus processes that are sustained over a period of seconds. Despite this limitation for assessing brief nociceptive responses, fMRI is well suited to studying neural processing during pain anticipation and the modulatory effects of pain expectancy on pain processing.

By contrast to dependency on haemodynamics, EEG directly detects currents generated by large columns of synchronously firing excitatory post-synaptic potentials, while MEG measures the magnetic fields induced by these cortical potentials [38]. E/MEG are useful for investigating the time course of neural events owing to their high temporal resolution (on the order of milliseconds). For this reason, the measurement of anticipatory neural responses is well also suited to EEG and MEG. Although the majority of E/MEG recordings reflect cortical neuronal activity, some investigators have taken scalp recordings of activity generated from the hippocampus, cerebellum and thalamus [39]. The major disadvantage of E/MEG is their more limited spatial resolution, which results from a much smaller number of spatial recordings as are typical with fMRI or PET. It is possible to use E/MEG scalp recordings to estimate the location of the current sources in the brain that contribute to those recordings by solving the 'inverse problem', i.e. creating a three-dimensional estimate of current sources from a two-dimensional array of scalp recordings. However, there is no unique mathematical solution to this problem, and the small number of scalp recordings relative to potential neuronal generators means that there are inherent uncertainties in solving it [39]. The recordings do not contain enough information about the generators to create an error-free localisation. As a result, E/MEG have a spatial resolution of 5–20 mm [40, 41]. Also, the accuracy of the source model may be affected by unknown/unmodelled concurrent neural responses.

Anticipatory event-related potentials (ERPs) or fields (ERFs) can be recorded with EEG and MEG, respectively, and are elicited following a pain-predictive cue [42, 43]. The best-characterised anticipatory ERP is the contingent negative variation (CNV), which is a negative waveform that is produced in anticipation of a stimulus when that stimulus requires a behavioural (motor) response [44]. The stimulus-preceding negativity (SPN) is very similar but does not require preparation for a motor response in order to be generated (Fig. 5). Hence, the CNV contains elements of the SPN, plus further motor preparation activity in the late phase of the waveform [44].

3 Differences in Brain Networks Supporting Pain and Anticipation

Studies have most commonly used neuroimaging methods including EEG, MEG, fMRI and PET to investigate the brain regions activated by pain anticipation and their temporal profiles. Here, an overview is provided of brain regions activated by nociceptive stimuli. As will become evident, brain regions activated during pain anticipation appear to explain many of the pain-related activations as measured with fMRI and can therefore be considered 'non-nociceptive'. E/MEG studies suggest a relatively more restricted network of regions are activated within the first few hundred milliseconds of pain stimulation, but have also been useful to track the temporal profile of anticipatory responses.

A meta-analysis examined English language studies until 2005 investigating human cerebral activity during acute and chronic pain states [45]. The authors concluded that the most commonly reported areas activated by pain stimuli in fMRI and PET studies were the anterior and midcingulate cortex (ACC and MCC, respectively), the primary somatosensory cortex (S1), the secondary somatosensory cortex (S2), the insular cortex (IC), the thalamus, basal ganglia and the prefrontal cortex (PFC). Regions that are less commonly activated by pain include the posterior parietal complex [46], supplementary motor area [47], amygdala [48] and nucleus accumbens [49]. However, as we shall see, many of these regions are more robustly activated during anticipation of pain.

Studies of pain provocation using EEG or MEG share consistent activation with fMRI/PET studies only in S1 and S2 cortices, the cingulate cortex (although a more posterior region on the border between the posterior MCC and the dorsal posterior cingulate cortex (PCC)—see Fig. 8) and the posterior insula and parietal operculum [45]. Activations found in fMRI/PET studies in the anterior cingulate, anterior insula and prefrontal cortex are rarely found in E/MEG studies [37] and normally only as part of context-dependent modulations of nociception such as via the induction of expectancies as we shall see in following sections of the chapter. Subcortical regions such as the thalamus and basal ganglia may not be expected in E/MEG studies due to their depth in the brain and differences in neuronal organisations compared to cortex [38]. It is noteworthy that the majority of research studies into cerebral activations in response to pain (including those in the meta-analysis of Apkarian et al. [45]) did not model anticipatory responses or expectancies as part of the experiment or analysis, which is problematic if stimuli are not presented in a way that is completely unpredictable. This means that many fMRI and PET studies recording 'pain' responses in anterior regions may in fact be measuring pain anticipation rather than nociception per se.

The neural correlates of pain anticipation are summarised in Fig. 6. Initial investigations used event-related fMRI, with early work showing that anticipation was associated with activity in more anterior regions of the insula and cingulate cortex than that modelled to occur post-stimulus [14]. However, subsequent work showed a greater overlap between anticipatory and pain-evoked responses, showing more widespread brain activity over broader areas of the 'pain matrix' during anticipation of pain [13]. In order to clarify which brain areas are the most consistently activated during anticipation of pain across different studies, Palermo et al. [12] used activation likelihood estimation meta-analysis to analyse pain anticipation responses in 19 fMRI studies. During anticipation (Fig. 4), activated foci were found in the dorsolateral prefrontal, MCC and anterior insula cortices, medial and inferior frontal gyri, inferior parietal lobule, middle and superior temporal gyrus, thalamus, and caudate. Deactivated foci were found in the ACC, superior frontal gyrus, parahippocampal gyrus and in the claustrum. These results highlight that, apart from notable exceptions that are in particular activated in E/MEG studies (posterior cingulate, somatosensory cortices, posterior insula, parietal operculum), regions commonly associated with pain-related activations in fMRI and PET studies are consistently activated during anticipation of pain. Another exception is the patterns of deactivation observed; for example a rostral region of the ACC is deactivated during anticipation but activated during pain.

Owing to their higher temporal resolution, E/MEG studies may give a finer picture as to the cortical responses occurring during nociception compared to anticipation, notwithstanding the limitations of these techniques in spatial resolution and their inability to map deeper subcortical sources. E/MEG studies have been used to map the temporal profile of anticipatory responses. To date, most studies have focussed on differentiating

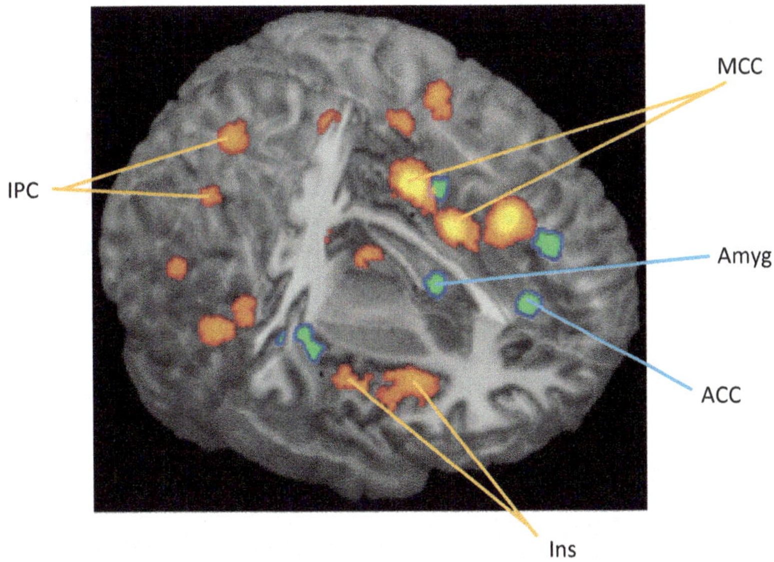

Fig. 4 Brain regions identified in a meta-analysis of pain anticipation neuroimaging studies, showing activations in *yellow/orange* and deactivations in *green/blue* on a 3D rendering of the brain. *ACC* Anterior cingulate cortex; *MCC* Midcingulate cortex; *Amyg* Amygdala; *IPC* Inferior parietal cortex. Adapted with permission from [12]

'early' (within a second after an anticipation cue) and 'late' (within a second prior to a pain stimulus) responses [42, 50–52]. As discussed by Seidel et al. [52], from a cognitive psychology perspective, one would expect early anticipation to be associated with processing of cue-information and basic attentional orienting processes, but not yet pain-related anticipatory processes. Hence, during late anticipation, one would expect pain-specific preparatory (e.g. motor) mechanisms and the expression of inhibitory or facilitatory mechanisms underlying expectancy modulation of pain. Also, comparing SPN responses to early and late phases of the CNV may be instructive for reflecting on to what extent the SPN might contain motor preparation components in the late phase. Such motor preparation might be expected as part of automatic withdrawal responses to stimuli that are expected to be intensity painful, even if such responses are not task-related as for the CNV [42].

Generally, the above view accords with empirical data; early and late phases of the SPN and CNV are associated with distinct sources, with some differences apparent between the SPN and CNV in the late phase. For both the CNV and SPN, the early part of the waveform has a frontal topography [53]. Using E/MEG, research has shown that the supplementary motor area and adjacent MCC might be involved in generating the early wave of the CNV [54, 55]. The early wave of the SPN (Fig. 5) may have much in common with the early wave of the CNV, showing a common broad fronto-central distribution [42], that co-occurs with activation of the SMA [52]. On the other hand, the latter part of the CNV is centro-parietal [53] and associated mainly with the preparation of motor execution in premotor and sensorimotor cortices [53, 55]. Late wave SPN also has a centro-parietal distribution [42] (Fig. 5). Sources of the late wave SPN have been localised to the MCC [43], although concurrent activations have been found using fMRI in the posterior insula [52]. In sum, the temporal dynamics of pain anticipation as revealed with E/MEG is characterised by early anterior sources (MCC and possibly supplementary motor areas) but late centro-posterior sources that likely derive from activity in MCC and insula regions. The MCC thus may participate as a 'hub' at all stages with a shift from initial attentional orienting responses in MCC/SMA to more pain-specific anticipation in MCC/insula just prior to pain (Fig. 6).

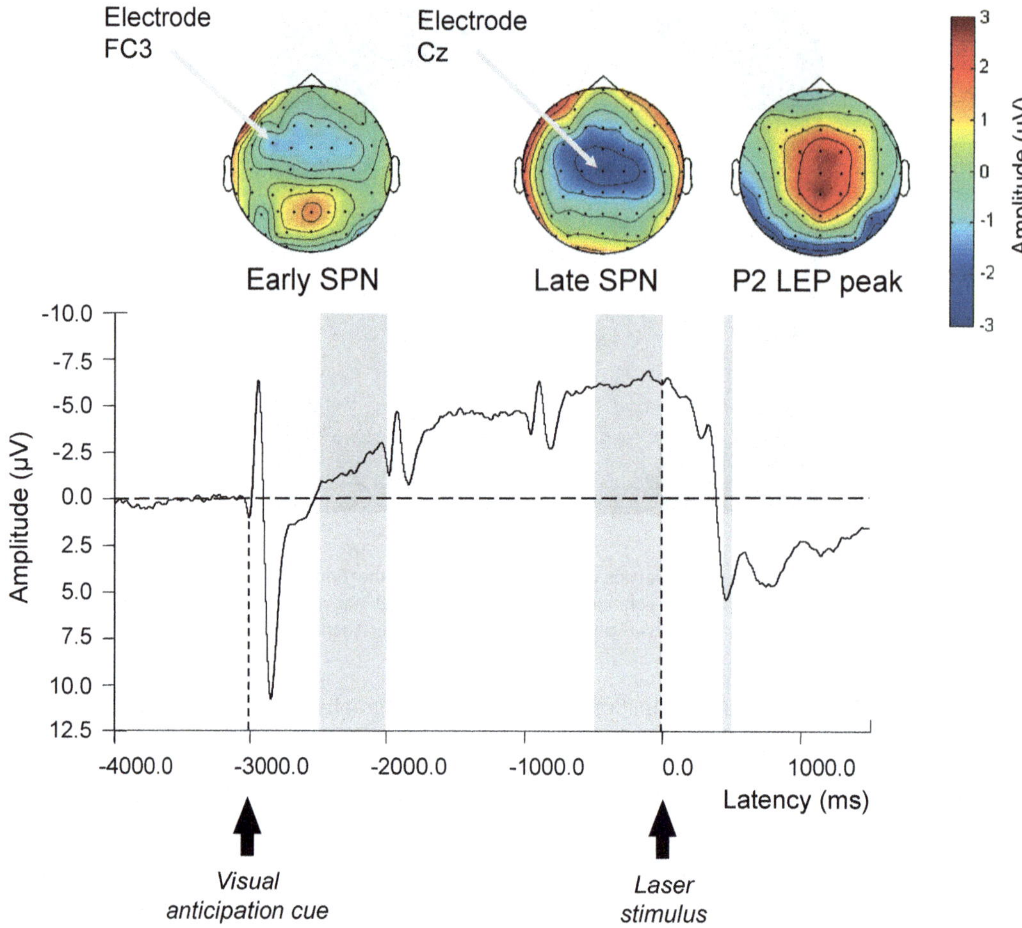

Fig. 5 Event-related potential (ERP) waveforms and topographic maps occurring during anticipation and experience of pain. The temporal range of early (−2500 to −2000 ms) and late (−500 to 0 ms) periods of the stimulus-preceding negativity (*SPN*) is shown, in addition to the P2 peak of the post-stimulus laser-evoked potential (*LEP*). The SPN is the gradually increasing negative potential occurring between −3000 and 0 ms. In the early phase, the SPN shows a broad fronto-central distribution. In the late phase, the SPN shows a more central distribution. Also evident are visual-evoked potentials caused by the visual anticipation cue and auditory-evoked potentials caused by predictive auditory tones. Figure taken with permission from [42]

4 Functional Attributes of the Anticipation Response

In this section, the results from recent neuroimaging studies will be discussed that have explicitly manipulated expectation in order to observe their effects on anticipatory and pain processing in the brain. Such studies provide insight into the functions of regions activated during anticipation of pain. As discussed, fMRI is well suited to studying how neural responses are modified by pain expectancy in order to identify potential or actual mediators of expectancy effects on pain perception [35, 56–59]. Further EEG studies have sought to dissect the temporal dynamics of these effects [15, 42, 52, 60]. Most studies have measured the effect of expectations on neuronal responses under either 'certain' or 'uncertain' conditions, but rarely have studies compared the two. Hence, in the following discussion, the relative (un)certainty of expectations will be highlighted and later, the few studies comparing certain and uncertainty

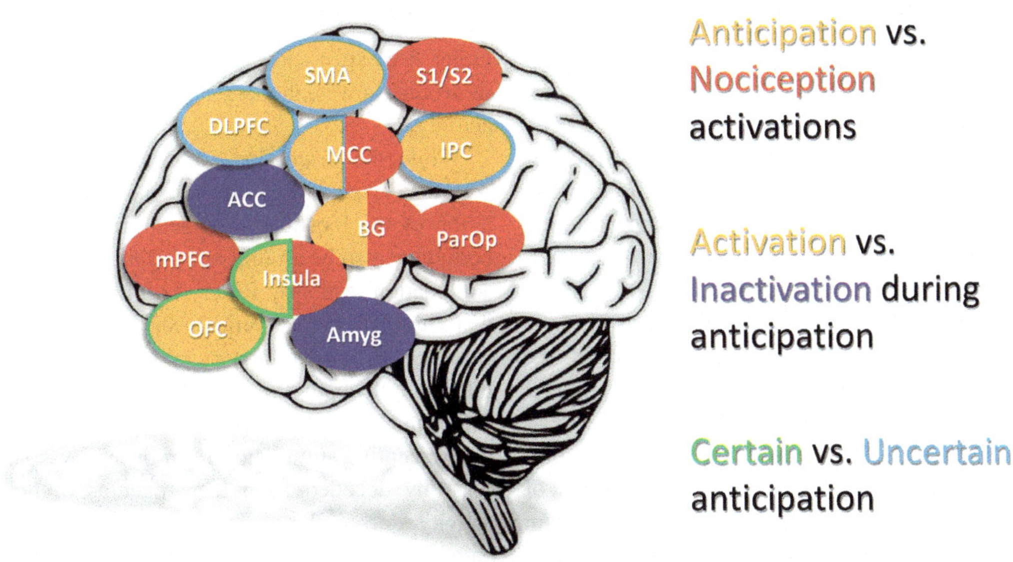

Fig. 6 Brain regions showing changes in activity during anticipation of pain, relative to nociception. Regions are also highlighted that are more active during either certain and uncertain expectations. *DLPFC* Dorsolateral prefrontal cortex. *SMA* Supplementary motor area. *S1/S2* Primary/secondary somatosensory cortices. *ACC* Anterior cingulate cortex. *MCC* Anterior Midcingulate cortex. *IPC* Inferior parietal cortex. *mPFC* Medial prefrontal cortex. *OFC* Orbitofrontal cortex. *Amyg* Amygdala. *BG* Basal ganglia. *ParOp* Parietal operculum

expectations will shed light on the contribution of relative (un)certainty to expectancy effects on anticipatory/pain processing. Finally, what we have discovered will be applied to the interpretation of recent studies of placebo analgesia, nocebo hyperalgesia, and hypnotic hyperalgesia, in order to better understand the mechanisms by which these phenomena might impact on acute and chronic pain treatments.

4.1 Modulation of 'Pain Processing' by Expectations

Brain imaging studies have highlighted the importance of a number of factors in determining precisely how pain processing is affected by expectations. As described here, the relative (un) certainty of expectancy cues, whether the cues predict high or low pain, the duration of the anticipation period, and the size of the difference between expected and actual stimulus intensities are all likely to impact on expectancy modulation of pain processing. Furthermore, as illustrated here, these factors differ across experiments and may account for some of the variability in results thus far observed.

An early fMRI study by Sawamoto et al. [56] showed that when subjects were cued to expect reduced pain with uncertainty, decreased pain responses occurred in the anterior portion of the midcingulate and the posterior insula/parietal operculum. Conversely, uncertain expectations of increased pain have been associated with greater activity in the left entorhinal cortex during pain [61]. However, it is not clear from these results what activity was related to expectation per se, and what was related to uncertainty.

Subsequent studies induced more certain expectations of pain by prior conditioning to predictive cues followed by subsequent miscuing during measurement of neuronal responses. A study by Koyama et al. [58] found that inducing expectations of lower pain reduced activity in the cerebellum, primary and secondary somatosensory cortex, MCC, thalamus, basal ganglia, DLPFC, PCC, IPC, SMA, and posterior and anterior insular cortex [58]. This agrees with the results of the study of uncertain expectations by Sawamoto et al. [56], but demonstrates modulation of a broader range of brain regions. However, in the Koyama et al. study, expectations of lower pain were signalled by increasingly shorter anticipation periods; furthermore, research has found that shorter (longer) anticipation periods result in less (greater) pain perception, because the anticipation of pain is thought to itself be unpleasant [62, 63]. It is therefore unclear from the study whether the reductions in brain activity resulted from 'certain' expectations or less anticipation time.

Subsequent fMRI work by Keltner et al. [59] looked at the effect of certain expectations of high intensity pain, which induced higher pain reports compared to an equivalent stimulus in a control condition in which there were expectations of a lower intensity stimulus. Expectation of high pain increased activation of the ipsilateral midcingulate cortex, the head of the caudate, cerebellum, and the contralateral nucleus cuneiformis (nCF). They proposed that pain intensity expectancy modulates activations produced by noxious stimuli through a distinct modulatory network that converges with afferent nociceptive input in the nCF. The study is notable for the lack of modulation of the somatosensory cortices or insula as found in the studies by Koyama et al. [58] and Sawamoto et al. [56]. Differences in methodology are likely to explain these differences. For example, in comparing two of the studies there was a higher probability of receiving a miscued stimulus (50%) in the Keltner et al. study compared to the Koyama et al. study (33%), which may have inadvertently increased the subjective uncertainty of the cues, and a smaller expected difference between pain stimuli being delivered (2 °C in the Koyama et al. study and 1 °C in the Keltner et al. study). A third possible factor is that, as Koyama et al. [58] noted in their paper, expectations of reduced pain produce bigger psychophysical effects than expectations of increased pain. Accordingly Koyama et al. studied changes in neural activity from expectations of decreased pain, while Keltner et al. analysed a contrast comparing high and low expectations which may have resulted in weaker effects. These points highlight that the larger the expectation of reduced pain, and the more certain it is, the larger the change in neural activity.

While the discussion so far has been with regard to expectations of pain intensity, expectation of the location of a pain stimulus in the body may also be an important component of the anticipation response. Ritter et al. [64] identified that higher order cortical structures of the descending pain modulatory system carry spatial information about the site of stimulation using fMRI and multivariate pattern analysis. The site of nociceptive stimulation (arm or leg) was successfully decoded from local patterns of brain activity during the anticipation and receipt of painful stimulation in the anterior and perigenual cingulate cortex, the dorsolateral prefrontal cortices, and the contralateral parietal operculum. Attempts to predict arm and leg stimulation from the peri-aqueductal grey, control regions (e.g. white matter) or the control time interval in the intertrial phase did not allow for classifications above chance level, suggesting spatial specificity of endogenous pain control limited to cortical areas.

4.2 Mediators of Expectancy Effects on 'Pain Processing'

Pain processing regions have been uncovered that are not only activated or deactivated by expectation, but rather *mediate* changes in pain perception. Mediation analysis is a statistical modelling procedure that has been available for some time but only recently applied to neuroimaging data [33]. In an experiment by Atlas et al. [16], anticipation cues and subjective reports of subsequent heat-evoked pain varied trial-by-trial. A mediation analysis tested whether responses in pain systems formally mediated cue effects on trial-by-trial pain reports when intensity did not vary. In this case, mediation implies that cue effects on a given brain region explain more variability in pain reports than cues themselves. The authors found that despite the fact that a broad array of regions were modulated by the changes in expectancy induced by cues (consistent with the regions discussed in the chapter thus far), only a subset of these regions formally mediated trial-by-trial cue effects on pain. These regions were the left anterior insula, MCC, right medial thalamus, left DLPFC, left basal ganglia, left amygdala and pons. Deactivations in the ACC also mediated expectancy effects. As noted previously in this chapter, despite the novelty of the analysis, it still cannot be determined from just this data what the precise timing of these activations were, whether just prior to or just after nociception.

A further mediation analysis sought to discover brain regions activated specifically during *anticipation* of the pain stimuli which might mediate the effect of expectancy on those regions mediating cue effects on pain perception [16]. It was found that anticipatory responses in the ventral striatum and medial OFC mediated the effects of expectancy cues on mediators of cue effects on pain. The authors noted that these regions have been linked to conditioning and associative learning in the context of appetitive and aversive stimuli, as discussed in Sect. 5.

4.3 Temporal Dynamics of Expectancy Effects During Anticipation and Pain

Although modulation of activity in the pain matrix from pain expectancy can be said to result from the 'top-down' effects of expectation, greater temporal resolution is required to identify the precise timing of these effects. In an MEG study, pain responses within a few hundred milliseconds of pain onset in contralateral secondary somatosensory cortex were both increased and decreased by expectations of higher or lower intensity pain, respectively [60]. However, expectations were found not to affect pain-evoked responses in cingulate cortex. This may suggest that earlier fMRI results were either recording modulation of anticipatory processing in this region, or modulation of responses taking place later than that recorded in MEG studies (e.g. a few seconds after pain). The anticipation hypothesis is supported by evidence from a study by Brown et al. [42] that used the high temporal resolution afforded by EEG to study the effects of expectation specifically on late anticipatory responses towards pain. It was found that the late SPN, previously found to derive from activity in anterior MCC [43], was increased and decreased by certain expectations of high and low pain relative to uncertain expectation [42]. However, there remains a possibility of modulation of further post-stimulus processing.

4.4 Temporal Dynamics of (Un) Certainty During Anticipation and Pain

The study by Brown et al. [42] was notable for comparing the effect on anticipatory and pain processing of certain versus uncertain expectations using the high temporal resolution of EEG. Subsequent analysis of the data using subjective confidence judgments (as a measure of uncertainty) shed further light on the topic [15], while subsequent work used similar EEG paradigms and also measured concurrent fMRI BOLD responses to improve understanding of the spatial localisation of anticipatory and pain responses [52]. Together, this work has begun to unravel the specific effects of uncertainty beyond that of expectation per se. Brown et al. [42] discovered that early anticipation was strongly affected by uncertainty (independently of the level of expectation). Uncertainty more greatly activated DLPFC, PCC, IPC and superior frontal gyrus (supplementary motor area). IPC and superior frontal gyrus were also more activated during uncertainty during late anticipation. The relationship of early IPC activations to uncertainty was further indicated by subsequent analyses showing that IPC activity, in addition to MCC activity, was inversely correlated with subjects' self-reported confidence in their expectations [15]. Seidel et al. [52] also found that early anticipatory responses are increased by uncertainty, reflected in a more pronounced frontal stimulus-preceding negativity (SPN). However, corresponding increased fMRI activations due to uncertainty were found in higher visual processing areas, suggesting that uncertainty captured more visual attention. In agreement with the results from Brown et al., Siedel et al. found that during late anticipation uncertainty was associated with fMRI activation in parietal areas indicating the recruitment of attentional processes.

Certain anticipation recruits a different set of brain regions from uncertain anticipation. During early anticipation, Brown et al. [42] discovered more greatly activated inferior frontal (vlPFC) and subgenual ACC activations relative to uncertain conditions. However, subjective confidence in expectations about pain positively correlated with activity in the anterior insula, which also predicted the extent to which expectations of high pain affected pain perception, suggesting a possible mediatory role for this region [15]. Siedel et al. [52] were able to delineate the contribution of certain expectation to changes in late anticipatory processing using EEG and fMRI. During late anticipation, stimulus-specific somatosensory preparation processes, as reflected in a centro-parietal SPN and posterior insula activation, were most pronounced for certain conditions. Certain anticipation was specifically associated with increased stimulus-specific preparatory activity in the anterior insula and motor preparation regions, which is in line with previous evidence [65, 66]. These responses were inversely related to activity in the MCC (the region more greatly activated during uncertainty), suggesting reduced 'conflict' [67] as subjects were more certain about their expectations.

The effects of uncertainty have also be examined with respect to nociceptive processing using EEG [52, 63, 68] and fMRI [52]. Uncertainty both due to unknown stimulus intensity [52] and unknown stimulus timing [63, 68] increases the amplitude of the mid-latency positive component of nociceptive-evoked potentials. Carlsson et al. [69] speculated that predictability may reduce the aversiveness of the painful stimulus by inducing feelings of control in comparison to an uncontrollable highly aversive condition during uncertain trials. fMRI data from their study showed that during stimulation, unexpected painful stimuli produced the strongest activation in affective pain processing regions: a large dorsomedial PFC cluster including SMA, bilateral lateral OFC, bilateral anterior insula, left cerebellum as well as right middle temporal gyrus. The findings of dorsomedial PFC/SMA concur with results showing activation of these regions during anticipation under uncertainty and suggest that these fMRI activations may reflect anticipatory processes rather than pain-evoked processes. On the other hand, activations of anterior insula, cerebellum and lateral OFC would be more consistent with certain anticipation, and hence these activations under uncertainty seem more likely to be related to nociceptive responses.

4.5 Placebo Analgesia: A Model for Positive Expectancy Effects

There is evidence that placebo analgesia is mediated by expectation and conditioning [10, 70], in particular positive expectancy [71], and is therefore a useful model to study expectancy effects. This modulation has been found to be related to the activation of μ-opioid receptors in the ACC and DLPFC [72]. Geuter et al. [73] also found that activity in ACC consistently scaled with increasing analgesic efficacy during both anticipation and pain processing. Interestingly, the placebo response is blocked by naloxone only if it is associated with a strong expectation of reduced pain; if no such expectation exists, the placebo response will not be sensitive to naloxone [72].

Some placebo studies are notable for identifying brain regions that predict placebo responses. *During pain,* Wager et al. [57] found that the individuals who reported the largest placebo effects on pain reports also showed the largest placebo effects on heat-evoked responses in insula, thalamus and MCC. Watson et al. [74] also found correlations between placebo analgesia and placebo-induced reductions during noxious stimulation in anterior MCC, but also in

PCC and post-central gyrus. Such inconsistent results are likely to reflect differences in study design such as the type of pain stimulus used and methods for modelling the imaging data. However, MCC was consistently showed to predict placebo responses across studies.

During anticipation, Wager et al. [57] found that typical pain regions were not the only regions to predict placebo analgesia. Placebo-related responses in dorsolateral prefrontal cortex (DLPFC) and orbitofrontal cortex (OFC) prior to noxious stimulation also predicted individual differences in placebo analgesia, such that the individuals who had the largest placebo effects on pain showed larger placebo-induced increases in these regions during pain anticipation. The findings were confirmed by Watson et al. [74], who found anterior MCC and DLPFC activation during anticipation predicting placebo-induced pain reduction. Kong et al. [75] also observed similar correlations between prefrontal activation and placebo responses in bilateral OFC, as well as ACC, lateral prefrontal cortex, cerebellum, parahippocampus and pons.

The prefrontal regions are associated with cognitive control and expected value computation, which might play key roles in shaping subsequent nociceptive responses. To address whether this is the case, a study [76] using machine learning and pattern analysis techniques aimed to find out whether individual differences in placebo analgesia were best predicted by patterns of activity within pain networks, emotion networks or networks involved in executive function and cognitive control. Responses in canonical pain regions were much less predictive of placebo responses than anticipatory responses in networks associated with emotion and affective value. They found that placebo effects on pain were best predicted by responses during both pain anticipation and during noxious stimulation in regions broadly associated with emotional appraisal, including orbitofrontal cortex, insula, amygdala, and other regions identified independently. Many of these regions do not directly respond to increases in noxious stimulus intensity. This is consistent with research that links placebo effects on pain to general reward processing [77, 78], and findings that placebo involves a reduction in anxiety [79, 80].

4.6 Nocebo Hyperalgesia: A Model for Negative Expectancy Effects

In the same manner as the study of placebo responses, nocebo-based studies provide insights into the underlying mechanisms by which negative expectancies change the pain experience. The term 'nocebo' refers to a negative verbal suggestion that a given treatment will result in an increase in symptoms such as pain [81]. Only a few studies have analysed the nocebo phenomenon and importantly, only one of them has described cortical–subcortical circuitries related to the nocebo response using fMRI [82]. Except for the hippocampus, these areas are in line with the ones found in a meta-analysis of neural activations during pain anticipation (see Fig. 4) where a special role is played by posterior MCC, anterior insula, and lateral and medial prefrontal cortices [12]. Finally, Kong et al. [82] found that nocebo-related increases in pain reports were positively correlated with heat-evoked responses in bilateral insula and left primary motor cortex regions, and inversely correlated with responses in bilateral DLPFC and OFC.

4.7 Hypnosis: An Experimental Model for Changing Certainty

Hypnosis or other suggestive techniques have been used to induce the physical sensation of pain even in the absence of a nociceptive stimulus [83–85]. It has been argued that hypnosis increases the subjective certainty in expectations [35]. In these studies, increased activation was found in areas of the pain matrix including ACC, MCC, insula, prefrontal cortex, thalamus and primary and secondary somatosensory cortices (with a different extent depending on the study), with the notable exception of the cerebellum and amygdala. An fMRI study [86] investigated whether hypnotic susceptibility (HS) explained inter-individual differences in the neural responses to placebo analgesia. While HS was not related to strength of placebo analgesia, subjects with higher HS had increased anticipatory activity in the right DLPFC and reduced functional connectivity of the right DLPFC with brain regions related to emotional and evaluative pain processing (anterior midcingulate cortex/medial prefrontal cortex). This highlights the potential importance of anticipatory responses in DLPFC during more 'certain' expectations of pain.

5 Hypothesised Anticipatory Functional Networks Mediating Expectancy Effects on Pain

How and why do expectations about pain influence sensory perceptions? There has been some debate among scientists as to whether expectations can change the sensory processing of pain in the brain, or whether the brain merely becomes biased in how it reports those sensations [87]. Brain imaging studies of pain anticipation are likely to be able to shed light on this issue by consideration of which functional networks are activated during anticipation of pain and predict subsequent modulation by expectations.

As pointed out by Palermo et al. [12], neural activations from brain imaging studies of pain anticipation are likely to encompass many independent and related processes. Thus far we have discussed brain regions activated during pain anticipation and which of those regions are modulated by expectations and uncertainty, mediate expectancy effects, and appear to be important in placebo, nocebo and hypnosis. In this section we focus on the most important brain regions and discuss them in terms of their known roles as part of functional networks. In short, the following discussion outlines four categories of brain functions that are hypothesised to play an important role in anticipation and pain (see also Fig. 7):

1. Prediction learning, valuation and conditioned responses. These are likely to be largely pre-attentive and include the generation of prediction error signals to encode cue value, coupled with the expression of Pavlovian motivational states including autonomic and other reflex responses.
2. Salience detection and conscious appraisal of cues and their affective consequences.
3. Evaluation of current and potential actions and the need for cognitive control if the current state of the organism is inconsistent with current goals.
4. Cognitive (top-down) control, including facilitation or inhibition of nociceptive networks and adjustments to cognition, emotion and behaviour.

In the following sections, these categories of brain functions are expounded with regard to empirical evidence. For the purpose of brevity, only neural regions involved in pain anticipation and expectancy effects on pain are discussed in detail, with the stipulation that many other regions likely contribute to these functions.

5.1 Prediction Learning, Valuation and Conditioned Responses

Neural processes associated with learning the contingencies between unconditioned painful stimuli and conditioned predictive stimuli have been studied and are important in understanding how expectations are generated and expressed during anticipation of pain. As detailed here, a network involving the OFC, amygdala and striatum are activated during anticipation of pain (see also Sect. 3) and form part of the 'limbic loop' [88, 89] associated with the evaluation of predictive cues and the learning of expectations via generation of dopamine-dependent prediction errors for rewarding and aversive stimuli [90, 26] (also see focus point 1). These prediction learning mechanisms are thought to underlie conditioned responses as a result of both Pavlovian and instrumental associations. The role of these regions in learning expectations may explain why, in one study, anticipatory responses in the OFC and ventral striatum statistically mediated the effects of anticipation cues on activity in a set of

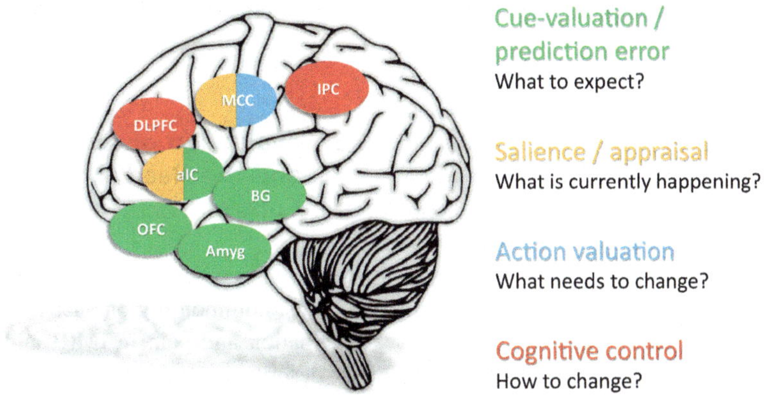

Fig. 7 Summary of main cognitive functions and their neural mappings corresponding to known brain responses during anticipation of pain. *DLPFC* Dorsolateral prefrontal cortex. *MCC* Midcingulate cortex. *IPC* Inferior parietal cortex. *aIC* Anterior insula cortex. *OFC* Orbitofrontal cortex. *Amyg* Amygdala. *BG* Basal ganglia

brain regions, including MCC, anterior insula and thalamus that appear to be important for the expression of expectancy effects on pain [16].

The medial OFC and vmPFC is a functionally complex region thought to be involved in computing and updating outcome values based on feedback or changes in state (e.g. predictive cues). It determines outcome value in the context of the *current* motivational state rather than relying on pre-computed values stored from previous experience [91]. The relationship of OFC and vmPFC activity to reward outcome values and subjective preferences have been evident from functional imaging studies (e.g. [92, 93]) but may be more lateralised with respect to aversive values [7].

Classical (Pavlovian) aversive conditioning paradigms have also revealed an important role for the amygdala in this learning process [2, 94, 95]. The basal nucleus of the amygdala is a point of convergence of information from the lateral nucleus of the amygdala, which initially receives sensory information from the temporal lobes and the orbitofrontal cortex (OFC) [96]. Whereas the mOFC is computing current outcome value, the function of this 'basolateral amygdala (BLA) complex' is thought to be to link value information with the sensory features of motivationally relevant cues. The OFC and BLA work closely together in encoding and retrieving value information (see [88], for a review) owing to their extensive anatomical and functional connectivity [97].

The ventral striatum is a region commonly associated with signalling rewarding outcomes and mediating reinforcement of appetitive behaviour [98]. However, this region is also commonly activated during anticipation of pain stimuli [99], particularly the caudate nucleus [12]. Another core region in the ventral striatum is the nucleus accumbens. Prediction error signals in the NAc have been specifically associated with Pavlovian conditioning while prediction errors required for instrumental conditioning involve the putamen [100]. The NAc is thought to bring evaluative information from the OFC and amygdala to bear on performance by selectively gating information projecting to basal ganglia output nuclei [101, 102].

5.2 Salience Detection and Affective Appraisal

The term 'salient' describes a stimulus or an aspect of a stimulus that stands out or that is set apart from others and that can be influenced by expectancies, goals and motivations [103]. We have seen that responses in the anterior insula occur in early anticipation of pain (as well as post-stimulus) (Sects. 3 and 4) and it is thought that this region is a centre of salience processing across multiple sensory and cognitive domains [103].

The insular cortex is located deep within the lateral sulcus of the cortex [104] and is highly connected to structures including the amygdala and OFC [105, 106]. The insula is believed to have a role in diverse functions of the body, not only pain, including emotional as well as homeostatic functions [107]. The roles of the anterior insula cortex (aIC) and the posterior insular cortex (pIC) differ, which is most clearly illustrated by their differential responses to pain stimuli. Stimulus intensity is linearly coded by activity in the pIC, while conscious evaluation (e.g. of stimulus intensity or emotion) is coded by the aIC [108–110]. The aIC becomes more active in response to cues that accurately predict pain and correlates with the error in prediction as subjects learn the value of predictive cues in addition to the ventral putamen [26], suggesting that the aIC receives Pavlovian prediction error signals from the ventral striatum for conscious appraisal. This may underlie the role of the insula in detecting novelty [111, 112], responses to action errors and other types of negative feedback [113].

In addition to coding perceptions (for example, pain and temperature), the aIC also receives interoceptive information to track emotions (for example, anxiety) associated with bodily states [114, 115]. Ascending interoceptive and visceromotor inputs, which track the moment-by-moment condition of the body, converge in the insular cortices [114]. The right aIC in particular has been shown to mediate the integration of autonomic nervous signals with consciousness [114, 116]. The activity of the insula, often together with amygdala activity, represents an individual's subjective and conscious emotional state, as well as the emotive value of external stimuli [117].

In sum, the evidence for anterior insula involvement in the anticipation of pain and the broader set of homeostatic, emotional and cognitive functions outlined here points to a role detecting the salience of environmental and internal cues that may predict pain. This is consistent with the view that the aIC integrates information about salience into perceptual decisions about pain [118], as well as with data showing that the subjective confidence in expectations about pain positively correlates with anticipatory activity in the anterior insula [15].

5.3 Evaluation of Actions and the Need for Cognitive Control

We have discussed that aMCC (Fig. 8) is one of the most common regions activated during both anticipation and pain in fMRI and PET studies. As described in Sect. 3, during anticipation, activity in aMCC is evident at all stages from early (post-cue) to late (pre-stimulus). However, we have also learnt that aMCC responses are not evident during the first few hundred milliseconds after delivery of a pain stimulus as measured in EEG/MEG studies, suggesting that the temporal profile of aMCC responses and their modulation by expectancy is weighted towards activity occurring outside this short-time window (either pre- or post-stimulus). This is consistent with aMCC responses reflecting slower, more deliberative and conscious aspects of anticipation and pain processing, which may follow the 'fast' evaluation of environmental cues in the amygdala and insula [2]. Indeed, the anticipation of pain increases functional connectivity between the aIC and aMCC [118]. Reciprocal connections between aMCC and the BLA have been established in animals [119] and are consistent with functional connectivity data in humans [120].

aMCC responses to pain stimuli are modulated by expectancy of pain (increased and decreased according to expectations of high and low pain—Sect. 4) and formally mediate expectancy effects on pain perception [16]. Consistent with this, the aMCC is commonly found to be modulated in studies of placebo analgesia (Sect. 4). However, aMCC responses during anticipation are greater under conditions of uncertainty, and there is evidence suggesting that although the level of aMCC activity is modulated by expectancy, greater overall anticipatory aMCC activity is associated with reduced expectancy effects on pain [15]. Although the functions of the MCC have been widely studied,

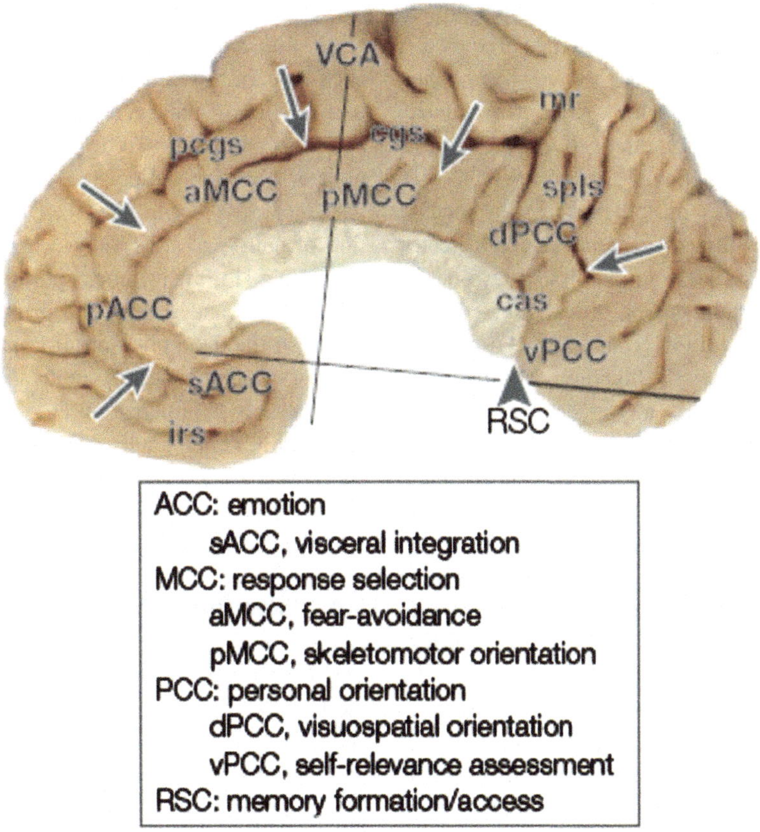

Fig. 8 Regions of the cingulate cortex and their functions. The cingulate cortex is divided into four regions. The anterior cingulate cortex (*ACC*) (24/32) has two subdivisions [subgenual (*sACC*) and pregenual (*pACC*)]. The midcingulate cortex (*MCC*) (24′/32′) which also has two subdivisions [anterior (*aMCC*) and posterior (*pMCC*)]. The posterior cingulate cortex (*PCC*) (23/31) subdivisions are dorsal (*dPCC*) and ventral (*vPCC*). Finally, the retrosplenial cortex (*RSC*) (23′/31′). The MCC is commonly referred to as 'dorsal' ACC, with the actual ACC is commonly labelled as 'rostral' ACC. Because of the use of ACC as a 'catch-all' terminology, this has led to inconsistent results and a heterogeneous plethora of functions labelled inaccurately as ACC [161]. In this chapter, the naming of regions of the cingulate cortex has been changed where appropriate with respect to the original publications, in order to maintain consistency in labelling as laid out in this figure. Figure taken with permission from [46]

they remain poorly understood. As discussed below, the aMCC has been investigated with respect to signalling the need for cognitive and behavioural *change*, and in this section this is translated into an understanding of the functioning of the aMCC during anticipation of pain.

MCC activation is associated with a number of general functions including conflict and action monitoring, motor inhibition and response selection [67, 121]. Its role in conflict monitoring is consistent with the greater activations of aMCC during anticipation in uncertain relative to certain expectancy contexts. The aMCC includes the motor areas of the cingulate cortex and has connections with the motor cortex [122]. The aMCC includes the rostral cingulate zone (RCZ), a premotor area that is somatotopically organised and projects to the spinal cord, dorsal (sensorimotor) striatum and primary motor, premotor and supplementary motor cortices [119]. The RCZ is thought to be sensitive to abstract aspects of action planning and inhibition [119] and is commonly activated by imaging studies of pain as well as studies of cognitive control [123].

An influential theory states that MCC monitors performance and signals the need for greater cognitive control in order to aid behavioural adaptation [124] (Fig. 9). *Cognitive control* indicates processes required to pursue a goal when distracting stimuli or competing (e.g. habitual) responses must be overcome because they are not sufficient to support goal-directed behaviour [125]. An example of such a situation is during uncertain and/or threatening contexts when the value of different courses of action needs to be considered. Uncertain anticipation of pain is therefore likely to trigger cognitive control mechanisms via aMCC.

The specific function of the aMCC in cognitive control is still debated. Shenhav et al. [126] recently proposed that the aMCC estimates the net value associated with allocating control to a given task. According to this theory, the primary representation of value, which feeds into aMCC decision-making mechanisms, is thought to be subserved by other structures that project to aMCC (e.g. insula, amygdala, and ventral/medial regions of PFC, basal ganglia and dopaminergic midbrain structures). Information about pain, or environmental cues that signal impending pain, could be passed to aMCC via ascending dopaminergic pathways [127]. It is also thought to receive inputs from all major divisions of the insula [128], enabling aMCC to track changes in emotional states as well as responses to errors and negative feedback [113] that may aid in evaluating the need for cognitive control. After receiving these inputs, the aMCC can then choose among competing tasks and allocate the appropriate amount of cognitive control to the one selected. This is likely to occur via reciprocal connections with fronto-parietal regions implicated in cognitive control [129]. This explains why the aMCC has been reported to play a role in directing (orienting) attention to a stimulus, whether it is painful or not [46, 121]. For example, during late anticipation of pain, it has been found that MCC and IPC responses predict the degree to which attention is interrupted by subsequent pain [130]. Interactions between the MCC and lateral prefrontal cortex are then required to implement behavioural changes [124].

In sum, the aMCC is well positioned to integrate information about unlearned reinforcers, such as pain, learned reinforcers such as predictive environmental cues, and information about current goals [123]. To do this, the aMCC must register both the anticipated value of outcomes ahead of their occurrence (e.g. during anticipation of pain) and their value when they actually occur (e.g. following nociception). The interaction of aIC and aMCC has been reasoned to allow for the integration of information about the value or significance of motivationally salient stimuli such as anticipation cues or pain stimuli (via the aIC) to engender adaptive adjustments to cognition, emotional states and behaviour (aMCC) [123].

5.4 Execution of Cognitive Control

The fronto-parietal attention systems, including the DLPFC and IPC, appear to be more robustly activated during anticipation of pain than during pain experience (Sect. 3). A current neurobiological model of attention posits a dorsal fronto-parietal network (intraparietal sulcus, superior parietal lobule and frontal eye fields) mediates goal-directed, top-down attention, contrasting with a ventral fronto-parietal network (temporoparietal junction, middle frontal gyrus and aIC) mediating stimulus-driven, bottom-up stimulation of attention [131, 132]. The ventral fronto-parietal network primarily responds to salient stimuli such as pain [133] or its anticipation [118]. However, the dorsal fronto-parietal network, implicated in cognitive control and the maintenance of goals (e.g. attentional sets and rules) [134], including DLPFC [129, 135], can be recruited via the aMCC signalling the need for cognitive control [124] (Fig. 9). Uncertainty activates fronto-parietal networks to a greater extent than certain anticipation [42], in which the aMCC can act as a conflict monitor to signal the need for fronto-parietal systems to resolve the conflict [67]. As described in previous sections of this chapter, there are theoretical reasons for hypothesising a stimulation of attention under conditions of uncertainty, especially when

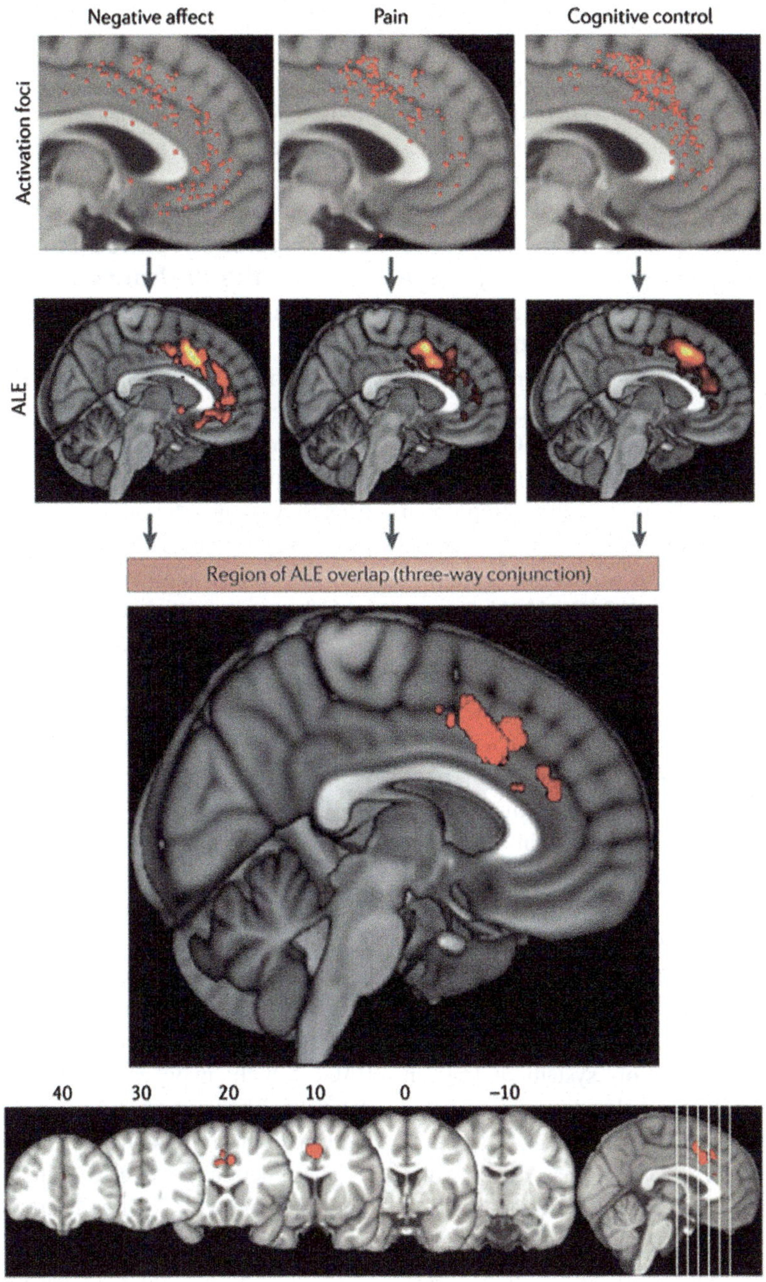

Fig. 9 Common activations of the anterior midcingulate cortex (aMCC) by pain, cognitive control and negative affect. The map depicts the results of a coordinate-based meta-analysis of 380 activation foci derived from 192 experiments involving more than 3000 participants. *Uppermost row* shows the spatially normalised foci for each domain. The *next row* shows thresholded activation likelihood estimate (*ALE*) maps for each domain considered in isolation. The *bottom two rows* depict the region of overlap across the three domains. The *red cluster* indicates the location of a three-way minimum significance conjunction of the three domains. The cluster lies in aMCC in the vicinity of areas 32′ and a24b′/c′. Numbers indicate mm from the anterior commissure. Figure taken with permission from [123]

uncertain cues are perceived as being ambiguous. Yet, responses in DLPFC prior to noxious stimulation also predict individual differences in placebo analgesia (Sect. 4). This suggests that DLPFC does not have a simple mediating function in pain expectancy but is rather activated when there is a need for cognitive control.

The lateral PFC is thought to have a regulative function in (1) supporting the active maintenance of task representations, and (2) biasing processing in pathways of posterior cortex that execute specific tasks [136]. This requires protection of these representations from interference by external or internal distractions. The PFC therefore has an important role in inhibiting actions that are inappropriate to the task at hand (top-down inhibition) [137]. Lateral PFC can be seen as implementing the control signal to support a given task, as specified by aMCC [126].

The DLPFC in particular may be important in maintaining or suppressing the conscious perception of pain via attention, showing negative correlations of activity here bilaterally with pain intensity and unpleasantness ratings [138]. This relates to findings that activity in bilateral DLPFC during anticipation of pain correlates with success in inducing placebo analgesia (Sect. 4). Pain catastrophising is associated with reduced activation of right DLPFC in response to pain [139], which may imply a deficit in the inhibition of ascending nociceptive input.

The IPC is one of the central components of the fronto-parietal attention system and is an output pathway for implementing cognitive control on sensory processing [132]. Tractography evidence shows that IPC has direct white matter connections to both the aIC and aMCC [140, 141]. The IPC is consistently associated with spatial attention [142]. For example, attention to the location of pain activates right IPC [121, 143], in addition to contralateral SI and SII/insula [143, 144].

In sum, the evidence strongly suggests recruitment of the fronto-parietal control systems when there is a need for cognitive control. Examples include conditions of uncertainty, when cues are ambiguous, or when prediction errors need to be minimised due to a mismatch between expectations and sensory input as in the case of placebo analgesia. Two functions of cognitive control are relevant to pain anticipation: modulation of nociceptive networks (discussed below) and goal-directed action.

5.5 Nociceptive Modulation by the Prefrontal and Cingulate Cortices

One of the outputs of the DLPFC appears to be the pregenual anterior cingulate cortex (pACC) which forms the frontal part of the cingulate cortex. Increases in activity of the pACC have been consistently found as a result of successful placebo analgesia [145]. Functional connectivity between the fronto-parietal network and the pACC during anticipation of pain has been shown to predict the effects of expectancies on pain perception [146]. Animal studies suggest that pACC activity coupled with OFC can modulate pain by altering the output properties of the peri-aqueductal grey/rostroventral medulla (PAG/RVM) system of the midbrain [147, 148]. Imaging studies have demonstrated the involvement of the PAG/RVM system in pain modulation produced by placebo analgesia [57, 145]. Top-down recruitment of PAG may occur via activation of DLPFC and pACC during anticipation of pain [57]. There is also evidence that MCC sends outputs to the lateral column of the PAG, which may enable the expression of defensive behaviours such as vigilance or fight-or-flight, in rats and cats [149].

The functions of the pACC include regulating autonomic activity such as heart rate and blood pressure [46]. The pACC has the largest number of opioid receptor binding sites in the cingulate cortex [46]. Placebo analgesia is likely to be related to the release of endogenous opioids in this region [145]. Deactivations in this area mediate expectancy effects on pain [16], consistent with a lack of opioidergic inhibition increasing perceived pain. The pACC is also activated during distraction from pain, in addition to the PAG, and is thought to act as an inhibitor of ascending nociceptive input [150, 151].

Normally, pACC is deactivated during anticipation of pain [12]. However, greater perception of anxiety during anticipation of pain is associated with increased activity in the ACC (specifically, the subgenual region) and OFC [152], suggesting that links between pain-related fear/anxiety and increased pain perception may involve this region. This provides a candidate mechanism for 'nocebo' responses, i.e. expectation-driven enhancement in pain perception.

In sum, the functions of the pACC with regard to pain and pain anticipation appear to be largely to do with outputs to the PAG/RVM system as a result of cognitive demands and are mediated by opioidergic mechanisms. The main nociceptive pathways that are affected by pain expectancy include the thalamus, somatosensory cortex, parietal operculum, posterior insula and PCC. However, it is not clear whether modulation of these areas is solely due to changes in brainstem outputs that affect downstream processing in subcortical and cortical areas, or whether there is some direct top-down modulation of nociceptive networks above the brainstem.

5.6 Other Brain Regions of Nociceptive Modulation by Expectancy

The thalamus is situated between the cerebral cortex and the midbrain. The ventral posterior nucleus of the thalamus receives input from afferent fibres arising from laminae IV–V (target for nociceptors), and then proceeds to the somatosensory cortices, which constitutes the main pathway for the nociceptive input reaching the brain [153]. Furthermore, the thalamus is known to encode pain intensity [154] and to send nociceptive information to the MCC and the ACC [155]. That said, thalamic activation is often bilateral, suggesting that it does not just reflect a sensory response which one would expect to be localised to the contralateral side of the pain stimulus [121]. The thalamus is commonly activated in studies of pain anticipation [12] and some studies have reported its modulation by pain expectancy, with one study showing that the medial thalamus formally mediates expectancy effects on pain [16]. Attentional processes and vigilance have been reported to enhance thalamic activation bilaterally, which have led researchers to believe that the thalamus plays a role in the cognitive modulation of pain [121].

The primary somatosensory cortex (S1) has repeatedly been found to encode the sensory-discriminative aspect of pain perception and non-painful tactile perception such as the intensity, localization and quality of the stimuli [45]. It is rarely activated in pain anticipation, but pain-evoked responses in S1 are modulated by expectancies, and hence anticipatory responses are likely to impact on S1 activity (Sects. 3 and 4). S1 is located in the post-central gyrus of the parietal lobe, and it is believed that the somatotopic organisation in S1 allows the discriminative localisation of pain [45]. Furthermore, S1 activation in response to noxious stimulation is greater in the brain hemisphere contralateral to the noxious stimulus [156]. However, tonic pain does not appear to result in activation in SI, despite notable activation in other areas such as PFC and cingulate cortex [157]. S1 may therefore be conceived of as having change-detection functions regarding the sensory-discriminative aspects of pain, particularly changes in location, and is therefore only activated either due to short-duration painful stimulation or to signal changes in ongoing pain. This is consistent with the general view that afferent sensory pathways transmit prediction error signals to update ongoing perceptual representations.

The secondary somatosensory cortex (S2) is located in the parietal operculum. S2 cortex during pain is modulated by expectancies of high and low pain, evident from both fMRI/PET and E/MEG studies (Sect. 4). Its role in anticipation of pain, however, is not well established. S2, together with the posterior insula, is one of the most commonly activated areas in acute pain studies, mainly contralateral to the stimulus but sometimes bilaterally, and is generally associated with the sensory discrimination of pain [45], but may be more closely related to its intensity coding rather than its location as it is in the case of S1 [158]. S2 activity substantially increases when intensity reaches painful levels [121]. The activation of S2 can also be triggered by non-nociceptive stimuli, including tactile, vibratory and olfacto-gustatory stimuli [45].

6 Conclusions and Future Directions

Improving our understanding of the mechanisms supporting pain anticipation and its impact on pain perception and behaviour would be expected to eventually relieve suffering associated with chronic pain. Research has discovered that much of the brain's response to expected pain is anticipatory and reflects a number of general brain functions rather than being specific to nociception. These anticipatory functions include those involved with associative learning, salience detection, action monitoring and cognitive control. These functions may underlie the biasing of pain perception by expectation.

This chapter has presented both methodologies for investigating pain anticipation and a summary of the most important research findings to date. Some shortcomings and gaps in knowledge have also been highlighted, which may inform future studies in the field. First, many studies of pain expectancy to date have not controlled for differing levels of uncertainty across their experimental conditions, such that the differential effects on neural activity of expectancy and certainty may be confounded in many experiments. This is consistent with a lack of recognition of these two factors as separate variables and a lack of clarity on the unique mechanisms underlying each. However, theoretical frameworks accounting for perception and behaviour under conditions of uncertainty (see focus point 2) can serve as a firm basis by which to further explore these effects.

Second, prefrontal cortical mechanisms are involved with making decisions about sensory perceptions in order to guide behaviour [159]. Recent evidence suggests that perceptual decision-making mechanisms may underlie at least some of the effect of expectations on pain perception [87]. Currently, the dominant conceptualisation of the role of the prefrontal and cingulate cortices in expectancy-related top-down pain modulation has been with regard to an inhibitory (likely opioid-mediated) pathway that changes the output properties of nociceptive modulatory systems in the brainstem. This top-down inhibition hypothesis, while well substantiated, may also be too narrow a view. Cognitive control is not simply inhibition, but also includes a broader set of functions including working memory and maintenance of goals (e.g. attentional sets and rules). The precise prefrontal mechanisms influencing expectancy effects on pain is likely to be a fruitful subject for further research.

Acknowledgements With gratitude to Lauren Atlas for commenting on a draft of this chapter.

References

1. Golshani AE, Kamdar AA, Spence SC, Beckmann NM. Congenital indifference to pain: an illustrated case report and literature review. J Radiol Case Rep. 2014;8(8):16–23.
2. LeDoux J. The emotional brain, fear, and the amygdala. Cell Mol Neurobiol. 2003;23(0272–4340):727–38.
3. Tom SM, Fox CR, Trepel C, Poldrack RA. The neural basis of loss aversion in decision-making under risk. Science. 2007;315(5811):515–8.
4. Melzack R, Wall PD. Pain mechanisms: a new theory. Science (80-.) 1965;150(0036–8075) (Print):971–9.
5. Craig AD, (Bud) Craig A. A new view of pain as a homeostatic emotion. Trends Neurosci. 2003;26 (0166–2236):303–7.
6. Legrain V, Iannetti GD, Plaghki L, Mouraux A. The pain matrix reloaded: a salience detection system for the body. Prog Neurobiol. 2011;93(1):111–24.
7. Seymour B, O'Doherty JP, Koltzenburg M, Wiech K, Frackowiak R, Friston K, Dolan R. Opponent appetitive-aversive neural processes underlie predictive learning of pain relief. Nat Neurosci. 2005;8(1097–6256):1234–40.
8. Price DD. Psychological mechanisms of pain and Analgesia, no. 15. IASP Press; 1999.
9. Fields HL. Pain modulation: expectation, opioid analgesia and virtual pain. Prog Brain Res. 2000;122(0079–6123) (Print):245–53.
10. Wager TD. Expectations and anxiety as mediators of placebo effects in pain. Pain. 2005;115(0304–3959):225–6.

11. Benedetti F, Carlino E, Pollo A. How placebos change the patient's brain. Neuropsychopharmacology. 2011;36(1):339–54.
12. Palermo S, Benedetti F, Costa T, Amanzio M. Pain anticipation: an activation likelihood estimation meta-analysis of brain imaging studies. Hum Brain Mapp. 2014.
13. Porro CA, Baraldi P, Pagnoni G, Serafini M, Facchin P, Maieron M, Nichelli P. Does anticipation of pain affect cortical nociceptive systems? J Neurosci. 2002;22(1529–2401):3206–14.
14. Ploghaus A, Tracey I, Gati JS, Clare S, Menon RS, Matthews PM, Rawlins JN. Dissociating pain from its anticipation in the human brain. Science (80-.) 1999;284(0036–8075):1979–81.
15. Brown CA, Seymour B, El-Deredy W, Jones AKP. Confidence in beliefs about pain predicts expectancy effects on pain perception and anticipatory processing in right anterior insula. Pain. 2008;139(2):324–32.
16. Atlas LY, Bolger N, Lindquist MA, Wager TD. Brain mediators of predictive cue effects on perceived pain. J Neurosci. 2010;30(39):12964–77.
17. Vlaeyen JW, Linton SJ. Fear-avoidance and its consequences in chronic musculoskeletal pain: a state of the art. Pain. 2000;85(0304–3959):317–32.
18. Hoy D, March L, Brooks P, Blyth F, Woolf A, Bain C, Williams G, Smith E, Vos T, Barendregt J, Murray C, Burstein R, Buchbinder R. The global burden of low back pain: estimates from the global burden of disease 2010 study. Ann Rheum Dis. 2014;73(6):968–74.
19. Woodworth RS. Reënforcement of perception. Am J Psychol. 1947;60(1):119–24.
20. Tolman EC. There is more than one kind of learning. Psychol Rev. 1949;56(3):144–55.
21. Bolles RC. Reinforcement, expectancy, and learning. Psychol Rev. 1972;79(5):394–409.
22. Rescorla RA, Wagner AR. A theory of Pavlovian conditioning: variations in the effectiveness of reinforcement and nonreinforcement. In: Black AH, Prokasy WF, editors. Classical Conditioning II. New York: Appleton-Century-Crofts; 1972. p. 64–99.
23. Reiss S. Pavlovian conditioning and human fear: an expectancy model. Behav Ther. 1980;11(3):380–96.
24. Sutton RS, Barto AG. Toward a modern theory of adaptive networks: expectation and prediction. Psychol Rev. 1981;88(2):135–70.
25. Niv Y. Reinforcement learning in the brain. J Math Psychol. 2009;53(3):139–54.
26. Seymour B, Doherty JPO, Dayan P, Koltzenburg M, Jones AK, Dolan RJ, Friston KJ, Frackowiak RS, O'Doherty JP. Temporal difference models describe higher-order learning in humans. Nature. 2004;429(1476–4687):664–7.
27. Friston K. Learning and inference in the brain. Neural Netw. 2003;16(0893–6080) (Print):1325–52.
28. Helmholtz H. Concerning the perceptions in general. In: Southall J, editor. Treatise on physiological optics, 3rd ed. New York: Dover.
29. Rao RP, Ballard DH. Predictive coding in the visual cortex: a functional interpretation of some extra-classical receptive-field effects. Nat Neurosci. 1999;2(1097–6256) (Print):79–87.
30. Friston K, Kilner J, Harrison L. A free energy principle for the brain. J Physiol Paris. 2006;100(1–3):70–87.
31. Friston K. The free-energy principle: a unified brain theory? Nat Rev Neurosci. 2010;11(2):127–38.
32. Dayan P, Abbott LF. Theoretical neuroscience: computational and mathematical modeling of neural systems, vol. 1. MIT Press; 2001.
33. Atlas LY, Wager TD. How expectations shape pain. Neurosci Lett. 2012;520(2):140–8.
34. Morton DL, El-Deredy W, Watson A, Jones AKP. Placebo analgesia as a case of a cognitive style driven by prior expectation. Brain Res. 2010;1359(C):137–41.
35. Ploghaus A, Becerra L, Borras C, Borsook D. Neural circuitry underlying pain modulation: expectation, hypnosis, placebo. Trends Cogn Sci. 2003;7(1364–6613):197–200.
36. Voudouris NJ, Peck CL, Coleman G. The role of conditioning and verbal expectancy in the placebo response. Pain. 1990;43(0304–3959):121–8.
37. Garcia-Larrea L, Frot M, Valeriani M. Brain generators of laser-evoked potentials: from dipoles to functional significance. Neurophysiol Clin. 2003;33(0987–7053) (Print):279–92.
38. Kirschstein T, Kohling R. What is the source of the EEG? Clin EEG Neurosci. 2009;40(3):146–9.
39. Baillet S, Mosher JC, Leahy RM. Electromagnetic brain mapping. Signal Process Mag IEEE. 2001;18(6):14–30.
40. Cuffin BN, Schomer DL, Ives JR, Blume H. Experimental tests of EEG source localization accuracy in realistically shaped head models. Clin Neurophysiol. 2001;112(1388–2457) (Print):2288–92.
41. Yao J, Dewald JPA. Evaluation of different cortical source localization methods using simulated and experimental EEG data. Neuroimage. 2005;25(1053–8119) (Print):369–82.
42. Brown CA, Seymour B, Boyle Y, El-Deredy W, Jones AKP. Modulation of pain perception by expectation and uncertainty: behavioral characteristics and anticipatory neural correlates. Pain. 2008;135(3):240–50.
43. Bocker KB, Baas JM, Kenemans JL, Verbaten MN. Stimulus-preceding negativity induced by fear: a manifestation of affective anticipation. Int J Psychophysiol. 2001;43(0167–8760):77–90.
44. Brunia CH, van Boxtel GJ. Wait and see. Int J Psychophysiol. 2001;43(0167–8760):59–75.
45. Apkarian AV, Bushnell MC, Treede R-DD, Zubieta J-KK. Human brain mechanisms of pain perception and regulation in health and disease. Eur J Pain. 2005;9(4):463–84.

46. Vogt BA. Pain and emotion interactions in subregions of the cingulate gyrus. Nat Rev Neurosci. 2005;6(1471–003X) (Print):533–44.
47. Coghill RC, Sang CN, Maisog JM, Iadarola MJ. Pain intensity processing within the human brain: a bilateral, distributed mechanism. J Neurophysiol. 1999;82(0022–3077) (Print):1934–43.
48. Veinante P, Yalcin I, Barrot M. The amygdala between sensation and affect: a role in pain. J Mol Psychiatry. 2013;1(1):9.
49. Baliki MN, Mansour A, Baria AT, Huang L, Berger SE, Fields HL, Apkarian AV. Parceling human accumbens into putative core and shell dissociates encoding of values for reward and pain. J Neurosci. 2013;33(41):16383–93.
50. Brown CA, El-Deredy W, Jones AK. When the brain expects pain: common neural responses to pain anticipation are related to clinical pain and distress in fibromyalgia and osteoarthritis. Eur J Neurosci. 2014;39(4):663–72.
51. Brown CA, Jones AKP. Psychobiological correlates of improved mental health in patients with musculoskeletal pain after a mindfulness-based pain management program. Clin J Pain. 2013;29 (3):233–44.
52. Seidel E-M, Pfabigan DM, Hahn A, Sladky R, Grahl A, Paul K, Kraus C, Küblböck M, Kranz GS, Hummer A, Lanzenberger R, Windischberger C, Lamm C. Uncertainty during pain anticipation: the adaptive value of preparatory processes. Hum Brain Mapp. 2015;36(2):744–55.
53. Bender S, Resch F, Weisbrod M, Oelkers-Ax R. Specific task anticipation versus unspecific orienting reaction during early contingent negative variation. Clin Neurophysiol. 2004;115(1388–2457):1836–45.
54. Cui RQ, Egkher A, Huter D, Lang W, Lindinger G, Deecke L. High resolution spatiotemporal analysis of the contingent negative variation in simple or complex motor tasks and a non-motor task. Clin Neurophysiol. 2000;111(1388–2457):1847–59.
55. Gomez CM, Marco J, Grau C. Preparatory visuo-motor cortical network of the contingent negative variation estimated by current density. Neuroimage. 2003;20(1053–8119):216–24.
56. Sawamoto N, Honda M, Okada T, Hanakawa T, Kanda M, Fukuyama H, Konishi J, Shibasaki H. Expectation of pain enhances responses to nonpainful somatosensory stimulation in the anterior cingulate cortex and parietal operculum/posterior insula: an event-related functional magnetic resonance imaging study. J Neurosci. 2000;20(1529–2401):7438–45.
57. Wager TD, Rilling JK, Smith EE, Sokolik A, Casey KL, Davidson RJ, Kosslyn SM, Rose RM, Cohen JD. Placebo-induced changes in fMRI in the anticipation and experience of pain. Science (80-.). 2004;303(1095–9203):1162–7.
58. Koyama T, McHaffie JD, Laurienti PJ, Coghill RC. The subjective experience of pain: where expectations become reality. Proc Natl Acad Sci U S A. 2005;102(0027–8424) (Print):12950–5.
59. Keltner JR, Furst A, Fan C, Redfern R, Inglis B, Fields HL. Isolating the modulatory effect of expectation on pain transmission: a functional magnetic resonance imaging study. J Neurosci. 2006;26(1529–2401) (Electronic):4437–43.
60. Lorenz J, Hauck M, Paur RC, Nakamura Y, Zimmermann R, Bromm B, Engel AK. Cortical correlates of false expectations during pain intensity judgments—a possible manifestation of placebo/nocebo cognitions. Brain Behav Immun. 2005;19(0889–1591) (Print):283–95.
61. Ploghaus A, Narain C, Beckmann CF, Clare S, Bantick S, Wise R, Matthews PM, Rawlins JNP, Tracey I. Exacerbation of pain by anxiety is associated with activity in a hippocampal network. J Neurosci. 2001;21(24):9896–903.
62. Berns GS, Chappelow J, Cekic M, Zink CF, Pagnoni G, Martin-Skurski ME. Neurobiological substrates of dread. Science (80-.). *2006*;312(1095–9203) (Electronic):754–8.
63. Clark JA, Brown CA, Jones AKP, El-Deredy W. Dissociating nociceptive modulation by the duration of pain anticipation from unpredictability in the timing of pain. Clin Neurophysiol. 2008;119 (12):2870–8.
64. Ritter C, Hebart MN, Wolbers T, Bingel U. Representation of spatial information in key areas of the descending pain modulatory system. J Neurosci. 2014;34(13):4634–9.
65. Drabant EM, Kuo JR, Ramel W, Blechert J, Edge MD, Cooper JR, Goldin PR, Hariri AR, Gross JJ. Experiential, autonomic, and neural responses during threat anticipation vary as a function of threat intensity and neuroticism. Neuroimage. 2011;55(1):401–10.
66. Schunck T, Erb G, Mathis A, Jacob N, Gilles C, Namer IJ, Meier D, Luthringer R. Test-retest reliability of a functional MRI anticipatory anxiety paradigm in healthy volunteers. J Magn Reson Imaging. 2008;27(3):459–68.
67. Bush G, Luu P, Posner MI. Cognitive and emotional influences in anterior cingulate cortex. Trends Cogn Sci. 2000;4(1364–6613):215–22.
68. Hauck M, Lorenz J, Zimmermann R, Debener S, Scharein E, Engel AK. Duration of the cue-to-pain delay increases pain intensity: a combined EEG and MEG study. Exp Brain Res. 2007;180(0014–4819) (Print):205–15.
69. Carlsson K, Andersson J, Petrovic P, Petersson KM, Ohman A, Ingvar M. Predictability modulates the affective and sensory-discriminative neural processing of pain. Neuroimage. 2006;32(1053–8119) (Print):1804–14.
70. Vase L, Robinson ME, Verne GN, Price DD. Increased placebo analgesia over time in irritable

bowel syndrome (IBS) patients is associated with desire and expectation but not endogenous opioid mechanisms. Pain. 2005;115(0304–3959) (Print):338–47.
71. Pollo A, Amanzio M, Arslanian A, Casadio C, Maggi G, Benedetti F. Response expectancies in placebo analgesia and their clinical relevance. Pain. 2001;93(0304–3959) (Print):77–84.
72. Zubieta JKK, Bueller JA, Jackson LR, Scott DJ, Xu Y, Koeppe RA, Nichols TE, Stohler CS. Placebo effects mediated by endogenous opioid activity on mu-opioid receptors. J Neurosci. 2005;25(1529–2401) (Electronic):7754–62.
73. Geuter S, Eippert F, Hindi Attar C, Büchel C. Cortical and subcortical responses to high and low effective placebo treatments. Neuroimage. 2013;67:227–36.
74. Watson A, El-Deredy W, Iannetti GD, Lloyd D, Tracey I, Vogt BA, Nadeau V, Jones AK. Placebo conditioning and placebo analgesia modulate a common brain network during pain anticipation and perception. Pain. 145(1872–6623) (Electronic):24–30.
75. Kong J, Gollub RL, Rosman IS, Webb JM, Vangel MG, Kirsch I, Kaptchuk YJ. Brain activity associated with expectancy-enhanced placebo analgesia as measured by functional magnetic resonance imaging. J Neurosci. 2006;26(1529–2401) (Electronic):381–88.
76. Wager TD, Atlas LY, Leotti LA, Rilling JK. Predicting individual differences in placebo analgesia: contributions of brain activity during anticipation and pain experience. J Neurosci. 2011;31(2):439–52.
77. Scott DJ, Stohler CS, Egnatuk CM, Wang H, Koeppe RA, Zubieta J-K. Individual differences in reward responding explain placebo-induced expectations and effects. Neuron. 2007;55(2):325–36.
78. Schweinhardt P, Seminowicz DA, Jaeger E, Duncan GH, Bushnell MC. The anatomy of the mesolimbic reward system: a link between personality and the placebo analgesic response. J Neurosci. 2009;29(15):4882–7.
79. Price DD, Milling LS, Kirsch I, Duff A, Montgomery GH, Nicholls SS. An analysis of factors that contribute to the magnitude of placebo analgesia in an experimental paradigm. Pain. 1999;83(0304–3959) (Print):147–56.
80. Morton DL, Watson A, El-Deredy W, Jones AKP. Reproducibility of placebo analgesia: effect of dispositional optimism. Pain. 2009;146(1–2):194–8.
81. Benedetti F, Lanotte M, Lopiano L, Colloca L. When words are painful: unraveling the mechanisms of the nocebo effect. Neuroscience. 2007;147 (2):260–71.
82. Kong J, Gollub RL, Polich G, Kirsch I, Laviolette P, Vangel M, Rosen B, Kaptchuk TJ. A functional magnetic resonance imaging study on the neural mechanisms of hyperalgesic nocebo effect. J Neurosci. 2008;28(49):13354–62.
83. Rainville P, Carrier B, Hofbauer RK, Bushnell MC, Duncan GH. Dissociation of sensory and affective dimensions of pain using hypnotic modulation. Pain. 1999;82(0304–3959) (Print):159–71.
84. Derbyshire SW, Whalley MG, Stenger VA, Oakley DA. Cerebral activation during hypnotically induced and imagined pain. Neuroimage. 2004;23 (1053–8119) (Print):392–401.
85. Raij TT, Numminen J, Narvanen S, Hiltunen J, Hari R. Brain correlates of subjective reality of physically and psychologically induced pain. Proc Natl Acad Sci U S A. 2005;102(0027–8424) (Print):2147–51.
86. Huber A, Lui F, Porro CA. Hypnotic susceptibility modulates brain activity related to experimental placebo analgesia. Pain. 2013;154(9):1509–18.
87. Wiech K, Vandekerckhove J, Zaman J, Tuerlinckx F, Vlaeyen JWS, Tracey I. Influence of prior information on pain involves biased perceptual decision-making. Curr Biol. 2014;24(15):R679–81.
88. Holland P, Gallagher M. Amygdala circuitry in attentional and representational processes. Trends Cogn Sci. 1999;3(2):65–73.
89. Cardinal RN, Parkinson JA, Hall J, Everitt BJ. Emotion and motivation: the role of the amygdala, ventral striatum, and prefrontal cortex. Neurosci Biobehav Rev. 2002;26(3):321–52.
90. O'Doherty JP, Dayan P, Friston K, Critchley H, Dolan RJ. Temporal difference models and reward-related learning in the human brain. Neuron. 2003;38(2):329–37.
91. Jones JL, Esber GR, McDannald MA, Gruber AJ, Hernandez A, Mirenzi A, Schoenbaum G. Orbitofrontal cortex supports behavior and learning using inferred but not cached values. Science. 2012;338 (6109):953–6.
92. O'Doherty JP, Deichmann R, Critchley HD, Dolan RJ. Neural responses during anticipation of a primary taste reward. Neuron. 2002;33(5):815–26.
93. Knutson B, Fong GW, Bennett SM, Adams CM, Hommer D. A region of mesial prefrontal cortex tracks monetarily rewarding outcomes: characterization with rapid event-related fMRI. Neuroimage. 2003;18(2):263–72.
94. Buchel C, Dolan RJ. Classical fear conditioning in functional neuroimaging. Curr Opin Neurobiol. 2000;10(0959–4388) (Print):219–23.
95. Calder AJ, Lawrence AD, Young AW. Neuropsychology of fear and loathing. Nat Rev Neurosci. 2001;2(1471–003X) (Print):352–63.
96. Freese JL, Amaral DG. No Title. In: Whalen PJ, Phelps EA, editors. The human Amygdala. New York: Guilford; 2009. p. 3–42.
97. Ghashghaei HT, Hilgetag CC, Barbas H. Sequence of information processing for emotions based on the anatomic dialogue between prefrontal cortex and amygdala. Neuroimage. 2007;34(3):905–23.
98. Dayan P, Balleine BW. Reward, motivation, and reinforcement learning. Neuron. 2002;36(2):285–98.

99. Delgado MR, Li J, Schiller D, Phelps EA. The role of the striatum in aversive learning and aversive prediction errors. Philos Trans R Soc Lond B Biol Sci. 2008;363(1511):3787–800.
100. O'Doherty J, Dayan P, Schultz J, Deichmann R, Friston K, Dolan RJ. Dissociable roles of ventral and dorsal striatum in instrumental conditioning. Science. 2004;304(5669):452–4.
101. Alheid GF, Heimer L. New perspectives in basal forebrain organization of special relevance for neuropsychiatric disorders: the striatopallidal, amygdaloid, and corticopetal components of substantia innominata. Neuroscience. 1988;27(1):1–39.
102. Groenewegen HJ, Wright CI, Beijer AV, Voorn P. Convergence and segregation of ventral striatal inputs and outputs. Ann N Y Acad Sci. 1999;877:49–63.
103. Uddin LQ. Salience processing and insular cortical function and dysfunction. Nat Rev Neurosci. 2014;16(1):55–61.
104. Treede R-D, Apkarian VA, Bromm B, Greenspan JD, Lenz FA. Cortical representation of pain: functional characterization of nociceptive areas near the lateral sulcus. Pain. 2000;87(2): 113–9.
105. Mesulam MM, Mufson EJ. Insula of the old world monkey. III: efferent cortical output and comments on function. J Comp Neurol. 1982;212(0021–9967) (Print):38–52.
106. Mufson EJ, Mesulam MM. Insula of the old world monkey. II: afferent cortical input and comments on the claustrum. J Comp Neurol. 1982;212(1):23–37.
107. Craig ADB. Significance of the insula for the evolution of human awareness of feelings from the body. Ann N Y Acad Sci. 2011;1225:72–82.
108. Craig AD, Chen K, Bandy D, Reiman EM. Thermosensory activation of insular cortex. Nat Neurosci. 2000;3(1097–6256) (Print):184–90.
109. Davis KD, Pope GE, Crawley AP, Mikulis DJ. Perceptual illusion of 'paradoxical heat' engages the insular cortex. J Neurophysiol. 2004;92(2): 1248–51.
110. Singer T, Seymour B, O'Doherty J, Kaube H, Dolan RJ, Frith CD. Empathy for pain involves the affective but not sensory components of pain. Science (80-.). 2004;303(1095–9203):1157–62.
111. Downar J, Crawley AP, Mikulis DJ, Davis KD. A multimodal cortical network for the detection of changes in the sensory environment. Nat Neurosci. 2000;3(3):277–83.
112. Linden DE, Prvulovic D, Formisano E, Völlinger M, Zanella FE, Goebel R, Dierks T. The functional neuroanatomy of target detection: an fMRI study of visual and auditory oddball tasks. Cereb Cortex. 1999;9(8):815–23.
113. Ullsperger M, Harsay HA, Wessel JR, Ridderinkhof KR. Conscious perception of errors and its relation to the anterior insula. Brain Struct Funct. 2010;214(5–6):629–43.
114. Craig AD. How do you feel? Interoception: the sense of the physiological condition of the body. Nat Rev Neurosci. 2002;3(1471–003X):655–66.
115. Zaki J, Davis JI, Ochsner KN. Overlapping activity in anterior insula during interoception and emotional experience. Neuroimage. 2012;62(1):493–9.
116. Critchley HD, Melmed RN, Featherstone E, Mathias CJ, Dolan RJ. Volitional control of autonomic arousal: a functional magnetic resonance study. Neuroimage. 2002;16(1053–8119):909–19.
117. Gray MA, Critchley HD. Interoceptive basis to craving. Neuron. 2007;54(2):183–6.
118. Wiech K, Lin C, Brodersen KH, Bingel U, Ploner M, Tracey I. Anterior insula integrates information about salience into perceptual decisions about pain. J Neurosci. 2010;30(48):16324–31.
119. Morecraft RJ, Tanji J. "Cingulofrontal interactions and the cingulate motor areas", in cingulate neurobiology and disease. Oxford: Oxford University Press; 2009. p. 114–37.
120. Yu C, Zhou Y, Liu Y, Jiang T, Dong H, Zhang Y, Walter M. Functional segregation of the human cingulate cortex is confirmed by functional connectivity based neuroanatomical parcellation. Neuroimage. 2011;54(4):2571–81.
121. Peyron R, Laurent B, Garcia-Larrea L. Functional imaging of brain responses to pain. A review and meta-analysis. Neurophysiol Clin. 2000;30(0987–7053):263–88.
122. Vogt BA, Berger GR, Derbyshire SW. Structural and functional dichotomy of human midcingulate cortex. Eur J Neurosci. 2003;18(0953–816X) (Print):3134–44.
123. Shackman AJ, Salomons TV, Slagter HA, Fox AS, Winter JJ, Davidson RJ. The integration of negative affect, pain and cognitive control in the cingulate cortex. Nat Rev Neurosci. 2011;12(3):154–67.
124. Ridderinkhof KR, van den Wildenberg WP, Segalowitz SJ, Carter CS. Neurocognitive mechanisms of cognitive control: the role of prefrontal cortex in action selection, response inhibition, performance monitoring, and reward-based learning. Brain Cogn. 2004;56(0278–2626):129–40.
125. Miller EK. The prefrontal cortex and cognitive control. Nat Rev Neurosci. 2000;1(1):59–65.
126. Shenhav A, Botvinick MM, Cohen JD. The expected value of control: an integrative theory of anterior cingulate cortex function. Neuron. 2013;79 (2):217–40.
127. Bromberg-Martin ES, Matsumoto M, Hikosaka O. Dopamine in motivational control: rewarding, aversive, and alerting. Neuron. 2010;68(5):815–34.
128. Cauda F, D'Agata F, Sacco K, Duca S, Geminiani G, Vercelli A. Functional connectivity of the insula in the resting brain. Neuroimage. 2011;55 (1):8–23.
129. Vogt BA, Pandya DN. Cingulate cortex of the rhesus monkey: II. Cortical afferents. J Comp Neurol. 1987;262(0021–9967) (Print):271–89.

130. Brown CA, Jones AKP. A role for midcingulate cortex in the interruptive effects of pain anticipation on attention. Clin Neurophysiol. 2008;119(10):2370–9.
131. Corbetta M, Kincade JM, Shulman GL. Neural systems for visual orienting and their relationships to spatial working memory. J Cogn Neurosci. 2002;14(3):508–23.
132. Corbetta M, Patel G, Shulman GL. The reorienting system of the human brain: from environment to theory of mind. Neuron. 2008;58(3):306–24.
133. Downar J, Mikulis DJ, Davis KD. Neural correlates of the prolonged salience of painful stimulation. Neuroimage. 2003;20(3):1540–51.
134. Corbetta M, Shulman GL. Control of goal-directed and stimulus-driven attention in the brain. Nat Rev Neurosci. 2002;3(1471–003X) (Print):201–15.
135. Hatanaka N, Tokuno H, Hamada I, Inase M, Ito Y, Imanishi M, Hasegawa N, Akazawa T, Nambu A, Takada M. Thalamocortical and intracortical connections of monkey cingulate motor areas. J Comp Neurol. 2003;462(1):121–38.
136. Miller EK, Cohen JD. An integrative theory of prefrontal cortex function. Annu Rev Neurosci. 2001;24:167–202.
137. Aron AR, Robbins TW, Poldrack RA. Inhibition and the right inferior frontal cortex. Trends Cogn Sci. 2004;8(4):170–7.
138. Lorenz J, Minoshima S, Casey KL. Keeping pain out of mind: the role of the dorsolateral prefrontal cortex in pain modulation. Brain. 2003;126(0006–8950):1079–91.
139. Seminowicz DA, Davis KD. Cortical responses to pain in healthy individuals depends on pain catastrophizing. Pain. 2006;120(0304–3959) (Print):297–306.
140. van den Heuvel MP, Mandl RCW, Kahn RS, Hulshoff Pol HE. Functionally linked resting-state networks reflect the underlying structural connectivity architecture of the human brain. Hum Brain Mapp. 2009;30(10):3127–41.
141. Uddin LQ, Supekar K, Amin H, Rykhlevskaia E, Nguyen DA, Greicius MD, Menon V. Dissociable connectivity within human angular gyrus and intraparietal sulcus: evidence from functional and structural connectivity. Cereb Cortex. 2010;20(11):2636–46.
142. Husain M, Nachev P. Space and the parietal cortex. Trends Cogn Sci. 2007;11(1):30–6.
143. Kulkarni B, Bentley DE, Elliott R, Youell P, Watson A, Derbyshire SWG, Frackowiak RSJ, Friston KJ, Jones AKP. Attention to pain localization and unpleasantness discriminates the functions of the medial and lateral pain systems. Eur J Neurosci. 2005;21(0953–816X):3133–42.
144. Bentley DE, Watson A, Treede RD, Barrett G, Youell PD, Kulkarni B, Jones AK. Differential effects on the laser evoked potential of selectively attending to pain localisation versus pain unpleasantness. Clin Neurophysiol. 2004;115(1388–2457):1846–56.
145. Petrovic P, Kalso E, Petersson KM, Ingvar M. Placebo and opioid analgesia—imaging a shared neuronal network. Science (80-.). 2002;295(1095–9203):1737–40.
146. Kong J, Jensen K, Loiotile R, Cheetham A, Wey H-Y, Tan Y, Rosen B, Smoller JW, Kaptchuk TJ, Gollub RL. Functional connectivity of the frontoparietal network predicts cognitive modulation of pain. Pain. 2013;154(3):459–67.
147. Hutchison WD, Harfa L, Dostrovsky JO. Ventrolateral orbital cortex and periaqueductal gray stimulation-induced effects on on- and off-cells in the rostral ventromedial medulla in the rat. Neuroscience. 1996;70(0306–4522) (Print):391–407.
148. Calejesan AA, Kim SJ, Zhuo M. Descending facilitatory modulation of a behavioral nociceptive response by stimulation in the adult rat anterior cingulate cortex. Eur J Pain. 2000;4(1090–3801) (Print):83–96.
149. An X, Bandler R, Ongür D, Price JL. Prefrontal cortical projections to longitudinal columns in the midbrain periaqueductal gray in macaque monkeys. J Comp Neurol. 1998;401(4):455–79.
150. Bantick SJ, Wise RG, Ploghaus A, Clare S, Smith SM, Tracey I. Imaging how attention modulates pain in humans using functional MRI. Brain. 2002;125(0006–8950):310–9.
151. Valet M, Sprenger T, Boecker H, Willoch F, Rummeny E, Conrad B, Erhard P, Tolle TR. Distraction modulates connectivity of the cingulo-frontal cortex and the midbrain during pain-an fMRI analysis. Pain. 2004;109(0304–3959):399–408.
152. Simpson Jr. JR, Snyder AZ, Gusnard DA, Raichle ME. Emotion-induced changes in human medial prefrontal cortex: I. During cognitive task performance. Proc Natl Acad Sci U S A. 2001;98(0027–8424):683–7.
153. Brooks J, Tracey I. From nociception to pain perception: imaging the spinal and supraspinal pathways. J Anat. 2005;207(1):19–33.
154. Lenz FA, Weiss N, Ohara S, Lawson C, Greenspan JD. The role of the thalamus in pain. Suppl Clin Neurophysiol. 2004;57:50–61.
155. Duerden EG, Albanese M-C. Localization of pain-related brain activation: a meta-analysis of neuroimaging data. Hum Brain Mapp. 2013;34(1):109–49.
156. Youell PD, Wise RG, Bentley DE, Dickinson MR, King TA, Tracey I, Jones AK. Lateralisation of nociceptive processing in the human brain: a functional magnetic resonance imaging study. Neuroimage. 2004;23(1053–8119):1068–77.
157. Derbyshire AW, Jones AK. Cerebral responses to a continual tonic pain stimulus measured using positron emission tomography. Pain. 1998;76(1–2):127–35.

158. Frot M, Magnin M, Mauguière F, Garcia-Larrea L. Human SII and posterior insula differently encode thermal laser stimuli. Cereb Cortex. 2007;17 (3):610–20.
159. Heekeren HR, Marrett S, Bandettini PA, Ungerleider LG. A general mechanism for perceptual decision-making in the human brain. Nature. 2004;431(1476–4687):859–62.
160. Seth AK, Suzuki K, Critchley HD. An interoceptive predictive coding model of conscious presence. Front Psychol. 2011;2:395.
161. Apps MAJ, Lockwood PL, Balsters JH. The role of the midcingulate cortex in monitoring others' decisions. Front Neurosci. 2013;7:251.

Author Biography

Dr. Christopher Brown is Lecturer in Psychology at the University of Liverpool in the United Kingdom. Dr. Brown's main areas of expertise are in the neuroimaging of pain anticipation in acute and chronic pain, and psychological phenotyping. He has conducted research at the University of Manchester (where he was awarded a Ph.D.) and completed post-doctoral research as the University of Cambridge.

Anticipation and Placebo Analgesia

7

Dominic E. Nathan, Taylor M. Ludman
and Luana Colloca

Abstract

The neural substrates of the anticipatory phase of an impending stimulus have been studied in the context of pain with neuromapping techniques, including both positron emission tomography (PET) and functional magnetic resonance imaging (fMRI). The neural response during the anticipatory phase is complex since it can be affected by many factors, including certainty and uncertainty, dispositional anxiety and pathology, which have been the topic of many past studies. Importantly, the anticipation of pain is affected by positive and negative expectancies, namely the placebo and nocebo phenomena which serve as great examples of how expectancies created during the anticipatory phase can modulate pain perception. Here we critically discuss the literature on the neural activity throughout the anticipatory phase preceding a noxious stimulus and during the delivery of a stimulus when placebo and nocebo effects modulate pain. Understanding the processes during the anticipatory phase and the placebo and nocebo effects can help increase knowledge of both acute and chronic pain mechanisms, identify biological predictors of variability in clinical pain phenotypes and ultimately contribute to new therapeutic approaches.

D.E. Nathan
Uniformed Services University of the Health Sciences, School of Nursing, Bethesda, MD, USA

T.M. Ludman · L. Colloca
Pain and Translational Symptom Science, University of Maryland Baltimore, School of Nursing, Baltimore, USA

T.M. Ludman · L. Colloca
UM Center to Advance Chronic Pain Research, Baltimore, MD, USA

L. Colloca (✉)
Department of Anesthesiology, University of Maryland Baltimore, School of Medicine, 655 W. Lombard Street, Baltimore, MD, USA
e-mail: colloca@son.umaryland.edu

L. Saba (ed.), *Neuroimaging of Pain*, DOI 10.1007/978-3-319-48046-6_7

Keywords
Anticipation · Expectancies · Placebo · Nocebo · Brain imaging

Abbreviations

Amy	Amygdala
ACC	Anterior cingulate cortex
rACC, sACC or pgACC (respectively)	Rostral or subgenual or pregenual ACC
INS	Insular cortex
MCC	Midcingulate cortex
NAcc	Nucleus accumbens
OFC	Orbitofrontal cortex
PAG	Periaqueductal gray
PFC	Prefrontal cortex
DLPFC	Dorsolateral prefrontal cortex
vmPFC	Ventral medial prefrontal cortex
SI	Primary somatosensory cortex
RVM	Rostral ventral medulla
Th	Thalamus
VTA	Ventral tegmental area

1 Introduction

According to Douglas Harper, a historian, and the "Online Etymology Dictionary", the English definition and use of the word 'anticipation' stems back to the Latin word 'anticipationem' which means preconception, or a preconceived notion. Moreover, it was considered to involve a noun of action from the past or, 'participlestem of anticipare' which is to take care of ahead of time. In a definition from the nineteenth century, the word evolved to literally mean an "action of looking forward to" [1]. In laboratory settings for brain mapping research on pain mechanisms and in this chapter, the term 'anticipation' is used to indicate the anticipatory phase that is the period of time stretching between the presentation of a cue associated with a painful stimulus and the delivery of the painful stimulus. The anticipatory cue alerts research participants about the occurrence of an impending painful stimulus that can be either predictable or unpredictable. In the context of pain modulation and placebo research, anticipation often pertains to expressions of expectations and desires for pain relief and benefits. For example, placebo and nocebo effects are examples of how positive and negative expectations during the anticipatory phase can lead to different pain outcomes and associated changes in neural responses when pain is experienced. Moreover, a positive attitude about a treatment can foster beliefs that pain can be eventually reduced, alleviated or nullified. By contrast, negative expectancies held about a treatment can drive nocebo effects and expectations of pain worsening can activate facilitatory mechanisms, including increased pain and neural modulation [2].

Many studies on the placebo and nocebo effects have focused on how positive and negative expectations of a treatment affect pain processing and experience. Ultimately the anticipation of treatment given for a noxious stimulus produces beliefs of analgesia (or hyperalgesia), affecting brain pain processing. Both positive and negative expectancies can be triggered by verbal instruction, conditioning, and social observation suggesting that past experience, and learning about a painful stimulus can shape cognitive interpretation and the outcome [3–7]. These top-down triggering experiences can induce hyperalgesia (nocebo) and hypoalgesia (placebo) by altering neural activity in brain areas associated with pain processing, emotions and pain valuation [8].

The first part of this chapter will elaborate on the neuromapping of the anticipatory response to impending experimental painful stimulations. The neural response during the anticipatory phase can be modulated by many factors, including the ability to predict the delivery of the painful stimulus and the existence of underlying chronic pain conditions or psychological traits (e.g., fear of pain, anxiety, catastrophizing traits) [9–11]. The second part of the chapter will cover the neuroscience behind the placebo and nocebo phenomena; which involves the associated activated and deactivated central nervous system regions and neurotransmitters involved in pain modulation.

2 Anticipation of an Impending Painful Stimulus

In the context of pain, anticipation is the action, including specific neural activities, which occurs before the presentation of a painful stimulus. The state of anticipation is considered to be orchestrated by many facets and aids in the unpleasant experience that accompanies pain [12–14]. The neural activity that occurs during the anticipatory period is thought to be shaped by cognitive factors and decisional processes about the intensity of the painful stimulus and the predictability of its delivery [11]. For example, a decision of how painful a stimulus will be is partially determined during the anticipatory phase. Ploner et al. [9] investigated the functional connectivity among brain regions during the anticipatory phase to explore the dynamics responsible for determining whether a stimulus will be painful even before the presentation of the painful stimulus. A simple near-threshold pain detection paradigm was used, meaning brief pulses of heat were consistently administered at an intensity close to the pain threshold throughout the experiment, so fluctuations in deeming whether a stimulus was painful or non-painful during different trials is likely attributable to changes in pain susceptibility. The brain activity three seconds before the presentation of the thermal stimulation was measured. The left anterior INS displayed strong activity during this short period even before the presentation of the noxious stimulus; therefore the anterior INS was responsible for *deciding* whether a stimulus could be painful. Based on this anticipatory phase and the decision that was eventually made, the anterior INS likely played a role in activating the PAG and pain modulation. Moreover, pre-stimulus connectivity among the anterior INS and brainstem, respectively determined whether a noxious event was perceived as painful. These responses co-varied with pain-relevant personality traits: more anxious and pain-attentive individuals displayed weaker descending connectivity with pain modulatory brain areas. These findings added on to prior studies that only assessed the regional activity during the anticipatory phase since pre-stimulus functional connectivity between regions also influences impending pain perception.

In a recent meta-analysis, Palermo et al. [14] used the activation likelihood estimation to analyze pain anticipation across several neuroimaging studies (19 functional magnetic resonance imaging (fMRI) studies published from 2003 to 2013) and showed that there were 21 activation and six deactivation clusters that appeared to be relevant for pain anticipatory activity. During anticipation, activated foci were found in the DLPFC, MCC, anterior INS, medial and inferior frontal gyri, inferior parietal lobule, middle and superior temporal gyrus, Th, and caudate. Deactivated foci were found in the ACC, superior frontal gyrus, parahippocampal gyrus and in the claustrum. A connectivity model was created that suggested the anterior INS and ACC select attentional, affective, and sensory resources when the model is activated by pain anticipation. These networks of co-activation overlapped and had a common origin of activation in the dorsolateral and medial PFC. Overall, these results suggest that self-regulation primes brain regions associated with emotions, action, and perception of a pain stimulus.

2.1 Certainty Versus Uncertainty in Anticipating Impending Noxious Stimuli Presentation

Compelling evidence from a set of studies that use fMRI and positron emission tomography (PET) suggests that knowing with certainty about how intense and when a painful stimulus will be presented is associated with increased activation of the rACC, anterior INS, and posterior cerebellum [11]. The elevated activity in these areas is most likely the result of the interaction between emotions (e.g., fear of pain) and pain processing. The anticipation leading up to pain stimulation stirs up fear, an emotion that prepares the body for a threatened aversive event [15]. In contrast, unpredictability about whether a stimulus will be noxious or salient causes increased activity in the vmPFC, MCC, and the primary SI. Moreover, uncertainty regarding the intensity of the impending painful stimulus is related to decreases in signaling in the vmPFC, MCC, and SI. Ploghaus et al. [11] concluded that unpredictability increases anxiety, by turning up the neural response in the entorhinal cortex of the hippocampus, which further amplifies the response to pain resulting in hyperalgesia [16].

2.2 More on Certain Anticipation

Certain expectations about the presentation of a painful stimulus are associated with distinct brain response patterns. Ploghaus et al. [17] measured the neural response to a thermal cue in study participants who knew when to expect the presentation of the stimulus. The results suggested that over the trials, the participants were increasingly anticipating the painful stimulus after the presentation of the associated visual cue. At the neural level, the presentation of the noxious stimulus associated cue induced increased activity in the rACC, INS, and posterior cerebellum but not when an innocuous stimulus associated cue was presented. During the actual experience of pain, different sites were activated including regions such as the MCC, mid-INS, and the vermis of the anterior cerebellum. These findings suggest that these are the main brain areas activated during pain anticipation and perception when expectations about a noxious stimulus are certain.

In a PET study by Chua et al. [18], the neural response to anticipatory anxiety, provoked by an impending painful stimulus, was also assessed. Electric shocks were delivered while participants completed a motor or learning task. We can assume that these tasks would have reduced the fear and anxiety related emotions that can be stirred up during the anticipatory phase when pain is expected by reducing attention to the impending painful stimulus. During the experiment, a red cue was displayed and the participant received a shock. The left INS displayed greater activation when anticipating a shock. Furthermore there was increased activation in the right superior temporal sulcus, left fusiform, and ACC. There were no significant distraction effects induced by the motor (low distraction) or learning task (high distraction) reflecting a relatively modest increase in anxiety with the shock and a simple distractor task. The main result was that the paralimbic structures played a role in anticipatory anxiety related to a painful shock.

Interestingly Fairhurst et al. [19], extended previous findings by assessing the activation of the brain stem in anticipating pain. FMRI data revealed brainstem activation in the PAG during the anticipation period. Before this study, the PAG was the only brain region from these to demonstrate significant increased activation during both the anticipatory and stimulation conditions [20]. When correlated with individual anticipation ratings, activation during anticipation included significant clusters within the entorhinal cortex and VTA. During pain stimulation, activation within the brainstem included the PAG, VTA, RVM, and the parabrachial nucleus, all elements of descending pain pathways. With a backward model approach, used for identifying functional links between the period of pain presentation and anticipation, the researchers further concluded that the activity in the posterior INS, during the painful stimulation, was predicted during the anticipatory condition by the PAG, VTA, and entorhinal cortex. This study concluded that in addition to the PAG increased activation, the VTA and entorhinal cortex showed increased activity during the anticipatory phase as well. All of these areas play a role in the cognitive–affective aspects of pain perception, which is thought to amplify pain perception [16, 21, 22].

2.3 More on Uncertain Anticipation

Using a PET approach, Hsieh et al. [23] studied regional cerebral blood flow (rCBF) in order to measure affective–motivational and cognitive–evaluative aspects during the anticipatory phase of a noxious stimulus. Participants in the anticipation group of the first experiment were told to expect a painful event at an unknown time during the scan session. In the first study, subjects were given either a saline or a noxious ethanol injection semi-randomly. Saline (20 μl, 0.9%) injection was given with explicit assurance that a non-pain control condition was studied (control). The subjects were then informed that during the subsequent study a noxious stimulation would sometimes be given during the scan without prior information. To maintain the tone of anticipation, a minute amount of ethanol (20 μl, 70%; immediate pain, peak intensity latency 7–10 s, duration 40–45 s) was semi-randomly injected during the course of the study. In a second study, subjects were instructed that an electric shock (intensity equal to $\geq$80% visual analogue scale or VAS) would be delivered sometime within the scanning period after the tracer administration and that the longer the delay, the more painful the shock would be. The anticipation of the unpredictable and unlearned pain stimulus activated the right ACC, vmPFC and PAG. By contrary, the anticipation of a learned pain-stimulus resulted in a decreased activity in the ACC and the vmPFC.

Anxiety levels appear to play a modulatory role on neural anticipatory responses. Simpson et al. [24] assessed the link between anxiety levels and neural responses during a phase when the participants were uncertain about how intense the painful stimulus would be. In this study, differences between baseline and experimental (anticipation) rCBF in the medial PFC, midbrain, and hypothalamus were identified. Those participants with lower anxiety ratings on the Spielberger State-Trait Anxiety Inventory (STAI), in response to the impending electrical shock to their finger during the anticipatory phase, presented with greater reductions in rCBF in the medial PFC (specifically the subgenual and a more anterior regions), midbrain, and hypothalamus. Those with greater anxiety ratings had either the same or slightly increased rCBF to these areas compared to baseline. The hypothalamus is highly interconnected with the vmPFC, a major cognitive processing region [25] The connection between the vmPFC and affective processing areas illuminates an interplaying relationship between cognitive and affective processes during the anticipatory stage of pain processing.

Interestingly, Porro et al. [26] captured the neural response to the anticipation of a painful subcutaneous injection of ascorbic acid into the foot. Participants were informed that a certain stimulus (subcutaneous injection of ascorbic acid) was painful or not. The participants had no prior experience of having the injection, and thus were unaware of the intensity and duration of the painful sensation that would occur. During the anticipation of noxious stimulation, the contralateral SI exhibited increased signaling which could mean that the activation of this area was the result of attentional mechanisms and not widespread arousal. Unlike prior studies that concluded no increased activity in the SI region, ascorbic acid was an entirely novel noxious stimulus to all participants which likely perpetuated the activity in the SI region. If the stimulus was not completely novel, endogenous inhibitory systems might have been triggered before SI activation could occur, which was the likely culprit in prior studies [23]. Increased activation of the SI was positively correlated with altered activity in the ACC, anterior INS, and vmPFC. The anteroventral cingulate and portions of the ipsilateral SI showed a decrease in activity.

2.4 Individual Variations and Subjectivity Throughout the Anticipatory Phase

Decisions about the painful nature of the stimulus are also highly individual. The expectations we have about anticipated pain are subjective, and thus result in varied neural activity during the anticipatory phase. Koyama et al. [27] manipulated expectancies of pain intensity; 1 s duration tone signaled increasingly intense stimulus temperatures (46, 48, and 50 °C) associated with longer expectation intervals (7.5, 15, and 30 s). To reinforce the association between the duration of the expectation phase and the stimulus temperatures, subjects participated in a training session before the fMRI session. During the fMRI session, 33% of both the 48 and 50 °C trials were falsely signaled. In the case of the 50 °C trials, expectations of decreased pain were created by using an expectation interval of 15 s (normally signaling a 48°C stimulus). In the case of 48 °C stimuli, expectations of increased pain were created by using an expectation interval of 30 s (normally signaling a 50 °C stimulus). Neural processes were identified accounting for the subjective predictions about the magnitude of the intensity of the pain experience and the expectation-induced modulation of pain. Activity, while participants were forming increased or decreased expectations of the three different noxious temperatures, was displayed in the PFC, INS, ACC, global pallidus/putamen, Th and cerebellum. Specifically, this activity was linked to the subjectivity of how painful a thermal stimulus would be. Recalling memories of past experiences, with the aid of the hippocampus and Amy, while creating a mental representation of the impending pain, could account for individual neural responses during the anticipatory phase. Overall, past experiences and expectations of pain may prime the brain when afferent noxious information is processed inevitably varying the nature and accuracy of pain processing [28, 29].

2.5 Anticipation in Chronic Pain Populations Further Illuminates Individual Differences

There is evidence suggesting that brainstem and other neural activity during the anticipatory phase is not the same in those who are healthy versus those with chronic pain conditions. Naliboff et al. [30] studied patients with irritable bowel syndrome (IBS) pathophysiology during exposure to repetitive stimuli leading to decreased salience of threat and reduction of hypervigilance. The authors sought to evaluate hypervigilance in IBS visceral hypersensitivity and associated brain activity and found altered brainstem activity during the anticipatory phase. Burgmer et al. [31] measured both state and trait catastrophizing characteristics and fMRI neural responses prior to exerting experimental painful pressure in patients with fibromyalgia syndrome (FMS). Patients were told to expect that they would experience mild, moderate, or a severe amount of pain. fMRI signaling in these patients was increased in the DLPFC, PAG, and posterior parietal cortex compared to healthy controls. FMS patients also reported engaging in more extreme catastrophizing behaviors and these traits could have been responsible for increased sensitization during anticipatory phases. Compared to healthy controls, FMS patients showed reduced gray matter volume in the DLPFC that is thought to play a role in reducing thoughts of pain, thus this reduction in gray matter density (GMD) could explain neural and behavioral responses when anticipating pain [32]. However, in a later study by Burgmer et al. [33] the FMS patients did not present with a different course of primary hyperalgesia but they presented with greater secondary hyperalgesia and an altered cerebral pattern corresponding to secondary hyperalgesia. The activity in the DLPFC was inversely correlated with secondary hyperalgesia in healthy controls but this correlation was disrupted in FMS, suggesting that alterations of pain processes can determine distinct patterns and the brain's representation of the anticipation of pain.

3 Neurochemistry and Neurophysiology of the Placebo Effect

There is mounting scientific evidence to confirm that there are many intricate psychoneurobiological events linked to pain modulation that take place when responding to a placebo or nocebo [7]. Results from neuromapping and pharmacological studies have advanced knowledge on which pain related and unrelated brain areas and neurotransmitters respond to placebo or nocebo effects.

3.1 Neurotransmitter Release During Placebo-Induced Analgesia

Endogenous opioids, cholecystokinin (CCK), and dopamine are all considered to be the neurochemical substrates of placebo hypoalgesia and nocebo hyperalgesia [7, 34]. Here we described some of the landmark studies advancing our knowledge about the analgesic and hyperalgesic neurochemical mechanisms linked to placebo administration or more simply the anticipation of positive and negative outcomes.

The findings from a hallmark study, conducted by Petrovic et al. [35], confirmed the involvement of the opioidergic system in placebo analgesia. Placebo analgesia and opioid analgesia (induced by remifentanil) produced analogous rCBF results in the rACC, an opioid rich area, therefore indicating that opioid release occurs in the rACC during the experience of placebo analgesia. In fact, administering a placebo along with a μ-opioid receptor-selective radiotracer can cause opioid ligand displacement which would confirm the release of endogenous opioids during placebo analgesia [36–38]. In these studies, μ-opioid receptor-mediated neurotransmission has been observed in the ACC, INS, DLPFC, OFC, Amy, PAG, and the Th during placebo effects [36–38]. Moreover, the dual administration of the opioid antagonist naloxone and a placebo treatment resulted in findings that supports that the opioidergic system plays a role in placebo analgesia and, therefore, pain modulation [39–41].

By using PET imaging and the selective μ-opioid receptor agonist [^{11}C]carfentanil, the effects of pain and a placebo were investigated at the level of the degree of μ-opioid receptor availability across distinct brain areas and the displacement of [^{11}C]carfentanil was taken as an index of the activity of the endogenous opioid system. Zubieta et al. [36] first explored the opioid reactivity to painful stimulation (without any placebo manipulation) and found that pain was associated with significant changes of opioid receptors occupancy in the dorsal ACC, medial PFC, rostral INS (contralaterally to the painful side), ventral basal ganglia, NAcc extending to the ventral pallidum, medial Th, right Amy, left subamygdalar temporal cortex, and PAG. When a placebo manipulation was introduced (e.g., given a saline intravenously, along with the information that the drug is either placebo or a strong painkiller) there were changes in the left DLPFC, rACC, ipsilateral NAcc, and right anterior INS. This report represents the first direct neurochemical evidence that a placebo procedure activates pain and stress inhibitory neurotransmitter systems, and the endogenous opioid system. Interestingly, the same group demonstrated that the placebo-induced activation of the endogenous opioid system is modulated by the internal affective state [42]. In fact, the emotional state of the subjects during pain, or the affective quality of experienced pain, were significantly associated with changes in placebo-induced endogenous opioid release measures, as gauged by using PET and [^{11}C]carfentanil, in the DLPFC, anterior INS, and NAcc [42]. Thus increases in endogenous opioid neurotransmission have been found in a number of key opioid-rich regions [36, 37] such as in PAG, rACC, pgACC, and multiple loci within OFC, anterior INS, Th, DLPFC and Amy. By contrast, in the anticipatory phase when a person expects a benefit from a given treatment, the brain activity linked to this time just before the treatment is given is characterized by a decrease in opioid activity in the right PFC [DLPFC, superior frontal sulcus (SFS) and inferior frontal junction (IFJ)], left Amy, left anterior INS, pgACC, dorsal PAG and caudate. Connectivity analyses on individual differences in opioid system activity extend on just structural or functional findings by revealing that placebo treatment increases functional connectivity between the PAG and rACC, as well as among a number of limbic and PFC regions [37].

More recently, Eippert et al. [41] explored the involvement of the opioid systems by using naloxone administration with fMRI. The administration of naloxone (0.15 mg/kg) blocked both behavioral and neural placebo effects. Indeed, naloxone reduced activity in pain-modulatory cortical structures, such as the rACC, and in key structures of the descending pain control system, including the hypothalamus, PAG, and RVM. Notably, naloxone abolished placebo-induced coupling between the rACC and PAG, which predicted both neural and behavioral placebo effects as well as activation of the RVM. These findings are in line with other research studies demonstrating a crucial involvement of the opioid system [37, 43]. Most importantly, these studies have indicated that placebo analgesia engages the DLPFC, which then turns up activity in the rACC influencing the PAG to induce opioid release, and thus, leading to pain reduction and further activation of the descending pain modulatory pathway [44]. The crucial role of the DLPFC as a sort of initiator of placebo responsiveness has been supported by different studies [32, 45].

In addition to opioids, the endogenous release of the neurotransmitter dopamine plays a leading role in stimulating pain modulation during the presentation of a placebo [34]. de la Fuente-Fernandez et al. [46] proposed a placebo-reward hypothesis that asserts the placebo response is mediated by the activation of the brain structures that release dopamine. Evidence from past research suggests that dopamine release in the NAcc, located in the ventral striatum, is signaled by the expectation of reward, or in the case of a placebo, pain relief [47, 48].

Scott et al. [38] examined the role of the NAcc, region centrally involved in the encoding of reward expectation, in the formation of placebo analgesic responses. Using functional molecular imaging, activation of NAcc dopamine release was observed during placebo administration and related to the anticipatory phase, perception-anticipation mismatches, and placebo effect development. Expectancy of monetary gain induced an increase in the NAcc synaptic activity and this change correlated with placebo-induced dopamine release accounting for 25% of the variance in the formation of placebo analgesia. The magnitude of increased activation of these receptors was positively correlated with positive anticipation ratings of how effective an administered placebo would be in reducing their pain symptoms. More specifically, dopamine receptor D2 activation drove the release of opioids in the NAcc which suggests that dopamine release is imperative for provoking opioid release in this area. In this same study, nocebo effects were associated with reduced dopaminergic and opioidergic activity.

Besides the pain arena, the dopaminergic system is obviously involved in modulating placebo responsiveness in patients with Parkinson's disease. For example, de la Fuente-Fernandez and colleagues [46, 49] detected a significant drop in [^{11}C]raclopride binding potential (BP) when Parkinson patients were injected with a saline solution and were verbally told that the solution would cause motor improvement. A reduction in [^{11}C]raclopride binding is an index of an increase in extracellular dopamine concentration. For this study, there was observed extracellular dopamine changes at the level the dorsal and ventral striatum. Those patients who experienced symptomatic benefit showed a larger release of dopamine in the dorsal striatum than those who did not, and the degree of placebo-induced dopamine release in the dorsal striatum correlated with perception of improvement reported by the patient [49]. By contrary, the level of placebo-dopamine release in the ventral striatum was independent of perception of clinical benefit [50] probably indicating that placebo-induced dopamine release might be related to expectation of reward (e.g., a clinical improvement).

The role of expectancy of therapeutic benefit was also investigated in the context of repetitive transmagnetic stimulation (rTMS) and associated changes in striatal [^{11}C]raclopride BP [51]. Placebo-rTMS induced a significant bilateral reduction in [^{11}C]raclopride BP in the dorsal and ventral striatum as compared to the baseline condition. With respect to previous studies [49, 50], they did not observe significant differences in [^{11}C]raclopride BP in the dorsal striatum between the group of patients who perceived the clinical benefit and the group who did not. In fact, placebo-rTMS induced a significant biochemical response in the striatum in all patients, although only four patients perceived a certain degree of clinical benefit. It has also been demonstrated for the first time that a placebo procedure affects specific neuronal populations by recording the activity from single neurons in the subthalamic region before and after a placebo administration following several preoperative administrations of apomorphine, according to a pharmacological conditioning procedure [52] . A saline solution was given in the operating room to Parkinsonian patients undergoing implantation for deep brain stimulation. Those patients who showed a clear-cut clinical placebo response, as assessed by means of both arm rigidity and well-being subjective reports, had a significant decrease of neural discharge in comparison with the baseline pre-placebo condition. The neural activity shifted significantly from a pattern of bursting activity to a pattern of non-bursting discharge. None of the placebo nonresponders showed such differences. These findings—decrease of frequency discharge and shift from bursting to no-bursting activity—were interpreted as a demonstration of conditioned drug-like effects and modulation of endogenous dopamine release.

Besides opioid and dopamine involvement, the CCK system plays a role in mediating both placebo and nocebo responses [34]. Benedetti et al. [53] in a recent study examined how CCK can counteract placebo analgesia. The use of pentagastrin, a CCK type-2 receptor agonist given to pharmacologically increase the activation of CCK type-2 receptor binding, disrupted the analgesic effects that were induced by a morphine preconditioning in an experimental human model of tonic pain via a tourniquet technique. This finding suggests that the critical balance between the levels of endogenous opioids and CCKs influences analgesic responses to a placebo [53–55]. In a prior study by Benedetti et al. [56], proglumide, a nonspecific CCK antagonist, prevented nocebo hyperalgesia and stimulated the placebo analgesic response during the ischemic pain induced by the tourniquet technique in placebo responders only. This finding in particular suggests that individual CCK-2 receptor activity could be responsible for why there are placebo responders and nonresponders [53, 57, 58].

3.2 Neural Responses Associated with Placebo Analgesia

Many of the pain processing associated brain regions that also respond to placebos have been studied most recently in brain imaging studies over the last decade. However, the pattern of involvement of these areas is not exactly the same in instances of placebo analgesia and pain processing.

In an fMRI study by Wager et al. [20] after a placebo treatment was administered for both an electrical shock and noxious thermal stimuli, brain areas within the pain network were found to show attenuated activity that was positively correlated with reported decreased pain ratings. More specifically, when pain was expected to be lessened, there was reduced activity in the rACC at the junction between the rostral and caudal ACC, contralateral INS, and the contralateral Th. Price et al. [59] replicated this finding of decreased neural activity in pain related brain areas in a population with a chronic health condition (those with IBS). Most importantly this study concluded that these areas signaled reduced activity not only after the painful stimulus was delivered, like in the Wager et al. [20] study, but also during the time the painful stimulus was being delivered. This finding reduces the likelihood that the relationship between the decreased pain ratings and activity within pain regions was solely a function of a report bias. In a most recent meta-analysis by

Atlas, Wager [60], not only were these findings supported, but additionally it was indicated that similarly designed studies reached the conclusion that placebo analgesia is associated with suppressed activity in notorious pain processing regions, including the dorsal ACC, Th, and INS, and also regions associated with affect and valuation including the Amy and striatum. Furthermore, expectations for reduced pain produced increased activation of the PFC (DLPFC, vmPFC, and OFC), PAG, and the rACC [60].

Indeed many studies agree that placebo analgesia reduces activity in both the ACC and the INS; however, these areas simultaneously exhibit increased activity [61]. In one such study by Kong et al. [62], expectations of a sham acupuncture needle led to a placebo effect that was positively correlated with increased activity in the bilateral rACC, specifically in the opioid rich pgACC and right anterior INS [35, 36, 63]. The authors concluded that the pgACC acts as a top down modulator by reducing the anxiety and negative emotions that surface when anticipating a painful stimulus [64, 65]. Prior research has suggested that the sensory representation of a painful stimulus is represented in the posterior INS before being represented again in the ipsilateral anterior INS, and then lastly in the right anterior INS which transforms a painful sensation into a cognition [66–68].

In addition to increased activity in certain areas of the ACC and INS, many studies have reported increased activity in the DLPFC during placebo analgesia. In a study by Watson et al. [69] the functional brain activity in the placebo conditioning phase was compared to activity during the placebo analgesia period. In the experimental group, anticipation of an analgesic response activated frontocingulate structures, including the DLPFC, along with the anterior MCC, mid-frontal cortex, and dorsal posterior cingulate cortex. The increased magnitude of DLPFC activity was correlated with pain responsiveness [20, 36, 45, 69, 70]. The DLPFC region stores memories of past placebo effectiveness and is thought to exert control over the ACC in order to modulate pain perception during placebo analgesia [20, 32, 71].

Lui et al. [70] performed a study to investigate acquisition and evocation phases of conditioned placebo analgesia in healthy participants receiving brief laser heat stimuli delivered to one foot (either right or left) and preceded by different visual cues, signaling either painful stimuli alone, or painful stimuli accompanied by a (sham) analgesic procedure (Fig. 1a). In the evocation phase, when all stimuli were surreptitiously set at the same painful intensity to test for placebo effects, participants showed robust behavioral conditioned placebo analgesic effects (Fig. 1b). During the first conditioning trials (acquisition phase) in which participants received red cues associated with painful stimuli and green cues paired with analgesic stimuli, progressive signal increases over time were found during the anticipatory phase of analgesia (*green* stimuli) compared to when pain was anticipated (*red* stimuli) in the medial prefrontal focus centered on medial area BA8, and in the bilateral lateral prefrontal foci (Fig. 1c). These frontal foci were adjacent to, and partially overlapped, those active during anticipation of analgesia in the evocation session. Signal changes observed in the frontal foci were related to the magnitude of the individual placebo analgesic response, and those foci active during placebo analgesia (see Fig. 1d). Specifically, a large focus in the right PFC showed activity related to analgesia, irrespective of the expected side of stimulation [70].

The study concluded that the PFC specifically the medial PFC and right DLPFC foci, are responsible for the development of placebo analgesia. In extension to the critical finding, replicated from Watson et al. [69], that frontal cortex activity occurred in both the conditioning and post-conditioning sessions, Lui et al. [70] concluded that the signals progressively increased over time in the conditioning trials which suggests long lasting placebo analgesia may intentionally be boosted over time. TMS studies have also concluded that the DLPFC plays a crucial role in the development of a placebo analgesic response. In one such study, repetitive TMS disrupted DLPFC function which, in turn, abolished the placebo effect to a heat pain paradigm stimulus [45]. This study's

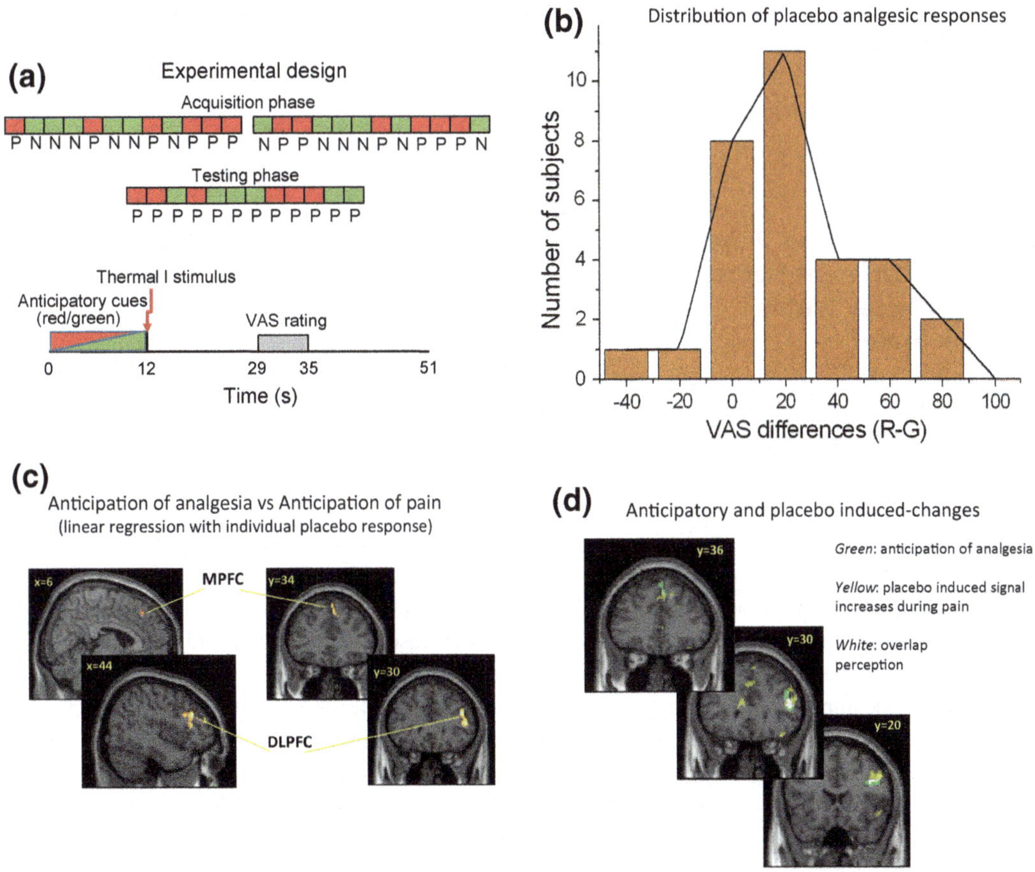

Fig. 1 Different visual cues signaled either painful stimuli alone, or painful stimuli accompanied by a (sham) analgesic procedure during the acquisition phase. In the evocation phase, all stimuli were surreptitiously set at the same painful intensity to test for placebo effects (**a**). For healthy participants receiving brief laser heat stimuli delivered to one foot (either *right* or *left*), pain was assessed using a visual analogue scale (VAS). Following the evocation (conditioning) phase, behavioral placebo analgesic effects were observed ranging from no changes to robust pain reductions (**b**). During the first conditioning trials (acquisition phase) in which participants received red cues associated with painful stimuli and green cues paired with analgesic stimuli, progressive signal increases over time were found during the anticipatory phase of analgesia (*green* stimuli) compared to when pain was anticipated (*red* stimuli) in the medial prefrontal focus centered on medial area BA8, and in the bilateral lateral prefrontal foci (**c**). Signal changes observed in the frontal foci during the anticipatory phase were related to the magnitude of the individual placebo analgesic response, and those foci active during placebo analgesia (**d**). Adapted from Lui et al. [70]

findings and other recent findings suggest that the altering of the excitability of the DLPFC can essentially turn off the placebo effect [72].

Huber et al. [73] retrospectively contacted the same subjects enrolled in the study by Lui et al. [70] to explore the connection between conditioning and hypnosis. Interestingly, high hypnotic susceptibility was associated with increased anticipatory activity in the right DLPFC and the ability to reduce functional connectivity of that focus with other brain regions such as the anterior MCC and medial PFC that are involved in emotional and evaluative pain processes. Subjects with low hypnotic susceptibility presented with a reverse pattern of fMRI changes and functional connectivity.

Finally, functional imaging studies have revealed that spinal cord changes occur during placebo analgesia. In one such study, spinal cord fMRI findings revealed a reduced response in the ipsilateral dorsal horn for those subjects assigned to the placebo conditioning group as compared to control nonconditioning group [74].

3.3 Do Nocebo and Placebo Effects Elicit Different Neural Responses Within the Brain?

Different behavioral studies have indicated that nocebo responses are formed in a faster manner than placebo ones and tend to not extinguish [5, 75–77].

Brain imaging studies indicated that during a nocebo response, the bilateral ACC, INS, and operculum signaled increased activity [78, 79]. The increased activation of these areas and the functional connectivity between the left frontal operculum and hippocampus with pain matrix regions suggests that the affective–cognitive pain pathway produces the nocebo effect. Other studies have replicated these findings, including a study by [80] that assessed the negative and positive expectancies associated with thermal pain by administering the μ-opioid agonist remifentanil. Negative expectancies of a thermal stimulus caused reduced activity in the sgACC. However, activity in the hippocampus and medial PFC increased, which could have resulted from the anxiety that is caused when a stimulus is anticipated to be painful [16]. In general most nocebo versus placebo studies conclude that nocebo hyperalgesia causes an increase in activity in pain associated regions and a decrease in opioid-sensitive brain regions and activation of the hippocampus [81, 82].

Spinal fMRI studies have also revealed spinal cord changes during a nocebo hyperalgesic response. In a study by Geuter, Buchel [83], negative expectations, initiated by nocebo conditioned verbal suggestion of an inert cream, led to increased exhibited ipsilateral dorsal horn activity indicating that psychological processes were related to negative expectations and nocebos act first on the spinal cord before regions within the brain.

Interestingly these responses can occur outside of conscious awareness. Neural pathways involved in nonconscious activation of conditioned nocebo pain responses include increased activation of the Th, Amy, and hippocampus. These results have a strong impact in clinical practice because they suggest that patients may unintentionally activate pathways that are responsible for making their pain worse [4].

3.4 Individual Differences in Placebo Responding

Not everyone responds to a placebo in the same way, in fact, some do not respond at all. Many placebo research studies conclude that the magnitude of brain and neurotransmitter activity is associated with pain rating scores and increased or decreased brain activity in other connected regions. A study by Wager et al. [84] reanalyzed previously collected fMRI data in order to predict, a priori, individual patterns of brain activity that occurred while participants were anticipating pain relief from a placebo. There was a substantial amount of individual variance in cortical anticipatory activity, including increased functioning in the frontopartietal network and decreased signaling in the posterior INS and temporal networks. Moreover, the decreased activity in the limbic and paralimbic regions, evident during placebo analgesic conditions, led to the conclusion that activity in areas involved in emotional appraisal circuity measure individual variation in placebo analgesia better than pain processing or cognitive control linked brain areas [84].

Differences in individual placebo responses have also been discovered by assessing chronic pain patients. In Hashmi et al. [85] the placebo response in patients with chronic back pain was measured. Functional connectivity between the left medial PFC and bilateral INS and high frequency oscillations in the DLPFC consistently predicted whether the participant would be classified as having post-treatment persistent chronic back pain or decreasing chronic back pain (placebo nonresponders vs. responders). The functional connectivity findings of this study indicate that prefrontal cognitive and pain processing regions interact to determine how those with chronic pain will respond to a placebo. In a later study by Hashmi et al. [86] individual differences in functional connectivity illuminated how well the brain converts positive expectations of chronic knee pain relief, stimulated by verbal suggestion, from an acupuncture treatment into analgesia.

Studies that assess individual differences in gray and white matter summarize additional individual ways of placebo processing. In one such study by Stein et al. [87], findings from a diffusion tensor MRI indicated individual differences in white matter. The fractional anisotropy (FA), an index of white matter integrity, of the right DLPFC, left rACC, and PAG was positivity correlated with individual placebo analgesic effects. More specifically, increased FA values in the white matter tracts that connect the PAG with the rACC and DLPFC predicted the magnitude of placebo analgesic responses. White matter integrity within and between regions of the descending pain modulatory pathways are linked with endogenous pain control [87].

As for gray matter, findings from a study by Schweinhardt et al. [88] that analyzed MRI voxel-based morphometry and pain ratings suggested that there is a relationship between GMD in the ventral striatum, INS, and PFC and the magnitude of placebo response. Moreover, since GMD in the ventral striatum and PFC is closely linked with dopamine related personality traits and reward anticipation, dopamine neurotransmission may play a key role in eliciting stronger placebo analgesic responses.

Connectivity analyses to illustrate individual opioid system differences, conducted by Wager et al. [37], showed increased connectivity between the PAG and rACC and increased functional coupling between various limbic and PFC regions. The placebo induces opioid activity across multiple regions, evident by the connectivity that varies based on the individual differences. An individual's response magnitude is consistent across the connected regions [37].

Another potential approach to estimate placebo responsiveness includes studies of brain imaging and genetics. Conditioning effects, evident by pain ratings, were associated with regional homogeneity (ReHo), which is a measure of neural coherence, in the ventral striatum. In addition to ReHo in the ventral striatum, the number of Met alleles at rs4680 in catechol-O-methyltransferase (*COMT*), and "Openness" personality scores all significantly fell onto a regression model that accounted for 59% of the variance for predicting conditioned placebo analgesic effects.

In addition to fMRI technology findings, results from PET scans displaying changes associated with endogenous opioid neurotransmission can potentially be used to predict placebo responsiveness. The magnitude of the μ-opioid system activation first evoked by placebo expectancies, in the DLPFC, pgACC, AI, and NAcc can be used as a biomarker of placebo responses. Indeed, NAcc activation correlated negatively with traits on the Positive Affect and Negative Affect scales (PANAS). Both the amount of painful stimulation and the affective characteristics of participants accounted for 40–68% of the individual variance of the neurochemical responses to the placebo effect, suggesting the importance of considering personality traits along with biochemical changes occurring in the brain [42].

3.5 Expectancies Modulate Brain Responses to Painkillers

The relationship between analgesic drug effects and individuals' expectancies is fundamental given that medications are not given in a vacuum but rather in a clinical encounter whereby anticipations of treatments and desires of benefit are pervasive.

Several studies have attempted to explore the relationship between drug and expectancies effects using the balanced placebo design [89, 5] in which instructions about the drug (told drug vs. told placebo) is one factor and the actual drug (given drug vs. given placebo) is the other factor, allowing testing of interactions between drug and expectancies of receiving a certain drug. Combining this design with neuroimaging helps advance understanding of how expectancies shape not only reported pain but also brain network associated changes [79].

Some recent brain imaging studies in the field of pain show the role of expectancies in modulating drug responses [90–92]. For example, Kong et al. (2009) combined verbal instruction (positive instruction vs. neutral instruction) with acupuncture treatment (real vs. sham) [90]. As hypothesized, pain ratings were significantly lower in the positive instruction groups compared with the neutral instruction groups, with no evidence of an interaction between instructions and treatment. Significant fMRI activity was associated with the main effect of instruction as well as an interaction with treatment in the bilateral inferior frontal gyrus and left medial frontal gyrus. Atlas et al. (2012) investigated behavioral pain ratings, using a balanced placebo design, and neural signals, using an open-hidden design, separating the effect of remifentanil and instruction with a pharmacokinetic model [91]. Remifentanil and instructions both reduced pain ratings, but the effect of remifentanil on pain reports and fMRI activity did not interact with expectancy effects. By contrary, a study by Schenk et al. (2014) provided some evidence of interactions between placebo and drug effects at the level of behavioral reports and neuronal changes while experiencing pain [92]. The authors investigated pain ratings and neural signals using fMRI and a within-subject balanced placebo design combining topical treatment (received lidocaine/prilocaine vs. received control cream) with distinct instructions (told lidocaine/prilocaine vs. told control cream). There was a significant treatment effect of lidocaine/prilocaine on pain ratings and in the anterior INS, as well as a significant interaction effect on pain ratings and in the rACC, anterior INS and the ventral striatum, suggesting that expectancy and drug effects may not be merely additive [92]. Clearly, additional research is necessary to further elucidate the neural mechanisms of interactions between drug and expectancy effects across different pain disorders.

4 Conclusion

The neural responses involved during the anticipatory phase and placebo and nocebo effects can greatly illuminate the human mechanisms of pain perception and decoding the experience of pain. For both the anticipatory phase and the nocebo and placebo effects, specific neural processes can be detected in the central nervous system at the level of cortical, subcortical and spinal activity. During the anticipatory phase of impending noxious stimulations, there is a variety of neural responses that can be attributed to prediction of pain delivery and intensity. Furthermore, expectancies, individual differences in psychological traits and past experiences, including a medical history of chronic pain, appear to shape anticipation and modulation of pain. The body of studies on neural correlates associated with the placebo and nocebo effects sheds light on inhibitory and facilitatory mechanisms of pain perception therefore helping to advance our knowledge of distinct pain phenotypes, and responses to painkillers and other non-pharmacological interventions.

Acknowledgements This project was supported by University of Maryland Baltimore (LC), the National Institute of Dental and Craniofacial Research (NIDCR; R01DE025946-01, LC) and International Association for Study of Pain (Early Research Grant [LC]).

References

1. Harper D. Anticipation. Online Etymology Dictionary.
2. Colagiuri B, Schenk LA, Kessler MD, Dorsey SG, Colloca L. The placebo effect: from concepts to genes. Neuroscience. 2015;307:171–90. doi:10.1016/j.neuroscience.2015.08.017.
3. Colloca L, Klinger R, Flor H, Bingel U. Placebo analgesia: psychological and neurobiological mechanisms. Pain. 2013;154(4):511–4. doi:10.1016/j.pain.2013.02.002.
4. Colloca L, Grillon C. Understanding placebo and nocebo responses for pain management. Curr Pain Headache Rep. 2014;18(6):419. doi:10.1007/s11916-014-0419-2.
5. Colloca L, Sigaudo M, Benedetti F. The role of learning in nocebo and placebo effects. Pain. 2008;136(1–2):211–8. doi:10.1016/j.pain.2008.02.006.

6. Colloca L, Benedetti F. How prior experience shapes placebo analgesia. Pain. 2006;124(1–2):126–33. doi:10.1016/j.pain.2006.04.005.
7. Colloca L, Benedetti F. Placebo analgesia induced by social observational learning. Pain. 2009;144(1–2):28–34. doi:10.1016/j.pain.2009.01.033.
8. Medoff ZM, Colloca L. Placebo analgesia: understanding the mechanisms. Pain Manag. 2015;5 (2):89–96. doi:10.2217/pmt.15.3.
9. Ploner M, Lee MC, Wiech K, Bingel U, Tracey I. Prestimulus functional connectivity determines pain perception in humans. Proc Natl Acad Sci U S A. 2010;107(1):355–60. doi:10.1073/pnas.0906186106.
10. Bushnell MC, Ceko M, Low LA. Cognitive and emotional control of pain and its disruption in chronic pain. Nat Rev Neurosci. 2013;14(7):502–11. doi:10.1038/nrn3516.
11. Ploghaus A, Becerra L, Borras C, Borsook D. Neural circuitry underlying pain modulation: expectation, hypnosis, placebo. Trends Cogn Sci. 2003;7(5):197–200.
12. Price DD, Milling LS, Kirsch I, Duff A, Montgomery GH, Nicholls SS. An analysis of factors that contribute to the magnitude of placebo analgesia in an experimental paradigm. Pain. 1999;83(2):147–56.
13. Staub E, Tursky B, Schwartz GE. Self-control and predictability: their effects on reactions to aversive stimulation. J Pers Soc Psychol. 1971;18(2):157–62.
14. Palermo S, Benedetti F, Costa T, Amanzio M. Pain anticipation: an activation likelihood estimation meta-analysis of brain imaging studies. Hum Brain Mapp. 2015;36 (5):1648–61. doi:10.1002/hbm.22727.
15. Barlow DH, Chorpita BF, Turovsky J. Fear, panic, anxiety, and disorders of emotion. Nebr Symp Motiv. 1996;43:251–328.
16. Ploghaus A, Narain C, Beckmann CF, Clare S, Bantick S, Wise R, Matthews PM, Rawlins JN, Tracey I. Exacerbation of pain by anxiety is associated with activity in a hippocampal network. J Neurosci. 2001;21(24):9896–903.
17. Ploghaus A, Tracey I, Gati JS, Clare S, Menon RS, Matthews PM, Rawlins JN. Dissociating pain from its anticipation in the human brain. Science. 1999;284(5422):1979–81.
18. Chua P, Krams M, Toni I, Passingham R, Dolan R. A functional anatomy of anticipatory anxiety. Neuroimage. 1999;9(6 Pt 1):563–71. doi:10.1006/nimg.1999.0407.
19. Fairhurst M, Wiech K, Dunckley P, Tracey I. Anticipatory brainstem activity predicts neural processing of pain in humans. Pain. 2007;128(1–2):101-10. doi:10.1016/j.pain.2006.09.001.
20. Wager TD, Rilling JK, Smith EE, Sokolik A, Casey KL, Davidson RJ, Kosslyn SM, Rose RM, Cohen JD. Placebo-induced changes in FMRI in the anticipation and experience of pain. Science. 2004;303 (5661):1162–7. doi:10.1126/science.1093065.
21. Ungless MA, Magill PJ, Bolam JP. Uniform inhibition of dopamine neurons in the ventral tegmental area by aversive stimuli. Science. 2004;303 (5666):2040–42. doi:10.1126/science.1093360.
22. Dunckley P, Wise RG, Fairhurst M, Hobden P, Aziz Q, Chang L, Tracey I. A comparison of visceral and somatic pain processing in the human brainstem using functional magnetic resonance imaging. J Neurosci. 2005;25(32):7333–41. doi:10.1523/JNEUROSCI.1100-05.2005.
23. Hsieh JC, Stone-Elander S, Ingvar M. Anticipatory coping of pain expressed in the human anterior cingulate cortex: a positron emission tomography study. Neurosci Lett. 1999;262(1):61–4.
24. Simpson JR, Jr., Drevets WC, Snyder AZ, Gusnard DA, Raichle ME. Emotion-induced changes in human medial prefrontal cortex: II. During anticipatory anxiety. Proc Natl Acad Sci U S A. 2001;98 (2):688–93. doi:10.1073/pnas.98.2.688.
25. Ongur D, An X, Price JL. Prefrontal cortical projections to the hypothalamus in macaque monkeys. J Comp Neurol. 1998;401(4):480–505.
26. Porro CA, Baraldi P, Pagnoni G, Serafini M, Facchin P, Maieron M, Nichelli P. Does anticipation of pain affect cortical nociceptive systems? J Neurosci. 2002;22(8):3206–14. doi:20026310.
27. Koyama T, McHaffie JG, Laurienti PJ, Coghill RC. The subjective experience of pain: where expectations become reality. Proc Natl Acad Sci U S A. 2005;102 (36):12950–55. doi:10.1073/pnas.0408576102.
28. Smith AP, Henson RN, Dolan RJ, Rugg MD. fMRI correlates of the episodic retrieval of emotional contexts. Neuroimage. 2004;22(2):868–78. doi:10.1016/j.neuroimage.2004.01.049.
29. Anderson AK, Phelps EA. Lesions of the human amygdala impair enhanced perception of emotionally salient events. Nature. 2001;411(6835):305–9. doi:10.1038/35077083.
30. Naliboff BD, Berman S, Suyenobu B, Labus JS, Chang L, Stains J, Mandelkern MA, Mayer EA (2006) Longitudinal change in perceptual and brain activation response to visceral stimuli in irritable bowel syndrome patients. Gastroenterology. 131 (2):352–365. doi:10.1053/j.gastro.2006.05.014.
31. Burgmer M, Petzke F, Giesecke T, Gaubitz M, Heuft G, Pfleiderer B. Cerebral activation and catastrophizing during pain anticipation in patients with fibromyalgia. Psychosom Med. 2011;73 (9):751–9. doi:10.1097/PSY.0b013e318236588a.
32. Lorenz J, Minoshima S, Casey KL. Keeping pain out of mind: the role of the dorsolateral prefrontal cortex in pain modulation. Brain. 2003;126(Pt 5):1079–91.
33. Burgmer M, Pfleiderer B, Maihofner C, Gaubitz M, Wessolleck E, Heuft G, Pogatzki-Zahn E. Cerebral mechanisms of experimental hyperalgesia in fibromyalgia. Eur J Pain. 2012;16(5):636–47. doi:10.1002/j.1532-2149.2011.00058.x.
34. Benedetti F, Amanzio M. Mechanisms of the placebo response. Pulm Pharmacol Ther. 2013;26 (5):520–3. doi:10.1016/j.pupt.2013.01.006.
35. Petrovic P, Kalso E, Petersson KM, Ingvar M. Placebo and opioid analgesia—imaging a shared

neuronal network. Science. 2002;295(5560):1737–40. doi:10.1126/science.1067176.
36. Zubieta JK, Bueller JA, Jackson LR, Scott DJ, Xu Y, Koeppe RA, Nichols TE, Stohler CS. Placebo effects mediated by endogenous opioid activity on mu-opioid receptors. J Neurosci. 2005;25(34):7754–62. doi:10.1523/JNEUROSCI.0439-05.2005.
37. Wager TD, Scott DJ, Zubieta JK. Placebo effects on human mu-opioid activity during pain. Proc Natl Acad Sci U S A. 2007;104(26):11056–61. doi:10.1073/pnas.0702413104.
38. Scott DJ, Stohler CS, Egnatuk CM, Wang H, Koeppe RA, Zubieta JK. Placebo and nocebo effects are defined by opposite opioid and dopaminergic responses. Arch Gen Psychiatry. 2008;65(2):220–31. doi:10.1001/archgenpsychiatry.2007.34.
39. Levine JD, Gordon NC, Fields HL. The mechanism of placebo analgesia. Lancet. 1978;2(8091):654–7.
40. Amanzio M, Benedetti F. Neuropharmacological dissection of placebo analgesia: expectation-activated opioid systems versus conditioning-activated specific subsystems. J Neurosci. 1999;19(1):484–94.
41. Eippert F, Bingel U, Schoell ED, Yacubian J, Klinger R, Lorenz J, Buchel C. Activation of the opioidergic descending pain control system underlies placebo analgesia. Neuron. 2009;63(4):533–43. doi:10.1016/j.neuron.2009.07.014.
42. Zubieta JK, Yau WY, Scott DJ, Stohler CS (2006) Belief or Need? Accounting for individual variations in the neurochemistry of the placebo effect. Brain Behav Immun. 20(1):15–26. doi:10.1016/j.bbi.2005.08.006.
43. Bingel U, Lorenz J, Schoell E, Weiller C, Buchel C. Mechanisms of placebo analgesia: rACC recruitment of a subcortical antinociceptive network. Pain. 2006;120(1–2):8-15. doi:10.1016/j.pain.2005.08.027.
44. Fields H. State-dependent opioid control of pain. Nat Rev Neurosci. 2004;5(7):565–75. doi:10.1038/nrn1431.
45. Krummenacher P, Candia V, Folkers G, Schedlowski M, Schonbachler G. Prefrontal cortex modulates placebo analgesia. Pain. 2010;148(3):368–74. doi:10.1016/j.pain.2009.09.033.
46. de la Fuente-Fernandez R, Schulzer M, Stoessl AJ. The placebo effect in neurological disorders. Lancet Neurol. 2002;1(2):85–91.
47. Garris PA, Kilpatrick M, Bunin MA, Michael D, Walker QD, Wightman RM. Dissociation of dopamine release in the nucleus accumbens from intracranial self-stimulation. Nature. 1999;398(6722):67–9. doi:10.1038/18019.
48. Fiorillo CD, Tobler PN, Schultz W. Discrete coding of reward probability and uncertainty by dopamine neurons. Science. 2003;299(5614):1898–902. doi:10.1126/science.1077349.
49. de la Fuente-Fernandez R, Ruth TJ, Sossi V, Schulzer M, Calne DB, Stoessl AJ. Expectation and dopamine release: mechanism of the placebo effect in Parkinson's disease. Science. 2001;293(5532):1164–6. doi:10.1126/science.1060937.
50. de la Fuente-Fernandez R, Stoessl AJ. The placebo effect in Parkinson's disease. Trends Neurosci. 2002;25(6):302–306.
51. Strafella AP, Ko JH, Monchi O. Therapeutic application of transcranial magnetic stimulation in Parkinson's disease: the contribution of expectation. Neuroimage. 2006;31(4):1666–72. doi:10.1016/j.neuroimage.2006.02.005.
52. Benedetti F, Colloca L, Torre E, Lanotte M, Melcarne A, Pesare M, Bergamasco B, Lopiano L. Placebo-responsive Parkinson patients show decreased activity in single neurons of subthalamic nucleus. Nat Neurosci. 2004;7(6):587–8. doi:10.1038/nn1250.
53. Benedetti F, Amanzio M, Thoen W. Disruption of opioid-induced placebo responses by activation of cholecystokinin type-2 receptors. Psychopharmacology (Berl). 2011;213(4):791–7. doi:10.1007/s00213-010-2037-y.
54. Benedetti F. Mechanisms of placebo and placebo-related effects across diseases and treatments. Annu Rev Pharmacol Toxicol. 2008;48:33–60. doi:10.1146/annurev.pharmtox.48.113006.094711.
55. Enck P, Benedetti F, Schedlowski M. New insights into the placebo and nocebo responses. Neuron. 2008;59(2):195–206. doi:10.1016/j.neuron.2008.06.030.
56. Benedetti F, Amanzio M, Maggi G. Potentiation of placebo analgesia by proglumide. Lancet. 1995;346(8984):1231.
57. Benedetti F, Amanzio M. The neurobiology of placebo analgesia: from endogenous opioids to cholecystokinin. Prog Neurobiol. 1997;52(2):109–25.
58. Benedetti F. Placebo analgesia. Neurol Sci. 2006;27(Suppl 2):S100–2. doi:10.1007/s10072-006-0580-4.
59. Price DD, Craggs J, Verne GN, Perlstein WM, Robinson ME. Placebo analgesia is accompanied by large reductions in pain-related brain activity in irritable bowel syndrome patients. Pain. 2007;127(1–2):63–72. doi:10.1016/j.pain.2006.08.001.
60. Atlas LY, Wager TD. A meta-analysis of brain mechanisms of placebo analgesia: consistent findings and unanswered questions. Handb Exp Pharmacol. 2014;225:37–69. doi:10.1007/978-3-662-44519-8_3.
61. Wager TD, Fields, H. Placebo analgesia. In: Textbook of pain. 2013. p 362–73.
62. Kong J, Gollub RL, Rosman IS, Webb JM, Vangel MG, Kirsch I, Kaptchuk TJ. Brain activity associated with expectancy-enhanced placebo analgesia as measured by functional magnetic resonance imaging. J Neurosci. 2006;26(2):381–8. doi:10.1523/JNEUROSCI.3556-05.2006.
63. Amanzio M, Benedetti F, Porro CA, Palermo S, Cauda F. Activation likelihood estimation meta-analysis of brain correlates of placebo analgesia

in human experimental pain. Hum Brain Mapp. 2013;34(3):738–52. doi:10.1002/hbm.21471.
64. Kalisch R, Wiech K, Critchley HD, Seymour B, O'Doherty JP, Oakley DA, Allen P, Dolan RJ. Anxiety reduction through detachment: subjective, physiological, and neural effects. J Cogn Neurosci. 2005;17 (6):874–83. doi:10.1162/0898929054021184.
65. Vase L, Robinson ME, Verne GN, Price DD. Increased placebo analgesia over time in irritable bowel syndrome (IBS) patients is associated with desire and expectation but not endogenous opioid mechanisms. Pain. 2005;115(3):338–47. doi:10.1016/j.pain.2005.03.014.
66. Craig AD. How do you feel? Interoception: the sense of the physiological condition of the body. Nat Rev Neurosci. 3(8):655–66. doi:10.1038/nrn894.
67. Craig AD. A new view of pain as a homeostatic emotion. Trends Neurosci. 2003;26(6):303–7.
68. Craig AD, Chen K, Bandy D, Reiman EM. Thermosensory activation of insular cortex. Nat Neurosci. 2000;3(2):184–90. doi:10.1038/72131.
69. Watson A, El-Deredy W, Iannetti GD, Lloyd D, Tracey I, Vogt BA, Nadeau V, Jones AK. Placebo conditioning and placebo analgesia modulate a common brain network during pain anticipation and perception. Pain. 2009;145(1–2):24–30. doi:10.1016/j.pain.2009.04.003.
70. Lui F, Colloca L, Duzzi D, Anchisi D, Benedetti F, Porro CA. Neural bases of conditioned placebo analgesia. Pain. 2010;151(3):816–24. doi:10.1016/j.pain.2010.09.021.
71. Craggs JG, Price DD, Verne GN, Perlstein WM, Robinson MM. Functional brain interactions that serve cognitive-affective processing during pain and placebo analgesia. Neuroimage. 2007;38(4):720–9. doi:10.1016/j.neuroimage.2007.07.057.
72. Egorova N, Yu R, Kaur N, Vangel M, Gollub RL, Dougherty DD, Kong J, Camprodon JA. Neuromodulation of conditioned placebo/nocebo in heat pain: anodal vs cathodal transcranial direct current stimulation to the right dorsolateral prefrontal cortex. Pain. 2015;156(7):1342–7. doi:10.1097/j.pain.0000000000000163.
73. Huber A, Lui F, Porro CA. Hypnotic susceptibility modulates brain activity related to experimental placebo analgesia. Pain. 2013;154(9):1509–18. doi:10.1016/j.pain.2013.03.031.
74. Eippert F, Finsterbusch J, Bingel U, Buchel C. Direct evidence for spinal cord involvement in placebo analgesia. Science. 2009;326(5951):404. doi:10.1126/science.1180142.
75. Colloca L, Petrovic P, Wager TD, Ingvar M, Benedetti F. How the number of learning trials affects placebo and nocebo responses. Pain. 2010;151(2):430–9. doi:10.1016/j.pain.2010.08.007.
76. Jensen K, Kirsch I, Odmalm S, Kaptchuk TJ, Ingvar M. Classical conditioning of analgesic and hyperalgesic pain responses without conscious awareness. Proc Natl Acad Sci U S A. 2015;112 (25):7863–7. doi:10.1073/pnas.1504567112.
77. Colagiuri B, Quinn VF, Colloca L. Nocebo hyperalgesia, partial reinforcement, and extinction. J Pain. 2015. doi:10.1016/j.jpain.2015.06.012.
78. Sawamoto N, Honda M, Okada T, Hanakawa T, Kanda M, Fukuyama H, Konishi J, Shibasaki H. Expectation of pain enhances responses to nonpainful somatosensory stimulation in the anterior cingulate cortex and parietal operculum/posterior insula: an event-related functional magnetic resonance imaging study. J Neurosci. 2000;20(19):7438–45.
79. Keltner JR, Furst A, Fan C, Redfern R, Inglis B, Fields HL. Isolating the modulatory effect of expectation on pain transmission: a functional magnetic resonance imaging study. J Neurosci. 2006;26 (16):4437–43. doi:10.1523/JNEUROSCI.4463-05.2006.
80. Bingel U, Wanigasekera V, Wiech K, Ni Mhuircheartaigh R, Lee MC, Ploner M, Tracey I. The effect of treatment expectation on drug efficacy: imaging the analgesic benefit of the opioid remifentanil. Sci Transl Med. 2011;3(70):70ra14. doi:10.1126/scitranslmed.3001244.
81. Colloca L, Finniss D. Nocebo effects, patient-clinician communication, and therapeutic outcomes. JAMA. 2012;307(6):567–8. doi:10.1001/jama.2012.115.
82. Jensen KB, Petrovic P, Kerr CE, Kirsch I, Raicek J, Cheetham A, Spaeth R, Cook A, Gollub RL, Kong J, Kaptchuk TJ. Sharing pain and relief: neural correlates of physicians during treatment of patients. Mol Psychiatry. 2014;19(3):392–8. doi:10.1038/mp.2012.195.
83. Geuter S, Buchel C. Facilitation of pain in the human spinal cord by nocebo treatment. J Neurosci. 2013;33 (34):13784–90. doi:10.1523/JNEUROSCI.2191-13.2013.
84. Wager TD, Atlas LY, Leotti LA, Rilling JK. Predicting individual differences in placebo analgesia: contributions of brain activity during anticipation and pain experience. J Neurosci. 2011;31(2):439–52. doi:10.1523/JNEUROSCI.3420-10.2011.
85. Hashmi JA, Baria AT, Baliki MN, Huang L, Schnitzer TJ, Apkarian AV. Brain networks predicting placebo analgesia in a clinical trial for chronic back pain. Pain. 2012;153(12):2393–402. doi:10.1016/j.pain.2012.08.008.
86. Hashmi JA, Kong J, Spaeth R, Khan S, Kaptchuk TJ, Gollub RL. Functional network architecture predicts psychologically mediated analgesia related to treatment in chronic knee pain patients. J Neurosci. 2014;34(11):3924–36. doi:10.1523/JNEUROSCI.3155-13.2014.
87. Stein N, Sprenger C, Scholz J, Wiech K, Bingel U. White matter integrity of the descending pain modulatory system is associated with interindividual differences in placebo analgesia. Pain. 2012;153 (11):2210–7. doi:10.1016/j.pain.2012.07.010.
88. Schweinhardt P, Seminowicz DA, Jaeger E, Duncan GH, Bushnell MC. The anatomy of the

mesolimbic reward system: a link between personality and the placebo analgesic response. J Neurosci. 2009;29(15):4882–7. doi:10.1523/JNEUROSCI.5634-08.2009.
89. Ross S, Krugman AD, Lyerly SB, Clyde ADJ. Drugs and placebos: a model design. Psychol Rep. 1962;10(2):383–92.
90. Kong J, Kaptchuk TJ, Polich G, Kirsch I, Vangel, Gollub R. Expectancy and treatment interactions: a dissociation between acupuncture analgesia and expectancy evoked placebo analgesia. Neuroimage. 2009;45(3): 940–9. doi:10.1016/j.neuroimage.2008.12.025.
91. Atlas LY, Wager TD. How expectations shape pain. Neurosci Letters. 2012;520(2):140–8. doi:10.1016/j.neulet.2012.03.03.
92. Schenk LA, Sprenger C, Geuter S, Büchel C. Expectation requires treatment to boost pain relief: an fMRI study. Pain. 2014;155(1):150–7. doi:10.1016/j.pain.2013.09.024.

Neuroimaging of Chronic Pain

8

Martin Diers and Christopher Milde

Abstract

In this chapter, we give an overview about the application of MRI-based neuroimaging approaches in the study of chronic pain. We emphasize the lack of neuroimaging studies investigating clinically relevant aspects of pain and how to overcome the need of finding biomarkers (brain signatures) for chronic pain conditions by cutting-edge imaging techniques. Moreover, we present techniques, which can help predict the outcome of pain treatments or the course of pain progression in the individual patient. We start with an introduction about the phenomenology of pain and give an overview about the technical and conceptual underpinnings of imaging techniques in the study of chronic pain. We emphasize methods that help to understand the transition of acute to chronic pain or to predict the development of chronic pain states in the context of therapeutic interventions. Further, we present selected experimental pain induction methods, which are important for the investigation of clinical pain signs (allodynia, hyperalgesia). We illustrate imaging acquisition and analysis methods enabling to assess spontaneous ongoing pain, which is the clinically most important aspect of chronic pain. Moreover, we accentuate that the perception of pain in chronic pain is often just weakly correlated with the temporal pattern of the stimulation and therefore needs suitable imaging methods. In the following sections,

M. Diers (✉)
Department of Psychosomatic Medicine and Psychotherapy, LWL University Hospital, Ruhr-University Bochum, Alexandrinenstrasse 1-3, 44791 Bochum, Germany
e-mail: martin.diers@ruhr-uni-bochum.de

C. Milde
Department of Cognitive and Clinical Neuroscience, Central Institute of Mental Health/Medical Faculty Mannheim, Heidelberg University, Square J5, 68159 Mannheim, Germany
e-mail: christopher.milde@zi-mannheim.de

L. Saba (ed.), *Neuroimaging of Pain*, DOI 10.1007/978-3-319-48046-6_8

the application of functional and structural imaging techniques are shown in selected chronic pain syndromes [chronic back pain, fibromyalgia syndrome, phantom limb pain and complex regional pain syndrome] and commonalities and peculiarities of functional and structural imaging correlates across different types of chronic pain will be discussed. We follow with a discussion of the current view on the pain matrix, which is discussed to be a putative brain signature of pain. In this context, we discuss the advantages and disadvantages of experimentally induced pain in the study of chronic pain. This chapter concludes with a presentation of selected studies using innovative imaging methods in chronic pain patients.

Keywords
Brain imaging · Spontaneous ongoing chronic pain · Chronic back pain · Fibromyalgia syndrome · Phantom limb pain · Complex regional pain syndrome · Brain biomarkers

1 Introduction

In this chapter, we want to give an overview about the application of MRI-based neuroimaging approaches in the study of chronic pain. We want to emphasize the lack of neuroimaging studies investigating clinically relevant aspects of pain and how to overcome the need of finding biomarkers (brain signatures) for chronic pain conditions by cutting-edge imaging techniques. The majority of imaging studies focus on brain activity in response to painful experimental stimuli often assessed in healthy subjects. However, there is a lack of studies investigating spontaneous (stimulus-independent) ongoing pain, the clinically most important aspect of chronic pain. The perception of pain is further often temporally incongruent with respect to the stimulation, especially in chronic pain patients. Contrasting brain circuitry in response to acute painful stimuli with clinically-relevant pains such as spontaneous ongoing pain, allodynia or hyperalgesia can shed light on factors transforming a "physiological" pain into a "pathophysiological" chronic pain. Therefore, novel imaging acquisition and analysis techniques are needed that allow the investigation of these clinically-relevant aspects of chronic pain. Moreover, we want to illustrate the usefulness of multivariate machine learning algorithms in neuroimaging of pain, which enables the prediction of pain treatments outcomes or the course of pain progression within the individual patient with considerably higher sensitivity and specificity than traditional imaging analyses.

We start with an introduction about the phenomenology of pain and will further give an overview about the technical and conceptual underpinnings of imaging techniques in the study of chronic pain. In this context, we present functional and structural imaging methods such as functional magnetic resonance imaging (fMRI), diffusion tensor imaging (DTI), voxel-based morphometry (VBM) as well as arterial spin labeling (ASL), percept-related fMRI, multivariate pattern analysis (MVPA), and graph analytical approaches as novel tools to acquire and analyze brain responses representing clinically relevant aspects of pain. We emphasize methods that help to understand the transition of acute to chronic pain or to predict the development of chronic pain states in the context of therapeutic interventions. Further, we present selected experimental pain induction methods, which are important for the investigation of clinical pain signs (allodynia, hyperalgesia) during acute stimulation or for the comparison between chronic pain patients and healthy controls.

In the following sections, the application of functional and structural imaging techniques will be shown in selected chronic pain syndromes [chronic back pain, fibromyalgia syndrome, phantom limb pain and complex regional pain syndrome] and commonalities and peculiarities of imaging correlates across different types of chronic pain will be discussed. We will proceed with a discussion of the current view on the pain matrix, which is discussed to be a putative brain signature of pain. In this context, we will discuss the advantages and disadvantages of experimentally induced pain in the study of chronic pain. This chapter concludes with a presentation of selected studies using innovative imaging methods in chronic pain patients.

2 Phenomenology of Pain

Pain is a multidimensional experience described in terms of its three dimensions: "sensory-discriminative" (sense of the intensity, location, quality and duration of pain), "affective-motivational" (unpleasantness and the urge to escape the unpleasantness), and "cognitive-evaluative" (cognitions such as distraction, appraisal, and cultural values). This suggests that pain is not only determined by the nociceptive input, the stimulus intensity and unpleasantness alone, and that cognitions can affect both, the sensory and the affective-motivational dimension [174]. This model corresponds with the neuroanatomical distinction between the lateral and the medial pain system [5]. The terminology of the pain system is deduced from the localization of the involved nuclei of the thalamus. The lateral system of the thalamus projects to the primary and secondary somatosensory cortex (S1 and S2) discussed to represent the sensory-discriminative component of pain. The affective-motivational component is discussed to be represented by the medial thalamic system, which projects to limbic (e.g., anterior cingulate cortex, ACC) and frontal structures [6]. However, the distinction of brain regions according to the medial and lateral branches of the spinothalamic tract fails for some regions known to be important for the processing of painful experimental stimuli, such as the insula [6]. The insula is connected to the limbic system, which has projections to the ACC [238]. The results from brain imaging studies suggest that instead of an isolated brain region being involved in pain, it is rather a network of several interconnected brain areas. These brain regions comprise somatosensory (S1, S2, insula), limbic (insula, ACC) and associative (prefrontal cortex, PFC) structures receiving parallel inputs from multiple nociceptive pathways [5]. But, different chronic pain syndromes seem to be characterized by unique functional and structural brain signatures [6, 15]. Some authors have stressed that chronic pain is characterized by decreased sensory processing and enhanced emotional/cognitive processing of pain [4, 5, 10]. It should be noted that nociception and pain are related but not the same. Nociception is the transduction of nociceptive information from the periphery to the central nervous system. Pain is a perceptual phenomenon integrating and modulating several neuronal, psychological and cultural processes and requires a conscious organism. However, nociceptive input into the brain does not necessarily lead to pain and pain is not necessarily accompanied by nociception [83, 236]. For instance, around 50 kg of weight is applied on 1 cm^2 of skin, when experienced ballerinas dance with point shoes for several hours. But, professional ballerinas are capable of dissociating nociception from perceiving pain [10].

2.1 Acute, Sub-acute, and Chronic Pain

Normally, acute pain can be treated and is limited to one site of the body. The function of acute pain is to warn of imminent danger and should encourage resting behavior of the affected body part. The *International Association for the Study of Pain* (IASP)-definition of pain is either based on duration of pain or as "*pain that extends beyond the expected period of healing*" [175]. According to the tautological timeframe definitions of pain, chronic pain is defined as lasting longer than 3 or 6 months (in some definitions even 12 months). It has lost its warning functions and attends with psychosocial changes. Normally, there is not only one triggering or maintaining cause but chronic pain is rather multi-causal emphasizing the psychosocial-model for the treatment of chronic pain [79]. Sub-acute pain lasts from 1 to 3–6 months. However, it should be noted that the timeframe definitions of pain rely upon arbitrary intervals of time from onset. Moreover, it is often unclear how the "expected period of healing" can be defined (e.g., when considering rheumatoid arthritis or trigeminal neuralgia) [83]. To address the complex phenomenon of chronic pain, Flor and Turk [83] suggested a 2-dimensional model for conceptualizing acute and chronic pain including a time dimension and a physical pathology dimension. In this model, cases with a short duration or high physical pathology would be viewed as acute pain, whereas cases with low physical pain and long duration would be viewed as chronic pain.

3 Imaging Methods in Chronic Pain Research

3.1 Neuroimaging in the Light of the Fundamental Organizational Principles of the Brain

The fundamental principles of how we understand the computational performance of the brain is also reflected by the various neuroimaging acquisition and analysis techniques currently available that promise us to provide an objective measure of chronic pain [29]. The different imaging methods vary in their degree of sophistication and also according to the demands they put on the experimental design and the inferences, which can be gained from them.

In the following section, we will give a basic overview about the models of the brain used in imaging neuroscience and the imaging modalities (structural and functional) currently available and subsume the different analysis methods used within those imaging modalities. We focus on structural and functional imaging techniques, however we acknowledge that neurochemical imaging [90, 107, 239] have also been used successfully to disentangle the neurobiological mechanism of chronic pain.

3.1.1 Functional Specialization and Integration

One of the oldest debates in the history of neuroscience deals with the question whether specific mental functions are localized to specific brain regions or rely on the processing of the entire brain [75]. In the nineteenth century, localizationism became prominent through the anatomical theories of Franz Gall and other phrenologists. The localizationalists accentuates that a certain mental function can be localized to a certain brain region and discounts interactions or functional integration among brain regions [86].

Nowadays, most neuroscientists agree on a more liberal version of localizationism, by admitting that there is a certain degree of localization of mental function, whereby this functional specialization is only meaningful in relation to functional integration among specialized brain regions to achieve coherent behavior and mental function. Thus a single function is ultimately processed in a segregated network of specialized brain modules whose context-dependent union is mediated via functional integration (effective connectivity between brain regions). So far, the neuroimaging literature mainly focuses on functional specialization. However, for a comprehensive understanding of brain function neuroimaging research must take the concept of functional integration seriously by incorporating different methods for analyzing brain connectivity [86, 224]. We and others propose that the neuronal underpinnings of chronic pain can only be understood in terms of functional specialization and integration [6, 148, 236].

3.1.2 Networks of the Brain: Structural-, Functional-, and Effective Connectivity

For a proper understanding of functional specialization and integration, one has to differentiate between the concepts of structural, functional, and effective connectivity. Brain networks can be derived from either structural or functional imaging methods [224]. Structural connectivity refers to the anatomical connections that connect a set of brain regions. These structural connections are white matter (WM) tracts. Changes in structural connectivity are thought of being slow with plastic changes taking place within hours to days [224]. The properties of WM tracts (structural connectivity) can be measured in vivo by diffusion tensor imaging (DTI) [2] (see Sect. 3.3).

In functional connectivity a "connection" refers to statistical dependencies (in its simplest version patterns of correlations) in the measured signals between brain regions. There is no commitment to how that connection was caused (no causal inference) [227]. Therefore, functional connectivity reflects only poorly functional integration because the mutual information between regions could in principle be realized via

1. A *direct influence* between region A and region B
2. An *indirect influence* of region A on B mediated by another region C
3. A *shared influence* via a common region C establishing coherence between region A and B.

It should be noted that only in the first case functional connectivity reflects functional integration, which is defined as causal influence between brain regions. Thus measures of functional connectivity are much more redundant as measures of effective connectivity. Effective connectivity is always directed and rests upon a parameterized (explicit) model of causal influences between brain regions [85]. The most prominent approach to infer causal relationships (functional integration) in the brain is called dynamic causal modeling (DCM) that was originally introduced by Friston [87]. Measures of functional connectivity rely on time-series data that can be derived from different modalities such as electroencephalography (EEG), magnetoencephalography (MEG), or fMRI [224]. While functional connectivity reflects the underlying structural connectivity quite well, strong functional coupling among certain brain regions often lack strong anatomical connectivity among those brain regions [224].

3.2 Functional Imaging Techniques

It has been shown that the different vascular measures assessed with brain imaging techniques such as MRI are tightly linked to changes in local neural activity. Thus vascular-based functional imaging techniques rely on the principle of neurovascular coupling to (indirectly) infer changes in neural activity [191]. Whereby the physiological basis of neurovascular coupling remains to be resolved, it is known that it includes coordinated activities in glial and neuronal cells as well as microvasculature [118]. In this context, we present functional imaging based on the blood oxygenation level dependent contrast (BOLD-fMRI) and perfusion-based arterial spin labeling (ASL).

3.2.1 Functional Magnetic Resonance Imaging Based on the Blood Oxygenation Level Dependent Contrast (BOLD-fMRI)

BOLD-fMRI is a noninvasive technique to assess brain function using the vascular MRI signal indirectly associated with neuronal activity. The BOLD-contrast relies on an increased hemodynamic response, that is mainly due to a rise in regional cerebral blood flow (rCBF) and regional cerebral volume (rCBV), resulting in an increase of the oxyhemoglobin–deoxyhemoglobin ratio, which in turn leads to a reduction of local magnetic inhomogeneity. An increase in magnetic inhomogeneity (in the presence of high concentrations of deoxyhemoglobin) is associated with a faster dephasing of protons, causing a decrease in the MRI signal of T2*-weighted images (usually echo-planar imaging (EPI) is used). fMRI allows the mapping of spatially segregated brain function with considerably high spatial resolution and moderate temporal resolution. However, little is known about the contributions of excitatory versus inhibitory neuronal signals to the fMRI-signals [154].

3.2.2 Task-Dependent BOLD-fMRI

Standard analysis of BOLD-fMRI data relies on statistical methods for separating noise from experimentally induced systematic fluctuations in the BOLD-signal [193]. An fMRI dataset can be described as a set of cuboid elements (i.e., voxels) with an associated time-series depicting the temporal course of the BOLD-signal. When using the most commonly used general linear model (GLM) approach, the time course associated with each voxel is explained by a weighted linear combination of one or more predictor variables, including experimental regressors (e.g., the temporal course of a painful stimulation), nuisance regressors (e.g., movement parameters) plus an error term [193]. The aim of the GLM-fitting procedure is to estimate if and to what extent the modeled regressors explain the variability observed in each individual voxel (the brain comprises tens of thousands of voxels). Thus, the fitting between the modeled responses and the actually observed BOLD-responses in each individual voxel of the brain (mass-univariate approach) determine the mapping of brain function based on BOLD-fMRI. The temporal sequence of on- and off-conditions is usually convoluted by a standardized hemodynamic response function, which accounts for the temporal delay between the neuronal and hemodynamic response of the brain to the conditions [154].

It is important to emphasize that the phenomenon under investigation must be temporally separable or interrupted by rest to be measurable by task-dependent BOLD-fMRI [86]. Thus task-dependent fMRI can only reveal brain responses in healthy subjects to experimental painful stimuli or stimulus-evoked pains such as allodynia or hyperalgesia in clinical pain states. By this, a spatially consistent pattern of brain responses "the pain matrix" to painful stimuli has been identified in healthy subjects [5] (see Sect. 6).

3.2.3 Non-task-dependent BOLD-fMRI: Resting-State Activity and Connectivity

A large portion of the spontaneous temporal fluctuations in activity in a certain brain area can be explained by activity in other brain regions that are either anatomically connected to that brain region or functionally related to that brain region, when the brain is at rest (i.e., not involved in an explicit task). It has been found, that during rest, the brain's BOLD-activity can be decomposed into distinct spatial maps of resting state, which reveal high resemblance to functional networks identified during task-dependent BOLD-fMRI [223, 224]. Different resting-state networks (components) have been consistently identified across individual subjects [50], scanning sessions [111, 218] and imaging centers [24]. Those resting-state networks include components that are composed of regions that can be regarded as sensory or motor and others that comprise modules that are involved in more complex multimodal stimuli and tasks, on the basis of their task-evoked responses. For instance, the default mode network (DMN) comprises a network of brain regions revealing temporally correlated BOLD-fluctuations when an individual is focusing on internal tasks [183]. It should be noted that in resting-state fMRI, it is in principle not possible to discern if brain areas were altered due to changes in the activity within the network (non-interactive co-activation) or due to changes between the connectivity between nodes of the network (interaction). The reason for this is that in resting-state fMRI it is conceptually impossible to model out shared main-effects [189].

Using seed-based functional connectivity, one can specifically interrogate patterns of functional connectivity between an a priori defined region of interest (e.g., the insula) and the rest of the imaged brain. Therefore the BOLD time-series from the seed region is extracted and correlated with the BOLD time-series of all other acquired voxels in the brain.

3.3 Structural Imaging Methods

As with the "functional brain signature" of chronic pain, it remains a matter of debate to which degree the "structural biomarkers" of chronic pain vary across different types of chronic pain [5, 6, 170]. Brain morphometry can be used to delineate "structural brain signatures" of chronic pain, which comprises techniques specifically suited to identify global, local, or architectonic properties of the brain under normal or pathologically altered states [153]. Compared to postmortem brain morphometry, structural MRI allows the investigation of morphometric parameters in vivo. Moreover, structural MRI allows [1] the selection of well-defined study samples, [2] a parallel assessment of fMRI data or neuropsychological test in the very same subjects and [4] the conduction of a longitudinal assessment, e.g., to evaluate treatment interventions. In most cases, structural brain imaging is based on the acquisition of high-resolution T1-weighted volumes capturing the whole brain. T1-weighted images are specifically suited to delineate the borders between gray matter (GM), WM, and out-of brain tissues and allow the size-estimation of cortical and subcortical regions [153]. In the following sections, we present different structural imaging methods and associated morphometric analyses that are specifically important for the imaging of chronic pain and discuss the potential of structural imaging in revealing brain signatures of chronic pain.

3.3.1 Diffusion Tensor Imaging (DTI)

DTI allows making spatial inferences about both micro- and macrostructure of the WM by measuring the disturbance (reduction) of random (Brownian) water molecule movements (diffusion) in vivo in the brain. In analogy to fMRI, in which neuronal population activity is indirectly inferred from the vascular signal, in DTI the WM-anatomy is indirectly measured by restrictions in random water molecule diffusion [2]. In an unconstrained medium, the movement of molecules is isotropic, i.e., it can be described by a Gaussian probability distribution across all spatial directions [18]. However, water diffusion in biological tissues is not free but rather constrained by the interactions with surrounding obstacles like other molecules, membranes, and fibers yielding a preferred spatial direction on molecular movement. This directionality of diffusion is called anisotropy. Especially the structural organization of the WM puts a directional constraint on free water diffusion so that water preferentially diffuses in parallel to the WM-fiber bundles [2].

Basser et al. [18] introduced the technique of diffusion tensor imaging into MRI by going beyond the scalar measurement of only measuring the average diffusion coefficient reflecting the interaction of water molecules with biological tissues as quantified in the apparent diffusion coefficient (ADC). In comparison to the ADC, as assessed within diffusion-weighted MRI (DWI), the diffusion tensor further captures the directionality of molecular movement in every volume element of the imaged space [2, 18].

Fractional anisotropy (FA) is the most commonly used measure derived from the diffusion tensor in DTI images (Fig. 1). FA describes the directionality of fiber tracts as derived from the anisotropy of water diffusion. In brain regions revealing a low degree of internal directional organization [like GM or cerebrospinal fluid (CSF)] the FA-values approach zero, while brain regions with well-organized WM-fiber bundles reveal high FA-values [2, 19]. FA-values are typically plotted in a color-coded FA-map depicting location and orientation of fiber tracts in brain areas revealing high anisotropy such as the corpus callosum (Fig. 1).

Moreover, using the in vivo multivariate spatial information of diffusion tensors it is also possible to perform DTI-tractography. DTI-tractography allows 3-dimensional reconstruction of WM-tracts connecting different cortical and subcortical regions and thus the investigation of structural connectivity in vivo [19] (Fig. 1). The basic idea of DTI-tractography is to follow the 3-dimensional trajectories of anisotropic structures in tissue by piecing together the voxel-wise estimates of the directions of the anisotropic diffusion tensors [129]. There are rapid developments in the analysis of DTI-based

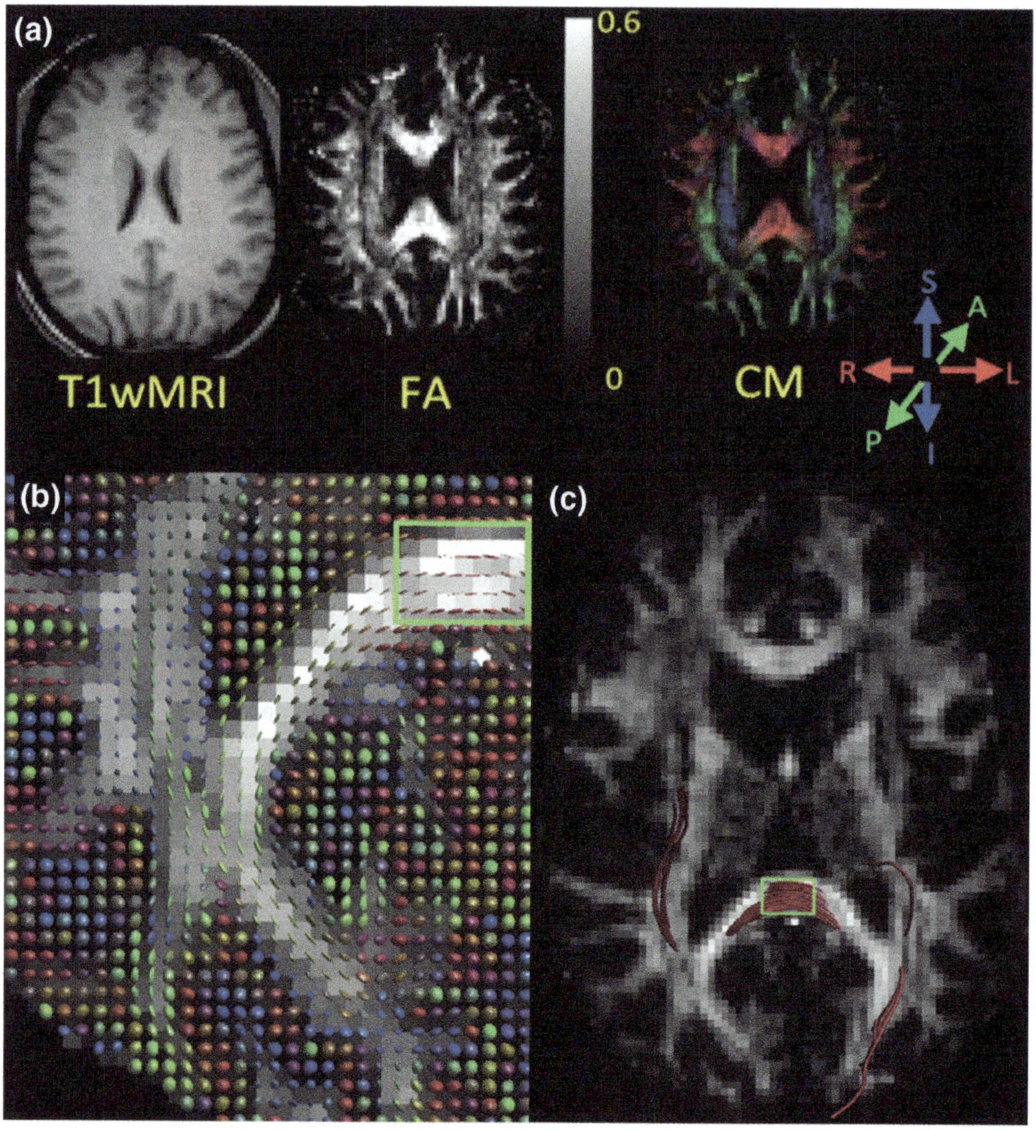

Fig. 1 **a** *left* T1-weighted structural magnetic resonance tomographic image (T1wMRI) depicting a transversal slice of a human brain spatially aligned to DTI-images presented as follows: *middle* Fractional anisotropy (FA) image: Brain regions showing high FA-values, thus revealing well-organized parallel axon arrays, show bright gray-scale values. *Right* DTI color map (CM) image depicting the spatial orientation of the principal eigenvector (principal direction of water molecule diffusion) across 3-dimensions: R-L (*red*): right-left; S-I (*blue*): superior-inferior; A-P (*green*): anterior-posterior. The intensity of the color is weighted by the FA-values to reduce the influence of non-white matter tissues. **b** FA-image with diffusion tensor ellipsoids overlaid and colored according to their orientation. The *green box* depicts a region of interest revealing high anisotropy as can be seen by the stretched shape of the diffusion tensor ellipsoid. **c** Results of a deterministic tractography showing fibers extending from the seed voxels of the region of interest. **a** adapted from Alger [2], **b**, **c** adapted from Jones [129]

tensor maps. However, the application of, e.g., probabilistic DTI-tractography in the study of chronic pain lags behind.

3.3.2 Voxel-Based Morphometry (VBM)

VBM is the most widely used automated post-processing method allowing the simultaneous identification of GM-changes in cortical and subcortical regions. VBM performs a region-wise volumetric comparison among groups of subjects. Preprocessing involves the images to be spatially normalized, segmented into three tissue types (GM, WM, CSF) and spatially smoothed, prior to performing statistical tests. The nonlinear registration into a reference space (in general the MNI-brain) serves to get rid of global differences in brain size and shape. The spatial normalization of VBM data has been optimized by only matching the individual GM maps into a GM reference to reduce the confounding effects of non-brain structural variability on the registration accuracy. Moreover, recent VBM approaches include an iterative procedure also including nonuniformities in image intensities for bias field correction. The circularity that registration needs an initial tissue classification and that tissue classification needs an initial registration is solved by alternating between tissue classification, bias correction, and registration to optimize these parameters for the generation of normalized gray matter maps [8]. The statistical inference is performed by voxel-wise comparisons between subject groups with statistical control for multiple comparison using Gaussian random field theory [27].

However, some authors pointed out that the distinction between global versus local effects is artificial and that the differences between subject groups found in VBM studies are mainly driven by differences in registration accuracy [27]. Despite some controversies according to the theoretical assumptions of VBM, the results from VBM studies have been replicated also by means of other GM morphometrics like within cortical thickness studies [117].

3.3.3 Possibilities and Limitations in Structural Imaging of Chronic Pain

Structural imaging methods can reveal alterations in GM (VBM, cortical thickness analysis) and WM (DTI and DTI tractography) associated with the chronification in various chronic pain syndromes. As opposed to adaptive changes often coming along with localized increases in GM (volume, density) such as increases in the visuomotor area MT when practicing juggling [170] most morphometric imaging studies report decreases in GM with chronification of pain ([170, 171]. However, it remains a matter of debate whether these decreases in GM reflect neurodegeneration, as proposed by some authors [3, 139] or tissue shrinkage [171]. Decreases in GM were not randomly distributed but mainly restricted to brain regions known to play a crucial role in pain modulation and pain processing [171], arguing against an undirected neurodegeneration. GM decreases are often observed in cingulate gyrus, insula, temporal lobe, and frontal/prefrontal regions. Schmidt-Wilcke [213] found GM decreases mainly in somatosensory cortex and brainstem and GM increases in the left thalamus and basal ganglia when comparing chronic back pain patients with controls. Notable, GM decreases in the brainstem did not correlate with disease duration but with intensity and unpleasantness of pain during scanning. These findings indicate that GM decreases might reflect impairments in antinociceptive circuitry rather than neurodegeneration [171].

Inferences can be drawn from structural changes with increasing chronification but also with normalization of brain structure, e.g., after therapeutic interventions. Studies investigating morphological brain changes in the course of treatment are particular powerful in shedding light on brain structures that might be relevant for the chronification of pain. Two studies on the effect of hip replacement therapy in osteoarthritis patients [99, 207] revealed that most of the GM decreases could be reversed, as revealed by a comparison to controls. For instance, Gwilym et al. [99] found decreased GM in ACC, insula and operculum, thalamus dorsolateral PFC, brainstem, and amygdala pre-surgery. These patients underwent surgery, which resolved pain in most of the patients post-surgery, while GM increases were observed in dorsolateral PFC, amygdala, and brainstem following hip replacement. However, chronic pain patients often suffer additionally from psychological comorbidities like anxiety or depression and reveal often sedentary behavior, both factors that affect brain structure as well. Pain medication such as the intake of opiates has further been shown to affect brain structure. Thus characteristic structural changes observed with pain chronification or pain relief might be also related to secondary factors like psychological comorbidity, motor behavior or pain medication [171].

3.4 Novel Approaches in Imaging of Chronic Pain

Pain is processed far more slowly by the nervous system than any other sensory modality and the perception of the time course for the subjective experience of pain is only weakly correlated with the time course of the stimulation [6]. There is a need to map the temporal course of the various perceptual aspects of painful experiences and not only of the stimulation itself [57]. Percept-related fMRI accounts for the temporal course of the perceived pain, which is often just weakly related to the temporal course of stimulation.

When patients come to the clinician, they rarely complain about allodynia or hyperalgesia in response to stimuli but rather about their spontaneous ongoing pain (pain in the absence of any external stimulation). But the measurement of spontaneous pain is far more challenging to assess by neuroimaging methods than the measurement of well-controlled brain activity in response to painful or innocuous (e.g., tactile allodynia) stimuli. So far, the majority of neuroimaging studies focus on brain responses to painful experimental stimuli in healthy subjects [101, 182], or on brain circuitry linked to allodynia or hyperalgesia in chronic pain populations [92, 156]. But, the most relevant aspect of chronic pain is spontaneous ongoing pain. Here, we present percept-related fMRI and ASL as novel imaging approaches allowing the investigation of brain responses specific for the perceptual and temporal characteristics of pain experiences or the measurement of spontaneous ongoing pain.

One important goal for the pain researcher is to find novel treatment approaches tailored to the individual patient. Thus, there is a need to predict the outcome of treatment approaches and to predict pain progression in the individual patient. Therefore novel analysis techniques are needed that help to infer a perceptual or cognitive state in the individual subject from their brain activity. In this context, we present multivariate pattern analysis (MVPA), a novel method based on machine learning algorithms. MVPA offers new avenues towards a personalized medicine with increased sensitivity and specificity compared to univariate analysis based on the general linear model (GLM) potentially reducing scanning time in vulnerable patient groups. Graph analysis allows the identification of relevant aspects of structural and functional connectedness among multiple brain regions and thus to go beyond a modern (brain-based) phrenology of pain towards a more mechanistic understanding of pain [6].

3.4.1 Percept-Related fMRI

Percept-related fMRI allows the exact identification of brain activity associated with the temporal pattern of a certain percept. It has been shown that the subjective experience of pain is highly variable across subjects even when using the same physical stimulus parameters [11, 12] (Fig. 2). Further there are marked differences in pain perception based on stimulus properties related to adaptations such as habituation or sensitization to repetitive stimuli [6] with a shift in the perceptual qualities across time [102]. This dissociation between the temporal features of different aspects of pain experiences and the time course of delivered stimuli seem to be even more pronounced in chronic pain populations [142, 212].

The group of Apkarian [84] have uncovered that spontaneous ongoing pain yields characteristic fluctuations in the scale of seconds to minutes. They found that these patterns of temporal fluctuations vary between different pain syndromes [84]. These fluctuations in spontaneous pain can be captured by continuous ratings by, e.g., assessing the intensity of pain. Davis et al. [57] originally introduced the concept of percept-related fMRI and by this the distinction between *stimulus-related* and *percept-related* fMRI. In the field of pain research, stimulus-related fMRI refers to the conventional

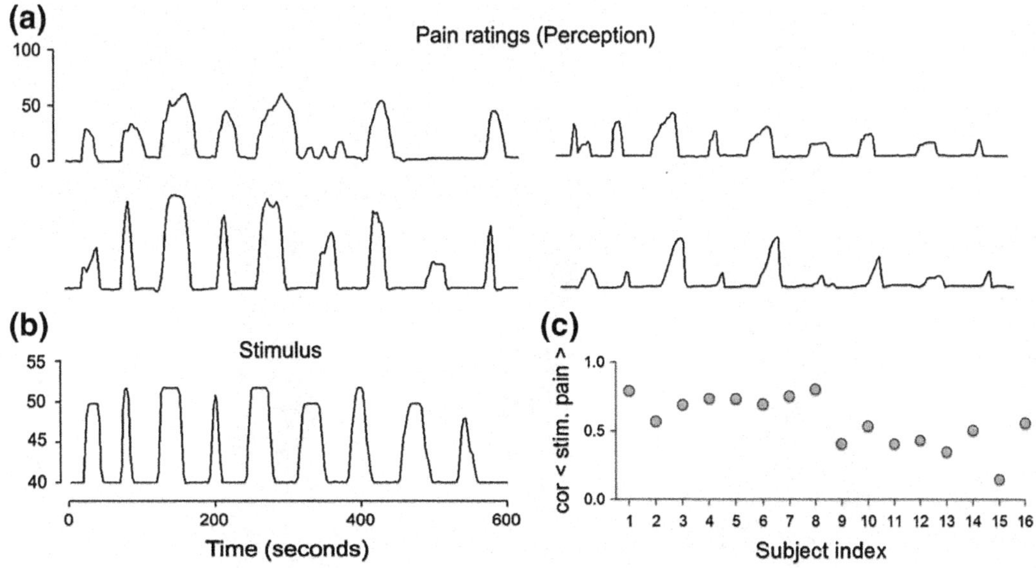

Fig. 2 Constant heat pain stimulation at the back of healthy subjects and perceived pain rated continuously. **a** Time courses of continuously rated perceived magnitudes of heat pain across 4 subjects revealing the inter-individual variability in subjective ratings on pain. **b** Time course and physical stimulus intensities of the constant thermal pain stimulus applied at the back. A comparison between the pain-ratings (**a**) and the time course of stimulation (**b**) reveals the delayed nature of pain percepts relative to the stimulation. **c** Variability in correlation strengths between stimulus and ratings across 16 subjects. Figure adapted from Apkarian et al. [6]: with data derived from Baliki et al. [12]

fMRI-approach in measuring (immediate) brain responses to nociceptive stimuli. Percept-related fMRI refers to the temporally delayed and changing experiences of different pain percepts relative to the nociceptive stimulus [245]. Based upon the finding of temporal fluctuations in spontaneous pain by Foss et al. [84], percept-related fMRI provides a tool to the pain researcher to map clinically relevant pain in the brain, even in the absence of a painful stimulus [6].

3.4.2 Arterial Spin Labeling (ASL)

ASL is a relatively novel perfusion-based MRI technique allowing the absolute quantification of rCBF, which is a surrogate measure of neuronal activity [244]. Like BOLD-fMRI, ASL is a noninvasive functional imaging method using magnetically labeled arterial blood water as an endogenous tracer to indirectly measure brain function. Like positron emission tomography (PET), it is a perfusion-based imaging technique. In contrast to PET, there is no need for an exogenous contrast agent, such as a radioactive $H_2^{15}O$, to measure rCBF [190]. The basic principles of ASL are shown in Fig. 3.

The advantages of ASL for pain research becomes apparent when considering the disadvantages of task-dependent BOLD-fMRI or resting-state fMRI. Spontaneous pain is difficult to measure with task-dependent BOLD-fMRI because the behavior must be separable into distinct epochs or interspersed by periods of rest, which is per definition not possible for spontaneous pain. ASL allows the assessment of non-stimulus driven brain activity, such as spontaneous ongoing pain, by detecting increased rCBF in specific brain areas associated with pain. In ASL, alterations in rCBF can be directly linked to fluctuations in e.g. spontaneous pain. In resting state fMRI, the mechanistic basis of the temporal coherence in low-frequency fluctuations between brain regions remains elusive [188]. Recent advantages in MRI such as ultra-high magnetic field strengths or the use of phased-array coils

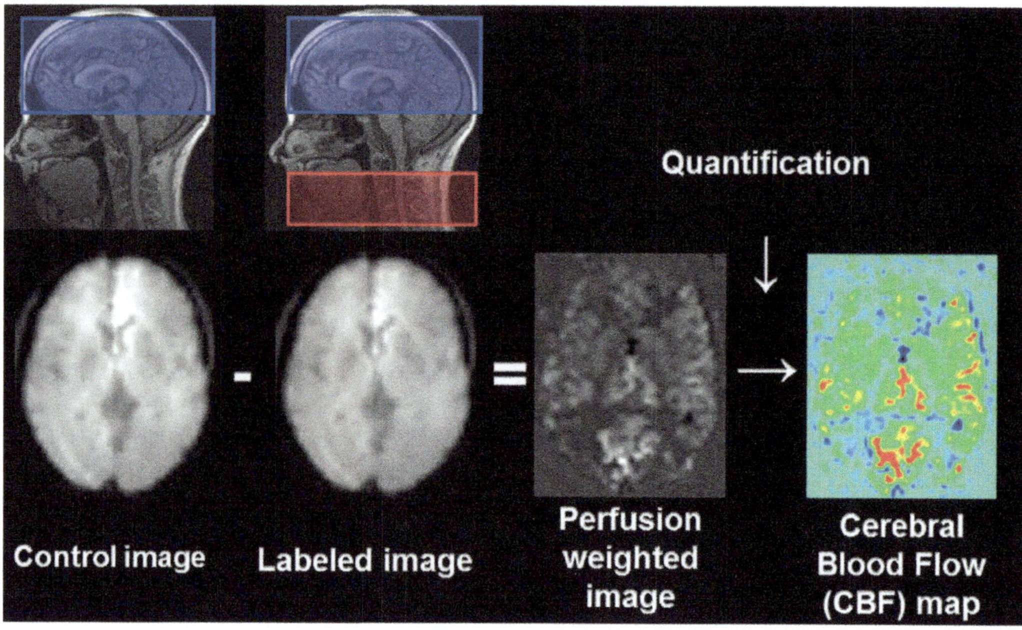

Fig. 3 Labeled and control acquisitions to generate perfusion weighted images to quantify cerebral blood flow (CBF). The labeled acquisition involves two basic steps (1) the magnetic labeling of the circulating blood protons upstream to the volume of interest (in general in the neck vessels) by radiofrequency pulses (*red box*), (2) the acquisition of images downstream to the labeled image (*blue box*). The control acquisition requires the imaging of the volume of interest without labeling of protons. The subtraction of the labeled from the control acquisitions provides the perfusion weighted image that suppresses the signal from the static tissues and allows the quantification of changes in regional CBF related to neuronal activity changes. Figure reprinted from Ferré et al. [73]

with pseudo-continuous ASL (PCASL) sequences significantly increased the signal-to-noise ratio and spatial and temporal resolution of ASL [73]. Additionally, task-dependent BOLD-fMRI is not well suited for the measurement of prolonged epochs of pain because with increasing periods of stimulus change the signal-to-noise ratio tremendously drops down due to low-frequency fluctuations in the fMRI-signal [1]. However, the duration of the subjective experience of pain has been shown to be often prolonged relative to the stimulation [92, 212]. ASL is not limited by task frequency and is therefore optimally suited for measuring sustained periods of pain [243].

BOLD-fMRI uses a composite measure of rCBF, rCBV, and oxygen metabolism to indirectly assess neuronal activity, while the ASL signal is solely reflecting the absolute measure of rCBF [190] (see Sect. 3.2). Often, alterations in basal CBF are in itself the variable of interest for the investigation of pathological states including chronic pain [113, 143].

3.4.3 Multivariate Pattern Analysis (MVPA)

Until now, the mainstay in the statistical analysis of imaging data are mass-univariate tests performed within the framework of the GLM [86]. One of the main advantages of the GLM approach is its great flexibility in terms of the multitude of different types of statistical analysis incorporated, including correlations, t-tests, analysis of variance and so forth [86, 178]. The basic logic of GLM analysis is the determination of model parameters (variables of interest, confounds, and error) that best describe the data set, while "best" means with, e.g., the lowest mean squared difference between the observed and fitted data points [178].

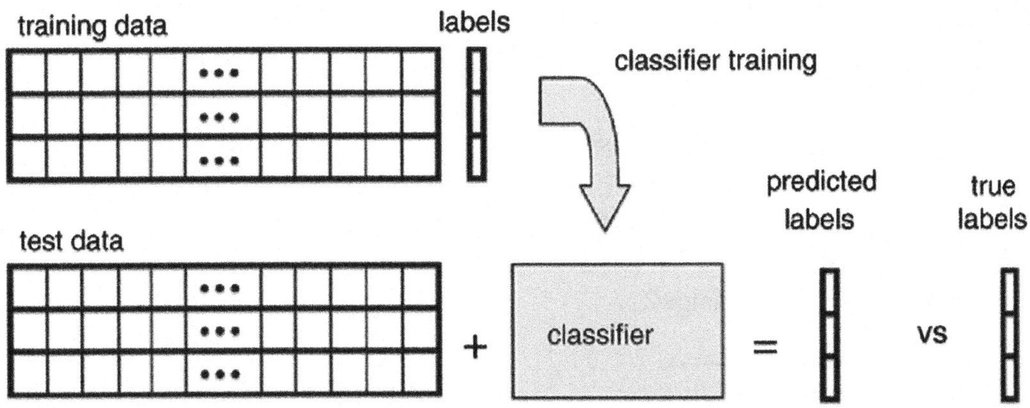

Fig. 4 Principle of training a classifier to discriminate brain signatures specifically linked with different class labels. A classifier is an algorithm learning the mapping (function) between features (voxels) and class labels (e.g. being a patient or healthy subject) based on a training dataset. The training dataset consists of multiple rows of examples were each case represents a feature (voxel). After training the classifier algorithm, the prediction (generalization) ability of the algorithm is evaluated based on its application to a test dataset. The test dataset consists of examples with rows representing multiple voxels, however, this time the algorithm is not informed about the labels the different examples belong to. Figure adapted from Pereira et al. [192]

In opposition, machine learning approaches, when applied to fMRI data often referred to as MVPA, seek to find model parameters that optimize the prediction ability about data that have not yet been observed (i.e., that create decision rules based on observed datasets that can efficiently categorize new observations) [197]. It should be noted, that the model that best fits the observed data (GLM) and the model that has the maximum ability to make predictions about data that have not yet been observed (MVPA) differ. MVPA approaches decode (and thereby predict) perceptual, psychological, or behavioral states related to individual MRI datasets based on a pattern-classification algorithm applied to multi-voxel MRI datasets [187]. By this, MVPA is a versatile tool of identifying brain signatures of acute and chronic pain. In this context, MVPA can help us to identify brain signatures to objectively identify individuals who suffer from chronic pain and thus improve diagnosis [38] or to differentiate between brain responses to painful versus innocuous stimulation [241]. Additionally, MVPA can be used to identify brain signatures that predict whether a patient will benefit from a certain treatment or not [151]. The different types of MVPA, including support vector machines, Gaussian process classifiers, sparse logistic regression or random forests are based on the principle of supervised learning that can be either used for classification or regression. The basic principles of how MVPA allows delineating brain-based biomarkers of pain perception can be summarized as follows.

The application of machine learning algorithms is typically applied to vectors of voxel activity values and can be summarized in four steps (see also Figs. 4 and 5):

1. ***Feature selection*** is used to decide which voxels will be included in the classification analysis. This is an important step to reduce the dimensionality of the dataset and thus improve the classifier performance [197].
2. ***Pattern assembly*** describes the labeling of brain activity patterns into discrete brain patterns corresponding to the experimental conditions, which generated the pattern [187].
3. ***Classifier training*** is the learning of a decision function that maps between patterns of activity across multiple voxels and experimental conditions based on the application of the classifier algorithm to a training dataset.

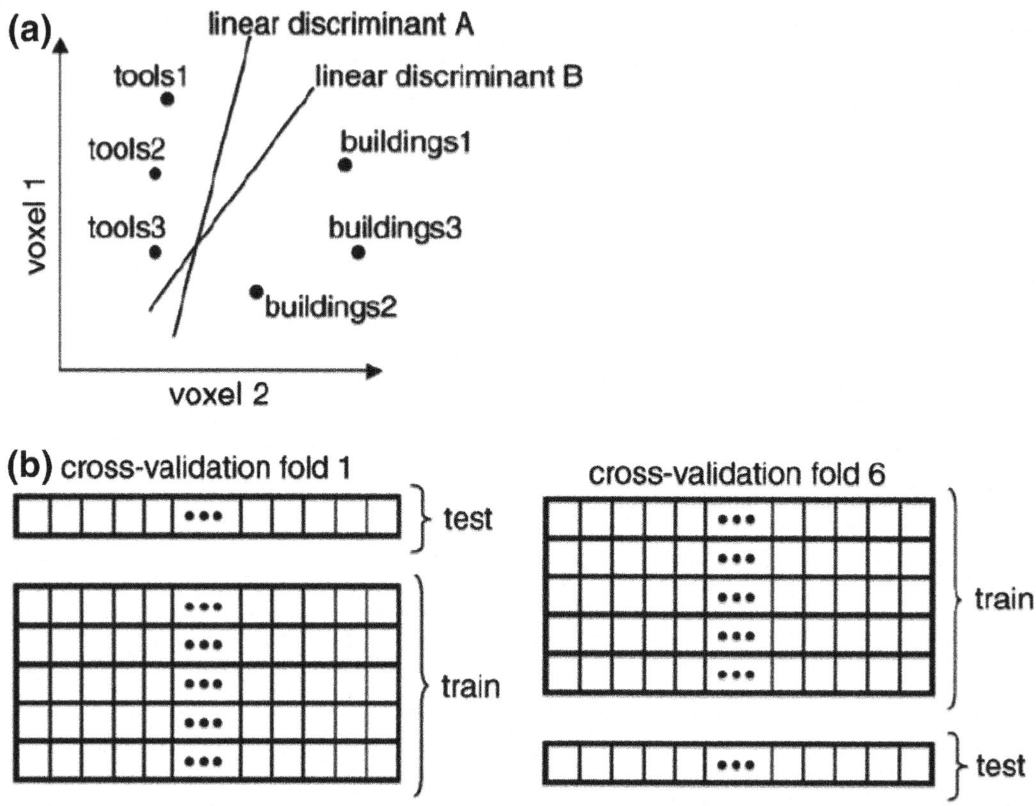

Fig. 5 The optimization of the classifier function and the principle of leave-one-run-out cross-validation. **a** Training the classifier algorithm is learning a line that bisects the feature space into two classes. The linear discriminants discern between multi-voxel brain signals (brain signatures) either belonging to class A (e.g. being a patient) or class B. (e.g. being a healthy subject). **b** Generalization performance of the classifier algorithm must be assessed in a statistically independent test dataset. However, most MRI-studies have limited sample sizes. Leave-one-run-out cross-validation circumvents this problem by taking each of the individual example rows once as a test dataset to check classification performance while the rest serves as a training set. Figure adapted from Pereira et al. [192]

Training is performed by iterative feeding in subsets of labeled patterns of brain activity into the classifier algorithm [197].

4. ***Generalization testing*** is performed on a statistically independent test dataset to assess the ability of the "trained" classifier to decode (predict) the category (label) in the examples of the test dataset [197]. To avoid sampling twice the number of observations for the training set and the test set, we can make use of *k*-fold-cross-validation. In *k*-fold-cross-validation, the dataset is split into k blocks of observations, and the classifier is trained on all blocks but one and then tested on the left-out block. This procedure is repeated for all blocks. If *k* is equal to the number of observations, we talk about *leave-one-out cross-validation* (Fig. 5).

Compared with conventional mass-univariate approaches, MVPA yields two main advantages: (1) it acknowledges the distributed (multi-voxel) nature of MRI datasets, offering greater sensitivity for the detection of spatially distributed MRI-effects (2) it can make predictions at the level of the individual subject [169, 187]. MVPA relies on changes in individual voxels, while mass-univariate methods focus on the detection of changes in overall activation in larger regions [187]. For instance, task-related effects on voxel values might be opposite (decrease vs. increase) in two adjacent voxels (which is, e.g., true for orientation-selective columns in the visual cortex) [105]. Classical univariate methods would discard this meaningful spatially fine-grained information, e.g., by applying spatial averaging (smoothing) across voxels to increase the signal-to-noise ratio to experimental conditions for group statistics. There is a necessary link between the principle of functional integration in the brain and multivariate analysis because multivariate analysis applied to MRI-data captures more than one voxel at once and can thus model interactions between distributed brain regions [86].

3.4.4 Graph Analytical Approaches

A new avenue to understand structure and function of the brain comes from the fusion of modern noninvasive imaging techniques with powerful network analysis tools originally developed to study social networks [196, 224]. The structural connectivity of the brain can be studied in vivo at a nearly millimeter scale by means of DTI [224]. These structural connections put a spatial constraint on large-scale neuronal dynamics, which in turn can be captured by functional and effective connectivity analysis [111]. Thus, the combined acquisition of DTI (structural connectivity) and fMRI data from a subject is complementary and can help to improve measures of functional and effective connectivity. A graph consists of nodes or vertices (a voxel or a region of interest) and edges (their mutual anatomical connections or statistical-dependencies between neural elements).

The conduction of brain network analysis with graph analytical tools can be summarized in four steps [209, 224] (Fig. 6):

(1) Definition of nodes based on parcelling GM cortical and subcortical regions. This parcellation can be performed based on anatomical borders or landmarks or by random parcellation of evenly spaced and sized clusters of voxels.
(2) Estimation of pairwise-couplings among all nodes and aggregation of structural or functional couplings within an adjacency matrix. A simple and common measure of adjacency is the Pearson correlation coefficient (r).
(3) Removal of weak or inconsistent connections within the adjacency matrix by statistical thresholding. Liberal thresholding involves the risk of preserving noninformative (noisy) edges, while highly conservative thresholding might split the graph into multiple isolated networks. Within a connection or adjacency matrix, which is already the simplest form of a graph, binary elements represent the presence or absence of edges between pairs of vertices [225] (Fig. 6).
(4) Following the extraction of relevant nodes and their respective edges, the resulting graph can be analyzed with graph analytical tools to detect, analyze, and visualize network topology (Fig. 7).

The variety of graph measures can be categorized into measures of *segregation* (to which degree elements of the network form separate clusters), *integration* (the capacity of a network as a whole to become interconnected and exchange information), and *influence* (quantifies the relevance of a given node according to the structural integrity or functional performance of a network). Measures of segregation include, e.g., cluster coefficients (a high cluster coefficient indicates a dense connection of a node to its neighbors forming a cluster or clique), measures of integration include, e.g., path length (the length of the shortest path corresponds to the topologically shortest distance between two nodes) and measures of influence, e.g., for example the nodes degree (the number of edges attached to a given node) (Fig. 7). The combination of high clustering and short path length is characteristic for small-world networks that allow efficient information sharing between nodes [209].

4 Methods for Experimental Pain Induction

The induction of experimental pain provides useful information for the neuronal processing of clinical pain signs such as allodynia or

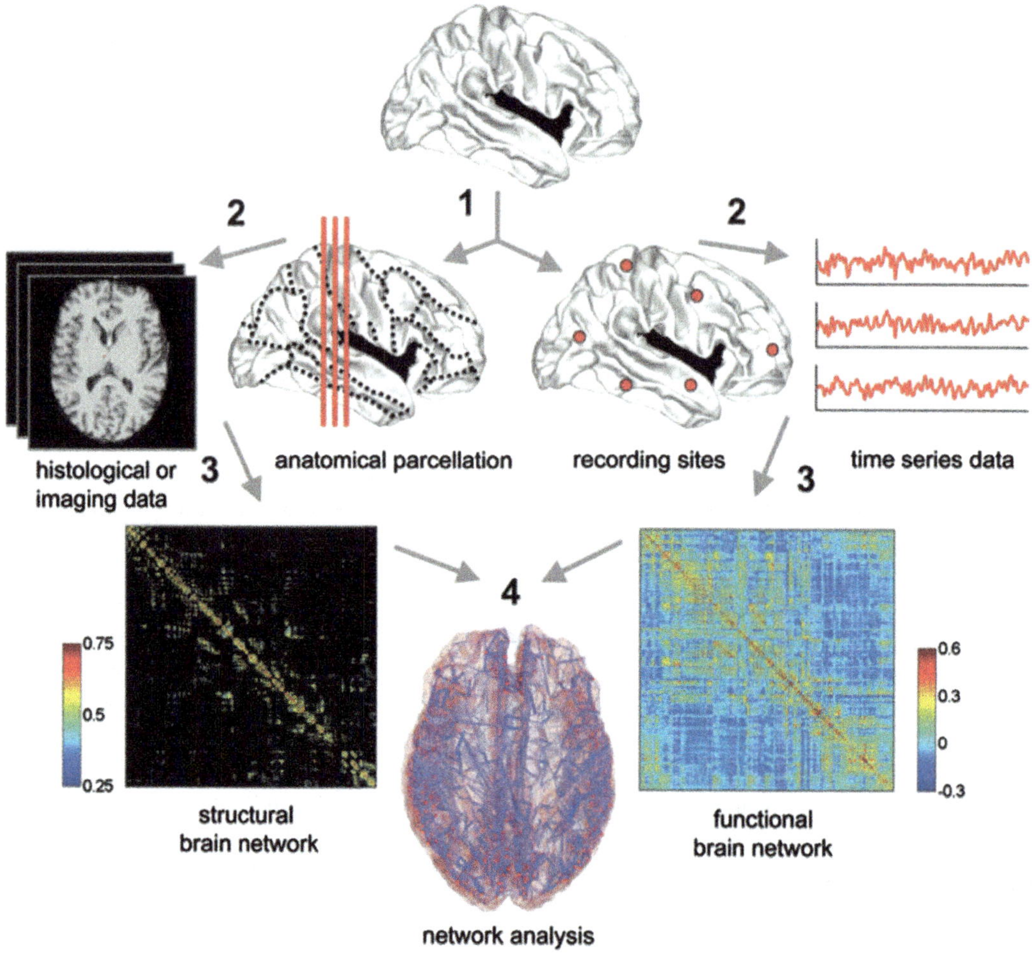

Fig. 6 Basic processing steps to generate a graph and study network topology. *1* Definition of network nodes by parcellation of the acquired brain volume (structural or functional MRI-volumes) or recording sites (for example positions of EEG-electrodes). *2* Definition of structural or functional network edges by estimating strength of structural connectivity (diffusion tensor imaging: DTI) or statistical dependencies (time-series data) among network nodes. *3* Construction of a network by aggregating nodes and edges into a connection matrix representing structural (*left*) or functional (*right*) connectivity. The example plots are from previously published data [111]. *4* Based on the thresholded connection matrix, different graph theoretical measures can be extracted. In many cases, the statistical evaluation of graph measures requires the formulation of an adequate null hypothesis, which is often a randomized global topology of a network, which preserves various subsets of structural parameters (e.g., node degrees). Figure reprinted from Sporns [224]

hyperalgesia. Additionally only experimental pain allows the direct comparison between chronic pain patients and healthy controls, thereby phasic as well as tonic pain states can be mimicked in healthy controls. To investigate the processing of experimental pain induction in healthy controls or chronic pain patients, several stimulation methods are available. Depending on the research question and the associated experimental paradigm and neuroimaging procedure used, the stimulation method has to be carefully chosen.

For the exact timing needed in EEG or MEG studies, electrical-, laser- or fast heat pain stimulations (e.g., Medoc Pathway Model CHEPS) are preferable. Other methods to induce experimental pain are chemical agents, pressure stimuli, an incision or thermal (heat, cold) pain. Additionally

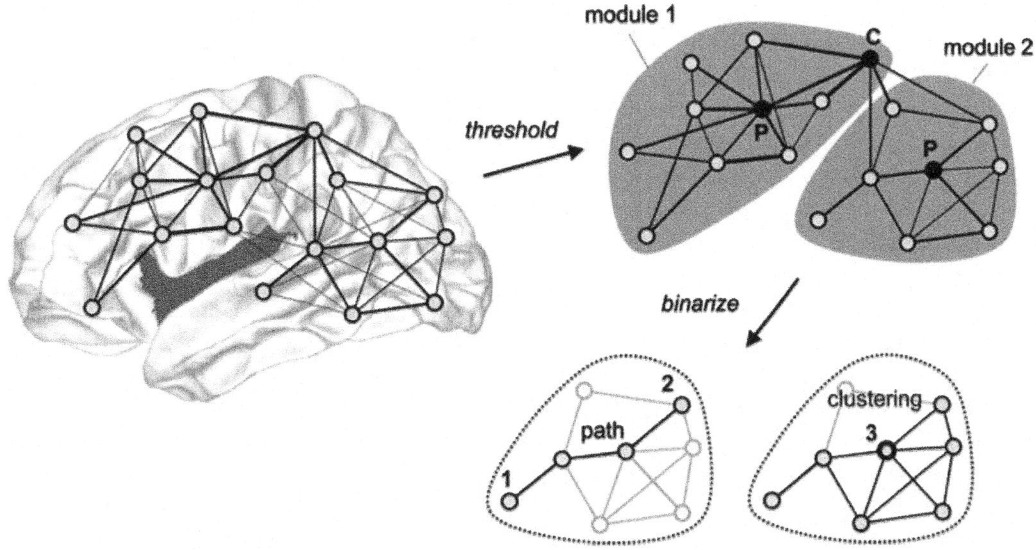

Fig. 7 Different graph theoretical measures and visualizations revealing the network topology can be extracted from the adjacency matrix. The diagram on the left reveals an undirected, weighted graph that might reveal the strength of structural connectivity between nodes as derived from diffusion tensor imaging (DTI). After thresholding, the graph reveals two prominent modules (network communities) that are interconnected by a single node, a so called 'connector hub' C. Each module further reveals a highly connected node with intramodular connections, a so called 'provincial hub' P. The diagram at the bottom shows the undirected, thresholded graph after binarizing (remaining connections are set to unity strength). One can see a typical graph measure of integration, which is path length (the number of intervening nodes that must be traversed to reach one node from the other (i.e. between nodes 1 and 2) and another, which is clustering (around node 3) reflecting segregation in the network. Figure reprinted from Sporns [224]

a psychological induction via empathy for pain [211] or hypnotic suggestion [202] was used. An incision, the insertion of a catheter for the application of a chemical agent or the application of intramuscular needle electrodes is invasive for the participants. Painful stimulation at the skin (transcutaneous or intracutaneous electrical stimulation, pressure, thermal pain, laser stimulation, incision, transcutaneous chemical stimulation) is different from painful stimulation directly applied to the muscle (intramuscular electrical stimulation, intramuscular chemical stimulation) which might be more comparable to a chronic musculoskeletal pain. Moreover these stimulation methods differ in their capability to excite A-delta, C-fibers, or even non-nociceptive nerve fibers. Psychological pain induction leads to the activation of similar networks, however, without nociceptive input [202, 222]. As examples of painful stimulation, we present brain responses to electrical stimulation with the differentiation between cutaneous and muscular stimulation, chemical muscular stimulation and close with a discussion of psychologically induced pain.

4.1 Electrically Induced Muscle Pain

In healthy subjects, electrical painful stimulation has been mainly applied transcutaneously [9, 64, 65] or intracutaneously (for review, see [32]). The comparison of transcutaneous and intracutaneous stimulation revealed common activations in S2, insula, medial cingulate cortex (MCC), thalamus, brainstem, vermis, and the anterior cerebellar hemisphere; transcutaneous stimulation activated additionally the supplementary motor area (SMA), motor thalamic nuclei, the ipsilateral insula, and the MCC, while the posterior cingulate cortex (PCC) was only activated by intracutaneous stimulation [210]. As a more specific model for muscle pain, fewer studies used intramuscular painful electrical stimulation [185, 220, 229] or chemically induced muscle pain (see below). The brain areas showing evoked activity through painful electrical intramuscular stimulation or painful electrical cutaneous stimulation are about the same, including the contralateral S1, S2 bilaterally, ACC, and PCC (for review see [33, 34]. Comparing intramuscular and intracutaneous stimulation, Shimojo et al. [220] found similar topographic patterns and source generators in the EEG in response to the non-painful and painful electrical stimulation. Furthermore, the authors found a positive relation between the stimulus intensity for both modalities and the magnitude of the P260 component measured at the vertex. The P260 component is thought to originate from the ACC [5, 9, 32]. This overlap between brain regions involved in the processing of cutaneous and intramuscular painful electrical stimulation was also reported by studies using PET. In addition to this substantial overlap of activation, Svensson et al. [229] reported larger ACC activity in response to the intramuscular stimulation. Muscle stimulation showed lower somatosensory evoked potential amplitudes and slightly prolonged latencies compared to skin stimulation at most recording sites [185].

4.2 Chemically Induced Muscle Pain

To assess the effect of experimentally induced tonic pain, which is more comparable to spontaneous pain in chronic pain patients, chemical agents were used. An advantage of such agents is that they can be applied directly into the muscle and thus are more comparable to habitual muscle pain. A disadvantage of this method is that it is invasive and often allows only one injection, which makes this model difficult to use with brain imaging methods needing multiple repetitions as in evoked potential or evoked field EEG or MEG studies or in BOLD-fMRI-studies (but see Sect. 3.4.2). Several studies investigated the neural correlates of the pain component associated with capsaicin-induced secondary hyperalgesia in healthy controls. Most of these studies were looking at changes in EEG frequency-bands, where one injection is sufficient [33–42], or used capsaicin to induce secondary mechanical or heat hyperalgesia and then stimulated this site with cold, static, or dynamic mechanical stimuli [17, 165, 177]. Maihöfner et al. [165] showed that brush-evoked allodynia compared to non-painful brushing led to significant increases in BOLD-signals in contralateral S1, PPC, inferior frontal cortex, and bilateral S2/insula. Repetitive stimulation with chemical agents is prone to tachyphylaxis, a habituation process that progressively diminishes responses to series of phasic stimulations. It is assumed that prostaglandin E_2 (PGE_2) reduces this tachyphylaxis [179, 226] and, therefore, in studies with chemical agents, it is a useful addition to have the possibility of applying repetitive injections of this agent. Otherwise only one injection of, e.g., hypertonic saline solution could be used [106, 146] with influence on the signal-to-noise ratio of the brain imaging data. We implement a fMRI-compatible model of chemically induced muscle pain by injecting a combination of prostaglandin and hypotonic saline solution [62].

4.3 Psychologically Induced Pain

We are able to discern between states of emotional distress and pain. With an undeniable certainty, we can make a distinction between how it feels to be socially rejected or seeing someone else in pain (empathy for pain) compared to how it feels to be physically hurt (physical pain). In recent years several authors found compelling evidence that "social pain" leads to similar patterns of brain activity than "physical pain" during nociceptive stimulation [160, 202, 222].

For instance, empathy for pain was psychologically induced. Persons empathizing with the pain of others can "feel" the pain by themselves [51, 52, 100, 124, 125, 145, 180, 181, 211, 222]. This subjective experience is mirrored in brain activations. Singer et al. [222] reported, that seeing loved ones receiving painful stimulation activates partly the same brain areas as receiving the painful stimulation oneself. The common activation was found bilaterally in the anterior insula, rACC, brainstem, and cerebellum. Activation in the anterior insula and the ACC was correlated with individual empathy scores, whereas activation in S1 an S2 was found only for painful stimulation to oneself.

A partial recruitment of the pain matrix was also observed, when the subjects receiving painful stimulation were strangers [181]. Psychological induction of empathy for pain was also done with pictures of persons in painful situations [125], painful faces [30], or faces of chronic pain patients expressing chronic and acute pain [211].

To identify the possible influence of the observer's own sensitivity to pain upon his or her perception of the others' pain, Danziger et al. [52] compared patients with congenital insensitivity to pain to controls. They showed that patients behave similarly to controls by rating verbally presented painful situations or by inferring pain from facial expression. But ratings of pain-inducing video clips in the absence of visible or audible pain-related behaviors showed more variable and significantly lower pain ratings as well as a reduction in aversive emotional responses in the patient group. Furthermore, only the patients showed a strong relation between pain judgments and emotional empathy scores suggesting that high empathy can substitute a lack of sensitivity to pain.

Based on similar patterns of brain activation, some social neuroscientists conclude that social rejection or empathy for pain really hurts [160]. However as summarized in Iannetti et al. [120] this interpretation is untenable based on logical and technical reasons as well as empirical evidences. Deducing that a subject is in pain merely based on a pattern of brain activity is an example of reverse inference, in which a mental state is inferred from a pattern of brain activity. The conclusion that a subject is in pain based on a pattern of brain activity is only valid when this pattern of brain activity is exclusively associated with that mental state. As will be discussed in Sect. 6, this is (not yet) true for the pain matrix, due to technical limitations in data acquisition and analysis of imaging data [120]. Irrespectively of this, it is important to interpret the high degree of overlap in brain regions between social and physical pain. According to Iannetti et al. [120] this overlap only reflects that both experiences trigger multimodal cognitive processing in response to behaviorally salient stimuli.

5 Neuroimaging of Pain in Selected Chronic Pain Syndromes

As pain is processed in a network of several brain areas, it is of interest how a chronic pain condition such as chronic back pain or fibromyalgia, phantom limb pain, complex regional pain syndrome influences this network during the processing of painful experimental stimuli.

5.1 Chronic Back Pain

Back pain with different amounts of pain intensity and triggered by different causes affects between 27 and 40% of the people. To observe these potential neuroplastic changes in the S1 representation of the back in chronic lower back pain patients, non-painful and painful electrical intracutaneous stimulation was used [80]. MEG recordings revealed a 25 mm medial shift of an equivalent current dipole source at about 70 ms after stimulus onset (Fig. 8). The extent of this shift was positively correlated with chronicity. Furthermore, within the early time window (100 ms) the root mean square amplitude of the somatosensory evoked field (SEF) was significantly higher in patients with chronic low back pain than in the healthy controls when painful back stimulation was used. In another study enhanced perceptual sensitization and enhanced processing of the sensory-discriminative aspect of pain, as expressed in the N80 component of the EEG could be reported [60]. Using fMRI, Giesecke et al. [96] reported that comparable levels of subjectively reported painful pressure stimulation to the left thumbnail resulted in activation patterns

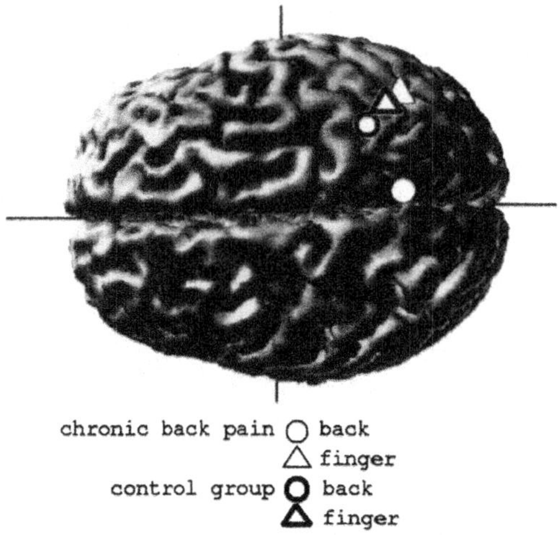

Fig. 8 Average location of the finger and back dipole for the chronic back pain patients and the healthy controls are superimposed schematically on a magnetic resonance image. The back position is shifted medial in the patients with chronic back pain. The amount of this shift is correlated with chronicity. Figure used from Flor et al. [80]

that were similar in chronic low back pain patients and healthy controls, whereas similar objective pressure intensities resulted in greater effects in patients in regions specific for pain processing. Although similar regions of brain activations were found for similar painful pressure stimulation, chronic low back pain patients showed a significantly reduced BOLD-signals in the periaqueductal gray (PAG) and significantly increased BOLD-responses in S1 and S2 and the lateral orbitofrontal cortex, supporting the hypothesis of a dysfunctional inhibitory system controlled by the PAG [95]. Similar results were found by stimulating 5 cm left to the fourth–fifth lumbar spinal interspace with patients showing augmented activation for subjectively identical stimuli compared to healthy controls specifically at the right insula, SMA, and the PCC [135]. In resting-state fMRI-analysis five meaningful resting-state networks were isolated of which only the DMN exhibited deviations from healthy controls [14]. The authors reported a decreased connectivity of medial PFC to the posterior constituents of the DMN, and increased connectivity to the insula in proportion to the intensity of pain. Similar results were found by Tagliazucchi et al. [230] demonstrating that chronic pain disrupts normal activity in the DMN even when the brain is in resting state.

If the described functional changes are caused by structural changes or cause structural alterations is still unclear. Apkarian et al. [7] reported 5–11% less neocortical gray matter (GM) volume in chronic back pain compared to control subjects. The magnitude of this decrease is equivalent to the GM volume lost in 10–20 years of normal aging. The decreased volume was related to pain duration, indicating a 1.3 cm^3 loss of GM for every year of chronic pain. GM density was reduced in bilateral dorsolateral PFC and right thalamus. The reduction of GM in the dorsolateral PFC could be replicated in two recent studies. Fritz et al. [88] found decreased GM volume in the ventrolateral and dorsolateral PFC, both the ventral and dorsal medial PFC and the anterior insula. Pain intensity showed a weak negative correlation with GM volume in the left dorsolateral and ventrolateral PFC, and ACC [88]. Ivo [123] showed decreased GM density in the dorsolateral PFC, the thalamus, and the MCC. Schmidt-Wilcke [213] found a significant decrease of GM in the brainstem and the S1. Correlation analysis of pain unpleasantness and intensity on the day of scanning revealed a strong negative correlation (i.e., a decrease in GM with increasing unpleasantness/increasing intensity of pain) in these areas. Additionally, a significant increase in GM bilaterally in the basal ganglia and the left thalamus was found.

5.2 Fibromyalgia Syndrome

Fibromyalgia is characterized by chronic widespread pain and tenderness at specific sites [246]. For the diagnosis of fibromyalgia 11 of 18 tender points have to be painful. Fibromyalgia is a chronic pain disorder with symptoms at the joints muscles and tendons on all four quadrants of the body. The pain is especially increasing under load. Additional symptoms include a general weakness, impaired concentration and cognitive dysfunction, sleep disturbance, chronic fatigue as well as a reduced mental and physical capacity. Physical, mental and also emotional load needs unnatural long recovery phases. In the population, the prevalence of fibromyalgia is between 0.6 and 4%, of these 85–90% are women. While the etiology of fibromyalgia remains unclear, the generalized hyperalgesia, widespread pain, and spontaneous pain in fibromyalgia cannot be explained by changes in peripheral tissues like muscle. To summarize fibromyalgia is a chronic pain condition with a sensitized pain perception but is not recognized or explained medically. The question arises if patients with fibromyalgia process experimental pain different from healthy controls?

Gibson et al. [94] showed that patients with fibromyalgia displayed a significant increase in the peak-to-peak amplitude of the cerebral potential in a time window 207 and 370 ms post stimulus evoked by painful CO_2 laser stimulation. De Tommaso [234] reported an increased amplitude of the vertex laser-evoked potential going along with an increased subjectively perceived laser pain intensity. Furthermore, patients with fibromyalgia had less habituation of perceived pain intensity and less habituation induced reduction of vertex laser-evoked potentials compared to controls. Similar results were found by Lorenz et al. [157] and Lorenz [155]. Compared to matched controls, patients with fibromyalgia exhibited significantly lower heat pain thresholds and had higher amplitudes of the laser-evoked potential components N170 and P390 [157]. The observation of amplitude enlargement of the N170 suggests enhanced nociceptive activation and neuronal synchronization in S2 by radiant heat. The P390 enhancement in patients with fibromyalgia might indicate greater attention to or cognitive appraisal of painful stimuli. In another study, lower electrical pain thresholds and a higher N80 amplitude of the EEG, both indicative of enhanced sensory processing were reported [61].

In an fMRI study, Gracely et al. [97] reported that only similar objective pressure intensities resulted in greater effects in patients with fibromyalgia in regions specific for pain processing whereas comparable levels of subjectively reported painful pressure stimulation resulted in activation patterns that were similar in fibromyalgia patients and healthy controls (Fig. 9). This applied to regions involved in the sensory-discriminative component of pain, such S1 and S2 as well as to those involved in the affective-motivational component, such as insula and ACC and could be replicated by Pujol et al. [201] using fMRI and by Maestu et al. [161] using MEG. Cook et al. [49] could replicate these results with heat stimuli and found differences between groups especially in the contralateral insula. Using an incision as an acute pain stimulus, it was shown that patients with fibromyalgia not only showed different brain activation responses compared to controls during painful stimulation but also during the anticipation of painful stimuli [36]. The authors conclude that central mechanisms of pain processing in the medial pain system alter cognitive/affective factors even during the anticipation of pain and may play an important role for pain processing in fibromyalgia. There is also evidence that the endogenous pain inhibitory system is changed in patients with fibromyalgia. Studies examining the descending modulation of pain could show the importance of the rostral (r)ACC in pain inhibition [22, 136, 240, 242]. Jensen et al. [126] reported that patients with fibromyalgia display significantly lower activation in the rACC than healthy subjects in response to unpredictable pressure pain stimulation [126] and that they have less functional connectivity between the rACC and hippocampi, amygdala, brainstem, and the rostral ventral medulla [127]. These findings support the hypothesis that fibromyalgia is characterized by cortical augmentation of pain processing. Resting-state data showed that patients with fibromyalgia had greater connectivity between the DMN and the insula. Furthermore, greater intensity of spontaneous pain at the time of measurement correlated with greater intrinsic connectivity between the insula and the DMN [184]. In another study patients with fibromyalgia showed decreased connectivity between thalamus and premotor areas, between the right insula and S1, and between supramarginal and PFC [77]. The authors suggest that

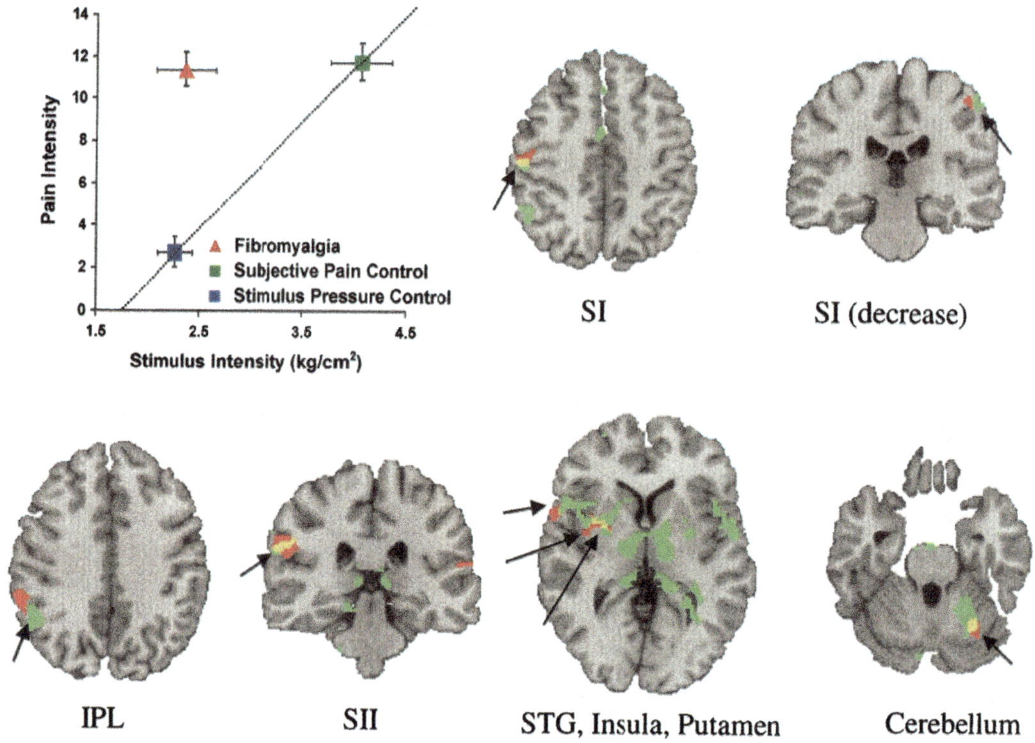

Fig. 9 *Upper left* In the stimulus pressure control condition (*blue*) the healthy controls rate the objective same stimulation intensity less intense compared to the patients with Fibromyalgia (*red*). To get a similar pain intensity (subjective pain control, *green*) higher objective stimulation intensities are need in healthy controls. On an MRI anatomic standard brain image common regions of activation in patients with fibromyalgia (*red*) and in the subjective pain control condition (*green*) are displayed. Overlapping activations are shown in *yellow*. The similar pain intensities, produced by significantly less pressure in patients, resulted in overlapping or adjacent activations. For objective same stimulation intensities, with higher pain ratings in Fibromyalgia, healthy controls show decreased brain activations compared to the patients (not displayed here). Figure reprinted from Gracely et al. [97]

abnormal connectivity patterns between pain-related regions and the remaining brain during rest reflect an impaired central mechanism of pain modulation in fibromyalgia. Weaker coupling between pain regions and prefrontal- and sensorimotor areas might indicate a less efficient system level control of pain circuits. In a recent study [121], state changes in resting brain connectivity following experimental pressure were investigated. The authors report that acute pressure pain stimulation increased the connectivity between the insula, the ACC, and the hippocampus compared to the measurement before the stimulation. Additionally, the authors found an increased thalamic connectivity to the precuneus/posterior cingulated cortex, which is part of the default mode network, in patients but not in controls. This connectivity was correlated with changes in clinical pain. The changes in resting-state brain activity following a noxious stimulus suggest that acute painful stimulation may contribute to the alteration of the neural signature of chronic pain.

There is conflicting evidence for the hypothesis that there is a general hypervigilance in fibromyalgia. Tiemann et al. [233] found no hypervigilance measured as an abnormal increase of attention to external stimuli in patients with fibromyalgia, Carrillo-de-la-Peña et al. [47] reported no differences in sensory gating of the P50 component, as indicated by P50 suppression rates to the second identical stimuli, and Lorenz

[155] found no hyper-reactivity in auditory evoked potentials. Contrary, McDermid et al. [172] reported lowered pain tolerance, pain threshold, and noise tolerance in fibromyalgia compared to rheumatoid arthritis and healthy controls and that both patient groups preferred lower levels of external stimulation. Corradi-Dell'Acqua [46] reported shorter N1 and P2 latencies and increased N1-P2 amplitudes in relation to loud tones suggesting that defects in an inhibitory system protecting against over-stimulation may be a crucial factor in the pathophysiology of fibromyalgia. In another study with 30 fibromyalgia patients 70% of the patients had a decreased noise tolerance [93].

If the described functional changes are caused by structural changes or cause structural changes is still unclear. In a VBM-study fibromyalgia patients had a significant decreased volume and a 3.3× increased age-related loss of the gray substance compared to healthy controls [139]. Additional the duration of the chronic pain correlates with the loss of GM, although one year of fibromyalgia is equivalent with 9.5 years of normal aging. Another study reported decreased GM volumes in the thalamus, but also increased volumes in the cerebellum and the striatum [214]. Voxel-based morphometry structural covariance network analysis showed more connections within the cerebellum in fibromyalgia and more connections in the frontal lobe in HC [134]. Spectral partitioning identified dense cerebellar connections to medial prefrontal/orbitofrontal cortex, medial temporal lobe, and right inferior parietal lobule in fibromyalgia. The GM volume of these regions was associated with severity of depressive symptoms. The number of fibers in these regions, measured by probabilistic DTI WM connectivity analyses, was associated with greater evoked pain hyperalgesia and clinical pain interference [134]. Reduced FA was reported in the bilateral thalamus, the thalamocortical tracts, and bilateral in the insula [159]. Additional reduced GM volumes were reported in pain processing areas going along with an increased FA. Sundgren et al. [228] reported a reduced FA in the right thalamus.

5.3 Phantom Limb Pain

The amputation or deafferentation of a limb or another body part is commonly followed by a global feeling that the missing limb is still present (phantom limb awareness), as well as specific sensory and kinesthetic sensations (phantom sensations) [116]. These non-painful phantom sensations may include a specific position, shape, or movement of the phantom, feelings of cold or warmth, tingling, itching, or electric sensations, and other paraesthesias [137], and are reported by almost all amputees. Phantom limb pain, or phantom pain, belongs to a group of neuropathic pain syndromes that is characterized by pain in the amputated limb or pain that follows partial or complete deafferentation. Residual limb (or stump) pain and non-painful residual limb phenomena are sensations in the still-present body part adjacent to the amputation or deafferentation line. Pain in the body part that is no longer present occurs in 60–80% of all amputees [128, 186]. Phantom pain or phantom sensations were reported after upper limb (e.g., [81]), lower limb [43, 215, 221], breast [114, 208], tooth [167, 168], internal organs [74], and penis [76, 205] amputations. The influence of peripheral, spinal, and central changes on phantom pain was reported previously (for a review see [78, 82]. Psychological factors do not seem to contribute to the causation but may instead affect the course and the severity of the pain [108, 219]. In persons with amputations it has been shown that the region of S1 that formerly received input from the now amputated limb reorganizes and receives input from neighboring regions (Fig. 10) [71, 81, 200, 247]. These changes are mirrored in MI [45, 131–133, 158]. Interestingly, reorganizational changes were only found in amputees with phantom limb pain after amputation, but not in amputees without pain [81]. This suggests that pain may contribute to the changes observed and that the persisting pain might also be a consequence of the plastic changes that occur. In several studies carried out on human upper-extremity amputee patients, displacement of the lip representation in the MI and S1 was positively correlated with the

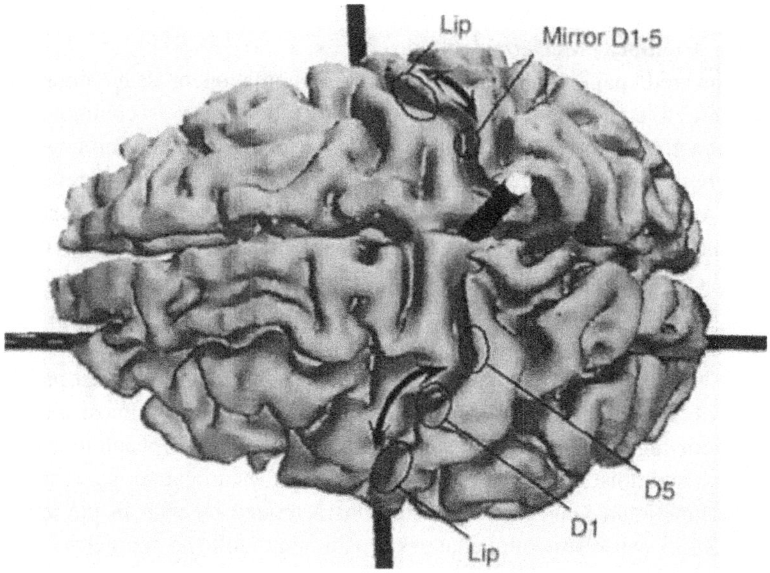

Fig. 10 Location of the left and right lip area, the thumb and little finger of the intact hand as well as the mirrored positions of the thumb and little finger to the side of the amputation in amputees with phantom limb pain superimposed schematically on a magnetic resonance image. At the side of amputation the lip area has shifted into the former hand area. This shift is called cortical reorganization and is correlated with the amount of phantom limb pain. Figure reprinted from Flor et al. [81]

intensity of phantom limb pain, and was not present in pain-free amputee patients or healthy control subjects (e.g., [81, 131]). In addition, in the patients with phantom limb pain, but not in the pain-free amputee patients, imagined movement of the phantom hand was shown to activate the neighboring face area [158]. This co-activation probably occurs due to the high overlap of the hand, arm, and mouth representations. Studies assessing brain activation of experimental pain in amputees are still missing. Resting-state functional connectivity values between the missing hand cortex and the sensorimotor network were reduced in amputees, and connectivity was weaker in individuals amputated for longer periods. Lower levels of functional coupling between the missing hand cortex and the sensorimotor network were also associated with emerged coupling of this cortex with the DMN [166].

There are only few studies reporting structural changes. Draganski et al. [68] reported a decrease in GM of the posterolateral thalamus contralateral to the side of the amputation. The thalamic GM differences were positively correlated with the time span after the amputation but not with the frequency or magnitude of coexisting phantom pain. Phantom limb pain was unrelated to thalamic structural variations, but was positively correlated to a decrease in brain areas related to the processing of pain. In a voxel-wise DTI analysis of the body of the corpus callosum Simoes et al. [221] found reduced FA values bilaterally in pain-free amputees compared with controls.

5.4 Complex Regional Pain Syndrome

As in phantom limb pain similar observations have been made in patients with complex regional pain syndrome. An external impact (e.g., trauma, operation, inflammation) leads to dystrophia or degeneration and atrophy of tissue. Symptoms of complex regional pain syndrome are circulatory disorders, edemata, alterations of the skin, pain, and restrictions of functions.

Frequently complex regional pain syndrome is following a distal radius fracture (7–37% of all cases). Accordingly, complex regional pain syndrome is more often on the upper than on the lower extremity. Changes in the somatosensory system were assessed in these patients too. Here the representation of the affected hand tended to be smaller compared with that of the unaffected hand and the individual digit representations had moved closer together [130, 163, 164, 232]. The extent of the pathological changes in the cortical representations correlated with the intensity of pain or motor dysfunction [162, 165, 232], but was additionally related to a degradation of sensibility in the affected hand. It was, however, unrelated to a loss of motor function [162]. It is so far not known how an expansion of adjacent representations and a shrinking of adjacent representations as observed in phantom limb pain and complex regional pain syndrome, respectively, can both be associated with pain. Similar to chronic back pain in resting-state fMRI analyses, five meaningful resting-state networks were isolated of which only the DMN exhibited deviations from healthy controls [14]. Again a decreased connectivity of medial PFC to the posterior constituents of the DMN, and increased connectivity to the insula in proportion to the intensity of pain was reported. In another study significantly greater reductions in functional default mode network connectivity were found in complex regional pain syndrome compared to controls [25]. The functional connectivity maps of S1/MI and intraparietal sulcus (IPS) in patients revealed greater and more diffuse connectivity with other brain regions, mainly with the cingulate cortex, precuneus, thalamus, and PFC. In contrast, controls showed greater intraregional connectivity within S1/MI and IPS.

Barad et al. [16] reported a decreased GM volume in several pain-affect regions, including the dorsal insula, left orbitofrontal cortex, and several aspects of the cingulate cortex. Greater GM volume was seen in the bilateral dorsal putamen and right hypothalamus. Pain duration was associated with decreased GM in the left dorsolateral PFC. Pain intensity was positively correlated with volume in the left posterior hippocampus and left amygdala, and negatively correlated with the bilateral dorsolateral PFC. Lee et al. [147] found thinner right dorsolateral PFC and left ventromedial PFC in complex regional pain syndrome patients compared to healthy controls. There were no correlations between cortical thickness and depression, although the Beck Depression Inventory and the Beck Anxiety Inventory differed significantly between the groups. WM changes measured by decreased FA were found in the left cingulum-callosal bundle [91].

6 Pain Matrix Reloaded: How Is Pain Represented in the Brain?

A large number of neuroimaging (fMRI, PET) as well as neurophysiological studies (EEG, MEG) have demonstrated a widely distributed network, comprising subcortical and cortical areas, which seems to be consistently activated during the processing of nociceptive stimuli and has been discussed to be an objective measure, "brain signature", representing painful experiences faithfully [5, 37, 89]. The most influential technique has been task-dependent fMRI due to its fairly high spatial and moderate temporal resolution and the fact that no exogenous tracer is needed to image the brain. This spatially consistent pattern of brain activity in response to nociceptive stimuli has been termed "pain matrix" as derived from the concept of the "neuromatrix" as originally proposed by [173] to describe one of the possible perceptual outputs of the neuromatrix.

The pain matrix comprises S1 and S2, the cingulate and insular cortices [5]. Some imaging studies also reported on brain responses to painful stimuli in other brain regions like the SMA, the motor cortex and the PFC and, sub-cortically, in the thalamus, basal ganglia and the brainstem [5, 69].

The specificity of the pain matrix for the perception of pain has been indicated by several studies showing linear and nonlinear relationships between physical stimulus intensity, magnitude of subjective pain and activity within various regions of the pain matrix [28, 35, 44, 59,

98]. Moreover, it has been shown that factors manipulating aspects of the painful experience were accompanied by altered brain responses in parts of the pain matrix [110, 204].

There are different perspectives on how the pain matrix codes the various aspects and dimensions of painful experiences [182]. According to the *population coding view*, painful experiences are the result of the "*flow and integration of information*" among specific brain regions, emphasizing the relevance of investigating functional integration between regions of the pain matrix [236]. Other authors advocate a *specialized subfunctions view*, accentuating that brain regions of the pain matrix subserve different roles corresponding to the sensory-discriminative, affective-motivational, and cognitive-evaluative dimensions of a painful experience [122, 199, 203]. Compliant with the specialized subfunctions perspective, it has not only been found that the intensity of the nociceptive stimulus affects the magnitude of brain responses in modules of the pain matrix [23, 242], but that manipulation of cognitive variables affecting pain experience lead to systematic alterations only in restricted regions of the pain matrix [110, 204]. According to the specialized subfunctions perspective, brain regions of the pain matrix are often subdivided according to the termination areas of the "sensory-discriminative" lateral and the "affective-motivational" medial pain system [6]. Therefore, S1 and S2 are typically considered to represent sensory-discriminative aspects of pain and the ACC affective-motivational aspects of pain (see Sect. 2). However, the "principle of labeled lines" interpretation of the pain matrix based on the anatomical distinction between the lateral and medial spinothalamic tract remains unclear for certain key structures like the insula [6]. The myriad of findings demonstrate a spatially consistent network of brain regions in response to various nociceptive stimuli. This network indicate that pain is not just associated with brain regions that are traditionally considered to be part of the nociceptive medial and lateral pain system, but also other brain regions known to be important for the processing of non-nociceptive stimuli, cognitive-evaluative, or motor functions [56]. These imaging findings indicate that nociceptive stimuli are accompanied by activity within a distributed network of brain region reflecting the multidimensionality of pain.

Nociception does not necessarily lead to pain and pain is not necessarily accompanied by nociception [83, 236]. Thus, an intriguing question is whether the pain matrix is also recruited when pain is induced in the absence of nociceptive input. In this context, it has been shown that "pain-like" experiences for example induced by suggestion-induced pain [202] or during empathy for others in pain [222] (see also Sect. 4.3) as well as during social or romantic rejection [70, 138] is also accompanied by activity in parts of the pain matrix. However, to infer that a subject suffers from pain merely based on similarities in brain activation between psychologically and physically induced pain is arguable. This is especially true when using image acquisition and statistical analysis techniques with limited spatiotemporal sensitivity and specificity [120] (see Sect. 4.3).

The view that the pain matrix provides a direct window to study the neuronal underpinnings of pain function and dysfunction has been fundamentally challenged [150, 182]. Currently, it is questionable that the pain matrix provides an objective and direct measure of the pain perception as stated by lots of researchers [29, 241]. There is growing evidence that activity in the pain matrix can be heavily influenced by factors independent of the intensity of the nociceptive drive, like salience [119], attention [149, 194, 215], response-inhibition or even stimulus modality [104] including a variety of innocuous stimuli [182] (Fig. 11). Neuroimaging studies revealed that brain responses towards experimental contexts that either evoke pain or are attention-demanding show actually a remarkable overlap in the brain [66, 150]. Indeed there are several studies now showing that those regions assigned to the pain matrix are actually involved in several other mental operations often invoking a nonspecific salience response [54, 58, 235]. But an evolutionary inherent feature of the experience of pain is its salience meaning [56].

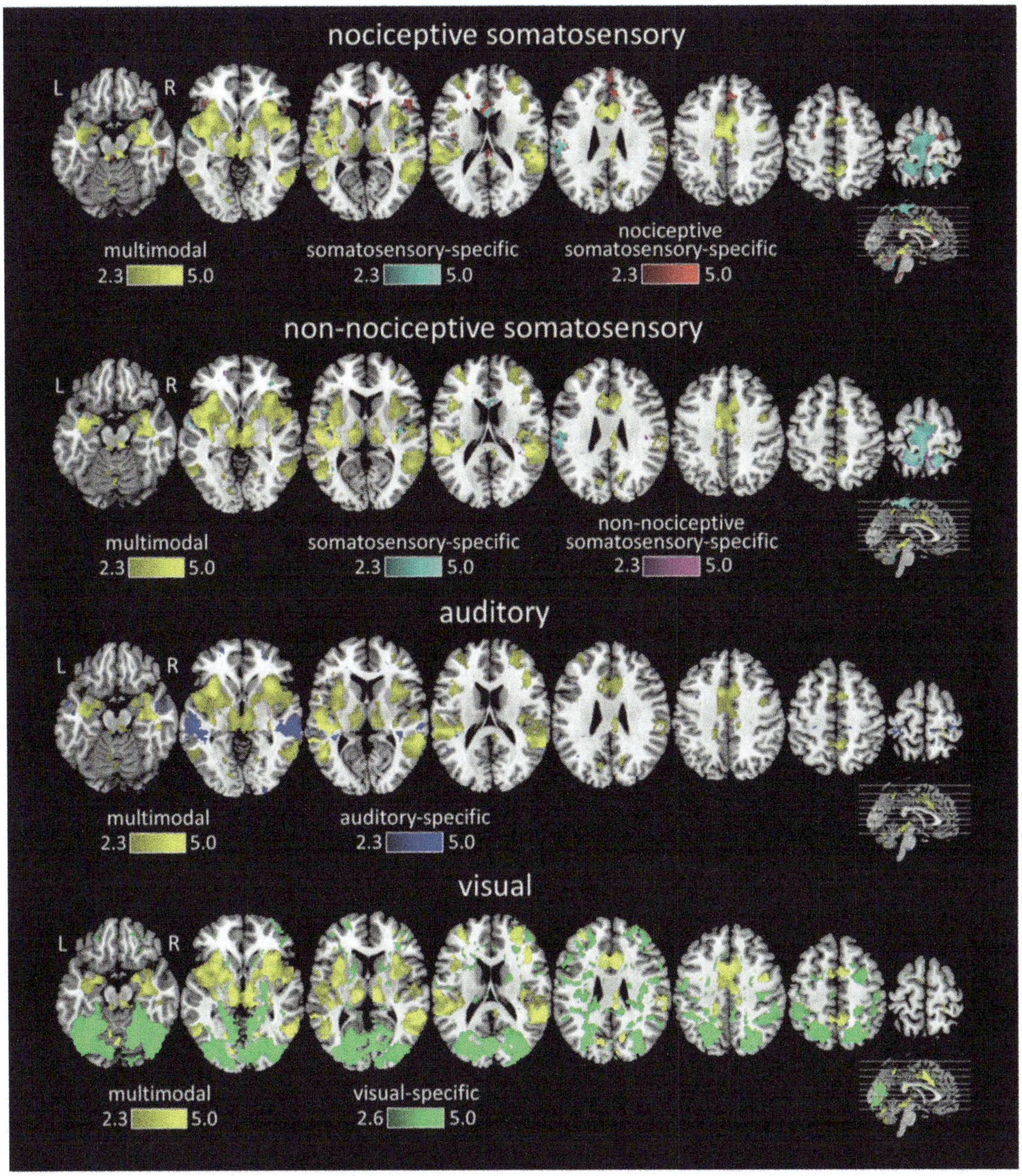

Fig. 11 Conjunction analysis showing high degree of overlap in brain responses to nociceptive, non-nociceptive and visual and auditory stimuli. Overlapping brain responses are shown in *yellow* (multimodal). Nociceptive and non-nociceptive stimuli as well as auditory and visual stimuli elicited comparable brain responses in anterior cingulate gyrus (ACC), insula and thalamus. Nociceptive and non-nociceptive stimuli elicited similar brain responses in primary somatosensory cortex (S1) and restricted parts of secondary somatosensory cortex (S2). Figure used from Mouraux et al. [182]

Abnormalities in brain networks processing the salience of stimuli might be also characteristic for pathological pain states [56]. Noteworthy, none of the brain regions being part of the pain matrix is exclusively linked to pain, but are rather involved in multiple other functions [6].

Further evidence for the nonspecificity of the pain matrix stems from studies analyzing brain correlates of nociception during states of reduced or altered states of consciousness. For example, in humans who are exposed to a high-dosage of propofol (which leads to loss of consciousness), some components of the pain matrix are still active [109]. Moreover, in an EEG-study by Bastuji et al. [20], nociceptive stimuli were applied to healthy-sleeping subjects. Even when the subjects were not awaked by arousing and alarming nociceptive stimuli, nociceptive stimuli elicited reproducible event-related potentials (ERPs) during both stage 2 and paradoxical sleep. Furthermore, activation in S1, S2, insula, and ACC was reported in subjects during states of minimal consciousness [26].

Apkarian et al. [6] criticizes the poor understanding of the underlying circuitry among brain regions of the pain matrix due to a lack of studies investigating functional integration during nociceptive stimulation. Thus neuroimaging remains at the level of modern (brain-based) phrenology [6, 56]. Furthermore, traditional imaging techniques often lack sensitivity and specificity to differentiate between similar cognitive states at the level of the brain. However, novel analysis methods such as MVPA have been used to show that brain activation pattern clearly differ between, for instance, "pain" due to romantic rejection versus pain during nociceptive stimulation [241]. Until now, there is no unique pattern of brain activity that allows undeniably the conclusion that a subject is in pain and how strong the pain is. There is accumulating evidence that the brain signatures of acute and spontaneous pain vary across different chronic pain syndromes (for a review see [6]). Pain is a multidimensional experience and different affective-motivational and cognitive variables can considerably affect brain circuitry during nociceptive stimulation. We want to emphasize that the investigation of brain signatures of pain must account for the multidimensionality of pain by effectively modeling the different components of pain. We propose that comparisons between chronic pain patients and healthy controls might reveal important differences in the neural processing of the different dimensions of pain either in presence or absence of nociceptive input. For instance, there is evidence showing systematic alterations in the neural processing of the affective-motivational dimension in chronic pain patients either in response to nociceptive stimulation [13] or during ongoing spontaneous pain [11].

7 Brain Signatures of Clinically Relevant Pain States Revealed by Novel Imaging Approaches

In conventional fMRI, the phenomenon under investigation must be separable into clear-cut on-and-off conditions or interrupted by rest, which is often contrary to the temporal characteristics of spontaneous ongoing pain, the defining feature of clinical pain states [190, 244]. However, it has been shown that the perception of pain to nociceptive or even non-noxious stimulation (i.e., allodynia) is often temporally dissociated from the stimulation in patients suffering from chronic pain [92, 142, 212]. Therefore, conventional fMRI is only poorly suited to capture certain aspects of pain that are only weakly stimulus-locked or show a high degree of variation with sustained or repetitive stimulation. In this section, we present selected studies using percept-related fMRI for the measurement of the temporally sluggish and variable nature of pain percepts relative to the stimulus and further to capture spontaneous fluctuations in spontaneous ongoing pain, in the absence of a stimulus.

Moreover, we want to present selected studies using machine learning algorithms that helped to predict the emergence or development of persistant pain or predict the therapeutic outcome in individual patients.

7.1 Percept-Related fMRI for the Mapping of the Temporal Dynamics of Spontaneous Pain

Baliki et al. [11] developed a finger-spanning device with which chronic back pain patients could continuously rate their subjective pain during an fMRI measurement. Thereby, two different temporal components were extracted from the ratings: (1) time periods with sustained spontaneous pain at a high intensity (subjective experience of chronic pain) and (2) time periods with transient increasing spontaneous pain (peripheral nociceptive input reaching the brain). Sustained pain was associated with BOLD-responses in the medial PFC in and around the rACC extending to the level of the genu. At a lower threshold, the authors additionally found bilateral activity in the posterior thalamus, ventral striatum, extended amygdala, and a large activity spread within the PFC. These results indicate that sustained spontaneous pain engages brain areas involved in emotion, cognition, and motivation [53, 63, 195] and suggesting the use of the second route of two proposed alternative routes mediating activity of medial PFC: (1) forward projections from parietal, insular, and cingulate regions encoding nociceptive information and (2) through midbrain and amygdala projections reflecting inputs from emotional and motivational circuitry [5, 115]. In contrast, in transient increasing habitual pain, Baliki et al. [11] found activation in the right anterior and posterior insula, S2, multiple portions of MCC, S1 region of the foot and cerebellum, corresponding to activations observed for acute pain containing regions of the sensory and affective dimension of pain [5, 44, 48, 231].

In a second experiment Baliki et al. [11], compared spontaneous pain with acute thermal heat pain. Brain activations for thermal heat pain were found in the bilateral insula, S2, cingulate cortex, and right dorsolateral PFC and were similar to the activations for transient increasing habitual pain. This data suggests a distinct network of brain areas responsible for habitual sustained spontaneous pain compared to transient increasing habitual and sustained acute thermal pain. In line with the results of Baliki et al. [11] the medial pain system is more active during the processing of habitual pain in patients with osteoarthritis whereas the lateral pain system is more active during acute experimental heat pain [140]. It seems plausible that other chronic pain disorders also process their clinical pain differently from experimental pain. However, it should be noted that fMRI does not optimally differentiate between anticipation and actual nociceptive processing and incomplete dissociation of these two aspects of processing could yield conflicting results in specific brain regions [140].

7.2 Segregating Pain Perception from Nociception in the Brain

The ACC and the insula have been commonly reported as typical nodes of the pain matrix. The ACC is discussed to process mainly affective components of the pain experience, while the insula is discussed to be either important for affective/emotional or sensory aspects of pain. Baliki et al. [12] used percept-related fMRI to disentangle their role for nociception versus pain perception respectively. Therefore the authors assessed the magnitude of the subjective experience of pain via a continuous rating within a group of healthy subjects. Participants rated the magnitude of pain in response to thermal stimulation using a finger-span device controlling the length of a bar which was visually presented. The authors found a segregated pattern of activity within the cingulate and insula differentiating between nociceptive- and pain-perceptive brain areas. The contrast between the stimulation and the perception regressor reveals that the ACC and the posterior insula better reflect pain perception and that large portions of the cingulate cortex and the insula reflect the thermal stimulus processing [6]. Moreover, the magnitude of pain was linearly correlated with activity in the posterior insula. Noteworthy, participants were also instructed to rate the magnitude of a visual stimulus also leading to activations in the posterior insula and the magnitude estimation of the visual stimulus also revealed a positive linear correlation with brain activity in this region. The authors conclude that the posterior insula might rather reflect a general magnitude estimator, which is independent of the stimulation-modality used. Indeed, inferring specificity of a brain region for pain processing based on pain magnitude ratings might be misleading since the coding of the intensity need not to occur for the basic perception of being in pain or not [55].

7.3 Segregating Pain Anticipation, Pain Perception and Pain Relief in Healthy and Chronic Back Pain Patients in the Brain

Baliki et al. [13] used percept-related fMRI to differentiate brain activity associated with pain anticipation, pain perception, or pain relief. Therefore the authors compared brain activity between healthy subjects and chronic back pain patients during randomly applied thermal heat pain stimuli. Most important, brain activity was contrasted between three epochs: (1) pain anticipation (brain activity during stimulus onset), pain perception (brain activity during the plateau) and pain relief (brain activity during stimulus offset). The anticipation epochs were related to brain activity in the nucleus accumbens (NAc), mid-cingulate and the anterior insula, the pain perception epochs were related to activity in the ACC, mid- and posterior insula and the dorsal striatum and the pain relief period epochs were linked to activity in the brainstem including the PAG. The authors were able to show that regions typically recruited in response to acute painful stimuli (nodes of the pain matrix) were recruited at different stages relative to the transformation of nociceptive information into subjective perception of pain.

A major challenge for pain researchers is the distinction of brain signatures linked to sensory-discriminative versus affective/emotional dimension of pain. The reason for this is the close correlation between the intensity of pain and the magnitude of unpleasantness [6]. Thus, it is hard to segregate sensory from affective responses to acute painful stimuli. Apkarian et al. [6] argues that the evidence for stating that a certain brain region of the pain matrix is sensory (e.g., S1) and another is affective (e.g., ACC) is often not validated or merely refers to anatomical distinctions (e.g., between the medial and lateral spinothalamic tracts). In the study by Baliki et al. [13], the sensory and affective dimension of pain was segregated in the brain by considering pain from the perspective of its motivational character. The motivational perspective accentuates the punishing and rewarding nature of pain forcing the subject to make decision with respect to the salient source of the painful stimulus. A painful stimulus involves a highly punishing effect during onset and maintenance of the stimulation but has a rewarding value during offset of the stimulation ("pain relief"). By investigating brain responses in relation to the different temporal components associated with a painful stimulation (pain anticipation, pain perception, and pain relief), the authors found evidence that phasic NAc-activity at stimulus offset represents an estimate of the magnitude of predicted pain relief in healthy subjects but predicted worsening of ongoing pain in chronic back pain patients. By considering pain from its motivational perspective, the authors were able to investigate the affective-motivational brain circuitry of pain independent from the sensory encoding of pain.

7.4 MVPA and ASL: Towards a Personalized Medicine

There is a huge effort but even a bigger need in effective pain management [141]. However, due to the absence of reliable biomarkers (e.g., brain signatures) of persistent pain the evaluation of treatment success is still bounded to subjective self-reports that often lack sensitivity, especially in highly vulnerable and cognitively impaired patient groups like older adults [21]. Pain researchers invest hopes in neuroimaging to improve pain phenotyping and to reveal more sensitive measures that complement subjective self-reports on pain [206].

Recent applications of machine learning algorithms techniques such as MVPA seem to be promising in delineating prognostic and diagnostic functional and structural imaging markers in pain [188, 241]. However, there is a lack of studies applying machine learning algorithms in clinically relevant pain states [188]. In a study by

O'Muircheartaigh et al. [188], multivariate machine learning algorithms were used in combination with an ASL-acquisition in patients with ongoing clinical pain (patients prior to and following surgical removal of molar teeth). Perfusion-based ASL allows the detection of changes in rCBF associated with ongoing, spontaneous pain, without the need of experimentally inducing pain in vulnerable patient populations [112, 237]. MVPA has two main advantages compared to conventional mass-univariate approaches: (1) MVPA acknowledges the spatially distributed (multi-voxel) nature of MRI datasets, offering greater sensitivity for the detection of spatially distributed MRI-effects, (2) it can make predictions at the level of the individual subject [187]. O'Muircheartaigh et al. [188] were able to accurately discriminate ongoing postsurgical pain from pain-free states within the same subjects without referring to subjective self-reports with a classification accuracy of 94.73%. Especially when dealing with vulnerable patients like chronic pain patients it is of crucial importance to reduce the total scanning time. Here, the greater sensitivity of MVPA compared to conventional fMRI measures allowed a tremendous reduction of scanning time [188].

7.5 Graph Analytical Approaches

Up so far, there is a lack of studies using graph theoretical approaches in chronic pain conditions to investigate structural (DTI tractography) or functional (e.g., resting state) connectivity. Graph theoretical measures allow the investigation of connectedness of not only two brain regions but to disentangle and prioritize relevant aspects of connectivity among large-scale networks of multiple brain regions.

In a study by Hashmi et al. [103] graph analytical measures revealing global and local network properties in resting-state networks were used to predict the efficacy of psychologically conditioned placebo analgesia effects to acupuncture treatment in chronic knee osteoarthritis patients. The authors hypothesized that greater network efficiency in resting-state networks allows better information transfer and enhance placebo induced analgesia by facilitating translation of psychological signals into pain relief. Analgesia was produced by building positive expectations towards acupuncture treatment with verbal suggestion and heat pain conditioning on a test site of the arm. The patients were naïve to the acupuncture treatment and were conditioned to believe that the needling procedure would produce analgesia only at the test site but not at the control site. Half of the patients were psychologically conditioned to the ulnar and the other half to the radial site of the arm. Additionally, sham acupuncture was performed at the test site with Streitberger needles that retract into the handle when pressed on the skin. The authors report significantly increased placebo analgesia at the test versus control site for both the acupuncture and the sham treatment, while there were no differences in analgesia between acupuncture and sham treatment.

The network topology and the efficiency of information transfer of the resulting graph were measured at the local and global level. Therefore a graph was constructed after parcellating the brains based on regions of interests mainly derived from a probabilistic anatomical atlas. The graph was thresholded at different levels. The statistical threshold was chosen to be not too liberal and not too conservative. Liberal thresholding involves the danger of preserving insignificant or inconsistent connections, while conservative thresholds might lead to isolated sub-graphs.

Clustering coefficients are measures of segregation assessing the amount of clustering in the network, which is expressed as the average proportion of connectivity between each node's neighbors. Thereby, the amount of clustering was compared to a random network. Characteristic path length is a measure of integration assessing the average number of edges that must be traversed to connect any two nodes of a network. Network efficiency measures the network's capacity for parallel information transfer between nodes via multiple series of edges. Global efficiency measures how efficiently information can

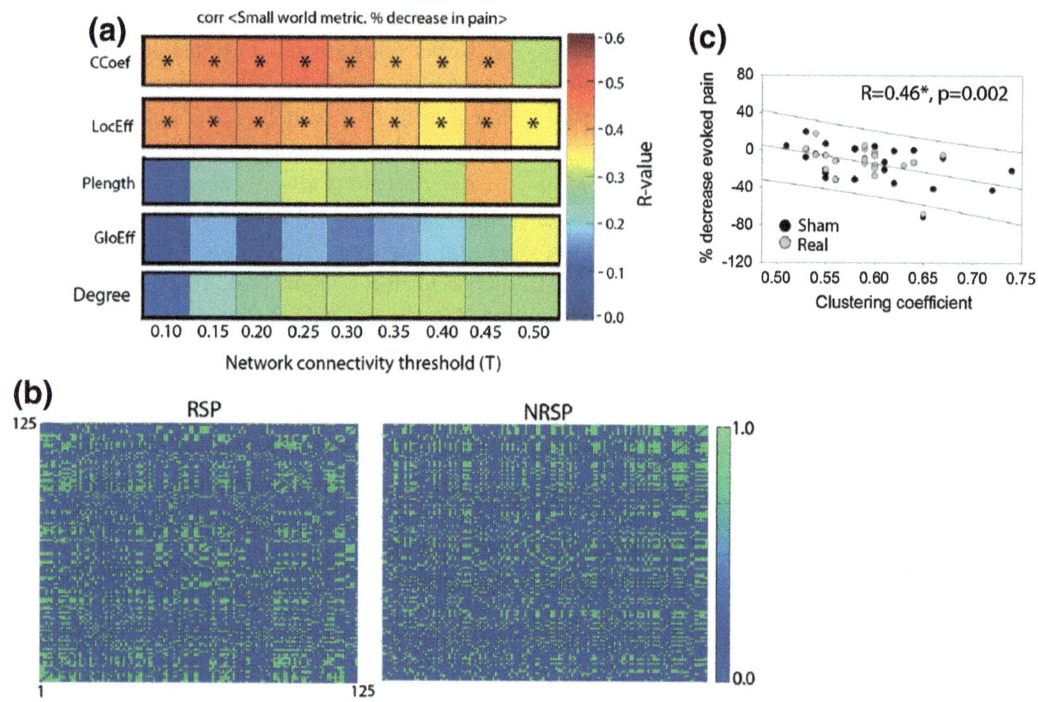

Fig. 12 Local aspects of the network topology of resting brains in osteoarthritis patients predict the amount of placebo analgesia. **a** Local (*CCoef* Clustering coefficient; *LocEff* Local efficiency) but not global (*Plenght* characteristic path length; *GloEff* Global efficiency, degree) measures of network topology and efficiency of information processing predict the percentage of decrease in pain to conditioned placebo treatment at different network connectivity thresholds. **b** NxN Correlation matrix of the 125 brain regions used for graph construction revealing greater clustering in responders (RSP) versus non-responders (NRSP). **c** Regression line with prediction intervals showing that the relationship between mean clustering coefficients and placebo-induce analgesia was not affected by the type of treatment (sham versus real acupuncture). **b** and **c** are plotted at a network connectivity threshold of $T = 0.25$. Figure adapted from Hashmi et al. [103]

be transferred across the entire network (is equal to the inverse of the characteristic path length), while local efficiency measures the efficiency of information transfer in local structures of the network. The node degree is a simple measure of influence/importance of a given node, simply reflecting the number of edges attached to a given node.

As depicted in Fig. 12, the authors showed that local but not global measures of the efficiency of local information processing (that is local efficiency and clustering coefficients) significantly predict the amount of conditioned placebo analgesia. Especially, clustering coefficients in regions associated with memory, motivation, and pain modulation were predictive for the amount of analgesia. The authors conclude that these alterations in efficiency of local networks might be a preparatory resource that facilitates subsequent performance of the brain in responding to adaptive environmental cues. This study reveals the usefulness of graph analysis in predicting treatment outcome for clinical use.

In a study by Liu et al. [152] resting-state fMRI was combined with graph theoretical approaches to compare changes in resting-state functional connectivity between patients suffering from migraine without aura and gender-matched healthy controls. To specifically address changes in patterns of functional connectivity with increasing duration of the disease, the patient group was subdivided into multiple groups varying in duration of the disease and compared to yoked controls. Liu et al. [152] found a systematic disruption in resting-state functional connectivity from local central nervous system to a more whole-brain network like topology with increasing disease duration in patients with long-term headache attacks compared to yoked controls. With increasing history of migraine the number of dysfunctional connections increased relative to the yoked controls. Notable, the size of the largest connected between-group differences network increased with disease progression with a size of 412 dysfunctional connections in the group with the longest disease history. Interestingly, the authors found that the disrupted connectivity did not follow a nonspecific generalized phenomenon of distribution, but rather transforms into a highly interconnected structure of mainly sensory-discriminative brain regions in patients with long-term headache attacks.

8 Conclusions

A central dogma of modern neuroscience is the plasticity of brain structure and function throughout lifetime [176, 198]. Thereby plastic changes can be adaptive as in musicians [72], learning for an exam [68], exercising juggling [67], or maladaptive to contexts such as ongoing nociceptive input, as it might be the case for the chronification of pain [80, 81]. Neuroimaging techniques promise to delineate the functional specialization and integration among brain regions specific for the encoding of experimental pain in health and disease. However, we emphasize the lack of studies investigating the clinically most striking aspect of chronic pain, which is (stimulus-independent) spontaneous ongoing pain.

We generally agree on the fact that pain is caused by bodily injury but, as a phenomenal experience, must emerge from neuronal activity in the brain [148]. Different neuroimaging approaches provide us the possibility to investigate the structural (e.g., DTI and volumetric imaging), functional (e.g., task-dependent and resting-state BOLD-fMRI, percept-related fMRI, ASL), and chemical (MR-spectroscopy) underpinnings of phenomenal experiences including pain. The use of neuroimaging techniques has revealed specific alterations in brain function, structure, and chemistry in different pain syndromes. These changes seem to be specific for different chronic pain syndromes arguing against a common brain signature of chronic pain. For instance, Baliki et al. [15] found unique structural brain signatures (GM) to be associated with different chronic pain populations. Contrarily the extent of chronicity of pain was localized to a common set of regions across all chronic pain conditions. Using percept-related fMRI in various chronic pain syndromes, Apkarian et al. [6] found varying patterns of brain activity correlating with spontaneous ongoing pain, but similar patterns of brain activity when comparing brain responses to various experimental painful stimuli. These results suggest that syndrome specific changes are mainly observed when assessing spontaneous pain and not experimental pain [6, 15].

We emphasized that morphological alteration in the course of pain chronification or pain relief (e.g., in response to a treatment) have to be interpreted with caution because they might be secondary to pain. We discussed that morphological signatures of worsening or relief of chronic pain might be also related to correlated changes in motor behavior and psychological comorbidities (anxiety or depression) that also lead to structural alterations in the brain. Accompanying changes in centrally acting pain medications have to be considered as source for structural brain plasticity too.

We (and others) criticize that most imaging studies on pain remain at the level of modern (brain-based) phrenologists [6, 148] since these studies focus only on localizing brain responses to painful stimuli but do not investigate the information flow between identified brain regions. As discussed in this chapter, the investigation of connectedness between brain regions allows the measurement of functional integration among brain regions and thus helps us to understand the mechanistic basis associated with the experience of pain. For instance, Sevel et al. [217] used dynamic causal modeling (DCM) to test whether differences in patterns of effective connectivity (that is causal relationships between brain regions) between pain-related brain regions can predict the amount of conditioned placebo analgesia in a group of healthy subjects. The authors found laterality specific differences in connection strengths with a right-hemispheric increase in modulatory dorsolateral PFC to PAG connectivity and a left-hemispheric endogenous connectivity between thalamus and dorsolateral PFC as being predictive for future placebo analgesia. An increasing number of imaging studies acknowledge that functioning of the brain can be understood from a network perspective. In this context, graph analytical approaches are a powerful tool for the investigation and extraction of relevant aspects of connectedness among multiple brain regions as derived from structural (e.g., DTI) or functional imaging (e.g., resting-state fMRI) data. Up so far, the application of graph analytical approaches in the field of chronic pain is limited [103, 144, 152].

We presented cutting-edge imaging approaches and associated statistical models that account for the inter-subject variability in functional and structural brain measures and offer new avenues towards a personalized medicine offering tailored treatments, thereby reducing side effects and increasing treatment efficacy. In this context, we introduced MVPA based on machine learning to infer whether a subject suffers from chronic pain or will benefit from a certain therapeutic intervention. These techniques might facilitate diagnosis, definition of brain-based biomarkers of relevant pain conditions, and help to improve the evaluation of pain treatments in the individual subject. Multivariate approaches show higher sensitivity and specificity in providing subject specific brain signatures associated with pain as compared to mass-univariate approaches [31]. The increased sensitivity of multivariate imaging approaches allows a substantial reduction of measurement time increasing the ethical acceptance for the measurement of chronic pain patients.

We conclude that it is still a matter of debate whether there is a common brain signature for chronic pain or if the different types of chronic pain can ultimately be described by their individual pathophysiology. Novel data acquisition (e.g., ASL) and analysis techniques (MVPA, graph analysis) offer us the opportunity to capture brain signatures of clinically relevant pain states with much higher spatiotemporal sensitivity and specificity, potentially reducing measurement time. Graph analytical approaches allow the identification of relevant network characteristics among complex patterns of connectivity within structural and functional imaging datasets. We strongly emphasize these techniques that have only been rarely applied in imaging of chronic pain.

References

1. Aguirre GK, Detre JA, Zarahn E, Alsop DC. Experimental design and the relative sensitivity of BOLD and perfusion fMRI. Neuroimage. 2002;15:488–500. doi:10.1006/nimg.2001.0990.
2. Alger JR. The diffusion tensor imaging toolbox. J Neurosci. 2012;32:7418–28. doi:10.1523/JNEUROSCI.4687-11.2012.
3. Apkarian AV. Chronic back pain as associated with decreased prefrontal and thalamic gray matter density. J Neurosci. 2004;24:10410–5.
4. Apkarian AV, Baliki MN, Geha PY. Towards a theory of chronic pain. Prog Neurobiol. 2009;87:81–97. doi:10.1016/j.pneurobio.2008.09.018.
5. Apkarian AV, Bushnell MC, Treede R-DD, Zubieta J-KK. Human brain mechanisms of pain perception and regulation in health and disease. Eur J Pain. 2005;9:463–84. doi:10.1016/j.ejpain.2004.11.001.
6. Apkarian AV, Hashmi JA, Baliki MN. Pain and the brain: specificity and plasticity of the brain in clinical chronic pain. Pain. 2011;152.
7. Apkarian AV, Sosa Y, Sonty S, Levy RM, Harden RN, Parrish TB, Gitelman DR. Chronic back pain is associated with decreased prefrontal and thalamic gray matter density. J Neurosci. 2004;24:10410–5.
8. Ashburner J, Friston KJ. Voxel-based morphometry—The methods. Neuroimage. 2000;11:805–21. doi:10.1006/nimg.2000.0582.
9. Babiloni C, Babiloni F, Carducci F, Cincotti F, Rosciarelli F, Rossini P, Arendt-Nielsen L, Chen A. Mapping of early and late human somatosensory evoked brain potentials to phasic galvanic painful stimulation. Hum Brain Mapp. 2001;12:168–79. doi:10.1002/1097-0193(200103)12:3<168::AID-HBM1013>3.0.CO;2-O.
10. Baliki MN, Apkarian AV. Nociception, pain, negative moods, and behavior selection. Neuron. 2015;87:474–91. doi:10.1016/j.neuron.2015.06.005.

11. Baliki MN, Chialvo DR, Geha PY, Levy RM, Harden RN, Parrish TB, Apkarian AV. Chronic pain and the emotional brain: specific brain activity associated with spontaneous fluctuations of intensity of chronic back pain. J Neurosci. 2006;26:12165–73. doi:10.1523/JNEUROSCI.3576-06.2006.
12. Baliki MN, Geha PY, Apkarian AV. Parsing pain perception between nociceptive representation and magnitude estimation. J Neurophysiol. 2009;101:875–87. doi:10.1152/jn.91100.2008.
13. Baliki MN, Geha PY, Fields HL, Apkarian AV. Predicting value of pain and analgesia: nucleus accumbens response to noxious stimuli changes in the presence of chronic pain. Neuron. 2010;66:149–60. doi:10.1016/j.neuron.2010.03.002.
14. Baliki MN, Mansour AR, Baria AT, Apkarian AV. Functional reorganization of the default mode network across chronic pain conditions. PLoS ONE. 2014;9:e106133. doi:10.1371/journal.pone.0106133.
15. Baliki MN, Schnitzer TJ, Bauer WR, Apkarian AV. Brain morphological signatures for chronic pain. PLoS ONE. 2011;6:e26010. doi:10.1371/journal.pone.0026010.
16. Barad MJ, Ueno T, Younger J, Chatterjee N, Mackey S. Complex regional pain syndrome is associated with structural abnormalities in painrelated regions of the human brain. J Pain. 2014;15:197–203.
17. Baron R, Baron Y, Disbrow E, Roberts TP. Brain processing of capsaicin-induced secondary hyperalgesia: a functional MRI study. Neurology. 1999;53:548–57.
18. Basser PJ, Mattiello J, LeBihan D. MR diffusion tensor spectroscopy and imaging. Biophys J. 1994;66:259–67. doi:10.1016/S0006-3495(94)80775-1.
19. Basser PJ, Pajevic S, Pierpaoli C, Duda J, Aldroubi A. In vivo fiber tractography using DT-MRI data. Magn Reson Med. 2000;44:625–32.
20. Bastuji H, Perchet C, Legrain V, Montes C, Garcia-Larrea L. Laser evoked responses to painful stimulation persist during sleep and predict subsequent arousals. Pain. 2008;137:589–99.
21. Bicket MC, Mao J. Chronic pain in older adults. Anesthesiol Clin. 2015;33:577–90. doi:10.1016/j.anclin.2015.05.011.
22. Bingel U, Lorenz J, Schoell E, Weiller C, Buchel C. Mechanisms of placebo analgesia: rACC recruitment of a subcortical antinociceptive network. Pain. 2006;120:8–15.
23. Bingel U, Schoell E, Herken W, Büchel C, May A. Habituation to painful stimulation involves the antinociceptive system. Pain. 2007;131:21–30. doi:10.1016/j.pain.2006.12.005.
24. Biswal BB, Mennes M, Zuo X-N, Gohel S, Kelly C, Smith SM, Beckmann CF, Adelstein JS, Buckner RL, Colcombe S, Dogonowski A-M, Ernst M, Fair D, Hampson M, Hoptman MJ, Hyde JS, Kiviniemi VJ, Kötter R, Li S-J, Lin C-P, Lowe MJ, Mackay C, Madden DJ, Madsen KH, Margulies DS, Mayberg HS, McMahon K, Monk CS, Mostofsky SH, Nagel BJ, Pekar JJ, Peltier SJ, Petersen SE, Riedl V, Rombouts SARB, Rypma B, Schlaggar BL, Schmidt S, Seidler RD, Siegle GJ, Sorg C, Teng G-J, Veijola J, Villringer A, Walter M, Wang L, Weng X-C, Whitfield-Gabrieli S, Williamson P, Windischberger C, Zang Y-F, Zhang H-Y, Castellanos FX, Milham MP. Toward discovery science of human brain function. Proc Natl Acad Sci USA. 2010;107:4734–9. doi:10.1073/pnas.0911855107.
25. Bolwerk A, Seifert F, Maihöfner C. Altered resting-state functional connectivity in complex regional pain syndrome. J Pain. 2013;14(1107–1115):e8. doi:10.1016/j.jpain.2013.04.007.
26. Boly M, Faymonville ME, Schnakers C, Peigneux P, Lambermont B, Phillips C, Lancellotti P, Luxen A, Lamy M, Moonen G, Maquet P, Laureys S. Perception of pain in the minimally conscious state with PET activation: an observational study. Lancet Neurol. 2008;7:1013–20.
27. Bookstein FL. "Voxel-based morphometry" should not be used with imperfectly registered images. Neuroimage. 2001;14:1454–62. doi:10.1006/nimg.2001.0770.
28. Bornhövd K, Quante M, Glauche V, Bromm B, Weiller C, Büchel C. Painful stimuli evoke different stimulus-response functions in the amygdala, prefrontal, insula and somatosensory cortex: a single-trial fMRI study. Brain. 2002;125:1326–36.
29. Borsook D, Sava S, Becerra L. The pain imaging revolution: advancing pain into the 21st century. Neuroscientist. 2010;16:171–85. doi:10.1177/1073858409349902.
30. Botvinick M, Jha AP, Bylsma LM, Fabian SA, Solomon PE, Prkachin KM. Viewing facial expressions of pain engages cortical areas involved in the direct experience of pain. Neuroimage. 2005;25:312–9. doi:10.1016/j.neuroimage.2004.11.043.
31. Brodersen KH, Wiech K, Lomakina EI, Lin C-S, Buhmann JM, Bingel U, et al. Decoding the perception of pain from fMRI using multivariate pattern analysis. Neuroimage. 2012;63:1162–70.
32. Bromm B. Brain images of pain. News Physiol Sci. 2001;16:244–9.
33. Bromm B. The involvement of the posterior cingulate gyrus in phasic pain processing of humans. Neurosci Lett. 2004;361:245–9.
34. Bromm B, Lorenz J. Neurophysiological evaluation of pain. Electroencephalogr Clin Neurophysiol. 1998;107:227–53.
35. Büchel C, Bornhovd K, Quante M, Glauche V, Bromm B, Weiller C. Dissociable neural responses related to pain intensity, stimulus intensity, and stimulus awareness within the anterior cingulate cortex: a parametric single-trial laser functional magnetic resonance imaging study. 2002.

36. Burgmer M, Pogatzki-Zahn E, Gaubitz M, Wessoleck E, Heuft G, Pfleiderer B. Altered brain activity during pain processing in fibromyalgia. Neuroimage. 2009;44:502–8. doi:10.1016/j.neuroimage.2008.09.008 S1053-8119(08)00999-3 [pii].
37. Bushnell MC, Duncan GH, Hofbauer RK, Ha B, Chen JI, Carrier B. Pain perception: is there a role for primary somatosensory cortex? Proc Natl Acad Sci USA. 1999;96:7705–9.
38. Callan D, Mills L, Nott C, England R, England S. A tool for classifying individuals with chronic back pain: using multivariate pattern analysis with functional magnetic resonance imaging data. PLoS ONE. 2014;9:e98007. doi:10.1371/journal.pone.0098007.
39. Chang PF, Arendt-Nielsen L, Graven-Nielsen T, Chen AC. Psychophysical and EEG responses to repeated experimental muscle pain in humans: pain intensity encodes EEG activity. Brain Res Bull. 2003;59:533–43.
40. Chang PF, Arendt-Nielsen L, Graven-Nielsen T, Svensson P, Chen AC. Topographic effects of tonic cutaneous nociceptive stimulation on human electroencephalograph. Neurosci Lett. 2001a;305:49–52. S030439400101802X [pii].
41. Chang PF, Arendt-Nielsen L, Graven-Nielsen T, Svensson P, Chen AC. Different EEG topographic effects of painful and non-painful intramuscular stimulation in man. Exp Brain Res. 2001;141:195–203. doi:10.1007/s002210100864.
42. Chang PF, Arendt-Nielsen L, Graven-Nielsen T, Svensson P, Chen AC. Comparative EEG activation to skin pain and muscle pain induced by capsaicin injection. Int J Psychophysiol. 2004;51:117–26.
43. Chen R, Corwell B, Yaseen Z, Hallett M, Cohen LG. Mechanisms of cortical reorganization in lower-limb amputees. J Neurosci. 1998;18:3443–50.
44. Coghill RC, Sang CN, Maisog JM, Iadarola MJ. Pain intensity processing within the human brain: a bilateral, distributed mechanism. J Neurophysiol. 1999;82:1934–43.
45. Cohen LG, Bandinelli S, Findley TW, Hallett M. Motor reorganization after upper limb amputation in man. Brain. 1991;114:615–27.
46. Corradi-Dell'Acqua C, Tomasino B, Fink GR. What is the position of an arm relative to the body? Neural correlates of body schema and body structural description. J Neurosci. 2009;29:4162–71.
47. Carrillo-de-la-Pena MT, Trinanes Y, Gonzalez-Villar A, Gomez-Perretta C, Garcia-Larrea L. Filtering out repetitive auditory stimuli in fibromyalgia: a study of P50 sensory gating. Eur J Pain. 2015;19:576–84.
48. Casey KL, Minoshima S, Morrow TJ, Koeppe RA. Comparison of human cerebral activation pattern during cutaneous warmth, heat pain, and deep cold pain. J Neurophysiol. 1996;76:571–81.
49. Cook DB, Lange G, Ciccone DS, Liu WC, Steffener J, Natelson BH. Functional imaging of pain in patients with primary fibromyalgia. J Rheumatol. 2004;31:364–78.
50. Damoiseaux JS, Rombouts SARB, Barkhof F, Scheltens P, Stam CJ, Smith SM, Beckmann CF. Consistent resting-state networks across healthy subjects. Proc Natl Acad Sci USA. 2006;103:13848–53. doi:10.1073/pnas.0601417103.
51. Danziger N, Faillenot I, Peyron R. Can we share a pain we never felt? Neural correlates of empathy in patients with congenital insensitivity to pain. Neuron. 2009;61:203–12.
52. Danziger N, Prkachin KM, Willer JC. Is pain the price of empathy? The perception of others' pain in patients with congenital insensitivity to pain. Brain. 2006;129:2494–507.
53. Davidson RJ. Anxiety and affective style: role of prefrontal cortex and amygdala. Biol Psychiatry. 2002;51:68–80.
54. Davis KD, Hutchison WD, Lozano AM, Tasker RR, Dostrovsky JO. Human anterior cingulate cortex neurons modulated by attention-demanding tasks. J Neurophysiol. 2000;83:3575–7.
55. Davis KD, Kucyi A, Moayedi M. The pain switch: an "ouch" detector. Pain. 2015;156:2164–6.
56. Davis KD, Moayedi M. Central mechanisms of pain revealed through functional and structural MRI. J Neuroimmune Pharmacol. 2013;8:518–34. doi:10.1007/s11481-012-9386-8.
57. Davis KD, Pope GE, Crawley AP, Mikulis DJ. Neural correlates of prickle sensation: a percept-related fMRI study. Nat Neurosci. 2002;5:1121–2. doi:10.1038/nn955.
58. Davis KD, Taylor KS, Hutchison WD, Dostrovsky JO, McAndrews MP, Richter EO, Lozano AM. Human anterior cingulate cortex neurons encode cognitive and emotional demands. J Neurosci. 2005;25:8402–6. doi:10.1523/JNEUROSCI.2315-05.2005.
59. Derbyshire SW, Jones AK, Gyulai F, Clark S, Townsend D, Firestone LL. Pain processing during three levels of noxious stimulation produces differential patterns of central activity. Pain. 1997;73:431–45. S0304395997001383 [pii].
60. Diers M, Koeppe C, Diesch E, Stolle AM, Hölzl R, Schiltenwolf M, van Ackern K, Flor H. Central processing of acute muscle pain in chronic low back pain patients: an EEG mapping study. J Clin Neurophysiol. 2007;24:76–83.
61. Diers M, Koeppe C, Yilmaz P, Thieme K, Markela-Lerenc J, Schiltenwolf M, van Ackern K, Flor H. Pain ratings and somatosensory evoked responses to repetitive intramuscular and intracutaneous stimulation in fibromyalgia syndrome. J Clin Neurophysiol. 2008;25:153–60.
62. Diers M, Schley MT, Rance M, Yilmaz P, Lauer L, Rukwied R, Schmelz M, Flor H. Differential central pain processing following repetitive intramuscular proton/prostaglandin E(2) injections in female

fibromyalgia patients and healthy controls. Eur J Pain. 2011;15:716–23. doi:10.1016/j.ejpain.2010.12.002.
63. Dolan RJ. Emotion, cognition, and behavior. Science (80-.) 2002;298:1191–94.
64. Dowman R. SEP topographies elicited by innocuous and noxious sural nerve stimulation. I. Identification of stable periods and individual differences. Electroencephalogr Clin Neurophysiol. 1994;92:291–302.
65. Dowman R, Darcey TM. SEP topographies elicited by innocuous and noxious sural nerve stimulation. III. Dipole source localization analysis. Electroencephalogr Clin Neurophysiol. 1994;92:373–91.
66. Downar J, Mikulis DJ, Davis KD. Neural correlates of the prolonged salience of painful stimulation. Neuroimage. 2003;20:1540–51.
67. Draganski B, Gaser C, Busch V, Schuierer G, Bogdahn U, May A. Neuroplasticity: changes in grey matter induced by training. Nature. 2004;427:311–2. doi:10.1038/427311a.
68. Draganski B, Gaser C, Kempermann G, Kuhn HG, Winkler J, Büchel C, May A. Temporal and spatial dynamics of brain structure changes during extensive learning. J Neurosci. 2006;26:6314–7. doi:10.1523/JNEUROSCI.4628-05.2006.
69. Duerden EG, Albanese M-C. Localization of pain-related brain activation: a meta-analysis of neuroimaging data. Hum Brain Mapp. 2013;34:109–49. doi:10.1002/hbm.21416.
70. Eisenberger NI, Lieberman MD, Williams KD. Does rejection hurt? An FMRI study of social exclusion. Science. 2003;302:290–2. doi:10.1126/science.1089134.
71. Elbert T, Flor H, Birbaumer N, Knecht S, Hampson S, Larbig W, Taub E. Extensive reorganization of the somatosensory cortex in adult humans after nervous system injury. NeuroReport. 1994;5:2593–7.
72. Elbert T, Pantev C, Wienbruch C, Rockstroh B, Taub E. Increased cortical representation of the fingers of the left hand in string players. Science. 1995;270(80–):305–7.
73. Ferré J-C, Bannier E, Raoult H, Mineur G, Carsin-Nicol B, Gauvrit J-Y. Arterial spin labeling (ASL) perfusion: techniques and clinical use. Diagn Interv Imaging. 2013;94:1211–23. doi:10.1016/j.diii.2013.06.010.
74. Fieldsen D, Wood S. Dealing with phantom limb pain after amputation. Nurs Times. 2011;107:21–3.
75. Finger S. Origins of neuroscience: a history of explorations into brain function. New York: Oxford University Press; 1994.
76. Fisher CM. Phantom erection after amputation of penis. Case description and review of the relevant literature on phantoms. Can J Neurol Sci. 1999;26:53–6.
77. Flodin P, Martinsen S, Löfgren M, Bileviciute-Ljungar I, Kosek E, Fransson P. Fibromyalgia is associated with decreased connectivity between pain- and sensorimotor brain areas. Brain Connect. 2014;4:587–94. doi:10.1089/brain.2014.0274.
78. Flor H. Phantom-limb pain: characteristics, causes, and treatment. Lancet Neurol. 2002;1:182–9.
79. Flor H, Birbaumer N, Turk DC. The psychobiology of chronic pain. Adv Behav Res Ther. 1990;12:47–85.
80. Flor H, Braun C, Elbert T, Birbaumer N. Extensive reorganization of primary somatosensory cortex in chronic back pain patients. Neurosci Lett. 1997;224:5–8.
81. Flor H, Elbert T, Knecht S, Wienbruch C, Pantev C, Birbaumer N, Larbig W, Taub E. Phantom-limb pain as a perceptual correlate of cortical reorganization following arm amputation. Nature. 1995;375:482–4.
82. Flor H, Nikolajsen L, Jensen T. Phantom limb pain: a case of maladaptive CNS plasticity? Nat Rev Neurosci. 2006;7:873–81.
83. Flor H, Turk DC. Basic concepts of pain. Chronic pain: an integrated biobehavioral approach. Seattle: IASP Press; 2011. p. 3–25.
84. Foss JM, Apkarian AV, Chialvo DR. Dynamics of pain: fractal dimension of temporal variability of spontaneous pain differentiates between pain states. J Neurophysiol. 2006;95:730–6. doi:10.1152/jn.00768.2005.
85. Friston K, Moran R, Seth AK. Analysing connectivity with Granger causality and dynamic causal modelling. Curr Opin Neurobiol. 2013;23:172–8. doi:10.1016/j.conb.2012.11.010.
86. Friston KJ. Models of brain function in neuroimaging. Annu Rev Psychol. 2005;56:57–87. doi:10.1146/annurev.psych.56.091103.070311.
87. Friston KJ, Harrison L, Penny W. Dynamic causal modelling. Neuroimage 2003;19:1273–302.
88. Fritz H-C, McAuley JH, Wittfeld K, Hegenscheid K, Schmidt CO, Langner S, Lotze M. Chronic back pain is associated with decreased prefrontal and anterior insular gray matter: results from a population-based cohort study. J Pain. 2015;17:111–8. doi:10.1016/j.jpain.2015.10.003.
89. Garcia-Larrea L, Frot M, Valeriani M. Brain generators of laser-evoked potentials: from dipoles to functional significance. Neurophysiol Clin. 2003;33:279–92.
90. Gear R, Becerra L, Upadhyay J, Bishop J, Wallin D, Pendse G, Levine J, Borsook D. Pain facilitation brain regions activated by nalbuphine are revealed by pharmacological fMRI. PLoS ONE. 2013;8:e50169. doi:10.1371/journal.pone.0050169.
91. Geha PY, Baliki MN, Harden RN, Bauer WR, Parrish TB, Apkarian AV. The brain in chronic CRPS pain: abnormal gray-white matter interactions in emotional and autonomic regions. Neuron. 2008;60:570–81. doi:10.1016/j.neuron.2008.08.022.
92. Geha PY, Baliki MN, Wang X, Harden RN, Paice JA, Apkarian AV. Brain dynamics for perception of tactile allodynia (touch-induced pain) in postherpetic neuralgia. Pain. 2008;138:641–56. doi:10.1016/j.pain.2008.02.021.

93. Gerster JC, Hadj-Djilani A. Hearing and vestibular abnormalities in primary fibrositis syndrome. J Rheumatol. 1984;11:678–80.
94. Gibson SJ, Littlejohn GO, Gorman MM, Helme RD, Granges G. Altered heat pain thresholds and cerebral event-related potentials following painful CO2 laser stimulation in subjects with fibromyalgia syndrome. Pain. 1994;58:185–93.
95. Giesecke T, Gracely RH, Clauw DJ, Nachemson A, Duck MH, Sabatowski R, Gerbershagen HJ, Williams DA, Petzke F. Central pain processing in chronic low back pain. Evidence for reduced pain inhibition. Schmerz. 2006;20(411–414):416–7. doi:10.1007/s00482-006-0473-8.
96. Giesecke T, Gracely RH, Grant MA, Nachemson A, Petzke F, Williams DA, Clauw DJ. Evidence of augmented central pain processing in idiopathic chronic low back pain. Arthritis Rheum. 2004;50:613–23.
97. Gracely RH, Petzke F, Wolf JM, Clauw DJ. Functional magnetic resonance imaging evidence of augmented pain processing in fibromyalgia. Arthritis Rheum. 2002;46:1333–43. doi:10.1002/art.10225.
98. Gross J, Schnitzler A, Timmermann L, Ploner M. Gamma oscillations in human primary somatosensory cortex reflect pain perception. PLoS Biol 2007;5:e133. doi:10.1371/journal.pbio.0050133 06-PLBI-RA-1755R3 [pii].
99. Gwilym SE, Filippini N, Douaud G, Carr AJ, Tracey I. Thalamic atrophy associated with painful osteoarthritis of the hip is reversible after arthroplasty: a longitudinal voxel-based morphometric study. Arthritis Rheum. 2010;62:2930–40.
100. Han S, Fan Y, Xu X, Qin J, Wu B, Wang X, Aglioti SM, Mao L. Empathic neural responses to others' pain are modulated by emotional contexts. Hum Brain Mapp. 2009;30:3227–37.
101. Hansen TM, Olesen AE, Graversen C, Drewes AM, Frøkjaer JB. The effect of oral morphine on pain-related brain activation—An experimental functional magnetic resonance imaging study. Basic Clin Pharmacol Toxicol. 2015;117:316–22. doi:10.1111/bcpt.12415.
102. Hashmi JA, Davis KD. Effect of static and dynamic heat pain stimulus profiles on the temporal dynamics and interdependence of pain qualities, intensity, and affect. J Neurophysiol. 2008;100:1706–15. doi:10.1152/jn.90500.2008.
103. Hashmi JA, Kong J, Spaeth R, Khan S, Kaptchuk TJ, Gollub RL. Functional network architecture predicts psychologically mediated analgesia related to treatment in chronic knee pain patients. J Neurosci. 2014;34:3924–36. doi:10.1523/JNEUROSCI.3155-13.2014.
104. Hatem SM, Plaghki L, Mouraux A. How response inhibition modulates nociceptive and non-nociceptive somatosensory brain-evoked potentials. Clin Neurophysiol. 2007;118:1503–16.
105. Haynes J-D, Rees G. Predicting the orientation of invisible stimuli from activity in human primary visual cortex. Nat Neurosci. 2005;8:686–91. doi:10.1038/nn1445.
106. Henderson LA, Gandevia SC, Macefield VG. Gender differences in brain activity evoked by muscle and cutaneous pain: a retrospective study of single-trial fMRI data. Neuroimage. 2008;39:1867–76. doi:10.1016/j.neuroimage.2007.10.045 S1053-8119(07)00996-2 [pii].
107. Henderson LA, Peck CC, Petersen ET, Rae CD, Youssef AM, Reeves JM, Wilcox SL, Akhter R, Murray GM, Gustin SM. Chronic pain: lost inhibition? J Neurosci. 2013;33:7574–82. doi:10.1523/JNEUROSCI.0174-13.2013.
108. Hill A. Phantom limb pain: a review of the literature on attributes and potential mechanisms. J Pain Symptom Manag. 1999;17:125–42.
109. Hofbauer RK, Fiset P, Plourde G, Backman SB, Bushnell MC. Dose-dependent effects of propofol on the central processing of thermal pain. 2004.
110. Hofbauer RK, Rainville P, Duncan GH, Bushnell MC. Cortical representation of the sensory dimension of pain. J Neurophysiol. 2001;86:402–11.
111. Honey CJ, Sporns O, Cammoun L, Gigandet X, Thiran JP, Meuli R, Hagmann P. Predicting human resting-state functional connectivity from structural connectivity. Proc Natl Acad Sci USA. 2009;106:2035–40. doi:10.1073/pnas.0811168106.
112. Howard MA, Krause K, Khawaja N, Massat N, Zelaya F, Schumann G, Huggins JP, Vennart W, Williams SCR, Renton TF. Beyond patient reported pain: perfusion magnetic resonance imaging demonstrates reproducible cerebral representation of ongoing post-surgical pain. PLoS ONE. 2011;6:e17096. doi:10.1371/journal.pone.0017096.
113. Hsieh JC, Belfrage M, Stone-Elander S, Hansson P, Ingvar M. Central representation of chronic ongoing neuropathic pain studied by positron emission tomography. Pain. 1995;63:225–36. doi:10.1016/0304-3959(95)00048-w.
114. Hsu C, Sliwa JA. Phantom breast pain as a source of functional loss. Am J Phys Med Rehabil. 2004;83:659–62.
115. Hunt SP, Mantyh PW. The molecular dynamics of pain control. Nat Rev Neurosci. 2001;2:83–91.
116. Hunter JP, Katz J, Davis KD. The effect of tactile and visual sensory inputs on phantom limb awareness. Brain. 2003;126:579–89.
117. Hutton C, Draganski B, Ashburner J, Weiskopf N. A comparison between voxel-based cortical thickness and voxel-based morphometry in normal aging. Neuroimage. 2009;48:371–80. doi:10.1016/j.neuroimage.2009.06.043.
118. Iadecola C, Nedergaard M. Glial regulation of the cerebral microvasculature. Nat Neurosci. 2007;10:1369–76. doi:10.1038/nn2003.
119. Iannetti GD, Hughes NP, Lee MC, Mouraux A. Determinants of laser-evoked EEG responses: pain perception or stimulus saliency? J Neurophysiol 2008;100:815–28.

120. Iannetti GD, Salomons TV, Moayedi M, Mouraux A, Davis KD. Beyond metaphor: contrasting mechanisms of social and physical pain. Trends Cogn Sci. 2013;17:371–8. doi:10.1016/j.tics.2013.06.002.
121. Ichesco E, Puiu T, Hampson JP, Kairys AE, Clauw DJ, Harte SE, Peltier SJ, Harris RE, Schmidt-Wilcke T. Altered fMRI resting-state connectivity in individuals with fibromyalgia on acute pain stimulation. Eur J Pain. 2016;. doi:10.1002/ejp.832.
122. Ingvar M. Pain and functional imaging. Philos Trans R Soc Lond B Biol Sci. 1999;354:1347–58. doi:10.1098/rstb.1999.0483.
123. Ivo R, Nicklas A, Dargel J, Sobottke R, Delank KS, Eysel P, et al. Brain structural and psychometric alterations in chronic low back pain. Eur Spine J. 2013;22:1958–64.
124. Jackson PL, Brunet E, Meltzoff AN, Decety J. Empathy examined through the neural mechanisms involved in imagining how I feel versus how you feel pain. Neuropsychologia. 2006;44:752–61.
125. Jackson PL, Meltzoff AN, Decety J. How do we perceive the pain of others? A window into the neural processes involved in empathy. Neuroimage. 2005;24:771–9.
126. Jensen KB, Kosek E, Petzke F, Carville S, Fransson P, Marcus H, Williams SC, Choy E, Giesecke T, Mainguy Y, Gracely R, Ingvar M. Evidence of dysfunctional pain inhibition in Fibromyalgia reflected in rACC during provoked pain. Pain. 2009;144:95–100. doi:10.1016/j.pain.2009.03.018 S0304-3959(09)00188-2 [pii].
127. Jensen KB, Loitoile R, Kosek E, Petzke F, Carville S, Fransson P, Marcus H, Williams SC, Choy E, Mainguy Y, Vitton O, Gracely RH, Gollub R, Ingvar M, Kong J. Patients with fibromyalgia display less functional connectivity in the brain's pain inhibitory network. Mol Pain. 2012;8:32. doi:10.1186/1744-8069-8-32 1744-8069-8-32 [pii].
128. Jensen TS, Krebs B, Nielsen J, Rasmussen P. Phantom limb, phantom pain and stump pain in amputees during the first 6 months following limb amputation. Pain. 1983;17:243–56.
129. Jones D. Studying connections in the living human brain with diffusion MRI. Cortex. 2008;44:936–52. doi:10.1016/j.cortex.2008.05.002.
130. Juottonen K, Gockel M, Silén T, Hurri H, Hari R, Forss N. Altered central sensorimotor processing in patients with complex regional pain syndrome. Pain. 2002;98:315–23.
131. Karl A, Birbaumer N, Lutzenberger W, Cohen LG, Flor H. Reorganization of motor and somatosensory cortex in upper extremity amputees with phantom limb pain. J Neurosci. 2001;21:3609–18.
132. Karl A, Mühlnickel W, Kurth R, Flor H. Neuroelectric source imaging of steady-state movement-related cortical potentials in human upper extremity amputees with and without phantom limb pain. Pain. 2004;110:90–102. doi:10.1016/j.pain.2004.03.013.
133. Kew JJ, Ridding MC, Rothwell JC, Passingham RE, Leigh PN, Sooriakumaran S, Frackowiak RS, Brooks DJ. Reorganization of cortical blood flow and transcranial magnetic stimulation maps in human subjects after upper limb amputation. J Neurophysiol. 1994;72:2517–24.
134. Kim H, Kim J, Loggia ML, Cahalan C, Garcia RG, Vangel MG, Wasan AD, Edwards RR, Napadow V. Fibromyalgia is characterized by altered frontal and cerebellar structural covariance brain networks. Neuroimage Clin. 2015;7:667–77. doi:10.1016/j.nicl.2015.02.022.
135. Kobayashi Y, Kurata J, Sekiguchi M, Kokubun M, Akaishizawa T, Chiba Y, Konno S, Kikuchi S. Augmented cerebral activation by lumbar mechanical stimulus in chronic low back pain patients: an FMRI study. Spine (Phila Pa 1976). 2009;34:2431–6. doi:10.1097/BRS.0b013e3181b1fb76.
136. Kong J, White NS, Kwong KK, Vangel MG, Rosman IS, Gracely RH, Gollub RL. Using fMRI to dissociate sensory encoding from cognitive evaluation of heat pain intensity. Hum Brain Mapp. 2006;27:715–21.
137. Kooijman CM, Dijkstra PU, Geertzen JH, Elzinga A, van der Schans CP. Phantom pain and phantom sensations in upper limb amputees: an epidemiological study. Pain. 2000;87:33–41.
138. Kross E, Berman MG, Mischel W, Smith EE, Wager TD. Social rejection shares somatosensory representations with physical pain. Proc Natl Acad Sci USA. 2011;108:6270–5. doi:10.1073/pnas.1102693108.
139. Kuchinad A, Schweinhardt P, Seminowicz DA, Wood PB, Chizh BA, Bushnell MC. Accelerated brain gray matter loss in fibromyalgia patients: premature aging of the brain? J Neurosci. 2007;27:4004–7.
140. Kulkarni B, Bentley DE, Elliott R, Julyan PJ, Boger E, Watson A, et al. Arthritic pain is processed in brain areas concerned with emotions and fear. Arthritis Rheum. 2007;56:1345–54.
141. Kupers R, Kehlet H. Brain imaging of clinical pain states: a critical review and strategies for future studies. Lancet Neurol. 2006;5:1033–44. doi:10.1016/S1474-4422(06)70624-X.
142. Kwan CL, Diamant NE, Mikula K, Davis KD. Characteristics of rectal perception are altered in irritable bowel syndrome. Pain. 2005;113:160–71. doi:10.1016/j.pain.2004.10.005.
143. Kwiatek R, Barnden L, Tedman R. Regional cerebral blood flow in fibromyalgia: Single-photon–emission computed tomography evidence of reduction in the pontine tegmentum and thalami. Arthritis Rheum. 2000;43:2823–33. doi:10.1002/1529-0131(200012)43:12<2823:AID-ANR24>3.0.CO;2-E.
144. Labus JS, Dinov ID, Jiang Z, Ashe-McNalley C, Zamanyan A, Shi Y, et al. Irritable bowel syndrome

in female patients is associated with alterations in structural brain networks. Pain. 2014;155:137–49.
145. Lamm C, Nusbaum HC, Meltzoff AN, Decety J. What are you feeling? Using functional magnetic resonance imaging to assess the modulation of sensory and affective responses during empathy for pain. PLoS ONE. 2007;2:e1292.
146. Le Pera D, Svensson P, Valeriani M, Watanabe I, Arendt-Nielsen L, Chen AC. Long-lasting effect evoked by tonic muscle pain on parietal EEG activity in humans. Clin Neurophysiol. 2000;111: 2130–7. S1388-2457(00)00474-0 [pii].
147. Lee DH, Lee KJ, Cho KI, Noh EC, Jang JH, Kim YC, et al. Brain alterations and neurocognitive dysfunction in patients with complex regional pain syndrome. J Pain. 2015;16:580–6.
148. Lee MC, Tracey I. Imaging pain: a potent means for investigating pain mechanisms in patients. Brit J Anaesth. 2013;111:64–72. doi:10.1093/bja/aet174.
149. Legrain V, Guérit JM, Bruyer R, Plaghki L. Attentional modulation of the nociceptive processing into the human brain: selective spatial attention, probability of stimulus occurrence, and target detection effects on laser evoked potentials. Pain. 2002;99: 21–39.
150. Legrain V, Iannetti GD, Plaghki L, Mouraux A. The pain matrix reloaded: a salience detection system for the body. Prog Neurobiol. 2011;93:111–24.
151. Liu F, Guo W, Yu D, Gao Q, Gao K, Xue Z, Du H, Zhang J, Tan C, Liu Z, Zhao J, Chen H. Classification of different therapeutic responses of major depressive disorder with multivariate pattern analysis method based on structural MR scans. PLoS ONE. 2012;7:e40968. doi:10.1371/journal.pone.0040968.
152. Liu J, Zhao L, Lei F, Zhang Y, Yuan K, Gong Q, Liang F, Tian J. Disrupted resting-state functional connectivity and its changing trend in migraine suffers. Hum Brain Mapp. 2015;36:1892–907. doi:10.1002/hbm.22744.
153. Lockhart SN, DeCarli C. Structural imaging measures of brain aging. Neuropsychol Rev. 2014;24:271–89. doi:10.1007/s11065-014-9268-3.
154. Logothetis NK. What we can do and what we cannot do with fMRI. Nature. 2008;453:869–78. doi:10.1038/nature06976 nature06976 [pii].
155. Lorenz J. Hyperalgesia or hypervigilance? An evoked potential approach to the study of fibromyalgia syndrome. Z Rheumatol. 1998;57(Suppl 2):19–22.
156. Lorenz J, Cross DJ, Minoshima S, Morrow TJ, Paulson PE, Casey KL. A unique representation of heat allodynia in the human brain. Neuron. 2002;35:383–93.
157. Lorenz J, Grasedyck K, Bromm B. Middle and long latency somatosensory evoked potentials after painful laser stimulation in patients with fibromyalgia syndrome. Electroencephalogr Clin Neurophysiol. 1996;100:165–8. 0013469495002596 [pii].
158. Lotze M, Flor H, Grodd W, Larbig W, Birbaumer N. Phantom movements and pain. an fMRI study in upper limb amputees. Brain. 2001;124:2268–77. doi:10.1093/brain/124.11.2268.
159. Lutz J, Jager L, de Quervain D, Krauseneck T, Padberg F, Wichnalek M, Beyer A, Stahl R, Zirngibl B, Morhard D, Reiser M, Schelling G. White and gray matter abnormalities in the brain of patients with fibromyalgia: a diffusion-tensor and volumetric imaging study. Arthritis Rheum. 2008;58:3960–9.
160. MacDonald G, Leary MR. Why does social exclusion hurt? The relationship between social and physical pain. Psychol Bull. 2005;131:202–23. doi:10.1037/0033-2909.131.2.202.
161. Maestu C, Cortes A, Vazquez JM, del Rio D, Gomez-Arguelles JM, del Pozo F, et al. Increased brain responses during subjectively-matched mechanical pain stimulation in fibromyalgia patients as evidenced by MEG. Clin Neurophysiol. 2013;124:752–60.
162. Maihöfner C, Baron R, DeCol R, Binder A, Birklein F, Deuschl G, Handwerker HO, Schattschneider J. The motor system shows adaptive changes in complex regional pain syndrome. Brain. 2007;130:2671–87. doi:10.1093/brain/awm131.
163. Maihöfner C, Handwerker HO, Birklein F. Functional imaging of allodynia in complex regional pain syndrome. Neurology. 2006;66:711–7. doi:10.1212/01.wnl.0000200961.49114.39.
164. Maihöfner C, Handwerker HO, Neundörfer B, Birklein F. Patterns of cortical reorganization in complex regional pain syndrome. Neurology. 2003;61:1707–15.
165. Maihöfner C, Schmelz M, Forster C, Neundorfer B, Handwerker HO. Neural activation during experimental allodynia: a functional magnetic resonance imaging study. Eur J Neurosci. 2004;19:3211–8.
166. Makin TR, Scholz J, Henderson Slater D, Johansen-Berg H, Tracey I. Reassessing cortical reorganization in the primary sensorimotor cortex following arm amputation. Brain. 2015;138:2140–6. doi:10.1093/brain/awv161.
167. Marbach JJ. Phantom tooth pain: differential diagnosis and treatment. J Mass Dent Soc. 1996;44:14–8.
168. Marbach JJ. Orofacial phantom pain: theory and phenomenology. J Am Dent Assoc. 1996;127:221–9.
169. Marquand AF, O'Daly OG, De Simoni S, Alsop DC, Maguire RP, Williams SCR, Zelaya FO, Mehta MA. Dissociable effects of methylphenidate, atomoxetine and placebo on regional cerebral blood flow in healthy volunteers at rest: a multi-class pattern recognition approach. Neuroimage. 2012;60:1015–24. doi:10.1016/j.neuroimage.2012.01.058.
170. May A. Chronic pain may change the structure of the brain. Pain. 2008;137:7–15.
171. May A. Structural brain imaging: a window into chronic pain. Neurosci. 2011;17:209–20.

172. McDermid AJ, Rollman GB, McCain GA. Generalized hypervigilance in fibromyalgia: evidence of perceptual amplification. Pain. 1996;66:133–44.
173. Melzack R. Labat lecture. Phantom limbs. Reg Anesth. 1989;14:208–11.
174. Melzack R, Casey KL. Sensory, motivational and central control determinants of chronic pain: a new conceptual model. In: Kenshalo DR, editor. The skin senses: proceedings of the first international symposium on the skin senses. Florida: Florida State University in Tallahassee; 1968. p. 432.
175. Merskey H, Bogduk N. IASP task force on taxonomy Part III: pain terms, a current list with definitions and notes on usage. In: IASP Task Force Taxon. http://www.iasp-pain.org/Content/NavigationMenu/GeneralResourceLinks/PainDefinitions/default.htm#Pain; 1994.
176. Merzenich MM, Nelson RJ, Stryker MP, Cynader MS, Schoppmann A, Zook JM. Somatosensory cortical map changes following digit amputation in adult monkeys. J Comp Neurol. 1984;224:591–605.
177. Mohr C, Leyendecker S, Mangels I, Machner B, Sander T, Helmchen C. Central representation of cold-evoked pain relief in capsaicin induced pain: an event-related fMRI study. Pain. 2008;139:416–30.
178. Monti MM. Statistical analysis of fMRI time-series: a critical review of the GLM approach. Front Hum Neurosci. 2011;5:28.
179. Moriyama T, Higashi T, Togashi K, Iida T, Segi E, Sugimoto Y, Tominaga T, Narumiya S, Tominaga M. Sensitization of TRPV1 by EP1 and IP reveals peripheral nociceptive mechanism of prostaglandins. Mol Pain. 2005;1:3.
180. Morrison I, Downing PE. Organization of felt and seen pain responses in anterior cingulate cortex. Neuroimage. 2007;37:642–51.
181. Morrison I, Lloyd D, di Pellegrino G, Roberts N. Vicarious responses to pain in anterior cingulate cortex: is empathy a multisensory issue? Cogn Affect Behav Neurosci. 2004;4:270–8.
182. Mouraux A, Diukova A, Lee MC, Wise RG, Iannetti GD. A multisensory investigation of the functional significance of the "pain matrix". Neuroimage. 2011;54:2237–49. doi:10.1016/j.neuroimage.2010.09.084.
183. Moussa MN, Steen MR, Laurienti PJ, Hayasaka S. Consistency of network modules in resting-state FMRI connectome data. PLoS ONE. 2012;7:e44428. doi:10.1371/journal.pone.0044428.
184. Napadow V, LaCount L, Park K, As-Sanie S, Clauw DJ, Harris RE. Intrinsic brain connectivity in fibromyalgia is associated with chronic pain intensity. Arthritis Rheum. 2010;62:2545–55. doi:10.1002/art.27497.
185. Niddam DM, Graven-Nielsen T, Arendt-Nielsen L, Chen AC. Non-painful and painful surface and intramuscular electrical stimulation at the thenar and hypothenar sites: differential cerebral dynamics of early to late latency SEPs. Brain Topogr. 2001;13:283–92.
186. Nikolajsen L, Ilkjaer S, Krøner K, Christensen JH, Jensen TS. The influence of preamputation pain on postamputation stump and phantom pain. Pain. 1997;72:393–405.
187. Norman KA, Polyn SM, Detre GJ, Haxby JV. Beyond mind-reading: multi-voxel pattern analysis of fMRI data. Trends Cogn Sci. 2006;10:424–30. doi:10.1016/j.tics.2006.07.005.
188. O'Muircheartaigh J, Marquand A, Hodkinson DJ, Krause K, Khawaja N, Renton TF, Huggins JP, Vennart W, Williams SCR, Howard MA. Multivariate decoding of cerebral blood flow measures in a clinical model of on-going postsurgical pain. Hum Brain Mapp. 2015;36:633–42. doi:10.1002/hbm.22652.
189. O'Reilly JX, Woolrich MW, Behrens TEJ, Smith SM, Johansen-Berg H. Tools of the trade: psychophysiological interactions and functional connectivity. Soc Cogn Affect Neurosci. 2012;7:604–9. doi:10.1093/scan/nss055.
190. Owen DG, Bureau Y, Thomas AW, Prato FS, St. Lawrence KS. Quantification of pain-induced changes in cerebral blood flow by perfusion MRI. Pain. 2008;136:85–96. doi:10.1016/j.pain.2007.06.021.
191. Pasley B, Freeman R. Neurovascular coupling. Scholarpedia. 2008;3:5340. doi:10.4249/scholarpedia.5340.
192. Pereira F, Mitchell T, Botvinick M. Machine learning classifiers and fMRI: a tutorial overview. Neuroimage. 2009;45:S199–209. doi:10.1016/j.neuroimage.2008.11.007.
193. Pernet CR. Misconceptions in the use of the general linear model applied to functional MRI: a tutorial for junior neuro-imagers. Front Neurosci. 2014;8:1. doi:10.3389/fnins.2014.00001.
194. Petrovic P, Petersson KM, Ghatan PH, Stone-Elander S, Ingvar M. Pain-related cerebral activation is altered by a distracting cognitive task. Pain. 2000;85:19–30.
195. Phelps EA, Delgado MR, Nearing KI, LeDoux JE. Extinction learning in humans: role of the amygdala and vmPFC. Neuron. 2004;43:897–905.
196. Poldrack RA. Modeling brain connectivity. Handbook of functional MRI data analysis. Cambridge: Cambridge University Press; 2011. p. 130–59.
197. Poldrack RA, Mumford JA, Nichols TE. Multivoxel pattern analysis and machine learning. Handbook of functional MRI data analysis. Cambridge: Cambridge University Press; 2011. p. 160–72.
198. Pons TP, Garraghty PE, Mishkin M. Lesion-induced plasticity in the second somatosensory cortex of adult macaques. Proc Natl Acad Sci USA. 1988;85:5279–81.
199. Porro CA. Functional imaging and pain: behavior, perception, and modulation. Neuroscientist. 2003;9:354–69.
200. Price DD, Verne GN, Schwartz JM. Plasticity in brain processing and modulation of pain. Prog Brain Res. 2006;157:333–52.

201. Pujol J, Lopez-Sola M, Ortiz H, Vilanova JC, Harrison BJ, Yucel M, et al. Mapping brain response to pain in fibromyalgia patients using temporal analysis of FMRI. PLoS One. 2009;4: e5224.
202. Raij TT, Numminen J, Närvänen S, Hiltunen J, Hari R. Brain correlates of subjective reality of physically and psychologically induced pain. Proc Natl Acad Sci USA. 2005;102:2147–51. doi:10.1073/pnas.0409542102.
203. Rainville P. Brain mechanisms of pain affect and pain modulation. Curr Opin Neurobiol. 2002;12:195–204.
204. Rainville P, Duncan GH, Price DD, Carrier B, Bushnell MC. Pain affect encoded in human anterior cingulate but not somatosensory cortex. Science. 1997;277(80–):968–71.
205. Ramachandran VS, McGeoch PD. Occurrence of phantom genitalia after gender reassignment surgery. Med Hypotheses. 2007;69:1001–3.
206. Robinson ME, Staud R, Price DD. Pain measurement and brain activity: will neuroimages replace pain ratings? J Pain. 2013;14:323–7. doi:10.1016/j.jpain.2012.05.007.
207. Rodriguez-Raecke R, Niemeier A, Ihle K, Ruether W, May A. Brain gray matter decrease in chronic pain is the consequence and not the cause of pain. J Neurosci. 2009;29:13746–50.
208. Rothemund Y, Grusser SM, Liebeskind U, Schlag PM, Flor H. Phantom phenomena in mastectomized patients and their relation to chronic and acute pre-mastectomy pain. Pain. 2004;107:140–6.
209. Rubinov M, Sporns O. Complex network measures of brain connectivity: uses and interpretations. Neuroimage. 2010;52:1059–69. doi:10.1016/j.neuroimage.2009.10.003.
210. Ruehle BS, Handwerker HO, Lennerz JK, Ringler R, Forster C. Brain activation during input from mechanoinsensitive versus polymodal C-nociceptors. J Neurosci. 2006;26:5492–9. doi:10.1523/JNEUROSCI.2059-05.2006 26/20/5492 [pii].
211. Saarela MV, Hlushchuk Y, Williams AC, Schurmann M, Kalso E, Hari R. The compassionate brain: humans detect intensity of pain from another's face. Cereb Cortex. 2007;17:230–7.
212. Samuelsson M, Leffler A-S, Hansson P. Dynamic mechanical allodynia: on the relationship between temporo-spatial stimulus parameters and evoked pain in patients with peripheral neuropathy. Pain. 2005;115:264–72. doi:10.1016/j.pain.2005.03.001.
213. Schmidt-Wilcke T, Leinisch E, Ganssbauer S, Draganski B, Bogdahn U, Altmeppen J, May A. Affective components and intensity of pain correlate with structural differences in gray matter in chronic back pain patients. Pain. 2006;125:89–97.
214. Schmidt-Wilcke T, Luerding R, Weigand T, Jurgens T, Schuierer G, Leinisch E, Bogdahn U. Striatal grey matter increase in patients suffering from fibromyalgia—A voxel-based morphometry study. 2007.
215. Schwenkreis P, Pleger B, Cornelius B, Weyen U, Dertwinkel R, Zenz M, Malin JP, Tegenthoff M. Reorganization in the ipsilateral motor cortex of patients with lower limb amputation. Neurosci Lett. 2003;349:187–90.
216. Seminowicz DA, Labus JS, Bueller JA, Tillisch K, Naliboff BD, Bushnell MC, et al. Regional gray matter density changes in brains of patients with irritable bowel syndrome. Gastroenterology. 2010;139(48–57):e2.
217. Sevel LS, O'Shea AM, Letzen JE, Craggs JG, Price DD, Robinson ME. Effective connectivity predicts future placebo analgesic response: a dynamic causal modeling study of pain processing in healthy controls. Neuroimage. 2015;110:87–94.
218. Shehzad Z, Kelly AMC, Reiss PT, Gee DG, Gotimer K, Uddin LQ, Lee SH, Margulies DS, Roy AK, Biswal BB, Petkova E, Castellanos FX, Milham MP. The resting brain: unconstrained yet reliable. Cereb Cortex. 2009;19:2209–29. doi:10.1093/cercor/bhn256.
219. Sherman RA, Sherman CJ, Bruno GM. Psychological factors influencing chronic phantom limb pain: an analysis of the literature. Pain. 1987;28:285–95. 0304-3959(87)90064-9 [pii].
220. Shimojo M, Svensson P, Arendt-Nielsen L, Chen AC. Dynamic brain topography of somatosensory evoked potentials and equivalent dipoles in response to graded painful skin and muscle stimulation. Brain Topogr. 2000;13:43–58.
221. Simoes EL, Bramati I, Rodrigues E, Franzoi A, Moll J, Lent R, Tovar-Moll F. Functional expansion of sensorimotor representation and structural reorganization of callosal connections in lower limb amputees. J Neurosci. 2012;32:3211–20. doi:10.1523/JNEUROSCI.4592-11.2012.
222. Singer T, Seymour B, O'Doherty J, Kaube H, Dolan RJ, Frith CD. Empathy for pain involves the affective but not sensory components of pain. Science. 2004;303:1157–62. doi:10.1126/science.1093535.
223. Smith SM, Miller KL, Moeller S, Xu J, Auerbach EJ, Woolrich MW, Beckmann CF, Jenkinson M, Andersson J, Glasser MF, Van Essen DC, Feinberg DA, Yacoub ES, Ugurbil K. Temporally-independent functional modes of spontaneous brain activity. Proc Natl Acad Sci USA. 2012;109:3131–6. doi:10.1073/pnas.1121329109.
224. Sporns O. Structure and function of complex brain networks. Dialogues Clin Neurosci. 2013;15:247–62. doi:10.1137/S003614450342480.
225. Sporns O. Brain connectivity. Scholarpedia. 2007;2:4695. doi:10.4249/scholarpedia.4695.
226. Steen AE, Reeh PW, Geisslinger G, Steen KH. Plasma levels after peroral and topical ibuprofen and effects upon low pH-induced cutaneous and muscle pain. Eur J Pain. 2000;4:195–209.

227. Stephan KE, Friston KJ. Analyzing effective connectivity with fMRI. Wiley Interdiscip Rev Cogn Sci. 2010;1:446–59. doi:10.1002/wcs.58.
228. Sundgren PC, Petrou M, Harris RE, Fan X, Foerster B, Mehrotra N, et al. Diffusion-weighted and diffusion tensor imaging in fibromyalgia patients: a prospective study of whole brain diffusivity, apparent diffusion coefficient, and fraction anisotropy in different regions of the brain and correlation with symptom severity. Acad Radiol. 2007;14:839–46.
229. Svensson P, Minoshima S, Beydoun A, Morrow TJ, Casey KL. Cerebral processing of acute skin and muscle pain in humans. J Neurophysiol. 1997;78:450–60.
230. Tagliazucchi E, Balenzuela P, Fraiman D, Chialvo DR. Brain resting state is disrupted in chronic back pain patients. Neurosci Lett. 2010;485:26–31. doi:10.1016/j.neulet.2010.08.053.
231. Talbot JD, Marrett S, Evans AC, Meyer E, Bushnell MC, Duncan GH. Multiple representations of pain in human cerebral cortex. Science (80-.) 1991;251:1355–58.
232. Tegenthoff M, Ragert P, Pleger B, Schwenkreis P, Förster A-F, Nicolas V, Dinse HR. Improvement of tactile discrimination performance and enlargement of cortical somatosensory maps after 5 Hz rTMS. PLoS Biol. 2005;3:e362. doi:10.1371/journal.pbio.0030362.
233. Tiemann L, Schulz E, Winkelmann A, Ronel J, Henningsen P, Ploner M. Behavioral and neuronal investigations of hypervigilance in patients with fibromyalgia syndrome. PLoS One. 2012;7:e35068.
234. de Tommaso M, Federici A, Santostasi R, Calabrese R, Vecchio E, Lapadula G, et al. Laser-evoked potentials habituation in fibromyalgia. J Pain. 2011;12:116–24.
235. Torta DM, Cauda F. Different functions in the cingulate cortex, a meta-analytic connectivity modeling study. Neuroimage. 2011;56:2157–72. doi:10.1016/j.neuroimage.2011.03.066.
236. Tracey I. Nociceptive processing in the human brain. Curr Opin Neurobiol. 2005;15:478–87. doi:10.1016/j.conb.2005.06.010.
237. Tracey I, Johns E. The pain matrix: reloaded or reborn as we image tonic pain using arterial spin labelling. Pain. 2010;148:359–60. doi:10.1016/j.pain.2009.11.009.
238. Treede RD, Kenshalo DR, Gracely RH, Jones AK. The cortical representation of pain. Pain. 1999;79:105–11.
239. Upadhyay J, Baker SJ, Rajagovindan R, Hart M, Chandran P, Hooker BA, Cassar S, Mikusa JP, Tovcimak A, Wald MJ, Joshi SK, Bannon A, Medema JK, Beaver J, Honore P, Kamath RV, Fox GB, Day M. Pharmacological modulation of brain activity in a preclinical model of osteoarthritis. Neuroimage. 2013;64:341–55. doi:10.1016/j.neuroimage.2012.08.084.
240. Valet M, Sprenger T, Boecker H, Willoch F, Rummeny E, Conrad B, Erhard P, Tolle TR. Distraction modulates connectivity of the cingulo-frontal cortex and the midbrain during pain—an fMRI analysis. Pain. 2004;109:399–408. doi:10.1016/j.pain.2004.02.033S0304395904000958.
241. Wager TD, Atlas LY, Lindquist MA, Roy M, Woo C-W, Kross E. An fMRI-based neurologic signature of physical pain. N Engl J Med. 2013;368:1388–97. doi:10.1056/NEJMoa1204471.
242. Wager TD, Rilling JK, Smith EE, Sokolik A, Casey KL, Davidson RJ, Kosslyn SM, Rose RM, Cohen JD. Placebo-induced changes in FMRI in the anticipation and experience of pain. Science. 2004;303:1162–7. doi:10.1126/science.1093065.
243. Wang J, Aguirre GK, Kimberg DY, Roc AC, Li L, Detre JA. Arterial spin labeling perfusion fMRI with very low task frequency. Magn Reson Med. 2003;49:796–802. doi:10.1002/mrm.10437.
244. Wasan AD, Loggia ML, Chen LQ, Napadow V, Kong J, Gollub RL. Neural correlates of chronic low back pain measured by arterial spin labeling. Anesthesiology. 2011;115:364–74. doi:10.1097/ALN.0b013e318220e880.
245. Wilcox CE, Mayer AR, Teshiba TM, Ling J, Smith BW, Wilcox GL, Mullins PG. The subjective experience of pain: an FMRI study of percept-related models and functional connectivity. Pain Med. 2015;. doi:10.1111/pme.12785.
246. Wolfe F, Clauw DJ, Fitzcharles MA, Goldenberg DL, Katz RS, Mease P, Russell AS, Russell IJ, Winfield JB, Yunus MB. The American College of Rheumatology preliminary diagnostic criteria for fibromyalgia and measurement of symptom severity. Arthritis Care Res. 2010;62:600–10. doi:10.1002/acr.20140.
247. Yang TT, Gallen C, Schwartz B, Bloom FE, Ramachandran VS, Cobb S. Sensory maps in the human brain. Nature. 1994;368:592–3.

Author Biographies

Martin Diers is Professor for Clinical and Experimental Behavioral Medicine at the Department of Psychosomatic Medicine and Psychotherapy, LWL University Hospital, Ruhr-University Bochum, Germany. He holds a PhD in Cognitive and Clinical Neuroscience from the Central Institute of Mental Health, Medical Faculty Mannheim, Heidelberg University. His scientific interests focus on psychobiological mechanisms in chronic musculoskeletal and neuropathic pain.

Christopher Milde is PhD candidate at the Department of Cognitive and Clinical Neuroscience, Central Institute of Mental Health, Medical Faculty Mannheim, Heidelberg University. His scientific interests focus on the application of surfaced-based morphometry on multimodal imaging datasets and the investigation of the neural representation of the body in illusory contexts and chronic pain.

Imaging of Pain in the Peripheral Nerves

9

Claudia Cejas and Diego Pineda

Abstract

The role in the assessment of the status of the plexus and peripheral nerves was traditionally yielded to the clinical evaluation and electrophysiological studies for a long time. High-resolution Magnetic Resonance nerve imaging (MR Neurography) has emerged as one of the most complete tools for the recognition of the anatomy and to categorize the abnormalities of the peripheral nervous system. Nowadays, Magnetic Resonance Neurography (MRN) with extended anatomical coverage is possible, improving the ability to discern between focal and nonfocal neuropathies. The aim of this chapter is to discuss the technical parameters of MRN, the anatomy of the brachial, lumbosacral plexus and the peripheral nerves, and the applications of MRN in the diagnosis of peripheral neuropathies and plexopathies. Additionally, the MRN features of the relevant causes (focal and diffuse neuropathies, vascular, traumatic, entrapment, and inflammatory radiation-induced and neoplastic diseases) are presented through illustrative cases.

Keywords

Neuropathies • Plexopathies • Protocol • Fascicular abnormality • Disease • Iatrogenic

1 Introduction

In the clinical setting the diagnosis of plexopathies and peripheral nerve disorders traditionally relies on information provided by physical examination and electrophysiological tests.

The main limitation of the electrophysiological testing is the failure on demonstration of the extent, exact location, type of nerve injury, and involvement of the perineural tissues [1].

C. Cejas (✉)
Department of Radiology, MRI Division, Foundation of Neurological Research, FLENI, Buenos Aires, Argentina
e-mail: ccejas@fleni.org.ar

D. Pineda
Department of Radiology, Neuroradiologist, FLENI, Montañeses 2325, CABA, C1428AQK Buenos Aires, Argentina
e-mail: dpineda@fleni.org.ar
URL: http://www.fleni.org.ar/

L. Saba (ed.), *Neuroimaging of Pain*, DOI 10.1007/978-3-319-48046-6_9

MRI has emerged as a useful complementary tool for the clinical routine. MRI of peripheral nerves also referred to as MR Neurography (MRN) by Chhabra et al. provides important information about nerve anatomy and pathology [2]. MRN has been successfully used to confirm the clinical diagnosis by direct visualization of the nerve abnormality and contributes to rule out other conditions such as compressive masses or inflammatory conditions [3, 4]. Moreover, it is helpful to evaluate the muscle denervation [5].

2 Technical Aspects Regarding Peripheral Nerve in MR Neurography

MRN examinations can be divided into anatomic (conventional T2-based) techniques and functional (diffusion-based) techniques. A complete MRN examination incorporates selective pulse sequences with high-resolution multiplanar isotropic images in order to visualize the nerve anatomy and lesions in a tailored plane.

The MRN imaging protocol includes both axial and longitudinal planes over the peripheral nerves using a combination of high-resolution 2D and 3D spin echo-type imaging techniques on a (ideally 3T) high-field scanner. The use of an iterative approach for lipid–water decomposition based upon the 2-point Dixon deconvolution technique has been introduced and applied to a series of MR imaging pulses including FSE imaging SPACE (Siemens, Erlangen, Germany), IDEAL (General Electric Healthcare, Milwaukee, WI), and VISTA (volumetric isotropic T2-weighted acquisition, Philips, Best, Netherlands) techniques [6, 7].

Sequence CUBE T2-weighted images acquires sub-millimeter isotropic 3D volume scan, without gaps, therefore without partial volume effect. These ultra thin slices help to visualize the nerve into any plane correctly [8].

Three-dimensional imaging is useful to obtain a longitudinal view of the nerves, especially through curved planar and MIP reformations. These images are particularly beneficial to delineate course deviations or focal changes in nerve caliber, and to demonstrate the relationship to the adjacent space-occupying lesions for presurgical planning. Fascicular abnormality is, however, best seen on axial planes, which is the pillar of diagnosis for neuropathy [9].

Diffusion tensor imaging (DTI) and tractography (3D fiber tracking) are useful techniques to evaluate the axonal tracts and structure. These methods allow to study the direction of axons and to quantify the tissue characteristics in order to distinguish diffusion anisotropy and other related parameters of water diffusion [10]. A *b*-value between 1000 and 1400 has been recently recommended [11].

Mean fractional anisotropy (FA) and apparent diffusion coefficient (ADC) map values may reflect microstructural changes and have been reported alterated in areas that may appear normal on anatomic MR images [12]. Tables 1, 2 and 3 summarize the models of the protocols used to study the brachial and lumbosacral plexus and, peripheral nerve, respectively, in author's institution.

Table 1 Protocol to study brachial plexus

Sequence	FOV	Thickness	TR/TE	Matrix	NEX
3D FIESTA axial	27	1.4/0.0	5/1.9	320 × 320	0.8
3D FIESTA coronal	35	0.6/0.0	4.3/1.8	320 × 320	0.8
IDEAL coronal T2	35	1.0/0.1	7080/92.7	320 × 256	3
e-DWI (b: 0–250–500–1000)	40	3	10,000/min	128 × 128	8
3D IDEAL coronal T1 (before/after contrast injection)	40	1.0/0.1	1320/10.2	320 × 224	3

FIESTA steady state free precession; *IDEAL* iterative decomposition of water and fat with echo asymmetry and least squares estimation; *e-DWI* diffusion

Table 2 Protocol to study lumbosacral plexus

Sequence	FOV (cm)	Thickness	TR/TE (ms)	Matrix	NEX
FSE FS T2 axial	44 × 44	4.0/0.5	4840/97.0	512 × 256	2
IDEAL T2 coronal	42 × 42	1.5/0.0	5240/86.5	320 × 256	3
CUBE T2 coronal	42 × 42	1.0/0.0	2060/182.8	288 × 288	1
e-DWI axial	40 × 28	2.4/0.0	6500/102.2	128 × 128	8
3D IDEAL T1 coronal (before/after contrast injection)	42 × 42	1.0/0.1	1320/10.2	320/224	3

FSE FS fast spin echo fat saturation; *IDEAL* iterative decomposition of water and fat with echo asymmetry and least squares estimation; *e-DWI* diffusion

Table 3 Protocol to study peripheral nerve

Sequence	FOV	Thickness	TR/TE	Matrix	NEX
3D FIESTA axial	18	1.0/0.0	5/1.9	320 × 320	1
FSE DP FS	18	3.0/0.5	2000/30	320 × 320	2
CUBE coronal T2	26	1.0/0.0	2300/255	320 × 320	1
IDEAL coronal T2	26	1.0/0.1	3620/90	320 × 256	3
e-DWI (b:0–250–500–1000)	26	3	10,000/min	128 × 128	8
3D IDEAL coronal T1 (before/after contrast injection)	26	1.0/0.1	1320/10.2	320 × 224	3

FIESTA steady state free precession; *FSE DP FS* fast spin echo proton density fat saturation; *IDEAL* iterative decomposition of water and fat with echo asymmetry and least squares estimation; *e-DWI* diffusion

3 Normal Appearance of Nerves in MR Neurography

MRN attempts to reflect the microscopic architecture of the nerve trunk. The smallest unit of the nerve is the axon enveloped by a layer of Schwann cells and connective tissue stroma called the endoneurium. Several axons joined together forms a fascicle covered by the perineurium, a connective tissue layer. Then the fascicular bundles are covered for an outer connective tissue layer, the epineurium. On imaging, the normal peripheral nerves more than 2–3 mm are easily visualized [13].

The normal nerves appear slightly hyperintense on T1-weighted and T2-weighted images compared to muscle and hypointense regarding the surrounding fat [2]. The endoneurial fluid can be seen as a slight hyperintensity in T2-weighted images. This findings could be better appreciated in the 3D inversion recovery images (SPAIR, STIR, TIRM, or T2-water IDEAL) than in the fat suppression sequences and be symmetric along the whole course of the nerve.

The size of the normal nerve is usually similar than the adjacent artery and gradually decreases along its distal course with minimal variations near the ramifications. The fascicular appearance, straight course and branching appearance allows the distinction of the normal peripheral nerves from the surrounding vessels (Fig. 1). Using the appropriate surface coils the thin epineurial layer, can also be seen in the largest nerves. The perineurial and epineurial layers are imperceptible, unless they are abnormally thickened [14–18].

4 Brachial Neuropathies

The brachial plexus (BP) is a set of nerve confluences and ramifications, which supply motor and sensory innervation to the upper extremity. Brachial plexus may be affected by a wide

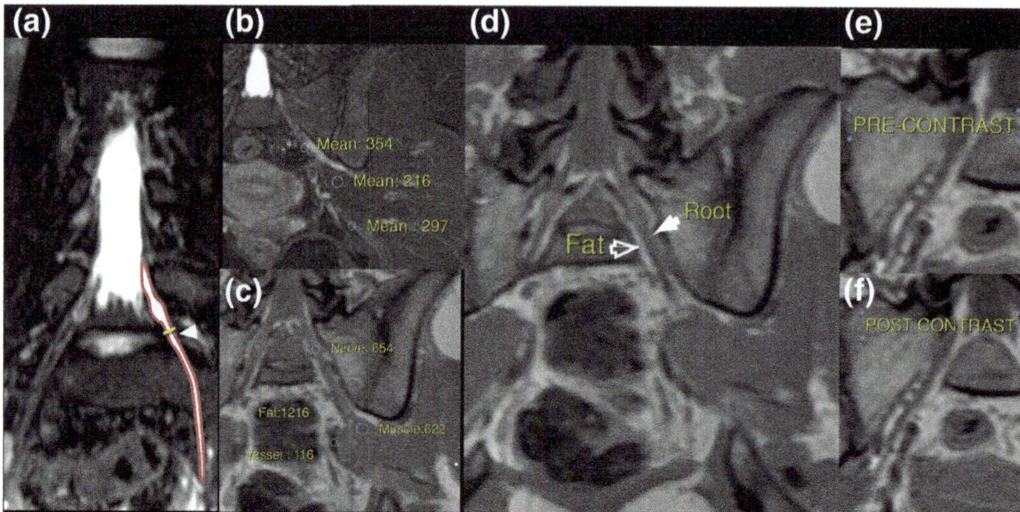

SIZE	SIGNAL INTENSITY	PERINEURAL FAT	COURSE	CONTRAST ENHANCEMENT
Similar to adjacent vessel. L5: widest root. L3: thinnest root.	T2-SAT: slightly hyperintense to muscle. T1-in phase: isointense to muscle. Vessel flow void is also a reference.	T1-in phase: hyperintense to root – nerve and clear margins. T2-SAT: hypointense to root-nerve.	Continuous and closely related to vessel.	No enhancement except in dorsal ganglia root.

Fig. 1 Normal appearance of the peripheral nerve in MR Neurography: SIZE (**a**). 3D IDEAL coronal FAT SAT T2-weighted image: symmetry of the normal L5 roots in shown (*arrowhead* in *red* drawing and *right* L5 root). A median of 5.6 mm was estimated in our normal population. SIGNAL INTENSITY (**b** and **c**). 3D IDEAL coronal FAT SAT T2-weighted image and 3D IDEAL coronal T1 in-phase: ROI's comparing signal intensity between muscle and nerve. In T2 weighted images nerve is slightly hyperintense to muscle, and in T1 in-phase weighted images nerve is isointense to muscle. PERINEURAL FAT (**d**). 3D IDEAL coronal T1 in-phase weighted images: normal perineural fat is hyperintense to the nerve (*white* and *void arrow* in **d**) and a sharp interphase can be recognized. Normally root or nerve trunk normally exhibits a straight and uninterrupted course. Contrast enhancement is not evident, except for the dorsal ganglia (**e**, **f**)

spectrum of conditions, such as trauma, entrapment, inflammation, and tumoral disease. Table 4 summarizes the causes of brachial plexopathy.

The main goal of the imaging diagnostic work-up in BP conditions is to determine the exact location of the injury and to differentiate between a mononeuropathy (single nerve), a multiple neuropathy (multiples nerves), or polyneuropathy based in them imaging features [19].

MRN offers an extended anatomical coverage of the BP and multiplanar high-resolution images of the nerves as well as the adjacent soft tissues. Also provides an objective assessment of the neuromuscular anatomy and related abnormalities. The purpose of this section is to review the appearance of the normal and abnormal BP with MRN.

4.1 Anatomy

Brachial plexus is divided into five segments: roots; trunks; divisions; cords; and terminal branches. The plexus is formed by the ventral rami of the lower four cervical (C5, C6, C7, C8) and the first thoracic (T1) roots. It may receive additional contributions from the C4 and T2 nerve roots. After exiting the neural foramina in

Table 4 Causes of brachial plexopathy

Trauma	Mechanical or entrapment	Inflammatory/vascular	Neoplastic and infiltrative	Iatrogenic
Vehicular Obstetric	Thoracic outlet syndrome	Diabetic Parsonage turner syndrome Chronic inflammatory demyelinating polyneuropathy Guillain–Barré syndrome Vasculitis Sarcoidosis	Nerve sheath tumors: schwannoma, neurofibroma, MPNST Metastases Lymphoma	Radiation, traction during anesthesia Shoulder surgery

MNST malignant peripheral nerve sheath tumors

the cervical spine, the C5 and C6 nerve roots constitute the upper trunk, C7 remain as the middle trunk, and C8 and T1 form the lower trunk (Fig. 2) [20, 21].

These three trunks run between the anterior and middle scalene muscles (interscalene space) and then separate into three anterior and three posterior divisions. The divisions run behind to the clavicule and joined together to form the lateral, posterior and medial cords, which name come from their spatial relationship with the subclavian artery [22]. The cords pass through the costoclavicular space. The three posterior divisions form the posterior cord, which separates into the thoracodorsal, subscapular, axillary and radial nerves. The three anterior divisions form two cords: the lateral cord, which receives contributions from the superior and medial trunks, and the medial cord, which is a direct continuation of the inferior trunk. The lateral cord then divides into two branches: the musculocutaneous nerve and other which joins with a similar branch from the medial cord to form the median nerve. The rest portion of the medial cord forms the medial brachial cutaneous, the medial antebrachial cutaneous and the ulnar nerves [23, 24]. Table 5 shows the innervation of BP.

4.2 Traumatic Injuries

The most frequent causes of injury of the BP are the motor vehicle accidents, sports related injuries, gunshot wounds, rucksack injuries, and iatrogenic traction during anesthesia. Traumatic injuries are divided into preganglionic, postganglionic, or mixed. Postganglionic injuries may demonstrate continuity or discontinuity of the nerve; making surgical repair of the nerve necessary in the last case [25].

From a pathological point of view, peripheral nerve injuries have been classified into neurapraxia, axonotmesis, and neurotmesis according to Seddon [26].

Neurapraxia, the mildest type of injury, involves the myelin sheath with axonal sparing. Muscle signal intensity on MRI could be normal or there is evidence of acute denervation. MRN demonstrates intraneural edema as an abnormal hyperintensity in T2 or STIR weighted images with mild enlargement of the nerve. The functional loss of nerve is temporary and the management is usually conservative. A complete recovery is expected within three weeks.

In axonotmesis, the axonal injury is complete with distal Wallerian degeneration but with perineurium and epineurium remaining unimpaired. This type of lesion causes acute muscle denervation appreciated on MRI as an increased signal in T2 or STIR weighted images as early as 4 days after insult. In the main of cases it is managed conservatively, but may require surgical exploration. MRN findings include effacement, enlargement, or disruption of individual fascicles in addition to the other imaging features seen in neurapraxia.

Neurotmesis is the most severe grade of nerve injury, with disruption of both the myelin sheath

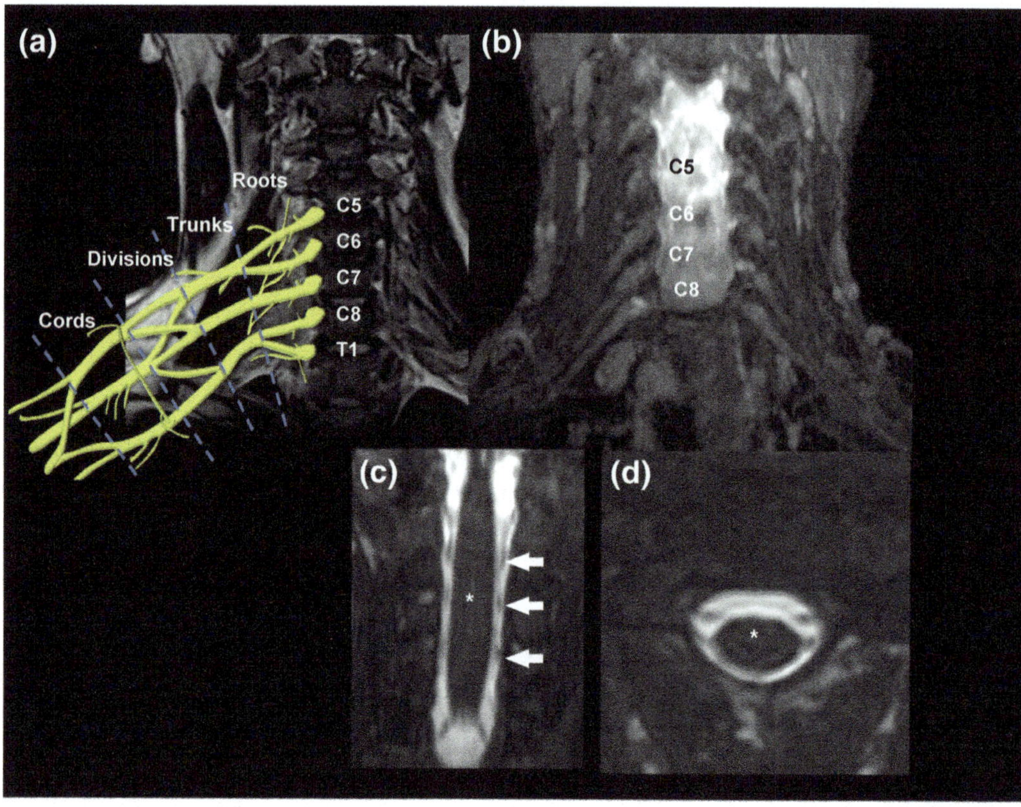

Fig. 2 Normal appearance of the brachial plexus in MR neurography: schema of the basic anatomy of the brachial plexus (**a**). The plexus is formed by the ventral rami of the *lower* four cervical (C5, C6, C7, C8) and the first thoracic (T1) roots. The C5 and C6 nerve roots constitute the *upper* trunk, C7 remain as the *middle* trunk, and C8 and T1 form the *lower* trunk. These three trunks running through the inter-scalene space, and then they are separate into three anterior and three posterior divisions. The divisions are posterior to clavicule and joined together to form the lateral, posterior, and medial cords, which name come from their spatial relationship with the subclavian artery. The cords pass through the costoclavicular space. The three posterior divisions form the posterior cord. The three anterior divisions form two cords: the lateral cord, which receives contributions from the superior and medial trunks, and the medial cord, which is a direct continuation of the inferior trunk. Sequence 3D IDEAL coronal FAT SAT T2-weighted image (**b**), shows bilateral C5 to C8 roots. Sequence 3D GRE coronal (**c**), and axial (**d**), show the spinal cord (*asterisk*) and, the anterior and posterior pre ganglionic roots (*arrows*). (Schema credited to Tatiana Escobar, MD)

and the axonal component with partial or complete discontinuity of the root (avulsion). Recovery cannot occur spontaneously and early surgical repair is required often with an interposed nerve graft.

It is essential to differentiate between complete and incomplete root injury or even from a rootlet avulsion.

A dural tear with pseudomeningocele formation, (an extradural fluid collection into the neural foramen) may be seen on conventional MR imaging and may be considered a pathognomonic finding of nerve root avulsion. However, approximately 20% of avulsions occur without evidence of pseudomeningocele formation [27].

MRN may also demonstrate loss of the normal fascicular appearance and discontinuity between both split nerve endings (Fig. 3). In almost 20% of patients with preganglionic injuries a spinal cord abnormality may appear as an increased signal in T2 or STIR weighted images representing edema. Hemorrhagic complications also may appear and commonly manifest as a

Table 5 Muscles innervated by the brachial plexus

Nerve	Roots	Motor innervation	Sensitive innervation
Phrenic N	C5	Diaphragm	
Dorsal scapulary N	C5	Levator scapulae, rhomboid	
Long thoracic N	C5, C6	Serratus anterior	
Subclavius N	C5, C6	Subclavius	
Suprascapular N	C5, C6	Supraspinatus, infraspinatus	
Lateral pectoral N	C5, C6, C7	Pectoralis major, minor	
Subscapular N	C5, C6	Subscapularis, teres major	
Thoracodorsal N	C6, C7, C8	Latissimus dorsi	
Axillary N	C5, C6	Deltoid, teres minor	Shoulder and lateral upper arm
Musculocutaneous N	C5, C6, C7	Biceps, brachialis, cracobrachialis	This continues us lateral cutaneous nerve of the forearm
Median N	C5, C6, C7, C8, T1	Flexor carpi radialis, pronator teres, palmaris longus, flexor digitorum superficialis and profunfus, flexor pollicis longus, abductor pollicis brevis, flexor pollicis brevis, 2–3 lumbricals	Region thenar and palm, elbow and wrist, thumb, index, middle and half of the ring
Radial N	C5, C6, C7, C8, T1	Triceps, supinator, anconeous, brachioradialis, extensor muscles of forearm	Posterior region of arm
Ulnar N	C8, T1	Flexor carpi ulnaris, flexor digitorum profundus, adductor pollicis, abductor digiti V, dorsal and palmar interosseous and, lumbricals to 4–5 fingers	Medial region of the arm, one and half finger of palmar side and two and half fingers of the dorsal side

decreased signal in T2 or T2* weighted images due to deposition of blood degradation products.

In the subacute and chronic stage the fibrosis is seen as a strandy hypointense soft tissue on T2-weighted imaging at the injury site, commonly with cord retraction. Terminal neuromas may develop as masses in continuity with the proximal stump of the injured nerve [28].

Muscles supplied by nerves distal to the site of injury may show denervation features. Hyperintensity in T2-weighted images of the paraspinal muscles due to edema, especially multifidus, has been shown to be an accurate indirect sign of root avulsion injury (Fig. 4) [29]. Imaging can be used to confirm the exact localization and, the degree of nerve injury and it is important to predict the functional outcome [30].

4.3 Mechanical and Entrapment Conditions: Thoracic Outlet Syndrome

The thoracic outlet syndrome (TOS) is an entrapment neuropathy where the BP may be

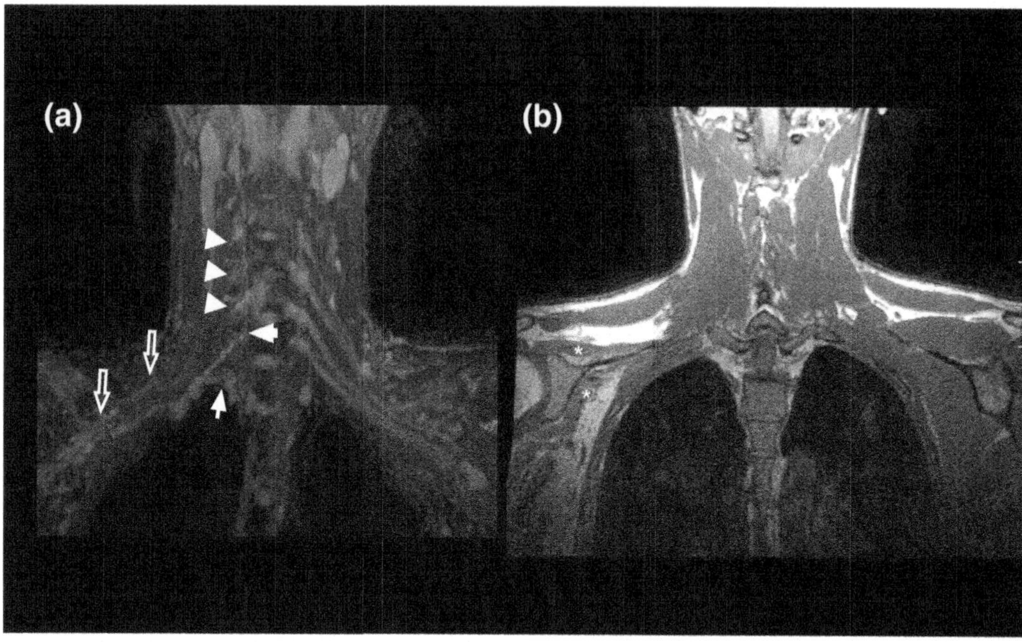

Fig. 3 Traumatic postganglionic injury: brachial MR neurography exam was obtained 6 months after motor vehicle crashes in a 25 years old man. Sequence 3D IDEAL coronal T2-weighted images (**a**), shows postganglionic injury of C5 to T1 roots, with nerve discontinuity of C5 to C7 and T1 (*arrowheads* in order superior to inferior) and retraction of nerves ends (*open arrows*), lesions in continuity of C8 root (*arrow*). The 3D IDEAL T1-weighted images (**b**) present chronic denervation in supraspinatus and subscapular (*white asterisk*) muscles

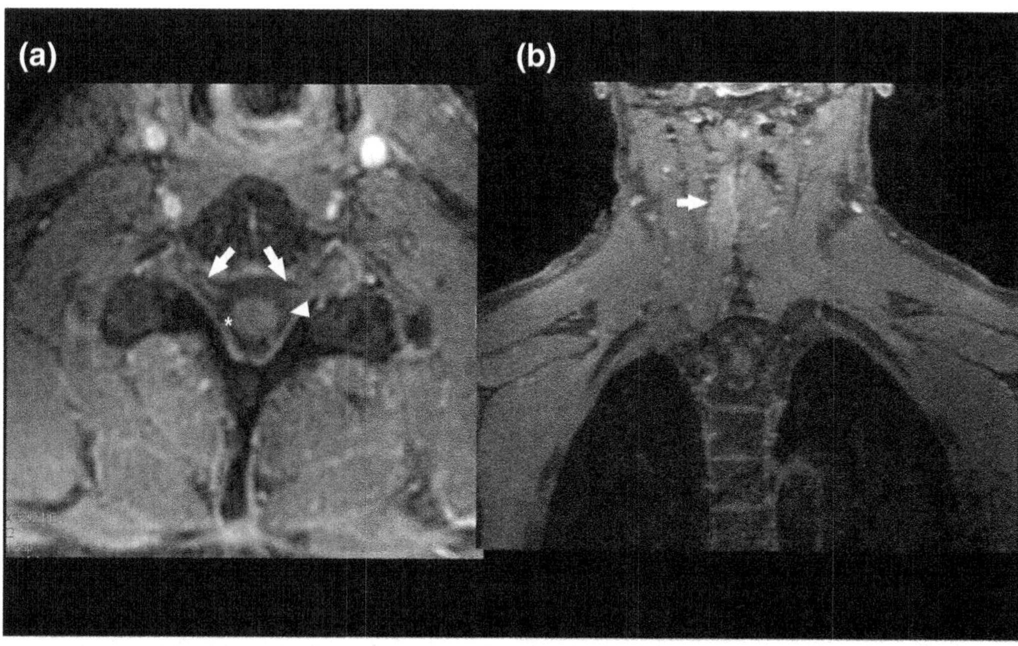

Fig. 4 Traumatic preganglionic injury: MR neurography examination was obtained 5 days after vehicle accident in a 53 years old man. Sequence 3D IDEAL T1-weighted images in axial planar reformation following intravenous gadolinium administration (**a**), shows avulsion of *right* preganglionic posterior root (*asterisk*) and, normal preganglionic *left* posterior root (*arrowhead*) and both anterior roots (*arrows*), visualized in C5 to C6 level. The coronal plane (**b**) shows acute denervation with enhancement of the *right* multifidus muscle (*white arrow*)

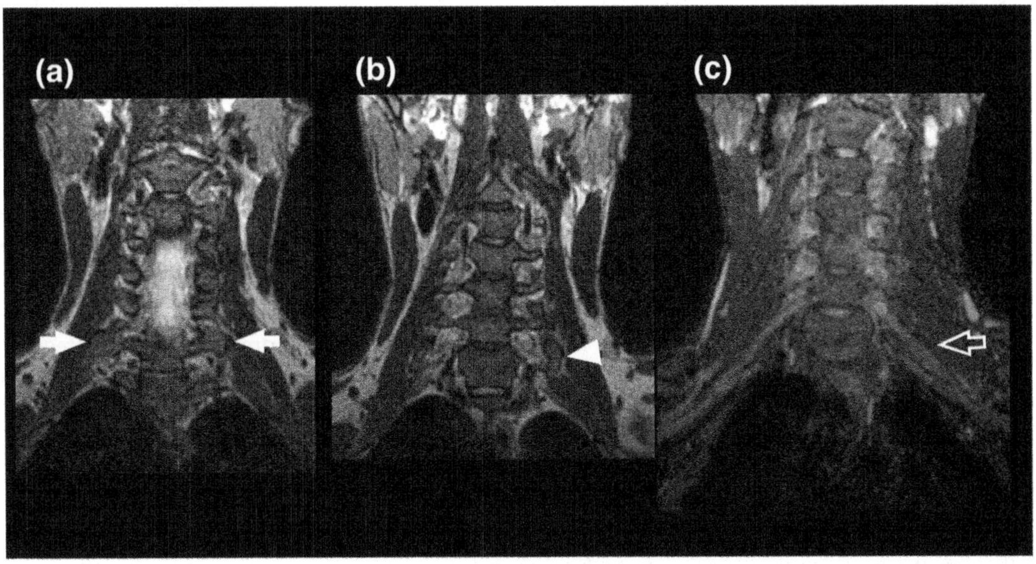

Fig. 5 Thoracic outlet syndrome: MRN in a 14 years old boy with paresthesia of the *left* superior arm during sports activity. Sequence 3D IDEAL coronal T2-weighted (**a**), shows bilateral C7 prominent transverse process (*arrows*). On the *left* side has a rudimentary cervical rib (*arrowhead* in **b**). Sequence 3D IDEAL coronal FAT SAT T2-weighted (**c**) exhibits enlargement of the *left middle* trunk (*open arrows*)

compressed in three levels: interscalene triangle, costoclavicular, and retropectoral spaces.

Usually, it is a dynamic condition that may occur due to arm elevation causing irritation or compression of the artery and/or subclavian vein and the three cords of BP when they pass through interscalene and costoclavicular spaces [31].

Entrapments can result from C7 prominent transverse process, cervical ribs, costoclavicular joint osteoarthritis, clavicule fractures, fibrous bands, scalene muscle accessories, muscle hypertrophy due to activities like swimming, weightlifting, and backpacking. Other causes of TOS are nerve sheath tumors, Pancoast tumors, or lipomas [32].

MRN is a useful tool to distinguish the abnormalities of the BP and to determine the different causes of entrapment (osseous, vascular, etc.). In Addition to the proper technical MRN sequences, the use of dynamic maneuvers may be beneficial in TOS. Contrast enhanced T1 3D SPGRE weighted images in both, elevated and resting arm position may contribute to determinate the relationship between BP, subclavian vessels and costoclavicular joint [31].

Common neurographic findings of the BP in TOS include displacement or enlargement of brachial cords and abnormal high signal in T2 weighted sequences of the affected cords. Moreover, the anatomic osseous variant may be demonstrated in basis of the conflict between osseous component and roots as well as the other causes of entrapment. It is necessary to search for denervatory changes in the regional muscles (Fig. 5) [33].

4.4 Inflammatory Diseases

The radiculoplexus neuropathies affect roots, plexus, and individual nerves to involving cervical, thoracic or lumbosacral segments. Occasionally, it is difficult to distinguish an acute plexitis from a cervical radiculopathy or even rotator cuff tears solely in clinical basis. MRN may explain virtually all cases [34].

Diabetes Neuropathy

Diabetes mellitus is the most common cause of neuropathy and can be associated with a wide spectrum of peripheral nerve disorders.

Table 6 Classification of the diabetic neuropathies

Impaired glucose tolerance and hyperglycemic neuropathy
• Generalized neuropathies
Sensorimotor
Acute painful (including treatment induced)
Autonomic
Acute motor
• Focal and multifocal neuropathies
Cranial
Thoracolumbar
Lumbosacral radiculoplexus (Bruns–Garland syndrome)
Focal limb (entrapment and compression)
• Superimposed chronic inflammatory demyelinating neuropathy
• Hypoglycemic neuropathy

Following the historical perspective from Dyck and Thomas it could be classified as shown in Table 6 [35, 36].

The most common is a slowly progressive distal sensorimotor diabetic polyneuropathy (DPN), with dominant sensory symptoms involving the lower extremities. The clinical picture usually includes pain, paresthesias, deep burning, or stabbing sensation at the onset with symmetrical distribution as a distinctive feature [37]

On the other hand the diabetic radiculoplexus neuropathies (DRPN) are uncommon conditions, mainly seen in older type 2 diabetics. DRPN could be classified into three subcategories: lumbosacral, thoracic, and cervical radiculoplexus neuropathies.

Although DRPN is most common in lumbosacral a significant proportion may be found in cranial or cervicothoracic location [38].

In a small proportion of diabetic patients, a multifocal neuropathy (MDN) is observed, with successive or simultaneous involvement of roots and nerves of the lower limbs, the trunk, and upper extremities over weeks or months, sometimes with a relapsing course [39].

Nerve biopsy detects axonal sensorimotor neuropathy with inflammation superimposed over ischemic nerve lesions. The hyalinization and thickening of the walls of small blood vessels suggests a role for nerve ischemia affecting unmyelinated fiber, while myelinated nerve fiber morphology remains normal in patients with early DPN [40]

On MRN the most frequent abnormality is the diffuse increase on T2 signal in the nerve, varying from mild–to–moderate in severity. In several diabetic patients, the BP trunks and cords are abnormal, even though the lesion is localized on a single nerve territory on neurological or electrophysiological examination.

Diffuse nerve hypertrophy is another frequent finding, commonly accompanied by contrast enhancement. The morphologic MRI sequences are useful to demonstrate the changes in the muscle signal.

Increased T2 signal may reveal miofibrilar edema in a subacute stage, while an abnormal T1 signal elevation may indicate fatty infiltration on the chronic setting.

Muscular signal changes are helpful in confirming the time course of disease. Variable degrees of axillary, mediastinal or suprascapular lymph node abnormalities such as enlargement and increased number may be seen in a subgroup of patients (Fig. 6) [41, 42].

Chronic Inflammatory Demyelinating Polyradiculoneuropathy

Chronic inflammatory demyelinating polyneuropathy (CIDP) is an acquired immunological peripheral neuropathy characterized by a progressive course with weakness and sensory loss in the legs and arms, and in some cases with cranial nerves affectation [43].

CIDP is classified into different clinical subtypes: typical CIDP with symmetric polyneuropathy involving both proximal and distal muscles, atypical CIDP, multifocal acquired demyelinating sensory and motor neuropathy (MADSAM or Lewis–Sumner syndrome) and demyelinating acquired distal symmetric polyneuropathy [44].

The diagnosis of CIDP is based on clinical presentation and electrophysiological findings consistent with demyelination. In clinical practice, CIDP is often difficult to diagnose [45].

MRN examination of the BP has become a helpful tool in the diagnosis work-up of CIDP. The most common findings are the

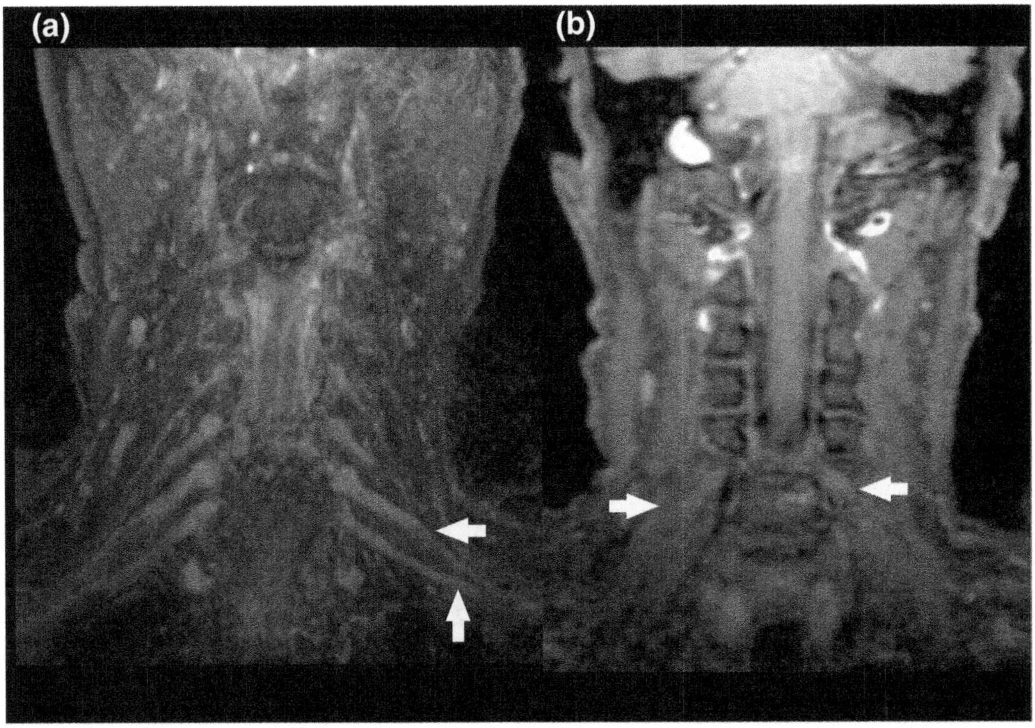

Fig. 6 Diabetic neuropathy: brachial MRN in 45 years old man, with uncontrolled diabetes type 2. Sequence 3D IDEAL coronal FAT SAT T2-weighted (**a**), shows diffusely hypertrophy and hyperintensity signal in both roots C6, C7 and C8. Note enlargement of *left middle* and *lower* trunk (*arrows*). The sequence IDEAL 3D coronal T1-weighted after contrast injection (**b**), show bilateral C7 roots enhanced (*arrows*)

enlargement and abnormal high signal intensity of nerves in T2-weighted images and DWI sequences (Fig. 7). High intensities on DWI and the ADC map, maybe due to T2 shine-trough.

Furthermore, an abnormal high signal weighted images may indicate the increased endoneurial collagen developing *"onion bulb"* seen on the histological analysis [46, 47].

In the typical form of CIDP a symmetric hypertrophy of the nerve roots is observed, while MADSAM subtype could exhibit a multifocal and fusiform hypertrophy specially in the peripheral nerve trunks [48, 49]. Gadolinium enhancement has been reported in some cases of CIDP, due increased permeability of the blood–nerve barrier [50].

Parsonage Turner or Neuralgic Amyotrophy

This is an idiopathic brachial plexitis that affects young and middle-aged patients, males more than females, and may be bilateral in up to 30% of patients. The etiology remains unknown, but proposed etiologies include immune and inflammatory.

Most commonly, the long thoracic, suprascapular, and anterior interosseous nerve are affected. The supraspinatus, infraspinatus, serratus anterior muscles are the most frequently affected. Multiple nerves may be involved, but the ulnar nerve is rarely affected.

As clinical features patients may note a constant and severe pain in neck, shoulder, or upper arm, followed by a profound weakness and atrophy of the muscles and in a few weeks [51].

Recurrences are seen in 25% of patients with idiopathic neuralgic amyotrophy, the sporadic subtype and in up 75% of patients with the hereditary variety ("HNA") [52].

MRN show mild or moderate diffuse enlargement of the nerves. T2 and STIR-weighted images may show a characteristic diffuse enlargement

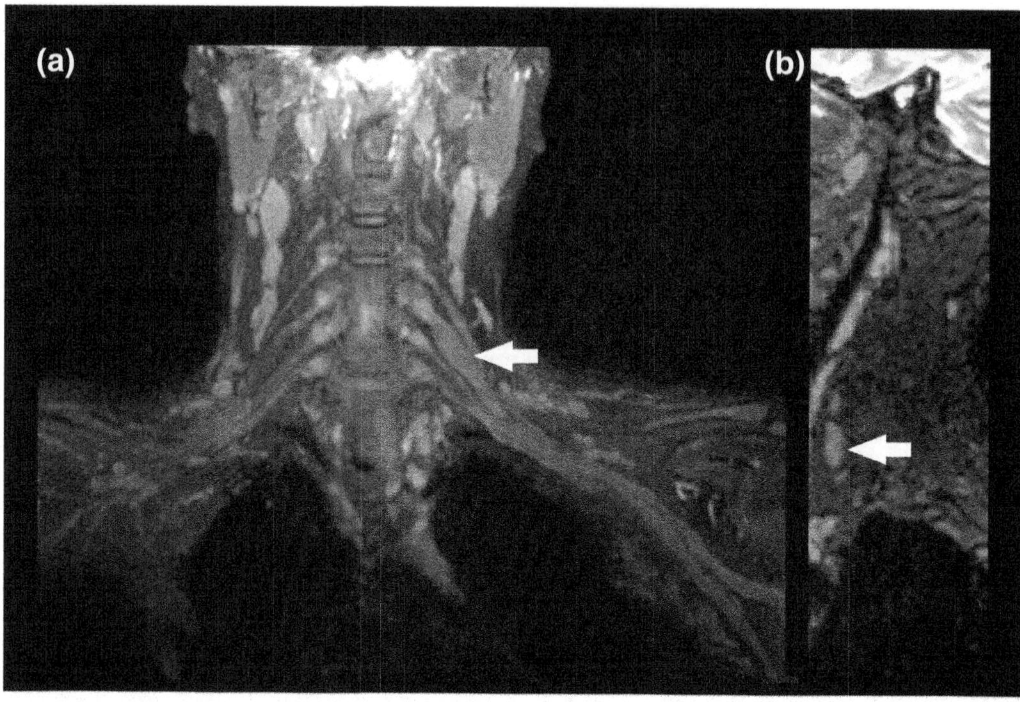

Fig. 7 Atypical form of chronic inflammatory demyelinating polyneuropathy: brachial MRN in 26 years old female with *left* biceps muscle atrophy last six months. Electrophysiological exam shows involvement of the *left* upper trunk. Sequence 3D IDEAL coronal FAT SAT T2-weighted images (**a**), shows bilateral asymmetric fusiform thickening of the roots, trunks, and divisions of brachial plexus, more evident in C6 root (*arrow*). The sagittal reconstruction (**b**) shows greater thickening of the *left* C6 root (*arrow*)

of nerves with abnormal high signal with alternating hypointense lineal foci along the nerves. Furthermore, muscle denervation changes is observed in affected regional muscles (Fig. 8) [41, 53].

4.5 Tumoral Disease: Primary and Secondary

Main features of a peripheral nerve sheath tumor (PNST) are expressed by clinical findings, the location of the lesion, and its radiological appearance. The most frequent tumors are summarizes in Table 7.

MRN is a valuable tool for the preoperative staging of the mass lesions involving the BP, facilitating the distinction between tumor and the nerve components [54].

MR imaging, especially whole-body imaging (WBMRI), is used in the assessment and follow-up of the disease burden in neurocutaneous syndromes, such as neurofibromatosis types 1 and 2 and schwannomatosis. Because WBMRI can detect even relatively small or asymptomatic tumors in all body

Table 7 Peripheral nerves tumors

Neurofibroma
Schwannoma
Malignant PNST
Neurolymphoma
Perineurioma
Lipoma
Intraneural epithelial or mesenchymal tumors

PNST peripheral neural sheath tumor

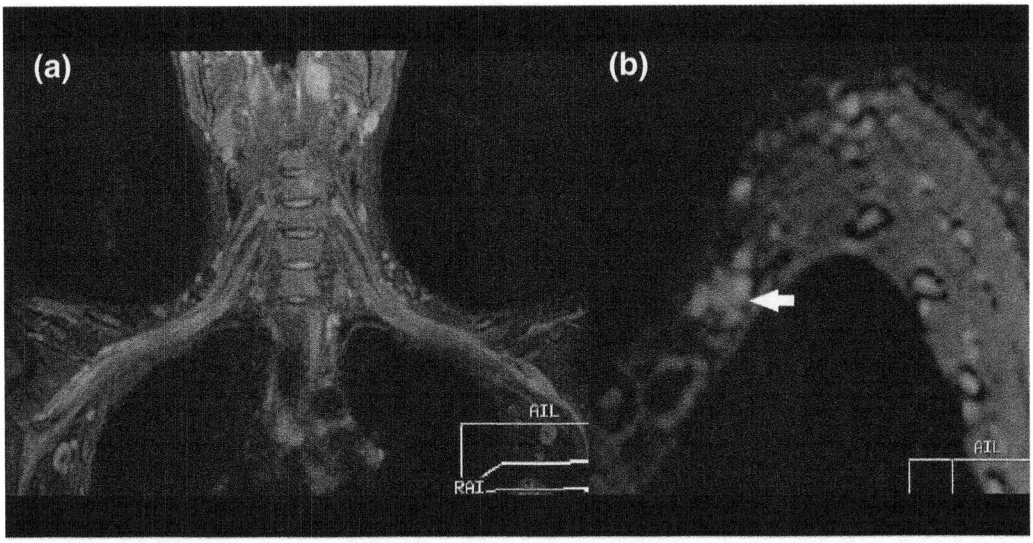

Fig. 8 Parsonage turner syndrome: brachial MRN in 19 years old man. Sequence 3D IDEAL coronal FAT SAT T2-weighted image (**a**), shows moderate diffusely enlarged with multiple hyperintense foci of roots, trunks, and divisions of brachial plexus. The sagittal view (**b**) of the division level (*arrow*). (Reprinted with permission of the publisher. Copyright © 2016 SERAM. Published by Elsevier España, S.L.U. All rights reserved)

Table 8 Neurocutaneous syndromes

NF-1	Cutaneous lesions Skeletal deformities Gliomas	Neurofibroma solitary/neurofibroma plexiform Malignant PNST
NF-2	Vestibular tumors Ependymomas, schwannomas or multiples meningiomas	Multiple benign PNST
Schwannomatosis	Multiple PNST Possible meningioma No vestibular tumors	Multiple benign PNST

NF neurofibromatosis; *PNST* peripheral neural sheath tumor

regions, it provides a more comprehensive view of tumor burden and, for the first time, allows for analysis of tumor distribution across body parts. The neurocutaneous syndromes are classify in Table 8 [55].

Benign Peripheral Nerve Sheath Tumors

The most common neurogenic tumors include neurofibroma, schwannoma, and perineurioma.

Neurofibroma

Neurofibromas are benign PNSTs arising from Schwann cells, with another mixed cell types including neuronal axons, fibroblasts, mast cells, macrophages, perineural cells, and collagen. There are two types of neurofibromas: solitary and plexiform [56, 57].

Solitary neurofibroma is the most common form of presentation. It is usually observed in younger individuals, between 20 and 30 years of age. Neurofibroma are fusiform well-defined lesions (<5 cm) with a nerve centrally located in the core of tumor. The tumor tissue is intimately imbricated with the nerve and cannot be separated from normal nerve fibers [57].

On MR images a solitary neurofibroma may be isointense or slightly hyperintense regarding muscle on T1-weighted images and markedly

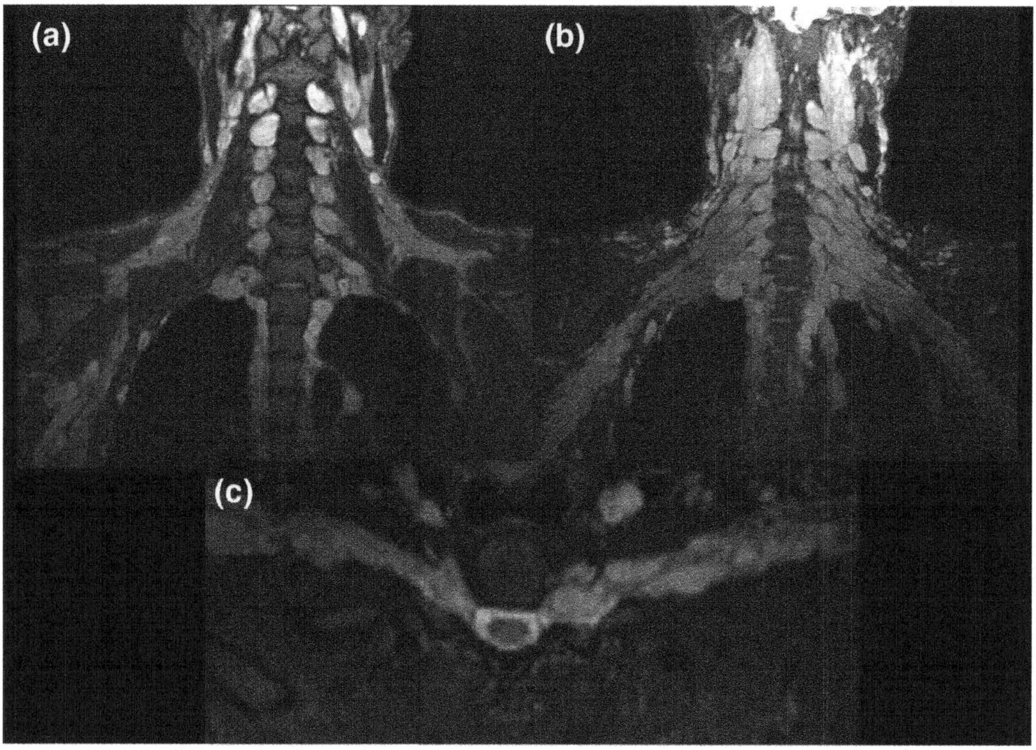

Fig. 9 Plexiform neurofibroma: brachial MRN in 16 year old man. Sequence 3D IDEAL coronal T2-weighted image (**a**), shows diffuse enlargement of roots. Sequence 3D IDEAL coronal FAT SAT T2-weighted (**b**), and curvilinear reconstruction of the trunk level (**c**), exhibits multiple tumors creating a "bag of worms" appearance. (Reprinted with permission of the publisher. Copyright © 2016 SERAM. Published by Elsevier España, S.L.U. All rights reserved)

hyperintense to fat on T2-weighted. When the intermuscular location occurs usually the neurofibromas are surrounded by fat ("split fat sign") on T1-weighted images. Lesions might exhibit high signal intensity in the periphery and low to intermediate signal intensity in the center on T2 weighted images ("target sign"). This sign was initially described to be pathognomonic of neurofibroma or at least distinctive of benign and malignant nerve sheath tumors, it has been observed in both neurofibromas and schwannomas, and occasionally even in MPNSTs [54].

Plexiform neurofibroma is another pathognomonic form of presentation of NF-1, usually observed in subjects between 10 and 20 years of age.

MR imaging shows a tortuous mass of irregularly expanded nerve branches creating a "bag of worms" appearance. That might more or less extensively invade adjacent muscles and connective tissues. Usually, plexiform neurofibromas have identical signal and enhancement pattern than a solitary lesion. Plexiform neurofibromas can be deep, superficial, or a combination of both. When involving an entire limb, might induce elephantiasis neuromatosa, a condition associated with enlargement of the affected extremity, hypertrophy of bone, and redundant skin (Fig. 9) [54]. The risk for malignant transformation is thought to be as high as 8–12% [58].

Schwannoma

Schwannoma, neurinoma or neurilemoma, is a slow-growing tumor arising from the Schwann cells of a nerve sheath. Generally, is depicted as a solitary lesion (<5 cm) developing eccentrically to the nerve fibers and is encapsulated by the perineurium [58].

Usually, they are incidentally discovered but, the large tumors exhibit neurological symptoms

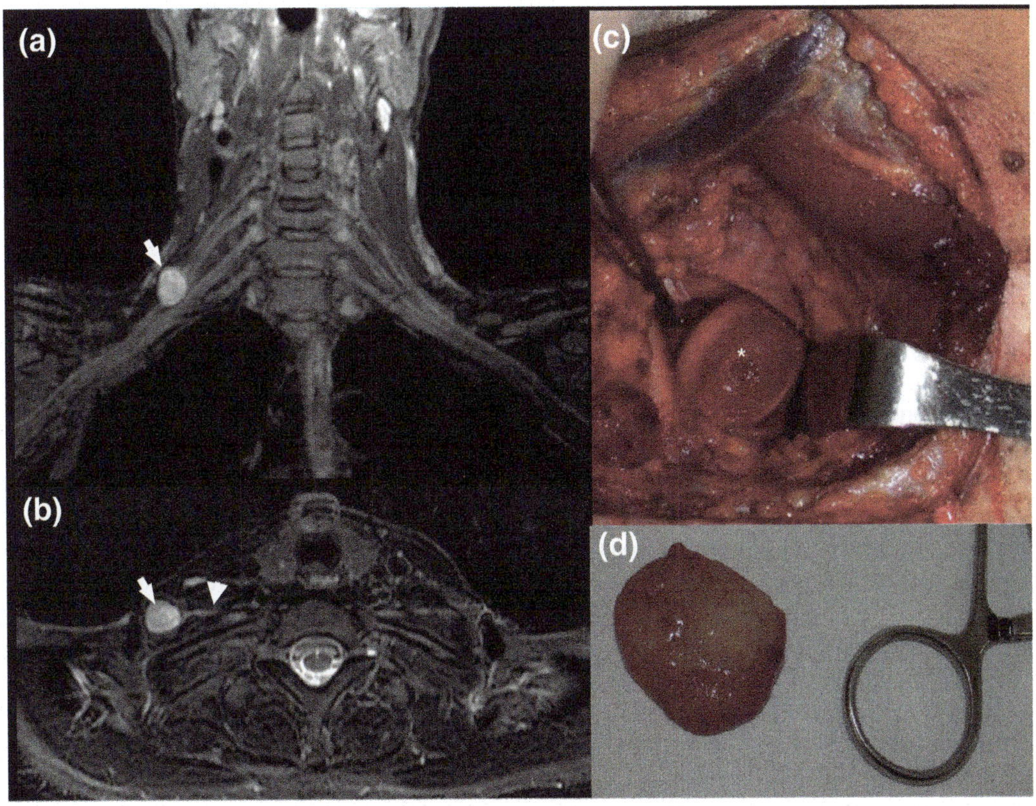

Fig. 10 Schwannoma: brachial MRN in 38 year old woman with a palpable mass in the *right* supraclavicular space and, paresthesias and pain in the hand. Sequence 3D IDEAL coronal FAT SAT T2-weighted image (**a**), and 3D IDEAL axial T2-weighted image (**b**), shows a well-defined, round soft tissue masses with high signal (*arrow*), in continuity with the *right* median trunk (*arrowhead* in **b**). Surgical specimen (**c**, **d**). (Figure credited to Mariano Socolovsky, MD from Foundation of Neurological Research—FLENI)

and pain. They can be seen at any age but are more commonly diagnosed in patients 20–30 years of age [54].

The mediastinum and retroperitoneum represent the most common sites of involvement, schwannomas can occur almost anywhere in the body. Large tumors may show degenerative change such as cystic areas, calcification, hemorrhage, and fibrosis, and are described as ancient schwannomas [59].

Schwannomas share MR imaging features with solitary neurofibromas. On MR imaging a hypointense capsule representing the epineurium is more commonly seen with schwannomas than with neurofibromas. Schwannomas are isointense to muscle on T1 weighted images and heterogeneously hyperintense on T2 weighted images and variable contrast enhancement (Fig. 10). Similar to neurofibromas, they can show target sign, fascicular sign, and split fat sign [60–62].

The distinction of schwannomas and neurofibromas by the position of the tumor relative to the nerve (eccentric versus central), is frequently impossible. On advanced DTI, these lesions also show lack of restricted diffusion with high ADC values. Just as neurofibromas can be plexiform, schwannomas can also be plexiform [63].

Malignant Peripheral Nerve Sheath Tumors

Malignant peripheral nerve sheath tumors (MPNST) are less common than their benign counterparts. MPNST most probably occurs in young people between 20 and 50 year old. without any gender predilection. Usually lesions larger than >5 cm present with dull pain or new neurological deficits [64].

MPNST shows a heterogeneous appearance on MRN, irregular boundaries and enhancement unlike benign schwannomas. However, the imaging findings may not accurately anticipate the histologic behavior of the lesions [58, 65]. Quantitative diffusion-weighted imaging has been reported to be of questionable aid for distinguishing benign and malignant soft tissue tumors including PNSTs [66].

In these context, Fluorine 18 (18F)–fluorodeoxyglucose (FDG) positron emission tomography (PET)/computed tomography (CT) can provide important information to discriminate between benign and malignant PNSTs. Increased glucose metabolism with a maximum standardized uptake value (SUV) greater than 3.5 has been proposed as a threshold for considering surgical excision because such increased activity is considered highly sensitive for the detection of malignant disease [67].

Neurolymphomatosis

Neurolymphoma (NL) is an extremely rare extranodal manifestation of non-Hodgkin lymphoma affecting peripheral nerves, roots or plexus or, cranial nerves. Although more frequently found in large B-cell non-Hodgkin lymphoma, it can also appear in the context of T-cell lymphoma. This is observed in both sexes and in a wide age range. Isolated neurolymphoma may progress to systemic disease in 7% of patients [68].

Frequently, the clinical features of neurolymphomatosis are pain and sensory motor involvement and may be clinically superimposed to other neuropathies [69].

MR Neurography findings are unspecific in NL. It frequently affects the BP plexus showing a nodular pattern or fusiform enlargement of the nerve roots, demonstrated in both T1 and T2-weighted images. Hyperintensity of the anterior and posterior roots, trunks and divisions can be shown in T2-weighted images. The nerves affected demonstrate mild to moderate enhancement with the gadolinium-based contrast agent. The nerve distal to the area infiltrated by lymphoma may demonstrate abnormal hyperintensity on T2-weighted images and enlargement associated with reactive neuritis. It is so difficult to distinguish between the limits of the infiltration and the irritative changes [70].

F-18 FDG PET/CT is a useful modality in patients with suspected NL. PET/CT typically shows a marked FDG uptake in lymphoma and it also allows to identify the spread of the disease. Moreover, F-18 FDG PET/CT is ideal for the evaluation of treatment response [71, 72].

Metastasis

Common malignancies affecting the BP include superior sulcus tumor (Pancoast tumor), breast and head and neck metastatic lymphadenopathy. Other less common tumors are sarcoma and melanoma. Extrinsic masses may cause irritation, compression, or invasion to the BP. In setting of invasion, neoplasms reach the plexus either by direct extension or metastasis through the lymphatics from the axilla. Focal masses or diffuse infiltration may onset with supraclavicular lower plexus (C8–T1 and lower trunk) involvement [73].

The most frequent symptom is severe pain starting in the shoulder girdle then radiating to the inner aspect of the upper limb [74].

MRN helps to determine tumoral position relative to the BP elements (intrinsic or extrinsic). Extrinsic masses are frequently irregular, with a perpendicular or vertical course regarding BP while nerve sheath neoplasms extends along the direction of the BP and may show diffuse contrast enhancement (Fig. 11) [22, 75]. In cases of breast cancer infiltration is difficult to discriminate from radiation plexopathy solely in terms of imaging (see later Radiation Neuropathy).

4.6 Iatrogenic and Therapeutic Related Disorders: Obstetric Palsy and Radiation Neuropathy

Obstetric Palsy

The incidence of brachial plexus birth palsies lies still between 0.4 and 4.6 per 1000 live births [76].

Erb–Duchenne palsy results from injury to the C5 and C6 roots or the upper trunk and accounts

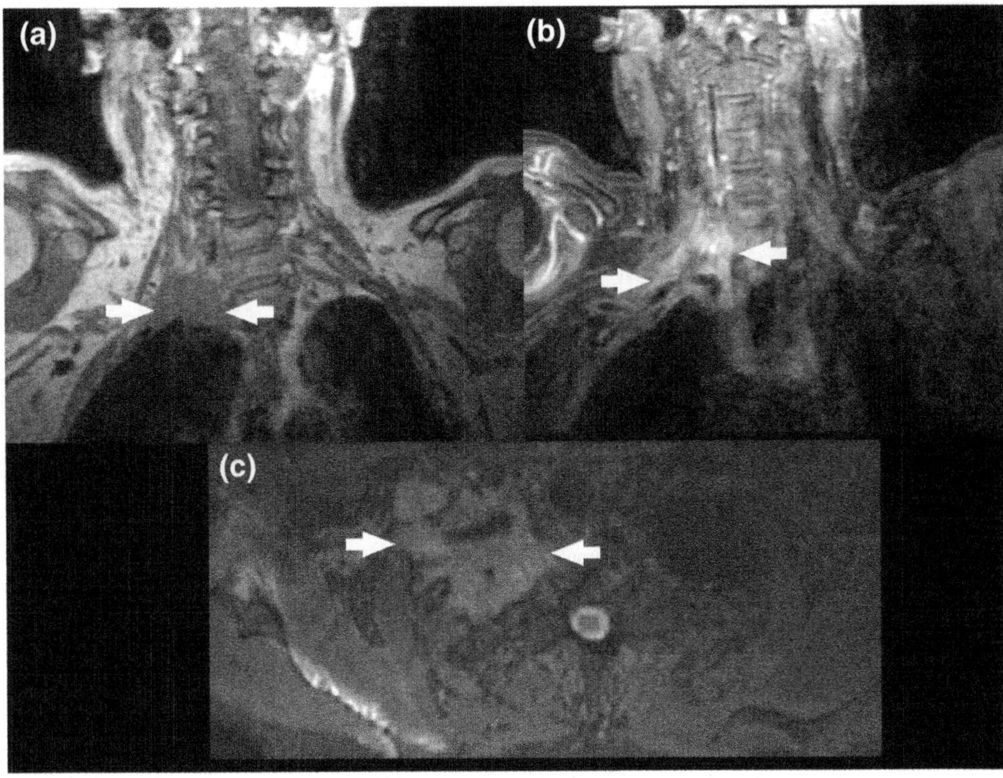

Fig. 11 Pancoast tumor: brachial MR neurography in 65 year old man with Horner syndrome (ptosis, miosis and anhidrosis of the ipsilateral face), paresthesias and shoulder pain. Sequence 3D IDEAL coronal T1-weighted image (**a**), 3D IDEAL coronal FAT SAT T1-weighted image after contrast injection (**b**), 3D IDEAL coronal FAT SAT T2-weighted image (**c**), demonstrate a large and irregular mass (*arrows*), with intermediate to low T1 intensity and increased heterogeneous T2 signal in midclavicular region encasing the brachial plexus. (Reprinted with permission of the publisher. Copyright © 2016 SERAM. Published by Elsevier España, S.L.U. All rights reserved)

for approximately 90% of obstetric brachial plexus injuries. Much less common is Dejerine–Klumpke palsy, which results from injury to the C8 and T1 roots or the lower trunk [77].

MRN is essential for early diagnosis to differentiate between preganglionar to postganglionar injuries (Fig. 12). Early diagnosis is crucial to offer surgical treatment in order to prevent major neurological deficits (see Traumatics Injuries).

Furthermore, conventional MRI may shows early abnormalities in glenohumeral joint, such as, a hypoplastic and flattened of glenoid fossa and, a blunt posterior labrum or, a flattened of humeral head and a blunt anterior labrum [78, 79].

Radiation Neuropathy

Brachial plexus is often involved during radiation treatment of axillary metastatic disease, often in the context of breast cancer. Patients presenting with neuropathy symptoms after radiation treatment may have tumor recurrence or inflammatory changes due to radiation plexopathy.

Neuropathy related to prior radiation therapy tends to occur between 5 and 30 months after treatment, with a peak of incidence between 10 and 20 months; however, there have been reports of symptoms first manifesting as long as decades after treatment [80].

The risk of developing radiation-induced plexopathy, and the degree of associated

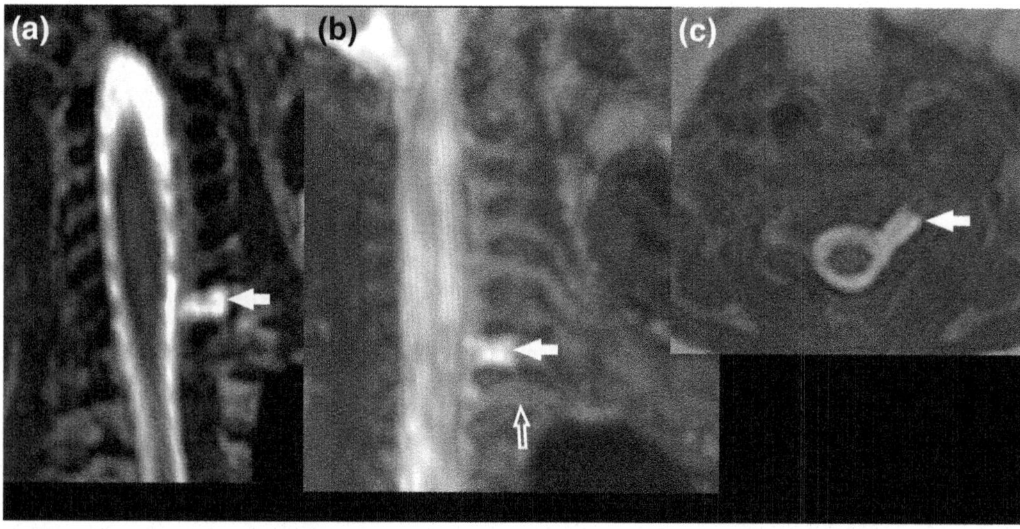

Fig. 12 Palsy obstetric: brachial MR neurography in 2 months child with Dejerine–Klumke palsy caused by trauma from vaginal birth. Sequence 3D GRE coronal (**a**), 3D IDEAL coronal FAT SAT T2-weighted image (**b**) and 3D GRE axial image (**c**), demonstrated the injury to the C8 and T1 roots or *lower* trunk. Presence of pseudomeningocele in C8 root (*white arrows*) from avulsion, and enlargement of the T1 root (*open arrow*)

damage, is directly proportional to radiation dose. Clinically evident damage is most likely to occur when at least 6000 cGy is administered. In addition, concurrent chemotherapy appears to potentiate the detrimental effects of radiation [81].

Tumor recurrences usually affect the lower supraclavicular plexus (C8, T1, and the lower trunk). Radiation plexopathy mostly involves the infraclavicular plexus. In cases of tumor recurrence, MRN shows a focal or diffuse asymmetric enlargement of the plexus rami with enhancing mass or nodular lesions.

Typical MR imaging findings in radiation-induced BP inflammation include relatively uniform, symmetric thickening, and hyperintensity in T2-weighted images and contrast enhancement of BP structures. Over the time actinic fibrosis manifests with a well-defined geographic pattern consistent with the radiation field. In addition, fat stranding with hypointensity in T1 and T2 weighted images could be noted. Also distortion and kinking of the nerve may coexist with diffuse and symmetric enlargement. Lack of a focal mass may contribute to differentiate fibrosis from recurrency in spite of contrast enhancement on MRI examinations seen in both condition (Fig. 13) [65, 82].

Distinguishing plexitis postirradiation from recurrent or metastatic tumor can be difficult at MR imaging, and PET/CT may serve in an adjunctive role [80, 83].

5 Upper Extremity Neuropathies

Electrodiagnostic test has been the traditional method for the evaluation of the upper limb neuropathies for years. Nowadays MRN with dedicated coils, 3D tailored pulse sequences in 3T scanners has emerged as a critical complementary tool in order to evaluate the nerve integrity and has become an important stair in the upper extremity work-up [84].

5.1 Anatomy

The upper extremity nerves originates from the BP are musculocutaneous, scapullary, axillary, radial, median, and ulnar nerves.

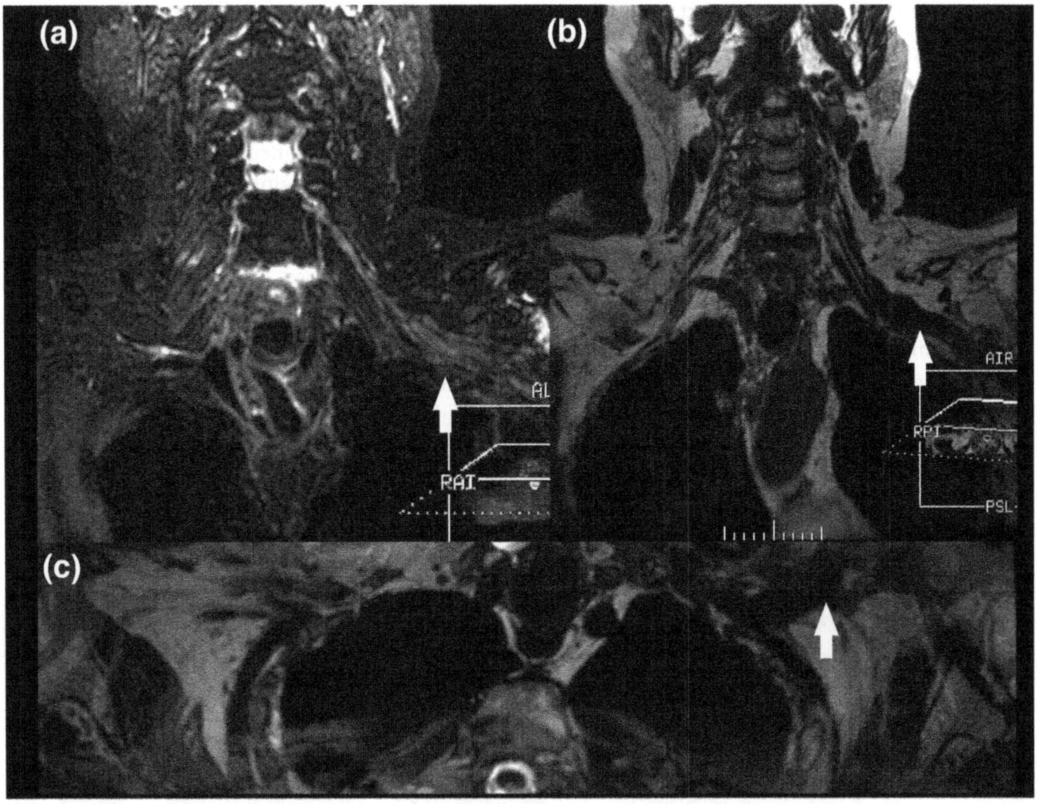

Fig. 13 Radiation neuropathy: brachial MR neurography in a 75 year old female with background of radiation due *left* breast cancer since 15 years ago. Sequence 3D IDEAL coronal FAT SAT T2-weighted image (**a**), 3D IDEAL coronal T2-weighted image (**b**) and 3D GRE axial image (**c**), exhibit a diffuse enlargement of divisions and cords of the *left* brachial plexus (*arrows*), with absence of soft tissue masses

Table 5 describes the muscles innervated by the BP.

5.2 Ulnar Neuropathies: Cubital Tunnel and Guyon Canal Syndromes

Ulnar nerve (UN; medial cord of C8–T1 of the BP), runs along medial posterior region of the upper arm. In the middle third the UN, penetrates into the intermuscular septum of triceps muscle and then enters into the cubital tunnel. This fiber osseous channel is bounded medially by the medial epicondyle, laterally by the olecranon, and the roof is the arcuate ligament. UN descends into the forearm in proximity to the ulnar artery, arriving into the Guyon tunnel in the hand. Then it divides into superficial sensory and deep motor branches. Guyon tunnel is formed by hipotenar muscles, hamate and flexor retinaculum and could be divided into three zones: zone 1 is proximally to the nerve bifurcation, while zones 2 and 3 comprise the deep and superficial branches, respectively [83–87].

The most common site of UN entrapment is the cubital tunnel at the elbow. Repetitive elbow flexion may be associated with UN friction [88]. Other causes of UN entrapment in the cubital tunnel include: thickened cubital tunnel retinaculum, accessory anconeous muscle, tumors (schwannoma, perineurioma, neurofibroma); and pseudotumors (fibrous bands, ganglion cysts, and osteophytes) [89]. After surgical decompression the nerve is positioned anterior to the medial humeral condyle. In some cases surrounding postsurgical fibrosis and fibrous bands may cause re-entrapment of the nerve [88].

The second site in frequency of the UN entrapment is the Guyon canal in the wrist. In activities with repetitive trauma (e.g., bicycle handle bars), the nerve is compressed by the pisohamate ligament. Other causes of UN entrapment include fractures, ganglion cysts, pisotriquetral joint osteoarthrosis, flexor carpi ulnar hypertrophy or tumors such as lipoma and schwannoma [90].

Symptoms include sensory complaints, numbness, and weakness of the fourth and fifth digits. Pain may be occurs in the elbow and, rarely in hand [91].

In the setting of the UN neuropathy, MRN offers exact depiction of the site of involvement. UN neuropathy diagnosis may be demonstrated by changes in the appearance of ulnar nerve. Anatomic boundaries and course are best visualized in T1-weighted images (Fig. 14). Typical findings of the UN neuropathy are defined by abnormal high intensity in T2-weighted images with diffuse enlargement and loss of the fascicular pattern. In addition scar tissue may preclude abnormal angulation of the nerve. Abnormal enhancement of the UN and the surrounding tissue may be seen after the contrast administration, probably due to inflammatory changes. In several cases muscle denervation changes may be seen in flexor carpi ulnaris, flexor digitorum profundus of 4th and 5th fingers, hypothenar and interosseous muscles of hand [16, 92, 93].

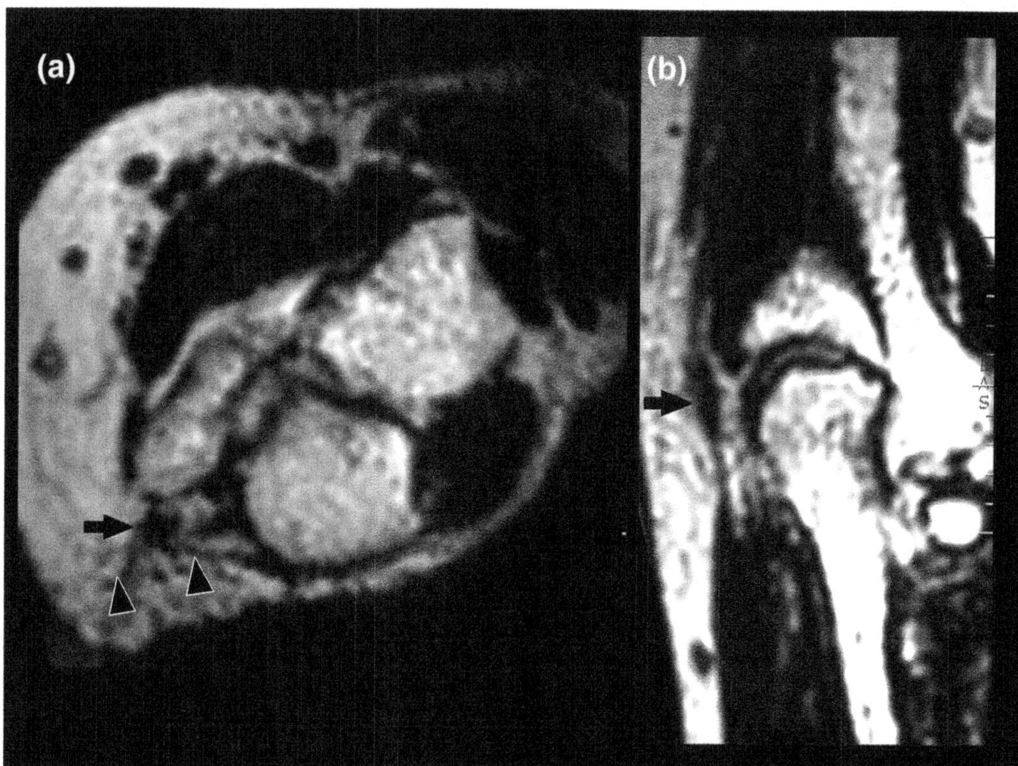

Fig. 14 Neuroma of ulnar nerve: sequence 3D CUBE; axial (**a**) and coronal planar reformations (**b**) in a 34 years old female with paresthesias in ulnar region. Abnormal enlargement of ulnar nerve (*arrow*) is present with fibrous bands surrounding the nerve (*arrowheads*)

5.3 Radial Neuropathies: Posterior Interosseous and Radial Tunnel Syndrome

The radial nerve (RN) is formed from the posterior cords of brachial plexus (C5–T1). RN descends across axilla passing by the inner side of the humerus and entering into the spiral groove of the humerus. Then, it pierces the intermuscular septum of triceps muscles and entering into the antecubital fossa [94].

The nerve passes lateral to the elbow joint capsule to reach the supinator muscle, where it divides into a superficial branch and a deep branch (the posterior interosseous nerve-PIN). The deep branch of radial nerve perforates the supinator muscle passing through the arcade of Frohse, which is formed by the superficial layer of the supinator muscle. After this stage passage, it is called as PIN [95].

RN neuropathy may be caused either by brachial inflammatory plexopathy or injury at the axillar region, often related with shoulder trauma or tumors.

The most common site of RN compression in the arm is the spiral groove of the humerus. This has been described in cases of anesthesia or even, during sleep, for a long period over the arm on a hard surface [96].

At the level or below the elbow, two neuropathic syndromes have been described: the radial tunnel syndrome and the PIN syndrome (also called supinator syndrome). The radial tunnel has an extension of 5 cm, starting from the radiohumeral joint to the supinator muscle. The deep branch radial nerve and PIN are susceptible to compression by thickening of superior or inferior edge of the supinator muscle (arcade of Frohse). Other causes of entrapment such as fibrous bands of the radiohumeral joint, vascular crossing and, tumors (i.e., schwannomas, neurofibromas, lipomas and ganglion cysts). Most frequent cause of the PIN neuropathy is repetitive pronation and supination microtrauma against a thickened arcade of Frohse [85, 97, 98].

The most frequent clinical presentation is wrist drop and paresthesia over the dorsum affecting the first to third fingers on physical examination [99].

MRN findings include diffuse and abnormal high intensity in T2-weighted images on radial nerve or PIN [90].

Also, muscle denervation changes may be seen in 50% of patients. Supinator is the most frequent affected muscle [97].

5.4 Median Neuropathies: Carpal Tunnel, Pronator, and Anterior Interosseous Nerve Syndrome

The median nerve (MN; medial and lateral cords of C6–C8, T1 of BP), runs together the brachial artery along the anteriomedial region of the upper arm between the biceps and brachialis muscles. When the Struther ligament exists, the MN nerve passes below it. Then, it goes through two superficial and deep heads of the pronator teres muscle arriving the antecubital fossa.

Before entering the anterior compartment of the forearm, MN runs under the flexor digitorum superficialis muscle then giving two branches in the forearm, i.e., the anterior interosseous and palmar cutaneous nerves.

At the carpal tunnel, MN is located between the flexor pollicis longus and flexor digitorum superficialis tendons and dorsally limited by the flexor retinaculum, the roof of the carpal tunnel. The Martin Gruber anastomosis is a communication between ulnar and medial nerves being the most common anatomical variant of the upper limb (30% of population) [85, 100].

There are three sites where the MN nerve can be compressed. The most common site of entrapment is the carpal tunnel. The so called carpal tunnel syndrome (CTS), results from compression by the flexor retinaculum, although the exact cause remains unknown. Systemic diseases, as diabetes, could be associated to CTS [35]. Other conditions as amyloid deposition, hypothyroidism, pregnancy; and masses (sinovial ganglion, fibrolipomatous hamartoma, schwannoma, has also been found (Fig. 15) [90].

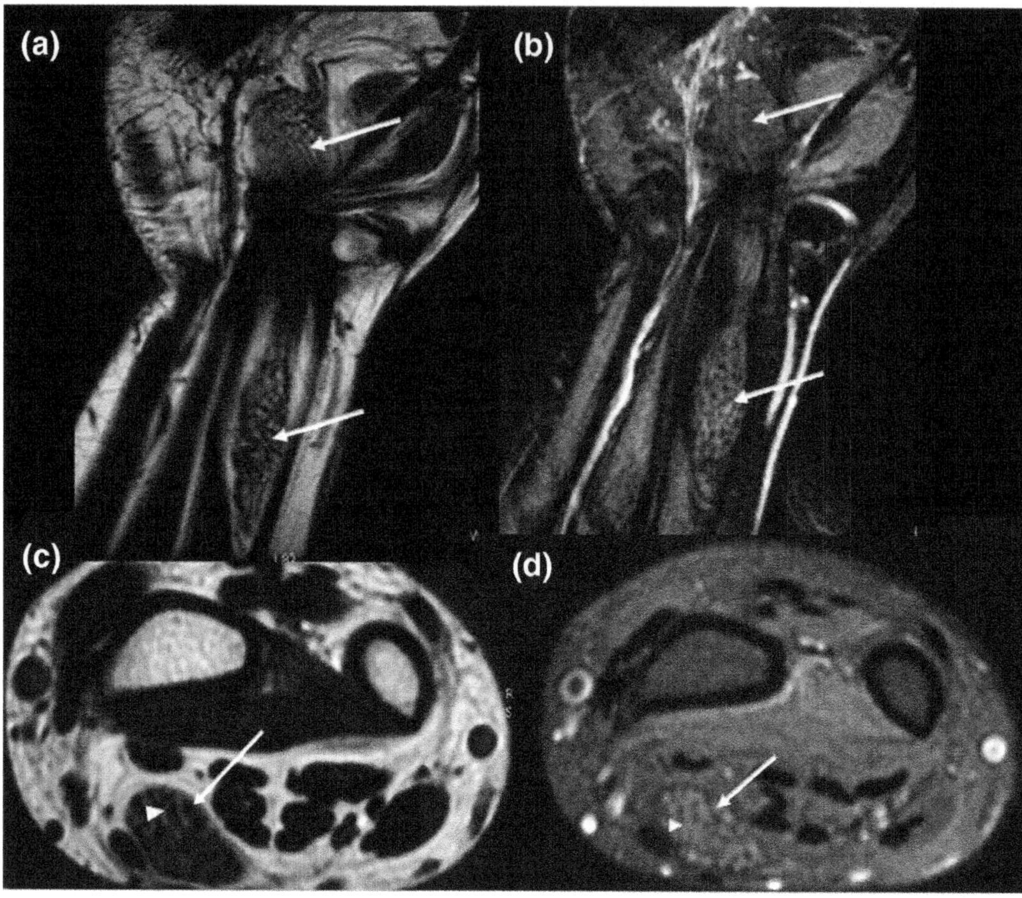

Fig. 15 Fibrolipomatous hamartoma of the median nerve: MRI of the wrist, in a teenage with macrodactyly. Sequence coronal T2-weighted image (**a**), FAT SAT T2-weighted image (**b**), axial T1-weighted image (**c**) and, FAT SAT T1-weighted image (**d**), reveal adipose tissue surrounded the nerve fascicles of the median nerve, causing thickened nerve and enlargement of the carpal tunnel (*arrows* in **a**, **b**). The images show the cylindric foci of low signal intensity corresponding to nerve fascicles (*arrow* in **c**, **d**) surrounded by fatty signal intensity (*arrowhead* in **c**, **d**). (Reprinted with permission of the publisher.

Other sites of MN compression include the inconstant Struthers' ligament (joining the supracondylar spur and medial humeral epicondyle), the entrapment of the MN as it passes between the two heads of the pronator teres (pronator syndrome) and when MN passes beneath of the anterior interosseous membrane (anterior interosseous nerve syndrome) [84]. The supracondylar spur, considered a vestigial structure in primitive mammals, is an anatomical variant present in 1% of population [101]. Pronator syndrome is the most common proximal neuropathy of the MN and may be related to pronator teres muscle hypertrophy, tumors, trauma, and fibrous bands along the bicipital aponeurosis [102, 103].

The anterior interosseous nerve, the largest branch of the MN, lies between the deep common flexor digitorum and the flexor pollicis longus. The most frequent causes of anterior interosseous nerve neuropathy are fibrous bands between the pronator teres and brachialis fascia, supracondylar fractures and local pressure after sleeping for a long period [104].

The MN innervates the flexor muscles of the forearm and digits, providing sensory innervation to the distal dorsal fingertips and the volar aspect of the first, second, third, and radial half of the fourth fingers. Injury to the MN results in an important loss of extremity function [105].

MRN shows abnormal high signal on T2-weighted images of MN at the site of entrapment. Hiperintensity may extend along the nerve. Conventional MRI sequences depict a supracondylar spur in cases of pronator syndrome, and thickening of flexor retinaculum with tenosynovitis of flexor tendons in cases of CTS.

Normally, the post denervation changes in the regional muscles can be used as a guide as far as they do not present Martin Gruber anastosis [106, 107]. Recently, DTI has been used as useful tool in patients with recurrent carpal tunnel syndrome. There is an ADC map reduction in these cases, indicating post surgical scar tissue [108].

6 Lumbosacral Neuropathies

Usually the LSP involvement is suspected when the additive effect of lower limb pain or numbness is followed by multicompartmental muscular weakness with an insidious onset over time. More often a LS plexopathy is found in association with thoracic and cervical neuropathies.

Assessment of the origin and extent of LSP plexopathies was formerly challenging for clinical and electrodiagnostic test due to variable muscle innervation and deep location of the nerves. MRN plays an important role on the recognition of abnormal patterns and determination of exact location and extension [109].

6.1 Anatomy

Often considered a single structure, LSP can be divided into two components, the "upper" lumbar plexus and the "lower" sacral plexus connected by the lumbosacral trunk (formed by fusion of a minor branch of L4 with the ventral ramus of L5) within the pelvis outlet [110].

Lumbar plexus lies within the paraspinal quadratus lumborum and psoas muscle and is formed by the ventral rami of T12 to L4 nerve roots. The six main nerves emerging from lumbar plexus provide sensitive innervation to lower abdomen (i.e., iliohypogastric, ilioinguinal, genitofemoral nerves) and anterolateral thigh (lateral femoral cutaneous). Motor innervation to pelvic muscles and thigh is provided by femoral and obturator nerves.

Sacral plexus originates from the ventral rami from L4 to S4 nerve roots and exits pelvic rim in independent cords with variable length. Sciatic, inferior gluteal, superior gluteal and pudendal nerves constitute the sacral branches of LSP arising form L4 to S4. A summary of the main nerves and roots is shown in Figs. 16 and 17.

A more detailed view for cutaneous and motor innervation is presented in Table 9 [111].

6.2 Lumbosacral Multifocal Diffuse Plexoradiculopathies

A simultaneous involvement of LSP roots and peripheral nerves can occur with systemic disease (mainly diabetes, sarcoidosis o amyloidopathy) or infiltrative conditions (bladder, prostate, rectal or cervical cancer dissemination and lymph nodes metastases).

Clinical picture is often attributable to multiple spinal levels without any recognizable pattern. The lumbar component of LSP may be compressed by retroperitoneal masses (i.e., psoas abscess, retroperitoneal fibrosis or hematoma). Also lymphoma or retroperitoneal lymph node metastases may infiltrate the lumbosacral plexus.

The sacral division of LSP may be affected by inflammatory or infectious conditions of the pelvic organs (i.e., pelvic abscess or ovarian masses) such as inflammatory arthritis of the sacroiliac joints,, or due to invasion of the sacrum and presacral space, by primary and secondary bone tumors (i.e., metastases, chordoma) or rectal and cervical carcinoma.

Some iatrogenic conditions may be implicated in LS pain syndromes. Radiation related plexopathy could have an insidious onset often painless, appearing on average 5 years after the initial exposition and becoming extremely painful. Labor and parturition also could lead LSP

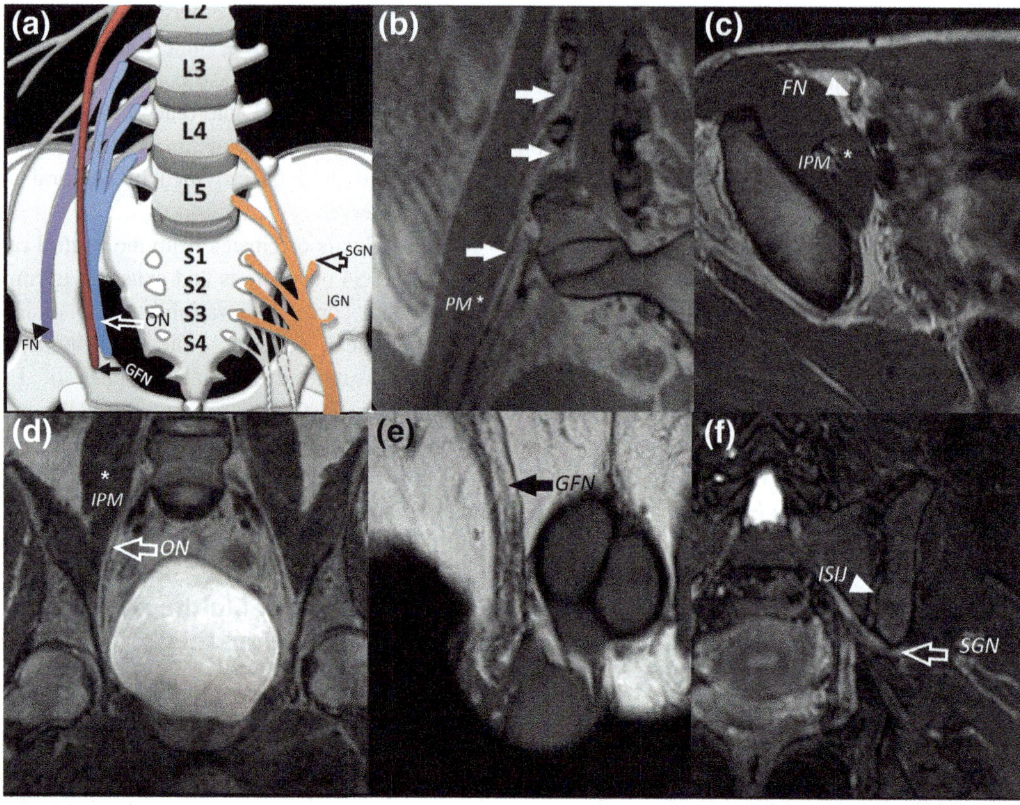

Fig. 16 Lumbar plexus (*LP*) and main pelvis nerves on MRN: (Abbreviations: *FN* femoral nerve, *GFN* genitofemoral nerve, *ON* obturator nerve, *SGN* superior gluteal nerve, *IGN* inferior gluteal nerve, *PM* psoas muscle, *IPM* iliopsoas muscle, *SC* spermatic chord, *ISIJ* inferior sacroiliac joint. Diagram of the LP showing nerve roots and anastomosis supplying pelvis (**a**): in the *right* side the FN in *purple* (formed by L2–L4 roots), GFN in *red* (formed by L1 and L2 roots) and ON in *light blue* originated from L3 to L5, are shown. The SGN and IGN in *orange* from (L4–S1 and L5–S2, respectively), are shown, in the *left* side. 3D IDEAL coronal oblique reformation T1 in-phase weighted images; (Schema credited to Tatiana Escobar, MD) (**b**): nerve roots L2, L3 and L4 (*white arrows*) emerging along the medial border of the **PM** (*asterisk*). **FN, GFN** and **ON** shares some radicular origins in close relationship with **PM** in retroperitoneal space. Sequence T1 in-phase weighted images (**c**) showing close relation of the FN and external iliac vessels before exiting pelvis and entering inguinal channel in proximity to IPM (*asterisk*). Sequence 3D IDEAL coronal T1 in-phase weighted images (**d**) showing ON (*void arrow*) easy to visualize by the surrounding perivescal fat. Sequence 3D IDEAL coronal T1 in-phase weighted images (**e**): in spermatic chord the GFN may be recognized medial to gonadal vessels (*black arrow*). The SGN (*void arrow*) is easy to identify due to close relation with the ISIJ in coronal FAT SAT T2 weighted image (**f**)

injury due to mechanical factors related to trauma as traction or nerve avulsions, less commonly [112]

Rarely there are hereditary neuropathies that may affect LSP. Commonly presents with sensory loss and muscle weakness starting from distal extremities. CMT or HMSN commonly involves the brachial and lumbosacral plexuses and exhibits a distinctive pattern characterized by enlargement of the nerve roots. A list of the causes of LSP plexopathy is shown in Table 10.

6.2.1 Inflammatory and Infectious Diseases

Diabetes Neuropathy

Diabetic neuropathies were discussed on Sect. 4.4. Regarding the multifocal varieties of diabetic lumbosacral radiculoplexus neuropathy

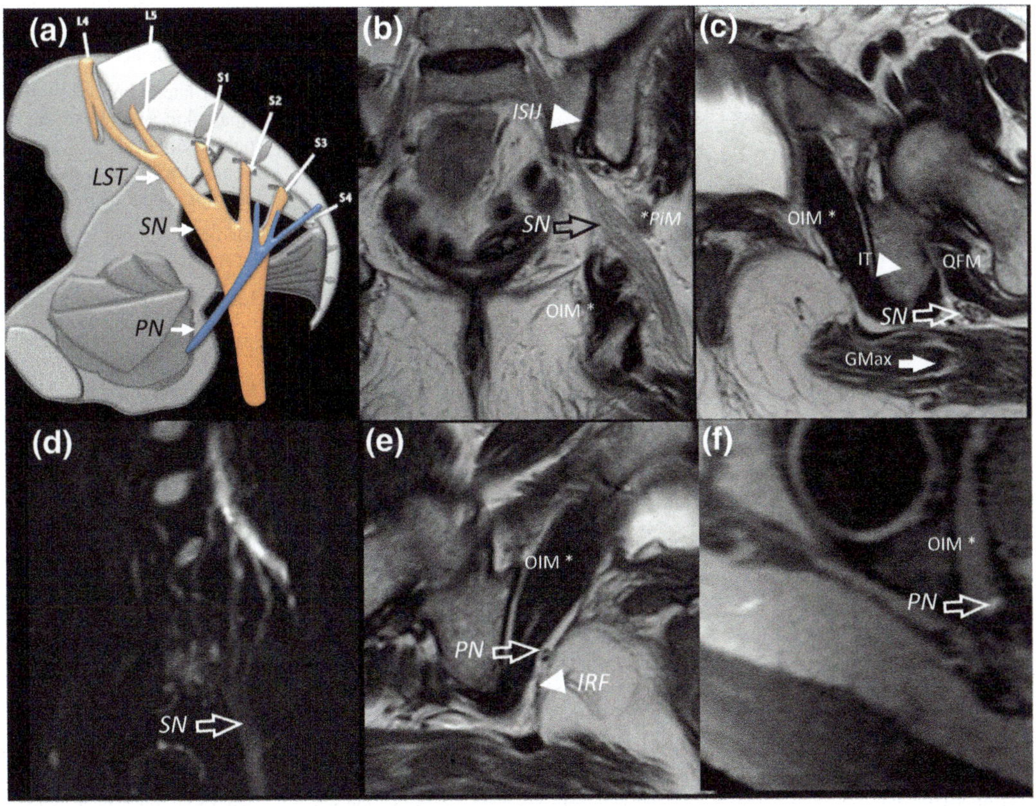

Fig. 17 Normal sacral plexus (*SP*) on MRN: (Abbreviations: **LST** lumbosacral trunk, **SN** sciatic nerve, **PN** pudendal nerve, **ISIJ** inferior sacroiliac joint, **PiM** piriformis muscle, **IPM** iliopsoas muscle, **OiM** obturator internus muscle, **IT** ischiatic tuberosity, **QFM** quadratus femoral muscle, **Gmax** gluteus maximus muscle, **IRF** ischiorectal fossa. Diagram (**a**) showing the spinal nerve roots origins of the SP (L4 to S4): **LST** formed by minor root of L4 and ventral L5, **SN** formed by L5, S1 and S2 and **PN** formed by S2, S3 and S4. Sequence T2 IDEAL in-phase weighted images in coronal oblique reformation; (Schema credited to Tatiana Escobar, MD) (**b**) **Pelvic SN** (*void black arrow*), slightly hypointense compared to fat passing below **ISIJ** (*white arrowhead*) and medial to **PiM** (*asterisk*). Sequence T2 IDEAL in-phase weighted images in axial plane (**c**) showing **SN** (*void black arrow*) in the subgluteal space. Note normal fascicular appearance of the **SN**. At this level **SN** is bounded by QFM, Gmax (*white arrow*) and IT *arrowhead*. Sequence DWI sagittal plane reformation (**d**) showing L5 to S1 spinal roots forming the **SN** (*white arrowhead*). Sequence 3D T2 IDEAL in-phase weighted images in axial plane (**e**) showing **PN** (*white void arrow*) within Alcock's canal (*AC*). AC is formed by medial border of OIM (*asterisk*), its aponeurosis and lateral wall of the IRF (*arrowhead*). Sequence 3D T2 IDEAL FAT SAT weighted images (**f**): **PN** is seen as a *bright dot* along the medial border of the OIM (*asterisk*)

(DLRPN) the LSP is the most frequently affected region with a lifelong incidence estimated in 1% among all diabetics [35].

DLRPN is a painful immune mediated neuropathy of the lower limbs often associated with severe weight loss. DLRPN could be differentiated from DPN by an acute or subacute onset, severe pain as dominant feature and asymmetry in the anatomic distribution [113].

Many denominations have been adopted possibly reflecting uncertainty about the anatomical involvement and pathophysiology of this condition (i.e., Bruns–Garland syndrome, diabetic amyotrophy, diabetic myelopathy, diabetic polyradiculopathy, femoral or femoral-sciatic neuropathy of diabetes, diabetic motor or paralytic neuropathy or proximal diabetic neuropathy [114].

Table 9 Motor and cutaneous innervation of the LS plexus

Nerve	Roots	Motor innervation	Sensitive innervation
Ilioinguinal N Iliohypogastric N Genitofemoral N	T12–L1	Transversus abdominis abdominal internal oblique	Upper lateral buttock, and a small skin area above the pubis symphysis Femoral triangle in the anterior aspect of the thigh
Femoral N	L2–L3–L4	Iliopsoas, pectineus, sartorius, quadriceps femoris	Medial lower leg
Lateral Femorocutaneous N	L2, L3		Lateral thigh
Obturator N	L3–L4–L5	Obturator externus, adductor longus, adductor brevis, gracilis, pectineus, adductor magnus	Upper medial thigh
Superior gluteal N Inferior gluteal N	L4–L5–S1–S2–S3	Gluteus medius, gluteus minimums, tensor fascia latae Gluteus maximus	Gluteal skin and lateral gluteus
Sciatic N	L4–L5 S1–S2–S3	Semitendinosus semimembranosus biceps femoris, long head, short head, adductor magnus	Provide indirect sensory innervation via its terminal branches
Tibial N	L4–L5, S1–S3	Deep and superficial posterior compartments of the leg including plantaris, gastrocnemius, popliteus, soleus, tibialis posterior, flexor digitorum longus and, flexor hallucis longus	Postero-lateral and antero-lateral leg, plantar surface of the foot
Common Peroneal N Superficial PN Deep PN		Short head of biceps femoris Peroneus longs, brevis Tibialis anterior, extensor digitorum longus, peroneus tertius and, extensor hallucis longus	Lateral leg and outer aspect of foot Antero-lateral aspect of the leg
Pudendal N	S2–S3–S4	Levator ani, external anal sphincter	Clitoris (in females) Penis (in males)

According a systematic revision by Chan et al. evidence suggests that DRNLP is caused by ischemic nerve injury secondary to immune mediated microvasculitis [115]. Increased expression of inflammatory mediators in different disease stages as: proinflammatory cytokines; tumor necrosis factor (TNF), interleukin (IL)-1, and IL-6 may be found [116].

MRI helps to determinate the pattern of involvement excluding other differential diagnosis and confirming results of the electrophysiological test. First clue to diagnosis is given so far by the distribution of muscle involvement. It usually starts with abnormal high signal in T2–FAT SAT weighted images demonstrating acute denervatory injury and segmental intrafascicular myoedema in the proximal tight muscles (most commonly in anterior compartment) following the distribution of the femoral or obturator nerves [117].

A pattern of muscle atrophy and fatty degeneration becomes the dominant features in later stages coincidentally with a significant weight loss (more than 4.5 kg in more than half of subjects). Fatty degeneration pattern demonstrates a confluent increase in T1-weighted signal intensity affecting more than 50% of the muscle (specially iliopsoas, pectineus, sartorius, quadriceps femoris, obturator externus, and adductor magnus) [118]

MRN abnormalities correspond to perivascular inflammatory changes detected in fascicular biopsies. Increased intraneural fluid and edema with ecstasies of surrounding small blood vessels of the epineurium, perivascular chronic inflammation in the subperineurial area and

Table 10 Causes of lumbosacral plexopathy

Immunitary or vascular	Inflammatory	Neoplastic and infiltrative	Mechanical or entrapment	Iatrogenic
Diabetic amiotrophy Diabetic radiculoplexopathy Sarcoidosis Amyloidosis Vasculitis	Guillain–Barré syndrome, chronic inflammatory polyradiculoneuritis, and multifocal mononeuritis	Lymphoma Local infiltration from: rectal, bladder, cervix or prostate cancer Peripheral Nerve sheath tumors: schwannoma, neurofibroma, MPNST Metastases	Trauma Pelvic rim fractures Retroperitoneal abcesses or hematomas Post-traumatic perineurioma	Labor and delivery Hip or knee surgery Injection Radiation plexoneuropathy

inflammatory cells infiltrating adjacent endoneurium may lead to intrinsic T2-weighted hyperintensities along the nerve [119].

Asymmetric and patchy nerve enlargement (best seen on DWI and T1 weighted images) is usually found affecting sciatic, femoral and obturator nerves. A segmental enhancement with gadolinium contrast could be found. An illustrative case for DLRPN is shown in Fig. 18 [120].

Vasculitis

Systemic (SVN) and no systemic vasculitides (NSVN) are known risk factors for peripheral neuropathy and been reported in up to 60–70% of patients [121].

Clinical course is determined by neuropathic pain and sensorimotor symptoms typically asymmetric at onset with multifocal and progressive spread resembling DLRPN or ILNRP in some point. Frequently patients with SVN caused by large vessels vasculitis develop constitutional symptoms such as weight loss, fatigue, fever, rash, or night sweats [122].

Large vessel vasculitis related to peripheral neuropathy includes Churg-Strauss syndrome, Wegener's granulomatosis, rheumatoid vasculitis, and polyarteritis nodosa. Imaging techniques are no further needed in the diagnostic algorithm of these conditions.

Idiopathic lumbosacral radiculoplexus neuropathy (ILRPN) is indistinguishable form its diabetic counterpart except for the impaired glucose metabolism in DLRPN [123].

Clinically both share the same pattern of distribution beginning with neuropathic pain starting at buttock, thigh, or leg. Lately it spreads from proximal to distal and finally becoming bilateral [116].

In histologic analyses there is multisegmental fiber loss, scar formation of the perineurium and epineurial neovascularization reflecting the vasculitic profile seen in DLNRP. Due to the restriction of the vasculitis to the peripheral nerves over time this entity is designated as a non-systemic vasculitic neuropathy (NSVN) [124].

Guillain Barré Syndrome

Guillain–Barré syndrome (GBS) is the most common acute immuno mediated postinfectious polyneuropathy, and constitutes one of the most serious emergencies in neurology [125]

Neurologic picture is usually preceded by an upper respiratory tract infection or diarrhea [126]. The main feature of GBS is a progressive bilateral and symmetric weakness of the limbs, followed by gait disturbations, progressing over 12 h to 28 days. After a plateau period patients typically experiment hyporeflexia or areflexia [127].

Neuropathic pain is reported in many patients, particularly children, when distal extremities are involved, commonly starting as back discomfort [128]. Some studies have suggested that pain may play a major role in the management strategy [128–131].

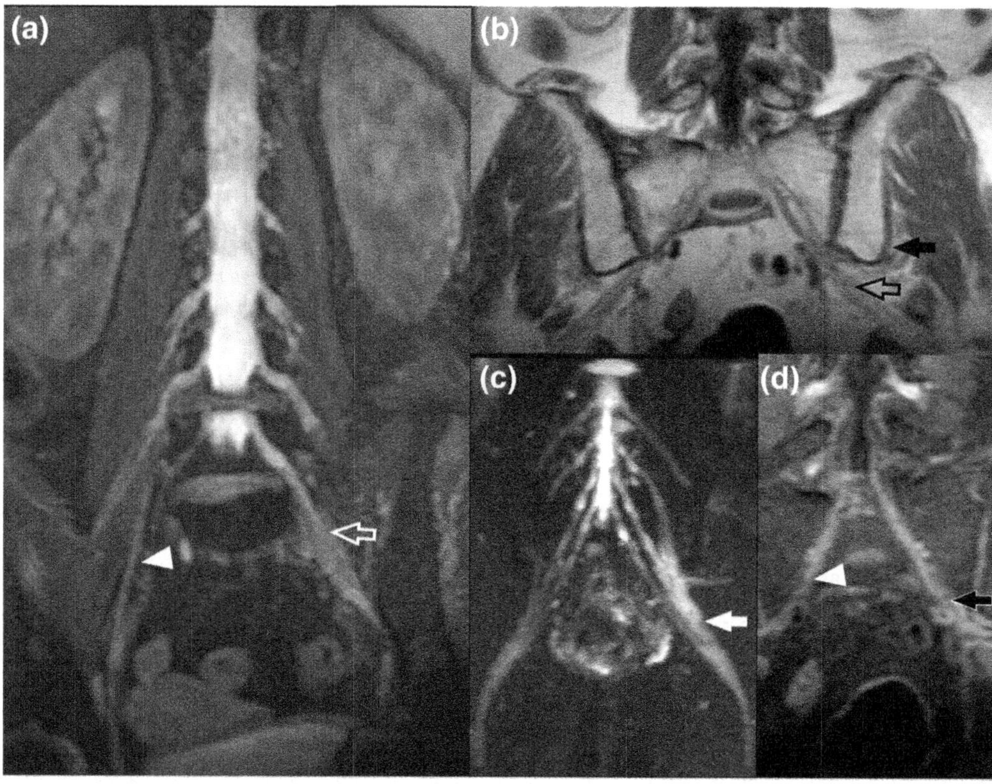

Fig. 18 Diabetic lumbosacral radiculoplexus neuropathy (*DLRPN*): 70 year old female with diabetes mellitus II. *Left* chronic hip and thigh pain progressing over two months. Weakness in both *lower* limbs coincident with previous 9 kg of weight loss. Saphenous nerve biopsy demonstrated vasculitis. Sequence 3D IDEAL coronal FAT SAT T2-weighted image (**a**): asymmetric enlargement of *left* L5 nerve and proximal sciatic nerve (*white arrow*) compared to contralateral (*white arrowhead*). Sequence T1 IN-PHASE weighted (**b**) images showing abnormal enlargement of *left* S1 and pelvic sciatic nerve (*void black arrow*) and superior gluteal nerve (*black arrow*). Sequence DWI (**c**) showing the abnormal enlargement of the main components of LSP exceeding a single. Sequence T1 IN-PHASE weighted images postgadolinium (**d**) showing enhancement of the abnormal *left* sciatic nerve

On MR images the most common reported pattern in classical GBS cases include diffuse enlargement and enhancement of the nerve roots. The most frequent location are the ventral roots and it is considered a distinctive feature [132, 133].

Nerve root enlargement has been postulated as reactive to lymphocytic and macrophagic infiltration around endoneural vessels that is associated with demyelination of the affected nerves [134].

Diffuse enhancement of the nerve roots and cauda equina are probably related to disruption of the blood–nerve barrier secondary to inflammation. This finding was confirmed by several authors, which tried to find a temporal correlation between the clinical features and the MR imaging characteristics. According to *Mulkey* et al., nerve roots can show post-gadolinium enhance as early as 2 days after the onset of clinical signs.

Other series have described a decrease on enhancement of the spinal nerve roots as the patient clinically improves. Complete resolution could be achieved after 6 months to 1 year away from the original clinical picture [133, 135]. Although nerve root enhancement appears to be a constant there is no consensus about the

Table 11 Differential diagnosis of the post-gadolinium spinal nerve and root enhancement

Infectious/inflammatory	Neoplastic/infiltrative	Entrapment or compressive	Iatrogenic	Other
Bacterial; meningitis (multiple organisms), mycobacterium tuberculosis, Borrelia burgdorferi (lyme disease), brucellosis, syphilis • Viral; Multiple—e.g., cytomegalovirus, herpes-zoster virus, herpes-simplex virus, HIV • Inflammatory polyradiculoneuropathies; Guillain–Barre syndrome, chronic inflammatory demyelinating polyneuropathy • Other; neurosarcoidosis, arachnoiditis (infectious, chemical, post-surgical, traumatic)	Leptomeningeal tumour/metastasis; primary central nervous system tumour (e.g., medulloblastoma, ependymoma), hematological malignancy, meningeal carcinomatosus • Nerve sheath tumors; schwannoma, neurofibroma	Space-occupying lesion • Disc herniation • Degenerative spinal disease	Lumbar puncture _ administration of intrathecal agent (e.g., anesthesia, methotrexate) • Spinal surgery • Radiation therapy	Radicular vessel enhancement‡ • Globoid cell
leukodystrophy				

relationship between disease severity and pattern of nerve root enhancement [136].

Additionally post-gadolinium nerve root enhancement is not specific for GBS and could be seen in a wide range of conditions that disrupt integrity of the blood–nerve root barrier as summarizes in Table 11. Despite lumbar puncture, it may be associated with meningeal enhancement, there is no large studies addressing the frequency of this or another iatrogenic causes of nerve root enhancement [137, 138].

Although nerve root enhancement appears to be a constant there is no consensus about the relationship between disease severity and pattern of nerve root enhancement [136]. A example of GBS in presented in Fig. 19.

Chronic Inflammatory Demyelinating Polyradiculoneuropathy

Chronic inflammatory demyelinating polyneuropathy (CIDP) as an acquired immuno mediated cause of neuropathic pain, affecting peripheral nerves and roots, was comprehensibly discussed in upper limbs section (Sect. 3). However some singularities about LSP and lower limb involvement will be highlighted in this apart [139].

Although CIDP commonly manifests with a wide symptomatic variety including motor deficit (83–94% of patients), sensory deficit (72–89%), facial palsy (4–15%) or oculomotor palsy (7%), a subset of cases presents with isolated sensory manifestations (ataxia, neuropathic pain, and paresthesia) especially in lower extremities. In these subjects a selective involvement of the LSP could precede brachial or cranial neuropathy. As far picture progresses a massive enlargement of the roots may appear without any predominance [140, 141].

Diffuse thickening in nerve roots and peripheral nerves extending from the lumbar to the brachial plexuses has been reported as the characteristic features of CIDP. These changes may reveal segmental demyelination, axonal degeneration, fiber loss, and reactive events (i.e., onion-bulb formation) [142].

Enhancement of the cauda equina, affecting both ventral roots has been reported. This feature has been postulated as a potential differentiator

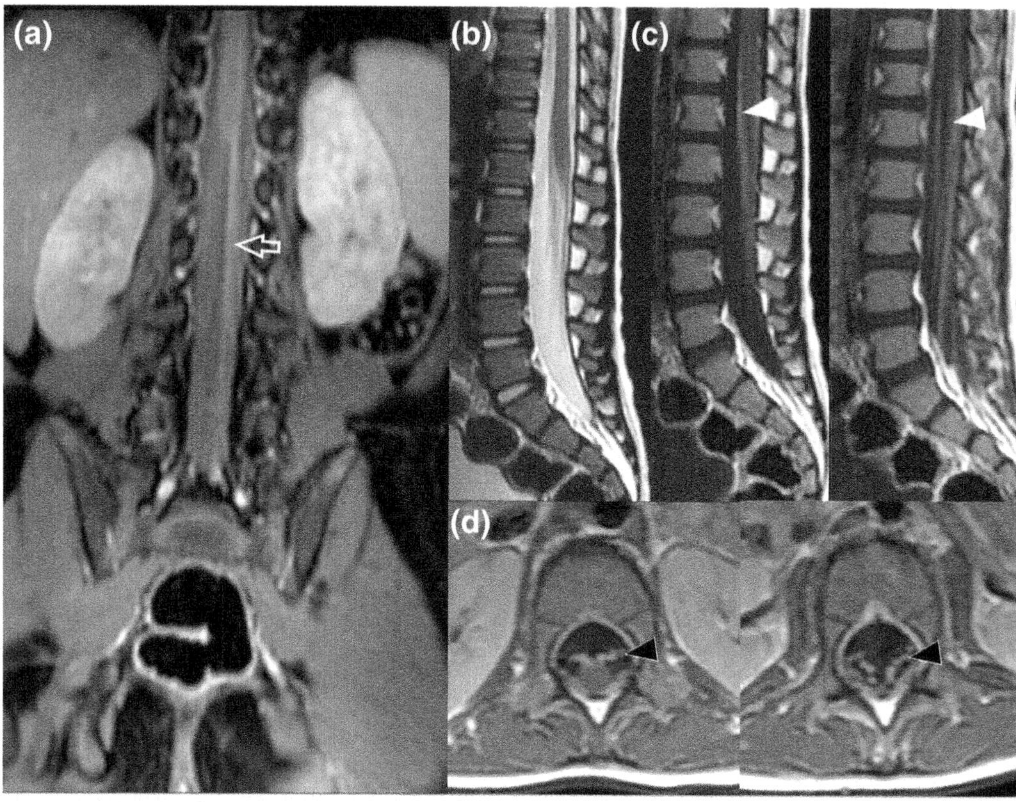

Fig. 19 Guillain Barré syndrome: 3 year old male with previous upper airway tract infection 9 days ago. Insidious onset of ataxia and thigh dull pain. Ascending motor weakness and hyporeflexia starting 2 days ago. Contrast enhanced T1 3D coronal WATER-IDEAL (**a**) showing diffuse enlargement and enhancement of cauda equina (*void arrow*). Sequence T2 sagittal FSE (**b**) showing symmetric enlargement of the cauda equina. Contrast enhanced T1 FSE sagittal plane (**c**) showing symmetric enhancement of the cauda equina (*white arrowheads*); same sequence that **c**, in axial plane (**d**) showing the symmetric involvement of the ventral roots (*black arrowheads*). This findings are considered a distinctive features of GBS

from GBS, alongside the relapsing or progressive course. A secondary involvement of the higher lumbar and thoracic nerve roots was reported, especially in children such as Ware et al. reported in 2014. Even cranial nerve thickening in CIDP has been reported in adult series, as well as rarely in children. Similar features are presented in a case of our series shown in Fig. 20 [143].

Enlargement and enhancement of nerve roots are not specific findings for CIDP and could be seen in other clinical conditions as neurofibromatosis type 1, hereditary polyneuropathies (i.e., Charcot–Marie–Tooth and Dejerine–Sottas diseases), metabolic diseases (i.e., metachromatic leukodystrophy, Krabbe disease) , tumors (i.e., leptomeningeal carcinomatosis, leukemia/ lymphoma), and toxic exposure (i.e., methotrexate or radiation) [144, 145].

6.2.2 Tumoral Disease: Primary and Secondary

Although neurofibromatosis type I is the commonest cause of primary neoplastic LS plexopathy is an unusual cause of neuropathic pain. Some case reports have demonstrated pain related to mechanical compression and mechanical factors in rare cases with plexiform neurofibromatosis [146].

In the peripheral nervous system (PNS), neoplastic neuropathic pain is almost exclusively derived from infiltration from pelvic tumors or the collateral effects of chemotherapy.

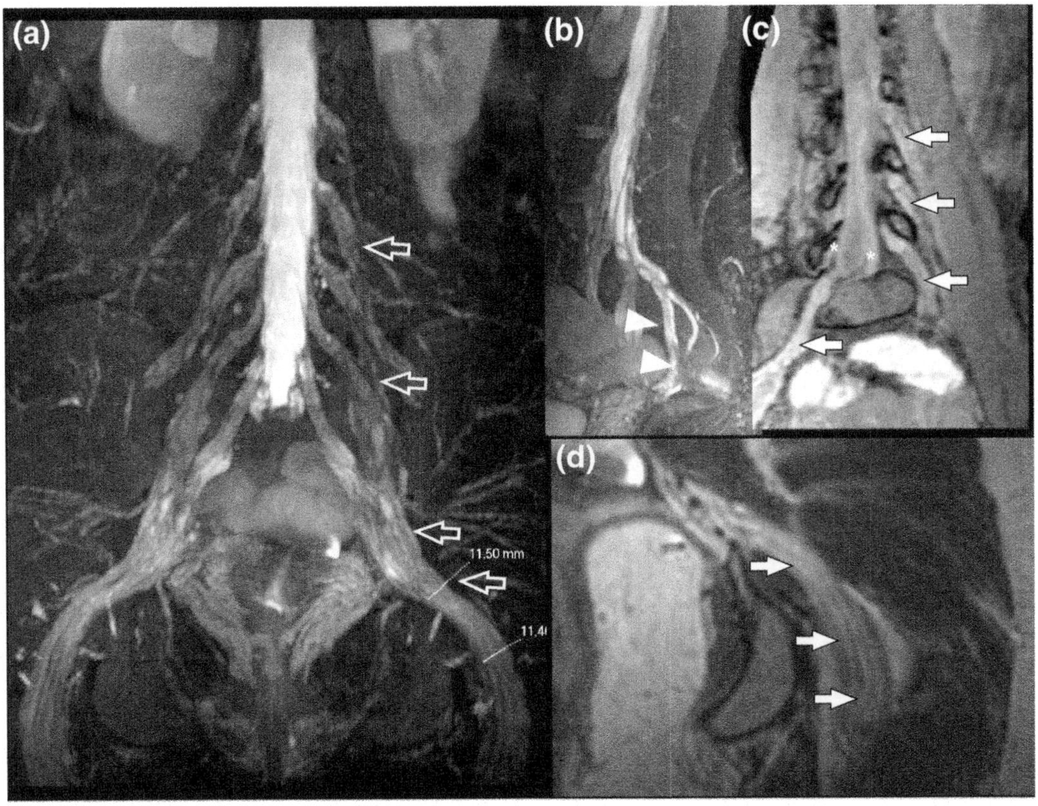

Fig. 20 Chronic inflammatory polyradiculoneuropathy (*CIDP*): 38 year old female with progressive hypoesthesia, dull pain in calf and walking impairment lasting six months. Weakness increased in the last month. Sequence T2 3D WATER IDEAL coronal (**a**), sagittal (**b**) demonstrated enlargement and abnormal T2 hyperintensity of nerve roots from L2 to S1 and lumbosacral trunk (*void arrows* in **a** and *white arrowheads* in **b**). Sequence T1 3D in-phase IDEAL post-gadolinium in coronal oblique reformation (**c**) showing diffuse and fusiform enlargement of roots of L3, L4 and L5, and S1 with mild contrast enhancement (*white arrows*). Note also, the cauda equina involvement (*asterisk*). In the same sequence that **c** (**d**), seen abnormal enlargement and losing of the fascicular pattern is seen in the sciatic nerve (*white arrows*)

Additionally hematologic malignancies (leukemias and lymphomas) can also involve the PNS by direct extension or compression. In a large retrospective review of neurologic complications of cancer the prevalence of lumbosacral plexopathy was 0.71% [147].

Colorectal, cervix, prostate, and bladder cancers accounts for the most common sources of local spreading to LSP. Retroperitoneal sarcomas and nodal metastasis from lymphoma may produce mechanical and compressive effects over nerve roots [148]. Identification of secondary invasion of LSP determinates an ominous prognosis for the survival since it has been estimated at 5.5 months according to the report of authors as Jaeckle et al. [147].

Neoplastic involvement of LSP usually starts with insidious pain followed by numbness and motor deficit in a protracted course over weeks. Bilateral involvement can be expected in 25% while incontinence and impotence became the dominant features [149]. Sympathetic nerve infiltration is closely related to "hot dry foot syndrome"; found in up to 30% of cases, and could be considered as an early feature of deafferentation [150].

According to the anatomical hypotheses raised by *Capek* et al. in 2015 and *Hébert-Blouin*

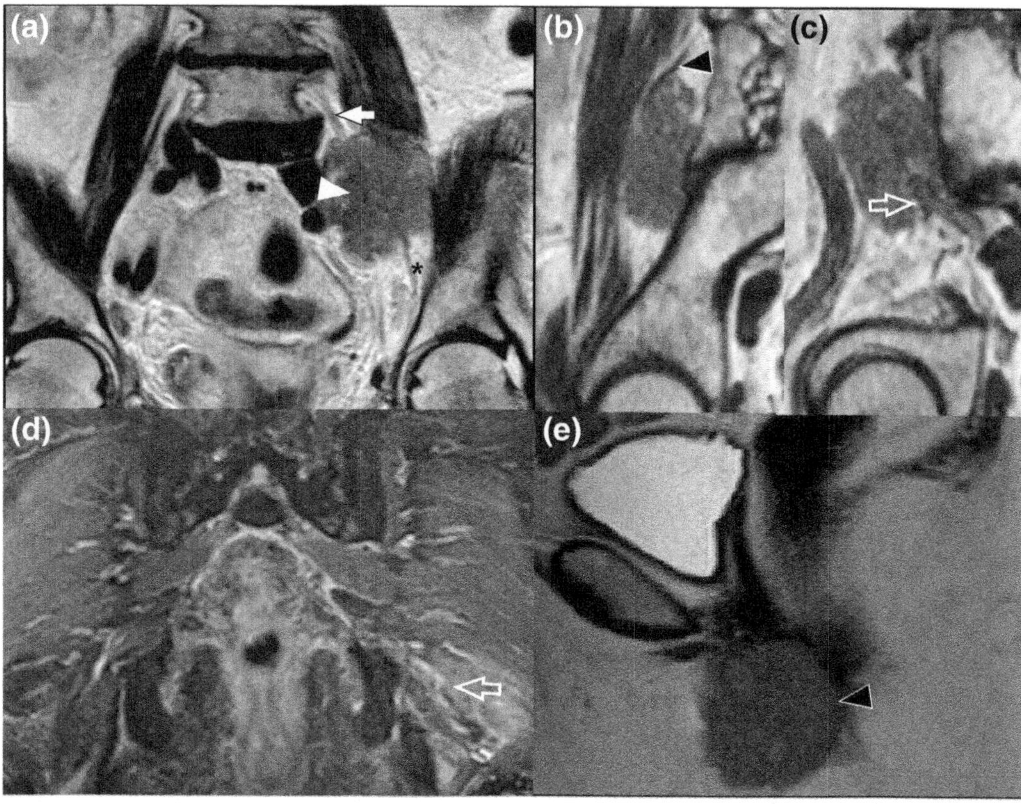

Fig. 21 Perineural spread of pelvic neoplasm: 80 year old female with recurrent rectal adenocarcinoma. Progressive leg and calf pain followed by limb enlargement and sphincter impairment 12 months after radiotherapy. Sequence 3D T2 in-phase dixon in coronal plane (**a**) and sagittal reformation (**b**, **c**) on pelvis showing a solid mass (*white arrowhead*) infiltrating the *left* L4 and L5 nerve roots (*white arrow* in **a** and *black arrowhead* in **b**). Obturator nerve is also involved (*asterisk* in **a**). Note also displacement of the psoas muscle. Vascular encasement may also be detected on MRN (*white void arrow* in **c**). **d**. Sequence 3D T2 STIR in coronal plane (**d**) showing edema of the *left* gluteus maximus (*void arrow*) due to denervatory changes. In T2-FSE sagittal plane (**e**) a solid rectal mass was detected in a former exam performed five months ago

et al. in 2010, routes of perineural spread (PNS) are defined by proximity. In recurrent prostate cancer retrograde infiltration of the LSP starts within the ipsilateral nerves of the affected lobe (S2 to S4), while in colorectal cancer begins with the inferior hypogastric plexus [151].

Subsequent spread continues through the parasympathetic pelvic and the sympathetic sacral nerves to finally attain the sacral plexus. Then invasion continues ascending to sacral spinal nerves, lumbar plexus, and lumbar spinal nerves via the lumbosacral trunk [152]. Several early reports suggest that PNS is most common than expected being observed even in 7–43% of biopsies of prostatectomy specimens.

MRI is the method of choice to demonstrate secondary neoplastic plexus involvement. Elevated signal on T2 weighted images, loss of normal fascicular pattern, and nodular enhancing pattern are the known features of perineural infiltration. It is important to consider the examination of the primary site of tumor in order to confirm invasion through the boundaries (e.g., mesorectal fascia in colorectal cancer, or parametrial fat in cervix cancer) [148].

Imaging features in a case of perineural extension form a lower rectum cancer is shown in Fig. 21.

Differential diagnosis in the oncology setting easily assessed by imaging includes bone

Table 12 Differential diagnosis between perineural spread (PNS) and radiation induced plexopathy

Features	PNS	Radiation induced plexopathy
MRI	• Adjacent mass (evidence of mesorectal fascial or parametrial invasion) • Low T1 and high T2 • Perineural fat effacement • Signal intensity similar to primary site tumor • Nodular enhancement • Focal (short segment) enlargement	• No mass evidence • Low T1 and low T2 • Perineural fat striking • Patchy or linear enhancement • Diffuse enlargement
Pain	• Frequent and severe • Unilateral or focal	• Mild or absent. Often preceded of numbness • Often bilateral
Onset	• Early or in late stages of cancer	• Months or years after exposition
Other clinical features	• Leg edema • Rectal, prostate or cervical mass • Warm and dry foot • Sphincter disfunction	• No edema • Non rectal o prostate mass- • Autonomic and sphincter function preserved
EMG	• Axonal loss/demyelination	• Myokymia

Based upon [148]

metastases, avascular necrosis of the hip, vertebral compression fractures, retroperitoneal hematoma, and radiation-induced plexopathy. Distinction between metastasis and radiotherapy induced lumbosacral plexopathy can be difficult. The former is more frequently associated with sphincter dysfunction, leg pain edema, and abnormal CSF while actinic LS plexopathy may be indolent and develops months or years after exposure. Clinical and imaging features useful to differentiate between metastatic infiltration and radiation induced injury are shown in Table 12. In Fig. 22 a case of radiation injury of LSP in a prostate cancer subject is presented [153].

6.3 Traumatic, Mechanical and Entrapment Conditions

Traumatic direct injuries of the LSP are rare. Lesions are commonly related to penetrating injury, pelvic rim fractures, or compression by hematomas. In a recently published series of 72 cases by Garozzo et al. [154] bone injuries were found in 85% of patients, internal lesions in 30% and vascular injuries in 8% [154].

Although paralysis could be the main symptom LSP injuries can cause pain in several degrees. The pattern of pain is commonly associated with the anatomic level of injury being more frequent in the sacral plexus injury and the lumbosacral trunk. When it is found associated with sphincter dysfunction, usually indicates a poor prognosis.

Nerve root and plexus injury are closely related to trauma biomechanics. Motorcycle accidents are associated with a higher incidence of pelvic crushing and sacroiliac joints separation. Suicidal jump and sacral transverse fracture also had a significantly higher risk for LSP injury according to a retrospective series published by *Sugimoto* et al. in 2010 [155].

Avulsion and compression of the spinal roots by a fracture through the sacral foramina may be the leading mechanism [156–158].

In plexus injuries, main role of the imaging is to rule out the presence of root avulsions. If multiple root avulsions are detected there are little chances for spontaneous recovery [154].

As in brachial plexus a root avulsion may be associated with the presence of meningocele. In CT myelography meningoceles are described as CSF pouches emerging from the foramina with complete filling by opaque contrast while in MRI they follows the signal of CSF in all the pulses sequences. Identification of remaining

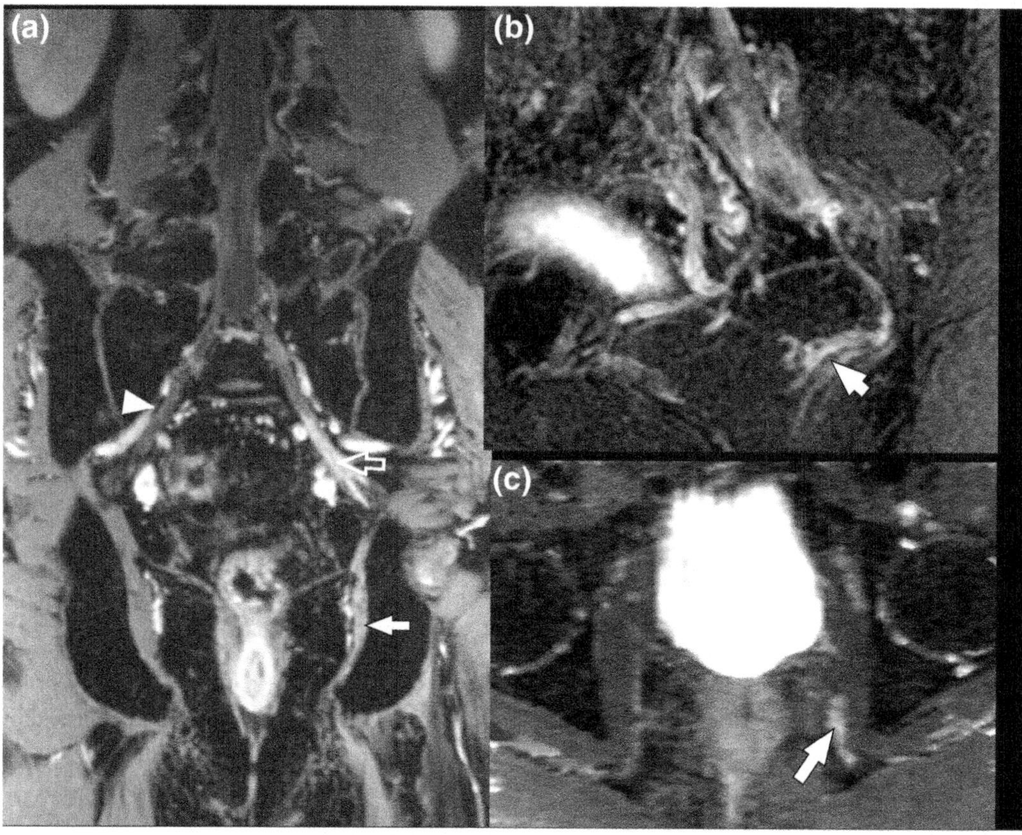

Fig. 22 Radiation injury of sacral plexus and pudendal nerve: 63 year old male with prostate cancer. Mild pain and dysesthesia in *left* buttock and hip. In the last 3 months a perineal pain onset without sphincterial involvement. Contrast enhanced T1 3D coronal WATER-IDEAL (**a**): focal enlargement and avid enhancement on the proximal *left* S1 root is shown (*void arrow*). Compare with the normal *right* side (*arrowhead*). Also, there is internal obturator muscle atrophy (*white arrow*). Sequence 3D T2 FAT SAT IDEAL coronal (**b**) and axial (**c**), showing enlargement of pudendal nerve (*white arrow*)

roots, feasible with high-resolution MRN, is mandatory because improves the opportunity for reinnervation surgery. Avulsions may be frequently found in sacral plexus with L5 and S1 being the roots more prone to this type of injuries. Additionally, the lumbosacral trunk exhibits a particular vulnerability in fractures of the posterior pelvic rim due to its little mobility, being relatively fixed to the sacral ala.

In the vast majority of LSP lesions the nerves are injured by compression due to a peri-fracture hematoma. In MRN the may appear focally enlarged with blurring of the perineural fat plane. Blood degradation products could be seen in the surrounding tissue. These types of injuries are related with spontaneous recovery as *Tonetti* et al. reported in the literature [159]. A case of LSP injury with posttraumatic neuroma formation is presented in Fig. 23.

LSP could be entrapped by retroperitoneal space-occupying masses due to the proximity with the emerging roots. Retroperitoneal hematoma extending over psoas or iliacus muscle secondary to common iliac or hypogastric ruptured aneurysm; anticoagulant therapy or hemophilia can compress the lumbar plexus or femoral nerve. LS trunk entrapment by fetal head during labor has also been reported.

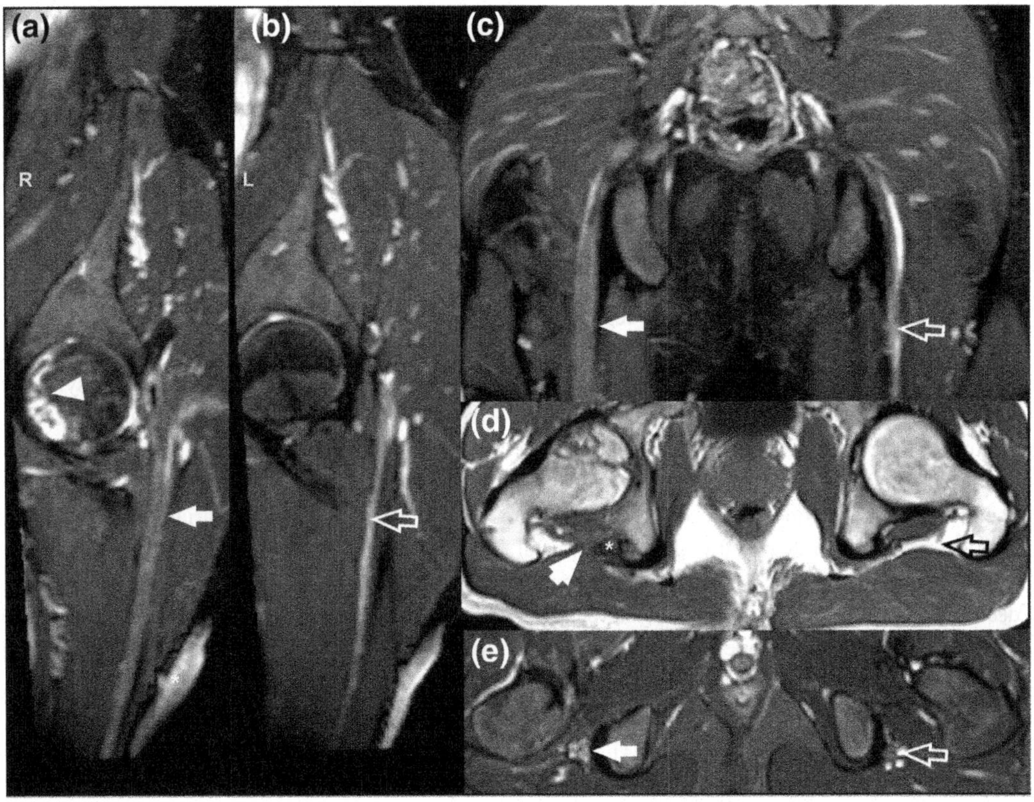

Fig. 23 **Traumatic** injury of the sciatic nerve: 18 year old male with drop foot after femoral neck fracture. Sequence 3D STIR SPACE sagittal (**a**, **b**) and coronal (**c**) reformation showing abnormal enlargement of the sciatic nerve (*SN*) along the subgluteal space behind to ischiatic tuberosity (*white arrow* in **a** and **c**). A fracture of the femur head is seen on the affected side (*arrowhead* in **a**). Compare with the normal *left* side (*void arrows* in **b** and **c**). Sequence T1 DIXON in-phase axial plane (**d**) showing blurring of the fat plane in the *right* subgluteal space due to fibrotic and scar tissue (*white arrow*). Note enlargement of the *right* SN (*asterisk*). Compare with the normal *left* side (*black void arrow*). Sequence 3D STIR SPACE axial plane (**e**) showing neuroma formation (*white arrow*), it is characterized by enlargement and losing of the fascicular pattern compared to normal side (*void white arrow*)

6.4 Pelvic Peripheral Neuropathies

Anatomic coverage by the pelvic rim prevents LSP and nerve root injuries lowering frequency in 1:3 or 1:4 compared to BP. However neuropathic pain may be more insidious and difficult to localize due superimposition of other symptoms related to variability of the components of pelvis.

6.4.1 Iliohypogastric, Ilioinguinal, and Genitofemoral Neuropathies

Ilioinguinal (L1), iliohypogastric (T12–L1), and genitofemoral (L1, L2) are predominantly sensory nerves sharing common origin from the L1 ventral roots, emerging through the lateral border of the psoas muscle. Iliohypogastric and genitofemoral nerves also receive contributions from the T12 and L2 anterior rami, respectively.

Ilioinguinal nerve travels along the superficial surface of the internal oblique muscle, passing on average 1.0 cm from the inguinal ligament. Then crosses the medial aspect of anterior superior iliac spine, and innervates a strip of skin along the inguinal canal giving sensory innervation to upper scrotum in male or upper major labia in women [160].

Iliohypogastric nerve innervates transversus abdominis and obliquus internus. Terminal branches provide sensory innervation to a small area of the upper lateral buttock, and a small skin area above the pubis symphysis.

Genitofemoral nerve provides cutaneous innervation to the femoral triangle in the anterior aspect of the thigh. A genital branch enters in spermatic cord in men supplying cremasteric muscle and the skin of the scrotum. In women runs along round ligament and provides sensory innervation to minor labia. Anatomic features are shown in Fig. 16a–c.

Genitofemoral neuralgia is the most common neuropathic pain syndrome along the cutaneous region of the groin, inner thigh, and lower abdomen. Injuries over these nerves are frequently related to surgical procedures. Iatrogenic injuries commonly associated with genitofemoral neuralgia may include caesarian section, appendectomy, lymph node biopsy, hysterectomy, vasectomy, and by far open and laparoscopic inguinal herniorrhaphy [161]. In fact post-herniorrhaphy chronic neuralgia have been reported as high as 63% of cases [162].

Mechanisms involved in pathogenesis of genitofemoral neuralgia may include entrapment within scar tissue, fibrous adhesions, and tethering of the nerve that cause the onset of nociceptive pain receptors.

Imaging and particularly ultrasound has an important role in image guided pain treatment allowing visualization of the nerve and the surrounding landmark structures [163]. MRN and conventional MR may detect extrinsic compressions by haematoma, tumoral, or direct trauma injury.

6.4.2 Obturator Neuropathies

The lumbar plexus (LP) is formed by L2, L3, and L4 ventral rami, and then ramificates into anterior and posterior divisions that form the obturator (ON) and femoral nerves (FN), respectively.

ON descends into pelvis along medial border of the psoas muscle, over the sacroiliac joint until reach the obturator canal. In MRN it can be visualized as a low signal band in T1 and T2 weighted images running parallel to medial psoas, surrounded by pelvic fat as shown in Fig. 16d.

ON exit pelvis trough the obturator foramen and divides into anterior and posterior branches. The anterior division supplies motor innervation to the hip, gracilis, adductor brevis, and longus muscles. The posterior branch innervates the obturator externus and part of the adductor magnus muscles. Sensory fibers innervate an area on the upper medial thigh.

Obturator neuropathies may result from trauma (specially pelvic and pubic symphysis fractures) or due to iatrogenic injuries (hip surgery, lithotomy position, aortofemoral bypass, or oophorectomy). Other authors found association with rectal or ovarian malignancies; endometriosis, tuboovarian abscesses or rarely schwannomas [164].

Obturator neuralgia commonly presents with pain, weakness of thigh adduction, sometimes associated with numbness in the medial thigh that worsen with exercise [165].

Adductor muscles, gracilis and obturator externus may show denervatory signal changes in the acute or subacute setting shown high signal in T2 weighted images or STIR. In the chronic phase fatty infiltration can be found. Displacement, interruption of the course of ON and enlargement may be seen in neoplastic infiltration or entrapment, respectively. Blurring of the perineural fat plane may also be recognizable in inflammatory conditions. A case of neoplastic invasion to ON from a recurrent cervical cancer is shown in Fig. 21.

6.4.3 Femoral Neuropathies

The femoral nerve (FN; L2, L3, L4 anterior rami) is formed for the posterior division of the anterior rami of the LP. Emerges from the lateral border of the psoas descending between the psoas and iliacus muscles. At the exit of pelvis passes beneath the inguinal ligament, lateral to artery, entering the proximal anterior thigh then dividing into terminal motor branches, which innervate the sartorius and quadriceps muscles. Nevertheless an early arborizing branching pattern as been proposed by authors as Lonchena et al. [166].

A cutaneous branch, the saphenous nerve supplies sensory innervation to the medial lower leg. Anatomic relationships and course are shown in Fig. 16a.

The main trunk of the FN may be vulnerable at least in two sites: the retroperitoneal pelvic space or beneath the inguinal ligament. Pelvic FN nerve may be injured in surgical procedures using retractor blades, by ischemic injury, compressed by retroperitoneal hematoma or in an inadvertent laceration [167].

Rarely mass lesions as lymphadenopathy, abscesses, cysts, lymphomas and malignancies of the colon or rectum may affect FN by contiguity. Primary femoral nerve neoplasms, such as schwannomas or neurofibromas, are extremely rare [168].

In inguinal channel injuries may be commonly iatrogenic, and result from nerve compression during lithotomy positioning or in hip replacement. Less frequent conditions as inguinal lymphadenopathies, femoral vessel catheterization or localized groin hematomas may appear at this level.

In MRN denervatory injuries may predict the level of harm. In pelvic compartment injuries edema may be seen in iliopsoas muscle and extrinsic compression could be found. Denervatory edema of the anterior compartment of the thigh could be seen in the subacute phase of FN injuries below inguinal ligament. FN is difficult to follow itself and could be recognized by its relationship with the femoral artery. In compression it may show focal enlargement or edema along the nerve. In chronic phase fibrotic changes could be seen.

6.4.4 Lateral Femoral Cutaneous Nerve

Lateral femoral cutaneous nerve (LFCN; L2 and L3 roots ventral rami) is a sensory branch providing innervation to lateral thigh. It travels between lower of the abdominal muscles and crosses through iliacus muscle. Leaves pelvis passing below inguinal ligament close to anterior superior iliac spine and enters the fascia latae to finally divide into its terminal branches.

Meralgia paresthetica (MP; a term derived from the Greek meros = thigh and algos = pain) is the most renowned neuropathy of the LFCN. The most common cause is entrapment beneath the inguinal ligament on near the anterior superior iliac spine (ASIP). Other commonly associated conditions could be pregnancy, obesity, and diabetes mellitus.

The main clinical features are thigh burning pain and numbness, usually unilateral. MP exhibits a higher prevalence in adult males and association with sports as gymnastics, baseball, soccer, or body buildings were found. Avulsion injuries or trauma involving ASIP also plays a major role in etiology [169].

In imaging, particularly in MRI is possible to determinate the level of the injury. In STIR pulse sequence edema may show bone bruise at AISP or tearing of the muscles of the high anterior compartment of thigh, especially sartorious. CT or conventional X-Ray may confirm the osseous injuries.

6.4.5 Superior and Inferior Gluteal Neuropathies

The superior gluteal nerve (SGN) is formed by L4, L5, and S1 ventral rami. Exits the pelvis through the sciatic notch, above the piriformis and passes between medius and minimus muscles, ending above the tensor fascia latae muscle. SGN provides motor innervation to the tensor fascia latae, gluteus medius, and gluteus minimus muscles. In the coronal plane it could be recognized by its close relationship with the inferior sacroiliac joints (ISIJ). Anatomy and relationships are shown in Fig. 16f.

The inferior gluteal nerve (IGN) is formed by L5, S1, and S2 ventral rami. It leaves the pelvis below the piriformis muscle very close to the sciatic and the posterior femoral cutaneous nerves. IGN provides motor innervation to the gluteus maximus muscle.

The clinical triad of gluteal nerve entrapment is characterized by buttock pain, weakness of abduction of the hip, and tenderness to palpation of the buttock. The isolated injury of the superior gluteal commonly manifests as abduction impairment of the hip causing a toddler gait. In the other hand buttock atrophy and weakness of hip extension are the common features of inferior gluteal neuropathies [170].

Lesions of the gluteal nerves are unusual and mostly related to iatrogenic injury. Excessive retraction or inadvertent section during total hip replacement is a well-known cause of SGN involvement.

The IGN may also be compressed by intrapelvic masses, such as colorectal malignancies or large iliac artery aneurysms and may be recognized as a complication of the posterior approach to hip arthroplasty [171].

Imaging features of the gluteal neuropathies may be recognized by signal abnormalities in the innervated muscles. Edema (shown as elevated signal in STIR or T2 weighted images) secondary to trauma or denervatory injury is found in gluteal muscles and fascia latae tensor. Normal SGN is visible in MR as a low signal strip on T1 or T2 weighted images surrounded by fat below to ISIJ in the coronal plane. In entrapment conditions nerve may look enlarged and perineural fat plane could be blurred. IGN nerve is usually not visible in MRN unless injured [172].

6.4.6 Pudendal Neuropathies

The pudendal nerve (PN) originates from S2, S3, and S4 sacral roots. It follows a straight course between the piriformis and coccygeus muscles through the greater sciatic foramen. PN together with the internal pudendal vessels enters in the pudendal canal. The lateral wall of the ischiorectal fossa, obturator internus muscle, and its aponeurosis forms Alcock's or pudendal canal, a fibrous cartilaginous structure prone to entrapment, compression or fibrosis [173].

In normal MRN the PN could be recognized on T2 or T1 weighted images as a low signal band posterior to the ischial spine and may be followed between the medial border of the obturator internus muscle and along lateral wall of the ischiorectal fossa. DWI or STIR images as a linear hyperintensity between the Alcock's canal (AC). These features are shown in Fig. 18e, f.

The pudendal nerve has three terminal branches, the perineal nerve, the dorsal nerve of the clitoris (in females) or the dorsal nerve of the penis (in males) and the inferior anal nerve [174].

Pudendal neuralgia is a neuropathic painful condition involving the vulva, vagina, clitoris, perineum, and rectum in females and glans, penis, scrotum, perineum, and rectum in males being either unilateral or bilateral [175].

Pudendal neuralgia can be caused by mechanical compression due to pelvic floor muscle spasm (levator ani or obturator internus), periradicular cysts, scar tissue or fibrosis from trauma, or surgeries involving the AC. Other less frequent causes may include viral infection, immunologic processes or, radiation injury (Fig. 22) [176].

Imaging (particularly MR and CT) could be contributive to the definitive diagnosis demonstrating infiltrative conditions (rectal, bladder and cervix carcinoma) involving PN trough the ischiorectal fossa. Other conditions as lymphoma or nodal metastasis may compress the adjacent structures. MRN may also show fibrotic or scar tissue within the Alcock's canal manifested as linear bands with low signal on T1 and T2-weighted images [111, 175].

6.4.7 Sciatic Pelvic Nerve: Sciatica and Piriformis Syndrome

The sciatic nerve (SN), the major nerve of body is formed from fusion of the L5, S1 and S2 ventral rami, with a little contribution from the L4 ventral ramus. Two independent trunks, the medial or tibial division and the lateral or fibular division constitute the sciatic nerve. With a wide range of variations the SN leave the pelvis through the greater sciatic foramen passing below the piriformis muscle (PM) in a common trunk. However in near of 30% of cases the peroneal division alone can pass through PM and tibial division may cross below [177].

In the buttock, it passes close to the posterior capsule of the hip joint then descends into the thigh between the adductor magnus and the gluteus maximus muscles. Anatomy is described in Fig. 17a–c.

In the pelvis, the sciatic nerve supplies the PM and femoral quadratus. In the thigh, the tibial nerve division innervates the long head of biceps femoris, semitendinosus, semimembranosus, and adductor magnus muscles. The peroneal division innervates the short head of the hamstring muscle [178].

Piriformis syndrome (introduced by Robin in 1947) is an ubiquitous term that encompasses a great number of conditions causing neuropathic pain, also commonly referred as non-discogenic sciatica, and not necessarily related to abnormalities in the course, muscle tone or size of the PM.

The PM originates from the anterior surface of the sacrum and inserts into the upper aspect of the greater trochanter, leaving out of the pelvis through the greater sciatic notch. Anatomical variations, such as a bipartite PM have been related to compression phenomena however similar variants have been found in healthy controls [179].

Other postulated causes are periarthritis involving the anterior sacroiliac ligament, stretching SN over the PM muscle or even a compression generated by the obturator internus [180].

Due to the fact that extraspinal sacral plexus and sciatic nerve entrapments may result from other pelvic conditions (specially within the subgluteal space) the term deep gluteal syndrome may be a more accurate description for the non-discogenic sciatica.

The subgluteal space is the cellular and fatty tissue located between the middle and deep gluteal aponeurosis layers as depicted in Fig. 17d–f. Vascular structures and SN may be entrapped by fibrovascular bands or conflict with muscular components resulting in neuropathic pain (piriformis hypertrophy, obturator internus impingement, quadratus femoris tears, hamstrings enthesopathy and gluteal contractures) [170].

Total hip joint replacement, intragluteal injection, tumoral infiltration (commonly by lymphoma, metastasis), or displacements by nerve sheath tumors are other relevant causes of sciatic neuropathy. A case of posttraumatic neuroma of the SN is shown in Fig. 23.

7 Lower Extremity Neuropathies

Neuropathies of the lower limb are less common but more disabling than their upper limb counterparts due to their greater impairment in locomotion. They also may be initially misdiagnosed because their clinical manifestations are easily superimposed or even confused with radiculopathies, vascular claudication, or trauma, which are more common by far. The typical clinical manifestation of nerve entrapment consists of neuropathic pain at rest commonly exacerbated by with continued exercise [181].

7.1 Common, Deep, and Superficial Peroneal Neuropathies

The common peroneal (CPN) and the tibial nerves (TN) come from a common trunk in the thigh, but they are functionally independent components of the SN [182].

At the proximal aspect of the popliteal fossa SN separates into two components: an antero medial cord, TN tibial nerve and a posterolateral the CPN.

CPN enters into a fat plane separating the lateral gastrocnemius muscle from biceps femoris.

The nerve descends to reach the posterior edge of the head of the fibula, encircling the fibular neck to get into the peroneal tunnel, a fiber osseous canal externally bordered by the neck of the fibula and peroneus longus muscle. At this point CPN is only covered by subcutaneous tissue and skin becoming vulnerable to trauma and compression injuries [183].

Caudal to the fibular neck CPN trifurcates branching into recurrent articular fascicle, superficial peroneal nerve (SPN), and deep peroneal nerve (DPN). SPN is located deep to peroneus longus muscle (PL) in the lateral compartment of leg entering across the anterior intermuscular septum. Then it continues descending to become superficial and emerging between peroneus brevis and extensor digitorum longus muscles (EDL). It pierces the crural fascia, approximately 12 cm above the lateral malleolus in most individuals and travels into subcutaneous fat as a sensory nerve. It may be a potential point of vulnerability for entrapment and trauma [184].

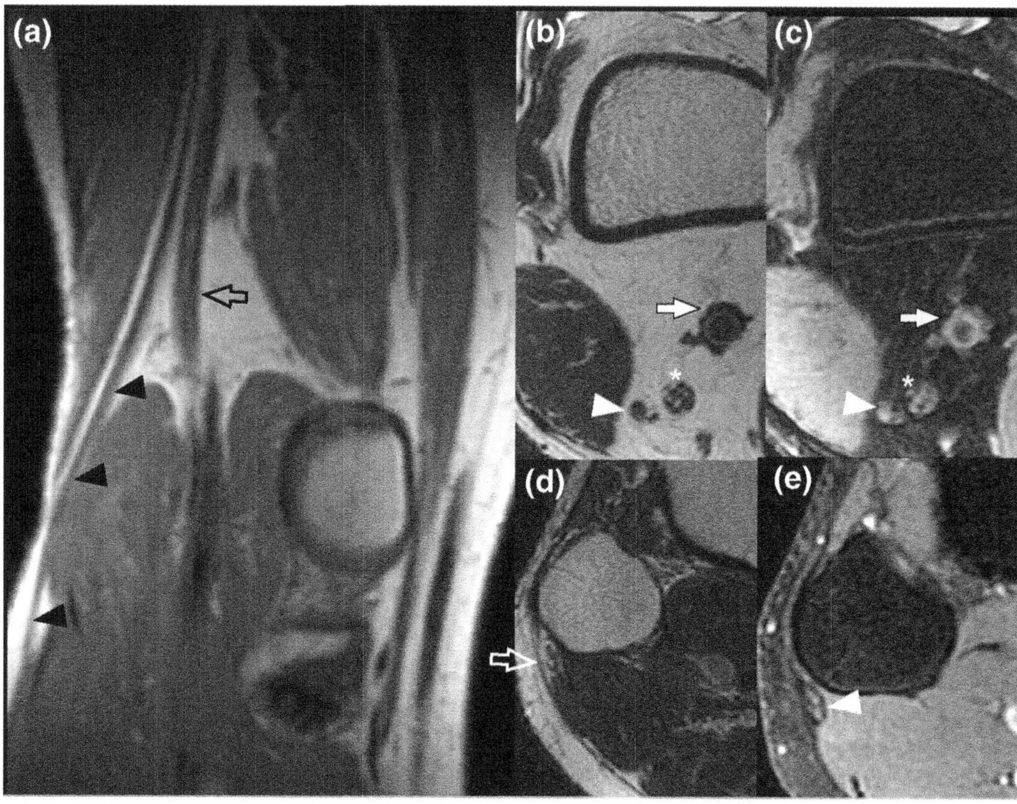

Fig. 24 Anatomic features of the common peroneal nerve (*PN*): coronal oblique T2 3D IDEAL in-phase reformation (**a**) showing the normal morphological division of the SN in TN (*void black arrow*) and CPN (*black arrowhead*). Sequence T2-FSE (**b**) and PD–FAT SAT (**c**) in axial plane showing the normal fascicular appearance of the TN (*asterisk*) and CPN (*white arrowhead*). Note the close relation to popliteal artery (*white arrow*). Sequence T2-FSE (**d**) and PD–FAT SAT (**e**) in axial plane. The CPN (*white arrowhead* in **e**) entering to the peroneal channel is mainly superficial, only covered by the posterior fascia of the leg (*void arrow* in **d**) and subcutaneous fat

SPN divides into terminal two branches: medial dorsal cutaneous nerve and intermediate dorsal cutaneous nerve [185].

The deep peroneal nerve (DPN) descends around the fibular neck and enters the anterior compartment through the intramuscular septum. In proximal leg it courses adjacent to the tibialis anterior artery ventrally to the interosseous membrane. In distal leg DPN passes beneath the extensor retinaculum between the extensor hallucis longus and extensor digitorum longus muscles, just lateral to the dorsalis pedis artery. Potential locations for entrapment of DPN are interosseous membrane and extensor retinaculum at the point where the extensor hallucis longus tendon crosses over it [183]. Anatomy of the CPN is detailed in Fig. 24.

Pain associated with common peroneal nerve entrapment is referred to the lateral leg and outer aspect of foot [186].

There is a wide spectrum of causes for peroneal neuropathy commonly resulting from external injuries as direct trauma mainly involving tibiofibular proximal joint.

Other causes of extrinsic compressions may be related to permanent leg crossing, squatting, certain surgery positions, casts, or braces. Intrinsic conditions resulting in nerve direct injury are neural sheath neoplasm, vascular abnormalities, exercise-induced compartment

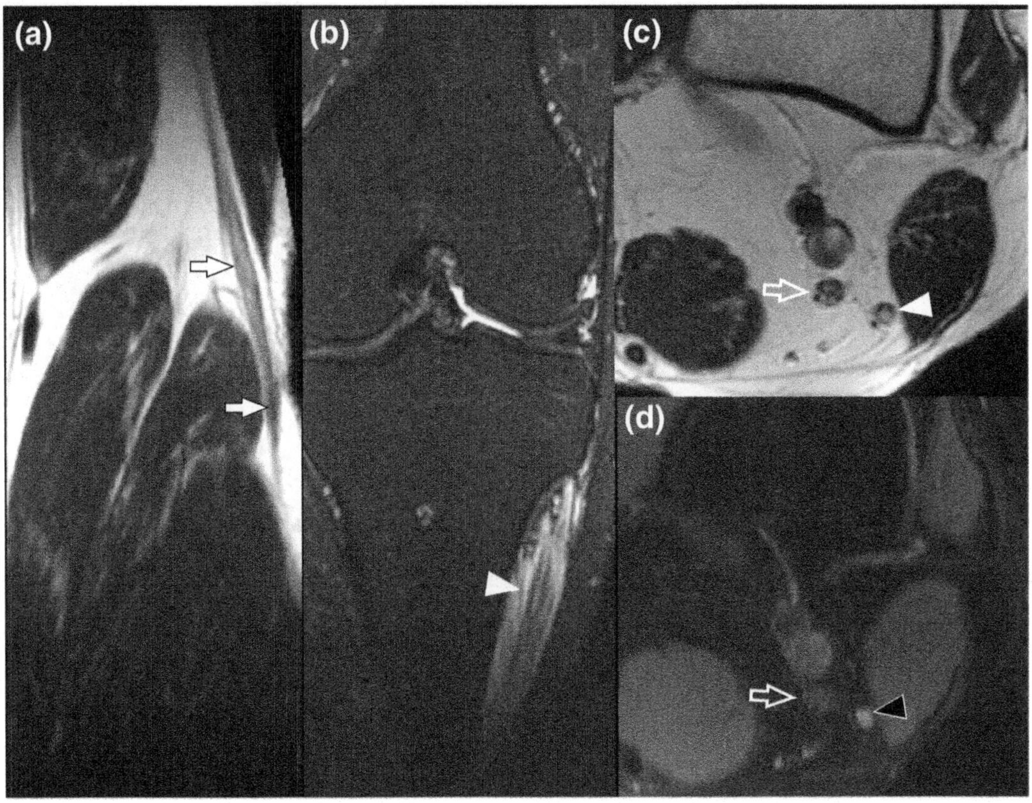

Fig. 25 Traumatic injury of the CPN: 33 year old male with previous posterolateral corner injury presented with foot drop and pain along CPN distribution several weeks after trauma. Sequence T2 3D IDEAL in-phase in coronal plane (**a**) shows a focal enlargement of the common peroneal nerve CPN (*white arrows*). Sequence T2 3D WATER IDEAL in coronal plane (**b**) demonstrated diffuse edema over tibialis anterior and extensor digitorum longus muscles (*white arrowhead*). Sequence T2 TSE in axial plane (**c**) through the distal portion of the popliteal fossa showing mild T2 hyperintensity of the CPN (*white arrowhead*). Note fascicular pattern disruption compared with the tibial nerve (*TN*) (*white void arrow*). In sequence PD FAT SAT in axial plane (**d**) the signal abnormalities are best appreciated. TN (*void arrow*) and CPN (*black arrowhead*)

syndrome, lacerations of the nerve, or postsurgical entrapment from sutures or hardware [187].

MRI in the acute setting of trauma may reveal edema, focal nerve enlargement, and perineural fat blurring seen as diffuse hyperintensity on T2W-imaging On chronic stage nerve scarring may produce neuroma, a fibrous scar tissue seen as a focal nodule in continuity with the injured nerve. Neuromas may have different appearances ranging from nodular hyperintensity on T2 weighted images related to surrounding fluid or low signal in all the pulse sequences due to fibrous collagenous tissue [183] Fig. 25.

In the nerve entrapment by cystic lesions at the peroneal channel Spinner et all described some reproducible MRI features in order to identify the joint connection (the tail sign) in para-articular cysts and to distinguish between peroneal intraneural and extraneural ganglia (the transverse limb sign and the signet ring sign) at the proximal tibiofibular joint [188].

Distinction between them has greatly influences in the treatment to prevent recurrences [189].

Tumor related peroneal neuropathy starts more frequently as drop foot than pain but in some cases sensitive symptoms as dysesthesias may precede the motor picture. More common peripheral nerve sheath tumors are schwannomas and neufibromas. MRI facilitates the

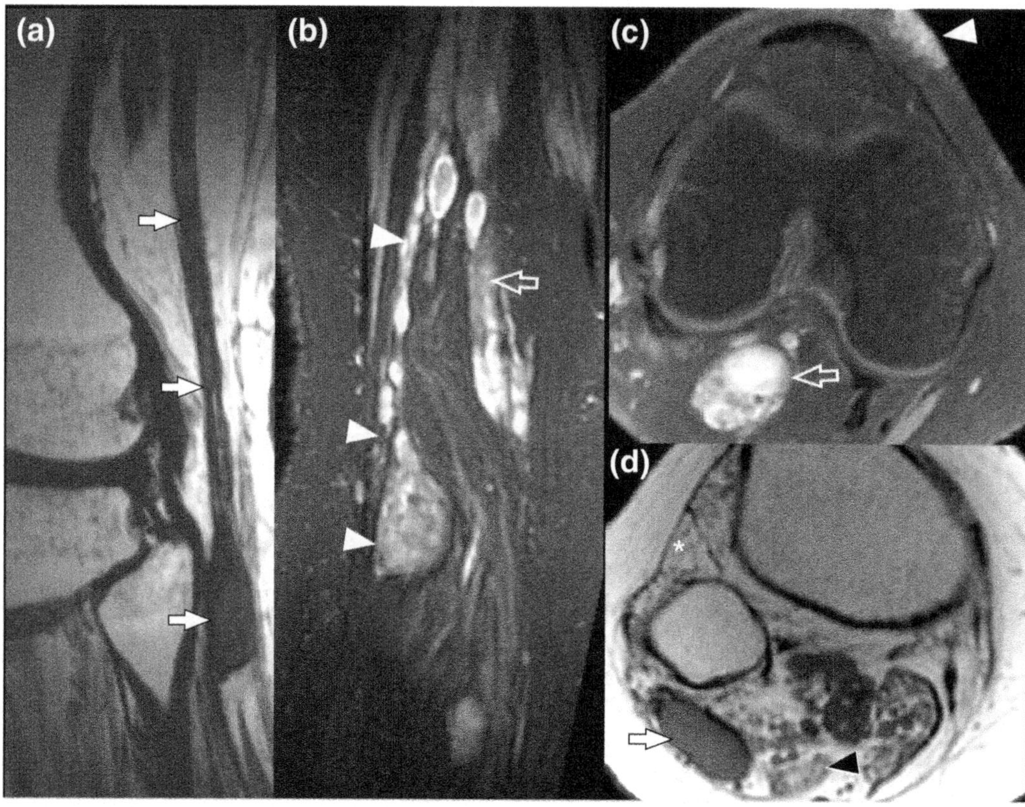

Fig. 26 Benign nerve sheath tumor of the *lower* limb (plexiform neurofibroma): 19 year old male with Von Recklinhausen disease and progressive *lower* limb paresthesias and dull pain. Sequence T1 3D IDEAL in-phase in sagittal oblique plane (**a**) showing a fusiform enlargement of the common peroneal nerve (*CPN*) extending through the peroneal channel (*white arrows*). Sequence T2 3D WATER IDEAL in coronal plane (**b**) demonstrated a multinodular and coalescent masses along the sciatic nerve (*SN*) and involving CPN (*white arrowheads*) and tibial nerve (*TN*, *void white arrow*). Sequence STIR in axial plane (**c**) at the level of the popliteal fossa showing a high signal mass with endolesional low signal strips representing the target sign (*white void arrow*). A neurofibroma is also seen in the cutaneous plane (*white arrowhead*). Sequence T1 3D IDEAL in-phase in axial plane (**d**) shows fatty infiltration of the muscular lateral compartment of the leg (*asterisk*): peroneus longus (*PL*), extensor digitorum longus (*EDL*) and tibialis anterior (*TA*) muscles due denervation changes. Denervatory chronic changes are also seen in posterior compartment of the leg

differentiation between them identifying eccentrically location at the nerve and encapsulation by the perineurium in schwannomas. It may improve the removal rate without damaging the nerve unlike neurofibromas because they are located in the center of the nerve [182] Fig. 26.

The characteristics of these tumors were explained in the chapter of brachial plexus (see Sect. 4.5 in this chapter).

Other causes of painful neuropathy include perineurioma, linfomatoses, endometriosis, radiation induced injuries, vascular malformations, and iatrogenic conditions [183].

Edema and denervatory changes on MRI involving the peroneus muscles suggests SPN lesions. The classic pattern SPN injury may result from traction of nerve in ankle strains, especially due to stretching or forced inversion. It may be useful to look for radiologic signs of lateral ligament insufficiency or instability [187].

In DPN lesions denervatory injuries may involve tibialis anterior and extensor digitorum muscles. It is commonly seen in patients with anterior tarsal tunnel syndrome defined by pain or burning sensations over the dorsum of the foot. Symptoms may be exacerbated with plantar flexion due to stretching of the nerve and contents of the anterior tarsal tunnel against the talonavicular joint [187].

7.2 Sural Neuropathies

The sural nerve (SN) is a sensory nerve supplying cutaneous innervations to the posterolateral aspect of the distal leg and the lateral border of the foot. SN is formed by union of the medial sural cutaneous branch of TN and the peroneal communicating branch of CPN.

SN travels along the distal popliteal fossa and continues descending next to lateral border of the Achilles tendon. Distally, the nerve travels posterior and inferiorly below to the peroneal tendons and lateral malleolus winding anteriorly to get the lateral aspect of foot. Finally SN bifurcates at the level of the fifth metatarsal base into its terminal branches. On MR imaging, SN could be localized as an isointense band relative the muscle in T1-weighted images adjacent to lateral the Achilles tendon next to the lesser saphenous vein.

Potential sites of compression of SN are the lateral aspect of the heel, the superficial sural aponeurosis at the junction of the Achilles tendon and the gastrocnemius and below the fifth metatarsal base. In the main cases neuropathic syndrome is closely related to the degree of tendinopathy.

MRI could be determining the level of entrapment or the concomitant conditions in the anatomic course of the nerve [190].

7.3 Saphenous Neuropathies

The saphenous nerve (SFN) originates distal to the inguinal ligament and descends through the femoral triangle, passing through adductor canal together to femoral vessels. Suprapatellar SFN travels down between sartorius and gracilis muscles and pierces the fascia latae. Then it continues descending next to greater saphenous vein. The infrapatellar SFN becomes superficial after exiting the adductor canal in the anteromedial aspect of the leg. On foot it descends ventrally to the medial malleolus and innervates the medial aspect of the foot.

In this long run SFN is vulnerable to entrapment especially distal to the adductor canal as it becomes subcutaneous, or injured in total knee replacement [190].

7.4 Tibial, Medial, Lateral, Calcaneal, and Plantar Digital Neuropathies

The tibial nerve (TN; L4–L5, S1–S3) emerges from the medial portion of the sciatic nerve. At the level of the popliteal fossa the TN travels through the posterior compartment of the leg, between the heads of the gastrocnemius muscle. At the ankle, the nerve becomes superficial and enters the tarsal tunnel passing below the flexor retinaculum. At this point TN originates the calcaneal nerve, which innervates the skin on the medial portion of the heel and calcaneum. In approximately 90% of the population the branching of the TN in its terminal branches for medial side of the foot (the medial and lateral plantar nerves) occurs within the tarsal tunnel. The distal branches of the plantar nerves give rise the interdigital nerves and the most medial branch of the medial plantar nerve, forms the plantar proper digital nerve distally [191].

On MRI the TN could be identified next to the tibial vessels aside the medial border of Achiles tendon. The flexor tendons of the foot, (posterior tibial, flexor digitorum longus, and flexor hallucis longus) borders TN behind the medial malleolus.

The upper tarsal tunnel (tibiotalar) is bordered by the deep fascia and the ventral aspects of tibia and talus. It contains the neurovascular structures being surrounded by homogeneous high signal in T1 weighted images.

The TN supplies motor innervation to the deep and superficial posterior compartments of

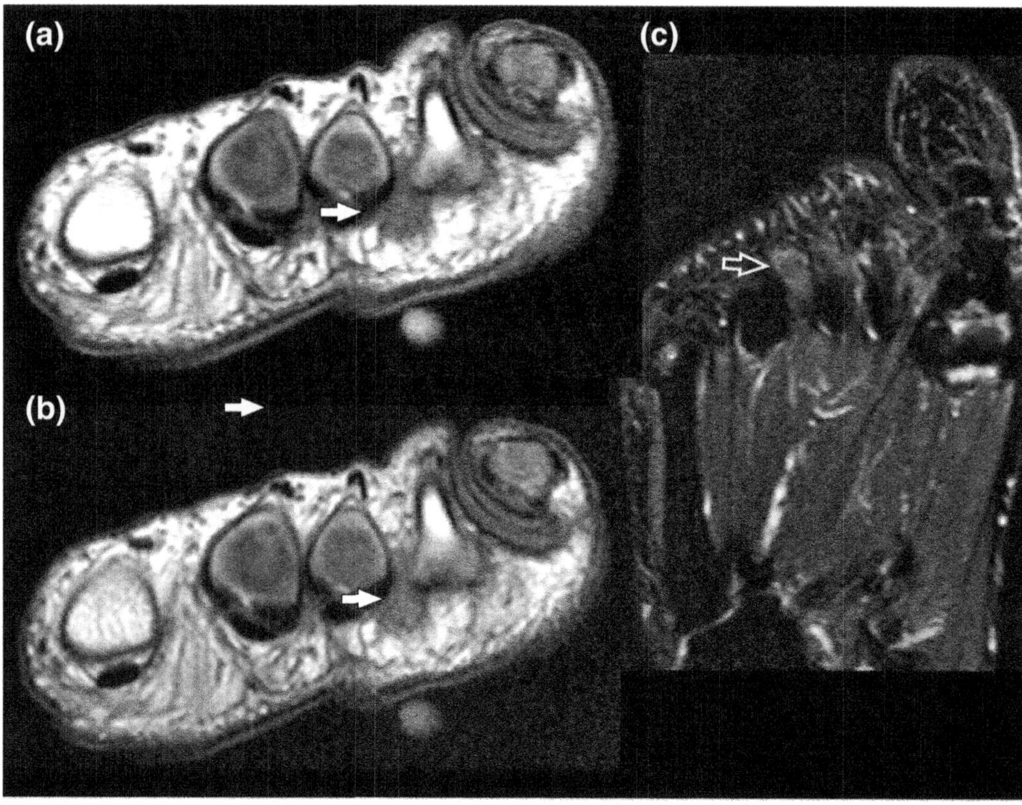

Fig. 27 Morton neuroma: 59 year old female with chronic plantar pain. Sequence DP (**a**) and T1 FSE (**b**) images in coronal plane at forefoot demonstrating focal low signal mass in the second interdigital space neuromas (*white arrows*). Sequence STIR weighted images in coronal plane (**c**) showing low signal probably due to collagen or fibrous tissue (*white void arrow*)

the leg including plantaris, gastrocnemius, popliteus, soleus, posterior tibialis, flexor digitorum longus, and flexor hallucis longus.

Proximal neuropathies of TN are infrequent are mainly caused by compression in distal popliteal fossa. Tarsal tunnel syndrome (TTS) is one most common causes of distal TN neuropathic pain and also may involve to the medial and lateral plantar nerves. MRI could depict the specific causes of entrapment and commonly include ganglion cysts, tenosynovitis of the flexor tendons, calcaneal fractures and accessory muscles [192].

The plantar digital nerves run between the transverse metatarsal ligament and the metatarsal heads. It can result compressed and perineural fibrotic changes may appear along the nerve. The third web space is the most commonly affected and pain at compression is useful clinical test. Morton's neuroma in MRI looks at an ovoid shaped or dumbbell lesion with low signal in both T1 and T2 weighted images and avid contrast enhancement [193] Fig. 27.

Acknowledgements We would like to thank Tatiana Escobar, MD, a radiologist at our institution, for the remarkable artworks provided for schemes in Figs. 2a, 16a and 17a, expressly made for this chapter, and to Mariano Socolovsky, MD, a peripheral nerve surgeon at our institution, for the surgical specimen in Fig. 10c, d.

References

1. Thompson PD, Thomas PK. Clinical patterns of peripheral neuropathy. In: Dyck PJ, Thomas PK, editors. peripheral neuropathy. 4th ed. Philadelphia, USA: Elsevier-Saunders; 2005. p. 1137–61.

2. Chhabra A, Williams EH, Wang KC, et al. MR neurography of neuromas related to nerve injury and entrapment with surgical correlation. AJNR Am J Neuroradiol. 2010;31:1363–8.
3. Grant GA, Britz GW, Goodkin R, et al. The utility of magnetic resonance imaging in evaluating peripheral nerve disorders. Muscle Nerve. 2002;25:314–31.
4. Andreisek G, Burg D, Studer A, et al. Upper extremity peripheral neuropathies: role and impact of MR imaging on patient management. Eur Radiol. 2008;18:1953–61.
5. Bendszus M, Koltzenburg M, Wessig C, et al. Sequential MR imaging of denervated muscle: experimental study. AJNR Am J Neuroradiol. 2002;23:1427–31.
6. Li Z, Graff Ch, Gmitro AF, et al. Rapid water and lipid imaging with T2 mapping using a radial IDEAL-GRASE technique. Magn Reson Med. 2009;61:1415–24.
7. Fuller S, Reeder S, Shimakawa A, et al. Iterative decomposition of water and fat with echo asymmetry and least-squares estimation (IDEAL) fast spin-echo imaging of the ankle: initial clinical experience. AJR Am J Roentgenol. 2006;187:1442–7.
8. Cejas C, Escobar I, Serra M, et al. High resolution neurography of the lumbosacral plexus on 3T magnetic resonance imaging. Radiologia. 2015;57:22–34.
9. Chhabra A. Peripheral MR neurography approach to interpretation. Neuroimag Clin N Am. 2014;24:79–89.
10. Hiltunen J, Suortti T, Arvela S, et al. Diffusion tensor imaging and tractography of distal peripheral nerves at 3 T. Clin Neurophysiol. 2005;116:2315–23.
11. Guggenberger R, Eppenberger P, Markovic D, et al. MR neurography of the median nerve at 3.0 T: optimization of diffusion tensor imaging and fiber tractography. Eur J Radiol. 2012;81:775–82.
12. Vargas M, Viallon M, Nguyen D, et al. Diffusion tensor imaging (DTI) and tractography of the brachial plexus: feasibility and initial experience in neoplastic conditions. Neuroradiology. 2010;52:237–45.
13. Filler AG, Maravilla KR, Tsuruda JS. MR neurography and muscle MR imaging for image diagnosis of disorders affecting the peripheral nerves and musculature. Neurol Clin. 2004;22:643–82.
14. Chhabra A, Andreisek G, Soldatos T, et al. MR neurography: past, present, and future. AJR Am J Roentgenol. 2011;197:583–91.
15. Kim S, Choi JY, Huh YM, et al. Role of magnetic resonance imaging in entrapment and compressive neuropathy: what, where, and how to see the peripherals nerves on the musculoskeletal magnetic resonance image: part 1. Overview and lower extremity. Eur Radiol. 2007;17:139–49.
16. Kim S, Choi J, Huh Y, et al. Role of magnetic resonance imaging in entrapment and compressive neuropathy: what, where, and how to see the peripherals nerves on the musculoskeletal magnetic resonance image: part 2. Upper extremity. Eur Radiol. 2007;17:509–22.
17. Cejas C, Aguilar M, Falcon L, et al. High resolution (3T) magnetic resonance neurography of sciatic nerve. Radiología. 2013;55:195–202.
18. Thawait SK, Wang K, Subhawong TK, et al. Peripheral nerve surgery: the role of high-resolution MR neurography. AJNR Am J Neuroradiol. 2012;33:203–10.
19. Pham M, Bäumer T, Bendszus M. Peripheral nerves and plexus: imaging by MR-neurography and high-resolution ultrasound. Curr Opin Neurol. 2014;27:370–9.
20. Johnson EQ, Vekris M, Demesticha T, et al. Neuroanatomy of the brachial plexus: normal and variant anatomy of its formation. Surg Radiol Anat. 2010;32:291–7.
21. Castillo M. Imaging the anatomy of the brachial plexus: review and self-assessment module. AJR Am J Roentgenoln. 2005;185:196–204.
22. Du R, Auguste KI, Chin CT, et al. Magnetic resonance neurography for the evaluation of peripheral nerve, brachial plexus, and nerve root disorders. J Neurosurg. 2010;112:362–71.
23. Martinoli C, Gandolfo N, Perez MM, et al. Brachial plexus and nerves about the shoulder. Semin Musculoskelet Radiol. 2010;14:523–46.
24. Truex R, Carpenter M. Segmental and peripheral innervation. In: Human neuroanatomy. 6th ed. Baltimore: Williams & Wilkins; 1969. 194 p.
25. Chhabra A, Thawait G, Soldatos T, et al. High-resolution 3T MR neurography of the brachial plexus and its branches, with emphasis on 3D imaging. AJNR Am J Neuroradiol. 2013;34:486–97.
26. Seddon H. Three types of nerve injury. Brain. 1943;66:237–88.
27. Silbermann-Hoffman O, Teboul F. Post-traumatic brachial plexus MRI in practice. Diag Interv Imag. 2013;94:925–43.
28. Upadhyaya V, Upadhyaya DN, Kumar A, et al. MR neurography in traumatic brachial plexopathy. Eur J Radiol. 2015;84:927–32.
29. Hayashi N, Masumoto T, Abe O, et al. Accuracy of abnormal paraspinal muscle findings on contrast—enhanced MR images as indirect signs of unilateral cervical root-avulsion injury. Radiology. 2002;223:397–402.
30. Kang A, Kliot M. Imaging of the peripheral nerves: clinical applications and diagnostic relevance. In: Wattjes M, Fisher D, editors. Neuromuscular imaging. New York: Springer; 2013.
31. Demondion X, Bacqueville E, Paul C, et al. Thoracic outlet: assessment with MR imaging in asymptomatic and symptomatic populations. Radiology. 2003;227:461–8.
32. Demondion X, Herbinet P, Van Sint Jan S, et al. Imaging assessment of thoracic outlet syndrome. Radiographics. 2006;26:1735–50.

33. Mallouhi A, Marik W, Prayer D, et al. 3 T MR tomography of the brachial plexus: structural and microstructural evaluation. Eur J Radiol. 2012;81:2231–45.
34. Sureka J, Cherian RA, Alexander M, et al. MRI of brachial plexopathies. Clin Radiol. 2009;64:208–18.
35. Dyck PJ, Kratz KM, Karnes JL, et al. The prevalence by staged severity of various types of diabetic neuropathy, retinopathy, and nephropathy in a population-based cohort: the Rochester Diabetic Neuropathy Study. Neurology. 1993;43:817–24.
36. Thomas PK. Classification, differential diagnosis and staging of diabetic peripheral neuropathy. Diabetes. 1997;46:54–7.
37. Kuntzer T, Ruiz J. Diabetic neuropathies: clinical sub-types, early detection and asking help from the specialist. Rev Med Suisse. 2014;30:950–3.
38. Dyck PJB, O'Brien PC, Bosch PE, et al. The multicenter double-blind controlled trial of IV methylprednisolone in diabetic lumbosacral radiculoplexus neuropathy. Neurology. 2006;66:37.
39. Said G, Lacroix C, Lozeron P, et al. Inflammatory vasculopathy in multifocal diabetic neuropathy. Brain. 2003;126:376–85.
40. Malik RA, Tesfaye S, Newrick PG, et al. Sural nerve pathology in diabetic patients with minimal but progressive neuropathy. Diabetologia. 2005;48:578–85.
41. Cejas C, Rollán C, Michelin G, et al. Neurografía de alta resolución en resonancia magnética 3 Tesla del plexo braquial. Radiologia. 2016;58(2):88–100 (2015).
42. Massie R, Mauermann ML, Staff NP, et al. Diabetic cervical radiculoplexus neuropathy: a distinct syndrome expanding the spectrum of diabetic radiculoplexus neuropathies. Brain. 2012;135:3074–88.
43. Dyck PJ, Lais AC, Ohta M, et al. Chronic inflammatory polyradiculoneuropathy. Mayo Clin Proc. 1975;50:621–37.
44. Van den Bergh PY, Hadden RD, Bouche P, et al. European Federation of Neurological Societies/Peripheral Nerve Society guideline on management of chronic inflammatory demyelinating polyradiculoneuropathy: report of a joint task force of the European Federation of Neurological Societies and the Peripheral Nerve Society—first revision. Eur J Neurol. 2010;17:356–63.
45. Ad Hoc Subcommittee of the American Academy of Neurology AIDS Task Force. Research criteria for the diagnosis of chronic inflammatory demyelinating polyradiculoneuropathy (CIDP). Neurology. 1991;41:617–8.
46. Adachi Y, Sato N, Okamoto T, et al. Brachial and lumbar plexuses in chronic inflammatory demyelinating polyradiculoneuropathy: MRI assessment including apparent diffusion coefficient. Neuroradiology. 2011;53:3–11.
47. Graham I, Lantos L. Greenfield's neuropathology, vol. 2. 7th ed. New York Oxford University Press; 2002. p. 617–20.
48. Shibuya K, Sugiyama A, Ito S, et al. Reconstruction magnetic resonance neurography in chronic inflammatory demyelinating polyneuropathy. Ann Neurol. 2015;77:333–7.
49. Tazawa K, Matzuda M, Yoshida T, et al. Spinal nerve root hypertrophy on MRI: clinical significance in the diagnosis of chronic inflammatory demyelinating polyradiculoneuropathy. Intern Med. 2008;47:2019–24.
50. Eurelings M, Notermans NC, Franssen H, et al. MRI of the brachial plexus in polyneuropathy associated with monoclonal gammopathy. Muscle Nerve. 2001;24:1312–8.
51. van Alfen N, van Engelen BG. The clinical spectrum of neuralgic amyotrophy in 246 cases. Brain. 2006;129:438–50.
52. Laccone F, Hannibal MC, Neesen J, et al. Dysmorphic syndrome of hereditary neuralgic amyotrophy associated with a SEPT9 gene mutation: a family study. Clin Genet. 2008;74:279–83.
53. Chalian M, Faridian-Aragh N, Soldatos T, et al. High-resolution 3T MRN of suprascapular neuropathy. Acad Radiol. 2011;18:1049–59.
54. Murphey MD, Smith WS, Smith SE, et al. From the archives of the AFIP. Imaging of musculoskeletal neurogenic tumors: radiologic-pathologic correlation. Radiographics. 1999;19:1253–80.
55. Cai W, Kassarjian A, Bredella MA, et al. Tumor burden in patients with neurofibromatosis types 1 and 2 and schwannomatosis: determination on whole-body MR images. Radiology. 2009;250:665–73.
56. Le LQ, Liu C, Shipman T, et al. Susceptible stages in Schwann cells for NF1 associated plexiform neurofibroma development. Cancer Res. 2011;71:4686–95.
57. Kransdorf MJ. Benign soft-tissue tumors in a large referral population: distribution of specific diagnoses by age, sex, and location. AJR Am J Roentgenol. 1995;164:395–402.
58. Kransdorf MJ. Malignant soft-tissue tumors in a large referral population: distribution of diagnoses by age, sex, and location. AJR Am J Roentgenol. 1995;164:129–34.
59. Isobe K, Shimizu T, Akahane T, et al. Imaging of ancient schwannoma. AJR Am J Roentgenol. 2004;183:331–6.
60. Vilanova JC, Woertler K, Narváez JA, et al. Soft-tissue tumors update: MR imaging features according to the WHO classification. Eur Radiol. 2007;17:125–38.
61. Kim DH, Murovic JA, Tiel RL, et al. A series of 146 peripheral non-neural sheath nerve tumors: 30-year experience at Louisiana State University Health Sciences Center. J Neurosurg. 2005;102:256–66.
62. Wu JS, Hochman MG. Soft-tissue tumors and tumorlike lesions: a systematic imaging approach. Radiology. 2009;253:297–316.
63. Woertler K. Tumors and tumor-like lesions of peripheral nerves. Semin Musculoskelet Radiol. 2010;14:547–58.

64. Ducatman BS, Scheithauer BW, Piepgras DG, et al. Malignant peripheral nerve sheath tumors. A clinicopathologic study of 120 cases. Cancer. 1986;57:2006–21.
65. Thawait SK, Chaudhry V, Thawait GK, et al. High-resolution MR neurography of diffuse peripheral nerve lesions. AJNR Am J Neuroradiol. 2011;32:1365–72.
66. Genovese E, Cani A, Rizzo S, et al. Comparison between MRI with spin-echo echo-planar diffusion-weighted sequence (DWI) and histology in the diagnosis of soft-tissue tumours. Radiol Med. 2011;116:644–56.
67. Warbey VS, Ferner RE, Dunn JT, et al. [18 F] FDG PET/CT in the diagnosis of malignant peripheral nerve sheath tumors in neurofibromatosis type-1. Eur J Nucl Med Mol Imag. 2009;36:751–7.
68. Baehring JM, Batchelor TT. Diagnosis of neurolymphomatosis. Cancer J. 2012;18:463–8.
69. Misdraji J, Ino Y, Louis DN, et al. Primary lymphoma of peripheral nerve: report of four cases. Am J Surg Pathol. 2000;24:1257–65.
70. Amrami KK, Felmlee JP, Spinner RJ. MRI of peripheral nerves. Neurosurg Clin N Am. 2008;19:559–72.
71. Trojan A, Jermann M, Taverna C, et al. Fusion PET-CT imaging of neurolymphomatosis. Ann Oncol. 2002;13:802–5.
72. Isasi CR, Lu P, Blaufox MD. A metaanalysis of 18F-2-deoxy-2-fluoro-D-glucose positron emission tomography in the staging and restaging of patients with lymphoma. Cancer. 2005;104:1066–74.
73. Kori SH, Foley KM, Posner JB. Brachial plexus lesions in patients with cancer: 100 cases. Neurology. 1981;31:45–50.
74. Khadilkar SV, Khade SS. Brachial plexopathy. Ann Indian Acad Neurol. 2013;16:12–8.
75. Parker EE, Glastonbury CM. MR imaging of the thoracic inlet. Magn Reson Imag Clin N Am. 2008;16:341–53.
76. Hoeksma AF, Wolf H, Oei SL. Obstetrical brachial plexus injuries: incidence, natural course and shoulder contracture. Clin Rehabil. 2000;14:523–6.
77. Schmitt C, Mehlman CT, Meiss AL. Hyphenated history: Erb-Duchenne brachial plexus palsy. Am J Orthop (Belle Mead NJ). 2008;37:356–8.
78. Gudinchet F, Maeder P, Oberson JC, et al. Magnetic resonance imaging of the shoulder in children with brachial plexus birth palsy. Pediatr Radiol. 1995;25:125–8.
79. Zafeiriou DI, Psychogiou K. Obstetrical brachial plexus palsy. Pediatr Neurol. 2008;38:235–42.
80. Crush AB, Howe BM, Spinner RJ, et al. Malignant involvement of the peripheral nervous system in patients with cancer: multimodality imaging and pathologic correlation. RadioGraphics. 2014;34:1987–2007.
81. Fathers E, Thrush D, Huson SM, Norman A, et al. Radiation-induced brachial plexopathy in women treated for carcinoma of the breast. Clin Rehabil. 2002;16:160–5.
82. Vargas MI, Viallon M, Nguyen D, et al. New approaches in imaging of the brachial plexus. Eur J Radiol. 2010;74:403–10.
83. Ho L, Henderson R, Luong T, et al. 18F-FDG PET/CT appearance of metastatic brachial plexopathy involving epidural space from breast carcinoma. Clin Nucl Med. 2012;37:263–4.
84. Chalian M, Behzadi AH, Williams EH, et al. High-resolution magnetic resonance neurography in upper extremity neuropathy. Neuroimag Clin N Am. 2014;24:109–25.
85. Miller TT, Reinus WR. Nerve entrapment syndromes of the elbow, forearm, and wrist. AJR Am J Roentgenol. 2010;195:585–94.
86. Bencardino JT, Rosenberg ZS. Entrapment neuropathies of the shoulder and elbow in the athlete. Clin Sports Med. 2006;25:465–87.
87. Blum AG, Zabel JP, Kohlmann R, et al. Pathologic conditions of the hypothenar eminence: evaluation with multidetector CT and MR imaging. Radiographics. 2006;26:1021–44.
88. Collier A, Burge P. Management of mechanical neuropathy of the ulnar nerve at the elbow. Curr Orthop. 2001;15:256–63.
89. Vucic S, Cordato DJ, Yiannikas C, et al. Utility of magnetic resonance imaging in diagnosing ulnar neuropathy at the elbow. Clin Neurophysiol. 2006;117:590–5.
90. Subhawong TK, Wang KC, Thawait SK, et al. High resolution imaging of tunnels by magnetic resonance neurography. Skelet Radiol. 2012;41:15–31.
91. Frost P, Johnsen B, Fuglsang-Frederiksen A, et al. Lifestyle risk factors for ulnar neuropathy and ulnar neuropathy-like symptoms. Muscle Nerve. 2013;48:507–15.
92. Andreisek G, Crook DW, Burg D, et al. Peripheral neuropathies of the median, radial, and ulnar nerves: MR imaging features. Radiographics. 2006;26: 1267–87.
93. Keen NN, Chin CT, Engstrom JW, et al. Diagnosing ulnar neuropathy at the elbow using magnetic resonance neurography. Skelet Radiol. 2012;41: 401–7.
94. Dang AC, Rodner CM. Unusual compression neuropathies of the forearm, part II: median nerve. J Hand Surg Am. 2009;34:1915–20.
95. Prasartritha T, Liupolvanich P, Rojanakit A. A study of the posterior interosseous nerve (PIN) and the radial tunnel in 30 Thai cadavers. J Hand Surg Am. 1993;18:107–12.
96. Brown WF, Watson BV. Acute retrohumeral radial neuropathies. Muscle Nerve. 1993;16:706–11.
97. Ferdinand BD, Rosenberg ZS, Schweitzer ME, et al. MR imaging features of radial tunnel syndrome: initial experience. Radiology. 2006;240: 161–8.
98. Bayramoglu M. Entrapment neuropathies of the upper extremity. Neuroanatomy. 2004;3:18–24.
99. Han BR, Cho YJ, Yang JS, et al. Clinical features of wrist drop caused by compressive radial neuropathy

and its anatomical considerations. J Korean Neurosurg Soc. 2014;55:148–51.
100. Lee MJ, LaStayo PC. Pronator syndrome and other nerve compressions that mimic carpal tunnel syndrome. J Orthop Sports Phys Ther. 2004;34:601–9.
101. Lordan J, Rauh P, Spinner RJ. The clinical anatomy of the supracondylar spur and the ligament of Struthers. Clin Anat. 2005;18:548–51.
102. Bordalo-Rodrigues M, Rosenberg ZS. MR imaging of entrapment neuropathies at the elbow. Magn Reson Imag Clin N Am. 2004;12:247–63.
103. Rehak DC. Pronator syndrome. Clin Sports Med. 2001;20:531–40.
104. Seror P. Anterior interosseous nerve lesions. Clinical and electrophysiological features. J Bone Joint Surg Br. 1996;78:238–41.
105. Shapiro BE, Preston DC. Entrapment and compressive neuropathies. Med Clin N Am. 2009;93:285–315.
106. Cudlip SA, Howe FA, Clifton A, et al. Magnetic resonance neurography studies of the median nerve before and after carpal tunnel decompression. J Neurosurg. 2002;96:1046–51.
107. Dunn AJ, Salonen DC, Anastakis DJ. MR imaging findings of anterior interosseous nerve lesions. Skelet Radiol. 2007;36:1155–62.
108. Brienza M, Pujia F, Colaiacomo MC, et al. 3T diffusion tensor imaging and electroneurography of peripheral nerve: a morphofunctional analysis in carpal tunnel syndrome. J Neuroradiol. 2014;41:124–30.
109. Soldatos T, Andreisek G, Thawait G, et al. High-resolution 3-T MR neurography of the lumbosacral plexus. RadioGraphics. 2013;33:967–87.
110. Laughlin R, Dyck P. Electrodiagnostic testing in lumbosacral plexopathies. Phys Med Rehabil Clin N Am. 2013;24:93–105.
111. Delaney H, Bencardino J, Rosenberg ZS. Magnetic resonance neurography of the pelvis and lumbosacral plexus. Neuroimag Clin N Am. 2014;24:127–50.
112. Tarulli A, Raynor E. Lumbosacral radiculopathy. Neurol Clin. 2007;25:387–405.
113. Said G, Goulon-Goeau C, Lacroix C, et al. Nerve biopsy findings in different patterns of proximal diabetic neuropathy. Ann Neurol. 1994;35:559–69.
114. Dyck PJ, Engelstad J, Norell J, et al. Microvasculitis in non-diabetic lumbosacral radiculoplexus neuropathy (LSRPN): similarity to the diabetic variety (DLSRPN). J Neuropathol Exp Neurol. 2000;59:525–38.
115. Chan YC, Lo YL, Chan ESY. Immunotherapy for diabetic amyotrophy. Cochrane Database Syst Rev. 2009;(3). Art. No.: CD006521.
116. Kawamura N, Dyck PJ, Schmeichel AM, et al. Inflammatory mediators in diabetic and non-diabetic lumbosacral radiculoplexus neuropathy. Acta Neuropathol. 2008;115:231–9.
117. Poliachik SL, Friedman SD, Carter GT, et al. Skeletal muscle edema in muscular dystrophy: clinical and diagnostic implications. Phys Med Rehabil Clin Am. 2012;23:107–22.
118. Wattjes MP, Kley RA, Fischer D. Neuromuscular imaging in inherited muscle diseases. Eur Radiol. 2010;20:2447–60.
119. Laughlin RS, Dyck PJ. Diabetic radiculoplexus neuropathies. Handb Clin Neurol. 2014;126:45–52.
120. Younger DS. Diabetic lumbosacral radiculoplexus neuropathy: a postmortem studied patient and review of the literature. J Neurol. 2011;258:1364–7.
121. Gwathmey KG, Burns TM, Collins MP, et al. Vasculitic neuropathies. Lancet Neurol. 2014;13:67–82.
122. Said G, Lacroix C. Primary and secondary vasculitic neuropathy. Neurol. 2005;252:633–41.
123. Dyck PJB, Windebank AJ. Diabetic and nondiabetic lumbosacral radiculoplexus neuropathies: new insights into pathophysiology and treatment. Muscle Nerve. 2002;25:477–91.
124. Collins M, Dyck PJ, Gronseth GS, et al. Peripheral Nerve Society guideline on the classification, diagnosis, investigation, and immunosuppressive therapy of non-systemic vasculitic neuropathy: executive summary. J Peripheral Nerv Syst. 2010;15:176–84.
125. Yuki N, Hartung HP. Guillain-Barré syndrome. N Engl J Med. 2012;366:2294–304.
126. Liu GF, Wu ZL, et al. A case-control study on children with Guillain-Barré syndrome in North China. Biomed Environ Sci. 2003;16:105–11.
127. Sejvar JJ, Baughman AL, Wise M, et al. Population incidence of Guillain-Barré syndrome: a systematic review and meta-analysis. Neuroepidemiology. 2011;36:123–33.
128. Martinez V, Fletcher D, Martin F, et al. Small fibre impairment predicts neuropathic pain in Guillain-Barré syndrome. Pain. 2010;151:53–60.
129. Kajimoto M, Koga M, Narumi H, et al. Successful control of radicular pain in a pediatric patient with Guillain-Barré síndrome. Brain Dev. 2015. doi:10.1016/j.braindev.2015.01.004.
130. Korinthenberg R, Schessl J, Kirschner J. Clinical presentation and course of childhood Guillain-Barre syndrome: a prospective multicentre study. Neuropediatrics. 2007;38:10–7.
131. Ruts L, Drenthen J, Jongen JL, et al. Dutch GBS Study Group. Pain in Guillain-Barre syndrome: a longterm follow-up study. Neurology. 2010;75:1439–47.
132. Byun WM, Park WK, Park BH, et al. Guillain-Barré syndrome: MR imaging findings of the spine in eight patients. Radiology. 1998;208:137–41.
133. Coskun A, Kumandas S, Pac A, et al. Childhood Guillain-Barre syndrome. MR imaging in diagnosis and follow-up. Acta Radiol. 2003;44:230–5.
134. Yikilmaz A, Doganay S, Gumus M, et al. Magnetic resonance imaging of childhood Guillain-Barre syndrome. Child's Nerv Syst. 2010;26:1103–8.
135. Mulkey SB, Glasier CM, El-Nabbout B, et al. Nerve root enhancement on spinal MRI in pediatric

Guillain-Barré syndrome. Pediatr Neurol. 2010;43:263–9.
136. Wilmshurst JM, Thomas NH, Robinson RO, et al. Lower limb and back pain in Guillain-Barré syndrome and associated contrast enhancement in MRI of the cauda equina. Acta Paediatr. 2001;90:691–703.
137. Mittl RL, Yousem DM. Frequency of unexplained meningeal enhancement in the brain after lumbar puncture. AJNR Am J Neuroradiol. 1994;15:633–8.
138. Forsberg A, Press R, Einarsson U, et al. Impairment in Guillain-Barré syndrome during the first 2 years after onset: a prospective study. J Neurol Sci. 2004;22:131–8.
139. McMillan HJ, Kang PB. Chapter 21—Chronic inflammatory demyelinating polyradiculoneuropathy. In: Jones HR, Darryl C, Darras BT, editors. Neuromuscular disorders of infancy, childhood, and adolescence. 2nd ed. San Diego: Academic Press; 2015. p. 398–417.
140. Said G, Krarup C. Chronic inflammatory demyelinative polyneuropathy. Handb Clin Neurol. 2013;115:403–13.
141. Abe Y, Terashima H, Hoshino H, et al. Characteristic MRI features of chronic inflammatory demyelinating polyradiculoneuropathy. Brain Dev. 2015. doi:10.1016/j.braindev.2015.01.006.
142. Duggins AJ, McLeod JG, Pollard JD, et al. Spinal root and plexus hypertrophy in chronic inflammatory demyelinating polyneuropathy. Brain. 1999;122:1383–90.
143. Ware TL, Kornberg AJ, Rodriguez-Casero MV, et al. Childhood chronic inflammatory demyelinating polyneuropathy. An overview of 10 cases in the modern era. J Child Neurol. 2014;29:43–8.
144. Rossi DP, Doria Lamba L, Pistorio A, et al. Chronic inflammatory demyelinating polyneuropathy of childhood: clinical and neuroradiological findings. Neuroradiology. 2013;55:1233–9.
145. De Smet K, De Maeseneer M, Talebian, et al. MRI in hypertrophic mono- and polyneuropathies. Clin Radiol. 2013;68:317–22.
146. DeFazio MV, Ter Louw R, Attinger CE, et al. Management of advanced plexiform neurofibromatosis of the foot presenting with skeletal deformation and intractable pain: an indication for proximal amputation. The Foot. 2015;25:30–5.
147. Jaeckle KA, Young DF, Foley KM. The natural history of lumbosacral plexopathy in cancer. Neurology. 1985;35:8–15.
148. Ramchandren S, Dalmau J. Metastases to the peripheral nervous system. J Neurooncol. 2005;75:101–10.
149. Hébert-Blouin MN, Amrami KK, Myers RP, et al. Adenocarcinoma of the prostate involving the lumbosacral plexus: MRI evidence to support direct perineural spread. Acta Neurochir (Wien). 2010;152:1567–76.
150. Evans RJ, Watson CP. The hot foot syndrome: Evans' sign and the old way. Pain Res Manag. 2012;17:31–4.
151. Capek S, Sullivan PS, Howe BM. Recurrent rectal cancer causing lumbosacral plexopathy with perineural spread to the spinal nerves and the sciatic nerve: an anatomic explanation. Clin Anat. 2015;28:136–43.
152. Sinaei M, Swallow C, Milot L, et al. Patterns and signal intensity characteristics of pelvic recurrence of rectal cancer at MR imaging. Radiographics. 2013;33:171–87.
153. Said G, Krarup C. Diagnosis of brachial and lumbosacral plexus lesions. Aminoff MJ, Boller F, Swaab DF, series editors. Peripheral nerve disorders: handbook of clinical neurology, vol. 115; 2013. p. 293.
154. Garozzo D, Zollino G, Ferraresi S. In lumbosacral plexus injuries can we identify indicators that predict spontaneous recovery or the need for surgical treatment? Results from a clinical study on 72 patients. J Brachial Plexus Peripheral Nerve Inj. 2014;9:1.
155. Sugimoto Y, Ito Y, Tomioka M, et al. 6 Risk factors for lumbosacral plexus palsy related to pelvic fracture. Spine (Phila Pa 1976). 2010;35(9):963–6.
156. Rai SK, Far RF, Ghovanlou B. Neurological deficits associated with sacral wing fractures. Orthopedics. 1990;13:1363–6.
157. Sabiston CP, Wing PC. Sacral fractures: classification and neurologic implications. J Trauma. 1986;6:197–202.
158. Lindahl J, Hirvensalo E. Outcome of operatively treated type-C injuries of the pelvic ring. Acta Orthop. 2005;76:667–78.
159. Tonetti J, Cazal C, Eid A. Neurological damage in pelvic injuries: a continuous prospective series of 50 pelvic injuries treated with an iliosacral lag screw. Rev Chir Orthop Reparatrice Appar Mot. 2004;90:122–31.
160. Ndiaye A, Diop M, Ndoye J, et al. Anatomical basis of neuropathies and damage to the ilioinguinal nerve during repairs of groin hernias (about 100 dissections). Surg Radiol Anat. 2007;29:675–81.
161. Cesmebasi A, Yadav A, Gielecki J, et al. Genitofemoral neuralgia: a review. Clin Anat. 2015;28:128–35.
162. Chen DC, Hiatt JR, Amid PK. Operative management of refractory neuropathic inguinodynia by a laparoscopic retroperitoneal approach. JAMA Surg. 2013;148:962–7.
163. Peng PW, Tumber PS. Ultrasound-guided interventional procedures for patients with chronic pelvic pain-a description of techniques and review of literature. Pain Physician. 2007;11:215–24.
164. Jendrzejewski F, Peltier J, Havet E, et al. The conflict between obturator nerve and ovary: a cadaveric and radioanatomic study. Morphologie: bulletin de l'Association des anatomistes 2013;97:54–8.
165. Sorenson EJ, Chen JJ, Daube JR. Obturator neuropathy: causes and outcome. Muscle Nerve. 2002;25:605–7.

166. Lonchena TK, McFadden K, Orebaugh SL. Correlation of ultrasound appearance, gross anatomy, and histology of the femoral nerve at the femoral triangle. Surg Radiol Anat. 2015:1–8.
167. van Alfen N, Malessy MJ. Diagnosis of brachial and lumbosacral plexus lesions. Handbook Clin Neurol. 2012;115:293–310.
168. Taira K, Jun S, Ryuta M, et al. Arthroscopic treatment of femoral nerve paresthesia caused by an acetabular paralabral cyst. Orthopedics. 2014;37:496–9.
169. Cheatham SW, Kolber MJ, Salamh PA. Meralgia paresthetica: a review of the literature. Int J Sports Phys Ther. 2013;8:883–93.
170. Hernando M, Cerezal L, Pérez-Carro L, et al. Deep gluteal syndrome: anatomy, imaging, and management of sciatic nerve entrapments in the subgluteal space. Skelet Radiol. 2015;44:919–34.
171. Petchprapa CN, Rosenberg ZS, Sconfienza LM, et al. MR imaging of entrapment neuropathies of the lower extremity: part 1. The pelvis and hip 1. Radiographics. 2010;30:983–1000.
172. Cass SP. Piriformis syndrome: a cause of nondiscogenic sciatica. Curr Sports Med Rep. 2015;14:41–4.
173. Oelhafen K, Shayota BJ, Muhleman M, et al. Benjamin Alcock (1801–?) and his canal. Clin Anat. 2013;26:662–6.
174. Insola A, Granata G, Padua L. Alcock canal syndrome due to obturator internus muscle fibrosis. Muscle Nerve. 2010;42:431–2.
175. Labat JJ, Riant T, Robert R, et al. Diagnostic criteria for pudendal neuralgia by pudendal nerve entrapment (Nantes criteria). Neurourol Urodyn. 2008;27:306–10.
176. Hibner M, Desai N, Robertson LJ, et al. Pudendal neuralgia. J Minim Invasive Gynecol. 2010;17:148–53.
177. Pais D, Casal D, Bettencourt P, et al. Sciatic nerve high division: two different anatomical variants. Acta Med Port. 2013;26:208–11.
178. Cejas C, Aguilar M, Falcón L, et al. Neurografía por resonancia magnética de alta resolución (3 Tesla) del nervio ciático. Radiologia. 2013;55:195–202.
179. Meknas K, Johansen O, Kartus J. Retro-trochanteric sciatica-like pain: current concept. Knee Surg Sports Traumatol Arthrosc. 2011;19:1971–85.
180. Meknas K, Christensen A, Johansen O. The internal obturator muscle may cause sciatic pain. Pain. 2003;104:375–80.
181. Edwards PH, Wright ML, Hartman JF. A practical approach for the differential diagnosis of chronic leg pain in the athlete. Am J Sports Med. 2005;33:1241–9.
182. Pineda D, Barroso F, Cháves H, et al. High resolution 3T magnetic resonance neurography of the peroneal nerve. Radiología (English Edition). 2014;56:107–17.
183. Chhabra A, Faridian-Aragh N, Chalian M, et al. High-resolution 3-T MR neurography of peroneal neuropathy. Skelet Radiol. 2012;41:257–71.
184. Canella C, Demondion X, Guillin R, et al. Anatomic study of the superficial peroneal nerve using sonography. AJR Am J Roentgenol. 2009;193:174–9.
185. Lopez-Ben R. Imaging of nerve entrapment in the foot and ankle. Foot Ankle Clin. 2011;16:213–24.
186. Ray WZ, Mackinnon SE. Nerve problems in the lower extremity. Foot Ankle Clin. 2011;16:243–54.
187. Flanigan RM, DiGiovanni BF. Peripheral nerve entrapments of the lower leg, ankle, and foot. Foot Ankle Clin. 2011;16:255–74.
188. Spinner RJ, Desy NM, Amrami KK. The cystic transverse limb of the articular branch: a pathognomonic sign for peroneal intraneural ganglia at the superior tibiofibular joint. Neurosurgery. 2006;59:157–66.
189. Spinner RJ, Desy NM, Amrami KK. Sequential tibial and peroneal intraneural ganglia arising from the superior tibiofibular joint. Skelet Radiol. 2008;37:79–84.
190. Burge AJ, Gold SL, Kuong S, Potter HG. High-resolution magnetic resonance imaging of the lower extremity nerves. Neuroimag Clin N Am. 2014;24:151–70.
191. Said G, Krarup C. Compression and entrapment neuropathies. Aminoff MJ, Boller F, Swaab DF, series editors. Peripheral nerve disorders: handbook of clinical neurology, vol. 115; 2013. p. 311.
192. Donovan A, Rosenberg ZS. MRI of ankle and lateral hindfoot impingement syndromes. AJR Am J Roentgenol. 2010;195:595–604.
193. Erickson SJ, Quinn SF, Kneeland JB, et al. MR imaging of the tarsal tunnel and related spaces: normal and abnormal findings with anatomic correlation. AJR Am J Roentgenol. 1990;155:323–8.

Author Biographies

Claudia Cejas, Ph, MD Specialist on the field of MRI in Neuromuscular Disorders.

Profesor of Radiology in Buenos Aires University.

Head, MRI Division in Foundation for Neurological Research "Dr. Raul Carrea" (FLENI). Buenos Aires, AR.

Editor in Chief in Revista Argentina de Radiologia.

Diego Pineda Ordoñez, MD Neuroradiology.

Senior radiologist in Foundation for Neurological Research "Dr. Raúl Carrea", (FLENI). Buenos Aires, AR.

Member of neuroimaging Chapter in Sociedad Argentina de Radiología.

Neuroimaging in Migraines

10

Antonio Russo, Alessandro Tessitore
and Gioacchino Tedeschi

Abstract

Neuroimaging techniques fall broadly into two great categories, examining either structure or function, but multiple methods can be employed in either approach. Structural imaging provides static anatomical information whereas functional imaging can be regarded as the method providing dynamic physiological information. However, the division between structural and functional imaging is difficult to make and arbitrary in some measure because structure and function can be often inextricably intertwined in the brain. Recent years have seen rapid growth of neuroimaging methodology which has provided new insights into functional brain organization of migraine patients. In particular, since migraine is regarded as a disorder of the brain, functional neuroimaging offers much in terms of understanding the physiological dysfunction that characterizes migraine. Furthermore, neuroimaging techniques are crucial for clinicians in order to further elucidate pathophysiological mechanisms underlying this complex and often disabling disease and to provide new therapeutical approaches for migraine patients. This chapter aim to focus on the results of structural and functional neuroimaging studies and attempts to synthesize the literature data to provide new pathophysiological concepts for understanding migraine mechanisms.

A. Russo · A. Tessitore · G. Tedeschi (✉)
Headache Center, Department of Medical, Surgical, Neurological, Metabolic and Aging Sciences, University of Campania "Luigi Vanvitelli", 80138 Naples, Italy
e-mail: gioacchino.tedeschi@unina2.it

A. Russo · A. Tessitore · G. Tedeschi
MRI Research Center SUN-FISM, University of Campania "Luigi Vanvitelli", Naples, Italy

L. Saba (ed.), *Neuroimaging of Pain*, DOI 10.1007/978-3-319-48046-6_10

Keywords
VBM: Voxel-based morphometry · CT: Cortical thickness · Tomography · Aura · Diffusivity · Neuronavigation

Abbreviations

MRI	Magnetic resonance imaging
NMR	Nuclear magnetic resonance
GM	Grey matter
WM	White matter
VBM	Voxel-based morphometry
CT	Cortical thickness
SB	Surface-based
DTI	Diffusion tensor imaging
HC	Healthy control
s-connectivity	Structural connectivity
rCBF	Regional cerebral blood flow
SPECT	Single photon emission computed tomography
PET	Positron emission tomography
fMRI	Functional magnetic resonance imaging
^{18}FDG	2-deoxy-2-[^{18}F] flour-D-glucose
BOLD	Blood oxygen level dependent
MwoA	Migraine without aura
MwA	Migraine with aura
WMH	White matter hyperintensities
PAG	Periacqueductal grey matter
dLP	Dorso lateral pons
IFG	Inferior frontal gyrus
PCG	Precentral gyrus
ACc	Anterior cingulated cortex
MFG	Middle frontal gyrus
PFc	Prefrontal cortex
OFc	Orbito frontal cortex
Vc	Visual cortex
Fc	Frontal cortex
SSc	Somatosensory cortex
IPG	Inferior parietal gyrus
PCc	Posterior cingulate cortex
CSD	Cortical spreading depression
CS	Cortical surface
AD	Axonal diffusivity
RD	Radial diffusivity
MD	Mean diffusivity
ROI	Region of interest
CC	Corpus callosum
TBSS	Tract-based spatial statistics
RS	Resting-state
f-connectivity	Functional connectivity

ADC	Apparent diffusion coefficient
dP	Dorsal pons
5-HT	5-hydroxytryptamine
μOR	μ-opioid receptor
lOR	L-opioid receptor
CA	Cutaneous allodynia
3D-IIN	3D immersive and interactive neuronavigation
NAcc	Nucleus accumbens
NCF	Nucleus cuneiformis
TP	Temporal pole
EC	Entorhinal cortex
MCc	Middle cingulated cortex
NAA	*N*–acetylaspartate
RP	Rostral pons
VM	Vestibular migraine
MRS	^{31}P-magnetic resonance spectroscopy
RSN	Resting-state networks
DMN	Default mode network
ReHo	Regional homogeneity
SMA	Supplementary motor area
FPN	Fronto-parietal networks
EF	Executive functions

1 Introduction

Neuroimaging techniques fall broadly into two great categories, examining either structure or function, but multiple methods can be employed in either approach. Structural imaging provides static anatomical information whereas functional imaging can be regarded as the method providing dynamic physiological information. However, the division between structural and functional imaging is difficult to make and arbitrary in some measure because structure and function can be often inextricably intertwined in the brain. Furthermore, although some neuroimaging techniques are based on structural high-resolution T1-weighted magnetic resonance imaging (MRI), applied statistical analyses methods are often used also for functional imaging data. The phenomenon of nuclear magnetic resonance (NMR) was first observed in 1945 [1, 2] but the first human in vivo MRI was produced by the end of 70th decade from the past century [3]. Compared with images from previous modalities, brain MRI provided excellent anatomical detail and strong grey matter (GM) and white matter (WM) contrast.

More recently, high-resolution structural MRI methods have been developed such as voxel-based morphometry (VBM), cortical thickness (CT) and other surface-based (SB) techniques and diffusion tensor imaging (DTI). VBM is a semiautomatic whole-brain method that enables comparisons of GM and WM between groups on a voxel basis, sensitive to subtle macroscopic and mesoscopic structural differences between groups of subjects that can be related to functional correlates and thus further understanding of disease pathophysiology in the brains of migraineurs and non-migraine subjects [4]. CT analysis is a categorical SB technique used in cohort studies, comparing the cortices of patients and healthy controls (HC) in vivo. DTI is specifically employed to assess WM microstructure and can potentially reveal even subtle anatomical abnormalities. Structural connectivity (s-connectivity) analysis is mostly performed on diffusion-derived data, and more recently in combination with volumetric measures.

An early advanced imaging approach for regional cerebral blood flow (rCBF) assessment has been provided by single-photon emission computed tomography (SPECT), using Xenon-133 during migraine attacks [5].

However, positron emission tomography (PET) and more recently functional MRI (fMRI) have superseded the older methods, as they enable the exploration of brain function with greater temporal and spatial resolution and are, today, the most frequently used techniques to attempt to clarify the complexity of migraine mechanisms [6, 7].

Many of the functional imaging studies in migraine research have applied PET to investigate brain activity and metabolism, as well as receptor neurochemistry [8] using different radiotracer such as respectively ^{15}O labelled water ($H_2^{15}O$), 2-deoxy-2-[^{18}F] fluoro-D-glucose (^{18}FDG) [9] or radioactively labeled ligands [10]. By means of PET it is possible to obtain useful insights into brain activation or functional patterns at rest in migraine [11]. Indeed, in the last few decades, PET studies have been extensively used to clarify the complex pathophysiology of migraine improving our understanding of pain processing [12].

Since the 1990s, the spectacular advent of fMRI revolutionized neuroimaging and improved tremendously our understanding of human brain processes to such an extent that in current practice, the definition of structural MRI seems to have shifted to mean "not functional" MRI.

Because migraine is mainly a disorder of brain function, brain fMRI studies are useful to study the underlying mechanisms of migraine. Since migraine is regarded as a disorder of the brain [13], functional neuroimaging offers much in terms of understanding the physiological dysfunction that characterizes migraine. fMRI is increasingly employed for its non-invasive nature, and by exploiting the so-called blood-oxygen-level dependent (BOLD) effect and the neurovascular coupling it has become a powerful tool.

For completeness of information, spectroscopy and chemical shift imaging could be cited. These techniques aim to measure chemical concentrations, and therefore should be considered separately from other MR techniques.

2 Structural Neuroimaging Changes in Gray Matter

In the past decade, VBM has been widely used in many types of headache conditions. However, in recent years, VBM studies have focused on migraine, but the results showed some contradictions. Although the physiological mechanisms underlying CT are not completely understood, thinning and thickening may reflect cytoarchitectural changes of neuronal density or synaptic pruning as well as the cortical hyperexcitability of the migraine brain.

An initial VBM study [14] explored 11 patients suffering from migraine with aura (MwA) and 17 patients with migraine without aura (MwoA), each patient's group compared with a HC group. The authors found no global or regional macroscopic structural difference in global GM or WM volumes between either patients with migraine (taken as homogenous groups) and HC or between patients with MwA and MwoA. The authors suggested that other methods of phenotyping migraine, such as by genotype or perhaps treatment response could help to better address the issue of subtle structural change in the brain of migraineurs. Rocca et al. [15] followed on this line, according with data from a population-based MRI study [16], and demonstrated that female patients with migraine have a high risk of developing WM hyperintensities (WMH), independently from the presence or the absence of aura. GM density abnormalities were investigated, by using a 3-T MRI scanner and an optimized version of VBM analysis, in seven patients with MwA and nine patients with MwoA (showing visible abnormalities on T2-weighted images) and HC. In these patients, characterized by a peculiar "neuroradiological phenotype", a reduced GM density, mainly located in the frontal and temporal lobes were observed when compared with HC. Moreover, an increased GM density of both periacqueductal (PAG) and dorsolateral pons (dLP), brain areas strictly related to the pathophysiological substrates of migraine, has been observed in patients with MwA when compared with patients with MwoA. Interestingly, reduced GM density was strongly related to age and disease duration in migraineurs.

A separate VBM study [17] aimed to evaluate the presence of global or focal GM or WM alterations in 27 migraineurs compared to HC and between 16 episodic and 11 chronic migraineurs confirmed that migraineurs are characterized by a significant GM reduction in several cortical areas involved in pain circuitry, independently from the presence of WMH. Episodic and chronic migraineurs (taken as homogenous group) presented a significant focal GM reduction in the right superior temporal gyrus (STG), right inferior frontal gyrus (IFG) and left precentral gyrus (PCG) when compared with HC. Furthermore, chronic migraineurs showed a focal GM decrease in the bilateral anterior cingulate cortex (ACc), left amygdala, left parietal operculum, bilateral insula, left middle frontal gyrus (MFG) and IFG when compared to episodic migraineurs. A significant correlation between GM reduction in ACc and frequency of migraine attacks was found in all the migraineurs. Similarly, Kim et al. [18] demonstrated a significant GM volume reductions in the bilateral insula, motor/premotor and prefrontal cortex (PFc), ACc, right posterior parietal cortex, and orbitofrontal cortex (OFc) in migraineurs (five with MwA and 15 with MwoA) when compared with HC. Observed GM volume changes were related to both increasing headache duration and frequency. A different study [19] has also supported a significantly GM volume reduction in the left medial PFc, dorsal ACc, right visual cortex (Vc), cerebellum and brainstem in 21 patients with MwoA compared with HC. The findings confirm previous observations of a significant correlation between GM reduction in ACc and the frequency of migraine attacks in migraineurs. All together, these VBM studies suggest the concept that migraine may be considered a progressive disorder. Indeed, frequent nociceptive inputs related to repeated migraine attacks in the course of migraineurs life could modify the structural patterns of specific brain regions involved in pain processing.

To clarify the role of repetitive noxious inputs as experienced by migraineurs and underlying GM changes, an elegant experimental paradigm has been conducted in HC receiving repetitive painful stimulation and innocuous thermal stimuli on the right forearm for 11 consecutive working days [20]. Behavioural data demonstrated that 14 HC were "sensitised", whereas the others 13 HC were "habituated" over the stimulation days. The VBM analysis has revealed in the group of "sensitisers" a significant reduction of GM density in several brain regions involved in pain processing such as the ACc, the insular cortex and the frontal cortex (Fc). By contrast, pain "habituaters" did not show any density changes in the GM. The repetitive application of painful stimuli changed the GM density in pain processing brain regions exclusively in those subjects who were characterized by the lack of habituation. Decrease GM density and increasing pain ratings over time observed in "sensitisers" HC are similar to findings observed in migraineurs and in consequence, an underlying sensitization phenomenon could be suggested also in migraineurs.

On the other hand, the presence of GM abnormalities early in the disease course, and the absence of correlation with patient clinical characteristics suggest that they may represent a phenotypic biomarker of migraine condition more than a consequence of repetitive nociceptive inputs experienced during migraine attacks. Indeed, using a 3.0 T scanner, significant GM atrophy of several regions of the frontal and temporal lobes and an increased volume of the right putamen have been observed in 12 paediatric migraineurs (7 with MwA and 5 with MwoA) when compared with paediatric HC [21]. Moreover, the left fusiform gyrus showed an increased volume in patients with MwA compared to patients with MwoA and HC, whereas it was significantly atrophied in patients with MwoA when compared to the other two groups.

Reduced regional GM was not correlated with disease duration and attack frequency, whereas a negative correlation was found between increased volume of the putamen and disease duration.

Nevertheless, in our studies [22–24], in which both functional and structural investigations have been conducted, no VBM abnormalities have been found in patients with MwoA compared to both patients with MwA and HC. To identify consistent results of VBM studies in migraineurs a recent meta-analysis [25] has been performed using activation likelihood estimation. A total of five studies were considered, comprising 126 migraineurs (including 23 patients with MwA, 41 patients with MwoA, 11 patients with episodic migraine and 16 with chronic migraine as well as 19 patients with menstrual migraine and 16 with not menstrual migraine) and 134 HC. The included studies have reported GM volume reduction at 84 coordinates as well as GM volume increase at two coordinates in migraine. There were significant reductions in middle Fc and the inferior Fc in migraineurs. However, due to difficulties related to VBM studies including migraineurs with different phenotypes in a single group (specifically both patients with MwoA and patients with MwA) or migraineurs with a single phenotype without comparison with other phenotypes (e.g. only patients with MwA) the authors were not able to perform a subgroup analysis and separate meta-analyses on each migraine phenotype. In consequence, whether VBM abnormalities are strictly related to specific subtypes of migraine or can distinguish the different subtypes of migraine is not defined.

Differences in CT have been reported by Hadjikhani and colleagues in two seminal studies in migraineurs [26, 27]. In the first study, the authors examining the motion-processing network in 24 migraine patients (12 with MwA and 12 MwoA) and HC founded that brain areas involved in motion processing were thickened in all migraineurs. Interestingly, one area of thickening corresponded to the region where previously was found the source of cortical spreading depression (CSD) during migraine aura [28] (area V3A) (see below). This finding raises the question as to whether a "silent" CSD develops as well in MwoA and structural abnormalities in the network of motion-processing areas could account for, or be the result of, the cortical hyperexcitability observed in migraineurs. The second study investigated morphologic changes in the somatosensory cortex (SSc) in 24 migraineurs (12 with MwA, 12 with MwoA) and 12 HC. The authors reported that migraine group had on average thicker SSc than the HC group. The most significant CT changes were observed in the caudal SSc, where the trigeminal area, including head and face, is somatotopically represented.

To delineate possible relationships between CT changes and clinical variables in migraineurs, cortical abnormalities have been investigated in an homogeneous group of 56 patient with MwoA (showing T2-visible WMH) compared with HC [29]. In these patients, cortical thickening in left rostral MFG and bilateral post-central gyri have been observed. The average CT of bilateral post-central gyri positively correlated with disease duration as well as estimated lifetime headache frequency.

Hougaard and colleagues [30] demonstrated difference in CT in the IFG comparing the typical migraine headache side of the patients to the contralateral side in 13 patients (within-subject comparisons) with frequent side-locked MwA (visual aura consistently occurring in the same hemifield). Interestingly, in the same work, the authors found no differences in GM structure with regard to aura, suggesting a structural reorganization of pain inhibitory circuits in response to the repeated intense nociceptive input due to the headache attacks. CT findings have been further elucidated by another study [31] conducted on 46 female migraineurs indicating that these patients show a lack of thinning in the insula by age in contrast to HC.

Recently, Schwedt and colleagues have conducted several investigations on CT changes in migraineurs. In a very interesting study [32], an atypical association between migraine and cortical aging has been demonstrated in 27 migraineurs (18 with MwoA, 9 with MwA) compared with HC. The authors demonstrated that, although both migraineurs and HC have expected age-related thinning in many regions along the cortical regions, migraineurs show structural alterations of temporal and parietal regions that become more pronounced over time. Moreover, CT-to-pain threshold correlations differed between migraineurs and HC for bilateral STG/inferior parietal gyrus (IPG), right PCG, posterior cingulate cortex (PCc)/precuneus, in 31 migraineurs (21 with MwoA, 10 with MwA) compared to HC [33]. In other terms, migraineurs exhibit a non-significant positive correlation between CT in STG/IPG with pain thresholds when compared with HC. Since this region participates in orienting and attention to painful stimuli, absence of the normal correlation might represent a peculiar inability to inhibit pain sensation via shifting attention away from the painful stimulus in migraineurs. Nevertheless, we cannot exclude that individual CT variability could be involved in pain perception as demonstrated by Erpelding and colleagues [34] using a high-resolution structural MRI in HC. In this study, brain GM analysis revealed a strong correlation between greater thermal and pain sensitivity and cortical thickening of the primary SSc. Additionally, greater sensitivity to cold stimuli correlated with CT in the paracentral lobule, and greater warm detection correlated with cortical thinning in the ACc. The authors also found that greater heat pain sensitivity correlated with thickening in the PCc and the OFc.

Furthermore, a study to identify the brain interregional CT correlations that most differed between migraineurs and HC has been conducted [35] on 64 migraineurs compared to HC.

CT was determined for 70 brain regions that cover the cerebral cortex and CT correlations amongst these regions were calculated. A model containing 15 interregional CT correlations differentiated groups of migraineurs from HC with high accuracy. Specifically, the right temporal pole was involved in 13 of the 15 interregional correlations while the right middle temporal cortex was involved in the other two, suggesting that these regions play an important role in migraine pathophysiology.

An alternative strategy to quantify GM morphometric abnormalities involves the use of SB methods, which produce measures of CT and cortical surface (CS). These two measures are thought to reflect different structural characteristics of the human cortex and to be driven by distinct cellular factors [36]. CS area increases dramatically during late foetal development as a consequence of cortical folding, while CT changes dynamically throughout the life span as a consequence of development and diseases [37, 38]. Cortical abnormalities, using the highly sensitive SB morphometry, have been explored in migraineurs (28 with MwA and 28 with MwoA) compared with HC [39]. No significant CT differences in SSc, cingulate gyrus, or V3A/MT+ were found between the groups, including analysis of specific subregions previously reported to be affected in migraineurs [39]. Interestingly, given the sample size, power analyses indicated that even a small difference in CT could have been detected between groups. In another study [40], CT and CS abnormalities have been investigated in patients with migraine to assess their correlation with clinical and radiologic manifestations of the disease. Both CT and CS areas were estimated in 63 migraineurs (31 with MwoA and 32 with MwA) compared with HC. Among patients with MwA, 25 experienced exclusively episodes of MwA, while seven experienced episodes either MwA or MwoA. Migraineurs showed reduced CT and CS area in regions involving in pain processing. Conversely, these two metrics were increased in regions involved in executive functions and visual motion processing. The anatomic overlap of CT and CS area abnormalities was only minimal, with CS area abnormalities being more pronounced and more widely distributed than CT abnormalities. CT and CS area abnormalities were related to aura and WMH but not to disease duration and attack frequency. These results shed a light on cortical abnormalities that could be observed in migraineurs, representing the results of a balance between an intrinsic predisposition, as suggested by CS area abnormalities, and disease-related processes, as indicated by CT abnormalities.

It is well known that the thalamus exerts a critical role in pain processing and cortical excitability control and to investigate thalamic microstructure an innovative multiparametric approach at high-field MRI has been used by Granziera and colleagues [41] in migraineurs (22 with MwoA and 15 patients with MwA) and HC. The authors found that patients with MwA exhibit broad changes in thalamic nuclei when compared with MwoA patients and HC. No structural differences in thalamic nuclei involved in pain processing (such as the ventro-postero-lateral nucleus and ventro-postero-medial nucleus) were observed in both patients with MwA and MwoA. Furthermore, by means of T2* relaxation times evaluation, a relatively higher iron accumulation in the thalamus of patients with MwA compared with patients with MwoA was demonstrated, suggesting a role in pathophysiological mechanisms underlying migraine attacks.

A very interesting functional and morphometric study [42], using high resolution MRI, aimed to investigate hippocampal morphometric changes in migraineurs with different frequency of headache attacks (ten patients suffering from migraine with a low frequency of attacks and ten patients suffering from migraine with a high frequency of attacks) compared with HC. By means of a segmentation approach a significant larger bilateral hippocampal volume was found in low frequency group compared with migraineurs with high frequency and HC. The observed alterations, suggesting an initial adaptive plasticity that may then become dysfunctional with increased frequency, support a hippocampus role in migraine. Indeed, structural (and functional) changes may be the result of repeated stress and, as a consequence, may alter biological responses (including the stress response) over time, as a negative cascade adding to the disease burden through allostatic overload. These responses would appear to be maladaptive, and lead to allostatic overload over time, and have significant implications for disease progression.

In conclusion, it is still not clear whether morphological changes are cause or consequence of abnormal pain processing, but it is well established that disease duration and frequency of migraine attacks correlate highly with such changes. On the other hand, results of several morphologic studies have consistently demonstrated structural abnormalities in brain regions that are part of the network subserving supraspinal nociceptive processing not only in migraine but also in other chronic painful conditions, including facial pain [43], post-traumatic headache [44] and fibromyalgia [45] supporting the notion that chronic stimulation of these areas might cause a loss of GM volume. According with this interpretation, the broad spectrum of individual characteristics in pain perception may be attributable to preceding vulnerabilities. Similarly, an inherited susceptibility for migraine may be responsible for a developmental change that leads to the structural differences in these areas [4]. Similarly, resilience or susceptibility to migraine might also be a consequence of inter-individual variations in brain structure.

3 Structural Neuroimaging Changes in White Matter

Although the clinical definition of migraine requires the brain of a patient to be normal and structural changes to be absent [46], an increasing number of studies support the association of migraine with an increased risk of MRI-detectable WMH [47, 48] probably associated with clinical parameters of disease severity, such as frequency of attacks, migraine duration as well as disease age and family history [49–51]. The mechanisms causing WM abnormalities and clinical implications for patients are not yet determined although several causes have been hypothesized [52, 53]. A longitudinal MRI study found clinically silent brain WMH to be predominantly progressive in nature [54], whereas other observations suggest no direct association between clinical features of migraine and WMH progression [55], supporting the hypothesis that this association is stable in older age and may be primarily attributable to changes occurring earlier in life [56].

Microvascular ischemic mechanisms, which in turn may be associated with ischemic stroke, have been suggested [57, 58], independently from aura symptoms [59] and a specifically increased risk of ischemic stroke as well as the risk for cognitive impairment due to WMH in migraineurs should not be assumed [60]. Probably WMH are markers of transient breakdown of the blood–brain resulting from intense but self-limited cerebral hyperperfusion [61] or decreased antioxidant response in migraineurs [62]. Very recently, an MRI study to determine the frequency of WMH and the relationship with both migraine characteristics and cardiovascular risk factors has been conducted in 90 migraineurs (70 with MwoA and 20 with MwA). Silent WMH were observed in 32% of migraineurs and were found more frequently in patients with chronic migraine. The majority of lesions were located in the supratentorial right hemisphere. Migraineurs with and without WMH did not show significant differences in cardiovascular risk factors, such as smoking, serum cholesterol, oral contraceptive pills use and body mass index. These results suggest that the relationship between migraine and WMH may be directly due to the effects of migraine itself (probably via a significant T cell accumulation, sphingomyelinase activation, increased oxidative stress and reduction of both GM and WM triggered by CSD)[63] rather than to cardiovascular risk factors [64]. A different possible explanation may rely on a peculiar vascular vulnerability of migraineurs that may contribute to the pathogenesis of migraine and, in the presence of some other unknown factors may also contribute, over time, to the development of both WMH and cardiovascular disease. At the moment, we can only consider migraine as a risk factor for WMH in the brain [65] but there are no reliable features that may indicate which subjects, across the overall migraine population, will develop vascular events [66], although a link to an increased risk of stroke, especially in patients with MwA, cannot be ruled out. The mechanisms underlying the relationship between migraine and WMH suffer from the lack of conclusive evidence and further research addressing this topic seems essential.

About the relation between migraine and dilated perivascular spaces a large, blinded, population-based study showed no differences in the number of visible perivascular spaces in the basal ganglia and hemispheric WM in both patients with MwA and MwoA compared with non-migrainous headache patients and HC [67].

4 Microstructural Neuroimaging Changes in White Matter

Newer imaging techniques (i.e. DTI) have provided more detailed information about microstructural brain changes in these patients. DTI allows visualization of the orientation and anisotropy of water diffusion characteristics, which are mainly influenced in the brain by tissue features and cellular membranes. It enables the reflection of the integrity of fibre bundles and to detect microstructural alterations in WM, which cannot be visualized on conventional MRI sequences [68]. Altered anisotropy can be the consequence of not only structural WMH, but also of alteration of myelination and axon density. Reduced fractional anisotropy (FA) may result from demyelination, axonal loss, gliosis and inflammation. Axonal diffusivity (AD) may help to detect axonal degeneration, whereas radial diffusivity (RD) may be affected by myelin loss [69]. Previous DTI studies in episodic migraineurs reported several alterations in the interictal phase. To assess the correlation between the extent of macroscopic T2-weighted abnormalities, specifically WMH, and "occult" tissue damage (pathological damage of normal appearing brain tissue), Rocca and colleagues [70] investigated, by means of DTI technique with a histogram-based analysis, 34 migraineurs and 17 HC. Migraineurs showed lower mean diffusivity (MD) histogram peak height of the normal appearing brain tissue compared with HC, whereas no differences were found in FA histogram-derived metrics between migraineurs and HC. Interestingly, no difference was found for any MD and FA histogram-derived metrics between migraineurs with and without brain MRI lesions, and between patients with MwA and MwoA. The authors concluded that although brain damage may extend beyond T_2-weighted abnormalities in migraineurs, the severity of these "occult" changes may result to have a mild impact. DTI approach [26] has been used also to explore motion-processing network, involving brain areas known as a CSD source involved in visual aura in patients with MwA and probably in "silent" CSD in MwoA, in 24 migraineurs (12 with MwA; 12 with MwoA) and HC. WM abnormalities in the areas subjacent visual motion-processing (MT+ and V3A) in superior colliculus and the lateral geniculate nucleus were found in migraineurs compared with HC. Another investigation [71] in 24 migraineurs (12 with MwoA; 12 with MwA) compared with HC demonstrated permanent interictal areas of lower FA in the ventrolateral PAG in patients with MwoA and in the ventral trigeminothalamic tract in patients with MwA, pointing to an effect of migraine on the trigeminal SSc and modulatory pain system.

Microstructural abnormalities have been investigated by means of a region-of-interest (ROI) approach, in the corpus callosum (CC) in 24 patients with MwoA (12 without depressive/anxious disorder; 12 with depressive/anxious disorder) compared with HC [72]. Significant differences in FA values at all locations of the CC among the three groups were observed. The FA values from both migraine groups were significantly lower than those from HC. The FA values from migraineurs with depressive/anxious disorder were significantly lower than those of the migraineurs without depressive/anxious disorder. There were negative correlations between FA value of genu of the CC and disease course as well as FA value of genu and body of the CC and headache frequency. However, negative correlations were also found between FA values at all locations of the CC and anxiety and depression severity, suggesting that microstructural changes in the CC could be a possible neuroanatomical basis of migraine complicated with depressive and anxious disorder.

Abnormalities in CC and microstructural WM changes related to depressive disorder have been independently explored also by means of a novel approach to detect microstructural WM integrity alterations using a diffusion-weighted imaging with a fine-tuned nonlinear registration and nonparametric permutation testing in an alignment-invariant tract representation (tract-based spatial statistics [TBSS]) in migraineurs [73]. Indeed, reduced FA values of the genu of CC has been demonstrated in 21 patients with MwoA compared with HC [74]. Furthermore, WM microstructural abnormalities seems to be correlated with interhemispheric functional connectivity (f-connectivity) changes in of ACc in these patients, suggesting the possibility that WM changes of the CC modulate the resting-state (RS) f-connectivity between defined and highly pain-related brain areas such as ACc.

Yu and colleagues (who had already previously demonstrated significant lower FA, MD and AD in multiple brain regions in 20 patients [75] with MwoA), investigated WM integrity in 40 patients with MwoA (20 with depressive symptoms and 20 without depressive symptoms) compared with HC. Patients with MwoA as a group showed several WM tracts abnormalities compared with HC. However, migraineurs with depression symptoms showed decreased FA and increased MD and RD, with conserved AD, in WM tracts including the genu, body and splenium of the CC, bilateral superior longitudinal fasciculi and the right anterior corona radiate compared with migraineurs without depression symptoms. These WM tracts changes correlated significantly with depressive severity. The results suggested that both depression symptoms (more sensitive as RD) and migraine (more sensitive as AD) could affected WM integrity. FA and apparent diffusion coefficient (ADC) values of red nuclei, PAG, thalami, posterior limbs of internal capsules and subcortical WM were explored, by [76] means of a ROI approach, in 14 patients with MwoA during a migraine attack compared with HC. WM abnormalities were found only in the red nuclei, where ADC showed higher values than in HC, without correlation with age, duration of disease, frequency of attacks and localization of pain in migraineurs.

Recently, in a DTI study [77] a comparison was made of FA and MD obtained from the analysis of migraine-recurrence-induced changes in the thalamus of 24 patients with MwoA both during (10 patients) and between (14 patients) attacks compared with HC. During the ictal phase (but not during interictal period) patients with MwoA showed a significantly higher FA and slightly lower MD values in bilateral thalami compared with HC. Furthermore, right thalamic FA was positively correlated with the number of days since the last attack in migraineurs. These findings support previous neurophysiological evidence of altered interictal thalamic activity in migraine probably related to plastic peri-ictal modifications in regional branching and crossing of fibres.

Obviously, it is not possible to confirm the exact underlying mechanism for the above mentioned observation and in particular whether WM microstructural changes are responsible for triggering an attack or if they are the consequents of the attack itself. In order to evaluate whether WM abnormalities in the first period of migraineurs' life, TBSS and DT probabilistic tractography analysis have been used in 15 paediatric migraineurs [78] and HC. A significant lower MD, AD and RD diffusivity of WM tracts located in the brainstem, thalamus and fronto-temporo-occipital lobes bilaterally has been shown in paediatric migraineurs compared to HC. Patients also exhibited increased FA of the optic radiations. No correlation was found between WM tract abnormalities and disease duration and attack frequency, suggesting that WM abnormalities could be interpreted as microstructural features of migraineurs from the earliest stages of life and independently from clinical parameters of disease severity.

Global probabilistic tractography was used to investigated the integrity of WM tracts that underlie regions of the "pain matrix" and to assess putative correlation with disease duration in 23 migraineurs and HC [21]. Migraineurs showed greater MD in the left and right anterior thalamic radiations, left corticospinal tract, and right inferior longitudinal fasciculus tract. Migraineurs also showed greater RD in the left anterior thalamic radiations, left corticospinal tract as well as left and right inferior longitudinal fasciculus tracts. A positive correlation between migraine duration and MD in the right anterior thalamic radiations and left corticospinal tract has been observed in these patients. By means of DT tractography structural changes in optic radiation were quantified in seven patients with MwA, eight patients with MwA (experiencing visual aura) and HC [79]. WM changes located in optic radiation and their relation to clinical manifestations and T_2-visible hyperintensities were investigated. No difference was found for any of the WM fibre bundles metrics between patients with MwoA and HC, whereas patients with MwA were characterized by a reduced average FA of both optic radiations compared with HC and reduced average FA of the right optic radiation compared with patients with MwoA. They also showed higher right optic radiation MD than HC. In this study, optic radiation metrics were not correlated with clinical parameters. More recently, DTI [80] data were analyzed using a TBSS approach and FA, MD, RD and AD were compared between 39 chronic migraineurs, 34 patients with episodic MwoA and HC as well as between migraineurs as a group and HC. In contrast to previous studies, the authors did not find alterations in DTI-derived metrics in episodic migraineurs compared with HC. Furthermore, no statistically significant differences in chronic migraineurs when compared with episodic migraineurs and HC were found. These data revolutionized previous insights about microstructural changes in both patients with episodic or chronic migraine. However, accordingly with Neeb and colleagues investigations, in our previous multiparametric studies [22–24] we have never observed DTI abnormalities (with both a whole-brain and ROI approach) in patients with episodic MwoA and MwA. This suggests that patients' sample homogeneity could be a critical factor to justify the different microstructural findings reported by different researchers in the same disease.

5 Functional Neuroimaging During Spontaneous Migraine Attacks

Because migraine is mainly a disorder of brain function, fMRI has been considered an appropriate tool to investigate the underlying mechanisms of migraine activation. In this context, the headache attack might be considered an obvious and "specific" stimulation paradigm, and BOLD changes during the headache attack could be contrasted to a baseline condition observed during the interictal period. Weiller and colleagues in a pioneering $H_2^{15}O$ PET study have shown significantly higher regional rCBF values in cingulate, Vc, auditory cortex and brainstem, specifically in the dorsal pons (dP) during a spontaneous attack [81] in nine patients with MwoA. The observed activations were abolished, after a therapeutic dose of sumatriptan, in cortical areas but not in the brainstem, suggesting that brainstem activation was unlikely to be the result of pain perception nor an increased activity of anti-nociceptive systems (because a persistent activation was present also after sumatriptan-related pain relief). Another $H_2^{15}O$ PET study, with a high-resolution PET, was conducted to test the hypothesis of brainstem activation during migraine attacks and to refine the anatomic brainstem localization. For this aim, five migraineurs (two patients with MwA and three patients with MwoA) underwent imaging both during spontaneous migraine attacks and interictal periods. A significant activation in the dP, lateralized to the left, was observed comparing the ictal with interictal states. The activation was also demonstrated in the right ACc, PCc, cerebellum, thalamus, insula, PFc and temporal lobes. Contextually, an area of deactivation during migraine phase was located in the pons, lateralized to the right [82].

A very interesting $H_2^{15}O$ PET study [83] was conducted in seven patients with MwoA in the early migraine phase, after headache relief by sumatriptan and during an attack-free period. The authors observed, during the headache phase, significant activations not only in the midbrain and pons but also in the hypothalamus and the activations were persistent even after successful treatment by sumatriptan. These findings support the concept that hypothalamic involvement may be not strictly related to trigemino-autonomic cephalalgias [83, 84] and corroborate clinical observation on the key role of the hypothalamus in the pathophysiological aspects of migraine attacks [85, 86] such as the trigger factors and the premonitory features. Premonitory phase of migraine and related neuronal correlates have been more recently explored by means of $H_2^{15}O$ PET imaging. Glyceryl trinitrate (nitroglycerin) has been used to trigger premonitory symptoms and migraine headache in eight patients with MwoA who habitually experienced premonitory symptoms during spontaneous attacks. In this case, the premonitory phase has been considered as the period following when the nitroglycerin-induced non-specific headache phase had completely ceased and patients started to experience symptoms warning them of an impending headache. Activations in the posterolateral hypothalamus, midbrain tegmental area, PAG, dP and Vc, temporal cortex and PFc have been found comparing the first premonitory scans to baseline scans in all migraineurs. In particular, hypothalamic activation observed in the premonitory phase of glyceryl trinitrate-triggered migraine attacks can explain many of the premonitory symptoms and provide insight into the migraine activation due to homeostasis changes [87]. Among premonitory symptoms nausea occurs in about a quarter of migraineurs, suggesting primary brain alterations unrelated to the experience of headache. To explore the neural correlates of nausea, a $H_2^{15}O$ PET study has been performed in the premonitory phase of nitroglycerin-induced migraine in ten patients with MwoA and then patients with and without nausea were compared (three patients had nausea and seven did not have nausea in the premonitory phase during the scanning session). The results showed activation in brain circuits mediating nausea such as rostral dorsal medulla (including the nucleus tractus solitarius, the dorsal motor nucleus of the vagus nerve and the nucleus ambiguous) and PAG only in the patients experiencing nausea. These structures were involved independently from pain and trigeminal activation, suggesting that nausea is a centrally driven symptom in migraine [88].

Recently, a ^{18}FDG-PET study [89] has been conducted to assess altered brain metabolism in vestibular migraine (VM), a disabling neurological disorder characterized by vestibular symptoms, such as vertigo, dizziness, or imbalance in at least 50% of migraine episodes in patients with MwoA or with aura MwA [46]. Two patients with VM were investigated during and between VM attacks in addition to detailed neurotological evaluation. During the attacks, both patients showed an activation of the bilateral cerebellum and frontal cortices, and deactivation of the bilateral posterior parietal and occipito-temporal areas. One patient also showed hypermetabolism in the dorsal pons and midbrain, right posterior insula,and right temporal cortex while the other patient had an additional activation of the left temporal cortex. Compared with interictal images, ictal PET showed increased metabolism in the bilateral cerebellum, frontal cortices, temporal cortex, posterior insula, and thalami. The findings of contemporary activation of the vestibulo-thalamo-cortical pathway and decreased metabolism in the Vc may represent a reciprocal inhibition between the visual and vestibular systems in patients with VM.

Experimental investigation by PET imaging has been further improved by the availability of suitable radiotracers targeting different neurochemical systems [90]. Since 5-hydroxytryptamine (5-HT) CA receptors were thought to be implicated in migraine pathogenesis, PET studies with specific radioligands have been conducted to investigate serotoninergic function in migraineurs. In an early study [91], PET with ^{18}F-fluorosetoperone (a 5-HT$_2$-specific radioligand) did not reveal differences of cortical 5-HT$_2$ receptors' distribution volumes in migraineurs (five patients with both MwA and MwoA and four patients with MwoA) when compared with HC. Another PET study using an α-[11C]methyl-l-tryptophan tracer was conducted to measure brain serotonin synthesis in 11 patients with MwoA during attacks, reporting an increased rate of brain serotonin synthesis in the acute phase [92]. These data have been recently confirmed using specific antagonist of serotonin receptors [93], and the authors advocate that increased 5-HT$_{1A}$ receptor availability is present early during migraine attacks in the pontine raphe of migraineurs [84, 91].

PET investigations during migraine attacks were also employed to investigate the effects of molecules known to be clinically effective. A $H_2^{15}O$ PET study has been conducted to investigate the effect on brain circulation of a 5-HT$_{1B/1D}$ receptor agonist (rizatriptan), which caused a 13% CBF and blood volume decrease possibly related to the effect of triptans on the large cerebral arteries or on arterioles [94]. In the same period, an interesting PET study using radioactive [carbonyl-11C] zolmitriptan [95] evaluated the uptake and distribution of triptans into the CNS supporting their central mode of action. Although PET imaging has offered much in terms of understanding the neural correlates of migraine and associated symptoms, and functional changes depending on pharmacological modulation, many questions about cerebral network functions in migraine are still open and the pending solution is dependent also on the refinement of technology. Today, there is an increasing interest in developing PET-radiotracers for specific receptors thought to be implicated in pain and headache pathogenesis such as glutamate and opioid receptors [12]. However, we still lack information regarding the impact of migraine attacks and its relief on the function of μ-opioid receptor (μOR) mediated neurotransmission, the primary target of opioid medications. This line of enquiry is of particular importance as this neurotransmitter system is arguably the endogenous brain mechanism most centrally involved in pain regulation, as well as in the effectiveness of opioid medications. Recently, a PET study using the selective l-opioid receptor (lOR) radiotracer [11C]carfentanil [96] has elucidate the allodynic response of the central l-opioid system during spontaneous migraine attack following a sustained pain threshold challenge on the trigeminal ophthalmic region. Six migraineurs showed ictal cutaneous allodynia (CA) during the thermal challenge that was concurrent and positively correlated with lOR activation in the midbrain, extending from red nucleus to ventrolateral PAG. These findings demonstrated for the first time in vivo the high lOR activation in the migraineurs' brains in response to their allodynic experience. The same research group using the same technique evaluated in vivo the μ-opioid system during spontaneous migraine attacks in seven migraineurs [97]. In the ictal phase, there was μOR activation in the medial PFc, which was strongly associated with the μOR availability level during the interictal phase. Furthermore, μ-opioid binding changes showed moderate negative correlation with the combined extension and severity of the attacks. These results indicated for the first time that there is high μOR activation in the brain during migraine attacks in response to pain. Similar PET data have been used to investigate, using a novel 3D immersive and interactive neuronavigation (3D-IIN) approach, the endogenous μ-opioid transmission in the ictal migraine phases in a patient with MwA who has been suffering with migraine for 10 years [98]. During the ictal PET session (spontaneous headache attack) there was a reduction in μOR BPND in the pain-modulatory regions of the endogenous μ-opioid system during the ictal phase, including the cingulate cortex, nucleus accumbens (NAcc), thalamus and PAG, indicating that μORs were already occupied by endogenous opioids released in response to the ongoing pain [98].

In MwA patients, early imaging studies have been performed to explore the theory suggesting that CSD was the electrophysiological correlate of visual aura [99, 100]. The seminal Olesen's SPECT study has shown that unilateral occipitoparietal oligemia during the aura was preceded by hyperemia, that oligemia may spread anteriorly and that severe headache could occur during this oligemic phase [101]. However, the work of Hadjikhani and colleagues [28] could be considered the most important fMRI study to better understand the pathophysiological mechanism underlying the aura phenomenon. The authors, using high-field fMRI with near-continuous recording during visual aura in three subjects, have initially observed a focal increase in BOLD signal (possibly reflecting vasodilation), developed within extrastriate cortex (area V3A). This BOLD changes progressed contiguously and slowly over the occipital cortex, congruent with the retinotopy of the visual percept. Following the same retinotopic progression, the BOLD signal then diminished (possibly reflecting vasoconstriction after the initial vasodilation), as did the BOLD response to visual activation. During periods with no visual stimulation, but while the subject was experiencing scintillations, BOLD signal followed the retinotopic progression of the visual percept. These data strongly suggest that an electrophysiological event such as CSD generates the aura in human visual cortex. Today, converging evidence suggests that the oligemia persists well into the pain phase supporting the concept that vasodilatation could not explain the pain during migraine attack [28, 102]. The role of CSD in MwA is well established, but its contribution to the pathophysiology of MwoA, which involves the new concept of a clinically silent CSD, is still an intriguing issue [103].

6 Functional Neuroimaging During Painful Stimuli

We previously wrote that migraine attack might be considered such an obvious and "specific" stimulation paradigm to investigate the underlying mechanisms of migraine activation. Nevertheless, the main limitation of this experimental approach lies in the capture of spontaneous and unpredictable attacks of relatively short duration, such as migraine, while imaging techniques require considerable planning [6, 104]. In the last few years, these factors have determined the selection of studies dominated by noxious stimulation paradigms designed to better explore abnormalities in sensory, adaptive, and affective components of pain processing network in migraineurs and HC. Since pioneer studies using nitroglycerine or capsaicin to elicit cranial pain in migraineurs, various noxious stimuli have been used in imaging studies. Among these, May and colleagues, to test the hypothesis that brainstem activation may represent the so-called 'migraine generator', performed a PET study [105] in seven HC. In these subjects, a small amount of capsaicin was administered subcutaneously in the right forehead to evoke a burning painful sensation in the first division of the trigeminal nerve. The authors found an increased rCBF bilaterally in the insula, in the ACc, the cavernous sinus and the cerebellum. Interestingly, using the same stereotactic space limits as in Weiller's study, no brainstem activation was observed in the acute pain state compared to the pain-free state. However, an increased activation was found in the region of the cavernous sinus, suggesting that this structure may be involved in trigeminal pain. In the last years, pain-inducing heat is applied with an MRI-compatible contact thermode with a predefined or individualized temperature to each patient to elicit pain of moderate or severe intensity. In particular, the contact thermode can offers an easy approach to explore trigeminal system using a painful stimulation. Indeed, the regions innervated by the three branches of the trigeminal nerve can be easily distinct and stimuli to activate the trigeminal system are well-identified. Moreover, the trigeminal system reflects a somatotopic brain representation, and functional changes in trigeminal system can be detected at multiple levels (from trigeminal ganglion to the trigeminal nucleus and even in higher brain centres). For these peculiar characteristics, experimental trigeminal pathway activation has been extensively used to explore neural mechanisms underlying migraine during both headache attack and interictal period.

The elegant study of Moulton and colleagues [106] could be considered one of the landmarks of migraine fMRI research in the course of a painful stimulation using the contact thermode. The authors determined the heat pain threshold as the average of three different evaluations in 12 migraineurs and HC. During BOLD-fMRI sessions, a non-painful stimulation (41 °C) and a noxious heat stimulus (pain threshold +1 °C) were applied to the side of the forehead involved during migraine attacks. Assuming a brainstem region of interest, during non-painful stimulation there was a significantly greater BOLD response in the dLP in HC than in migraineurs. Conversely, during the painful stimulation a significant activation of the nucleus cuneiformis (NCF), a dLP structure involved in descending pain modulation, was observed. Interestingly, perception of painful stimuli did not show differences between patients with migraine and HC. Clinical and fMRI findings suggested that a central sensitization during attacks may be related to NCF "hypo-function" in patients with migraine experiencing CA. The same research group has lately conducted a BOLD-fMRI study using a painful trigeminal stimulation [107] in 11 migraineurs during the interictal period and HC. Moreover, eight migraineurs were tested by means of the same experimental stimulation during both the ictal and interictal periods. The authors demonstrated, using a ROI-based approach, an increased BOLD response to trigeminal painful stimulation in temporal pole (TP) and parahippocampal gyrus, centred on the entorhinal cortex (EC) in migraineurs, during the interictal period compared with HC and during migraine attack compared with the interictal period. Microstructural connectivity analysis, by means of DTI, revealed that TP and EC showed an enhanced connectivity with different brain structures involved in pain processing. These findings shed some light on migraine mechanisms, suggesting that hyperexcitability of associative multisensory areas, such as TP and EC (during both migraine attack and the interictal period), may be related to pain circuits.

Our group has explored the functional reorganization of pain-related pathways during trigeminal painful stimulation, using a whole-brain analysis approach, in 16 drug-naïve patients with MwoA during the interictal period [108]. By means of the contact thermode, a severe noxious (53 °C), a moderate noxious (51 °C) and a control (41 °C) stimulus were applied randomly to the maxillary skin. During the control trigeminal stimulus no differences in activation were observed between patients with MwoA and HC, whereas a significantly greater activation to the moderately painful heat stimulus was observed in the perigenual part of the ACc, and a significantly decreased activation to the severe painful heat stimulus was observed bilaterally in the secondary SSc. A group-by-stimulus whole-brain interaction analysis revealed a significant BOLD response in the pons which was associated with higher headache-related disability, intensity of pain in the course of a migraine attack and frequency of migraine. Similarly to the behavioural findings observed in the Moulton's study [106], patients and HC did not show any significant difference in perception at any level of experimental stimulation. In our opinion, the functional reorganization of pain-related cortical areas in patients with MwoA could represent a compensatory or adaptive mechanism to reduce painful input to the cortex by increasing cerebral anti-nociceptive activity.

A new experimental stimulation has been developed by Stankewitz et a. [109] based on the intranasal administration of low concentration of gaseous ammonia, producing a trigeminal nerve irritation, which can be well-implemented within an event-related BOLD-fMRI study. The authors, for the first time, have explored processing, perception and modulation of pain by means of BOLD-fMRI in the course of repeated trigeminal painful stimulation over several days in 15 migraineurs [110] compared with HC. Migraineurs and HC were stimulated for eight consecutive days. BOLD-fMRI was assessed in the course of trigemino-nociceptive stimuli (ammonia) and no-noxious control stimuli (air puffs) on days 1, 8 and 90 in migraineurs. PFc, ACc, red nucleus and ventral medulla exhibited an increased activity in HC and a decreased response in migraineurs, from the first to the eighth day. These divergent BOLD responses did not correlate with pain perception (i.e. migraineurs and HC showed a gradual decrease of pain ratings from day 1 to day 8, which only marginally increased again on day 90). The findings suggested that altered pain processing networks may explain the dysfunctional neuronal filters of sensory input in migraineurs, likely due to repetitive migraine attacks.

The role of recurring headache attacks in migraineurs has been further explored in association to the migraine cycle in 20 migraineurs [111] (ten patients experienced a migraine attack in the next 72 h after scanning and were therefore in the preictal phase and 13 patients were scanned during acute headache attacks). During painful trigeminal stimulation using ammonia gas, the authors observed a robust activation in cortical and subcortical areas involved in pain processing in migraine patient exclusively within the interictal period and in HC. However, a lower activation in a brainstem area corresponding to the spinal trigeminal nucleus was detected in migraineurs compared with HC. Interestingly, the BOLD response increased during the pain-free migraine cycle toward the migraine attack, and it was down-regulated just before or immediately at the beginning of a migraine attack. In our opinion, beyond the putative role of spinal trigeminal nucleus as "migraine modulator", this event-related BOLD-fMRI study highlights two important concepts. The first is a phenomenological concept, which is necessary to better understand the neurobiological significance of periodic functional changes of migraineous brain. Migraine cycle spans over several days during different phases (prodromic, aura, headache, resolution and recovery), and trigeminal activity in migraineurs is not constant but strongly variable. The second is a methodological concept, which underlines the importance of taking the time to the next attack into account when investigating migraineurs. Another recent study [112] has investigated the "migraine cycle" and its relation with pain-induced activation of specific brain regions in 24 adult migraineurs (who were at least 48 h pain free) and HC. There were no significant correlations between brain activation and time to next migraine attack. However, a greater pain-induced activation of lentiform nucleus, fusiform gyrus, subthalamic nucleus, hippocampus, middle cingulate cortex (MCc), premotor cortex, SSc, dorsolateral PFc, and a reduced activation in PCG and STG have been observed in migraineurs compared to HC. Moreover, there were significant correlations between BOLD response and headache frequency for MCc, right dorsolateral PFc, left fusiform gyrus, left PCG and left hippocampus and with disease duration for left fusiform gyrus. It is evident that the majority of regions with enhanced pain-induced activation in headache-free migraineurs participate in cognitive aspects of pain perception such as expectation of pain and pain memory. Enhanced cognitive pain processing by migraineurs might reflect cerebral hypersensitivity related to high expectations and hypervigilance for pain. Indeed, pain perception is a complex sensory experience that is processed in a network of distributed cortical areas and within this network (the so-called "pain matrix" or, more recently, "neurolimbic pain network") the encoding and evaluation of painful events depend crucially on the functional interplay of these regions [113].

7 Functional Neuroimaging During Visual Stimuli

Around 45% of migraineurs report symptoms of light hypersensitivity in the interictal state, and about 90% during a migraine attack [114–116]. Significant evidences to understand migraine mechanisms were provided by $H_2^{15}O$ studies using luminous stimulations, which demonstrated a multisensory integration between light perception and trigeminal nociception.

Cao and colleagues [117] investigated Vc activation in the early phase of the visually triggered migraine attack in 12 migraineurs (10 with MwA and visual symptoms, 2 with MwoA). Visually triggered headache and visual change in migraineurs were correlated with spreading suppression of the initial neuronal activation and increased Vc oxygenation. The authors suggest that this spreading suppression may be associated with initial activation of a migraine attack, independent of whether there are associated aura symptoms. Some years later, Boulloche and colleagues [118] have demonstrated that luminous stimulations activated the Vc bilaterally in seven migraineurs also during interictal period. A concomitant heat pain stimulation (applied in the territory of the ophthalmic branch of the right trigeminal nerve) potentiated cortical activation in these patients and induced Vc activation in HC. The authors hypothesized that Vc hyperexcitability could be related to brainstem modulation of cortical excitability characterized by integration mechanisms with trigeminal structures.

Brainstem activation has been explored in 26 migraineurs (23 with MwA; 3 with MwoA) during repeated visual stimulation [119]. Repetitive visual stimulation triggered migraine symptoms in 12 patients: four with MwA developed both visual symptoms and headaches, and six with MwA and two with MwoA experienced headaches only. Four patients who had MwA experienced the onset of their usual aura or onset of their typical headache either during the experiment or immediately after. In the remaining 10 migraineurs, and all HC, visual stimulation failed to trigger symptoms at any time. A significant BOLD response has been observed in red nucleus and substantia nigra in association with visually triggered symptoms of migraine, suggesting that these brainstem structures are a part of a neuronal network activated during an attack. Denuelle and colleagues [120] have investigated photophobic mechanism during spontaneous migraine attacks, after headache relief by sumatriptan and during attack-free interval in eight migraineurs. The authors found that low luminous stimulation could activate the Vc during migraine attacks and after headache relief but not during the attack-free interval. The Vc activation was statistically stronger during migraine headache than after pain relief.

By ^{1}H-magnetic resonance spectroscopy (MRS) [121] changes in brain metabolites due to Vc activation during visual stimulus have been investigated in 22 patients with MwA and 22 patients with MwoA. In the Vc, photic stimulation is linked with a consistent decrease of the N-acetylaspartate (NAA) signal and a parallel increase of the lactate peak in patients with MwA when compared with MwoA and HC. NAA loss might result from a decrease in NAA formation subsequent to ATP depletion, and these data could be related to a transient dynamic uncoupling following a rapid recoupling of oxidative metabolism after stimulation, due to a less efficient mitochondrial functioning in patients with MwA. In view of these associations, several studies have used fMRI to investigate responses to visual stimuli in migraineurs. Some of these studies specifically investigated patients with MwA, because Vc hyperexcitability might predispose the brain to visual hypersensitivity and visual aura.

Using fMRI-BOLD approach [122], light sensitivity and photophobia have been assessed exploring the response of the Vc to light stimuli in 19 patients with migraine (7 with MwA; 12 with MwoA) compared to HC. This study showed a significant hyperexcitability of the Vc with a wider photoresponsive area in migraineurs during interictal period. The authors suggested that the underlying mechanism of cortical reactivity in migraineurs is probably dual and may be part of a constitutional (defensive) mechanism or represents an acquired (sensitization) phenomenon.

Beyond primary visual areas, the dynamics of the basic interictal state with regard to extrastriate, motion-responsive middle temporal area (MT-complex) has been explored with BOLD-fMRI at 3 T using coherent/incoherent moving dot stimuli in 24 migraineurs (12 with MwA, 12 with MwoA) in the interictal period [123]. A weaker bilateral activation has been found in the MT-complex in both patients with MwA and MwoA compared with HC, whereas a significant stronger activation mainly at the left side in response to visual stimulation in the MT-complex was found in patients with MwoA and MwA compared with HC.

Cortical response to a visual stimulus during the interictal period has been compared also in another study investigating 25 patients with MWA and 25 patients with MwoA [124]. Despite similar interictal symptoms of visual discomfort, BOLD-fMRI response to visual stimulation within primary Vc and lateral geniculate nuclei were greater in patients with MwA compared to patients with MwoA and HC suggesting a direct connection between cortical hyperresponsiveness and migraine aura. Based on both altered visual motion processing in striate and extrastriate areas and optokinetic stimulation inducing symptoms associated with migraine in migraineurs, activation patterns and the hemodynamic response to optokinetic stimulation have been explored in 18 patients with MwA [125] using a novel approach based on a structural (by fMRI approach) and temporal (by functional transcranial Doppler) resolution. In this way, the activation pattern of the Vc (V1–V5) as well as the vasomotor reactivity of the posterior cerebral artery have been investigated. The authors found attenuation of the physiological right lateralization with a significantly increased activation in the left V5 complex, the left area V3, and the right V5 complex. Furthermore, the analysis of the visually evoked flow response of the rCBF in the posterior cerebral artery showed a larger side-difference of the offset latency and a reduced steepness of the decreasing slope on the left side, supporting the concept of an interictal motion-processing deficit in migraine. Recently [126], functional interhemispheric differences in responses to visual stimulation between symptomatic and asymptomatic hemispheres during the interictal phase has been evaluated in 20 patients with frequent side-fixed visual aura attacks ($\geq 90\%$ of auras occurring in the same visual hemifield). BOLD responses were selectively increased in the symptomatic hemispheres in the IPG, the IFG and the superior parietal lobule. The migraineurs also showed a significantly increased response in the same cortical areas when compared to HC. These findings suggest a hyperexcitability of the visual network (involved in oculomotor control, guidance of movement, motion perception, visual attention and visual spatial memory system) in the interictal phase of migraine with visual aura.

All together the reported data confirm that migraineurs, during visually stimulating patterns, have high activation in the primary and extrastriate Vc likely correlated to a cortex hyperexcitability that could not be explained only by trigeminal nociception because it persisted also during interictal period.

8 Functional Neuroimaging During Olfactory Stimuli

It is well known that, although the osmophobia is not reported in IHS classification criteria, migraineurs are hypersensitive to odours during and between migraine attacks. Furthermore, half of migraineurs report that certain odours can trigger migraine attacks [127]. PET studies also provided important insights into the neural mechanisms underlying associated migraine symptoms, such as photophobia, phonophobia and osmophobia, the latter being very specific to this form of headache [128]. During olfactory stimulation, migraineurs subjects exhibited a significantly higher activation in piriform and temporal cortices when compared with HC. Demarquay and colleagues [129], using voxel-based and ROI analyses, evaluated olfactory processing in 11 migraineurs experiencing olfactory hypersensitivity and investigated whether rCBF associated with olfactory stimulation was modified in patients compared with HC. During both olfactory and non-olfactory conditions, a higher rCBF in the left piriform cortex and antero-STG in has been found in migraineurs. During odour stimulation, migraineurs also showed significantly higher activation in the left temporal pole and significantly lower activation in the frontal (left IFG as well as left and right MFG) and temporo-parietal (left and right angular, and right posterior-STG) regions, PCc and right locus coeruleus. These results could reflect a particular role of both the piriform cortex and antero-STG in migraineurs experiencing olfactory hypersensitivity and odour-triggered migraine. The abnormal cerebral activation patterns during olfactory stimulation might reflect altered cerebrovascular response to olfactory stimulation due to the migraine disease, or an abnormal top-down regulation process related to olfactory hypersensitivity. More recently, migraine neuronal processing in response to olfactory stimulation (rose odour) has been investigated during interictal (in 20 migraineurs) and ictal period (13 of the 20 patients were scanned within 6 h after the onset of a spontaneous migraine attack) [130]. Imaging data showed that migraineurs during interictal period did not differ from control subjects. However, during spontaneous and untreated attacks, migraineurs showed significantly higher BOLD response in brain areas including limbic structures (amygdala and insular cortices). Interestingly, in response to olfactory stimulation, a significant activation has been observed also in the rostral pons (RP). The findings suggest that the activity level of this structure can be triggered by olfactory input and thus points to the strong physiologic relationship between the olfactory and the trigemino-nociceptive pathway in the migraine pathophysiology. Specifically, odour-induced activation of the RP might be a mechanism by which could odours trigger migraine attacks.

9 Functional Neuroimaging During Vestibular Stimuli

Recently our group has conducted a BOLD-fMRI study [131] in patients with VM (according to ICHD-III, beta version) [46] during the interictal period. The functional response of vestibular neural pathways during caloric vestibular stimulation in 12 patients with VM, 12 patients with MwoA and HC has been explored. Electronystagmography evaluation was performed to exclude vestibular disorders and to verify that caloric stimulus induced vestibular nystagmus. In all subjects, caloric vestibular stimulation elicited a statistically significant activation in bilateral insular cortex, thalamus, cerebellum and brainstem. Interestingly, a discrete PAG activation was observed, suggesting a peculiar relationship between vestibular stimulation and activation of a brain area which plays a key role in pain processing [132]. This finding could suggest that reciprocal connections between brainstem vestibular nuclei and structures involved in modulation of trigeminal nociceptive inputs may have some role in VM pathophysiology [133]. Furthermore, the analysis of difference between groups showed a significant divergent response in the mediodorsal thalamus in patients with VM relative to both patients with MwoA and HC. It is noteworthy that the thalamus represents a key structure in transmitting sensory input from the brainstem to the cortex, exerting a pivotal function in pain processing and cortical excitability control. This observation could clarify the VM pathophysiological mechanism, suggesting a dys-modulation in the multimodal sensory integration and processing of both vestibular and nociceptive information, resulting in a vestibulo-thalamo-cortical dysfunction. Furthermore, thalamic functional abnormalities exhibited a positive correlation with the frequency of VM attacks. Nevertheless, it is not possible to establish whether thalamic findings are a primary phenomenon due to the hereditary liability resulting in VM attacks or a secondary phenomenon as a result of repetitive VM attacks.

10 Resting Brain in Migraine

^{18}FDG-PET is widely used to measure glucose uptake into tissue including the brain and in the last decades, several ^{18}FDG-PET studies have been conducted to compare brain metabolism between migraineurs and HC, demonstrating substantial differences in brain metabolism between the two subject groups. Among these, Kassab and colleagues have explored resting glucose uptake in posterior supratentorial and infratentorial WM in migraineurs during the interictal period in 11 migraineurs compared with HC. The authors identified two regions of significant increase in glucose uptake mapped predominantly to the posterior WM of the cerebrum and cerebellum in migraineurs relative to the HC. These findings suggested a primary metabolic disturbance in the posterior WM of the brain in migraineurs. This point of view has been supported by Montagna and colleagues that investigated 22 patients with MwoA in headache-free periods by means of ^{31}P-magnetic resonance spectroscopy (MRS) of brain and muscle [134]. Brain ^{31}P-MRS showed significantly low phosphocreatine, increased adenosine diphosphate, and decreased phosphorylation potential demonstrating an abnormal energy metabolism in MwoA, as previously demonstrated in patients with migraine stroke and MwA [135]. To compare metabolism in the brain of migraineurs during headache-free periods with those obtained from HC a recent ^{18}FDG-PET study [136] has been conducted to evaluate interictal metabolic differences between 20 episodic migraineurs (four with MwA; 16 with MwoA) and HC. A significant hypometabolism in several regions known to be involved in central pain processing, such as bilateral insula, bilateral ACc and PCc, left premotor and PFc and left primary SSc has been found. Moreover, regional metabolism of both the insula and the ACc showed significant negative correlations with disease duration and lifetime headache frequency, suggesting that repeated migraine attacks over time could lead to metabolic abnormalities of selective brain regions belonging to the central pain matrix. These findings may be interpreted as a primary metabolic brain deficit related to migraine disorder or, alternatively, could suggest that a phenotypic trait could play a role in secondary metabolic abnormalities of brain regions involved in pain processing.

Recently, a novel tool which explores connectivity between functionally linked, but anatomically separated, brain regions has been developed. The use of this technique, called RS-fMRI, has allowed the identification, at rest, of the main brain functional networks without requiring subjects to perform specific active tasks. Methodologically, [137] several approaches can be applied for the analysis of RS-fMRI, including seed-based, independent component analysis based and/or cluster-based methods. The seed approach is the simplest to investigate spatial patterns, based on the direct correlations with time courses of signal change from a seed measurement. This technique is widely used in f-connectivity mainly due to its ease of interpretation and good sensitivity, however, its main limitation is the dependence on the a priori definition of a seed region, which prevents the method from studying multiple systems simultaneously. To overcome this limitation, blind source separation algorithms, such as independent component analysis (ICA), have become popular in f-connectivity analysis of BOLD-fMRI data. Indeed, ICA transforms individual patient RS-fMRI data sets into series of networks maps, allowing for a voxel-based population analysis of whole-brain f-connectivity without the need to specify the ROI constituting the layout of the neural network. RS-fMRI allows to identify a set of biologically meaningful spatial maps of independent components that are topographically organized in highly reproducible functional networks with biological relevance, called RS networks (RSN).

Analysis of f-connectivity investigates the functional organization of the brain based on temporal correlations in BOLD signal fluctuations in different brain regions. Most f-connectivity analyses are done when the brain is at rest, whit the person being studied is not performing any task and is not being stimulated. In the RS there is continuous low frequency fluctuation in the BOLD signal throughout the brain. Brain regions with temporal correlations in BOLD signal are deemed to be functionally connected or functionally communicating. The most commonly reported RSN are the default mode network (DMN), the FPN (or executive network), the sensorimotor network and the visual and auditory networks. The presence of functional connections and the strength of such functional connections can be atypical in the presence of neurological diseases including migraine. Using RS-fMRI, several studies have identified f-connectivity abnormalities in migraineurs, mainly located at the level of the pain processing network. Along this research line, Mainero and colleagues [138] have analyzed the baseline functional interaction within the networks of PAG and a subset of brain areas involved in nociceptive and somatosensory processing and as well as in pain modulation. The study, conducted in 17 migraineurs (eight with MwA; nine with MwoA) during the interictal period compared with HC. The authors demonstrated a stronger f-connectivity between the PAG and several brain areas within the nociceptive and somatosensory processing pathways in migraineurs compared to HC. In addition, as the monthly frequency of migraine attacks worsens, the strength of the f-connectivity in some areas within these pathways increased, whereas a significant decrease in RS f-connectivity between the PAG and brain regions with a predominant role in pain modulation (PFc, ACc, amygdala) was evidenced. Interestingly, migraineurs with a history of CA exhibited significantly reduced f-connectivity between PAG, PFc, and ACc compared to migraineurs without CA. These data revealed on interictal dysfunctional dynamics within pain pathways in migraine manifested as an impairment of the descending pain modulatory circuits, likely leading to loss of pain inhibition, and hyperexcitability primarily in nociceptive areas. Yu and colleagues [139] have applied regional homogeneity (ReHo) method to analyze local temporal homogeneity of intrinsic fluctuation, and investigated the f-connectivity alterations of regions showing morphometric deficits during rest condition in 26 patients with MwoA compared with HC. Migraineurs showed a significant decrease in interval ReHo values in the right rostral ACc, PFc, OFc and the supplementary motor area (SMA) when compared with HC. In addition, ReHo values were negatively correlated with the duration of disease in the right ACc and PFc. The results suggested that the RS abnormalities of these regions may be associated with functional impairments in pain processing in patients with MwoA.

Our group in a series of RS-fMRI studies, has demonstrated f-connectivity abnormalities in several RSN. To explore DMN f-connectivity in patients with MwoA and investigate its clinical significance, the RS f-connectivity of the DMN in 20 patients with MwoA, during the interictal period, and 20 HC have been compared [22]. Patients with MwoA showed decreased f-connectivity in prefrontal and temporal regions of the DMN when compared to HC. Observed functional abnormalities were unrelated to detectable structural abnormalities or clinical and neuropsychological features of migraineurs. Similarly, based on converging neuropsychological evidence suggesting executive difficulties in migraine during interictal periods, we have evaluated the f-connectivity of the fronto-parietal networks (FPN) known to be associated with executive functions (EF), in 14 patients with MwoA, in the interictal period, in comparison to HC [140]. Our data showed that MwoA patients, compared to HC, had significant f-connectivity reduction within the right FPN and specifically in the MFG and the dorsal ACc. In addition, we found that MFG reduced f-connectivity was negatively correlated with the pain intensity of migraine attacks. There were no structural differences between the two groups. Surprisingly, neuropsychological data revealed no significant executive dysfunction in MwoA patients. This observation has been supported by a recent study confirming a disrupted executive control network f-connectivity not only in patients with MwoA but also in patients with MwA, in the interictal period. Although f-connectivity abnormalities are present in the absence of clinically relevant executive deficits, they may reflect a vulnerability to high-demanding conditions of daily living activities in patients with migraine. Indeed, a putative explanation of these clinical, neuropsychological and functional findings is that the observed connectivity dysfunction in DMN and FPN could underlie or be related to a maladaptive brain response to repeated stress which seems to characterize patients with migraine [141, 142]. Indeed, according to recent studies, recurrent migraine attacks alter both f-connectivity and s-connectivity [4], and these changes may disrupt mechanisms of stress response [141, 142]. When behavioural or physiological stressors are frequent or severe, allostatic responses can become maladaptive, leading, in a vicious cycle, to further allostatic load. Moreover, due to a high energetic demand, the observed DMN dysfunction may be associated with an impaired brain energy metabolism which has been demonstrated in previous MR spectroscopy studies in migraineurs, likely due to an imbalance between ATP production and ATP use [134, 135].

In the last decades, based on the prominent role played by the Vc in migraine aura pathophysiology, visual pathways have been extensively explored in patients with MwA both during the aura phase and the interictal period. Visual evoked potentials studies have revealed, although with conflicting results, abnormal visual information processing in migraineurs, subtending abnormalities in habituation and sensitization mechanisms [143]. Similarly, conflicting results has been observed in f-connectivity findings related to visual network in patients with MwA. More recently, the RS-fMRI approach has been used to evaluate the f-connectivity in 20 patients with MwoA and 20 MwoA during the interictal period, compared with HC [24]. A significant increased f-connectivity in the right lingual gyrus within the RS visual network has been demonstrated in patients with MwA, compared to both patients with MwoA and HC. These abnormalities were present in the absence of structural or microstructural abnormalities and not related to migraine severity. The results support the hypothesis of an extrastriate cortex involvement, centred in the lingual gyrus, a brain region related to mechanisms underlying the initiation and propagation of the migraine aura. This RS-fMRI finding may represent a functional biomarker that could differentiate patients experiencing the aura phenomenon from patients with MwoA, even between migraine attacks. Niddam and colleagues [144], have recently investigated f-connectivity abnormalities in intrinsic cognitive networks in 26 patients with MWA and 26 patients with MwoA during interictal period and HC. The authors used a whole-brain approach with seeds placed in the anterior insula and the MFG, key nodes of the salience and dorsal attention networks, respectively. In opposite to our observations, a reduced f-connectivity has been observed Vc (area V3A) in patients with MwA during interictal period when compared with patients with MwoA and HC. Another RS-fMRI study exploring RS f-connectivity in patients with MwA outside of attacks showed different results. Indeed, no differences of f-connectivity were found between 40 patients with MwA and HC, suggesting an abnormal f-connectivity, during interictal period, only during exposure to external stimuli, as reported by previous studies demonstrating increased cortical hyperresponsivity in the interictal phase of MwA [145].

The possible mechanisms underlying the disrupted f-connectivity in both patients with MwA and MwoA are currently unknown and it is not possible to answer the question whether observed functional abnormalities are acquired through a genetic predisposition or as a result of migraine burden. Nevertheless, longitudinal changes in brain activity between repeated observations within a short time ReHo and interregional f-connectivity were assessed in 19 female patients with MwoA and 20 HC [146]. All patients reported that their headache activity increased over time. Abnormal ReHo changes in the patient group relative to the HC were found in the putamen, OFc, secondary SSc, brainstem, and thalamus. Moreover, these brain regions exhibited longitudinal ReHo changes at the 6-week follow-up examination. These headache activity changes were accompanied by disproportionately dysfunctional connectivity in the putamen in the migraineurs, as revealed by f-connectivity analysis, suggesting that the putamen plays an important role in integrating diverse information among other migraine-related brain regions. The results obtained in this study suggest that progressive brain aberrations in migraine progress as a result of increased headache attacks. The same research group [147], examined RS abnormalities in patients with MwoA to explore the relationship between neuroimaging markers and disease duration (long-term and short-term) in 40 patients with MwoA compared with HC. MwoA patients with long-term disease duration showed comprehensive neuronal dysfunction than patients with short-term disease duration when compared with HC, In addition, increased average ReHo values in the thalamus, brain stem, and temporal pole showed significantly positive correlations with the disease duration. On the contrary, ReHo values were negatively correlated with the duration of disease in the ACc, insula, PCc and superior occipital gyrus. These findings of progressive brain damage in relation to increasing disease duration suggest that MwoA is a progressive central nervous disease, and the length of the disease duration was one of the key reasons to cause brain dysfunction in migraineurs. The repeated migraine attacks over time result in RS abnormalities of selective brain regions belonging to the pain processing and cognition.

Based on the hypothalamus implications in the autonomic symptoms of migraine as well as neuroimaging evidence of hypothalamic activation during attacks, RS-fMRI techniques have been used to explore, using a seed approach, f-connectivity changes between the hypothalamus and the rest of the brain in in 12 patients with MwoA compared to HC [148]. Patients with MwoA showed an increased hypothalamic FC with a number of brain regions involved in regulation of autonomic functions, including the locus coeruleus, caudate, parahippocampal gyrus, cerebellum, and the temporal pole. Stronger functional connections between the hypothalamus and brain areas that regulate sympathetic and parasympathetic functions may explain some of the hypothalamic-mediated autonomic symptoms that accompany or precede migraine attacks. A peculiar f-connectivity between pain-modulating circuits and the limbic system have been also investigated comparing f-connectivity between the amygdala and the cortex in 11 patients with MwA and 11 with MwoA as well as in HC and in two other chronic pain conditions not associated with CSD: trigeminal neuralgia (nine patients) and carpal tunnel syndrome (11 patients) [149]. The authors evidenced that amygdala f-connectivity to the visceroceptive insula was increased in both patients with MwA and MwoA when compared to all other patient groups. These data reinforce the evidence of a neurolimbic pain network dysfunction and may likely reflect repetitive episodes of CSD leading to the development of migraine pain.

RSN of the BG in patients with MWoA has been investigate in 40 patients with MwoA [150]. Increased f-connectivity between the BG and several brain regions within nociceptive and somatosensory processing pathways was observed in migraineurs compared with HC. Correlation analysis revealed significant correlations between the volume of the bilateral caudate and right NAcc and disease duration. In addition, an increased monthly frequency of migraine attack was associated with increased f-connectivity between the bilateral caudate and left insula, and longer disease duration was correlated with increased f-connectivity between the right NAcc and bilateral ACc. BG dysfunctional dynamics during interictal RS-fMRI within pain pathways have been also correlated with reduced volume of the BG in MwoA. The findings support the hypothesis that impaired pain processing and modulatory processes in MwoA may be associated with abnormal structure and function within the pain-related pathways of the BG.

The conflicting data we are facing probably reflect the low numbers and clinical heterogeneity of subjects involved in the available neuroimaging studies (i.e. inclusion of patients suffering from MwA and MwoA; very different frequency or intensity of attack; different current pharmacological treatments) as well as the modality of imaging used. We believe that combining functional and structural techniques will prompt a new and more effective way of looking into migraneous brain organization and function. Recruiting patients for studies aimed at exploring the acute or prodromal phase of migraine is certainly not easy, but more specifically designed studies, looking at different stages of the migraine attack, will clarify the involvement of several brain structures, especially the real role of brainstem. It is noteworthy that the main difficulty with neuroimaging studies, in an episodic disorder such as migraine, is to capture spontaneous attacks when the imaging techniques require considerable planning. We are aware that experimental pain is absolutely different from spontaneous migraine and that by doing so we cannot draw any firm conclusions about real migraine mechanisms. However, we believe that further refinements of this kind of approach will improve our capabilities of exploring pain processing-related cerebral activity in migrainous subjects. Functional techniques suitable to study increasingly smaller brain structures and to detect even subtle abnormalities are rapidly evolving and will take us closer to the key structures and mechanisms of migraine. However, the interpretation of the biological significance of these various functional changes could remain incomplete without a combination of expanding genomic information about neurochemical pathways and genetic polymorphisms linked to specific migraine phenotypes or subtypes. Finally, there is no doubt that the clinicians' expertise is of pivotal importance, as only an headache expert may have the basic and clinical clues to plan a specific functional neuroimaging protocol to address a still rather complicated open issues.

References

1. Bloch F. Nuclear induction. Phys Rev. 1946;70:460–74.
2. Purcell EM, Torrey HC, Pound RV. Resonance absorption by nuclear magnetic moments in a solid. Phys Rev. 1946;69:37–8.
3. Symms M, Jäger HR, Schmierer K, Yousry TA. A review of structural magnetic resonance neuroimaging. J Neurol Neurosurg Psychiatry. 2004;75 (9):1235–44.
4. Ellerbrock I, Engel AK, May A. Microstructural and network abnormalities in headache. Curr Opin Neurol. 2013;26(4):353–9.
5. Olesen J, Larsen B, Lauritzen M. Focal hyperemia followed by spreading oligemia and impaired activation of rCBF in classic migraine. Ann Neurol. 1981;9(4):344–52.
6. Tedeschi G, Russo A, Tessitore A. Functional neuroimaging in migraine: usefulness for the clinical neurologist. Neurol Sci. 2012;33(Suppl 1):S91–4.
7. May A. A review of diagnostic and functional imaging in headache. J Headache Pain. 2006;7 (4):174–84.
8. Sprenger T, Henriksen G, Valet M, et al. Positron emission tomography in pain research. From the structure to the activity of the opiate receptor system. Schmerz. 2007;21:503–13.
9. Sprenger T, Ruether KV, Boecker H, et al. Altered metabolism in frontal brain circuits in cluster headache. Cephalalgia. 2007;27:1033–42.
10. Sprenger T, Willoch F, Miederer M, et al. Opioidergic changes in the pineal gland and hypothalamus in cluster headache: a ligand PET study. Neurology. 2006;66:1108–10.
11. Grupen C, Buvat I. PET imaging: basics and new trends. Handbook of particle detection and imaging. Springer; 2012. p. 935–971.
12. Tedeschi G, Russo A, Tessitore A. Relevance of functional neuroimaging studies for understanding migraine mechanisms. Expert Rev Neurother. 2013;13(3):275–85.
13. Goadsby PJ, Lipton RB, Ferrari MD. Migraine–current understanding and treatment. N Engl J Med. 2002;346:257–70.
14. Matharu MS, Good CD, May A, et al. No change in the structure of the brain in migraine: a voxel-based morphometric study. Eur J Neurol. 2003;10(1):53–7.
15. Rocca MA, Ceccarelli A, Falini A, et al. Brain gray matter changes in migraine patients with T2-visible lesions: a 3-T MRI study. Stroke. 2006;37(7):1765–70.
16. Kruit MC, Van Buchem MA, Hofman PA, et al. Migraine as a risk factor for subclinical brain lesions. J Am Med Assoc. 2004;291:427–34.

17. Valfrè W, Rainero I, Bergui M, et al. Voxel-based morphometry reveals gray matter abnormalities in migraine. Headache. 2008;48(1):109–17.
18. Kim JH, Suh SI, Seol HY, et al. Regional grey matter changes in patients with migraine: a voxel-based morphometry study. Cephalalgia. 2008;28:598–604.
19. Jin C, Yuan K, Zhao L, et al. Structural and functional abnormalities in migraine patients without aura. NMR Biomed. 2013;26(1):58–64.
20. Stankewitz A, Valet M, Schulz E. Pain sensitizers exhibit grey matter changes after repetitive pain exposure: a longitudinal voxel-based morphometry study. Pain. 2013;154(9):1732–7.
21. Messina R, Rocca MA, Colombo B, et al. White matter microstructure abnormalities in pediatric migraine patients. Cephalalgia. 2015.
22. Tessitore A, Russo A, Giordano A, et al. Disrupted default mode network connectivity in migraine without aura. J Headache Pain. 2013;8(14):89.
23. Tessitore A, Russo A, Conte F, et al. Abnormal connectivity within executive resting-state network in migraine with Aura. Headache. 2015;55(6):794–805.
24. Tedeschi G, Russo A, Conte F, et al. Increased interictal visual network connectivity in patients with migraine with aura. Cephalalgia. 2015.
25. Hu W, Guo J, Chen N, et al. A meta-analysis of voxel-based morphometric studies on migraine. Int J Clin Exp Med. 2015;8(3):4311–9.
26. Granziera C, DaSilva AF, Snyder J, et al. Anatomical alterations of the visual motion processing network in migraine with and without aura. PLoS Med. 2006;3(10):e402.
27. DaSilva AF, Granziera C, Snyder J, et al. Thickening in the somatosensory cortex of patients with migraine. Neurology. 2007;69(21):1990–5.
28. Hadjikhani N, Sanchez del Rio M, Wu O, et al. Mechanisms of migraine aura revealed by functional MRI in human visual cortex. Proc Natl Acad Sci U S A. 2001;98(8):4687–92.
29. Kim JH, Kim JB, Suh SI, et al. Thickening of the somatosensory cortex in migraine without aura. Cephalalgia. 2014;34(14):1125–33.
30. Hougaard A, Amin FM, Hoffmann MB, et al. Structural gray matter abnormalities in migraine relate to headache lateralization, but not aura. Cephalalgia. 2015;35(1):3–9.
31. Maleki N, Barmettler G, Moulton EA, et al. Female migraineurs show lack of insular thinning with age. Pain. 2015;156(7):1232–9.
32. Chong CD, Dodick DW, Schlaggar BL, et al. Atypical age-related cortical thinning in episodic migraine. Cephalalgia. 2014;34(14):1115–24.
33. Schwedt TJ, Chong CD. Correlations between brain cortical thickness and cutaneous pain thresholds are atypical in adults with migraine. PLoS One. 2014;9(6):e99791.
34. Erpelding N, Moayedi M, Davis KD. Cortical thickness correlates of pain and temperature sensitivity. Pain. 2012;153:1602–9.
35. Schwedt TJ, Berisha V, Chong CD. Temporal lobe cortical thickness correlations differentiate the migraine brain from the healthy brain. PLoS One. 2015;10(2):e0116687.
36. Rakic P. Specification of cerebral cortical areas. Science. 1988;241(4862):170–6.
37. Frye RE, Liederman J, Malmberg B, et al. Surface area accounts for the relation of gray matter volume to reading-related skills and history of dyslexia. Cereb Cortex. 2010;20(11):2625–35.
38. Kapellou O, Counsell SJ, Kennea N, et al. Abnormal cortical development after premature birth shown by altered allometric scaling of brain growth. PLoS Med. 2006;3(8):e265.
39. Datta R, Detre JA, Aguirre GK, et al. Absence of changes in cortical thickness in patients with migraine. Cephalalgia. 2011;31(14):1452–8.
40. Messina R, Rocca MA, Colombo B, et al. Cortical abnormalities in patients with migraine: a surface-based analysis. Radiology. 2013;268(1):170–80.
41. Granziera C, Daducci A, Romascano D, et al. Structural abnormalities in the thalamus of migraineurs with aura: a multiparametric study at 3 T. Hum Brain Mapp. 2014;35(4):1461–8.
42. Maleki N, Becerra L, Brawn J, et al. Common hippocampal structural and functional changes in migraine. Brain Struct Funct. 2013;218(4):903–12.
43. Schmidt-Wilcke T, Hierlmeier S, Leinisch E. Altered regional brain morphology in patients with chronic facial pain. Headache. 2010;50(8):1278–85.
44. Obermann M, Nebel K, Schumann C, et al. Gray matter changes related to chronic posttraumatic headache. Neurology. 2009;73(12):978–83.
45. Burgmer M, Gaubitz M, Konrad C, et al. Decreased gray matter volumes in the cingulo-frontal cortex and the amygdala in patients with fibromyalgia. Psychosom Med. 2009;71(5):566–73.
46. Headache Classification Committee of the International Headache Society (IHS). The international classification of headache disorders. 3rd edn (beta version). Cephalalgia. 2013;33(9):629–808.
47. Swartz RH, Kern RZ. Migraine is associated with magnetic resonance imaging white matter abnormalities: a meta-analysis. Arch Neurol. 2004;61:1366–8.
48. Honningsvåg LM, Hagen K, Håberg A, et al. Intracranial abnormalities and headache: a population-based imaging study (HUNT MRI). Cephalalgia. 2015.
49. Seneviratne U, Chong W, Billimoria PH. Brain white matter hyperintensities in migraine: clinical and radiological correlates. Clin Neurol Neurosurg. 2013;115(7):1040–3.
50. Dinia L, Bonzano L, Albano B, et al. White matter lesions progression in migraine with aura: a clinical and MRI longitudinal study. J Neuroimag. 2013;23:47–52.
51. Trauninger A, Leé l-Ossy E, Kamson DO, et al. Risk factors of migraine-related brain white matter

hyperintensities: an investigation of 186 patients. J Headache Pain. 2011;12:97–103.
52. Kruit MC, Van Buchem MA, Launer LJ, et al. Migraine is associated with an increased risk of deep white matter lesions, subclinical posterior circulation infarcts and brain iron accumulation: the population-based MRI CAMERA study. Cephalalgia. 2010;30:129–36.
53. Evans RW. Clinical neuroimaging of migraine. In: Borsook D, May A, Goadsby P, Hargreaves R, editors. The migraine brain. New York: Oxford University Press; 2012. p. 85–95.
54. Erdélyi-Bótor S, Aradi M, Kamson DO, et al. Changes of migraine-related white matter hyperintensities after 3 years: a longitudinal MRI study. Headache. 2015;55(1):55–70.
55. Palm-Meinders IH, Koppen H, Terwindt GM, et al. Structural brain changes in migraine. JAMA. 2012;308:1889–97.
56. Hamedani AG, Rose KM, Peterlin BL, et al. Migraine and white matter hyperintensities: the ARIC MRI study. Neurology. 2013;81(15):1308–13.
57. Aradi M, Schwarcz A, Perlaki G, et al. Quantitative MRI studies of chronic brain white matter hyperintensities in migraine patients. Headache. 2013;53(5):752–63.
58. Kurth T, Diener H-C. Migraine and stroke: perspectives for stroke physicians. Stroke. 2012;43:3421–6.
59. Bashir A, Lipton RB, Ashina S, et al. Migraine and structural changes in the brain: a systematic review and meta-analysis. Neurology. 2013;81(14):1260–8.
60. Friedman DI, Dodick DW. White matter hyperintensities in migraine: reason for optimism. JAMA. 2012;308:1920–1.
61. Gupta VK. White matter hyperintensities: pearls and pitfalls in interpretation of MRI abnormalities. Stroke. 2004;35(10):2239–41.
62. Aytaç B, Coşkun Ö, Alioğlu B, et al. Decreased antioxidant status in migraine patients with brain white matter hyperintensities. Neurol Sci. 2014;35(12):1925–9.
63. Pusic AD, Mitchell HM, Kunkler PE, et al. Spreading depression transiently disrupts myelin via interferon-gamma signaling. Exp Neurol. 2015;264:43–54.
64. Toghae M, Rahimian E, Abdollahi M, et al. The prevalence of magnetic resonance imaging hyperintensity in migraine patients and its association with migraine headache characteristics and cardiovascular risk factors. Oman Med J. 2015;30(3):203–7.
65. Hougaard A, Amin FM, Ashina M. Migraine and structural abnormalities in the brain. Curr Opin Neurol. 2014;27(3):309–14.
66. Sacco S, Kurth T. Migraine and the risk for stroke and cardiovascular disease. Curr Cardiol Rep. 2014;16(9):524.
67. Husøy AK, Indergaard MK, Honningsvåg LM, et al. Perivascular spaces and headache: a population-based imaging study (HUNT-MRI). Cephalalgia. 2015.
68. Basser PJ, Pierpaoli C. Microstructural and physiological features of tissues elucidated by quantitative-diffusion-tensor MRI. J Magn Reson B. 1996;111:209–19.
69. Alexander AL, Lee JE, Lazar M, Field AS. Diffusion tensor imaging of the brain. Neurotherapeutics. 2007;4:316–29.
70. Rocca MA, Colombo B, Inglese M, et al. A diffusion tensor magnetic resonance imaging study of brain tissue from patients with migraine. J Neurol Neurosurg Psychiatry. 2003;74(4):501–3.
71. DaSilva AF, Granziera C, Tuch DS, et al. Interictal alterations of the trigeminal somatosensory pathway and periaqueductal gray matter in migraine. NeuroReport. 2007;18(4):301–5.
72. Li XL, Fang YN, Gao QC, et al. A diffusion tensor magnetic resonance imaging study of corpus callosum from adult patients with migraine complicated with depressive/anxious disorder. Headache. 2011;51(2):237–45.
73. Smith SM. Fast robust automated brain extraction. Hum Brain Mapp. 2002;17:143–55.
74. Yuan K, Qin W, Liu P, et al. Reduced fractional anisotropy of corpus callosum modulates inter-hemispheric resting state functional connectivity in migraine patients without aura. PLoS One. 2012;7(9):e45476.
75. Yu D, Yuan K, Qin W, et al. Axonal loss of white matter in migraine without aura: a tract-based spatial statistics study. Cephalalgia. 2013;33(1):34–42.
76. Kara B, Kiyat Atamer A, Onat L, et al. DTI findings during spontaneous migraine attacks. Clin Neuroradiol. 2013;23(1):31–6.
77. Coppola G, Tinelli E, Lepre C, et al. Dynamic changes in thalamic microstructure of migraine without aura patients: a diffusion tensor magnetic resonance imaging study. Eur J Neurol. 2014;21(2):287-e13.
78. Chong CD, Schwedt TJ. Migraine affects white-matter tract integrity: a diffusion-tensor imaging study. Cephalalgia. 2015.
79. Rocca MA, Pagani E, Colombo B, et al. Selective diffusion changes of the visual pathways in patients with migraine: a 3-T tractography study. Cephalalgia. 2008;28(10):1061–8.
80. Neeb L, Bastian K, Villringer K, et al. No microstructural white matter alterations in chronic and episodic migraineurs: a case-control diffusion tensor magnetic resonance imaging study. Headache. 2015;55(2):241–51.
81. Weiller C, May A, Limmroth V, et al. Brainstem activation in spontaneous human migraine attacks. Nat Med. 1995;1(7):658–60.
82. Afridi SK, Giffin NJ, Kaube H, et al. A positron emission tomographic study in spontaneous migraine. Arch Neurol. 2005;62(8):1270–5.
83. Denuelle M, Fabre N, Payoux P, Chollet F, Geraud G. Hypothalamic activation in spontaneous migraine attacks. Headache. 2007;47(10):1418–26.

84. Russo A, Tessitore A, Giordano A, et al. The pain in migraine beyond the pain of migraine. Neurol. Sci. 2012;33(Suppl 1):S103–6.
85. Aurora SK, Barrodale PM, Tipton RL, et al. Brainstem dysfunction in chronic migraine as evidenced by neurophysiological and positron emission tomography studies. Headache. 2007;47 (7):996–1003; discussion 1004.
86. Pavlovic JM, Buse DC, Sollars CM, et al. Trigger factors and premonitory features of migraine attacks: summary of studies. Headache. 2014;54 (10):1670–9.
87. Maniyar FH, Sprenger T, Monteith T, et al. Brain activations in the premonitory phase of nitroglycerin-triggered migraine attacks. Brain. 2014;137(Pt 1):232–41.
88. Maniyar FH, Sprenger T, Schankin C, et al. The origin of nausea in migraine-a PET study. J Headache Pain. 2014;3(15):84.
89. Shin JH, Kim YK, Kim HJ, Kim JS. Altered brain metabolism in vestibular migraine: comparison of interictal and ictal findings. Cephalalgia. 2014;34 (1):58–67.
90. Aurora SK, Barrodale PM, Tipton RL, et al. Brainstem dysfunction in chronic migraine as evidenced by neurophysiological and positron emission tomography studies. Headache. 2007;47 (7):996–1003.
91. Chabriat H, Tehindrazanarivelo A, Vera P, et al. 5HT2 receptors in cerebral cortex of migraineurs studied using PET and 18F-fluorosetoperone. Cephalalgia. 1995;15(2):104–8.
92. Chugani DC, Niimura K, Chaturvedi S, et al. Increased brain serotonin synthesis in migraine. Neurology. 1999;53(7):1473–9.
93. Demarquay G, Lothe A, Royet JP, et al. Brainstem changes in 5-HT1A receptor availability during migraine attack. Cephalalgia. 2011;31(1):84–94.
94. Okazawa H, Tsuchida T, Pagani M, et al. Effects of 5-HT1B/1D receptor agonist rizatriptan on cerebral blood low and blood volume in normal circulation. J Cereb Blood Flow Metab. 2006;26(1):92–8.
95. Wall A, Kågedal M, Bergström M, et al. Distribution of zolmitriptan into the CNS in healthy volunteers: a positron emission tomography study. Drugs R&D. 2005;6(3):139–47.
96. Nascimento TD, DosSantos MF, Lucas S, et al. μ-Opioid activation in the midbrain during migraine allodynia—brief report II. Ann Clin Transl Neurol. 2014; 1(6):445–50.
97. DaSilva AF, Nascimento TD, DosSantos MF, et al. Association of μ-opioid activation in the prefrontal cortex with spontaneous migraine attacks—brief report I. Ann Clin Transl Neurol. 2014;1(6):439–44.
98. DaSilva AF, Nascimento TD, Love T, et al. 3D-neuronavigation in vivo through a patient's brain during a spontaneous migraine headache. J Vis Exp. 2014;(88).
99. Leao AAP. Spreading depression of activity in cerebral cortex. J Neurophysiol. 1944;7:359–90.
100. Smith JM, Bradley DP, James MF, Huang CL. Physiological studies of cortical spreading depression. Biol Rev Camb Philos Soc. 2006;81(4):457–81.
101. Olesen J, Friberg L, Olsen TS, et al. Timing and topography of cerebral blood flow, aura, and headache during migraine attacks. Ann Neurol. 1990;28(6):791–8.
102. Ayata C. Cortical spreading depression triggers migraine attack: pro. Headache. 2010;50(4):725–30.
103. Sprenger T, Goadsby PJ. What has functional neuroimaging done for primary headache and for the clinical neurologist? J Clin Neurosci. 2010;17 (5):547–53.
104. Russo A, Tessitore A, Tedeschi G. Migraine and trigeminal system-I can feel it coming…. Curr Pain Headache Rep. 2013;17(10):367.
105. May A, Kaube H, Büchel C, et al. Experimental cranial pain elicited by capsaicin: a PET study. Pain. 1998;74(1):61–6.
106. Moulton EA, Burstein R, Tully S, et al. Interictal dysfunction of a brainstem descending modulatory center in migraine patients. PLoS One. 2008;3: e3799.
107. Moulton EA, Becerra L, Maleki N, et al. Painful heat reveals hyperexcitability of the temporal pole in interictal and ictal migraine states. Cereb Cortex. 2011;21:435–48.
108. Russo A, Tessitore A, Esposito F, et al. Pain processing in patients with migraine: an event-related fMRI study during trigeminal nociceptive stimulation. J Neurol. 2012;259:1903–12.
109. Stankewitz A, Voit HL, Bingel U, Peschke C, May A. A new trigemino-nociceptive stimulation model for event-related fMRI. Cephalalgia. 2010;30:475–85.
110. Aderjan D, Stankewitz A, May A. Neuronal mechanisms during repetitive trigemino-nociceptive stimulation in migraine patients. Pain. 2010;151:97–103.
111. Stankewitz A, Aderjan D, Eippert F, May A. Trigeminal nociceptive transmission in migraineurs predicts migraine attacks. J Neurosci. 2011;31:1937–43.
112. Schwedt TJ, Chong CD, Chiang CC, et al. Enhanced pain-induced activity of pain-processing regions in a case-control study of episodic migraine. Cephalalgia. 2014;34(12):947–58.
113. Maizels M, Aurora S, Heinricher M. Beyond neurovascular: migraine as a dysfunctional neurolimbic pain network. Headache. 2012;52 (10):1553–65.
114. Main A, Dowson A, Gross M. Photophobia and phonophobia in migraineurs between attacks. Headache. 1997;37:492–5.
115. Vanagaite J, Pareja JA, Storen O, et al. Light-induced discomfort and pain in migraine. Cephalalgia. 1997;17:733–41.
116. Wober-Bingol C, Wober C, Karwautz A, et al. Clinical features of migraine: a cross-sectional study

in patients aged three to sixty-nine. Cephalalgia. 2004;24:12–7.
117. Cao Y, Welch KM, Aurora S, et al. Functional MRI-BOLD of visually triggered headache in patients with migraine. Arch Neurol. 1999;56 (5):548–54.
118. Boulloche N, Denuelle M, Payoux P, Fabre N, Trotter Y, Géraud G. Photophobia in migraine: an interictal PET study of cortical hyperexcitability and its modulation by pain. J Neurol Neurosurg Psychiatry. 2010;81(9):978–84.
119. Cao Y, Aurora SK, Nagesh V, et al. Functional MRI-BOLD of brainstem structures during visually triggered migraine. Neurology. 2002;59(1):72–8.
120. Denuelle M, Boulloche N, Payoux P, et al. A PET study of photophobia during spontaneous migraine attacks. Neurology. 2011;76(3):213–8.
121. Sarchielli P, Tarducci R, Presciutti O, et al. Functional 1H-MRS findings in migraine patients with and without aura assessed interictally. Neuroimage. 2005;24(4):1025–31.
122. Martín H, Sánchez del Río M, de Silanes CL, et al. Photoreactivity of the occipital cortex measured by functional magnetic resonance imaging-blood oxygenation level dependent in migraine patients and healthy volunteers: pathophysiological implications. Headache. 2011;51(10):1520–8.
123. Antal A, Polania R, Saller K, et al. Differential activation of the middle-temporal complex to visual stimulation in migraineurs. Cephalalgia. 2011;31 (3):338–45.
124. Datta R, Aguirre GK, Hu S, et al. Interictal cortical hyperresponsiveness in migraine is directly related to the presence of aura. Cephalalgia. 2013;33:365–74.
125. Griebe M, Flux F, Wolf ME, et al. Multimodal assessment of optokinetic visual stimulation response in migraine with aura. Headache. 2014;54(1):131–41.
126. Hougaard A, Amin FM, Hoffmann MB, et al. Interhemispheric differences of fMRI responses to visual stimuli in patients with side-fixed migraine aura. Hum Brain Mapp. 2014;35(6):2714–23.
127. Zanchin G, Dainese F, Trucco M, et al. Osmophobia in migraine and tension-type headache and its clinical features in patients with migraine. Cephalalgia. 2007;27(9):1061–8.
128. De Carlo D, Toldo I, Dal Zotto L, et al. Osmophobia as an early marker of migraine: a follow-up study in juvenile patients. Cephalalgia. 2012;32 (5):401–6.
129. Demarquay G, Royet JP, Mick G, et al. Olfactory hypersensitivity in migraineurs: a H(2)(15)O-PET study. Cephalalgia. 2008;28(10):1069–80.
130. Stankewitz A, May A. Increased limbic and brainstem activity during migraine attacks following olfactory stimulation. Neurology. 2011;77(5):476–82.
131. Russo A, Marcelli V, Esposito F, et al. Abnormal thalamic function in patients with vestibular migraine. Neurology. 2014;82(23):2120–6.
132. Borsook D, Burstein R. The enigma of the dorsolateral pons as a migraine generator. Cephalalgia. 2012;32(11):803–12.
133. Espinosa-Sanchez JM, Lopez-Escamez JA. New insights into pathophysiology of vestibular migraine. Front Neurol. 2015;6(6):12.
134. Montagna P, Cortelli P, Monari L, et al. 31P-magnetic resonance spectroscopy in migraine without aura. Neurology. 1994;44(4):666–9.
135. Barbiroli B, Montagna P, Cortelli P, et al. Abnormal brain and muscle energy metabolism shown by 31P magnetic resonance spectroscopy in patients affected by migraine with aura. Neurology. 1992;42(6):1209–14.
136. Kim JH, Kim S, Suh SI, et al. Interictal metabolic changes in episodic migraine: a voxel-based FDG-PET study. Cephalalgia. 2010;30(1):53–61.
137. Colombo B, Rocca MA, Messina R, et al. Resting-state fMRI functional connectivity: a new perspective to evaluate pain modulation in migraine? Neurol Sci. 2015;36(Suppl 1):41–5.
138. Mainero C, Boshyan J, Hadjikhani N. Altered functional magnetic resonance imaging resting-state connectivity in periaqueductal gray networks in migraine. Ann Neurol. 2011;70(5):838–45.
139. Yu DH, Yuan K, Zhao L, et al. Regional homogeneity abnormalities in patients with interictal migraine without aura: a resting-state study. NMR Biomed. 2012;25(5):806–12.
140. Russo A, Tessitore A, Giordano A, et al. Executive resting-state network connectivity in migraine without aura. Cephalalgia. 2012;32(14):1041–8.
141. Maleki N, Becerra L, Borsook D. Migraine: maladaptive brain responses to stress. Headache. 2012;52(Suppl 2):102–6.
142. Borsook D, Maleki N, Becerra L, et al. Understanding migraine through the lens of maladaptive stress responses: a model disease of allostatic load. Neuron. 2012;73(2):219–34.
143. Aurora SK, Wilkinson F. The brain is hyperexcitable in migraine. Cephalalgia. 2007;27:1442–53.
144. Niddam DM, Lai KL, Fuh JL, et al. Reduced functional connectivity between salience and visual networks in migraine with aura. Cephalalgia. 2015.
145. Hougaard A, Amin FM, Magon S, et al. No abnormalities of intrinsic brain connectivity in the interictal phase of migraine with aura. Eur J Neurol. 2015;22(4):702-e46.
146. Zhao L, Liu J, Yan X, et al. Abnormal brain activity changes in patients with migraine: a short-term longitudinal study. J Clin Neurol. 2014;10(3):229–35.
147. Zhao L, Liu J, Dong X, et al. Alterations in regional homogeneity assessed by fMRI in patients with migraine without aura stratified by disease duration. J Headache Pain. 2013;17(14):85.
148. Moulton EA, Becerra L, Johnson A, et al. Altered hypothalamic functional connectivity with autonomic circuits and the locus coeruleus in migraine. PLoS One. 2014;9(4):e95508.

149. Hadjikhani N, Ward N, Boshyan J, et al. The missing link: enhanced functional connectivity between amygdala and visceroceptive cortex in migraine. Cephalalgia. 2013;33(15):1264–8.

150. Yuan K, Zhao L, Cheng P, et al. Altered structure and resting-state functional connectivity of the basal ganglia in migraine patients without aura. J Pain. 2013;14(8):836–44.

Author Biographies

Antonio Russo graduated in Medicine and Surgery in 2003 and specialized in Neurology in 2009 at the Second University of Naples (SUN). In 2013, he earned his Ph.D. in Experimental Physiopathology and Neurosciences at the Second University of Naples under the supervision of Prof. Gioacchino Tedeschi. He became assistant professor of Neurology at SUN in 2014. His researches are focused on the study of the migraine pathophysiology by means of advanced neuroimaging techniques.

Alessandro Tessitore earned his medical degree from the Second University of Naples (SUN). In November 2004, he completed his neurology training with honours and then undertook a visiting fellowship at the Clinical Brain Disorders Branch at the National Institutes of Health, Bethesda (USA) under the supervision of Professor Daniel Weinberger. In 2008, he earned his Ph.D. in Neuroscience at the SUN under the supervision of Prof. Gioacchino Tedeschi. He has co-authored approximately 65 peer-reviewed articles. He is currently Associate Professor of Neurology at the University of Campania "Luigi Vanvitelli", Naples.

Gioacchino Tedeschi graduated in Medicine and Surgery in 1977 and specialized in Neurology in 1981 at the Second University of Naples (SUN). Since then, he has been Research Fellow at Synthelabo in Paris (for 2 years), Assistant Professor at the University of Galles-UK (for 2 years) and Visiting Scientist at the National Institutes of Health, Bethesda-USA (for 4 years). He became professor of Neurology at SUN in 2000 and since then he is Chief of the Department of Neurology. His past research interests covered clinical pharmacology, epilepsy and eye movements. For the last few years, his researches are concentrated on the study of the CNS neural networks in a number of disorders, including headache. He is author of more than 250 papers, many of whom have appeared in the most prestigious neurological journals.

Rheumatic Pain

11

Debbie L. Morton and Anthony K.P. Jones

Abstract

Rheumatic pain describes pain of the joints and their connective tissues which is commonly associated with osteoarthritis (OA), which is a degenerative disorder of the joints and rheumatoid arthritis (RA), which is a systemic inflammatory disease. Until relatively recently it was assumed that the patient's pain was solely derived from peripheral mechanisms in affected joints. However, evidence now shows that a poor relationship exists between radiographic evidence of joint damage and pain (Bedson and Croft in BMC Musculoskelet Disord 9:116, [12]) with many patients reporting pain that does not correspond with the extent of joint pathology and pain that occurs adjacent to or at sites without tissue damage (Kean et al. in Inflammopharmacology 12(1):3–31, [94]; Gwilym et al. in J Bone Joint Surg Br 90(3):280–287, [66]). Treating the problem surgically by replacing the damaged joint does not always alleviate pain (Wylde et al. in Pain 152 (3):566–572, [186]) whilst in some patients sham surgery (placebo) does (Moseley et al. in N Engl J Med 347(2):81–88, [121]). The extent of the pain experienced by rheumatic pain patients can be highly variable and in some cases there may be pain and tenderness present over much of the body [chronic widespread pain or fibromyalgia (FM)] in association with other somatic symptom pain disorders. Before non-invasive neuroimaging was available neuroscientists had to rely on studies of patients with brain or spinal cord lesions, or patients responses during neurosurgical procedures to understand the neural basis of human pain perception. Functional brain neuroimaging comprises a number of non-invasive brain imaging techniques which have improved our understanding of the underlying mechanisms involved in acute and chronic pain and pain therapy. These

D.L. Morton · A.K.P. Jones (✉)
Human Pain Research Group, Institute of Brain, Behaviour and Mental Health, University of Manchester, Salford, UK
e-mail: anthony.jones@manchester.ac.uk

L. Saba (ed.), *Neuroimaging of Pain*, DOI 10.1007/978-3-319-48046-6_11

techniques are now beginning to influence the development of future treatments for pain.

Keywords
Neuroimaging · Rheumatic pain · EEG · fMRI · PET

1 Introduction

Significant advances in functional and structural neuroimaging techniques [166] mean that the discrepancies between clinical pathology and pain self-report are now beginning to be understood [66]. Neuroimaging allows for objective measurement of the cortical and sub-cortical processes that contribute to the experience of pain. Neuroimaging methods can capture brain activity non-invasively and permit the study of the cerebral processing of pain without interfering with neurophysiological processing. Advances in functional neuroimaging over the past two and a half decades have led to the definition of the neural substrates of human pain processing. Pain processing is reflected in a matrix of neuronal structures incorporating motor function, attention, emotional and sensory processing [43] with the pattern of response strongly influenced by the psychological context of pain [41]. Identifying differences in cognitive and emotional processing of pain has been a major focus of neuroimaging studies of patients with chronic pain compared in healthy volunteers over the last 25 years. The areas of the brain that consistently respond to noxious stimuli include the thalamus, brainstem and amygdala, the insula cortex (IC), primary and secondary somatosensory cortices (SI and SII), the anterior cingulate cortex (ACC) and the prefrontal cortex (PFC) [4, 131, 162, 165, 168]. These are all areas which are believed to play an important role in the sensory-discriminative and cognitive aspects of pain processing. The network of brain areas correlating with pain is dynamic [4, 79, 100, 105, 166] and is dependent on variables such as the type and duration of pain stimulus [100, 128, 152], the type of patients studied [10], the psychological context of the pain and the imaging method used.

The possible function of the different areas of the pain matrix in normal volunteers has been well reviewed elsewhere [90, 114, 166] and will not be detailed here. However, in general terms the pain matrix is activated whatever the type of pain whether clinical or experimental. The pattern of response is as strongly (sometimes more so) influenced by the psychological and cognitive context of the pain (top-down influences) as the sensory-discriminative components (bottom-up), as will be illustrated in the following sections.

The focus of this chapter is to summarise the current knowledge of brain mechanisms in rheumatic pain as revealed by neuroimaging with a focus on RA, OA and FM. Whilst chronic widespread pain and FM are often treated as though they are separate from other rheumatic disorders, there is no physiological reason for this and in this chapter they will be considered as part of a range of musculoskeletal pain disorders.

2 Methods of Neuroimaging

There are a number of different methods of neuroimaging used in the study of pain which we will discuss here. One of the most commonly used imaging methods in both clinical and experimental pain research is functional magnetic resonance imaging (fMRI). fMRI is based on the magnetisation difference between oxygenated blood travelling to active cells and the resultant deoxygenated blood (blood oxygen level dependant (BOLD) response). As such it is an indirect, non-invasive way to measure neuronal activity using cerebral blood flow (CBF) in response to a stimulus. The BOLD technique whilst ideal for investigating experimental pain is not always suited for the study of chronic pain. For the ongoing pain as seen in chronic pain, arterial spin labelling (ASL) perfusion contrast scanning is more appropriate. ASL uses magnetically labelled arterial blood water to create an image of changes in perfusion in the area of interest. This is then compared to a control image taken without magnetically labelling the arterial blood to create a perfusion image reflecting the amount of arterial blood delivered to each voxel within the area of interest within the transit time [45]. ASL has been used to show changes in regional CBF in brain regions previously associated with pain perception [126].

Structural MRI can be useful to assess brain plasticity due to chronic pain and its treatments [115] by providing detailed information about the grey and white matter of the brain and spinal cord. Using voxel-based morphometry (VBM) to analyse structural MRI data, changes in volume of brain tissue, often seen in chronic pain, can be measured. Another MRI based technique is diffusion tensor imaging (DTI). DTI uses the diffusion of water in biological tissues to create patterns that can characterise microstructural changes in the brain. Water diffuses more rapidly in the direction aligned with the internal structure and so DTI can elucidate the orientation of white matter tracts as well as imaging functionally localised brain regions to increase our understanding of brain networks [2, 36].

PET is used to image the changes in metabolism or chemical events at receptor and neurotransmitter reuptake sites in living tissues including the brain. Using radionucleotide labelling, PET is able to measure changes in regional CBF (rCBF), blood volume, and oxygen uptake and glucose metabolism (i.e. FDG-PET). Both rCBF and glucose metabolism allow for the indirect measurement of neuronal activity in response to a painful stimulus. rCBF can be measured directly by assessing changes in uptake of $H_2^{15}O$, inhaled $C^{15}O_2$, or ^{15}O-butanol see [81]. For investigation of the neuro-anatomy of chronic ongoing neuropathic pain see [74] and for experimentally induced acute pain see [30, 87]. PET studies using radio-labelled ligands allow the evaluation of receptor occupancy and density [53, 104] and metabolism within certain neurochemical pathways such as the dopaminergic system. Studies using ^{11}C-carfentanil have shown that sustained pain triggers release of endogenous opioids in region specific manor and that a reduction in severity of pain correlates with increased occupation of mu-opioid receptors by endogenous opioids [188]. Studies using ^{11}C-diprenorphine have shown reduced density of opioid receptors mainly within the medial pain system in patients with central neuropathic pain due to post-stroke pain [91, 183] but not in conditions of peripheral neuropathic pain [111].

Whilst both fMRI and PET are techniques that offer high spatial resolution they are hindered by their performance in the temporal domain, whereas electroencephalography (EEG) has much better temporal resolution. The EEG signal represents the discharge from post-synaptic potentials of cell bodies and large dendrites of pyramidal cells in layers 3–5 of the neocortex which is measured using an array of electrodes on the scalp. The application of a brief painful stimulus (usually from a laser pulse) creates a time locked response called the laser evoked potential (LEP) which has been widely used to investigate pain processing in the brain. EEG provides millisecond accuracy in the timing of the pain-related neuronal event but suffers from a lack of spatial resolution meaning there is uncertainty regarding the source of the event. This can, to an extent, be overcome by combining EEG with other imaging techniques such as fMRI or using mathematical modelling to estimate source localisation (for a review see [62]).

3 Abnormalities of the Pain System Seen in Chronic Pain Patients

Neuroimaging has elucidated distinct networks of brain regions involved in pain processing commonly referred to as the 'pain matrix'. This includes somatosensory, insular, cingulate, prefrontal and inferior parietal cortices and as well

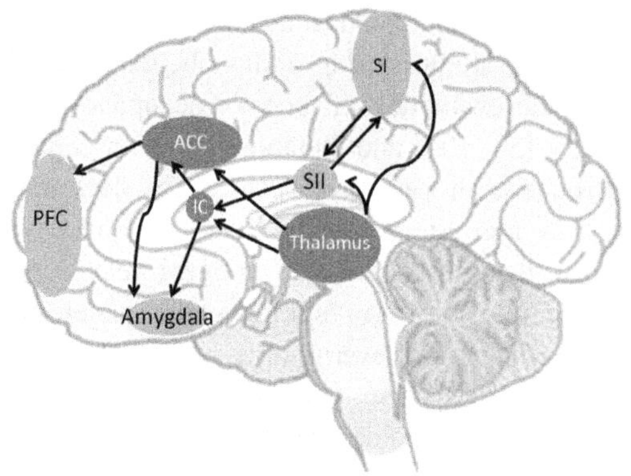

Fig. 1 A schematic representation of the pain matrix. In the pain matrix there are two complementary pathways through which pain processing takes place. The medial pathway (*dark grey*) projects from medial thalamus to the anterior cingulate (*ACC*) and insula cortex (*IC*) and processes the affective-motivational component of pain (i.e. unpleasantness). The lateral pathway (*light grey*) projects from the lateral thalamus to the primary and secondary (*SI* and *SII*) and insula cortex (*IC*) and processes the sensory-discriminative aspect of pain (i.e. location and intensity)

as the thalamus (Fig. 1). The magnitude of perceived pain can be predicted by the magnitude of the hemodynamic responses in SI, SII, the IC and the ACC using fMRI [16, 25] and PET [34, 43, 164]. Distraction from the painful stimulus can result in decreases in pain ratings and decreases in the magnitude of the elicited brain responses [29, 130, 132, 170]. Responses to pain are also dependant on the psychological context. For example, expectations of painful stimuli leads not only to increased magnitude of reported pain but also to increases in brain responses to the stimuli [134, 137, 167]. Being uncertain about expected pain intensity has been shown to involve a cortical network often associated with attention, whilst being relatively certain about pain intensity involves areas associated with semantic and prospective memory [23]. Altering the emotional context of the painful stimuli modulates the pattern of brain activity within the pain matrix [47, 101]. For example, Kulkarni et al. asked participants to concentrate on either location or unpleasantness of the painful stimuli which was maintained at the same intensity. They showed that these tasks significantly increased relative activity within the lateral and medial pain systems, suggesting that the two systems were mainly focused on sensory-discriminative and emotional aspects of processing respectively [101].

Apkarian et al. in their meta-analysis comparing acute pain in healthy subjects with acute pain in clinical pain patients, demonstrated that there were distinct but overlapping brain activation patterns during acute pain stimulation. Pain perception in healthy subjects was conveyed as afferent nociceptive information through the thalamus to S1, S2, IC and ACC. These are areas involved in the sensory (SI, S2) and emotional (IC and ACC) aspects of pain processing and were activated in 82% of healthy volunteer studies whilst in chronic pain patients the activation of these areas was seen only in 42% of studies. Activity of the PFC was increased in clinical pain conditions (81% in patients vs. 55% in healthy controls) [4]. The PFC is involved in processing the ascending nociceptive input from the spinothalamic tract but is also a source of descending modulation via its connections to the brainstems' descending pain modulatory system [176]. It is an area that plays an important part in interoception, cognition and the processing of negative emotions [48, 177] and is also involved in executive control of attention [23, 138]. The increased activation of the PFC and decreased activation of S1, S2 and the ACC seen in patients may imply that they have an increase in affective and cognitive processing of pain and a decrease in the sensory aspects of pain processing respectively. This may be a result of clinical pain having a stronger emotional value [139]. Another possible explanation could be the ongoing chronic pain experienced by patients may lead to more generalised changes which affect the baseline state. This change in baseline would lead to an altered response to evoked pain [4]. This explanation fits well with a study by Baliki et al. who reported changes in the default activity of areas of the brain known to be active at rest in chronic back pain patients compared to healthy controls [8, 9]. They found that patients brains were 'altered' by their persistent pain leading to them process non-pain-related information in a different manner to healthy controls. The authors suggest that the cognitive and behavioural changes associated with chronic pain may be due to disruptions of the default mode network (DMN), a network of brain regions active when the brain is a rest.

The manipulation of aspects of pain experience such as intensity vs. unpleasantness modulates the responses in specific sub-regions of the pain network or matrix [72, 141] known as the medial and lateral pain pathways [18]. The emotional aspects of pain such as pain unpleasantness, and cognitive responses to pain such as attention and anticipation are associated with activity within the medial pain pathway. The latter includes areas such as the medial nucleus of the thalamus and projects to the anterior insula and BA24 of the ACC [164, 101]. Sensory aspects of pain such as its location and duration, is the responsibility of the lateral pain pathway which incorporates SI and SII, parietal operculum (BA7b) and posterior insula [101, 136]. Repeated episodes of arthritic pain over long periods of time can lead to changes in behaviour and pain processing in the CNS which facilitate disability and reduce life quality. Chronic pain patients demonstrate a heightened affective-motivational response and many of the abnormalities of nociceptive processing observed during chronic pain have been in the medial pain system. During experimental and arthritic pain of the same intensity there is activation of both the medial and lateral pain pathways in OA patients. However during arthritic pain alone, there is increased activation of the medial system associated with increased pain unpleasantness. This suggests that arthritic pain has more emotional salience for OA patients than experimental pain [100]. In RA, the pain the patient reports cannot be explained by the pathology of the peripheral joint [163].

Using PET, Jones and Derbyshire investigated the affective-motivational brain response to an experimental pain stimulus in RA patients. There was a significant reduction of PFC and ACC activation and an additional decrease in insula and inferior parietal lobe seen in patients when compared to healthy controls [89]. It was suggested that the ongoing inflammatory pain observed in RA led to cortical adaptation in the central nociceptive system which may be related to patients coping and attentional strategies. Further studies in atypical facial pain, which is a type of somatoform pain disorder, showed reduced PFC responses to an acute pain stimulus compared to pain-free volunteers possibly related to abnormal coping [42].

The psychological state of the individual has significant bearing on prefrontal responses to pain. Brown et al. [21] showed that individuals with meditation experience perceived pain as less unpleasant than controls and during pain anticipation meditators showed a reversal of the normal positive correlation between medial PFC activity and pain unpleasantness. Both OA and FM patients had reduced activation of the dorsolateral prefrontal cortex (dlPFC) during the anticipation of a painful stimulus. This related to poorer psychological coping across both patient groups [20] and was associated with increased processing during anticipation in the IC. The activity seen in IC was correlated with the extent of pain and tenderness in both groups. Activity in the dlPFC negatively correlates with perceived pain intensity [109]. The dlPFC is associated with the top-down modulation of pain by the descending pain inhibitory system [19] and the modulation of pain due to placebo analgesia [98, 171, 174]. The dlPFC was one of the prefrontal cortical regions whose activity negatively correlated with pain catastrophising scores in healthy volunteers subjected to intense pain stimulation [154]. This suggests that individuals who display catastrophising behaviour may have difficulty disengaging from intense pain through a lack of top-down control. Using fMRI to investigate the neural circuitry of depressive mood on RA pain Schweinhardt et al. [152] linked depressive symptoms to increased cerebral processing of provoked joint pain in the mPFC, an area of the limbic system shown to mediate fear and anxiety during pain processing [125]. The increased activity of the mPFC correlated with scores on the Beck depression inventory (BDI) [11] and tender-to-swollen joint ratio ratings. This suggests that the mPFC may mediate depressive symptoms resulting in amplification of clinical pain severity. The degree of depressive symptoms in FM patients also correlates with the magnitude of pressure evoked pain responses in medial structures including the ACC and medial temporal lobe [58].

There is widespread activation of the cingulate cortex and greater activity in the amygdala, orbitofrontal cortex (OFC), and putamen during periods of OA pain compared to periods of experimental pain [65, 100, 128]. The amygdala, OFC and putamen are areas previously associated with aversive conditioning, reward and fear. The activation of these areas in OA implies fear of additional injury and disability by patients [3, 16]. In comparison with experimental pain, OA pain is also associated with increased activation of the PFC and the inferior posterior parietal cortex [100, 128]. These regions are associated with the supervision of attention [138] and it is hypothesised that descending fibres from the PFC inhibit neuronal coupling along the ascending midbrain-thalamic-cingulate pathway, modulating pain in a 'top-down' fashion [67, 107, 109]. These findings suggest that OA patients have enhanced affective pain processing. They may attach more emotional significance to their pain [100] and this is what may cause the variance in perceived pain that is seen in OA.

4 Central Sensitisation and Inhibition

A large number of patients with rheumatic pain also report feeling pain at other non-injured sites of the body, with localised pain (pain from only one site) being rare [93, 124, 133]. This implies that rheumatic pain patients may activate an alternative mechanism which is responsible for this widespread pain experience. Central sensitisation is one such candidate mechanism, defined as an 'amplification of neural signalling within the central nervous system that elicits pain hypersensitivity' [185]. Central sensitisation has two main characteristics, allodynia (pain in response to non-painful stimuli) and hyperalgesia (increased pain response to normally painful stimuli). The inflamed and damaged joints seen in RA are widely thought to result in peripheral sensitisation, but this cannot be the source of pain that is reported at sites without inflammation [147].

Meeus et al. [118] in their extensive literature review concluded that RA patients show many features symptomatic of central sensitisation such as hyperalgesia and this may account for the poor relationship between disease activity and symptoms. For example, EEG has revealed that RA patients have significantly larger pain-related potentials compared to controls when subjected to standardised, repetitive nociceptive stimuli of equal intensity, and significantly larger responses to the first stimulus in the chain. Differences were only seen between patients and controls when the interstimulus interval was less than 8 s. Additionally patients also habituated less to the painful stimuli than controls [76, 77, 175]. These results suggest that the chronic inflammatory joint pain seen in RA results in central sensitisation at least partly due to temporal summation of noxious stimuli leading to amplification of chronic pain.

Patients with symptomatic OA report diffuse alteration of pain perception in response to various stimuli with subjects having increased pain intensity and significantly larger referred and radiating pain areas than matched controls [6]. Localised changes in peripheral nociceptive activity are unlikely to account for these findings and point to the presence of enhanced central pain processing in OA. Other studies of OA patients have revealed additional features associated with central sensitisation such as secondary hyperalgesia [65, 80, 187]. For instance, OA, RA and FM patients have increased pain response to repeated pain stimulation with a small interstimulus-interval compared to healthy controls [5, 54, 76, 77]. fMRI has revealed that OA patients experience significantly more activation of the brainstem, ACC and dlPFC than controls during pressure pain stimulation. The ACC is thought to be a component of descending facilitation of pain [96] as well as descending inhibition and the increased activation seen in OA patients has been interpreted as reflecting hyperactive descending facilitation pathways which could also cause sensitisation of the CNS. In addition, activation of the periaqueductal grey (PAG) in patients was associated with stimulation of the skin in referred pain areas, highlighting the role of the PAG in central sensitisation [65].

Results from functional imaging studies have led to a growing consensus that the enhanced pain processing that is seen in FM is caused by central sensitisation. FM patients have reduced pain thresholds [37, 57, 61, 108, 161], which are unlikely to be due to peripheral sensitisation [86, 112, 146] as FM patients show no evidence of peripheral damage, inflammation or other abnormal changes of the tissue [159]. Additionally, FM patients experience widespread pain that is similar to secondary hyperalgesia [160] and an increase in pain intensity over time when a second non-painful stimulus is repeatedly delivered (wind up) [120, 158, 160]. As with OA, the causes of central sensitisation in FM may originate from enhanced descending facilitation of sensory inputs. An fMRI study of 29 FM patients undergoing experimental pressure stimulation found a positive correlation between catastrophising behaviour and pain-evoked activation of brain areas associated with anticipation of pain, attention to pain, emotional aspects of pain and motor control [60].

Pain catastrophising has been linked to hypervigilance to pain which is thought to contribute to enhanced descending facilitation of pain in FM [38]. fMRI has shown increased activity in the IC, ACC and basal ganglia in FM patients compared to controls even when subjective ratings for the painful stimulus were equal [140]. The insula is involved with affective pain processing and interoceptive awareness and the ACC is associated with affective pain processing [151]. These findings are therefore consistent with FM patients demonstrating hypervigilance and attaching more emotional significance to painful stimuli [60, 140].

Studies of OA patients show abnormalities in the descending inhibitory pathways consistent with a failure in diffuse noxious inhibitory control (DNIC) [5, 97]. DNIC is a process by which a painful stimulus is inhibited by a secondary painful stimulus and can be thought of as pain inhibiting pain. In healthy subjects, counter-irritation results in activation of DNIC and this can be measured by the modulation of the RIII component of the nociceptive long-loop flexion withdrawal reflex (LLFR). The RIII reflex is often, but not always, correlated with stimulus intensity and is regarded as an objective measure of descending influences on experimental pain in humans [103, 180–182]. Lautenbacher and Rollman's [102] study points to a failure of DNIC in FM as concurrent thermal and electrical stimulation increased pain thresholds in healthy volunteers but not in FM patients. In a later counter-irritation study the pain of an electrical stimulus was reduced in FM patients if they believed a concurrent distant cold stimulus would reduce their pain [59]. Conversely, the amplitude of the patients withdrawal reflex (LLFR) did not decrease during counter-irritation as would be expected with reduced pain reports but increased. This suggests that descending influences on spinal-brain stem reflexes is abnormal in FM. PET studies of FM patients often report reduced resting rCBF in the thalamus and basal ganglia, a finding which is also seen in neuropathic pain and cancer patients. It has been suggested that this may reflect reduced background inhibition of nociceptive processing [92]. Reduced thalamic activity may also reflect suppression of background pain signalling [179] but there is no direct evidence to implicate either mechanism. OA patients process their clinical pain differently to how they process experimental pain, with PET studies showing increased activation of the cingulate cortex, thalamus and amygdala; areas associated with fear and emotion [100]. The same may be true for FM although this has not been directly addressed and would be very difficult to accomplish experimentally. Most fMRI studies of FM use some kind of evoked pain usually involving either evoked pressure or thermal pain. fMRI findings regarding the function of pain modulatory areas such as the ACC and frontal cortical areas are often contradictory. For example, whilst Jensen et al. [82] found reduced activity of the ACC and thalamus in FM patients compared to controls Pujol et al. [140]

found increased ACC activity. fMRI studies that utilise frequent repetition of stimuli cannot unequivocally differentiate between actual pain processing and the anticipation of pain due to limitations of temporal resolution. The conflicting results seen may be more to do with the incomplete dissociation of these two processes. EEG is an ideal tool for separating pain anticipation from actual pain processing and has been used to image experimental pain in FM for this reason [57, 59, 108, 120, 161]. Brown et al. used EEG to look at the differences in pain anticipation between FM, OA and healthy controls. They found increased insula activity in FM compared to OA and controls, that correlated with extent of pain and tenderness, and decreased activation of the dlPFC in both FM and OA during pain anticipation compared to controls that correlated with poor coping (Figs. 2 and 3) [20]. Quantitative EEG (qEEG) has been used in FM to analyse patients' clinical pain. A study of 85 FM patients at rest demonstrated EEG abnormalities including reduced alpha, delta or theta power and increased beta or hypercoherence in patients when compared to healthy controls. Unfortunately it's difficult to draw conclusions as the control group were not tested in the same setting and the FM patients' pain severity was negatively correlated with degree of EEG abnormality [69]. Stevens et al. [161] have also found reduced alpha power in FM patients at rest. This is thought to be related to thalamic over-activity and may indicate enhanced afferent pain processing. In other chronic pain conditions studies have also shown reduced alpha power with increased theta [144, 145]. These findings are interesting in relation to the finding of increase frontal alpha power during placebo conditioning associated with expectation of analgesia [78].

Neuroimaging studies of OA, FM and RA patients suggest that altered neural processing, consistent with central sensitisation, does occur in rheumatic pain. Currently there is no direct evidence for these changes predisposing patients to chronic pain or maintaining their chronic pain. However, the correlation between the abnormalities in cognitive components of pain processing with the extent of clinical symptoms and signs flags these as potential mechanisms. The amenability of anticipatory responses to cognitive therapy highlights the potential to target these potential mechanisms for improved cognitive therapies and neurofeedback approaches to individualised treatment strategies.

The findings from these neuroimaging studies are particularly relevant to arthritis patients as they demonstrate the potential for distinguishing between different types of chronic musculoskeletal pain phenotypes within the spectrum of RA, OA and FM, on the basis of condition-specific neural pain signatures [27, 31]. The work by Brown et al. [20] underlines common brain mechanisms of OA and FM that may drive the extent of pain and tenderness. This suggests that perhaps we should be working towards an approach more geared to physiological phenotyping than to disease-related phenotyping which may be less relevant to both aetiology and the development of new therapies.

5 Structural Changes Associated with Rheumatic Pain

As we can see from the previous two sections, neural processing is modified by experience. There is evidence for morphological brain changes taking place in less than two weeks [56] as a result of changes in pain experience. The structural changes seen in the brains of rheumatic pain patients can be assessed by structural MRI voxel-based morphometry (VBM). In RA, OA and FM there is evidence for structural reorganisation which might relate to functional changes seen in the medial pain pathway and central sensitisation. Using VBM, Wartolowska et al. [172] found an increase in grey matter in the basal ganglia and smaller intracranial volume in RA patients compared to age-matched controls. As the basal ganglia are involved in motor control, aversive conditioning [156] pain processing and modulating behaviour in response to aversive stimuli, the changes seen here may occur as a result of altered motor control and/or prolonged pain processing. This hypothesis is supported by studies suggesting that a change in basal ganglia

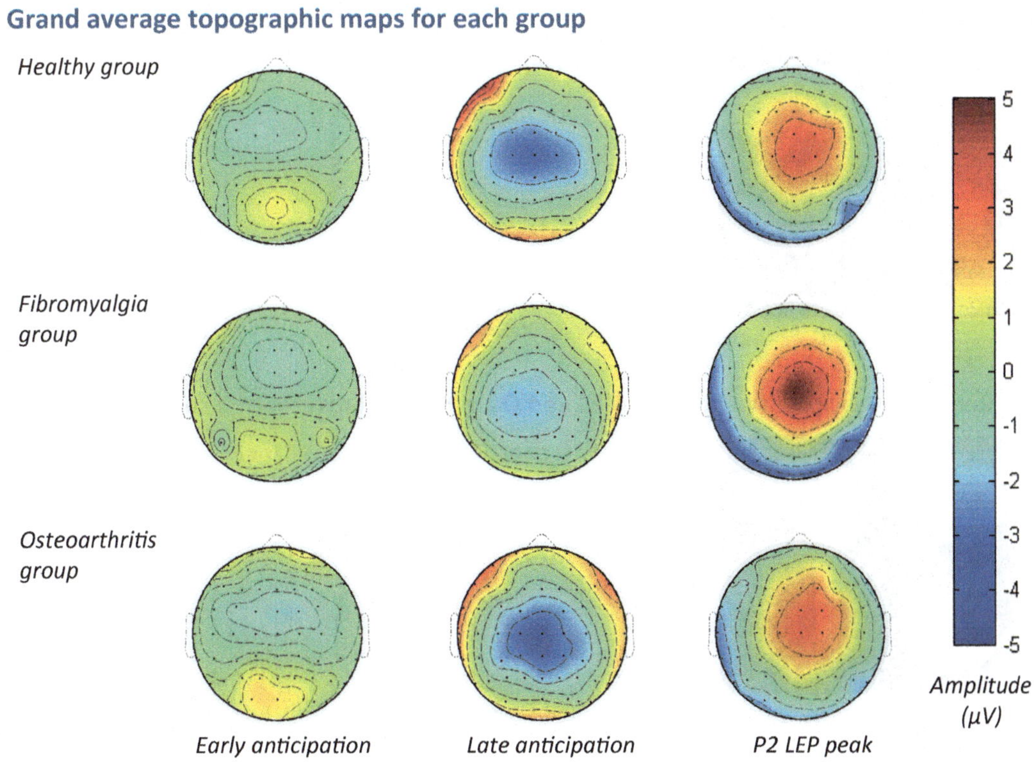

Fig. 2 **a** ERP topography plots. Topography plots represent the average of each group, with ERP data corrected to the pre-anticipation baseline. The anticipatory response had the highest amplitude in the OA group, and the lowest amplitude in the FM group, with HP group amplitudes being in between but nearer to the those in the OA group

connectivity may be important in the maintenance of chronic pain [17].

In chronic pain patients, grey matter volume loss occurs primarily in those regions related to pain and pain processing [115]. When OA patients are compared to healthy controls reductions in grey matter in the thalamus [64], insula, cingulate cortex, S1 and S2 are seen [10] with reduced grey matter in the thalamus positively correlating with pain severity and disability. The thalamus has previously been implicated in chronic pain conditions e.g. decreased thalamic blood flow contralateral to the site of pain is seen in cancer pain patients, suggesting that chronic pain may be associated with a decrease of synaptic activity in the thalamus [46]. Administering pain relief led to thalamic reperfusion in a study on a single non-cancer patient who also had reduction in thalamic perfusion contralateral to the site of pain [55]. A common mechanistic explanation for these observations may be a reduction in interneuronal inhibition in the thalamus that is restored by therapy.

FM patients also have reduced grey matter volume. Structural neuroimaging has pointed to grey matter volume reductions in the postcentral gyrus, amygdala, hippocampus, superior frontal gyrus, and ACC [110] and the PFC, amygdala and ACC [28]. Kuchinad et al. [99] found reductions of grey matter in the cingulate, insular and medial frontal cortices and parahippocampal gyri of female FM patients compared to controls. Interestingly, the longer the individual had suffered with FM, the greater was the loss of grey matter volume, with each year of FM being corresponding to 9.5 times the loss of grey matter seen with normal ageing. In another study using VBM, Schmidt-Wilcke et al. [148] found reduced grey matter in the right superior temporal gyrus, left posterior thalamus and increases

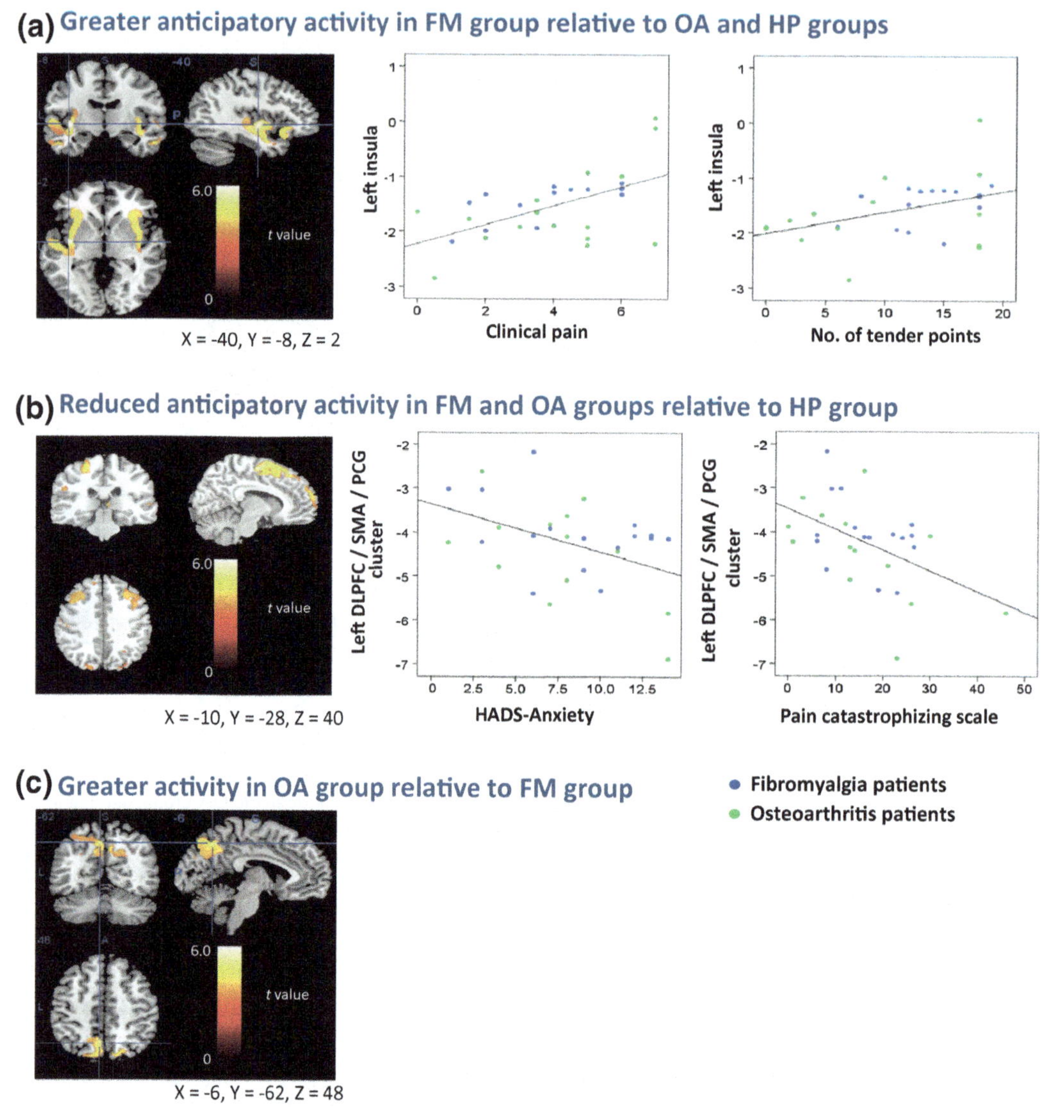

Fig. 3 ERP source estimates and their correlates during late anticipation (average activity over −500 to 0 ms). **a** Group effects on ERP sources, showing regions of greater activity in the FM group than in the other two groups [i.e. FM patients vs. each of the other two groups: (FM > OA) + (FM > HP)]. The fixation cross shows the most activated voxel. *y*-axis units represent the log-transformed current source density. Greater activations are seen in the FM group during late anticipation in the bilateral insula cortices, and in the right inferior temporal gyrus. **b** Group effects on ERP sources showing regions of reduced activity in the patient groups relative to HPs [i.e. HPs vs. each patient group: (HP > FM) + (HP > OA)]. Coordinates of the image are chosen to allow the visualisation of all activated clusters, and do not denote a region of activation. *y*-axis units represent the log-transformed current source density. Areas with significantly reduced activations in the patients during late anticipation, include the frontal and parietal brain regions consisting of the left (contralateral) postcentral gyrus, the left superior frontal gyrus (supplementary motor area), and the left middle frontal gyrus (dlPFC). **c** Greater activity in the OA group than in the FM group (OA > FM). The fixation cross shows the most activated voxel. *HP* Healthy participants; *HADS* Hospital anxiety and depression scale; *PCG* postcentral gyrus; *SMA* supplementary motor area. Reproduced with permission from Brown et al. [20]

in grey matter in the left OFC, left cerebellum and bilaterally in the striatum, areas known to be both part of the somatosensory system and motor system.

The structural changes observed in chronic pain may not be exclusively pain-specific. Lifestyle changes which result from living with pain, and pain associated mood disorders such as anxiety and depression may be a common factor contributing to the observed changes in grey matter [13, 117]. When affective mood disorders such as current depressive episode, generalised anxiety disorder, bipolar and dysthymia were accounted for, Hsu et al. found no difference between grey matter volume between FM patients without mood disorders and healthy controls. Grey matter volume loss, particularly in the insula was only seen in FM patients with affective mood disorders and was inversely correlated with trait anxiety [75]. There is substantial evidence to show that loss of grey matter volume is reversible when pain is successfully treated. In OA patients treated with surgery to repair their hip damage, increases in grey matter volume were seen in the thalamus [64]. In another study, patients with hip OA suffering from chronic pain exhibited grey matter decreases in the ACC, right insular cortex and operculum, dlPFC, amygdala, and brainstem when compared with healthy controls. The subgroup of patients who, after total hip replacement surgery, were completely pain-free showed grey matter increases in the ACC, dlPFC, amygdala, and brainstem at 6 weeks and 4 months after surgery [142]. These studies suggest that the changes in grey matter observed in chronic pain conditions area consequence of chronic pain rather than the cause of chronic pain. The exact significance of these findings is still not entirely clear.

6 Functional Responses Indicating 'Brain State'

Most of the functional imaging studies in patients with rheumatic pain compare the neural response to a task versus a resting or baseline state. Resting-state BOLD fMRI records resting state-fluctuations over time by measuring the spontaneous slow (<0.1 Hz) fluctuations that occur in the brain [15] (for a review see [51]). It can be used to identify the correlates of the default mode network (DMN). The DMN is a network of brain regions that is active when the brain is at 'rest' and is involved in self-referential orientation and monitoring [52]. Resting-state fMRI (R-fMRI) is increasingly being used to identify large-scale connectivity patterns in 'resting-state networks' which show overlap with patterns of task-induced activity [157] and are consistently found across participants and over time [39]. The altered resting state of large-scale networks correlates with individual differences in behavioural performance [51] and in disease [63, 153]. This type of approach is particularly useful when the neural mechanism is not well understood or cannot be accurate predicted which is often the case in the ongoing physical process of disease onset and/or development [26]. For example, patients with FM demonstrate less connectivity within the brain's pain inhibitory network during calibrated pressure pain than healthy controls suggesting that the dysfunction of the descending pain modulatory network plays an important role in maintenance of FM pain [83, 84]. Napadow et al. [123] provided evidence for linkage between resting-state brain connectivity and spontaneous pain in FM. The authors found greater intrinsic connectivity between the insula and the DMN which correlated with increased reports of spontaneous pain. In a further study reductions in spontaneous clinical pain correlated with a decrease in insula-DMN activity [122]. The hypervigilance to pain, which is seen in FM and other chronic pain disorders, may be due to the increased connectivity of insula-DMN. What is still unclear however, is whether the alterations seen in the DMN are causation or correlation in relation to pain and cognition.

EEG can also measure resting brain 'state'. It is often used in the study of sleep-wake interactions by identifying the frequency oscillations associated with the different sleep stages. For example, RA patients have been documented as having abnormal alpha sleep oscillations and show correlations which are suggestive of a relationship between sleep disturbances, morning stiffness and pain [49, 119].

To date, only one study has used ASL to look at arthritic pain. Howard et al. found increases in rCBF in pain processing areas including S1 and S2, IC, cingulate cortex, amygdala and thalamus associated with ongoing OA pain at rest. When they repeated the study between 7 and 21 days later they showed that the variations seen in rCBF in these areas were related to changes in patients ongoing pain [73].

7 Functional Brain Imaging in Drug Development and Pain Treatments

Functional neuroimaging, particularly fMRI, can be used to study the pharmacokinetics and pharmacodynamics of analgesics and monitor the effectiveness and time-course of treatment effects in rheumatic pain [150, 173, 184]. They can also be used to improve our understanding of the mechanism of action of many commonly prescribed analgesics. For example, administration of a selective cyclooxygenase-2 inhibitor (COX-2i) to a patient suffering from psoriatic arthritis resulted in significantly decreased reported pain intensity that correlated with a decrease in activity in the anterior insula and SII [7]. In a slightly larger study of 6 patients suffering from chronic knee OA, brain activity was observed with fMRI in a two-week, repeated measure study of treatment effects due to a COX-2i. Spontaneous pain and clinical characteristics of OA decreased whilst increased blood and cerebral spinal fluid levels of the drug correlated positively with prefrontal-limbic brain activity [128]. This finding is part of a consistent pattern of increases in prefrontal cortical activity associated with therapeutic benefit whether as a result of mindfulness [20–22], placebo analgesia [171, 174] or COX-2 inhibition. What is not clear from the COX studies is to what extent the effects are centrally or peripherally generated.

A first of its kind, double blind placebo controlled study investigated the effects of naproxen on OA pain using fMRI. They found reductions in brain activity in the bilateral S1, thalamus, and amygdala with naproxen compared to placebo. They also showed significant correlations between changes in reported pain intensity and changes in brain activity in brain regions previously associated with the lateral pain pathway [143]. Inhibition of anti-tumour necrosis factor (TNF)-α is a successful, fast acting treatment for arthritic pain [50, 116]. Hess et al. [70] used fMRI to show that within 24 h after the inhibition of TNF-α, pain-related brain activity in S2, the thalamus and limbic areas were blocked *before* clinical and laboratory markers of inflammation were altered. This is important because it shows that TNF-α blockade in RA may modulate pain processing brain areas before the anti-inflammatory effects are seen in the joints. Baliki et al. [8, 9] compared the effects of the local anaesthetic Lidocaine on chronic low back pain (CBP) and evoked knee OA pain. Whilst they found significantly decreased brain activity in both groups, in CBP the reduction in BOLD was found in the mPFC/rACC and OFC which are primarily emotional areas whereas in OA only activity in the thalamus was reduced. As the PFC and ACC are areas involved in mediating placebo analgesia [14, 171, 174] this could indicate that in CBP, lidocaine is acting mainly through placebo mechanisms whereas in OA, the nociceptive processing is blocked, hence the reduced thalamic activity.

The discrepancy between the high cost of drug discovery and development and the small number of effective compounds for chronic pain reaching the market is high [53, 95]. And, despite an array of pharmacological treatment options available, for many chronic pain patients only adequate pain relief is achievable [35, 68]. For many popular analgesic drugs there is also much discussion regarding the site of action where the analgesic effect is mediated [88, 135, 149]. In addition, there are many problems in the translation of pre-clinical research into the pain clinic as in vivo pre-clinical studies are (1) undertaken on homogenous groups of animals, (2) animal models cannot reliably measure spontaneous on ongoing pain, and (3) efficacy studies are often very short whilst conversely patients often suffer from progressively deteriorating conditions over several years. Human brain imaging has the potential to circumvent some of these problems and fMRI, EEG and PET all have their respective strengths and can be used to image different aspects therapeutic interventions [1, 53, 106]. EEG and FDG-PET provide direct measures of neuronal activity and are therefore less susceptible to drug effects on the vascular response. However, the most common imaging modality has been fMRI combined with careful modelling of vascular direct effects. Combining fMRI and the administration of a drug is termed pharmacological fMRI (phFMRI). phFMRI is a sensitive tool able to assess potential drug effects by measuring local changes in cerebral blood flow, cerebral blood volume and blood oxygenation resulting from neuronal activity, giving information about the local drug action on neuronal activity [150]. Pharmacodynamic data from phFMRI can be used as a measure of drug efficacy at an early stage in drug development providing an objective marker of brain responses during treatment in phase 1 and phase 2 clinical trials. Functional brain imaging techniques provide potential for monitoring modulation of the pain matrix during therapeutic intervention and better evaluation of whether there is a credible biological effect of therapy at an early stage of development. The logical use of these techniques during early phases of development is likely to make drug development more cost-effective.

Cognitive-based treatments have also been assed using neuroimaging and have helped us understand how psychological interventions affect the brains of chronic pain patients. Hypnotic induction in FM shows activations in the cerebellum, anterior midcingulate, anterior and posterior insula and inferior parietal cortex with fMRI [44] and increases in CBF orbitofrontal cortex, right thalamus and inferior parietal cortex and decreases in the cingulate cortex [178]. Brown et al. gave an eight week course of mindfulness-based cognitive therapy to a group of chronic pain patients. Mindfulness meditation reduced the unpleasantness of their clinical pain and in the experimental pain condition, anticipatory and pain-evoked potentials decreased in the IC, an area responsible for modulation of emotional responses, and increased processing in executive control areas such as the dlPFC. This suggests that greater perceived control of pain may result from improved regulation of the emotional response to pain as result of increased dlPFC processing [22]. The effects of cognitive behavioural therapy (CBT) have also been assessed using fMRI. These studies have shown increased pain-evoked activation of the PFC in FM patients after 12 weeks of treatment [83, 84] and increases in grey matter volume of the PFC have been seen in chronic pain patients after 11 weeks [155]. Increased activity of the PFC is seen during placebo induced analgesia [14, 129] and during expectation of pain relief before the placebo response occurs [171, 174]. From this is can be concluded that psychological therapies such as hypnosis, mindfulness meditation and CBT may share common therapeutic mechanisms with placebo analgesia. This suggests that treatment of rheumatic pain using placebo might be as effective as more traditional therapeutic interventions and that we should be exploring their routine integration into clinical practice more rigorously.

Functional brain imaging has also been used to develop neurofeedback to reduce experimental pain based on fMRI responses in forebrain areas [40]. This promising approach has also been applied to other non-invasive techniques such as EEG [85] and may provide an important adjunct to more standard cognitive therapeutic approaches in the future.

8 Measuring Pain Quantitatively Using Neuroimaging

Pain is currently measured using the subjective report of the patient using one-dimensional rating scales. Whilst various pain scales have been developed to measure patient subjective experience of pain [71] functional imaging can be used to objectively measure some of the brain responses that underpin this experience. Their provision of candidate mechanisms for chronic pain provides the potential to target these potential mechanisms for therapeutic intervention [32, 90]. Neuroimaging however, is an unbiased, quantitative method to measure the pain-related brain activation [33]. Gaussian process multivariate regression models are able to predict self-reported, thermally induced pain from whole brain fMRI volumes [113]. In another study, patterns of brain activity were used to predict from fMRI data analysed using vector machine learning whether an individual was experiencing evoked pain or just heat (with accuracy of 81%) [24]. Recently these mathematical modelling techniques have been used to classify chronic low back pain patients from healthy controls with a prediction accuracy of 76% [169].

These studies suggest that, with adequate analytical techniques, functional brain imaging can be used to accurately measure the brain substrates of different aspects of pain experience. However, philosophically it seems unlikely that brain imaging can be the final arbiter of what a subject may be experiencing. In other words an objective imaging technique cannot make a subjective experience objective.

9 Conclusion

Neuroimaging is a non-invasive method of measuring the processes that lead to and comprise pain experience. Modern neuroimaging techniques have led to rapid progress in our understanding of brain networks involved in pain processing in health and disease. These techniques allow us to understand how sensory-discriminative and the cognitive-evaluative and emotional components of pain are processed and how these are integrated. More importantly, functional imaging techniques, in particular EEG, have allowed us to have a much more detailed understanding of how expectations influence pain experience and how cognitive therapies and placebo effects influence these processes. Recent studies using EEG suggest that expectations and abnormalities in how these are processed may be as important to the understanding of chronic pain as events during the pain experience. This is likely to broaden the number of potential physiological targets in the brain for which we can develop new pharmacological and non-pharmacological therapies. New advances in neuroimaging such as the development of arterial spin labelling which can be used to monitor tonic rather than phasic neural events [73, 126, 127], and the use of high power 7-T scanners for anatomical and functional imaging which will lead to increased temporal and spatial resolution and will give us even greater ability to understand the mechanisms pain and pain therapy.

Despite a wealth of neuroimaging data, the mechanisms of how chronic pain is maintained are still not fully understood. The challenge is to use combinations of functional brain imaging to better understand the neurophysiology of chronic pain, optimize the use of existing pain therapies and develop new therapies targeting some of the candidate new brain mechanisms outlined in this chapter. In the longer term, it may be possible to identify patients who would respond to treatment and identify new treatment approaches using cheaper and more readily available brain imaging techniques such as EEG. The use of functional brain imaging as a therapy in the form of neurofeedback has shown some initial promise [40]. Again the challenge is to now develop this further at lower cost.

Functional brain imaging has led to the concept of chronic pain as a brain problem and has led to a shift in the focus of how we think of this common clinical problem and how we should be explaining pain to patients and health-care professionals. It has also pointed to how plastic many of the processes involved in pain perception are. The development of new pain therapies has been slow over the last 50 years. The challenge for the next 50 years is to use this knowledge to develop better and individually tailored therapies. The management of these patients with greater compassion and encouragement would be a good first step. Just the empowerment of the brain to do what it is designed to do with greater access to cognitive therapy and other lifestyle adjustments would be an excellent second step. A greater understanding of the mechanisms of pain chronicity should also be encouraging us to explore strategies to prevent pain chronicity.

References

1. Biomarkers Definitions Working Group, Bethesda MD. Biomarkers and surrogate endpoints: preferred definitions and conceptual framework. Clin Pharmacol Ther. 2001;69(3):89–95.
2. Alexander AL, Lee JE, Lazar M, Field AS. Diffusion tensor imaging of the brain. Neurotherapeutics: J Am Soc Exp NeuroTher. 2007;4(3):316–29.
3. Anderson AK, Phelps EA. Lesions of the human amygdala impair enhanced perception of emotionally salient events. Nature. 2001;411(6835):305–9.
4. Apkarian AV, Bushnell MC, Treede RD, Zubieta JK. Human brain mechanisms of pain perception and regulation in health and disease. Eur J Pain. 2005;9(4):463–84.
5. Arendt-Nielsen L, Nie H, Laursen MB, Laursen BS, Madeleine P, Simonsen OH, Graven-Nielsen T. Sensitization in patients with painful knee osteoarthritis. Pain. 2010;149(3):573–81.
6. Bajaj P, Bajaj P, Graven-Nielsen T, Arendt-Nielsen L. Osteoarthritis and its association with muscle hyperalgesia: an experimental controlled study. Pain. 2001;93(2):107–14.
7. Baliki M, Katz J, Chialvo DR, Apkarian AV. Single subject pharmacological-MRI (phMRI) study: modulation of brain activity of psoriatic arthritis pain by cyclooxygenase-2 inhibitor. Mol Pain. 2005;1:32.
8. Baliki MN, Geha PY, Apkarian AV, Chialvo DR. Beyond feeling: chronic pain hurts the brain, disrupting the default-mode network dynamics. J Neurosci. 2008;28(6):1398–403.
9. Baliki MN, Geha PY, Jabakhanji R, Harden N, Schnitzer TJ, Apkarian AV. A preliminary fMRI study of analgesic treatment in chronic back pain and knee osteoarthritis. Mol Pain. 2008;4:47.
10. Baliki MN, Schnitzer TJ, Bauer WR, Apkarian AV. Brain morphological signatures for chronic pain. PLoS ONE. 2011;6(10):e26010.
11. Beck AT, Ward CH, Mendelson M, Mock J, Erbaugh J. An inventory for measuring depression. Arch Gen Psychiatry. 1961;4:561–71.
12. Bedson J, Croft PR. The discordance between clinical and radiographic knee osteoarthritis: a systematic search and summary of the literature. BMC Musculoskelet Disord. 2008;9:116.
13. Beyer JL, Krishnan KR. Volumetric brain imaging findings in mood disorders. Bipolar Disord. 2002;4(2):89–104.
14. Bingel U, Lorenz J, Schoell E, Weiller C, Buchel C. Mechanisms of placebo analgesia: rACC recruitment of a subcortical antinociceptive network. Pain. 2006;120(1–2):8–15.
15. Biswal B, Yetkin FZ, Haughton VM, Hyde JS. Functional connectivity in the motor cortex of resting human brain using echo-planar MRI. Magn Reson Med. 1995;34(4):537–41.
16. Bornhovd K, Quante M, Glauche V, Bromm B, Weiller C, Buchel C. Painful stimuli evoke different stimulus-response functions in the amygdala, prefrontal, insula and somatosensory cortex: a single-trial fMRI study. Brain. 2002;125(Pt 6):1326–36.
17. Borsook D, Upadhyay J, Chudler EH, Becerra L. A key role of the basal ganglia in pain and analgesia–insights gained through human functional imaging. Mol Pain. 2010;6:27.
18. Bowsher D. Termination of the central pain pathway in man: the conscious appreciation of pain. Brain. 1957;80:607–22.
19. Brighina F, Piazza A, Vitello G, Aloisio A, Palermo A, Daniele O, Fierro B. rTMS of the prefrontal cortex in the treatment of chronic migraine: a pilot study. J Neurol Sci. 2004;227(1):67–71.
20. Brown CA, El-Deredy W, Jones AK. When the brain expects pain: common neural responses to pain anticipation are related to clinical pain and distress in fibromyalgia and osteoarthritis. Eur J Neurosci. 2014;39(4):663–72.
21. Brown CA, Jones AK. Meditation experience predicts less negative appraisal of pain: electrophysiological evidence for the involvement of anticipatory neural responses. Pain. 2010;150(3):428–38.
22. Brown CA, Jones AK. Psychobiological correlates of improved mental health in patients with musculoskeletal pain after a mindfulness-based pain management program. Clin J Pain. 2013;29(3):233–44.
23. Brown CA, Seymour B, Boyle Y, El-Deredy W, Jones AK. Modulation of pain ratings by expectation and uncertainty: Behavioral characteristics and anticipatory neural correlates. Pain. 2008;135(3):240–50.
24. Brown JE, Chatterjee N, Younger J, Mackey S. Towards a physiology-based measure of pain: patterns of human brain activity distinguish painful from non-painful thermal stimulation. PLoS ONE. 2011;6(9):e24124.
25. Buchel C, Bornhovd K, Quante M, Glauche V, Bromm B, Weiller C. Dissociable neural responses related to pain intensity, stimulus intensity, and stimulus awareness within the anterior cingulate cortex: a parametric single-trial laser functional magnetic resonance imaging study. J Neurosci. 2002;22(3):970–6.
26. Buckner RL, Andrews-Hanna JR, Schacter DL. The brain's default network: anatomy, function, and relevance to disease. Ann N Y Acad Sci. 2008;1124:1–38.
27. Burgmer M, Pogatzki-Zahn E, Gaubitz M, Stüber C, Wessoleck E, Heuft G, Pfleiderer B. Fibromyalgia unique temporal brain activation during experimental pain: a controlled fMRI Study. J Neural Transm. 2010;117(1):123–31.
28. Burgmer M, Pogatzki-Zahn E, Gaubitz M, Wessoleck E, Heuft G, Pfleiderer B. Altered brain

activity during pain processing in fibromyalgia. Neuroimage. 2009;44(2):502–8.
29. Bushnell MC, Duncan GH, Hofbauer RK, Ha B, Chen JI, Carrier B. Pain perception: is there a role for primary somatosensory cortex? Proc Natl Acad Sci U S A. 1999;96(14):7705–9.
30. Casey KL, Minoshima S, Berger KL, Koeppe RA, Morrow TJ, Frey KA. Positron emission tomographic analysis of cerebral structures activated specifically by repetitive noxious heat stimuli. J Neurophysiol. 1994;71(2):802–7.
31. Chen JJ, Wang JY, Chang YM, Su SY, Chang CT, Sun SS, Kao CH, Lee CC. Regional cerebral blood flow between primary and concomitant fibromyalgia patients: a possible way to differentiate concomitant fibromyalgia from the primary disease. Scand J Rheumatol. 2007;36(3):226–32.
32. Chizh BA, Greenspan JD, Casey KL, Nemenov MI, Treede RD. Identifying biological markers of activity in human nociceptive pathways to facilitate analgesic drug development. Pain. 2008;140 (2):249–53.
33. Coghill RC, McHaffie JG, Yen YF. Neural correlates of interindividual differences in the subjective experience of pain. Proc Natl Acad Sci U S A. 2003;100(14):8538–42.
34. Coghill RC, Sang C. Pain intensity processing within the human brain a bilateral distributed mechanism. Am Phys Soc. 1999;1934–43.
35. Collins SD, Chessell IP. Emerging therapies for neuropathic pain. Expert Opin Emerg Drugs. 2005;10(1):95–108.
36. Conturo TE, Lori NF, Cull TS, Akbudak E, Snyder AZ, Shimony JS, McKinstry RC, Burton H, Raichle ME. Tracking neuronal fiber pathways in the living human brain. Proc Natl Acad Sci U S A. 1999;96(18):10422–7.
37. Cook DB, Lange G, Ciccone DS, Liu WC, Steffener J, Natelson BH. Functional imaging of pain in patients with primary fibromyalgia. J Rheumatol. 2004;31(2):364–78.
38. Crombez G, Eccleston C, Van den BA, Goubert L, Van HB. Hypervigilance to pain in fibromyalgia: the mediating role of pain intensity and catastrophic thinking about pain. Clin J Pain. 2004;20(2):98–102.
39. Damoiseaux JS, Rombouts SA, Barkhof F, Scheltens P, Stam CJ, Smith SM, Beckmann CF. Consistent resting-state networks across healthy subjects. Proc Natl Acad Sci U S A. 2006;103 (37):13848–53.
40. deCharms RC, Maeda F, Glover GH, Ludlow D, Pauly JM, Soneji D, Gabrieli JD, Mackey SC. Control over brain activation and pain learned by using real-time functional MRI. Proc Natl Acad Sci U S A. 2005;102(51):18626–31.
41. Derbyshire SW. Meta-analysis of thirty-four independent samples studied using PET reveals a significantly attenuated central response to noxious stimulation in clinical pain patients. Curr Rev Pain. 1999;3(4):265–80.
42. Derbyshire SW, Jones AK, Devani P, Friston KJ, Feinmann C, Harris M, Pearce S, Watson JD, Frackowiak RS. Cerebral responses to pain in patients with atypical facial pain measured by positron emission tomography. J Neurol Neurosurg Psychiatry. 1994;57(10):1166–72.
43. Derbyshire SW, Jones AK, Gyulai F, Clark S, Townsend D, Firestone LL. Pain processing during three levels of noxious stimulation produces differential patterns of central activity. Pain. 1997;73 (3):431–45.
44. Derbyshire SW, Whalley MG, Oakley DA. Fibromyalgia pain and its modulation by hypnotic and non-hypnotic suggestion: an fMRI analysis. Eur J Pain. 2009;13(5):542–50.
45. Detre JA, Leigh JS, Williams DS, Koretsky AP. Perfusion imaging. Magn Reson Med. 1992;23(1):37–45.
46. Di Piero V, Jones AK, Iannotti F, Powell M, Perani D, Lenzi GL, Frackowiak RS. Chronic pain: a PET study of the central effects of percutaneous high cervical cordotomy. Pain. 1991;46(1):9–12.
47. Dillmann J, Miltner WHR, Weiss T. The influence of semantic priming on event-related potentials to painful laser-heat stimuli in humans. Neurosci Lett. 2000;284(1–2):53–6.
48. Dolan RJ. Emotion, cognition, and behavior. Science. 2002;298(5596):1191–4.
49. Drewes AM, Svendsen L, Taagholt SJ, Bjerregard K, Nielsen KD, Hansen B. Sleep in rheumatoid arthritis: a comparison with healthy subjects and studies of sleep/wake interactions. Br J Rheumatol. 1998;37(1):71–81.
50. Firestein GS. Evolving concepts of rheumatoid arthritis. Nature. 2003;423(6937):356–61.
51. Fox MD, Raichle ME. Spontaneous fluctuations in brain activity observed with functional magnetic resonance imaging. Nat Rev Neurosci. 2007;8 (9):700–11.
52. Fox MD, Snyder AZ, Vincent JL, Corbetta M, Van Essen DC, Raichle ME. The human brain is intrinsically organized into dynamic, anticorrelated functional networks. Proc Natl Acad Sci U S A. 2005;102(27):9673–8.
53. Frank R, Hargreaves R. Clinical biomarkers in drug discovery and development. Nat Rev Drug Discov. 2003;2(7):566–80.
54. García-Fernández E, Godoy-Izquierdo D, Pérez--García M, Jiménez-Alonso J, López-Chicheri I, Godoy JF. Differences in pressure-pain threshold between healthy women and patients with fibromyalgia syndrome, systemic lupus erythematosus, and rheumatoid arthritis. J Musculoskelet Pain. 2009;17(2):139–54.
55. Garcia-Larrea L, Maarrawi J, Peyron R, Costes N, Mertens P, Magnin M, Laurent B. On the relation between sensory deafferentation, pain and thalamic

activity in Wallenberg's syndrome: a PET-scan study before and after motor cortex stimulation. Eur J Pain. 2006;10(8):677–88.
56. Gauthier LV, Taub E, Perkins C, Ortmann M, Mark VW, Uswatte G. Remodeling the brain plastic structural brain changes produced by different motor therapies after stroke. Stroke; J Cereb Circ. 2008;39(5):1520–5.
57. Gibson SJ, Littlejohn GO, Gorman MM, Helme RD, Granges G. Altered heat pain thresholds and cerebral event-related potentials following painful CO_2 laser stimulation in subjects with fibromyalgia syndrome. Pain. 1994;58(2):185–93.
58. Giesecke T, Gracely RH, Williams DA, Geisser ME, Petzke FW, Clauw DJ. The relationship between depression, clinical pain, and experimental pain in a chronic pain cohort. Arthritis Rheum. 2005;52 (5):1577–84.
59. Goffaux P, de Souza JB, Potvin S, Marchand S. Pain relief through expectation supersedes descending inhibitory deficits in fibromyalgia patients. Pain. 2009;145(1–2):18–23.
60. Gracely RH, Geisser ME, Giesecke T, Grant MA, Petzke F, Williams DA, Clauw DJ. Pain catastrophizing and neural responses to pain among persons with fibromyalgia. Brain. 2004;127(Pt 4):835–43.
61. Gracely RH, Petzke F, Wolf JM, Clauw DJ. Functional magnetic resonance imaging evidence of augmented pain processing in fibromyalgia. Arthritis Rheum. 2002;46(5):1333–43.
62. Grech R, Cassar T, Muscat J, Camilleri KP, Fabri SG, Zervakis M, Xanthopoulos P, Sakkalis V, Vanrumste B. Review on solving the inverse problem in EEG source analysis. J NeuroEng Rehabil. 2008;5:25–7.
63. Greicius MD, Krasnow B, Reiss AL, Menon V. Functional connectivity in the resting brain: a network analysis of the default mode hypothesis. Proc Natl Acad Sci U S A. 2003;100(1):253–8.
64. Gwilym SE, Filippini N, Douaud G, Carr AJ, Tracey I. Thalamic atrophy associated with painful osteoarthritis of the hip is reversible after arthroplasty: a longitudinal voxel-based morphometric study. Arthritis Rheum. 2010;62(10):2930–40.
65. Gwilym SE, Keltner JR, Warnaby CE, Carr AJ, Chizh B, Chessell I, Tracey I. Psychophysical and functional imaging evidence supporting the presence of central sensitization in a cohort of osteoarthritis patients. Arthritis Rheum. 2009;61 (9):1226–34.
66. Gwilym SE, Pollard TC, Carr AJ. Understanding pain in osteoarthritis. J Bone Joint Surg Br. 2008;90 (3):280–7.
67. Hadjipavlou G, Dunckley P, Behrens TE, Tracey I. Determining anatomical connectivities between cortical and brainstem pain processing regions in humans: a diffusion tensor imaging study in healthy controls. Pain. 2006;123(1–2):169–78.
68. Harden N, Cohen M. Unmet needs in the management of neuropathic pain. J Pain Symptom Manage. 2003;25(5 Suppl):S12–7.
69. Hargrove JB, Bennett RM, Simons DG, Smith SJ, Nagpal S, Deering DE. Quantitative electroencephalographic abnormalities in fibromyalgia patients. Clin EEG Neurosci. 2010;41(3):132–9.
70. Hess A, Axmann R, Rech J, Finzel S, Heindl C, Kreitz S, Sergeeva M, Saake M, Garcia M, Kollias G, Straub RH, Sporns O, Doerfler A, Brune K, Schett G. Blockade of TNF-alpha rapidly inhibits pain responses in the central nervous system. Proc Natl Acad Sci U S A. 2011;108 (9):3731–6.
71. Hjermstad MJ, Fayers PM, Haugen DF, Caraceni A, Hanks GW, Loge JH, Fainsinger R, Aass N, Kaasa S. Studies comparing numerical rating scales, verbal rating scales, and visual analogue scales for assessment of pain intensity in adults: a systematic literature review. J Pain Symptom Manage. 2011;41 (6):1073–93.
72. Hofbauer RK, Rainville P, Duncan GH, Bushnell MC. Cortical representation of the sensory dimension of pain. J Neurophysiol. 2001;86 (1):402–11.
73. Howard MA, Sanders D, Krause K, O'Muircheartaigh J, Fotopoulou A, Zelaya F, Thacker M, Massat N, Huggins JP, Vennart W, Choy E, Daniels M, Williams SC. Alterations in resting-state regional cerebral blood flow demonstrate ongoing pain in osteoarthritis: an arterial spin-labeled magnetic resonance imaging study. Arthritis Rheum. 2012;64(12):3936–46.
74. Hsieh JC, Belfrage M, Stone-Elander S, Hansson P, Ingvar M. Central representation of chronic ongoing neuropathic pain studied by positron emission tomography. Pain. 1995;63(2):225–36.
75. Hsu MC, Harris RE, Sundgren PC, Welsh RC, Fernandes CR, Clauw DJ, Williams DA. No consistent difference in gray matter volume between individuals with fibromyalgia and age-matched healthy subjects when controlling for affective disorder. Pain. 2009;143(3):262–7.
76. Hummel T, Schiessl C, Wendler J, Kobal G. Peripheral and central nervous changes in patients with rheumatoid arthritis in response to repetitive painful stimulation. Int J Psychophysiol. 2000;37 (2):177–83.
77. Hummel T, Schiessl C, Wendler J, Kobal G. Peripheral and central nervous changes in patients with rheumatoid arthritis in response to repetitive painful stimulation. Int J Psychophysiol. 2000;37:177–83.
78. Huneke NT, Brown CA, Burford E, Watson A, Trujillo-Barreto NJ, El-Deredy W, Jones AK. Experimental placebo analgesia changes resting-state alpha oscillations. PLoS ONE. 2013;8 (10):e78278.

79. Iannetti GD, Mouraux A. From the neuromatrix to the pain matrix (and back). Exp Brain Res. 2010;205(1):1–12.
80. Imamura M, Imamura ST, Kaziyama HH, Targino RA, Hsing WT, de Souza LP, Cutait MM, Fregni F, Camanho GL. Impact of nervous system hyperalgesia on pain, disability, and quality of life in patients with knee osteoarthritis: a controlled analysis. Arthritis Rheum. 2008;59 (10):1424–31.
81. Ingvar M, Eriksson L, Greitz T, Stone-Elander S, Dahlbom M, Rosenqvist G, af Trampe P, von Euler C. Methodological aspects of brain activation studies: cerebral blood flow determined with [15O] butanol and positron emission tomography. J Cereb Blood Flow Metab. 1994;14(4):628–38.
82. Jensen KB, Kosek E, Petzke F, Carville S, Fransson P, Marcus H, Williams SCR, Choy E, Giesecke T, Mainguy Y, Gracely R, Ingvar M. Evidence of dysfunctional pain inhibition in Fibromyalgia reflected in rACC during provoked pain. Pain®. 2009;144(1–2):95–100.
83. Jensen KB, Kosek E, Wicksell R, Kemani M, Olsson G, Merle JV, Kadetoff D, Ingvar M. Cognitive behavioral therapy increases pain-evoked activation of the prefrontal cortex in patients with fibromyalgia. Pain. 2012;153(7):1495–503.
84. Jensen KB, Loitoile R, Kosek E, Petzke F, Carville S, Fransson P, Marcus H, Williams SC, Choy E, Mainguy Y, Vitton O, Gracely RH, Gollub R, Ingvar M, Kong J. Patients with fibromyalgia display less functional connectivity in the brain's pain inhibitory network. Mol Pain. 2012;8:32.
85. Jensen MP, Gertz KJ, Kupper AE, Braden AL, Howe JD, Hakimian S, Sherlin LH. Steps toward developing an EEG biofeedback treatment for chronic pain. Appl Psychophysiol Biofeedback. 2013;38(2):101–8.
86. Ji RR, Kohno T, Moore KA, Woolf CJ. Central sensitization and LTP: do pain and memory share similar mechanisms? Trends Neurosci. 2003;26 (12):696–705.
87. Jones AK, Brown WD, Friston KJ, Qi LY, Frackowiak RS. Cortical and subcortical localization of response to pain in man using positron emission tomography. Proc R Soc Lond B Biol Sci. 1991;244 (1309):39–44.
88. Jones AK, Derbyshire SW. Cerebral mechanisms operating in the presence and absence of inflammatory pain. Ann Rheum Dis. 1996;55(7):411–20.
89. Jones AK, Derbyshire SWG. Reduced cortical responses to noxious heat in patients with rheumatoid arthritis. Ann Rheum Dis. 1997;56:601–7.
90. Jones AK, Huneke NT, Lloyd DM, Brown CA, Watson A. Role of functional brain imaging in understanding rheumatic pain. Curr Rheumatol Rep. 2012;14(6):557–67.
91. Jones AK, Watabe H, Cunningham VJ, Jones T. Cerebral decreases in opioid receptor binding in patients with central neuropathic pain measured by [11C]diprenorphine binding and PET. Eur J Pain. 2004;8(5):479–85.
92. Jones AKP, Kulkarni B. The role of cingulate cortex in central neuropathic pain: functional imaging and cortical model of allodynia. In: Cingulate neurobiology and disease. B.A. Vogt. New York: Oxford University Press; 2009.
93. Kamaleri Y, Natvig B, Ihlebaek CM, Bruusgaard D. Localized or widespread musculoskeletal pain: does it matter? Pain. 2008;138(1):41–6.
94. Kean WF, Kean R, Buchanan WW. Osteoarthritis: symptoms, signs and source of pain. Inflammopharmacology. 2004;12(1):3–31.
95. Kola I, Landis J. Can the pharmaceutical industry reduce attrition rates? Nat Rev Drug Discov. 2004;3 (8):711–5.
96. Kong J, Gollub RL, Polich G, Kirsch I, Laviolette P, Vangel M, Rosen B, Kaptchuk TJ. A functional magnetic resonance imaging study on the neural mechanisms of hyperalgesic nocebo effect. J Neurosci. 2008;28(49):13354–62.
97. Kosek E, Ordeberg G. Lack of pressure pain modulation by heterotopic noxious conditioning stimulation in patients with painful osteoarthritis before, but not following, surgical pain relief. Pain. 2000;88(1):69–78.
98. Krummenacher P, Candia V, Folkers G, Schedlowski M, Schonbachler G. Prefrontal cortex modulates placebo analgesia. Pain. 2010;148 (3):368–74.
99. Kuchinad A, Schweinhardt P, Seminowicz DA, Wood PB, Chizh BA, Bushnell MC. Accelerated brain gray matter loss in fibromyalgia patients: premature aging of the brain? J Neurosci. 2007;27 (15):4004–7.
100. Kulkarni B, Bentley DE, Elliott R, Julyan PJ, Boger E, Watson A, Boyle Y, El-Deredy W, Jones AK. Arthritic pain is processed in brain areas concerned with emotions and fear. Arthritis Rheum. 2007;56(4):1345–54.
101. Kulkarni B, Bentley DE, Elliott R, Youell P, Watson A, Derbyshire SW, Frackowiak RS, Friston KJ, Jones AK. Attention to pain localization and unpleasantness discriminates the functions of the medial and lateral pain systems. Eur J Neurosci. 2005;21(11):3133–42.
102. Lautenbacher S, Rollman GB. Possible deficiencies of pain modulation in fibromyalgia. Clin J Pain. 1997;13(3):189–96.
103. Le Bars D, Dickenson AH, Besson JM. Diffuse noxious inhibitory controls (DNIC). I. Effects on dorsal horn convergent neurones in the rat. Pain. 1979;6(3):283–304.
104. Lee CM, Farde L. Using positron emission tomography to facilitate CNS drug development. Trends Pharmacol Sci. 2006;27(6):310–6.
105. Legrain V, Iannetti GD, Plaghki L, Mouraux A. The pain matrix reloaded: a salience detection system for the body. Prog Neurobiol. 2011;93(1):111–24.

106. Lesko LJ, Atkinson AJ Jr. Use of biomarkers and surrogate endpoints in drug development and regulatory decision making: criteria, validation, strategies. Annu Rev Pharmacol Toxicol. 2001;41:347–66.
107. Lorenz J, Cross DJ, Minoshima S, Morrow TJ, Paulson PE, Casey KL. A unique representation of heat allodynia in the human brain. Neuron. 2002;35 (2):383–93.
108. Lorenz J, Grasedyck K, Bromm B. Middle and long latency somatosensory evoked potentials after painful laser stimulation in patients with fibromyalgia syndrome. Electroencephalogr Clin Neurophysiol/Evoked Potentials Sect. 1996;100 (2):165–8.
109. Lorenz J, Minoshima S, Casey KL. Keeping pain out of mind: the role of the dorsolateral prefrontal cortex in pain modulation. Brain. 2003;126(Pt 5):1079–91.
110. Lutz J, Jager L, de Quervain D, Krauseneck T, Padberg F, Wichnalek M, Beyer A, Stahl R, Zirngibl B, Morhard D, Reiser M, Schelling G. White and gray matter abnormalities in the brain of patients with fibromyalgia: a diffusion-tensor and volumetric imaging study. Arthritis Rheum. 2008;58(12):3960–9.
111. Maarrawi J, Peyron R, Mertens P, Costes N, Magnin M, Sindou M, Laurent B, Garcia-Larrea L. Differential brain opioid receptor availability in central and peripheral neuropathic pain. Pain. 2007;127(1–2):183–94.
112. Magerl W, Fuchs PN, Meyer RA, Treede RD. Roles of capsaicin-insensitive nociceptors in cutaneous pain and secondary hyperalgesia. Brain. 2001;124 (Pt 9):1754–64.
113. Marquand A, Howard M, Brammer M, Chu C, Coen S, Mourao-Miranda J. Quantitative prediction of subjective pain intensity from whole-brain fMRI data using Gaussian processes. Neuroimage. 2010;49(3):2178–89.
114. May A. Neuroimaging: visualising the brain in pain. Neurol Sci. 2007;28(Suppl 2):S101–7.
115. May A. Chronic pain may change the structure of the brain. Pain. 2008;137(1):7–15.
116. McInnes IB, Schett G. Cytokines in the pathogenesis of rheumatoid arthritis. Nat Rev Immunol. 2007;7(6):429–42.
117. Mease P. Fibromyalgia syndrome: review of clinical presentation, pathogenesis, outcome measures, and treatment. J Rheumatol Suppl. 2005;75:6–21.
118. Meeus M, Vervisch S, De Clerck LS, Moorkens G, Hans G, Nijs J. Central sensitization in patients with rheumatoid arthritis: a systematic literature review. Semin Arthritis Rheum. 2012;41(4):556–67.
119. Moldofsky H, Lue FA, Smythe HA. Alpha EEG sleep and morning symptoms in rheumatoid arthritis. J Rheumatol. 1983;10(3):373–9.
120. Montoya P, Pauli P, Batra A, Wiedemann G. Altered processing of pain-related information in patients with fibromyalgia. Eur J Pain. 2005;9 (3):293–303.
121. Moseley JB, O'Malley K, Petersen NJ, Menke TJ, Brody BA, Kuykendall DH, Hollingsworth JC, Ashton CM, Wray NP. A controlled trial of arthroscopic surgery for osteoarthritis of the knee. N Engl J Med. 2002;347(2):81–8.
122. Napadow V, Harris RE. What has functional connectivity and chemical neuroimaging in fibromyalgia taught us about the mechanisms and management of 'centralized' pain? Arthritis Res Ther. 2014;16(5):425.
123. Napadow V, LaCount L, Park K, As-Sanie S, Clauw DJ, Harris RE. Intrinsic brain connectivity in fibromyalgia is associated with chronic pain intensity. Arthritis Rheum. 2010;62(8):2545–55.
124. Natvig B, Bruusgaard D, Eriksen W. Localized low back pain and low back pain as part of widespread musculoskeletal pain: two different disorders? A cross-sectional population study. J Rehabil Med. 2001;33(1):21–5.
125. Ochsner KN, Ludlow DH, Knierim K, Hanelin J, Ramachandran T, Glover GC, Mackey SC. Neural correlates of individual differences in pain-related fear and anxiety. Pain. 2006;120(1–2):69–77.
126. Owen DG, Bureau Y, Thomas AW, Prato FS, St Lawrence KS. Quantification of pain-induced changes in cerebral blood flow by perfusion MRI. Pain. 2008;136(1–2):85–96.
127. Owen DG, Clarke CF, Ganapathy S, Prato FS, St Lawrence KS. Using perfusion MRI to measure the dynamic changes in neural activation associated with tonic muscular pain. Pain. 2010;148(3):375–86.
128. Parks EL, Geha PY, Baliki MN, Katz J, Schnitzer TJ, Apkarian AV. Brain activity for chronic knee osteoarthritis: dissociating evoked pain from spontaneous pain. Eur J Pain. 2011;15(8):843. e841–814.
129. Petrovic P, Ingvar M. Placebo and opioid analgesia imaging a shared neuronal network. Science. 2002;295:1737–40.
130. Petrovic P, Petersson KM, Ghatan PH, Stone-Elander S, Ingvar M. Pain-related cerebral activation is altered by a distracting cognitive task. Pain. 2000;85(1–2):19–30.
131. Peyron R, Garcia-Larrea L, Gregoire MC, Convers P, Richard A, Lavenne F, Barral FG, Mauguiere F, Michel D, Laurent B. Parietal and cingulate processes in central pain. A combined positron emission tomography (PET) and functional magnetic resonance imaging (fMRI) study of an unusual case. Pain. 2000;84(1):77–87.
132. Peyron R, Garcia-Larrea L, Gregoire MC, Costes N, Convers P, Lavenne F, Mauguiere F, Michel D, Laurent B. Haemodynamic brain responses to acute pain in humans: sensory and attentional networks. Brain. 1999;122(Pt 9):1765–80.
133. Picavet HS, Schouten JS. Musculoskeletal pain in the Netherlands: prevalences, consequences and risk groups, the DMC(3)-study. Pain. 2003;102 (1–2):167–78.

134. Porro CA, Baraldi P, Pagnoni G, Serafini M, Facchin P, Maieron M, Nichelli P. Does anticipation of pain affect cortical nociceptive systems? J Neurosci. 2002;22(8):3206–14.
135. Porro CA, Cavazzuti M. Functional imaging studies of the pain system in man and animals. Prog Brain Res. 1996;110:47–62.
136. Porro CA, Cettolo V, Francescato MP, Baraldi P. Temporal and intensity coding of pain in human cortex. J Neurophysiol. 1998;80(6):3312–20.
137. Porro CA, Cettolo V, Francescato MP, Baraldi P. Functional activity mapping of the mesial hemispheric wall during anticipation of pain. NeuroImage. 2003;19(4):1738–47.
138. Posner MI, Petersen SE. The attention system of the human brain. Annu Rev Neurosci. 1990;13:25–42.
139. Price DD. Psychological and neural mechanisms of the affective dimension of pain. Science. 2000;288 (5472):1769–72.
140. Pujol J, López-Solà M, Ortiz H, Vilanova JC, Harrison BJ, Yücel M, Soriano-Mas C, Cardoner N, Deus J. Mapping brain response to pain in fibromyalgia patients using temporal analysis of fMRI. PLoS ONE. 2009;4(4):e5224.
141. Rainville P, Duncan GH, Price DD, Carrier B, Bushnell MC. Pain affect encoded in human anterior cingulate but not somatosensory cortex. Science. 1997;277(5328):968–71.
142. Rodriguez-Raecke R, Niemeier A, Ihle K, Ruether W, May A. Brain gray matter decrease in chronic pain is the consequence and not the cause of pain. J Neurosci. 2009;29(44):13746–50.
143. Sanders D, Krause K, O'Muircheartaigh J, Thacker MA, Huggins JP, Vennart W, Massat NJ, Choy E, Williams SC, Howard MA. Pharmacologic modulation of hand pain in osteoarthritis: a double-blind placebo-controlled functional magnetic resonance imaging study using naproxen. Arthritis Rheumatol. 2015;67(3):741–51.
144. Sarnthein J, Jeanmonod D. High thalamocortical theta coherence in patients with neurogenic pain. Neuroimage. 2008;39(4):1910–7.
145. Sarnthein J, Stern J, Aufenberg C, Rousson V, Jeanmonod D. Increased EEG power and slowed dominant frequency in patients with neurogenic pain. Brain. 2006;129(Pt 1):55–64.
146. Schaible H-G, Ebersberger A, Natura G. Update on peripheral mechanisms of pain: beyond prostaglandins and cytokines. Arthritis Res Ther. 2011;13 (2):210.
147. Schaible HG, Grubb BD. Afferent and spinal mechanisms of joint pain. Pain. 1993;55(1):5–54.
148. Schmidt-Wilcke T, Luerding R, Weigand T, Jurgens T, Schuierer G, Leinisch E, Bogdahn U. Striatal grey matter increase in patients suffering from fibromyalgia—a voxel-based morphometry study. Pain. 2007;132(Suppl 1):S109–16.
149. Scholz J, Woolf CJ. Can we conquer pain? Nat Neurosci. 2002;5(Suppl):1062–7.
150. Schweinhardt P, Bountra C, Tracey I. Pharmacological FMRI in the development of new analgesic compounds. NMR Biomed. 2006;19(6): 702–11.
151. Schweinhardt P, Bushnell MC. Pain imaging in health and disease—how far have we come? J Clin Investig. 2010;120(11):3788–97.
152. Schweinhardt P, Kalk N, Wartolowska K, Chessell I, Wordsworth P, Tracey I. Investigation into the neural correlates of emotional augmentation of clinical pain. Neuroimage. 2008;40(2):759–66.
153. Seeley WW, Crawford RK, Zhou J, Miller BL, Greicius MD. Neurodegenerative diseases target large-scale human brain networks. Neuron. 2009;62 (1):42–52.
154. Seminowicz DA, Davis KD. Cortical responses to pain in healthy individuals depends on pain catastrophizing. Pain. 2006;120(3):297–306.
155. Seminowicz DA, Shpaner M, Keaser ML, Krauthamer GM, Mantegna J, Dumas JA, Newhouse PA, Filippi CG, Keefe FJ, Naylor MR. Cognitive-behavioral therapy increases prefrontal cortex gray matter in patients with chronic pain. J Pain. 2013;14(12):1573–84.
156. Seymour B, O'Doherty JP, Koltzenburg M, Wiech K, Frackowiak R, Friston K, Dolan R. Opponent appetitive-aversive neural processes underlie predictive learning of pain relief. Nat Neurosci. 2005;8(9):1234–40.
157. Smith SM, Fox PT, Miller KL, Glahn DC, Fox PM, Mackay CE, Filippini N, Watkins KE, Toro R, Laird AR, Beckmann CF. Correspondence of the brain's functional architecture during activation and rest. Proc Natl Acad Sci U S A. 2009;106 (31):13040–5.
158. Staud R, Craggs JG, Perlstein WM, Robinson ME, Price DD. Brain activity associated with slow temporal summation of C-fiber evoked pain in fibromyalgia patients and healthy controls. Eur J Pain. 2008;12(8):1078–89.
159. Staud R, Smitherman ML. Peripheral and central sensitization in fibromyalgia: pathogenetic role. Curr Pain Headache Rep. 2002;6(4):259–66.
160. Staud R, Vierck CJ, Cannon RL, Mauderli AP, Price DD. Abnormal sensitization and temporal summation of second pain (wind-up) in patients with fibromyalgia syndrome. Pain. 2001;91(1–2):165–75.
161. Stevens A, Batra A, Kotter I, Bartels M, Schwarz J. Both pain and EEG response to cold pressor stimulation occurs faster in fibromyalgia patients than in control subjects. Psychiatry Res. 2000;97(2–3):237–47.
162. Talbot JD, Marrett S, Evans AC, Meyer E, Bushnell MC, Duncan GH. Multiple representations of pain in human cerebral cortex. Science. 1991;251 (4999):1355–8.
163. Thompson PW, Carr AJ. Pain in the rheumatic diseases. Ann Rheum Dis. 1997;56(6):395.

164. Tolle TR, Kaufmann T, Siessmeier T, Lautenbacher S, Berthele A, Munz F, Zieglgansberger W, Willoch F, Schwaiger M, Conrad B, Bartenstein P. Region-specific encoding of sensory and affective components of pain in the human brain: a positron emission tomography correlation analysis. Ann Neurol. 1999;45(1):40–7.
165. Tracey I. Nociceptive processing in the human brain. Curr Opin Neurobiol. 2005;15(4):478–87.
166. Tracey I, Mantyh PW. The cerebral signature for pain perception and its modulation. Neuron. 2007;55(3):377–91.
167. Tracey I, Ploghaus A, Gati JS, Clare S, Smith S, Menon RS, Matthews PM. Imaging attentional modulation of pain in the periaqueductal gray in humans. J Neurosci. 2002;22(7):2748–52.
168. Treede RD. Transduction and transmission properties of primary nociceptive afferents. Ross Fiziol Zh Im IM Sechenova. 1999;85(1):205–11.
169. Ung H, Brown JE, Johnson KA, Younger J, Hush J, Mackey S. Multivariate classification of structural MRI data detects chronic low back pain. Cereb Cortex. 2014;24(4):1037–44.
170. Valet M, Sprenger T, Boecker H, Willoch F, Rummeny E, Conrad B, Erhard P, Tolle TR. Distraction modulates connectivity of the cingulo-frontal cortex and the midbrain during pain—an fMRI analysis. Pain. 2004;109(3):399–408.
171. Wager TD, Rilling JK, Smith EE, Sokolik A, Casey KL, Davidson RJ, Kosslyn SM, Rose RM, Cohen JD. Placebo-induced changes in FMRI in the anticipation and experience of pain. Science. 2004;303(5661):1162–7.
172. Wartolowska K, Hough MG, Jenkinson M, Andersson J, Wordsworth BP, Tracey I. Structural changes of the brain in rheumatoid arthritis. Arthritis Rheum. 2012;64(2):371–9.
173. Wartolowska K, Tracey I. Neuroimaging as a tool for pain diagnosis and analgesic development. Neurotherapeutics. 2009;6(4):755–60.
174. Watson A, El-Deredy W, Iannetti GD, Lloyd D, Tracey I, Vogt BA, Nadeau V, Jones AK. Placebo conditioning and placebo analgesia modulate a common brain network during pain anticipation and perception. Pain. 2009;145(1–2):24–30.
175. Wendler J, Hummel T, Reissinger M, Manger B, Pauli E, Kalden JR, Kobal G. Patients with rheumatoid arthritis adapt differently to repetitive painful stimuli compared to healthy controls. J Clin Neurosci. 2001;8(3):272–7.
176. Wiech K, Ploner M, Tracey I. Neurocognitive aspects of pain perception. Trends Cogn Sci. 2008;12(8):306–13.
177. Wiech K, Tracey I. The influence of negative emotions on pain: behavioral effects and neural mechanisms. Neuroimage. 2009;47(3):987–94.
178. Wik G, Fischer H, Bragée B, Finer B, Fredrikson M. Functional anatomy of hypnotic analgesia: a PET study of patients with fibromyalgia. Eur J Pain. 1999;3(1):7–12.
179. Wik G, Fischer H, Finer B, Bragee B, Kristianson M, Fredrikson M. Retrospenial cortical deactivation during painful stimulation of fibromyalgic patients. Int J Neurosci. 2006;116(1):1–8.
180. Willer JC, Bathien N. Pharmacological modulations on the nociceptive flexion reflex in man. Pain. 1977;3(2):111–9.
181. Willer JC, De Broucker T, Le Bars D. Encoding of nociceptive thermal stimuli by diffuse noxious inhibitory controls in humans. J Neurophysiol. 1989;62(5):1028–38.
182. Willer JC, Roby A, Le Bars D. Psychophysical and electrophysiological approaches to the pain-relieving effects of heterotopic nociceptive stimuli. Brain. 1984;107(Pt 4):1095–112.
183. Willoch F, Schindler F, Wester HJ, Empl M, Straube A, Schwaiger M, Conrad B, Tolle TR. Central poststroke pain and reduced opioid receptor binding within pain processing circuitries: a [11C] diprenorphine PET study. Pain. 2004;108(3):213–20.
184. Wise RG, Preston C. What is the value of human FMRI in CNS drug development? Drug Discov Today. 2010;15(21–22):973–80.
185. Woolf CJ. Central sensitization: implications for the diagnosis and treatment of pain. Pain. 2011;152(3 Suppl):S2–15.
186. Wylde V, Hewlett S, Learmonth ID, Dieppe P. Persistent pain after joint replacement: prevalence, sensory qualities, and postoperative determinants. Pain. 2011;152(3):566–72.
187. Zambreanu L, Wise RG, Brooks JC, Iannetti GD, Tracey I. A role for the brainstem in central sensitisation in humans. Evidence from functional magnetic resonance imaging. Pain. 2005;114(3):397–407.
188. Zubieta JK, Smith YR, Bueller JA, Xu Y, Kilbourn MR, Jewett DM, Meyer CR, Koeppe RA, Stohler CS. Regional Mu opioid receptor regulation of sensory and affective dimensions of pain. Science 293(5528 July 13th):311–15.

12 Neuroimaging Studies of Somatoform Pain Disorder: How Far Have We Come?

Atsuo Yoshino, Yasumasa Okamoto and Shigeto Yamawaki

Abstract

We review neuroimaging studies for somatoform pain disorder. It is important for studies of somatoform pain disorder to assess various psychosocial elements as well as somatic complaints per se. Multidimensional understandings are crucial for development of useful etiological models and treatment approaches. The brain regions altered in somatoform pain disorder appear to include the ACC, insula, amygdala, hippocampus, parahippocampus, SI, SII, basal ganglia, and PFC. We believe that negative psychosocial factors and restricted emotional responses are linked to such brain-based findings, and various studies have supported such relationships. However, somatoform pain disorder neuroimaging studies remain few and far between, and further study is needed to elucidate the relationship between pathophysiology in somatoform pain disorder and the associated brain mechanisms.

Keywords

Insula · Behavior · Pain perception · Fibromyalgia · Parahippocampus · Hemodynamic · Psychosocial

1 Introduction

1.1 Clinical Features of Somatoform Pain Disorder

Somatization is defined as a tendency to experience somatic symptoms in response to psychosocial stress, and to consult professionals in clinic or hospital settings in response to these symptoms [29]. Somatic symptoms are often reported by patients in primary care and in many

A. Yoshino (✉) · Y. Okamoto · S. Yamawaki
Department of Psychiatry and Neurosciences, Division of Frontier Graduate School of Biomedical Sciences, Hiroshima University, 1-2-3 Kasumi, 734-8551 Minami-ku, Hiroshima, Japan
e-mail: yoshino@hiroshima-u.ac.jp

L. Saba (ed.), *Neuroimaging of Pain*, DOI 10.1007/978-3-319-48046-6_12

Table 1 Diagnostic criteria for somatoform pain disorder and somatic symptom disorder based on DSM

[Somatoform pain disorder] DSM-IV-TR – Pain perception in one or more anatomical body parts center on clinical feature, and is serious to concern clinical treatment – The pain brings about clinically grave distress or dysfunction in social, occupational, or other important field – Psychological factors are inferred to play an important role in the onset, severity, worsening, or persistence of the pain – The symptom or defect is not intentionally represented or forged – The symptom cannot be explained well by the existence of affective disorder, anxiety disorder, and is not fulfilled with the criteria of dyspareunia
[Somatic symptom disorder] DSM-5 – One or more distressful somatic symptoms that arise significant confusion on daily life – Extreme thought, emotion, or behavior related to the somatic symptoms or followed health apprehension. The at least following one is included: (1) Mismatched or persistent thought about the seriousness to one's own symptoms (2) Persistent and strong anxiety about one's own health or symptoms (3) Excessive effort that is spent to the concern to these symptoms or health – These symptoms are persistent (usually more than six months), although any one somatic symptom may not be lasting
Somatoform pain disorder is involved with pain as somatic symptoms, if it is considered from the view of somatic symptom disorder

cases remain medically unexplained even after detailed assessment. One study has reported that at least 33% of somatic symptoms cannot be medically explained [26]. Somatoform pain disorder is defined as the occurrence of one or more physical complaints for which appropriate medical evaluation reveals no explanatory physical pathology or pathophysiologic mechanism, or when such a pathology is present, the physical complaints or resulting impairment are grossly in excess of what would be expected from the physical findings, according to the Diagnostic and Statistical Manual of Mental Disorders IV (DSM-IV) [1] (Table 1). Individuals with somatoform pain disorder are characterized by pain symptoms that cannot be fully explained by a general medical condition or other mental disorders such as depression. In DSM-IV and ICD-10, it is stated that the symptoms must cause clinically substantial distress or impairment in social, occupational, or other areas of functioning. Of 294 internal medical inpatients, 20.2% had somatoform pain disorder [18]. Thus, somatoform pain disorder is thought to be fairly prevalent. In the Diagnostic and Statistical Manual of Mental Disorders, 5th Edition (DSM-5), somatoform pain disorder has been subsumed into a new diagnosis, somatic symptom disorder, together with somatization disorder, undifferentiated somatoform disorder, and hypochondriasis [2]. Somatic symptom disorder is diagnosed when three criteria are met. Three criteria are (1) occurrence of somatic symptoms that are distressing or result in significant disruption of daily life, (2) the symptoms are persistent, and (3) the symptoms are associated with dysfunctional thoughts, feelings, or behaviors [2]. Both somatoform pain disorder and somatic symptom disorder are associated with specific pain-related dysfunctional beliefs, feelings, and behaviors, as well as a relative inability to control and self-manage pain, rather than dysfunctional pain perception per se. The cognitive features of somatoform pain disorder include excessive attention directed towards pain perception, concerns about disease, catastrophic thoughts such as associating normal physical sensations with physical disease, and fear of one's own body being damaged by various physical activities [2]. Behavior features of somatoform pain disorder include seeking repeated confirmation that one's own body is not in an abnormal state, repeated help-seeking in the form of medical support and obtaining guarantees, and avoidance of physical activities [2]. This disorder diminishes quality of life and is associated with increased depression and anxiety [2]. For example, patients with somatoform pain disorder are resistant to medication, they might feel helpless, and they might harbor dysfunctional beliefs such as "*It's terrible and it's never going to get any better*". The medical and other economic costs, including medical care utilization and overall healthcare costs, associated with somatization are considerable [47].

1.2 Importance of Neuroimaging Studies for Somatoform Pain Disorder

Although its etiopathology remains unclear, cultures, learning, memory, genetic or biological vulnerabilities, early traumatic events, and so on all seem to have an influence on the presentation of somatoform pain disorder [29], and it is important to assess various psychosocial elements as well as somatic complaints per se [2]. With regard to treatment, both pharmacological (e.g. tricyclic compounds such as amitriptyline) and non-pharmacological treatments (including psychotherapy) are important for somatoform pain disorder [27], which is different from the treatment of peripheral pain such as that due to damage and inflammation.

Methodologies such as neuroimaging can help to clarify the psychophysiology, neuroanatomy, and neurochemistry underlying somatoform pain disorder, and thus can help to clarify etiology and treatment. Although the specific mechanisms underlying somatoform pain disorder require further clarification, progress in the development of neuroimaging methods make it possible to study the neural mechanisms underlying somatoform pain disorder. Neuroimaging approaches include functional magnetic resonance imaging (fMRI), magnetoencephalography (MEG), positron emission tomography (PET), and single photon emission computed tomography (SPECT), and have been used to investigate abnormal neural mechanisms in somatoform pain disorder, although compared with other mental disorders, such knowledge remains quite limited, and neuroimaging of somatoform disorders can still be regarded as a developing field [3].

In this review, we will first review existing neuroimaging studies of somatoform pain disorder, documenting evidence of distinctive neural functioning or structures that are specifically associated with somatoform pain disorder. Studies examining other similar disorders, including fibromyalgia, chronic fatigue syndrome, and irritable bowel syndrome, were excluded to maintain a focus on somatoform pain disorder. However, we did not exclude studies of "somatoform disorders" per se. We also review evidence from treatment-related neuroimaging studies of somatoform pain disorder, where evidence of distinctive neural function or structure is noted.

2 Neuroimaging Studies of Somatoform Pain Disorder

Various researchers have compared somatoform pain disorder patients with healthy control subjects. First, we review task-based fMRI studies. Second, we examine resting-state-based studies including MRI, SPECT, and PET methods. Furthermore, we reviewed structural MRI, magnetic resonance spectroscopy (MRS), and electroencephalography (EEG) studies, and finally, treatment-related neuroimaging studies of somatoform pain disorder are examined.

2.1 Neuroimaging Studies Examing Abnormal Neural Mechanisms Underlying Somatoform Pain Disorder

2.1.1 Task-Based fMRI Studies (Table 2)

Some fMRI studies have used a pain stimulus task. Stoeter et al. have investigated cerebral activation induced by pain stimuli, cognitive stress, and emotional stress. Seventeen patients with somatoform pain disorder and an age-matched control group were included in their study. They found greater activation of brain regions such as the thalamus, anterior insula, hippocampus, and prefrontal cortex during exposure to pain stimuli in the patient group [49]. These regions are generally regarded as

Table 2 Studies conducted during task-based paradigms

Authors	Year	*N* (patients)	Participant groups	Stimuli	Positively related regions	Negatively related regions
Stoeter et al.	2007	17	SPD versus control	Pain stimuli Cognitive and emotional stress	Th, aIC, Hippo, Parietal cortex VLPFC, MPFC, Putamen	Motor cortex
Gündel et al.	2008	12	SPD versus control	Pain stimuli	SI, SII, Parietal cortex, Amy Parahippo, IC	VMPFC, OFC
Noll-Hussong et al.	2010	Total 15	8 abused SPD versus 7 non-abused SPD	Empathy for pain	Lateral and medial PFC	Hippo
de Greck et al.	2012	20	Somatoform disorder (5 SPD) versus control	Emotion recognition by using facial stimuli	–	Postcentral gyrus, STG Parahippo, pIC, Amy
Noll-Hussong et al.	2013	21	SPD versus control	Empathy for pain	–	ACC
Yoshino et al.	2013	11	SPD versus control	Pain stimuli modulated by emotion (sadness)	aIC, pIC, Hippo	–

SPD Somatoform Pain Disorder, *Th* Thalamus, *aIC* anterior Insular Cortex, *pIC* posterior Insular Cortex, *Hippo* Hippocampus, *VLPFC* Ventrolateral Prefrontal Cortex, *MPFC* Medial Prefrontal Cortex, *SI* Primary Somatosensory Cortex, *SII* Secondary Somatosensory Cortex, *Amy* Amygdala, *Parahippo* Parahippocampus, *BG* Basal Ganglion

pain-related brain regions. For example, the insula is frequently identified in neuroimaging studies of pain. The insula is probably a critical region for pain perception along with the anterior cingulate cortex (ACC) [41]. The insula has been classically subdivided into anterior and posterior regions [41]. The anterior insula is connected to the limbic system and has been linked with affectively and motivationally valenced information related with pain perception, which has been ascertained using various methods including pain anticipation tasks and pain empathy processing [3, 11]. Negative emotions such as anxiety may amplify pain perception, and the anterior insula is central for these understanding [11]. This involvement of the anterior insula in subjective pain perception, as modulated by negative emotion, may coincide with a role in interoceptive processing, including the monitoring of internal body states as an innate characteristic of an organism, like thirst or hunger [11]. Stoeter et al.'s results appear to show that the emotional modulation of pain perception is disturbed in somatoform pain disorder. In the face of cognitive stress induced using a task involving the continuation of a given sequence of symbols under time pressure, patients showed increased activation of the parieto-occipital cortex, precuneus, parahippocampal gyrus, posterior insula, and thalamus. During emotional stress induced by pictures and audio of physical violence, the left insula was more strongly activated in patients as compared to healthy controls. Patient-specific brain activations common to pain and cognitive stress were identified in the anterior part of the superior temporal gyri, bilaterally. Stoeter et al. suggest that both central neural processing in pain perception and the experience of cognitive stress are disturbed in patients. Another study has examined cerebral pain processing using noxious heat stimuli [19]. They performed fMRI on twelve right handed women with somatoform pain disorder and 13 age-matched healthy controls. Experimentally induced pain-related brain activations were more strongly shown in the primary somatosensory cortex (SI), the secondary somatosensory cortex (SII), the inferior parietal cortex, the contralateral amygdala, ipsilateral parahippocampus and the contralateral insula in patients. Conversely, in comparison to the patients, healthy control groups showed increased activation in the ventromedial prefrontal cortex (VMPFC) and the orbitofrontal cortex (OFC). There is accumulating evidence from neuroimaging studies that the frontal area, including the VMPFC, OFC, and MPFC, in involved in the processing and modulation of pain [3]. It was thought that the PFC controls pain-related brain regions such as the ACC, insula, amygdala, hippocampus, and PAG in terms of implementing emotional and cognitive modulation of pain perception, with roles played by attention, expectation, and learning [30]. Gündel et al.'s study suggests that dysfunctional activation of brain regions including the PFC is linked to the persistence of somatoform pain disorder.

Emotion plays as important modulatory role in pain perception, within a context of various mental disorders [3], and we examined the neural mechanism that is engaged in response to pain stimuli modulated by negative emotion, using a group of 11 somatoform pain disorder patients (40.9 ± 6.5 years) and an age-matched control group (40.6 ± 6.1 years) [60]. A stimulation method was used to induce pain at the superficial skin level, and we set the intensity to either moderate or low. Negative emotionality was induced using pictures of faces, with sad and neutral facial stimuli (control) used. Participants received the pain stimulus while sad or neutral facial expressions were displayed, and they rated the average intensity of the pain stimuli using a numerical rating scale (NRS) during an MRI scan. We investigated how sadness affected subjective pain perception and brain mechanisms in patients with somatoform pain disorder. For both patients and controls, pain intensities were significantly higher in the sad emotional condition than in the neutral emotional condition, and greater sensitivity to low levels of pain in an emotional context of sadness in the patients was speculated. Differences in cerebral processing during pain stimuli between the groups were shown in the anterior insula, posterior insula, and hippocampus. Activation of the anterior and posterior insula during low-pain stimuli in the sad condition was significantly greater in patients than in controls. Posterior insula activations have previously been reported during pain perception and appear to provide a subjective evaluation of bodily states such as somatotopy [41]. Furthermore, we found increased functional connectivity between the anterior insula and parahippocampus during the sad-specific low-pain condition. The parahippocampus has been reported to play a key role in the recollection of autobiographical memories, sending information from the hippocampus to the association areas, in particular during the retrieval of emotional memories [12]. The hippocampus is activated in cases of mismatch between expected and actually received painful stimuli, and a role of the hippocampus in learning and verbal memory involved in pain processing has been identified, such that the hippocampus can amplify aversive events during negative emotions such as anxiety [3]. Our fMRI results suggest that a high sensitivity to pain perception modulated by negative emotion may be an important distinctive feature in the development and persistence of somatoform pain disorder, and we have speculated that stronger sensitivity to low levels of pain during a negative emotion such as sadness might be linked to the psychopathology of somatoform pain disorder.

In contrast to directly perceived pain stimuli such as those used in thermal pain experiments, one study has investigated an alteration of neural response to reveal emotion recognition deficits in somatoform disorder patients [14]. This fMRI study compared neural activity across 20 somatoform disorder patients (12 females; mean age = 42.5) including five somatoform pain disorder patients, and 20 healthy controls, using a test of emotion recognition abilities that featured facial stimuli depicting anger, joy, disgust, and neutral emotional states, as well as a control condition that featured unrecognizable stimuli. During the emotion recognition test, patients with somatoform disorder had significantly more difficulties of response accuracy compared to healthy controls. Comparing hemodynamic responses of patients with somatoform disorders and healthy controls, these researchers found decreased hemodynamic responses at the right parahippocampus and left amygdala for the contrast "all emotions including anger, disgust, joy, and neutral expression"—"control (smoothed pictures with unrecognizable contents)". Furthermore, just for the angry emotional contrast ("anger"—"control"), there were significantly decreased hemodynamic responses at the left postcentral gyrus, left superior temporal gyrus, bilateral parahippocampus, left posterior insula, left amygdala, and left cerebellum in patients compared to controls. Furthermore, there was a significant negative correlation between somatization scores on the Symptom Checklist-90 (SCL-90-R) and hemodynamic responses for the "anger"—"control" contrast at the left superior temporal gyrus. The authors speculated that these results might explain the development of somatoform disorder in terms of emotion suppression (anger in particular) to avoid interpersonal conflict, and that decreased brain activity during emotional recognition might be linked to less reactivity of various emotional states. Another fMRI study has also examined neural responses during empathy for pain, both in 21 somatoform pain disorder patients (17 women; mean age = 46.2 years) and 19 healthy controls (12 women; mean age = 48.8 years), to ascertain the ability to imagine how one would feel in a particular painful situation [35]. These authors used pictures which presented human feet and hands in painful and non-painful situations in daily life, to evaluate pain empathy. Healthy controls tended to provide higher pain intensities. fMRI data identified pain-related activations of the somatosensory cortex, insula, and ACC for the "Pain > No Pain" contrast, particularly for the healthy controls. Moreover, there was significantly stronger activation of the posterior ACC in healthy controls than in patients with somatoform pain disorder for the "Pain > No Pain" contrast. Pain stimuli elicit robust activations of the ACC, and the ACC is an important region for pain perception, at both cognitive and emotional levels. This region is involved in various forms of cognitive processing such as response selection, negative expectation bias, and negative affect modulation during pain processing [41], and is more directly concerned with the evaluation of pain intensity and with the affective evaluation of pain than with the sensory component per se. Previous work has shown that pain-related activations of the ACC overlap with ACC activations related to emotion [54], and it has been suggested that the activation of the ACC expressed during pain perception might largely reflect the affective components of pain processing. In particular, Vogt et al. suggested that the posterior ACC may activate during affective responses to noxious perception rather than somatic component [54], and may play a role in the processing of affective information. Abnormal ACC activity has been reported in various mental disorders such as anxiety disorders or depression [3], and neuroimaging studies of somatoform pain disorder have also provided evidence for abnormal ACC activation. Patients with somatoform pain disorder are often reported to have difficulty identifying emotional signals within themselves, thus perceiving emotional stress as mere somatic symptoms [51], and Noll-Hussong et al. suggested that these results may indicate that, in patients with somatoform pain disorder, the experience of pain generated via visual paradigms is not recognized as an emotional signal as it might be in controls, because of difficulties with emotional perception and associated problems with pain empathy. These researchers also examined the neural responses in both eight somatoform pain disorder patients with a history of early sexually aversive life experiences (7 women; mean age = 44.8 years) and eight patients with a non-abusive history (7 women; mean age = 46.1 years), using a pain stimulus picture empathy task [34]. Greater brain activation in abused somatoform pain disorder patients was shown in the left lateral and medial superior frontal gyrus, with greater activation for non-abused patients shown at the left hippocampus. They speculated that left frontal gyrus activation might be associated with enhanced recollection of emotionally aversive events such as pain stimuli, which might be strengthened by life stress such as abuse.

Although pain stimuli elicited by an experimental method may differ from clinical pain, there appear to be distinctive neural activities that can be distinguished from those of healthy controls in patients with somatoform pain disorder, and that brain areas associated with cognitive-emotional components such as the ACC, insula, hippocampus, and parahippocampus may be important for the psychopathology of somatoform pain disorder. However, there is only scarce evidence from task-based studies, and to our knowledge, emotional pain modulation has been reported only for pain perception during sadness in somatoform pain disorder, although the psychopathology of somatoform pain disorder might be related to various emotional states such as anger, worry, or even positive emotions. Furthermore, many neuroimaging studies have estimated intensity of pain perception, and the methods used to measure the affective dimension of pain or somatic symptoms in neuroimaging studies (e.g., during an fMRI trial) might be also limited. Neuroimaging studies using adequate psychometric methods are important to extract brain regions. Future various task-based research needs should be added to further clarify the neural mechanisms that underlie somatoform pain disorder.

2.1.2 Resting-State Studies (fMRI, PET, and SPECT) (Table 3)

In contrast with the task-based fMRI approach, resting-state fMRI (R-fMRI) studies observe the blood oxygen level-dependent signal in the absence of an overt task or other form of stimulation, and are crucial for the understanding of intrinsic brain functional architecture under both normal and pathological conditions [61]. Resting-state neuroimaging studies can extract resting-state functional connectivities including the default mode network (DMN) [such as the posterior cingulate cortex (PCC), precuneus, lateral parietal cortex, and medial prefrontal cortex (MPFC)], which are thought to be involved in the processes of self-referential mental activity, autobiographical memory, and social cognition [61]. Otti et al. have tested resting-state functional connectivity for somatoform pain disorder [37]. They revealed that the fronto-insular network (FIN) and anterior DMN in patients with somatoform pain disorder had significantly higher power at high frequencies (0.20–0.24 Hz) in comparison with healthy controls. They speculate that this result might reflect impaired subjective emotional awareness, affective meaning construction and social understanding of patients with somatoform pain disorder, considering the role of FIN and anterior DMN. We investigated resting-state functional connectivity in patients with somatoform pain disorder using another analytical methodology [57]. The participants were nine patients with somatoform pain disorder (mean age = 39.0 ± 5.1 years) and 20 control subjects (mean age = 42.4 ± 6.8 years). Compared with healthy controls, neural activity was significantly higher in the left precentral gyrus for the patients. The precentral gyrus has been implicated in haemodynamic changes in brain regions related to motor functioning during the experience of pain, such that the possibility of interactions between pain and motor functioning has been raised [41]. Moreover, rTMS of the precentral gyrus appears to have been shown as an important brain region for treatment of chronic pain [20]. Our results have shown potentially important left precentral gyrus disturbance in patients with somatoform pain disorder, and that such resting-state abnormalities may be associated with structural and functional impairments in the pain processing of these patients. A resting-state fMRI study has researched regional neural activity of the brain in 26 medication-naive somatization disorder patients [50]. Somatization disorder is a subtype of somatoform disorder. The core feature of somatization disorder is a pattern of recurring, multiple, clinically significant somatic complaints that cannot be fully explained by any general medical condition or the direct effects of

Table 3 Studies conducted during resting-state

Authors	Year	Scan type	*N* (patients)	Participant groups	Stimuli	Positively related regions	Negatively related regions
Garcia-Campayo et al.	2001	SPECT	7	SD versus control	^{99m}TC-HMPAO and ^{99m}Tc-Bicisate	–	Frontal, temporoparietal, and cerebellar regions
Hagelberg et al.	2003	PET	8	Atypical facial pain versus control	[^{18}F] FDOPA, [^{11}C] NNC 756, and [^{11}C] raclopride	Putamen	–
Hakala et al.	2006	PET	10	Somatoform disorder versus control	2-[18]-fluoro 2 deoxy-D-glucose	–	Caudate, putamen
Karibe et al.	2010	SPECT	10	SPD versus control	rCBF	Th, Caudate, ACC Brain stem, PCC	PFC, Occipital lobe Temporal lobe
Koh et al.	2012	SPECT	32	Somatoform disorder (15 MIS and 17 LIS) versus control	Phytohemagglutinin and 99m-Tc-ethyl cysteinate dimer	Decreased CBF of parietal lobule and supramarginal gyrus in the MIS group	
Otti et al.	2013	R-fMRI	21	Multisomatoform disorder versus control	Patients-controls	FIN, aDMN	–
Su et al.	2014	R-fMRI	26	SD versus control	Patients-controls	MPFC	Precuneus
Yoshino et al.	2014	R-fMRI	9	SPD versus control	Patients-controls	Precentral gyrus	–

SPD Somatoform Pain Disorder, *SD* Somatization disorder, *MIS* More Immune-Suppressed, *LIS* Less Immune-Suppressed, *rCBF* regional Cerebral Blood Flow
Th Thalamus, *aIC* anterior Insular Cortex, *pIC* posterior Insular Cortex, *Hippo* Hippocampus, *VLPFC* Ventrolateral Prefrontal Cortex, *MPFC* Medial Prefrontal Cortex, *SI* Primary Somatosensory Cortex, *SII* Secondary Somatosensory Cortex, *Amy* Amygdala, *Parahippo* Parahippocampus, *BG* Basal Ganglion, *PCC* Posterior Cingulate Cortex, *aDMN* anterior Default Mode

a substance [1]. The somatization disorder group presented with significantly stronger neural activities in the bilateral superior MPFC (BA8, 9) compared with controls, and reduced neural activities in the left precuneus compared with controls. The MPFC receives sensory information from the body and the external environment via affective-limbic areas such as the amygdala, hippocampus, and hypothalamus, and supports unity of the processing of emotional aspects from information gathered from the internal and external environments [5]. It has also been reported that the MPFC might be important for intensity encoding in chronic pain and in the developmental pathway by which pain becomes chronic [5]. The MPFC also plays a role in the modulation of emotional processing, including attention to emotional stimuli and emotion identification [43]. Dysfunction in this region might be causally associated with emotional and cognitive changes in somatoform pain disorder.

Some resting-state studies have examined the brain neurochemistry of patients with somatoform pain disorder, using PET and SPECT. A SPECT study has described the regional cerebral blood flow (rCBF) of patients with orally localized somatoform pain disorder and a control group [24]. Ten patients (9 females; mean age 55.0 years) and 12 healthy individuals (7 females; mean age 61.8 years) participated in a *N*-isopropyl-4-[^{123}I] iodoamphetamine (^{123}I-IMP) SPECT scan. Increased rCBF was shown in the bilateral posterior cingulate gyri, brain stem, caudate nucleus, left thalamus, and right anterior cingulate gyrus in the patients with somatoform pain disorder compared with the control group. Furthermore, there was decreased rCBF in the bilateral frontal lobes, the bilateral occipital lobes, and the left temporal lobe in the patient group. In this study, changes in brain perfusion that represented increased rCBF were shown mainly in the limbic regions, whereas decreases were seen mainly in the cortex. The authors concluded that these findings may be responsible for the maintenance of somatoform pain disorder as a result of an abnormal neural mechanism in both cortical and limbic structures. Garcia-Campayo et al. have reported a preliminary SPECT study in somatization disorder [21]. Eleven patients with somatization disorder were investigated using ^{99m}TC-HMPAO and 99mTc-Bicisate, and hypoperfusion was seen primarily in the non-dominant frontal, temporoparietal, and cerebellar regions. This study may contribute to the discussion about an association between asymmetric brain activation and somatization related to emotional disturbance.

It has been reported that cytokine activity in the immune system is likely to be altered in somatization disorder [45], and Koh et al. have examined the relationship between neural activity and immune function in 32 patients with a somatoform disorder (12 men; mean age 35.3 years) and 42 healthy control subjects (23 men; mean age 34.9 years) [25]. This association has been illustrated by using blastogenic response to phytohemagglutinin (PHA) as a measure of cell-mediated immunity along with 99m-Tc-ethyl cysteinate dimer SPECT. Patients were classified into a relatively more immune-suppressed (MIS) subgroup ($N = 15$) and a relatively less immune-suppressed (LIS) subgroup ($N = 17$) via a median split of the blastogenic responses to PHA. The results have shown that: [1] compared with healthy controls, blastogenic responses to PHA was reduced in patients, [2] decreased cerebral blood flow was significant at the left inferior parietal lobule and the left supramarginal gyrus in the MIS subgroup as compare to LIS subgroup patients, and [3] there were the positive relationships between cerebral perfusion in the left interior parietal lobule and left supramarginal gyrus and blastogenic responses to PHA in patients. These researchers speculated that the left inferior parietal lobule and left supramarginal gyrus might play an important immunomodulating role in patients with a somatoform disorder. Another study has compared glucose metabolism of 10 female somatoform disorder patients and 12 female healthy volunteers via PET using 2-[18]-fluoro 2 deoxy-D-glucose (FDG) [23]. Participants also completed the Temperament and Character Inventory (TCI), which evaluates personality in terms of a 7-factor psychobiological model comprising four temperament dimensions (novelty-seeking, harm avoidance, reward dependence and persistence) and three character dimensions (self-directedness, cooperativeness and self-transcendence). Lower novelty-seeking and higher harm avoidance scores on the TCI were significantly related to reduced glucose metabolism in the caudate and putamen of patients with somatoform disorders. Harm avoidance in particular might be important in the development of somatoform disorders and these metabolic results might underlie various clinical characteristics such patients, including fatigue, low activity levels, and avoidant behavior.

The development of reliable resting-state biomarkers for somatoform pain disorder might offer a meaningful contribution in terms of efforts to diagnose and treat this condition.

2.1.3 Structural MRI Studies (Table 4)

There has been more interest in structural changes of the brain associated with somatoform pain disorder, although only a small number of studies have been conducted to date. Valet et al. have investigated whether the symptoms in somatoform pain disorder might be influenced by structural brain changes, using voxel-based morphometric (VBM) analyses [53]. Fourteen female patients (mean age = 51.1 years) who met the DSM-IV criteria for somatoform pain disorder and 25 healthy age-matched women were selected. There were no significant differences in global gray matter volumes between patients with somatoform pain disorder and healthy controls, but the somatoform pain disorder patient group showed significant decreases of gray matter density in prefrontal structures such as the ventromedial prefrontal cortex (VMPFC) and orbitofrontal cortex (OFC). Decreases were also detected for the ACC, insula, parahippocampal cortex, and the cerebellum. On the other hand, no significant increase in gray matter density was found. Another somatoform disorder study examined MRI volumes of the caudate, putamen, and hippocampus [22]. Ten female patients who met the criteria for somatization disorder (n = 6) or undifferentiated somatoform disorder (n = 4) according to the DSM-IV diagnostic classification and 16 healthy female controls participated. Significantly larger volumes for the patients in comparison to the controls were found for the bilateral caudate.

Atmaca et al. have evaluated the hippocampus and amygdala using structural MRI in patients with somatization disorder [4]. Twenty patients with somatization disorder (mean age = 43.6 years; all females) and 20 healthy controls (mean age = 40.0 years; all females) were recruited. There were significantly smaller mean volumes of the bilateral amygdala in patients with somatization disorder in comparison to healthy controls but no differences in terms of the whole brain or total gray and white matter hippocampus volumes. The amygdala is an important brain region, like other limbic areas, and has been suggested to play a facilitative and inhibitory role in the modulation of emotional pain behavior or emotional-affective and cognitive dimensions of pain, namely, attention, conditioning-based learning, and memory retrieval [3, 41]. This region also contributes to convergence and integration in sensitivity of pain perception. The connection from the amygdala to sensory cortical regions such as the ACC can facilitate attention or perception, and as a result, can work to enhance pain perception. For example, our previous analysis revealed that the amygdala is directly connected to the ACC during the experience of pain in a sad experimental condition, and that afferent nociceptive processing in the amygdala is associated with pain in the sad condition [59]. These data suggest that it is important for the amygdala to modulate pain perception as a function of emotional context.

In summary, some morphological studies have debated the question of whether somatoform pain disorder is associated with structural changes in the brain. So far, it has been proposed that decreased cortical thickness have been observed in brain regions such as the PFC, hippocampus, and amygdala, and that somatoform pain disorder might be related to specific morphological alterations in structures understood to

Table 4 Structural MRI studies related to somatoform pain disorder

Authors	Year	*N* (patients)	Participant groups	Stimuli	Positively related regions	Negatively related regions
Hakala et al.	2004	10	6 SD and 4undifferentiated somatoform disorder versus control	Patients-controls	Caudate	–
Valet et al.	2009	14	SPD versus control	Patients-controls	–	VMPFC, OFC, ACC IC, Parahippo, Cerebellum
Atmaca et al.	2011	20	SD versus control	Patients-controls	–	Amy

SPD Somatoform Pain Disorder, *SD* Somatization disorder, *Th* Thalamus, *IC* Insular Cortex, *Hippo* Hippocampus, *VLPFC* Ventrolateral Prefrontal Cortex, *Amy* Amygdala, *Parahippo* Parahippocampus

play a role in antinociception. Regional decreases in gray matter volume suggest the possibility of contributing to a neural degenerative process via excitotoxicity of neurons [7]. However, there are only a scarce number of structural MRI studies of somatoform pain disorder, and results have not been particularly robust so far. These studies do suggest a need for more work to identify potential structural brain alterations in this condition.

2.1.4 Other Studies

Magnetic Resonance Spectroscopy (MRS) Studies

MRS is able to measure chemical and cellular features in human beings. In particular, it can assess the chemical constitution of tissues, metabolic processes of tissues, and distinctive chemical or metabolic factors in diseases. In brain regions, the density and mobility of chemicals are measured as spectral peaks via MRS [17]. Previous MRS studies have measured changes in glutamate (Glu) levels, an excitatory neurotransmitter, and glutamine (Gln) levels. Fayed et al. have evaluated brain metabolite patterns in patients with somatization disorder ($n = 10$), fibromyalgia ($n = 10$), which most closely resembles somatoform pain disorder in many aspects, and healthy controls ($n = 10$) using spectroscopy techniques [16]. They also used the Pain Catastrophizing Scale (PCS) to assess the extent to which a person engages in catastrophic thinking about pain perception. This study specifically emphasized pain-related brain regions such as the insula, hippocampus and posterior cingulate cortex (PCC). The results from this study demonstrated that levels of glutamate + glutamine (Glx), a combined measure of glutamate (Glu) and glutamine (Gln) within the PCC, were significantly increased in patients with fibromyalgia and somatization disorder compared with healthy controls, with levels significantly higher in patients with fibromyalgia. Furthermore, Glx levels in the PCC were positively correlated with PCS scores in both patient groups. Glu has been reported to be an important mediator in neurotransmission and to be associated with sensitization in pain perception [15]. Glu/Gln cycling shows energy production that occurs in the brain and is directly related to increased energy demands with neural activation, and increases in excitatory neurotransmitters such as Glx might result in neuronal hyperexcitability in the PCC associated with fibromyalgia and somatization disorder. This pathway might be associated with augmented pain processing in fibromyalgia and somatization disorder. Furthermore, extreme Glx levels can damaged neurons via excitotoxicity [7], and it has been suggested that such structural damage or significant atrophy might be a maintaining factor in fibromyalgia and somatization disorder.

EEG Studies

We found one EEG study of neural mechanisms in patients with somatoform pain disorder. Stefanie et al. elucidated cortical activation (spectral power) in eight frequency bands (1.5–6, 6–8, 8–10, 10–12, 12–18, 18–21, and 21–30 Hz) in 15 patients (mean age = 50.1 years; 7 women) and controls (mean age = 50 years; 7 women) with somatoform pain disorder at rest [48]. They found significant spectral power reductions for many brain regions including the SI, SII, ACC, insula, amygdala, PFC, and hippocampus within most frequency bands for patients with somatoform pain disorder compared to controls. After Bonferroni correction, spectral power reductions within 21–30 Hz for many brain regions including the SI, SII, ACC, insula, amygdala, PFC, and hippocampus remained significant. A power reduction of this frequency range can be regarded as indicative of less neural activation due to the positive association between a power of this range and cortical activities. The pathophysiology of somatoform pain disorder might be related to a hypoactive anti-nociceptive system.

2.2 Neuroimaging Studies Related to Treatment

Treatment of somatoform pain disorder typically includes medication and psychotherapy as is the case for other mental disorders, and some efficacy date are available [27]. In Kroenke's study, somatoform pain disorder did not respond well to conventional pain therapy, and antidepressants such as fluoxetine, citalopram, and fluvoxamine might be efficacious for somatoform pain disorder although the number of antidepressant trials is small and further studies is needed. Antidepressants that inhibit the reuptake of serotonin or noradrenaline can reduce nociceptive afferent transmission in the ascending spinal pain pathways by increasing monoamine neurotransmission in the descending inhibitory spinal pathways, and the efficacy of antidepressants for pain is independent of their effects on depression [39]. Regarding psychotherapy, it has been claimed that cognitive behavior therapy (CBT) should be efficacious and that empirical evidence supports using CBT for the treatment of somatoform pain disorder [27]. This disorder has many psychological as well as physiological aspects [27], including negative emotionality and impaired social functioning, such that multimodal treatment is essential. In many patients with somatoform pain disorder, the improvement of negative emotionality or thought patterns are critically linked to changes in pain perception [56]. Although to some extent the clinical efficacy of treatment for somatoform pain disorder has been characterized, the neural mechanisms underlying pharmacological or nonpharmacological treatments (including CBT) for somatoform pain disorder remain poorly understood.

De Greck et al. have investigated changes in brain activities of patients with somatoform pain disorder during an emotional empathy task, pre-treatment and post-treatment (psychodynamic psychotherapy; [13]). The participants were 15 patients with a somatoform disorder (8 women, mean age = 42.6 years), diagnosed according to the DSM-IV-TR criteria, and 15 control subjects (8 women, mean age = 37.0 years). The empathy task used facial stimuli, and participants were supposed to rate their subjective impressions of empathy capability for emotion during the evaluation period. Psychotherapy was associated with decreases in somatization, alexithymia, and depressive symptoms. Changes in brain activation were assessed using a region of interest (ROI) and a voxel-wise whole brain analysis. Results revealed that brain activity in the left postcentral gyrus, left superior temporal gyrus, left posterior insula, left amygdala, left cerebellum, and the bilateral parahippocampal gyrus during empathy with anger all increased after psychotherapy. The importance of changes to the parahippocampal gyrus' activities through the treatment process were emphasized, on the basis of previous results (refer to the Sect. 2.1.1 Task-based fMRI studies session) and the previously reported role of the parahippocampal gyrus in pain perception (e.g., recall of autobiographical memories and recollection of autobiographical memories) [12].

Using neuroimaging in combination with a treatment intervention may allow one to determine medication types and dosages, and to discriminate case-specific effects by referring to distinct brain systems. Although to our knowledge, no somatoform pain disorder neuroimaging studies have examined the effects of medication treatment, one fMRI study did report neural changes after amitriptyline treatment in patients with irritable bowel syndrome [31]. These researchers performed painful rectal stimulation which included either stressful acoustic noise or relaxing music in different trials with 19 patients (all females, mean age = 39 years), and cerebral activation during rectal distension was compared between placebo and amitriptyline groups using MRI. Amitriptyline reduced pain-related brain activations in the ACC and the left posterior parietal cortex occurred only during rectal distension in the stressful state. This study would seem to be relevant for prediction of responsiveness to treatment in somatoform pain disorder patients. However, little is yet known about neuroimaging studies applied to better treatment strategies, and it may be useful to combine advances in neuroimaging techniques with various clinical effects parameters for both pharmacological and non-pharmacological interventions. More precise neural changes or assessment of treatment effects could be provided via transcranial magnetic stimulation (TMS) or transcranial direct current stimulation (tDCS).

2.3 Summary of Neuroimaging Studies for Somatoform Pain Disorder

Based on our search, we identified a total of 18 neuroimaging studies including six task-based fMRI, seven resting-state (fMRI and SPECT/PET), three structural MRI, one MRS, and one EEG study, all of which examined the neural basis of somatoform pain disorder. The task-based fMRI studies mainly examined specific brain activations during pain perception, emotionally modulated pain perception, and presentation of emotional stimuli for recognition or empathy. R-fMRI, structural MRI, and other methods including SPECT/PET and MRI studies mainly drew comparisons between patients and controls. We could not find neuroimaging studies related to pharmacotherapy in somatoform pain disorder. There was one fMRI study of somatoform pain disorder that assessed patients pre- and post-psychotherapy (psychodynamic psychotherapy). However, there are no neuroimaging studies of other forms of psychotherapy, including CBT. The brain regions altered in somatoform pain disorder appear to include the ACC, insula, amygdala, hippocampus, parahippocampus, SI, SII, basal ganglia, and PFC, along with other regions already know to be associated with chronic pain states. We will discuss the specific features or pathophysiology of somatoform pain disorder, based on these neuroimaging studies.

3 Understanding Somatoform Pain Disorder from Pain-Related Studies

As mentioned above, various methodologies including task-based fMRI, resting-state fMRI, structural MRI, PET, SPECT, and EEG have been used to study somatoform pain disorder. Brain alterations have been shown for the ACC, insula, amygdala, hippocampus, parahippocampus, SI, SII, basal ganglia, and PFC, and neuroimaging data have contributed to our understanding of the etiology and pathology of somatoform pain disorder.

3.1 Property of Pain Perception Modulated by Emotion and Cognition

It has been argued that the development and persistence of somatoform pain disorder is influenced by the quality of pain perception itself. Pain has many physiological as well as psychological aspects. Pain can be divided into sensory-discriminative and emotional-affective dimensions, and neuroimaging studies have helped to elucidate these dimensions [44], and neuroimaging and psychological studies have revealed that both dimensions are influenced by various emotional and cognitive elements [43, 46, 55, 58, 59]. For example, we examined the association between emotion (mainly during a sad emotional condition) and pain perception within the context of a neuroimaging study, using fMRI and MEG to examine neural responses to electrical pain-inducing stimuli [58, 59]. We found that subjective pain ratings were higher in the sad emotional context than in the happy and neutral contexts, and that pain-related brain activation of regions such as the ACC and amygdala was more pronounced in the sad context relative to the happy and neutral contexts. Sawamoto et al. have demonstrated that anticipation can increase ACC and insula activities that underlie pain perception [46]. Various cognitive processes such as focus of attention, expectation (e.g. catastrophic thought), learning, reappraisal and social interactions have been also shown to influence pain perception and biased pain-related brain activation [55]. For example, twelve healthy men rated esophageal stimulation, ranging from non-painful to painful, during which they completed a task (1-back working memory task) aimed at distracting them from the esophageal stimulus [10]. When pain was delivered during distraction, there was a significant positive relationship between intensity of pain ratings and brain activity in the right ACC and right prefrontal cortex. In terms of this latter region, the ventrolateral prefrontal cortex appears to be engaged when pain perception is experienced as less strong during the expression of religious beliefs [55]. Such results suggest an effect of distraction in terms of attenuating pain and altering associated brain activity. The right mid-ACC might be involved in attentional aspects of pain processing, and there appears to be a modulatory effect of attention on the PFC, concerning a pathway that descends from the PFC to the amygdala, PAG, RVM, and onto the spinal dorsal horn. These conclusions have been supported by previous studies that included distraction tasks, pain intensity-related expectation cues, placebo-induced analgesia, etc. [55]. Placebo-induced analgesia should be associated with cognitive factors such as learning, attention, expectation and reappraisal, and decreased neural responses to pain stimuli in brain regions such as the ACC, insula, thalamus, and PAG, which would be related to activation of the descending pain modulatory system. Thus, many experimental studies have shown altered pain processing in relevant brain regions as a function of emotional or cognitive factors, indicating that pain perception is easily modulated. The brain regions which play an important role in pain modulation constitute the prefrontal, ACC, insula, amygdala, and brainstem structures such as PAG and RVM, and pain perception is influenced by both bottom-up and top-down mechanisms, which can be modulated by emotional or cognitive mechanisms. For example, the brainstem structures, including the PAG and RVM, play an important role in the descending pain neuromodulatory system which is closely connected to the thalamus, amygdala, and PFC, and might either inhibit or amplify pain processing within the spinal cord [41]. Moreover, brain imaging studies have revealed that pain-related cortical and subcortical networks such as the ACC-fronto-PAG-RVM circuitry serve as a pain modulatory system [8]. Bushnell et al. have also demonstrated that chronicity of pain is related to abnormalities in these pain modulatory circuits. Thus, various factors, including cognition and social environment, are associated with pain perception, and the emotional-affective dimensions of pain perception also contribute to such a feeling. As to causes, pain perception is considered to be more intense than other somatic symptoms [44]. The different dimensions (e.g., emotion or cognition) of pain do not result from isolated brain regions, but are thought to be constructed via multiple inputs from various regions. Moreover, pain is not only modulated by the above dimensions but also by social factors such as adverse childhood experiences [34]. Thus, pain perception is not necessarily directly linked to noxious input but is instead affected by psychological and social environmental variables. These properties of pain perception are considered to influence the pathophysiology of somatoform pain disorder. In fact, extant neuroimaging studies of somatoform pain disorder have shown ACC, insula, amygdala, hippocampus, parahippocampus, SI, SII, basal ganglia, and PFC activation to be associated with psychological variables in pain perception, which is consistent with earlier findings from laboratory studies of pain perception.

3.2 Affected Factors to Somatosensory Amplification

Somatosensory amplification has been advocated as playing a critical role in the pathophysiology of somatization [6], and might be useful in terms of understanding somatoform pain disorder. This term refers to a tendency to recognize perception of bodily sensations as disruptive, noxious, and intrusive. For example, depression, more serious disease, and gender (with females complaining of more discomfort) were related to stronger sensitivity to somatic symptoms [6]. To understand the mechanisms underlying somatosensory amplification, Barsky et al. [6] claimed three factors: (1) a hypersensitive attention to somatic sensations, (2) excessive extraction of small and relatively minor sensations, and (3) the tendency to recognize bodily sensations as alarming and distressing. Somatosensory amplification is not a somatic symptom itself but rather represents a sense of discomfort or unpleasantness regarding somatic sensations. In addition to pain, various other sensory experiences (e.g., auditory and visual) are also amplified, and it has been suggested that these dispositions are associated with a fundamental problem of sensory processing rather than an abnormality of the part of body associated with the physical experiences in question. Amplification is influenced by various factors, including emotion, level of arousal, and social environment. Negative emotions, including anxiety, depression, and hostility are associated with high levels of many somatic symptoms, and it has been suggested that negative emotions lead to higher levels of attention being paid to one's own body [6]. For example, anxiety is reported to amplify visceral-somatic processing, with greater sensitization at higher anxiety levels [36]. Healthy participants experienced alternating blocks of painful heat and non-painful warmth stimulation during the collection of fMRI data. Activation in two pain processing regions, the anterior and posterior cingulate, was correlated with individual pain-related fear. A regression analysis also revealed a significant correlation between anxiety sensitivity questionnaire scores or fear to pain and ACC and PFC regions activated by pain stimuli. Thus, pain perception is influenced by many complex conditions, and the mechanisms are engaged by various brain regions. Perez et al. have proposed a neurocircuitry framework of somatosensory amplification that includes frontolimbic, subcortical, and brainstem structures [40]. Neuroimaging studies concerned with somatosensory amplification and with the relationship between cognitive-affective processes and pain perception would contribute to the elucidation of distinctive neural mechanisms associated with somatoform pain disorder. We believe that the psychopathology of somatoform pain disorder as well as other chronic pain disorders is clearly influenced by such pain properties and is linked to emotional dysregulation.

3.3 Psychopathology of Somatoform Pain Disorder Including Psychological Factors

It has been reported that abnormal psychological complaints tend to be expressed as somatic symptoms in patients with somatoform pain disorder. Some studies have suggested that somatization may be associated with lower levels of emotional awareness (including empathy) or rather experiences as bodily sensation, and that patients with somatoform pain disorder may experience emotional responses only as passive implicit manifestations of emotion [14, 51]. It was also reported that patients with somatoform pain disorder have difficulty detecting emotional distress within themselves and tended to experience and report such distress as bodily sensations [51]. Other studies have reported that alexithymia, which is associated with emotional dysfunction, could be an underlying psychological basis of somatoform pain disorder, and emotional awareness or diminished emotional recognition seems to be impaired in somatoform pain disorder patients, who recognize these symptoms in the form of physical perceptions [38]. In fact, it seems that some patients with somatoform pain disorder recognize themselves as having no mental problems and tend to dissimulate emotional distress. Subic-Wrana et al. consider that the ability to experience an emotion such as sadness, anxiety, or anger may be addressed by helping patients to replace unexplained somatic symptoms with a mental representation, such that establishing such a representation may lead to the improvement of symptoms in patients with somatoform disorders [51]. Such deficits in emotional awareness might underlie the phenomenon of somatization, and focused training in emotional awareness may be needed for such patients. Furthermore, patients who have a stronger somatization tendency perform less well during semantic memory or verbal episodic memory tasks in comparison to healthy controls [33]. Thus, psychological elements are thought to be central to the onset, exacerbation, and maintenance of the complaint for patients with somatoform pain disorder. However, few neuroimaging studies of emotional deficits in somatoform pain disorder have been undertaken so far. De Greck et al. have tested their views using an emotion recognition paradigm [14]. They found that there were significantly decreased hemodynamic responses in brain regions such as the left postcentral gyrus, left superior temporal gyrus, bilateral parahippocampus, left posterior insula, left amygdala, and left cerebellum in patients when compared to controls in an angry emotional context. They also showed a significant correlation between several brain regions active during an emotion recognition test and somatization scores (on the SCL-90-R), and they proposed a strong association between impaired emotion recognition and somatization. Another fMRI study has also examined neural responses associated with empathy for pain in somatoform pain disorder patients, to ascertain the ability to imagine how one would feel in a particular painful situation [35]. Behavioral and neural responses showed that patients often report difficulty with identifying emotional signals within themselves and that the functioning of empathy in somatoform pain disorder is disturbed (discussed in detail, Sect. 1.1). However, few neuroimaging studies have examined the relationship between emotion recognition ability or emotional awareness or alexithymia and the neural mechanisms that underlie somatoform pain disorder. It is important to investigate the pathophysiology of somatoform pain disorder using neuroimaging methods to assess regulation of emotion and emotional awareness, given that somatoform pain disorder is strongly related to the disturbance in the regulation of emotion.

3.4 Summary of Pain-Related Neuroimaging Studies Associated with Psychopathology of Somatoform Pain Disorder

First, we described neuroimaging studies of the specific characteristics of pain itself that are easily modulated by emotional and cognitive factors, and these studies have shown that negative psychosocial factors are linked to higher pain sensitivity. We then reviewed the idea of somatosensory amplification, which is a sense of discomfort or unpleasantness and is affected by various factors including emotion, level of arousal, and environment. Finally, we reviewed deficits in emotional awareness or alexithymia as a possible contributing factor. These studies have identified abnormal pain-related brain activations in regions such as the insula, ACC, amygdala, parahippocampus, and PFC, and it might prove useful in understanding the psychopathology of somatoform pain disorder.

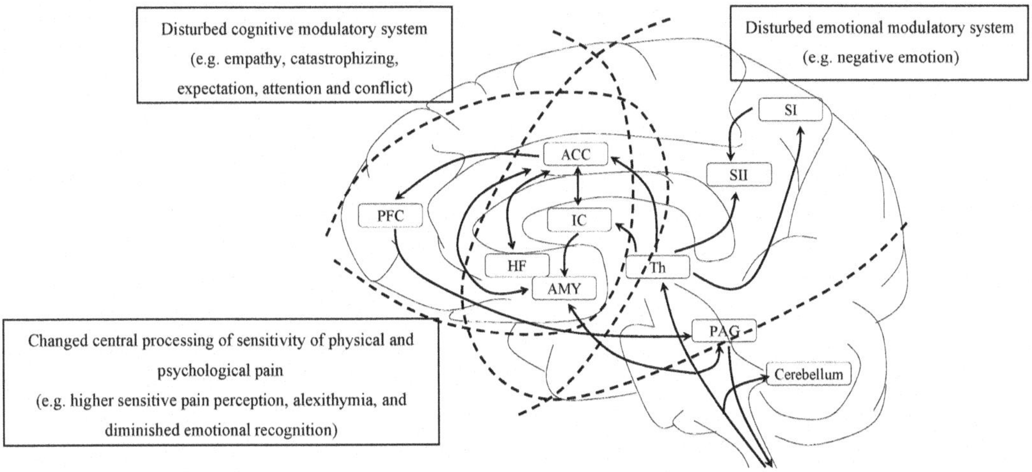

Fig. 1 Distinctive neural circuits of somatoform pain disorder. Disturbed cognitive modulatory system (e.g., empathy, catastrophizing, expectation, attention and conflict) is mainly related with high-order cortical areas, such as PFC and ACC. Disturbed emotional modulatory system (e.g., negative emotion) is largely associated with the brain regions in the affective dimension of pain perception such as ACC, insula, and amygdala. Changed central processing of sensitivity of physical and psychological pain is related with extensive pain-related brain regions. *ACC* anterior cingulate cortex, *AMY* amygdala, *HF* hippocampal formation (including the parahippocampus), *IC* insular cortex, *PAG* periaqueductal gray, *PFC* prefrontal cortex, *SI* primary somatosensory cortex, *SII* secondary somatosensory cortex, *Th* thalamus

4 Neural Mechanisms of Somatoform Pain Disorder

Considering the above described neuroimaging studies of somatoform pain disorder, and that pain perception is mediated by various cognitive and emotional processes, we suggest that the dysfunction in pain perception modulated by various psychosocial factors such as emotion and environmental or expressed emotion deficits influence the complicated etiology of somatoform pain disorder, and, to some extent, that neuroimaging studies also revealed its pathophysiology (Fig. 1). A distinctive neural network involved in somatoform pain disorder includes the SI, SII, amygdala, hippocampus, insula, the ACC, and the PFC, and altered pain perception in somatoform pain disorder is linked to interactions between the pain-related brain network and the pain modulatory system. Studies of patients with somatoform pain disorder have mostly showed increased activation in pain-related brain regions and decreased activation in the PFC. The PFC appears to have a substantial role in the cognitive control of pain perception, and impaired prefrontal functioning has been postulated to underlie a pathological neural basis of somatoform pain disorder [19, 24, 53], and it is assumed that such altered PFC activity is accompanied by changes in other pain-related brain regions. Increased ACC and insula activity in response to pain stimuli have been consistent both with the role of these areas in modulating the negative emotional response to pain stimuli and with cognitive factors of pain stimuli that play an important role in increased personal salience [55]. It has been proposed that pain-related somatosensory or limbic regions which process pain-related sensory-affective factors interact with the PFC that subserves attentional control, producing a sensitization (or desensitization) to pain perception [55]. We predict a raised salience of pain perception in these patients and this state may arise from abnormalities in higher-order modulation of pain perception. However, the evidence of neural abnormalities associated with this condition is not yet strong. Further experimental approaches which may more directly evaluate questions of pathology are expected to augment knowledge to date.

5 Future Research Directions

Many extant neuroimaging studies of somatoform pain disorder suffer from small sample sizes. Moreover, although possible explanations for this disorder include distinctive neural activities such as those observed in pain-related brain regions, few studies have examined possible abnormal neural networks characterized by functional connectivities among brain regions. For example, the ACC has functional connectivities with various brain regions such as the OFC, amygdala, PAG, and insula, and plays an important role in the integration of sensory, motor, cognitive, and emotional aspects of pain [3]. We need to elucidate further brain mechanisms for somatoform pain disorder. We also noted that there were few biological studies (including neuroimaging studies) examining the persistence or intractability of somatic symptoms, including therapeutic response. Furthermore, more detailed analysis of causal models is needed. For example, neurotransmitter, neurochemistry, psychosocial factors such as life history and personal relations, immune system, autonomic responses, and personality characteristic patterns such as highly anxious temperament may be considered for studies to further elucidate somatoform pain disorder. Other neuroimaging techniques including spinal cord imaging, optical imaging, magnetoencephalography (MEG) and near infrared spectroscopy may also be available to investigate brain regions of somatoform pain disorder. SI and SII functioning correlate with somatotopy, intensity of sensory information, and temporal attributes of pain, and are thought to play an important role in the sensory-discriminatory aspect of pain processing [44]. However, our MEG study suggests that the SI may also be involved in the processing of affective components, such as sadness [58], and it is necessary to examine SI and SII activation using various neuroimaging techniques to more clearly to examine the pathophysiology of somatoform pain disorder.

Pain perception is a multidimensional experience that includes affective and behavioral factors, and a disconnection between subjective affectivity or perception and objective behavior exists in many patients with somatoform pain disorder. Neuroimaging techniques might be useful to decrease this gap in the future, and are hoped to play an important part in endophenotypes for pain perception [52]. Somatoform pain disorder may be arbitrarily diagnosed on occasion because there is no gold standard to diagnosis. In the future, these advances will hopefully lead to an assurance of somatoform pain disorder being a definite "disease".

Moreover, in the future, neuroimaging studies may be directly more useful as indicators of treatment for patients with somatoform pain disorder. For example, it has been demonstrated that participants can acquire an established brain activation by themselves, using real-time functional MRI (rtfMRI) as the training [9]. A study showed that healthy subjects could learn activation of the rostral ACC (rACC) which is involved in pain perception and regulation via specifically guided training and that such activation was related to decreased intensity of pain perception during presentation of pain stimuli [9]. Chronic pain patients also showed approximately 50% reduced pain ratings after training with rtfMRI information from the rACC. In the future, this intervention might be applied for clinical use for patients with chronic pain states including somatoform pain disorder, although further evaluation is required.

Somatoform pain disorder has been reported to have a high comorbidity with other mental disorders including depression and anxiety disorders [32]. On the other hand, there appears to be a separate relationship between medically unexplained somatic complaints and depression [28]. It has been also reported that syndromes such as fibromyalgia and irritable bowel (which closely resemble somatoform pain disorder in many aspects) are separable from major depressive disorder and other psychiatric disorders in terms of having a strong familial or genetic predisposition [42]. However, there is little direct evidence to propose that any of the abnormalities found in neuroimaging studies are particular to somatoform pain disorder compared with other psychiatric diagnoses such as depression, which may present with similar symptoms. It would be necessary to further clarify differences and similarities in the cognitive, emotional, and behavioral responses to somatic symptoms or neural mechanisms between patients with somatoform pain disorder and other mental disorders such as depression or anxiety disorders. Similarly, it may be important to examine the relationship of psychiatric symptoms such as depression and anxiety between both disorders. It appears that it may be useful for neuroimaging techniques to distinguish somatoform pain disorder from other mental disorders by exploring the etiology of the disorders. In the future, further neuroimaging techniques could help to identify highly detailed neurobiological substrates of mental disorders related to somatoform pain disorder. It would be critical to develop the neurobiology of somatic symptoms including pain perception in the future. Neuroimaging studies are one of the most helpful approaches to contribute to such developments.

6 Conclusion

We have reviewed neuroimaging studies for somatoform pain disorder. Various psychosocial factors such as cultures, learning, memory, genetic or biological vulnerabilities, and early traumatic events seem to have an influence on the presentation of somatoform pain disorder. Multidimensional understandings are crucial for development of useful etiological models and treatment approaches, and we proposed the distinctive neural mechanisms of somatoform pain disorder in view of disturbed cognitive modulatory system, disturbed emotional modulatory system, and changed central processing of sensitivity of physical and psychological pain in reference to previous various studies. However, somatoform pain disorder neuroimaging studies remain few and far between, and further study is needed to elucidate the relationship between pathophysiology in somatoform pain disorder and the associated brain mechanisms.

References

1. APA. Diagnostic and statistical manual of mental disorders: DSM-IV. Washington, DC: American Psychiatric Publishing Inc.; 1994.
2. APA. Diagnostic and statistical manual of mental disorders, (DSM-5®). American Psychiatric Publishing; 2013.
3. Apkarian AV, Bushnell MC, Treede RD, Zubieta JK. Human brain mechanisms of pain perception and regulation in health and disease. Eur J Pain. 2005;9:463–84.
4. Atmaca M, Sirlier B, Yildirim H, Kayali A. Hippocampus and amygdalar volumes in patients with somatization disorder. Prog Neuropsychopharmacol Biol Psychiatry. 2011;35(7):1699–703.
5. Baliki MN, Petre B, Torbey S, Herrmann KM, Huang L, Schnitzer TJ, et al. Corticostriatal functional connectivity predicts transition to chronic back pain. Nat Neurosci. 2012;15(8):1117–9.
6. Barsky AJ, Goodson JD, Lane RS, Cleary PD. The amplification of somatic symptoms. Psychosom Med. 1988;50(5):510–9.
7. Bleich S, Römer K, Wiltfang J, Kornhuber J. Glutamate and the glutamate receptor system: a target for drug action. Int J Geriatr Psychiatry. 2003;18(S1): S33–40.
8. Bushnell MC, Čeko M, Low LA. Cognitive and emotional control of pain and its disruption in chronic pain. Nat Rev Neurosci. 2013;14(7):502–11.
9. Christopher deCharms R, Maeda F, Glover GH, Ludlow D, Pauly JM, Soneji D et al. Control over brain activation and pain learned by using real-time functional MRI. Proc Nat Acad Sci USA. 2005;102 (51):18626–31.
10. Coen SJ, Aziz Q, Yágüez L, Brammer M, Williams SC, Gregory LJ. Effects of attention on visceral stimulus intensity encoding in the male human brain. Gastroenterology. 2008;135(6):2065–74. e1.
11. Craig AB. How do you feel–now? The anterior insula and human awareness. Nat Rev Neurosci. 2009;10:59–70.
12. Damasio AR, Grabowski TJ, Bechara A, Damasio H, Ponto LL, Parvizi J, et al. Subcortical and cortical brain activity during the feeling of self-generated emotions. Nat Neurosci. 2000;3(10):1049–56.
13. de Greck M, Bölter AF, Lehmann L, Ulrich C, Stockum E, Enzi B et al. Changes in brain activity of somatoform disorder patients during emotional empathy after multimodal psychodynamic psychotherapy. Front Hum Neurosci. 2013;7.

14. de Greck M, Scheidt L, Bölter AF, Frommer J, Ulrich C, Stockum E, et al. Altered brain activity during emotional empathy in somatoform disorder. Hum Brain Mapp. 2012;33(11):2666–85.
15. Dickenson AH, Matthews EA, Suzuki R. Neurobiology of neuropathic pain: mode of action of anticonvulsants. Eur J Pain. 2002;6(SA):51–60.
16. Fayed N, Andres E, Rojas G, Moreno S, Serrano-Blanco A, Roca M, et al. Brain dysfunction in fibromyalgia and somatization disorder using proton magnetic resonance spectroscopy: a controlled study. Acta Psychiatr Scand. 2012;126 (2):115–25.
17. Fayed N, Olmos S, Morales H, Modrego PJ. Physical basis of magnetic resonance spectroscopy and its application to central nervous system diseases. Am J Appl Sci. 2006;3(5):1836.
18. Fink P, Hansen MS, Oxhoj M-L. The prevalence of somatoform disorders among internal medical inpatients. J Psychosom Res. 2004;56(4):413–8.
19. Gündel H, Valet M, Sorg C, Huber D, Zimmer C, Sprenger T, et al. Altered cerebral response to noxious heat stimulation in patients with somatoform pain disorder. Pain. 2008;137(2):413–21.
20. Galhardoni R, Correia GS, Araujo H, Yeng LT, Fernandes DT, Kaziyama HH, et al. Repetitive transcranial magnetic stimulation in chronic pain: a review of the literature. Arch Phys Med Rehabil. 2015;96(4):S156–72.
21. Garcia-Campayo J, Sanz-Carrillo C, Baringo T, Ceballos C. SPECT scan in somatization disorder patients: an exploratory study of eleven cases. Aust NZJ Psychiatry. 2001;35(3):359–63.
22. Hakala M, Karlsson H, Kurki T, Aalto S, Koponen S, Vahlberg T, et al. Volumes of the caudate nuclei in women with somatization disorder and healthy women. Psychiatry Res Neuroimaging. 2004;131 (1):71–8.
23. Hakala M, Vahlberg T. NIEMI PM, Karlsson H. Brain glucose metabolism and temperament in relation to severe somatization. Psychiatry Clin Neurosci. 2006; 60(6):669–75.
24. Karibe H, Arakawa R, Tateno A, Mizumura S, Okada T, Ishii T, et al. Regional cerebral blood flow in patients with orally localized somatoform pain disorder: a single photon emission computed tomography study. Psychiatry Clin Neurosci. 2010;64 (5):476–82.
25. Koh KB, Sohn S-H, Kang JI. Lee Y-j, Lee JD. Relationship between neural activity and immunity in patients with undifferentiated somatoform disorder. Psychiatry Res Neuroimaging. 2012;202(3):252–6.
26. Kroenke K. Patients presenting with somatic complaints: epidemiology, psychiatric co-morbidity and management. Int J Methods Psychiatr Res. 2003;12 (1):34–43.
27. Kroenke K. Efficacy of treatment for somatoform disorders: a review of randomized controlled trials. Psychosom Med. 2007;69(9):881–8.
28. Lane RD, Carmichael C, Reis HT. Differentiation in the momentary rating of somatic symptoms covaries with trait emotional awareness in patients at risk for sudden cardiac death. Psychosom Med. 2011;73 (2):185–92.
29. Lipowski ZJ. Somatization: the concept and its clinical application. Am J Psychiatry. 1988;145 (11):1358–68.
30. Mohr C, Leyendecker S, Mangels I, Machner B, Sander T, Helmchen C. Central representation of cold-evoked pain relief in capsaicin induced pain: an event-related fMRI study. Pain. 2008;139(2):416–30.
31. Morgan V, Pickens D, Gautam S, Kessler R, Mertz H. Amitriptyline reduces rectal pain related activation of the anterior cingulate cortex in patients with irritable bowel syndrome. Gut. 2005;54(5):601–7.
32. Muller JE, Wentzel I, Nel DG, Stein DJ. Depression and anxiety in multisomatoform disorder: Prevalence and clinical predictors in primary care. SAMJ S Afr Med J. 2008;98(6):473–6.
33. Niemi P, Portin R, Aalto S, Hakala M, Karlsson H. Cognitive functioning in severe somatization—a pilot study. Acta Psychiatr Scand. 2002;106 (6):461–3.
34. Noll-Hussong M, Otti A, Laeer L, Wohlschlaeger A, Zimmer C, Lahmann C, et al. Aftermath of sexual abuse history on adult patients suffering from chronic functional pain syndromes: an fMRI pilot study. J Psychosom Res. 2010;68(5):483–7.
35. Noll-Hussong M, Otti A, Wohlschlaeger AM, Zimmer C, Henningsen P, Lahmann C, et al. Neural correlates of deficits in pain-related affective meaning construction in patients with chronic pain disorder. Psychosom Med. 2013;75(2):124–36.
36. Ochsner KN, Ludlow DH, Knierim K, Hanelin J, Ramachandran T, Glover GC, et al. Neural correlates of individual differences in pain-related fear and anxiety. Pain. 2006;120:69–77.
37. Otti A, Guendel H, Wohlschläger A, Zimmer C, Noll-Hussong M. Frequency shifts in the anterior default mode network and the salience network in chronic pain disorder. BMC Psychiatry. 2013;13 (1):84–92.
38. Pedrosa Gil F, Ridout N, Kessler H, Neuffer M, Schoechlin C, Traue HC et al. Facial emotion recognition and alexithymia in adults with somatoform disorders. Depress Anxiety. 2009;26(1):E26–E33.
39. Perahia DG, Pritchett YL, Desaiah D, Raskin J. Efficacy of duloxetine in painful symptoms: an analgesic or antidepressant effect? Int Clin Psychopharmacol. 2006;21(6):311–7.
40. Perez DL, Barsky AJ, Vago DR, Baslet G, Silbersweig DA. A neural circuit framework for somatosensory amplification in somatoform disorders. J Neuropsychiatry Clinic Neurosci. 2015.
41. Peyron R, Laurent B, Garcia-Larrea L. Functional imaging of brain responses to pain. A review and

meta-analysis. Clinic Neurophysiol. 2000;30:263–88. S0987-7053(00)00227-6 [pii].
42. Phillips K, Clauw DJ. Central pain mechanisms in chronic pain states–maybe it is all in their head. Best Pract Res Clin Rheumatol. 2011;25(2):141–54.
43. Phillips ML, Drevets WC, Rauch SL, Lane R. Neurobiology of emotion perception I: The neural basis of normal emotion perception. Biol Psychiat. 2003;54(5):504–14.
44. Price DD. Central neural mechanisms that interrelate sensory and affective dimensions of pain. Mol Interv. 2002;2:392–403.
45. Sadock BJ, Sadock VA. Synopsis of psychiatry. 2003.
46. Sawamoto N, Honda M, Okada T, Hanakawa T, Kanda M, Fukuyama H, et al. Expectation of pain enhances responses to nonpainful somatosensory stimulation in the anterior cingulate cortex and parietal operculum/posterior insula: an event-related functional magnetic resonance imaging study. J Neurosci. 2000;20(19):7438–45.
47. Smith GR Jr, Monson RA, Ray DC. Patients with multiple unexplained symptoms: their characteristics, functional health, and health care utilization. Arch Intern Med. 1986;146(1):69–72.
48. Stefanie K, Peter A, Gerda S-Z, Marion F, Bernd S, Wolfgang P, et al. Dysfunctional pain modulation in somatoform pain disorder patients. Eur Arch Psychiatry Clin Neurosci. 2011;261(4):267–75.
49. Stoeter P, Bauermann T, Nickel R, Corluka L, Gawehn J, Vucurevic G, et al. Cerebral activation in patients with somatoform pain disorder exposed to pain and stress: an fMRI study. Neuroimage. 2007;36(2):418–30.
50. Su Q, Yao D, Jiang M, Liu F, Jiang J, Xu C, et al. Dissociation of regional activity in default mode network in medication-naive, first-episode somatization disorder. PLoS ONE. 2014;9(7):e99273.
51. Subic-Wrana C, Beutel ME, Knebel A, Lane RD. Theory of mind and emotional awareness deficits in patients with somatoform disorders. Psychosom Med. 2010;72(4):404–11.
52. Tracey I. Can neuroimaging studies identify pain endophenotypes in humans? Nat Rev Neurol. 2011;7(3):173–81.
53. Valet M, Gündel H, Sprenger T, Sorg C, Mühlau M, Zimmer C, et al. Patients with pain disorder show gray-matter loss in pain-processing structures: a voxel-based morphometric study. Psychosom Med. 2009;71(1):49–56.
54. Vogt BA. Pain and emotion interactions in subregions of the cingulate gyrus. Nat Rev Neurosci. 2005;6(7):533–44.
55. Wiech K, Ploner M, Tracey I. Neurocognitive aspects of pain perception. Trends Cognit Sci. 2008;12(8):306–13.
56. Yoshino A, Okamoto Y, Horikoshi M, Oshita K, Nakamura R, Otsuru N et al. Effectiveness of group cognitive behavioral therapy (GCBT) for somatoform pain disorder patients in Japan: a preliminary non-case-control study. Psychiatry and clinical neurosciences. 2015; in press.
57. Yoshino A, Okamoto Y, Kunisato Y, Yoshimura S, Jinnin R, Hayashi Y, et al. Distinctive spontaneous regional neural activity in patients with somatoform pain disorder: a preliminary resting-state fMRI study. Psychiatry Res Neuroimaging. 2014;221:246–8.
58. Yoshino A, Okamoto Y, Onoda K, Shishida K, Yoshimura S, Kunisato Y, et al. Sadness enhances the experience of pain and affects pain-evoked cortical activities: An MEG study. J Pain. 2012;13:628–35.
59. Yoshino A, Okamoto Y, Onoda K, Yoshimura S, Kunisato Y, Demoto Y et al. Sadness enhances the experience of pain via neural activation in the anterior cingulate cortex and amygdala: an fMRI study. Neuroimage. 2010;50:1194–201. doi:10.1016/j.neuroimage.2009.11.079 S1053-8119(09)01265-8 [pii].
60. Yoshino A, Okamoto Y, Yoshimura S, Shishida K, Toki S, Doi M et al. Distinctive neural responses to pain stimuli during induced sadness in patients with somatoform pain disorder: an fMRI study. Neuroimage: Clinic. 2013;2:782–9.
61. Zhang D, Raichle ME. Disease and the brain's dark energy. Nat Rev Neurol. 2010;6(1):15–28.

Neuroimaging of Visceral Pain

13

James K. Ruffle, Jens B. Frokjaer and Adam D. Farmer

Abstract

Visceral pain is a complex and multidimensional experience associated with a plethora of underlying pathologies. Chronic visceral pain remains poorly understood and sub-optimally managed in the clinic. Thus, it exerts a significant socioeconomic burden and reduction in health-related quality of life. In excess of 100 studies over the past 20 years have provided insights into the supraspinal processing of visceral pain both in health and disease, with many utilising neuroimaging. However, there remains considerable controversy due to marked variability in stimulation paradigms and participant selection. In this chapter, we provide a focused review of the literature concerning the neuroimaging of visceral pain and also propose some new directions. Such new directions may provide further insights into the pathophysiology of visceral pain syndromes and facilitate the objective evaluation of novel treatment strategies.

Keywords

Interoceptive · Brain–gut axis · Viscerotopic · Oesophageal · Irritable bowel syndrome

J.K. Ruffle · A.D. Farmer (✉)
Centre for Digestive Diseases, Blizard Institute, Wingate Institute of Neurogastroenterology, Barts and The London School of Medicine & Dentistry, Queen Mary University of London, London, UK
e-mail: a.farmer@qmul.ac.uk

J.K. Ruffle
Acute Medical Assessment Unit, Barts Health NHS Trust, Whipps Cross University Hospital, Whipps Cross Rd, Leytonstone, London E11 1NR, UK

J.B. Frokjaer · A.D. Farmer
Mech-Sense, Aalborg University Hospital, Aalborg, Denmark

J.B. Frokjaer · A.D. Farmer
Department of Clinical Medicine, Aalborg University, Aalborg, Denmark

A.D. Farmer
Department of Gastroenterology, University Hospitals of North Midlands NHS Trust, Royal Stoke University Hospital, Newcastle Road, Stoke on Trent ST4 6QG, UK

A.D. Farmer
University Hospitals of North Midlands, County Hospital, Weston Road, Staffordshire ST16 3RS, UK

L. Saba (ed.), *Neuroimaging of Pain*, DOI 10.1007/978-3-319-48046-6_13

1 Introduction to Visceral Pain

The brain receives a vast number of sensory signals from visceral afferents, which encode a variety of modalities. A large proportion of these signals are responsible for the regulation of gastrointestinal (GI) homeostasis, which include the balancing of GI motility and secretion. Furthermore, these sensory signals can be subdivided into subconscious and conscious. Subconscious signals include information from baroreceptors, stretch receptors and GI secretor–motor reflex control, whilst conscious signals include the viscerosensations of satiety, nausea and, most importantly, pain. Cumulatively, a complex GI sensory spectrum is achieved, all of which invoke activity in numerous cortical and subcortical structures. Visceral pain refers to nociception from any of the large interior organs from bodily cavities, for example the GI tract, and is a central defining feature of numerous disorders such as irritable bowel syndrome (IBS), a prevalent functional GI disorder (FGID) [1–3].

Perceptions and sensory inputs are central components to the definition of the bodily senses. Importantly, senses are often divided into 'exteroceptive' and 'interoceptive'. 'Exteroception' refers to the perception of stimuli that originate from outside of the body, and include visual, acoustic or tactile modalities. In contrast however, and of upmost relevance to this chapter, the perception of internally originating sensory signals, such as visceral pain, and the subsequent brain processing of these is often referred to as 'interoception'. Interoception may be defined as the neurophysiology comprising the perception of one's internal bodily state [2]. To that end, much research over the last two or three decades has focused on identifying the functional neurophysiology implicated in this. Whilst multiple stimulus modalities have been investigated [4], a large proportion of interoceptive research has regarded pain. This chapter will focus on the important interoceptive dimension that is visceral pain. The functional neurophysiology underpinning visceral pain will be described, along with advances in knowledge gained from functional brain imaging studies in humans. The 'brain–gut axis' [5, 6], a construct denoting the bidirectional intercommunication between the gut and the brain, will be discussed with reference to the numerous neuroimaging studies that have investigated it. The experimental paradigms in numerous neuroimaging studies investigating visceral pain will be briefly detailed, including investigations of oesophageal pain and rectal pain. The neuroimaging modalities for exploring painful visceral disorders will also be described, including that of IBS and inflammatory bowel disease (IBD). Furthermore, much visceral pain neuroimaging research has been to characterise the inter-individual factors that influence the cognitive interpretation pain merits. These factors include physiological, anatomical and psychophysiological, and will be discussed throughout this chapter. The neuroimaging of visceral pain is a highly topical area of modern research, thus we will conclude with the clinical utility and suggest future potential directions for the field.

2 Functional Neuroanatomy of Visceral Pain

2.1 Introduction to the 'Brain–Gut Axis'

The brain–gut axis can be defined as a bidirectional intercommunication between the gut and the brain [5, 6]. Research from a diverse array of disciplines, coupled with technological advances, particularly in functional neuroimaging, has provided an increasing body of evidence for the proposal of a brain–gut axis over the last decade, such that it is now arguably a central component to research comprising the neuroimaging of visceral pain [7–9]. Although disparate experimental methods have been used to instigate pain at various portions of the viscera, and indeed different neuroimaging modalities utilised, a more or less consistent 'visceral sensory neuromatrix' has been elucidated. Lastly, it should be noted that the brain–gut axis does not only encompass neural components. In particular, the autonomic nervous system (ANS), neuroendocrine [hypothalamo-pituitary-adrenal (HPA)] and neuroimmune systems are also important aspects as they mediate the stress response to pain [6].

This proposed communication system between gut and brain has gained widespread acceptance as the germane construct for providing an explanation of both the normal, acute and chronic alterations in digestive tract function. Moreover, this model of circuitous communication has provided a biological construct to underpin the biopsychosocial concept of numerous GI disorders by facilitating the integration of many contributing factors whether they be biological, psychological or social in nature. One such disorder where research of the brain–gut axis is well studied is that of IBS [3], and will be discussed in further detail later.

2.2 Differences Between Visceral and Somatic Pain

Numerous differences are apparent when comparing the neuroanatomy of both the somatic and visceral nervous systems. Pain originating from the viscera is characteristically difficult to localise [1], and frequently is rated more unpleasant than the somatic counterpart [10]. For somatic pain, cutaneous nociceptive afferents transmit impulses to the respective level of the spinal cord dorsal horn, following which the spinothalamic tract projects to the somatic pain processing nuclei, in particular the lateral and medial thalamic, before further transmission to other cortical regions [11]. For the case of visceral pain however, key differences are exhibited both at the periphery and centrally [12]. In order to describe the neuroimaging of visceral pain, its underpinning neuroanatomy and neurophysiology must firstly be understood, and hence will be discussed below, spanning from the ascending spinal pathways, central brain processing and lastly its descending modulation.

2.3 Ascending Pathways

The ascending neural pathways of visceral sensation, and indeed pain, are comprised of GI afferents, referred to as vagal and spinal pathways [13–16]. These two afferent types converge at multiple supraspinal levels, but notably include the nucleus tractus solitarii (NTS), a cluster of brainstem nuclei associated with multiple inputs including baroreceptors, GI and pulmonary afferents [17, 18]. The NTS subsequently projects principally to the parabrachial nucleus (PB), prior to higher transmission. However, it is important to note that the PB also receives its own inputs from visceral dorsal horn laminae. This highlights a complex interlinking system of both autonomic and sensory afferents.

Spinal and vagal afferents indirectly project to a diverse group of both cortical and subcortical targets. These secondary projections include the thalamus (in particular the ventral posterior lateral, medial dorsal and ventral medial posterior nuclei), insula, amygdala, prefrontal cortex (PFC), primary somatosensory cortex (SI), secondary somatosensory cortex (SII) and cingulate cortices [including the anterior cingulate cortex (ACC)] [6]. It is thought that these brain regions are, by large, viscerotopically organised, that is to say specific aspects or coordinates of these brain regions are thought to correspond to specific components of the viscera [19–21]. By elaborate connections, a visceral sensory neuromatrix exists, both for innocuous and nociceptive sensation [22].

2.4 The Brain Processing of Visceral Pain

Historically, most prior literature investigating the functional roles of brain regions implicated in pain has concerned somatic sensation. However, over the last decade, there has been a significant number of functional brain imaging studies investigating visceral sensation. Consequentially, the wealth of visceral pain neuroimaging research has permitted a more or less robust definition of the brain–gut axis, especially with regards to pain processing. Examples of sites for visceral sensation are both the oesophageal and rectal anatomical regions and, although somatic and visceral pain show some similarities with regards to their central neural processing, differences of brain activity are apparent as a consequence of differing topographic representation at both individual sites and indeed the nervous system (peripheral and central) more generally [20, 21, 23–25]. The brain regions perhaps best implicated in visceral pain processing, thus far, are the SI, SII, cingulate cortex, insula, PFC and thalamus [6, 26–29]. The understood roles of these individual brain regions will be discussed herein.

First, the SI and SII principally act to encode intensity and localisation of visceral pain [21, 30]. These sensory regions are often referred to as the 'lateral pain system' [6]. Second, the cingulate cortex is implicated in the emotional interpretation of the stimulus [31]. Regarded as the 'medial pain system', the cingulate cortex has been shown to involve two distinct pain dimensions: the affective-motivational (pain unpleasantness and related anxiety) and cognitive-evaluative (anticipation and attention of pain) [32–36]. In addition, the cingulate cortex is anatomically separated into the ACC and mid cingulate cortex (MCC). This is then further subdivided into the pregenual/subgenual ACC and anterior/posterior MCC [6].

The insula is a complex region well implicated in neural processing of visceral sensation (Fig. 1) [37–40]. Specifically, the right anterior insula has been regarded by some as the 'interoceptive cortex' [2, 37], with a central role in the subjective awareness and perception of the material or bodily self as a sentient (or feeling) entity, thus it is thought that the right anterior insula engenders emotional awareness [41]. Whilst its role is not specific to painful sensation, it is thought that the insula is implicated in the processing of the affective dimension of pain, integrating it with emotional information [6, 41–43]. In addition, the insula encompasses efferent outputs to regions such as the amygdala, hypothalamus and periaqueductal grey (PAG), and thus serves an additional role in incorporating additional brain regions in visceral pain processing [20, 35, 40, 44–46]. These brain regions are understood to have important and unique functions of their own in how we process pain. Importantly, the amygdala has a critical role in the attribution of an affective dimension, in particular fear, to the painful sensation (and percepts in general) [32, 45]. Furthermore, the PAG, along with the amygdala, yields an important role in the descending modulation of pain [44]; a neurophysiological response to pain that will be discussed further later.

Located at superoanterior aspect of the brain is the PFC, a large region implicated in one's cognitive influence on pain [47]. A major subdivision of the PFC is the orbitofrontal cortex (OFC), which functions to process sensory information and encode its affective, motivational and hedonic dimensions [48, 49]. The OFC is additionally implicated in the decisions of an autonomic and behavioural response to the stimulus [6], as well as the complex interaction between cognition (such as anticipation) and emotion to pain [50, 51]. Another subdivision of the PFC is the dorsolateral PFC, which is mainly implicated in cognition, including the attention to and anticipation of visceral pain [21].

One of the best-implicated regions in pain processing is the thalamus. Comprised of six nuclear groups per thalamic hemisphere (anterior, medial, lateral, intralaminar, midline and reticular), its functionality includes relay nuclei (which interconnect cortical regions to convey information such as perception and emotion) and projection nuclei (which receive numerous inputs, and output to various higher brain regions). It is well implicated in pain sensation and the arousal response, to which its outputs include the described insula, SI and PFC [28, 32, 52]. However, in visceral pain neuroimaging studies, its activation is not universally reported, and thus some questions remain regarding the extent of its role in visceral pain when compared to somatosensory [12].

2.5 Comparing the Brain Processing of Visceral and Somatic Pain

Whilst there are numerous similarities in brain regions activated in those exposed to either visceral or somatic pain, such as the insula, thalamus, ACC and PAG [53], differences have been observed by multiple research groups. When comparing the brain processing of somatic pain, a recent review by Johns et al, highlights the main differences to be the preferential activation of the perigenual ACC and

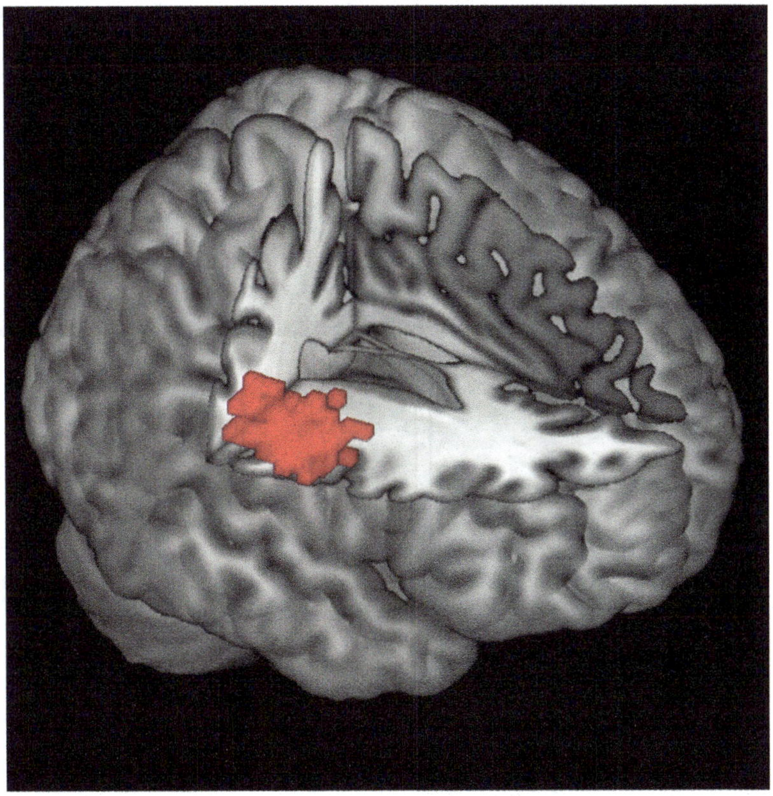

Fig. 1 Activity in the right insula in subjects undergoing visceral pain, by means of oesophageal distension. Three-dimensional render prepared utilising fMRI data from [67]

the occasional absence of activation in the thalamus for visceral pain [12, 54, 55]. Whilst some studies do report thalamic activity following painful visceral stimulation, this is not universal, a surprising finding given its central role in integration of information from spinal cord to brain, not least that of somatic pain. A lesser extent of activity in the primary somatosensory cortex is also reported in visceral pain studies when compared to somatic [56]. Lastly, Dunckley et al, report greater activity in the PAG and nucleus cuneiformis, suggested by the authors to represent a greater 'nocifensive' response and, furthermore, greater emotional salience following painful visceral stimulation compared to somatosensory [25]. Needless to say, critical disparities between the brain processing of somatic and visceral pain processing exist [21], although further research is required to tease these apart further by means of studies whose primary aims are to investigate this, as opposed to crudely comparing neuroimaging studies which often have markedly different paradigms or analysis techniques.

Eickhoff et al., conducted an single-study comparison of somatosensory and visceral pain [38]. The group harnessed functional magnetic resonance imaging (fMRI) and ano-rectal balloon distension, combined with the prior knowledge that, whilst the anal canal is innervated by somatosensory afferents, the rectum is innervated by the enteric nervous system. The group found key differences between rectal (visceral) and anal (somatosensory) pain processing. Whilst many comparable brain regions were activated under both stimuli, distinct clusters of activity within regions were also apparent. For example, disparate aspects of the insula were activated and frontal–parietal operculum between somatosensory or visceral pain. Furthermore, anal pain evoked activity in the SII area primarily, whilst rectal pain evoked brain activity more anteriorly on the precentral operculum, leading the group to conclude that an element of segregation is apparent between visceral and somatosensory pain (Fig. 2) [38].

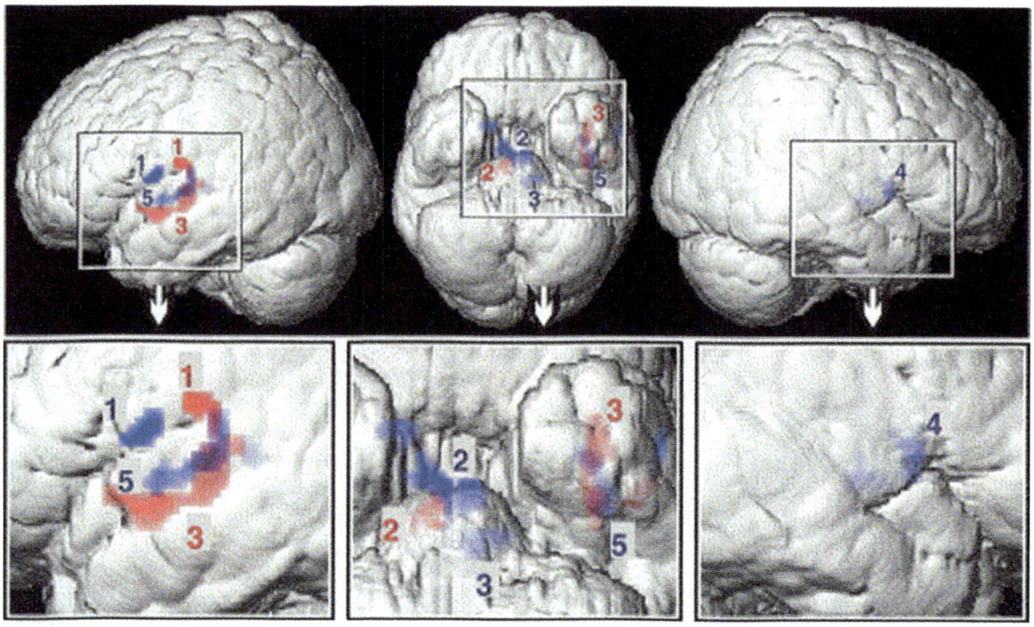

Fig. 2 (*Upper row*) Foci associated with the processing of anal (*red*) and rectal (*blue*) sensations. Areas of significant increase in blood oxygen level dependent signal are shown superimposed on a surface rendering of the subject template. (*Lower row*) Detail views on these renderings. The *black* and *white boxes* in the *upper row* of images denote the location of the detail images. Figure from [38]

2.6 Descending Modulation

Thus far, we have described the sensory projections from gut to brain and subsequent cortical processing. To complete the description of the aforementioned bidirectional brain–gut axis however, one must also discuss the projections from the brain caudally to the gut. Most brain structures receiving visceral sensory input, be it nociceptive or innocuous, do indeed project caudally in order to modulate ongoing transmission of visceral afferents, in particular to the spinal cord dorsal horn [9, 57]. It is thought that the ACC is central to this descending modulation, with additional projections to the amygdala and PAG, to which numerous functional imaging studies support this [25, 44, 58]. By processing at the ACC, it is thought that both cognitive and affective factors may therefore modulate and hence influence visceral pain transmission [58]. The PAG and amygdala subsequently project to the locus coeruleus (LC), which subsequently projects caudally to the spinal cord dorsal horn to mediate nociceptive afferents, termed 'gate control theory' [22]. Different neurotransmitters have been implicated including, but not limited to, endogenous opioids, serotonin and noradrenaline [58]. Importantly, descending modulation of visceral pain is considered to be influenced by psychological factors (and indeed 'pain endophenotypes', which shall be discussed later) and hence numerous groups have sought to investigate this, many of which have utilised neuroimaging [59–61]. This notion of psychophysiological influence is not unfounded; moreover the majority of the brain regions that process visceral sensation have been shown to additionally have a role in emotional perception, generation and regulation [6].

3 Brain Networks of the Visceral Pain Neuromatrix

Recent functional brain imaging studies have focused on elucidating potential brain 'connectivity networks' implicated in visceral pain processing [62, 63]. This aspect of research focuses on the interaction, or networking, between several brain regions, as opposed to independent brain region analysis [64]. Based upon multiple visceral pain neuroimaging studies, consensus has been largely reached that concludes three key networks formulate the visceral pain neuromatrix. These are as follows: (1) the homeostatic-afferent network, comprised of the thalamus, insula, OFC and dorsal ACC, (2) the emotional arousal network, comprised of the amygdala, ACC subregions and LC complex, and lastly (3) the cortical modulatory network, comprised of the ACC, amygdala, insula and medial OFC [33, 45, 62, 63, 65, 66]. The homeostatic-afferent network is thought to be predominantly responsible for processing the interoceptive input in a physiological sense via the PB [63]. The emotional arousal network, as name would suggest, is thought more responsible for how we emotionally interpret and perceive visceral pain [66], and lastly the role of the cortical modulatory network is the modifying of pain perception, also referring to the caudally travelling aspect of the brain–gut axis [13].

For the case of FGIDs, a term characterising numerous disorders but perhaps most often associated with IBS, connectivity network research has also played an important role. In particular, the homeostatic-afferent and emotional arousal networks have been a major focus. These disorders and their associated findings apparent on functional neuroimaging will be discussed in greater detail later, but many of these disorders by definition are chronic visceral pain syndromes. It is consequently thought that in these disorders, increased engagement of the homeostatic-afferent and emotional arousal network occurs [63].

4 Methods and Sites of Experimentally Induced Visceral Pain

In investigating acute visceral pain using neuroimaging, numerous experimental stimuli have been utilised to generate or evoke it. Furthermore, different bodily sites have also been investigated, and include aspects of the GI tract such as the oesophagus, stomach, duodenum and rectum, but also other GI such as the pancreas, or indeed other visceral systems entirely such as those urological or gynaecological. Investigating these different visceral sites is most beneficial for research progression, as it eliminates any assumption that the brain processes the entire visceral sensory system equivocally.

Frequently adopted paradigms to investigate GI pain are that of balloon distension and electrical stimulation, whereby different sites have been tested for both. For balloon distension, this method that has been used in sites such as the oesophagus [21, 28, 67, 68], stomach [53] and rectum [38, 69, 70] typically involves the use of an elasticated balloon that is either automated to inflate by air, or is manually done so by means of attached tubing, such as nasogastric tube and an extrinsic syringe permitting insufflation. Disparate to balloon distension, electrical stimulation has also been utilised, for example, at the oesophagus [71], stomach [72], duodenum [72] and rectum [25], whereby electrical stimulating catheters are typically utilised (Fig. 3).

5 Functional Neuroimaging

In evaluating the neural correlates of visceral pain, be that utilising structural or functional neuroimaging, it must be noted that the 'viscera' encompasses a vast amount of bodily tissue with the capacity for nociception. Furthermore, a presumption that the brain processing of all visceral pain is completely identical would likely be erroneous. For example, is the central processing of oesophageal and rectal pain identical? Phrasing this clinically, is the brain processing of pain occurring in IBS equivocal to that of gastric oesophageal reflux disease? There is a necessity to explore the regional visceral neuroimaging to answer these questions. Therefore, we will discuss region-specific neuroimaging of visceral pain sites where such studies exist, be it with experimentally induced pain or resultant from a GI disorder. In particular, studies utilising functional imaging, such as fMRI with blood oxygenation level dependent (BOLD) technique, along with positron emission tomography (PET) will be focused upon herein.

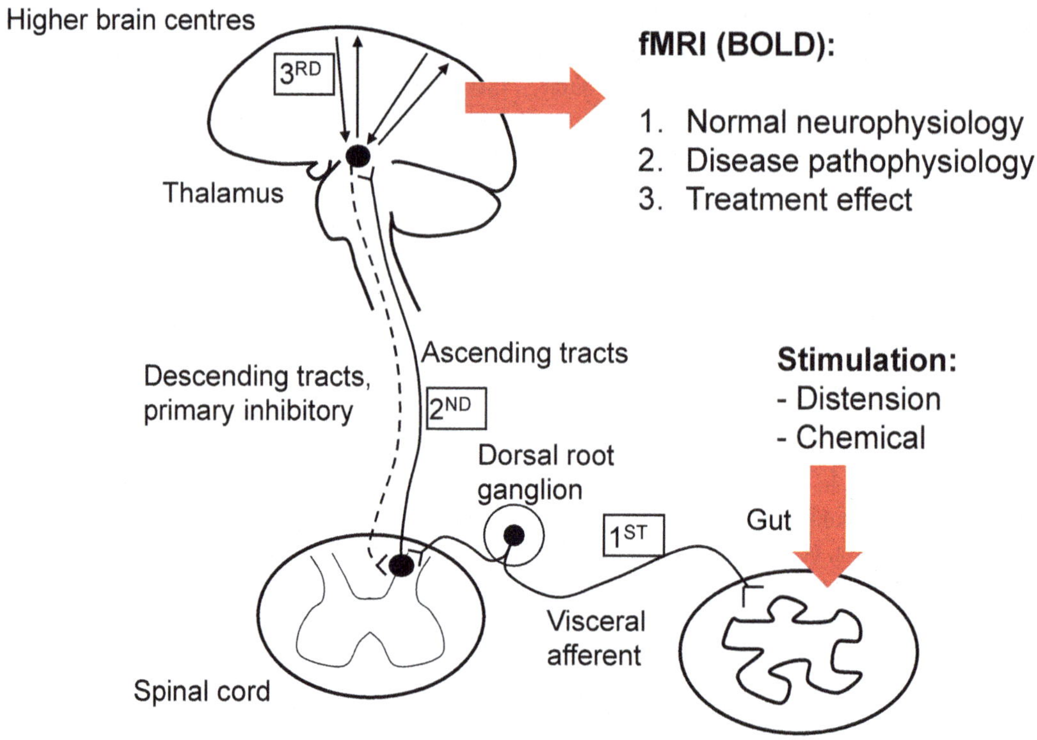

Fig. 3 Pathway from experimentally induced visceral pain (such as by distension of chemical stimulation) to fMRI data capture

5.1 Irritable Bowel Syndrome

IBS is one of the most frequently studied 'functional' GI disorders with regards to the neuroimaging of pain, given its incomplete understanding and additional high prevalence of 5–20% [3, 73]. The disorder is also more prevalent in women, and thus its associated brain processing of pain ties into the influence of gender that we shall discuss in detail later. IBS has been described as one of the most common persistent pain syndromes, whereby dysregulation of the brain–gut axis is central to the pathology, encompassing the development of visceral hypersensitivity [3]. The disorder is defined by the third iteration of the Rome criteria which centres around the presence of recurrent abdominal pain that improves with defecation [74]. As an incompletely understood disorder with minimal efficacious therapies, IBS is well associated with a significant socioeconomic impact through a reduction in health-related quality of life and lastly enormous healthcare costs (for example from over investigation and recurrent consultations) [3].

In the neuroimaging of IBS patients, much, if not all, of the visceral pain neuromatrix is typically activated, as would be expected (Fig. 4). A 2011 meta-analysis of fMRI studies utilising painful rectal inflation in both IBS patients and healthy controls highlights many of the apparent disparities [75]. Typically, in individuals exposed to painful rectal distension, activated brain regions include the thalamus, insula (occasionally bilateral, but in particular the right anterior aspect, as this side is well implicated in the processing of visceral nociception as opposed to somatic [2]), pregenual ACC, MCC and amygdala [75]. Furthermore, dependent on the study, other regions may also be shown to activate, see Mayer et al. [70].

Connectivity studies have also been utilised in IBS research. When areas of brain activation

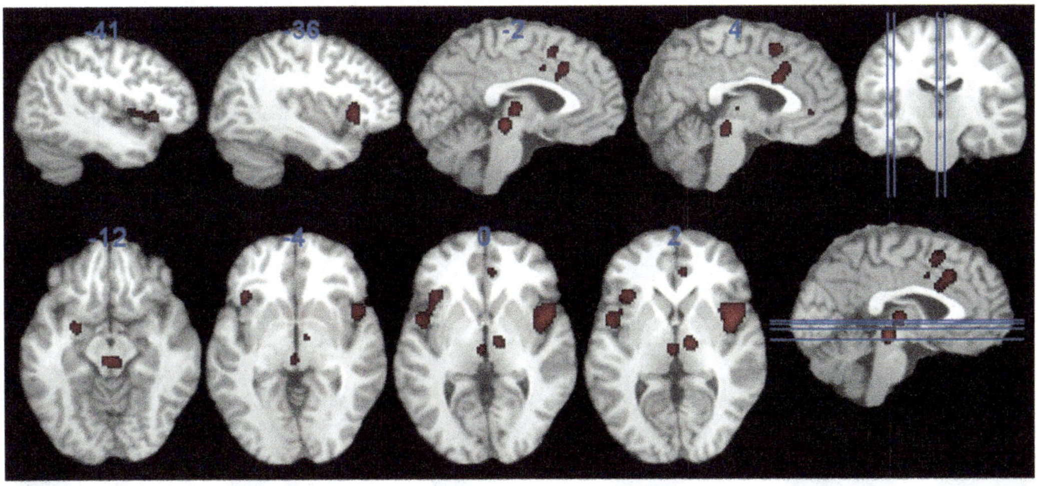

Fig. 4 Regions showing consistent and reliable activation across in patients with irritable bowel syndrome. Data and figure from [75]

are compared between the healthy and IBS populations, IBS patients typically show a greater degree of activity in the pregenual ACC and amygdala, suggested by the authors of the meta-analysis to reflect greater activity of the emotional arousal neural network, and additionally midbrain clusters, a brain area described earlier associated with nociceptive descending modulation (Fig. 5) [70, 75–77]. In comparison, healthy controls appear to display greater activity in regions of the cognitive-modulatory neural network, in particular the medial and lateral PFC [75]. Furthermore, when IBS was co-varied with gender, disparate connectivity strengths of the emotional arousal circuitry is apparent, as opposed to any visceral processing units [62, 78].

5.2 Inflammatory Bowel Disease

Another GI disorder well associated with visceral pain is that of IBD, an umbrella term encompassing both Crohn's disease (CD) and ulcerative colitis (UC). Both CD and UC are characterised by colonic inflammation and pain, although numerous GI pathophysiological differences differentiate them apart, see Abraham et al, for a comprehensive review [79]. The brain processing of pain in IBD patients has been studied utilising neuroimaging. In a ^{15}O-water PET study, Mayer et al., sought to compare brain activity following rectal distention in healthy controls, UC and IBS patients (Fig. 6) [70]. First, UC patients displayed regional activations as would be expected following a painful visceral stimulus, including the insula, ACC and PAG [6, 70]. Furthermore, when comparing UC and IBS patients, the UC cohort showed greater activity in the bilateral dorsal pons and PAG, whilst IBS patients displayed greater activity in the amygdala and rostroventral ACC and dorsomedial PFC. These findings are particularly interesting when the understood roles of these individuals are taken account of. For example, is there greater activity of corticopontine pain inhibitory circuits, including descending modulation from the PAG, in UC-afflicted patients? Furthermore, the regions of greater activity in IBS patients are well-documented emotional arousal network regions. Therefore, does IBS encompass greater activity in the emotional arousal visceral pain neuromatrix? The research group also investigated differences in IBS, UC and healthy controls with connectivity analysis, whose findings are of greater activity in frontal cortex-PAG pain inhibitory circuitry for UC patients and healthy controls [70].

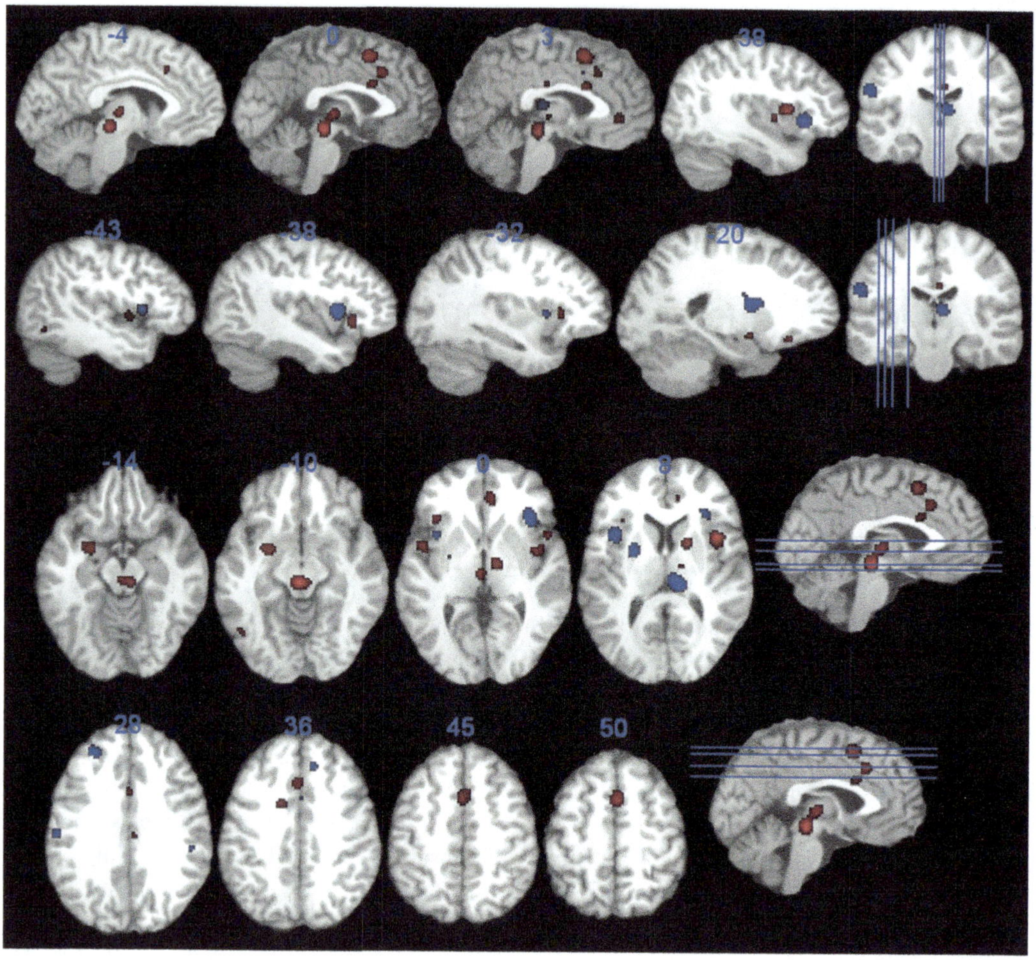

Fig. 5 Selected axial and sagittal slices representing brain areas demonstrating difference greater activation in IBS (*red*) and greater activation in controls (*blue*) across all studies. Data and figure from [75]

6 Structural Neuroimaging of Chronic Visceral Pain

Tangential to functional imaging, structural neuroimaging studies have also been used to investigate changes apparent in syndromes characterised by chronic visceral pain [47]. Many of these studies are comprised of 'resting state' scans and hence do not investigate acute pain syndromes, but moreover focus on any chronic neuroplastic changes that may be apparent over time. The neuroimaging of chronic pain is an interesting subject field and reflects many differences when compared to acute pain, such that it has been suggested by some that chronic pain be considered an individual disease entity [80]. These structural changes will be discussed herein, with the caveat that limitations have been suggested for these types of studies in chronic pain disorders. Notably, the authors of a 2011 review argue that, with utilising structural imaging studies, it is not known whether any changes apparent are pre-existing risk factors for the initial development of the pain syndrome, or secondary changes resultant from it [63]. It is possible that repeated scans in patients at risk of developing chronic pain may help to answer this question, for example in patients both before and after surgery. Furthermore, the underlying

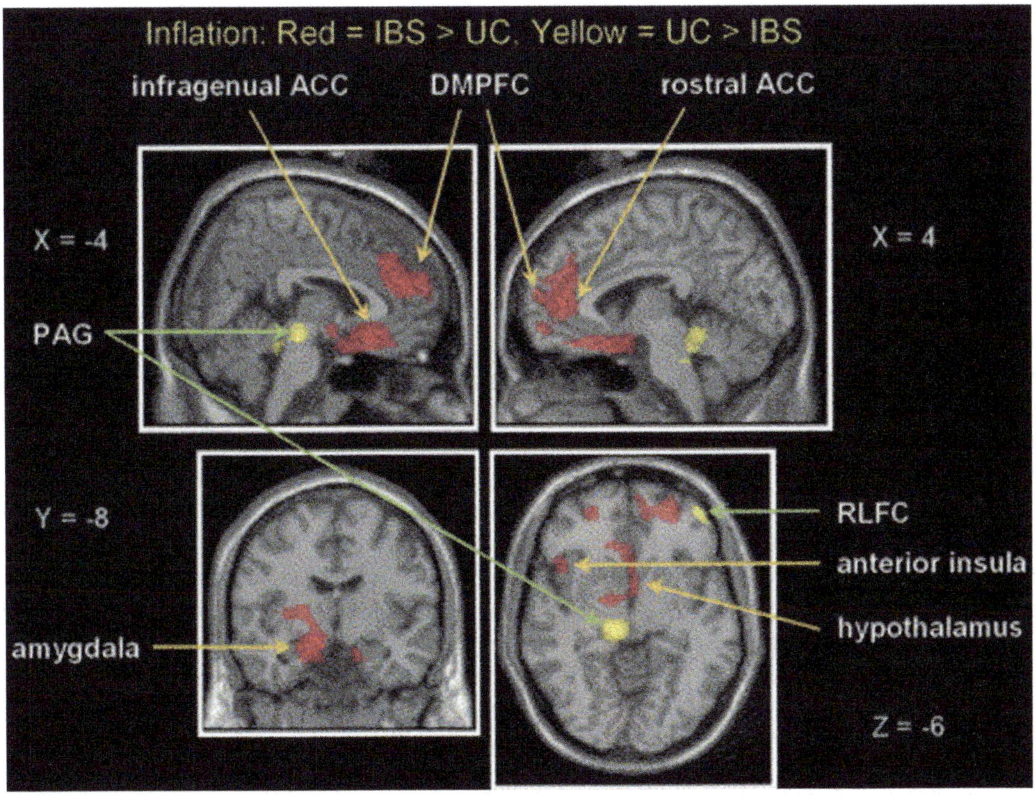

Fig. 6 Regions showing a significant difference between IBS and UC patients in regional cerebral blood flow response during rectal balloon inflation. Right brain is represented by positive *x* values (sagittal views) and is shown on the *right side* in coronal and axial views of a template brain that is representative of Montreal Neurological Institute (MNI) space. Figure from [146]. Abbreviations: *ACC* anterior cingulate cortex; *DMPFC* dorsomedial prefrontal cortex; *PAG* periaqueductal grey; *RLFC* right lateral frontal cortex

processes that may lead to changes in grey matter density (GMD) are not yet known [63].

6.1 Cortical Thickness and Grey Matter Density

Utilising voxel-based morphometry (VBM) and cortical thickness analysis, Seminowicz et al., have compared structural brain differences between a large cohort of both IBS patients and healthy controls (Fig. 7) [81]. The group showed that IBS was associated with a decreased GMD in numerous brain regions implicated in visceral pain processing. Namely, a decreased GMD was found in the thalamus, PFC, posterior parietal cortex and ventral striatum of IBS patients. However, IBS patients also showed an increase in GMD in the ACC and OFC, with non-significant trends to increased GMD in the posterior insula and SII. Based upon these findings, the group suggested IBS to be associated with a decrease in GMD in brain regions typically described to have increased responsiveness to rectal distension. Second, the group proposes that IBS is associated with increased GMD in brain regions implicated in the cognitive and attentional aspects of visceral sensation (including pain). Third, the group suggests that anxiety and depression account for a degree of the structural changes between IBS and healthy controls, as shown with further analyses whereby subject anxiety and depression were added as covariates [81].

One limitation with the interpretation of these aforementioned structural changes shown in IBS is that of reproducibility across studies. The study described above was comprised of more than 50 IBS patients and almost 50 controls, a large cohort by neuroimaging standards, and thus the reported findings arguably hold a significant weight [81]. However, two earlier studies investigating structural changes between IBS patients and healthy controls show contrasted findings, for example in that IBS patients showed cortical thinning in the ACC and anterior insula [82, 83]. Whilst one study cohort only comprised nine patients with IBS, it should not be dismissed and instead should prompt further studies to investigate the paradigm of IBS and chronic pain for reproducibility. For the case of insula thickness, Blankstein et al., report cortical thinning in short-term IBS, but normal thickness in long-term IBS [83]. It is perhaps possible that duration of disease, along with other inter-individual factors, compounds these findings and hence convolutes comparison between the studies.

6.2 Diffusion Tensor Imaging

Diffusion tensor imaging (DTI) is a structural neuroimaging modality that permits the evaluation of white matter integrity, and also allows one to investigate the interplay between grey and white matter. Specific white matter imaging modalities include fractional anisotropy (FA) and quantitative fibre tracking (or probabilistic tractography) [63]. As an example centralised around a disorder of chronic visceral pain, a pilot study of amygdala emotional arousal circuitry in IBS patients and healthy controls, utilising probabilistic tractography, showed that there was significantly reduced structural connectivity between the amygdala and the dorsolateral PFC [84]. Between controls and patients, no other differences were found in amygdala emotional arousal connectivity. This led the authors to suggest therefore that, with decreased connectivity between the amygdala and dorsolateral PFC in IBS patients, decreased or inadequate inhibition of emotional arousal pain circuitry may subsequently occur and therefore lead to amplification of experienced visceral pain [84].

DTI has also been utilised in experimentally induced visceral pain models. Moisset et al., used FA and quantitative fibre tracking to investigate healthy controls undergoing non-painful and painful rectal distensions (Fig. 8) [85]. First, fMRI was used to compare the brain processing of non-painful and painful stimuli, to which they showed expected findings whereby primarily visceral somatosensory regions were activated for the former, but for the latter included pain processing regions such as the thalamus and insula. Subsequently, using DTI, the authors showed that during both non-painful and painful rectal distension, a neural network encompassing the insula, thalamus, ACC, PFC and somatosensory cortices was evident [85]. These are all regions well associated with both visceral pain individually, however it is interesting that research has progressed to reveal the degree of connectivity between them, and how they may interact to process the complex visceroceptive sensation that is pain. Furthermore, the group describes that, although all these aforementioned brain region connections were apparent across the total cohort, a large degree of variability in the degree of connectivity was apparent between individuals [85]. This likely shows further inter-individual variability in the brain processing of visceral pain, encompassing the structural level. The degree of functional connectivity between these visceral pain regions must be an area further investigated with research.

6.3 Electroencephalography

Although fMRI modalities such as BOLD or PET yield excellent spatial resolution, and indeed have significantly furthered our understanding of the neuroimaging of visceral pain, a main limitation is that of their temporal resolution, which is typically in the region of several seconds [86]. As of this limitation, a role for electroencephalography (EEG) is apparent, a modality known for its high temporal resolution (thus permitting the investigation of changes on a millisecond time scale [64]), albeit poor spatial resolution [5, 87, 88]. EEG measures neuronal

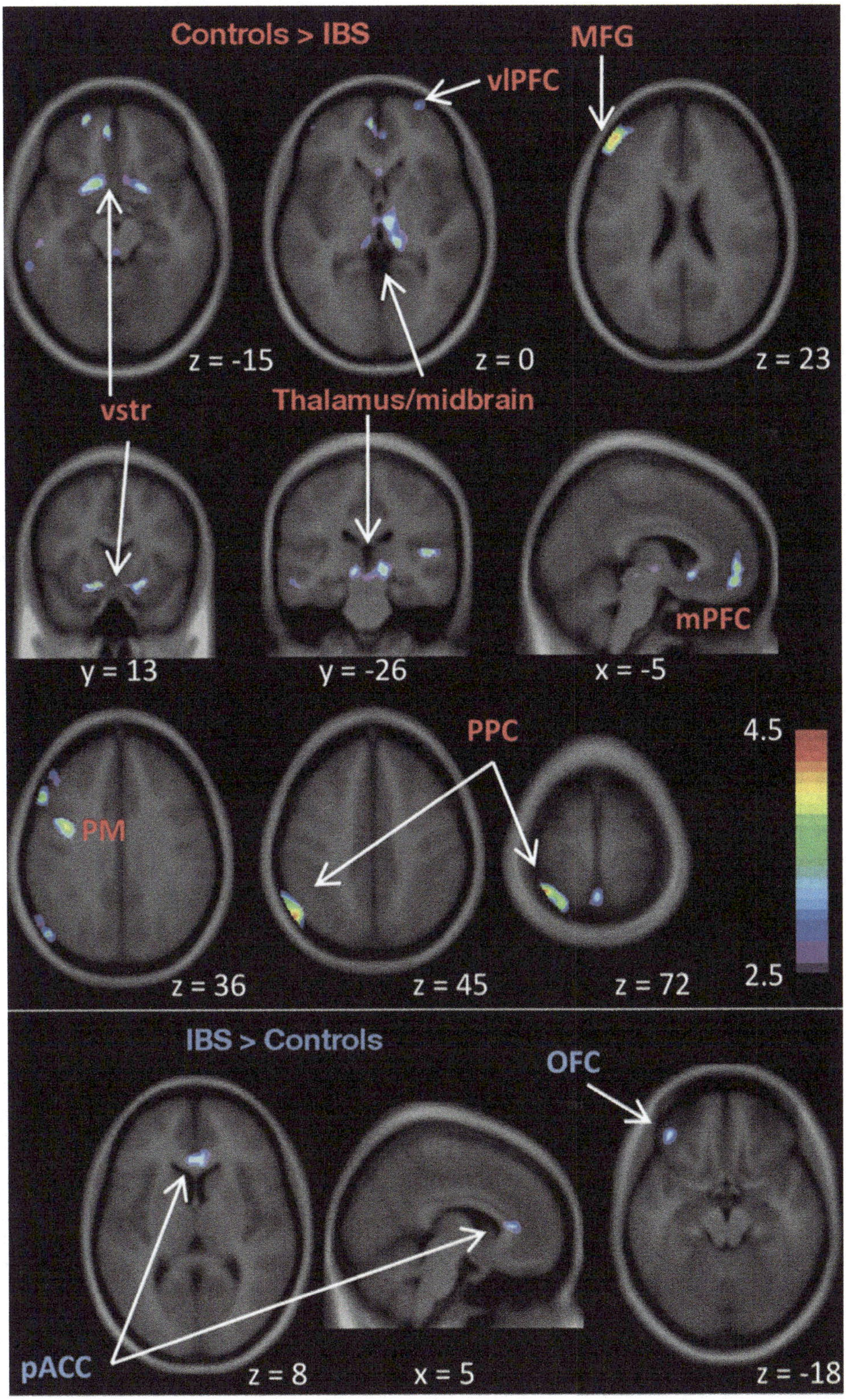
Controls > IBS
MFG
vlPFC
z = -15
z = 0
z = 23
vstr
Thalamus/midbrain
mPFC
y = 13
y = -26
x = -5
PPC
4.5
PM
z = 36
z = 45
z = 72
2.5
IBS > Controls
OFC
pACC
z = 8
x = 5
z = -18

◀ **Fig. 7** VBM results. Significant GMD clusters from GLM comparing patients with IBS and controls, with age as a covariate. Results are displayed on a group average brain in stereotaxic (MNI) space. *Colour bar* shows *t* value. *Left side* of image is left side of brain. Figure from [81]. Abbreviations: *IBS* irritable bowel syndrome; *MFG* *mPFC* medial prefrontal cortex; *OFC* orbitofrontal cortex; *pACC* pregenual anterior cingulate cortex; *PM* premotor area; *PPC* posterior parietal cortex; *vlPFC* ventrolateral prefrontal cortex; *vstr* ventral striatum

activity in a direct manner, and its low cost and ease of use strengthens the proposition of its use in a clinical setting [87, 88]. EEG has been utilised in numerous GI disorders, including rectal hyposensitivity, whereby a defect in sensory neural function (but normal cerebral processing) has been shown [89], diabetes, whereby differences in the operculum-cingulate network were found following painful electrical stimulation of the rectum [90] and also IBS, whereby significantly lower alpha power and higher beta power during baseline have been reported [91].

As an excellent visceral pain example, EEG has frequently been used in numerous studies of chronic pancreatitis (CP) [87, 92]. In this disorder, progressive destruction of the pancreatic glands occurs, leading to the development of significant visceral pain. Furthermore, it has been suggested by research that aberrant CNS pain processing occurs in this disorder, and hence may contribute to the intense pain experienced (for example by central sensitisation) [87, 92]. For that reason, the use of neuroimaging has been warranted in order to further investigate this disorder.

EEG has been utilised whereby experimentally induced visceral pain has been investigated [93, 94]. Comparing a cohort of patients with CP with healthy controls, Dimcevski et al., electrically stimulated the oesophagus, duodenum and stomach in order to induce pain (Fig. 9) [72]. An EEG

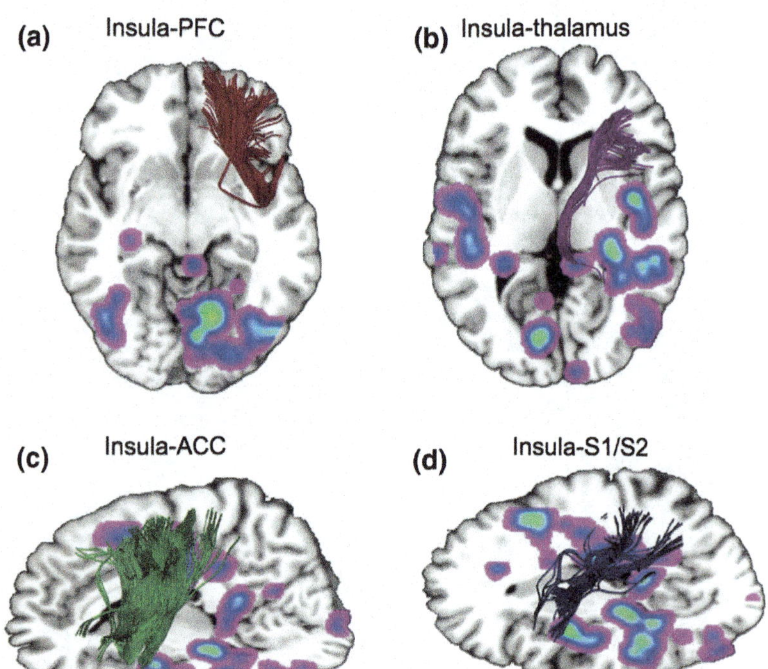

Fig. 8 Connections between the main areas activated during visceral perception. **a**, **b** coronal sections showing the bundles of fibres between the insula and the prefrontal cortex (PFC) (*red*), and the insula and the thalamus (*purple*). **c**, **d** oblique sagittal sections showing the bundles of fibres between the insula and the cingulate cortex (ACC) (*green*) and the insula and the primary and secondary somatosensory cortices (S1/S2) (*blue*). Figure from [85]

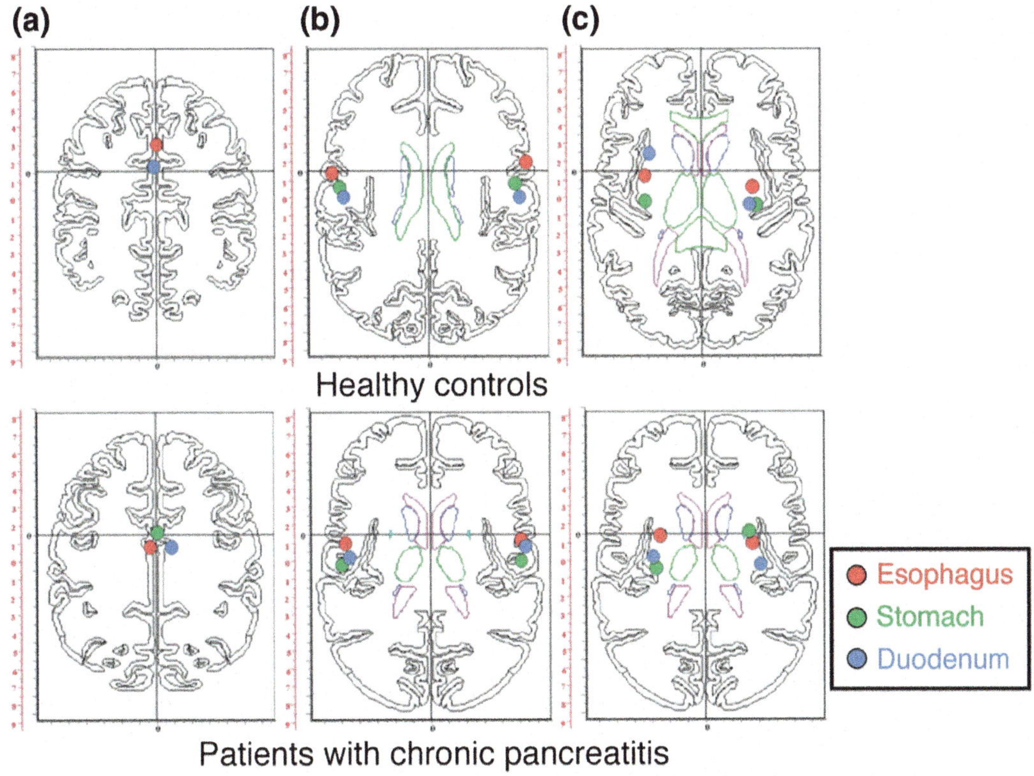

Fig. 9 Locations of dipoles fitting evoked brain potentials to painful oesophagus, stomach and duodenum stimulation in patients with chronic pancreatitis and in healthy controls. The dipolar coordinates were projected onto brain representations issued from the Talairach and Tournoux atlas. **a** Dipole sources localised in the anterior cingulate gyrus. Note that in healthy subjects the cingulate dipoles to stomach and duodenum stimulations superimpose each other. **b** Middle bilateral generators in the secondary somatosensory area. **c** Bilateral dipoles corresponding to the insular regions. The insular dipoles differed between the groups for stimulation of all gut segments. Figure from [72]

was also recorded and event-related potentials were (EP) obtained. Interestingly, the CP patient group showed a decrease in early EP component latencies, which was shown by source analysis to localise in the bilateral insula, anterior cingulate gyrus and bilateral SII. Furthermore, the exact localisations of activity were disparate between healthy controls and CP patients, for example, insular dipoles were more medial, and the cingulate cortex was more posterior in CP patients, compared to controls [72]. This led the group to suggest that CP is associated with changes in the cortical projections of the visceral pain system (including both hyperexcitability and reorganisation), a similar finding when compared to neuropathic pain, thus suggesting the pain in CP to hold a neuropathic component [72]. In this same cohort, it was also found that significant differences were apparent in delta and theta bands between patients and controls [95]. The group suggested that the demonstrated increased theta activity in CP patients may reflect thalamocortical dysrhythmia, and adds weight to a neuropathic dimension of the pain in this disorder [95, 96].

7 The Non-gastrointestinal Aspects of Visceral Pain

The vast majority of studied visceral pain sites are that of the GI tract, such as oesophagus, stomach or rectum. However, it is important not to neglect other aspects of the viscera, although this has frequently been the case in research. Some studies have investigated non-GI visceral pain, and these include examples such as gynaecological or urological pain [97, 98]. However, compared to the GI tract, pelvic and urological pain is hugely understudied in a neuroimaging aspect. Although a paucity of data is available, some research groups have investigated specific visceral pain disorders, such as urologic chronic pelvic pain syndrome (CPPS) [98–100] and vulvar vestibulitis syndrome (VVS) [97]. Needless to say, the data that is available on these neglected aspects of visceral pain neuroimaging will be discussed herein.

CPPS is a condition affecting the pelvic floor muscles, and is characterised by pelvic or perineal pain for more than 3 months in the absence of pathology such as a urinary tract infection [101]. In neuroimaging, resting state functional connectivity analyses have revealed that, when compared to healthy controls, patients afflicted with CPPS show altered connectivity between the posterior insula and superior temporal gyrus [100]. Furthermore, in a follow-up study conducted by the same group, they showed that the posterior medial cortex is detached from the default mode network in CPPS [98]. These are interesting findings, and suggest that further research is required into these other visceral pain conditions. It is likely that we have only scratched the surface in investigating some of these disorders, particularly for CPPS, whereby authors of a 2015 published paper have described it as the first connectivity study of the disorder [100]. Such novelty is rare to find with neuroimaging of the GI tract, and this reflects the paucity of research in certain aspects of neuroimaging research concerning visceral pain.

Moving to gynaecological causes of pain, VVS is a cause of dyspareunia (pain during sexual intercourse) that may occur in pre-menopausal women [102]. Pukall et al, reported in a 2005 fMRI study utilising vulvar vestibule pressure as a painful stimulus that, when compared to healthy controls, patients with VVS display significantly greater brain activity in the insula and frontal cortical regions [97]. Moreover, the group noted significant similarities in the findings when compared to other chronic pain syndromes such as IBS and fibromyalgia [97]. It would be interesting for these disorders of chronic visceral pain, such as IBS, fibromyalgia and VVS, to be compared with regards to their brain processing utilising modern neuroimaging techniques.

8 Inter-individual Variability Influences Visceral Pain Processing

An interesting aspect of neurogastroenterology research over the last decade has been to characterise the inter-individual factors that influence the brain processing of pain. Thus far, the influential parameters characterised are numerous, but include genetic, physiological, neuroanatomical and psychophysiological, such that some groups have moved to characterise visceral pain 'endophenotype' clusters [60, 61]. For parsimony and relevance to this chapter, only factors which have been reported with neuroimaging will be described herein, but for further comprehensive review of these in general, see Farmer et al. [103].

8.1 Gender

One difference between studied individuals for visceral pain is frequently gender [104–106], such that many modern studies will control for it entirely or only study either female or male subjects [107]. It is frequently documented that women demonstrate a higher pain sensitivity and prevalence of chronic painful visceral conditions than males, however until recently, very few visceral pain neuroimaging studies have implicitly investigated a prospective influence of gender. Kano et al, recently utilised an oesophageal pain model with BOLD fMRI, by means of distal oesophageal distension, to investigate the differences in the brain processing of visceral pain amongst males and females (Fig. 10) [108]. The group showed that females displayed greater brain activity in the MCC, anterior insula and premotor cortex, whilst males showed greater activity in the supplementary motor area (SMA) [108]. Notably, the regions female subjects showed greater activity in are well associated with the emotional arousal aspect of the visceral pain neuromatrix, leading the group to suggest that females may attribute more emotional importance to painful stimuli than in males [108]. Other groups have also reported these findings of greater emotional arousal region activity in female subjects, for example, a connectivity analysis study conducted by Labus et al. [62].

8.2 Genetic

Genetic differences between individuals have been proposed as influential in the brain processing of visceral pain. In particular, differences in the 5-hydroxytryptamine signalling system are thought to be implicated, a brain signalling system which holds an important role in the processing of interoceptive signals, the stress response and emotional regulation [76, 109–111]. Furthermore, the 5-hydroxytryptamine transporter gene-linked polymorphic region (5-HTTLPR) has been associated with disparate brain responses to visceral pain, when neuroimaging has been utilised. One recent connectivity study has shown how, in healthy male subjects undergoing painful rectal distension, the *S/S* 5-HTTLPR genotype corresponded to significantly greater influence of the hippocampus on the amygdala, in comparison to alternate genotypes (*l*-carriers) [109]. This disparity in brain region connectivity illustrates how differences in the 5-HTTLPR genotype influence the brain processing of visceral pain, with particular localisation to regions involved in the stress response and emotional regulation [109]. The 5-HTTLPR has also been studied in the context of IBS. Camilleri et al, showed that, in a study of IBS patients, the 5-HTTLPR *SS* genotype was associated with decreased IBS symptoms, whilst the *LS/SS* genotype corresponded to increased rectal compliance and increased pain ratings [112]. In a more recent study evaluating pain severity in IBS, the *LS/SS* 5-HTTLPR genotype was again correlated with severity of symptoms experienced [113].

Neuroimaging has been utilised in the context of studying the 5-HTTLPR [66]. In a PET study of healthy controls undergoing colonic distension, Fukudo et al, have shown that, in the *S/S* genotype individuals, a significantly larger increase in regional cerebral blood flow occurs, compared to the individuals with the *L* allele (Fig. 11) [114, 115]. In particular, those with the *S/S* genotype displayed significantly greater activity in the ACC (Fig. 11a), hippocampus (Fig. 11b) and OFC (Fig. 11c), leading the group to suggest that the emotional regulation of visceral pain is influenced by variability in the 5-HTTLPR and, furthermore, the functional gene polymorphism may, in part, predict the effect of a selective serotonin reuptake inhibitor in the use of visceral pain [114]. In a BOLD fMRI study of visceral pain, by means of oesophageal distension, healthy volunteers with the *S/S* genotype showed significantly greater brain activity in the insula, inferior frontal gyrus, SMA, precentral gyrus and cerebellum [116]. In visceral pain endophenotypes studies, the inclusion of the 5-HTTLPR genotype has also yielded interesting results, see 'Visceral Pain Endophenotypes' and [60]. Further studies investigating the neuroimaging of visceral pain and the influence of genetic factors are warranted, in effort to elucidate the full picture of this inter-individual variability factor.

8.3 Personality

Of the psychophysiological factors, personality has historically been thought of as highly influential in the cognitive interpretation of pain, long before the advent of functional neuroimaging [117–119]. However, the vast majority of early studies utilised somatic pain as the experimental stimulus, and thus it is only within the last decade or so that the influence of personality and mood on visceral pain has begun to be evaluated [120–122]. Using an oesophageal pain model, by means of a distensible nasogastric tube positioned in the distal oesophagus, we have

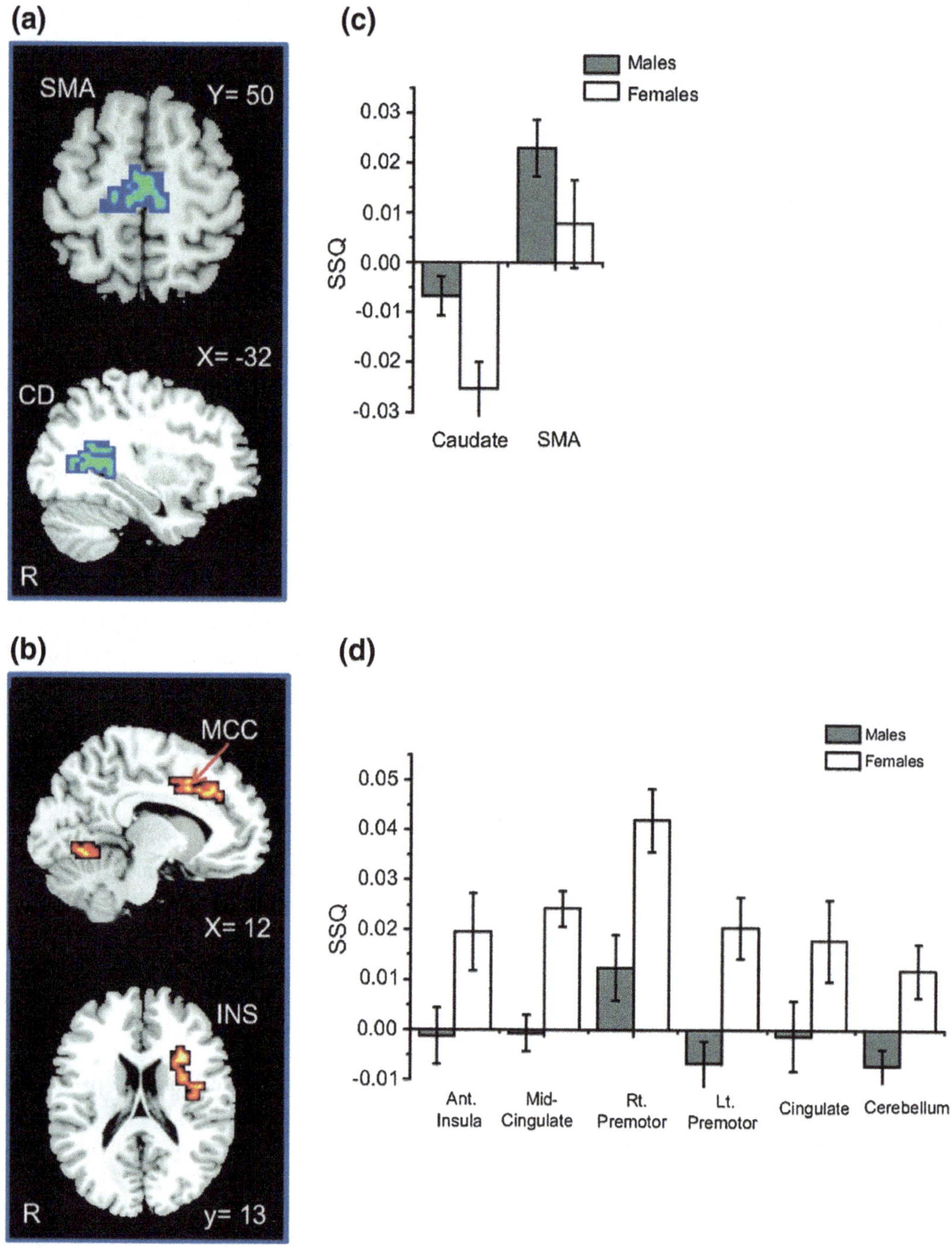

Fig. 10 Brain regions where there were significant differences between men and women during pain. **a** Brain areas where men demonstrated greater brain activity compared with women in the SMA (BA6) and left caudate. **b** Brain areas where women produced greater brain activity than men in the midcingulate cortex (BA32) and left insula. **c** Plot of mean blood oxygenation level dependent signal extracted from each significant cluster in (**a**). **d** Plot of mean blood oxygenation level dependent signal extracted from each significant cluster where women produced greater brain activity than men. *Ant.* anterior; *CD* caudate; *INS* insula; *MMC* midcingulate cortex; *SMA* supplementary motor area; *SSQ* sum of squares. Figure from [108]

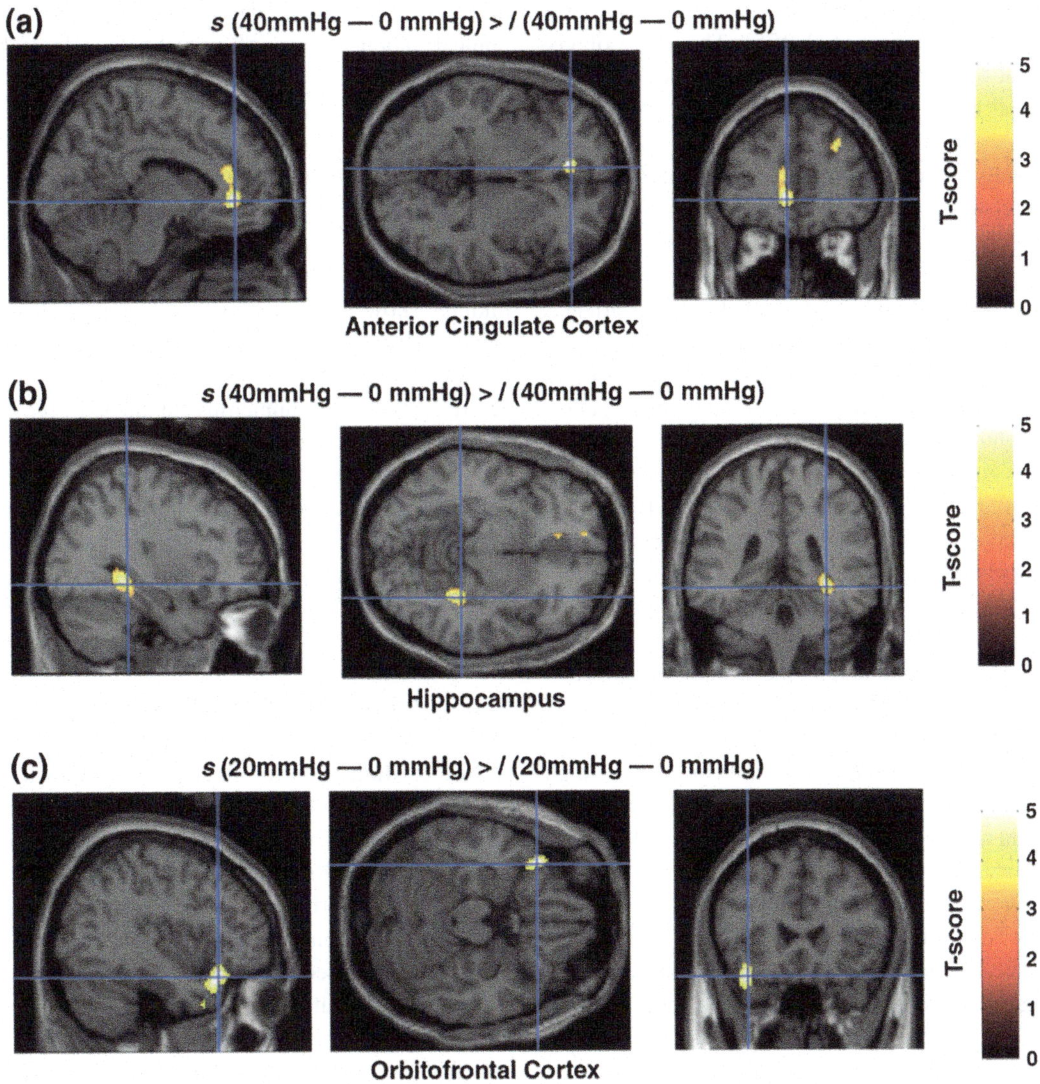

Fig. 11 **a** Moderate colorectal distention in the S allele group significantly activated more the left anterior cingulate cortex than that in the L group. The image with 40 mm Hg was subtracted by that with 0 mm Hg. BA 32, *x*, *y*, *z* = −8, 40, −2. Figure from [114]. **b** Mild colorectal distention in the S allele group significantly activated more the right hippocampus than that in the L group. The image with 20 mm Hg was subtracted by that with 0 mm Hg. BA 47, *x*, *y*, *z* = −38, 24, −20. Figure from [114]. **c** Mild colorectal distention in the S allele group significantly activated more the left orbitofrontal cortex than that in the L group. The image with 20 mm Hg was subtracted by that with 0 mm Hg. BA 47, *x*, *y*, *z* = − 38, 24, − 20. Figure from [114]

previously shown that the degree of an individual's neuroticism (defined by the tendency to experience negative emotions or anxiety) or extraversion (defined by the tendency to be optimistic, outgoing and sociable) influences the brain processing of visceral pain, using BOLD fMRI [67, 68]. Furthermore, an added dimension to these studies was also to investigate the brain response to visceral pain anticipation, that is to say the subsequent brain region activity when volunteers were told they would imminently experience visceral pain. The functional neuroimaging of pain anticipation is an interesting sub-dimension of visceral pain research; given the nature of pain as a stressful event that merits emotional arousal [123–125]. The study of visceral pain anticipation will be discussed in more detail later.

In our study of neuroticism, it was found that during visceral pain anticipation, higher neuroticism corresponded to greater activity in brain regions attributed to emotional and cognitive appraisal, but that the degree of activity in these regions decreased during actual pain experience. In particular, this was shown for the thalamus, parahippocampal gyrus and ACC (Fig. 12). Coen et al, suggest that this finding may reflect heightened arousal during pain anticipation but an avoidance coping mechanism during actual pain in the high neuroticism subjects [68].

In a further study, high extraversion was associated with greater brain activity in the right insula during both visceral pain anticipation and pain experience, a brain region central to both the sensory discriminatory and emotional affective dimensions of the visceral pain neuromatrix (Fig. 13) [6, 28]. This finding during anticipation which may support the theory that high extraversion subjects show greater change in brain activity (from low baseline cortical arousal) in brain regions involved in emotional and cognitive processing subsequent to an emotionally and cognitively salient stimulus [65, 126]. Interestingly, insula activity during visceral pain has been correlated with sympathetic nervous system (SNS), one dimension of the ANS [127]. Furthermore, extraversion has been associated with predominance for a SNS response during visceral pain, a finding in contrast to neuroticism [60]. Coalescence of these findings may suggest therefore that greater insula activity may correspond to greater a SNS response to visceral pain in high extraversion individuals [67]. Thus, it is likely that an individual's personality influences visceral pain processing at the brain level, and indeed is a notion supported now by numerous groups [128].

Interestingly, IBS has strong ties to the personality trait neuroticism [129]. Notably, neuroticism is found to be higher in patients with IBS, and is also a risk factor for chronic unexplained abdominal pain in those afflicted with the disorder [130, 131]. These findings link the clinical applicability of psychophysiological research, finding that those with chronic visceral pain syndromes may tend to occupy specific psychological traits. In a recent DTI study, comparing the white matter of IBS and healthy controls, it was shown that neuroticism correlated with FA in aspects of the thalamus in IBS patients, but not healthy controls [132]. These findings show the importance of factors such as personality in visceral pain, in both health and disease, but also in both research and clinical practice. With the numerous findings of how personality influences chronic pain, including visceral in nature, it has been proposed by some that an assessment of personality structure in a hospital setting will aid in pain intervention programmes and predict treatment outcomes [133].

8.4 Visceral Pain Endophenotypes

A further dimension of inter-individual factor research is that of visceral pain 'endophenotypes', defined as measurable components of a disease/condition, such as neuroanatomical or cognitive characteristics, that have simpler ties to the genetic underpinnings than the disorder itself [59, 134]. It has been suggested that pain-cluster endophenotype characterisation may help to identify patients at risk for developing chronic pain, but also reduce variability in neuroimaging studies [60, 61]. Utilising a healthy cohort, it has been shown that both pain sensitive and more resistant phenotypes exist, of which these endophenotypes encompass numerous psychophysiological, physiological and genetic qualities [60, 61]. For sensitive pain-cluster individuals, they have been shown experimentally to exhibit high neuroticism and anxiety, a greater sympathetic tone and a significantly higher cortisol level at baseline, whilst during a painful stimulus they display a greater increase in parasympathetic tone [60]. The converse has been shown apparent for pain clusters whereby individuals were typically more pain resistant, as well as this phenotype of individuals having the

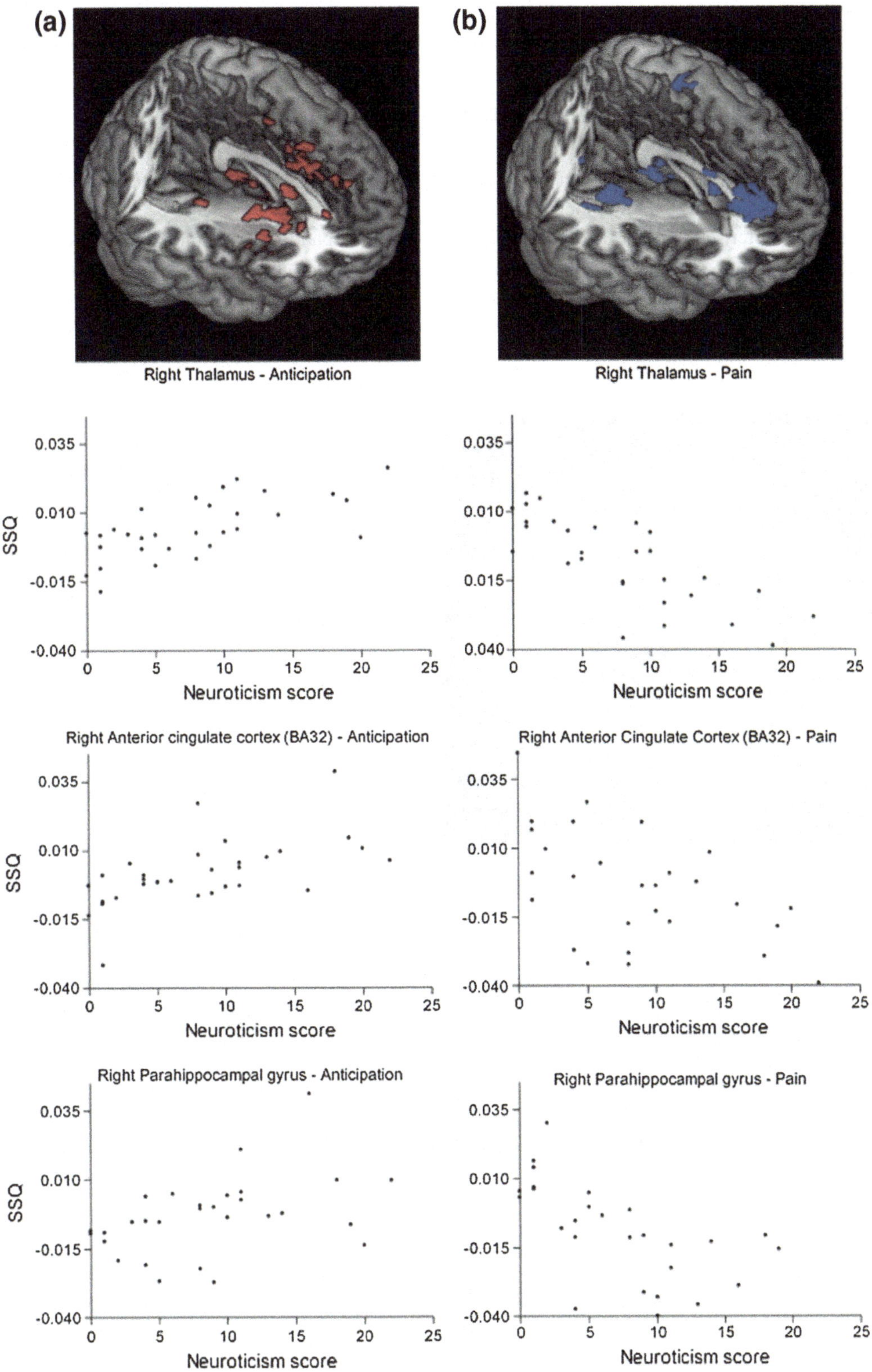
(a)
(b)
Right Thalamus - Anticipation
Right Thalamus - Pain
SSQ
0.035
0.010
-0.015
-0.040
0.035
0.010
0.015
0.040
0 5 10 15 20 25
Neuroticism score
Right Anterior cingulate cortex (BA32) - Anticipation
Right Anterior Cingulate Cortex (BA32) - Pain
Right Parahippocampal gyrus - Anticipation
Right Parahippocampal gyrus - Pain

◀ **Fig. 12** Representative graphical examples of correlations (positive, *red*; negative, *blue*) found between neuroticism and level of brain activity (SSQ) during **a** anticipation and **b** pain. This figure shows several brain regions, including the thalamus, parahippocampal gyrus, and ACC, where brain activity increases with higher neuroticism during anticipation but decreases during pain processing. Figure from [68]. Abbreviations: *ACC* anterior cingulate cortex; *SSQ* sum of squares

tendency to be high extraversion scoring (Fig. 14). Furthermore, the endophenotypes pain clusters characterised are associated with disparate brain activity when using neuroimaging, including in the insula, putamen, inferior frontal gyrus and caudate (Fig. 15) [60].

8.5 Anticipation of Visceral Pain

The anticipation of visceral pain, that is to say the knowing and apprehension when told that pain is imminent, is processed at the neural level very differently from when actual pain is experienced.

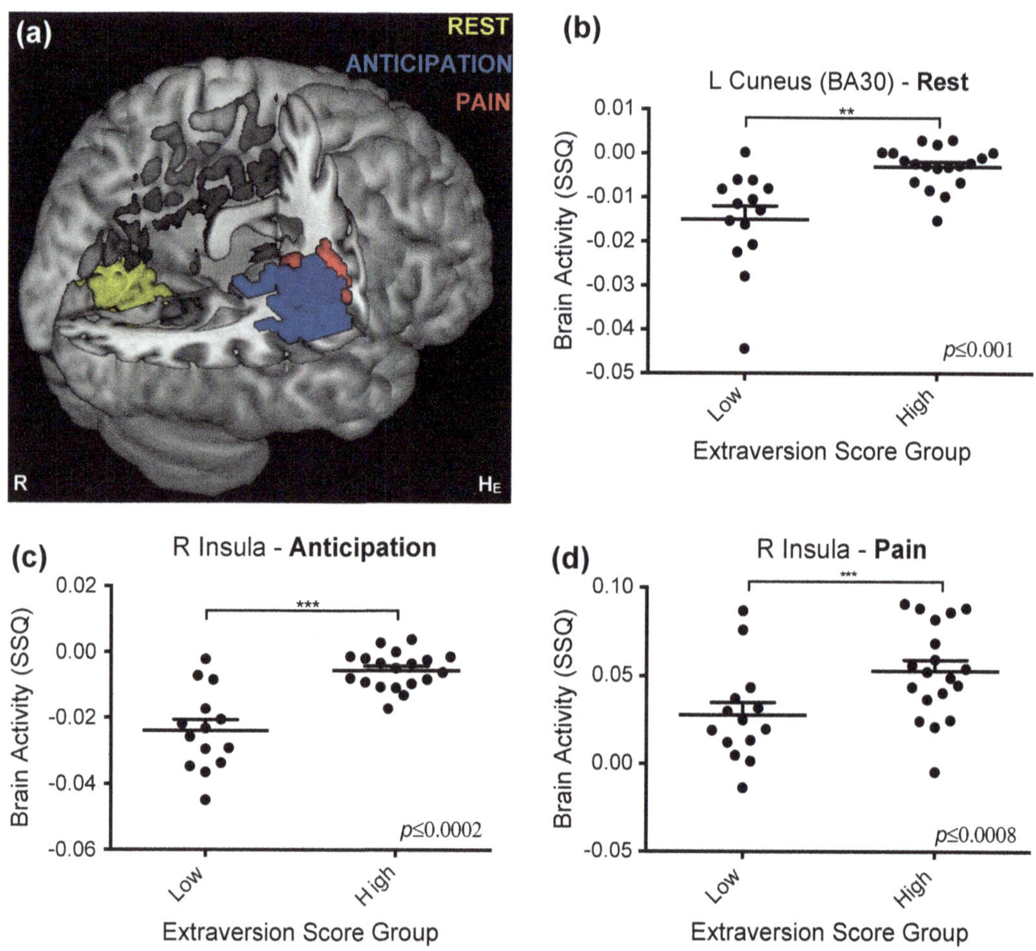

Fig. 13 Brain regions significantly more active in the high extraversion group during rest, pain anticipation and pain. **a** 3D render of all regions more active in high extraversion participants. Yellow clusters signify greater activity during rest, blue during anticipation, and red during pain. SSQ extractions (the statistical analyses used in the XBAM fMRI analysis package) showing greater activity for the high extraversion group in the left cuneus during rest (**b**) and the right insula during both anticipation (**c**) and pain (**d**) displaying mean ± SEM. Figure from [67]. Abbreviations: *BA* Brodmann area; *HE* high extraversion; *L* left; *R* right; *SSQ* sum of squares

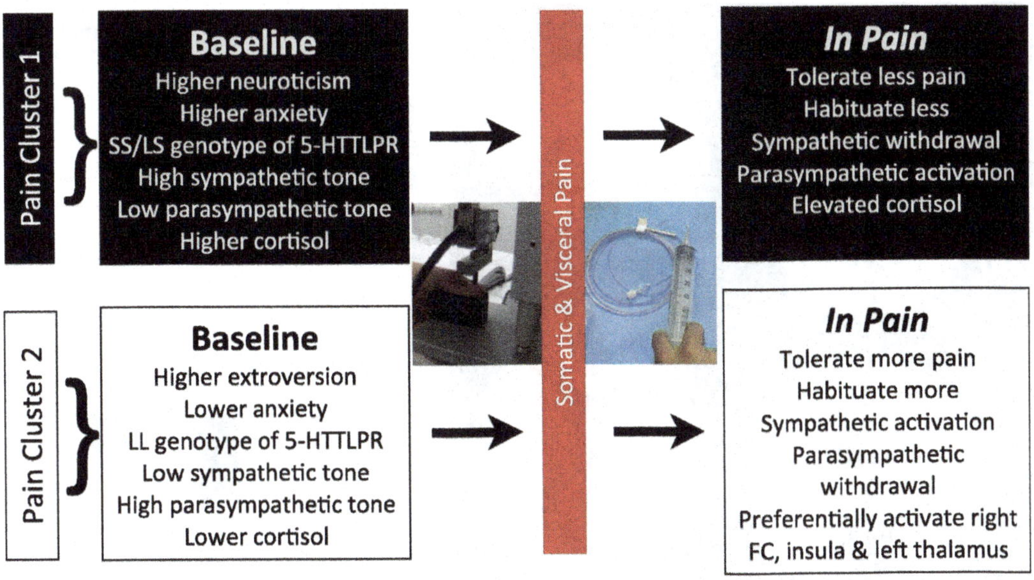

Fig. 14 A schematic summary of the salient differences between the PCs at baseline and in response to visceral and somatic pain. Figure and data from [60]. Abbreviations: *FC* frontal cortex

When we anticipate a threat meriting emotional arousal, such as the threat of pain, our anticipation modulates the subsequent cognitive response to the impending stimuli, such as degree of stress and anxiety accompanying it [135]. As of this complex role, a rationale for its study is apparent and indeed has been an interesting focus of visceral pain neuroimaging research over the last decade. Many have studied pain anticipation in effort to describe brain regions implicated, and perhaps even an anticipatory neural network, if such a matrix exists.

Early functional neuroimaging studies have focused on elucidating brain regions that activate in a situation of pain anticipation. Given its well-established role in the perception of pain, one frequently proposed brain region to be implicated in such modulation in anticipation is that of the anterior insula [2, 21, 67, 135, 136]. Other proposed regions to be implicated include the ACC, PAG, PFC, medial frontal lobe, amygdala and OFC (Fig. 16) [21, 136–140]. It should be noted that these regions are well implicated in the processing of sensing visceral pain also, and thus an established role in pain anticipation is interesting.

The study of the phenomenon that is visceral pain anticipation is a developing field, one that has made considerable advance utilising functional neuroimaging [123]. More recent studies have added a dimension of complexity to the research, that is to say investigating visceral pain anticipation in both health and disease. A frequently investigated example of visceral pain in disease is that of IBS, due to its strong associations with both anxiety and the personality trait neuroticism [130, 140, 141]. Berman et al. conducted an fMRI study of rectal distension in IBS patients and healthy controls, adding an anticipatory cue prior to said distension in order to study how the brain in fact anticipates this threat [140]. The group showed a significantly positive correlation in patients between neuroticism and degree of startle responses during the anticipation of threat. Moreover, whilst healthy controls were found to inactivate certain brain regions in visceral pain anticipation, including the insula, supragenual ACC and amygdala, this was shown not to be apparent to the same degree as in the IBS patients [140]. The 'normal' inactivation of these brain regions is thought of as a cognitive coping strategy to threats that merit emotional arousal, such as visceral pain [142], leading to the proposition that these IBS patients may therefore encompass poorer pain coping resources at the cognitive level, be that resultant from IBS or secondary to the

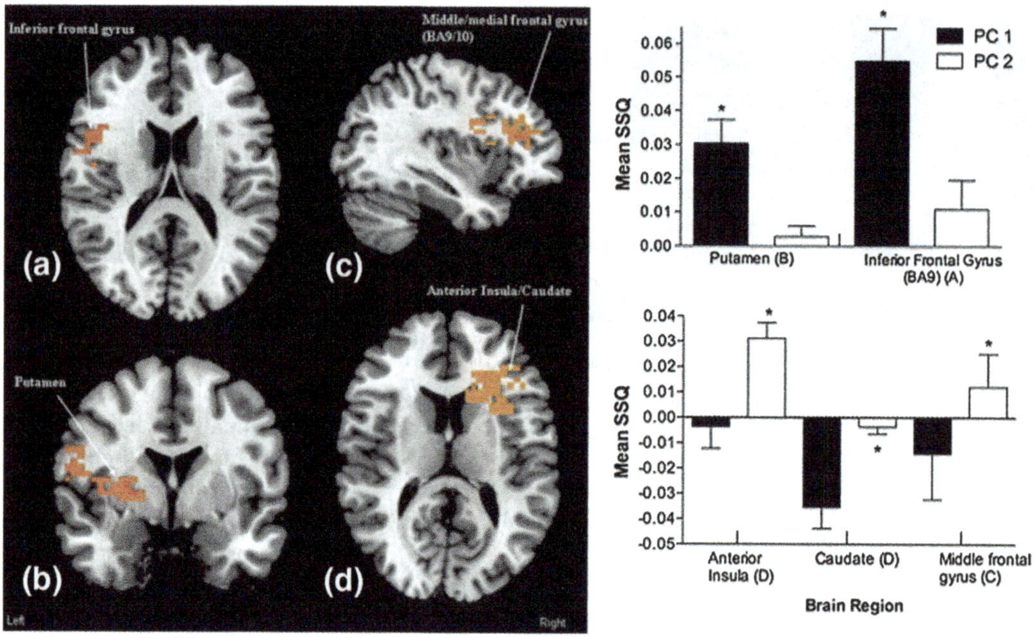

Fig. 15 Representative images of the groups' brain activation maps, summarising regions showing significant differences in brain activity between endophenotype pain clusters and graphs showing relative differences in activity between groups (PC1 [*shaded black*], PC2 [*white*]) in these regions. PC1 showed increased brain activity (SSQ) compared with PC2 in **a** the left inferior frontal gyrus; and **b** the left putamen. PC2 showed significantly more brain activity compared with PC1 in **c** the right middle/medial frontal gyrus and **d** the right anterior insula and caudate. Figure from [60]. Abbreviations: *PC* [endophenotype] pain cluster

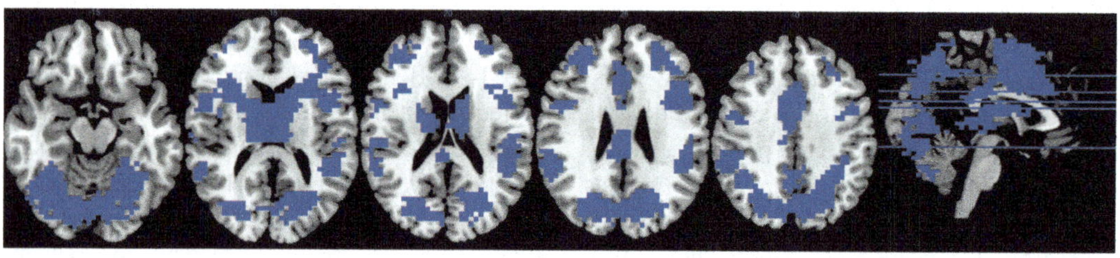

Fig. 16 Brain regions activated during the anticipation of visceral pain, in a healthy cohort. Data and figure obtained from [67, 147]

psychophysiological characteristics which frequently coexist with IBS [68, 140].

9 Pain Habituation and Cognitive Modulation

Commonly, when individuals experience a painful condition chronically (for example over months or years), their perception or interpretation of this pain will change. The cognitive interpretation and attribution of visceral pain shows some degree of fluidity, that is to say it is not rigid and fixed in the same way that a study may illustrate its processing at one specific time point [27]. For example, in individuals suffering with chronic painful disorders, they will often find ways to 'cope with' or 'deal' with the pain. This difference in how one interprets pain over time is interesting to study using functional neuroimaging. Labus et al, sought to investigate this in IBS patients using PET, whereby over a 1-year period they would repeatedly attend testing sessions of rectal balloon distension and accompanying neuroimaging every 3 months [78]. The group found that, over time with repeated rectal distensions sessions, there was reduced connectivity within the selective attention to threat network, and a reduction of amygdala-related interference with the cognitive attentional mechanisms [78].

In a further interesting example, Coen et al, sought to investigate the reproducibility of brain activity following visceral pain by means of a BOLD fMRI study investigating repeated oesophageal pain [27]. By use of a dilatable oesophageal balloon, volunteers underwent numerous dilatations, and furthermore this was repeated for a total of three separate sessions, each approximately 4 weeks apart. Brain regions activated during the painful stimulus were highly comparable to that of the described visceral pain neuromatrix, however, another interesting finding was also observed over the three sessions [27]. Notably, a significant decrease in the strength of activity was found in the ACC, SI and SMA, which was significantly correlated with a decrease in volunteer pain ratings, despite no change to the actual stimulus delivered. This interesting finding led the authors to conclude that a habituation effect may occur with the repeated experience of visceral pain. Furthermore, this finding highlights a limitation of studies whereby repeated painful visceral, or indeed perhaps somatic, sensation is studied [27].

It is likely that this described experimental habituation to pain ties in with how individuals perceive pain, such that over time they may 'fear' it to a lesser degree, therefore meriting less emotional arousal [47]. This possibility would be consistent with the numerous studies investigating how one's emotion may influence visceral pain processing, for example, in studies whereby the extent of an individual's 'fear of pain' has been shown to correlate with brain activity in the insula, ACC and thalamus and indeed vice versa (Fig. 17) [51].

As an alternative suggestion to a decrease in emotional fear, with continual visceral pain it is possible that it may merit less and less attention by the individual. This construct whereby, if in pain, you may take the attitude to 'ignore it' is not uncommon, and studies have also shown how decreased attention focuses on the visceral pain being experienced correlates with disparate brain activity in numerous visceral pain processing regions, such as the anterior insula [23, 35, 135]. It is important to note that this also ties in strongly with the aforementioned habituation effect shown by Coen et al. [27]. Furthermore, Bantick et al, report in an fMRI study of somatic pain that, when distracted by the counting Stroop task, decreased activity in the thalamus, insula and ACC is found, which may in fact be suggestive of reduced pain perception at the brain level (Fig. 18) [50]. It is therefore possible that when the brain is distracted, or indeed the pain is ignored, this can be reflected by a reduction in brain activity in the visceral pain neuromatrix.

In a clinical setting, it is not uncommon for medical staff to attempt to modulate how a patient may interpret and attribute their pain. Furthermore, the findings of Labus et al, show just how dynamic the brain processing of visceral pain can be [78]. A good formal example of this would be cognitive behavioural therapy, but this notion also includes that of simply encouraging more optimistic or positive thinking (tying in strongly to the influence of personality [67, 68]). It has also been shown that the perception of pain may be modulated by the behavioural context in which it is experienced. Although studying somatic pain, Ploner et al, show that, in an fMRI study of healthy controls, increased attention to pain and a negative emotional context (instigated by emotionally negative or positive images being displayed during the experiment) increases pain perception, and in particular negative emotion significantly correlated with activity in the 'interoceptive cortex' that is the anterior insula [2, 43]. With further connectivity analysis, the group also showed that the insula selectively connected to emotional arousal pain processing regions with this negative emotional context [43]. For the future, a study to compare

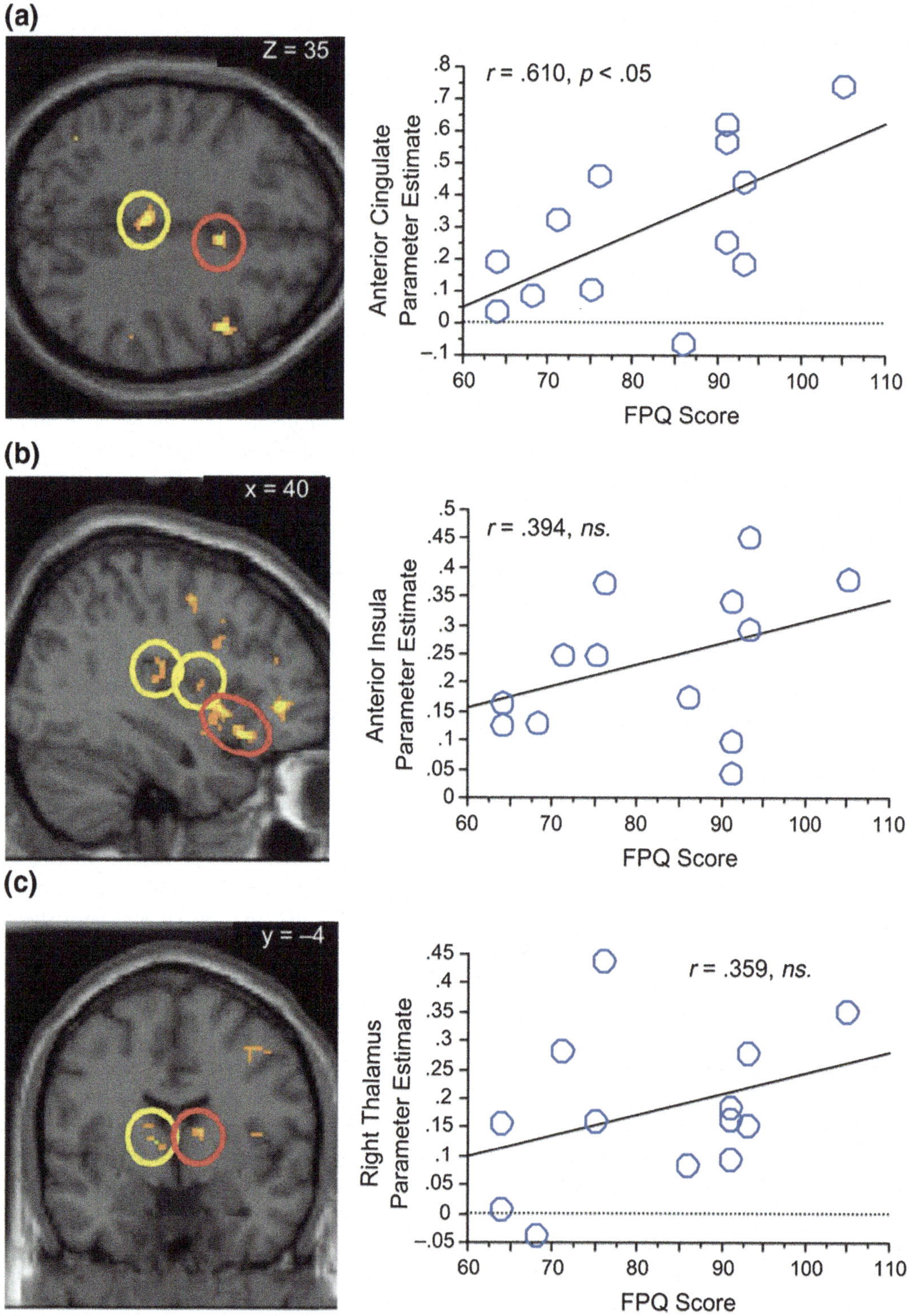

Fig. 17 Regions of anterior and posterior cingulate cortex **a**, posterior, mid and anterior insular cortex **b**, and bilateral thalamus **c** activated in the pain > warmth contrast. Cingulate, insula and thalamic regions activated in this contrast were hypothesised on a priori grounds to be related to ASI or FPQ scores. Those highlighted in *red circles* showed the strongest correlations, but only with FPQ scores. Scatter plots illustrate these correlations. Figure from [51]. Abbreviations: *ASI* anxiety sensitivity index; *FPQ* fear of pain questionnaire

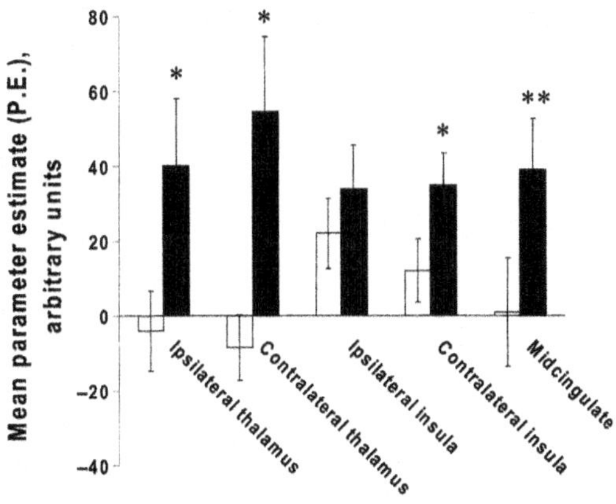

Fig. 18 Mean parameter estimates within key areas of the pain matrix during painful stimulation in the interference task (*white*) and the neutral task (*black*). The group results show mean ± 1 SEM regional activation, and significance levels are indicated for $P < 0.05$ (*) and $P < 0.005$ (**). The parameter estimate is the factor by which the linear signal model is scaled to best fit the fMRI time-course data. Parameter estimates are measured in arbitrary units and are proportional to fMRI signal changes. The ipsilateral thalamus, contralateral thalamus, contralateral insula and midcingulate all showed a significant drop in pain activation during the distracting interference condition compared with the neutral condition. Figure from [50]. Abbreviations: SEM, standard error of the mean

these findings with a visceral pain paradigm would be an interesting one.

10 Translating Neuroimaging Research into Clinical Practice

Throughout this chapter, we have highlighted numerous studies whereby the use of neuroimaging has provided interesting insight as to the brain processing of visceral pain. However, despite the publication of a vast number of visceral pain neuroimaging studies over the last decade, this has not fundamentally made an impact on clinical practice. Translational medicine, that is to say the application of research findings into advancement in a clinical setting, must be employed, and is a core focus of the rationale behind many of the discussed visceral pain neuroimaging studies, particularly in the context of painful GI disorders such as IBS.

There are multiple possible reasons for the current lack of translation to clinical practice. First, as we discussed earlier in this chapter, see 'Inter-individual variability effects visceral pain processing', the vast heterogeneity in studied subjects has shown to have profound effects on neuroimaging findings. These factors include genetic, physiological, psychophysiological and neuroanatomical [60, 67, 68, 108]. Moreover, disease heterogeneity is another limiting factor for translating these findings, for example the case of IBS, a disorder separated into numerous subcategories, but also where pain experienced is highly variable between patients [3]. Second, another possible reason for the lack of influence of clinical practice is that of the study methodologies employed, see 'Methods and sites of experimentally induced visceral pain'. In this chapter, we have discussed studies whereby common painful stimuli have been that of balloon distension (for example oesophageal or rectal), but also used methods are that of electrical stimulation and indeed acid infusion. It is important to note that these are not 'normal' physiological sensations per se, that is to say that it is somewhat a 'leap of faith' when inducing visceral pain by balloon distension to extrapolate this to the pain experienced in GI disorders, such as IBS. Furthermore, these are experimentally induced models of acute visceral pain, and therefore unlikely reflect the pathophysiologically distinct entity of chronic pain that is apparent in many clinical disorders, such as IBS. Lastly, an important limitation exists when studying healthy controls against patients with painful visceral conditions. This is that, for the case of healthy subjects, many individuals are often used as part of the healthy controls cohort for multiple different studies. As of this, they may have been in the scanner many different times, leading to a familiarity in this environment that the patient cohort simply do not have. Furthermore, this yields the problem of studying a 'supernormal' cohort, whereby the healthy control group utilised may not be representative of the typical healthy population [143, 144]. Needless to say, these limitations of neuroimaging studies must be addressed as to enable the translation of these findings to clinical practice.

11 Future Directions

For future directions in the neuroimaging of visceral pain, we will draw upon three specific areas: (1) 'multi-centre' studies, (2) 'multi-modal' imaging methods and (3) the use of the methods and obtained knowledge on disease pathophysiology in order to propose and test new treatment approaches. First, the combination of datasets across numerous research centres has become a possibility in visceral pain neuroimaging research with the recent creation of data archives such as the 'Pain and Interoception Imaging Network (PAIN) Repository' (https://www.painrepository.org). This initiative enables research groups to upload neuroimaging datasets of both health and disease, where visceral pain is a central element. Furthermore, neuroimaging datasets can be downloaded by groups to analyse with significantly larger cohorts than has previously been possible in single-centre studies [5]. The aim of the PAIN Repository is to facilitate discovery in chronic pain states, with particular emphasis in brain endophenotypes and biomarker elucidation, in hope that brain responses may be linked to genetic and biological data sets. A second future direction to discuss is the advent of multi-modal imaging. Historically, early neuroimaging studies utilised one sole imaging methodology, such as PET or BOLD contrast fMRI. However, many more recent studies have utilised more than one neuroimaging method, for example, the combination of EEG and connectivity analysis [64], functional EEG with structural MRI, or even BOLD fMRI with both DTI and VBM. More combinations, also including metabolic spectroscopy, will likely appear in the future as computer power increases. The rationale is to obtain more knowledge of the changes in brain activation when also taking the neuroplastic or structural changes into consideration in the same model. Lastly, concerning the use of neuroimaging in elucidating new treatment approaches, it has been proposed that neuroimaging methods in visceral pain may hold a role as a biomarker, for example, when elucidating the presence or absence of a response to treatment in a painful visceral disorder [145]. If it is the case whereby the degree of visceral pain positively correlates with the degree of brain activity in pain processing regions, it therefore may be possible to investigate for the effect of drugs in decreasing region activity.

12 Conclusions

With significant advances in imaging modalities over the last two or three decades, vast advances as to the neuroimaging of visceral pain have been made. Early studies utilised neuroimaging modalities to describe the brain regions activated during visceral pain, and have led to the characterisation of the visceral pain neuromatrix. Subsequently, with the advent of functional neuroimaging, the degree of activity of visceral pain processing regions has been compared in both health and disease. Moreover, GI disorders whereby the brain is understood to play an important role have been significantly investigated, most notably IBS. This has led to significant advances in our understanding of the brain–gut axis, that is to say the bidirectional communication between the gut and the brain. Furthermore, the top-down influence of the brain on the experience of visceral pain has been thoroughly investigated, for example, the influence of psychophysiological factors such as personality, even leading to the proposition of visceral pain endophenotypes. More recent studies of both health and disease have utilised connectivity analysis to investigate the strength of connection between these different brain regions comprising the visceral pain neuromatrix. With all that has been accomplished in the last two decades in the neuroimaging of visceral pain, what may be accomplished in the coming years is a particularly exciting prospect. The translation of neuroimaging research into clinical practice may stand to benefit patients suffering from chronic painful conditions, particularly those that are not fully understood. Furthermore, it is hoped that the future directions of neuroimaging in visceral pain could lead to new and efficacious treatments of visceral pain.

References

1. Knowles CH, Aziz Q. Basic and clinical aspects of gastrointestinal pain. Pain. 2009;141:191–209.
2. Craig AD. How do you feel? Interoception: the sense of the physiological condition of the body. Nat Rev Neurosci. 2002;3:655–66.
3. Farmer AD, Ruffle JK. Irritable bowel syndrome. Hamdan Med J. 2015;8.
4. Farmer AD, Ban VF, Coen SJ, Sanger GJ, Barker GJ, Gresty MA, et al. Visually induced nausea causes characteristic changes in cerebral, autonomic and endocrine function in humans. J Physiol. 2015.
5. Mayer EA, Aziz Q, Coen S, Kern M, Labus JS, Lane R, et al. Brain imaging approaches to the study of functional GI disorders: a Rome working team report. Neurogastroenterol Motil. 2009;21:579–96.
6. Van Oudenhove L, Coen SJ, Aziz Q. Functional brain imaging of gastrointestinal sensation in health and disease. World J Gastroenterol. 2007;13:3438–45.
7. Aziz Q, Thompson DG. Brain-gut axis in health and disease. Gastroenterology. 1998;114:559–78.
8. Van Oudenhove L, Demyttenaere K, Tack J, Aziz Q. Central nervous system involvement in functional gastrointestinal disorders. Best Pract Res Clin Gastroenterol. 2004;18:663–80.
9. Jones MP, Dilley JB, Drossman D, Crowell MD. Brain-gut connections in functional GI disorders: anatomic and physiologic relationships. Neurogastroenterol Motil. 2006;18:91–103.
10. Strigo IA, Bushnell MC, Boivin M, Duncan GH. Psychophysical analysis of visceral and cutaneous pain in human subjects. Pain. 2002;97:235–46.
11. Tracey I. Nociceptive processing in the human brain. Curr Opin Neurobiol. 2005;15:478–87.
12. Johns E, Tracey I. Neuroimaging of visceral pain. Rev Pain. 2009;3:2–5.
13. Coen SJ, Hobson AR, Aziz Q. Processing of gastrointestinal sensory signals in the brain. In: Johnson LR, editor. Physiology of the gastronintestinal tract. Oxford: Academic Press; 2012. p. 689–702.
14. Berntson GG, Sarter M, Cacioppo JT. Ascending visceral regulation of cortical affective information processing. Euro J Neurosci. 2003;18:2103–9.
15. Henry TR. Therapeutic mechanisms of vagus nerve stimulation. Neurology. 2002;59:S3–14.
16. Willis WD, Westlund KN. Neuroanatomy of the pain system and of the pathways that modulate pain. J Clin Neurophysiol. 1997;14:2–31.
17. Sugiyama Y, Suzuki T, DeStefino VJ, Yates BJ. Integrative responses of neurons in nucleus tractus solitarius to visceral afferent stimulation and vestibular stimulation in vertical planes. Am J Physiol Regul Integr Comp Physiol. 2011;301:R1380–90.
18. Altschuler SM, Bao XM, Bieger D, Hopkins DA, Miselis RR. Viscerotopic representation of the upper alimentary tract in the rat: sensory ganglia and nuclei of the solitary and spinal trigeminal tracts. J Comp Neurol. 1989;283:248–68.
19. Cechetto DF, Saper CB. Evidence for a viscerotopic sensory representation in the cortex and thalamus in the rat. J Comp Neurol. 1987;262:27–45.
20. Carrive P, Bandler R. Viscerotopic organization of neurons subserving hypotensive reactions within the midbrain periaqueductal grey: a correlative functional and anatomical study. Brain Res. 1991;541:206–15.
21. Aziz Q, Thompson DG, Ng VW, Hamdy S, Sarkar S, Brammer MJ, et al. Cortical processing of human somatic and visceral sensation. J Neurosci. 2000;20:2657–63.
22. Melzack R, Wall PD. Pain mechanisms: a new theory. Science. 1965;150:971–9.
23. Dunckley P, Aziz Q, Wise RG, Brooks J, Tracey I, Chang L. Attentional modulation of visceral and somatic pain. Neurogastroenterol Motil. 2007;19:569–77.
24. Dunckley P, Wise RG, Aziz Q, Painter D, Brooks J, Tracey I, et al. Cortical processing of visceral and somatic stimulation: differentiating pain intensity from unpleasantness. Neuroscience. 2005;133:533–42.
25. Dunckley P, Wise RG, Fairhurst M, Hobden P, Aziz Q, Chang L, et al. A comparison of visceral and somatic pain processing in the human brainstem using functional magnetic resonance imaging. J Neurosci. 2005;25:7333–41.
26. Peyron R, Laurent B, Garcia-Larrea L. Functional imaging of brain responses to pain. A review and meta-analysis. Neurophysiologie clinique Clin Neurophysiol. 2000;30:263–88.
27. Coen SJ, Gregory LJ, Yaguez L, Amaro E Jr, Brammer M, Williams SC, et al. Reproducibility of human brain activity evoked by esophageal stimulation using functional magnetic resonance imaging. Am J Physiol Gastrointest Liver Physiol. 2007;293:G188–97.
28. Aziz Q, Andersson JL, Valind S, Sundin A, Hamdy S, Jones AK, et al. Identification of human brain loci processing esophageal sensation using positron emission tomography. Gastroenterology. 1997;113:50–9.
29. Derbyshire SW. A systematic review of neuroimaging data during visceral stimulation. Am J Gastroenterol. 2003;98:12–20.
30. Binkofski F, Schnitzler A, Enck P, Frieling T, Posse S, Seitz RJ, et al. Somatic and limbic cortex activation in esophageal distention: a functional magnetic resonance imaging study. Ann Neurol. 1998;44:811–5.
31. Tolle TR, Kaufmann T, Siessmeier T, Lautenbacher S, Berthele A, Munz F, et al. Region-specific encoding of sensory and affective components of pain in the human brain: a positron emission tomography correlation analysis. Ann Neurol. 1999;45:40–7.
32. Price DD. Psychological and neural mechanisms of the affective dimension of pain. Science. 2000;288:1769–72.

33. Kulkarni B, Bentley DE, Elliott R, Youell P, Watson A, Derbyshire SW, et al. Attention to pain localization and unpleasantness discriminates the functions of the medial and lateral pain systems. Euro J of Neurosci. 2005;21:3133–42.
34. Porro CA, Cettolo V, Francescato MP, Baraldi P. Functional activity mapping of the mesial hemispheric wall during anticipation of pain. Neuroimage. 2003;19:1738–47.
35. Gregory LJ, Yaguez L, Williams SC, Altmann C, Coen SJ, Ng V, et al. Cognitive modulation of the cerebral processing of human oesophageal sensation using functional magnetic resonance imaging. Gut. 2003;52:1671–7.
36. Vogt BA, Derbyshire S, Jones AK. Pain processing in four regions of human cingulate cortex localized with co-registered PET and MR imaging. Eur J Neurosci. 1996;8:1461–73.
37. Craig AD. Interoception: the sense of the physiological condition of the body. Curr Opin Neurobiol. 2003;13:500–5.
38. Eickhoff SB, Lotze M, Wietek B, Amunts K, Enck P, Zilles K. Segregation of visceral and somatosensory afferents: an fMRI and cytoarchitectonic mapping study. Neuroimage. 2006;31:1004–14.
39. Wiech K, Lin CS, Brodersen KH, Bingel U, Ploner M, Tracey I. Anterior insula integrates information about salience into perceptual decisions about pain. J Neurosci. 2010;30:16324–31.
40. Augustine JR. Circuitry and functional aspects of the insular lobe in primates including humans. Brain Res Brain Res Rev. 1996;22:229–44.
41. Zaki J, Davis JI, Ochsner KN. Overlapping activity in anterior insula during interoception and emotional experience. Neuroimage. 2012;62:493–9.
42. Tracey I, Mantyh PW. The cerebral signature for pain perception and its modulation. Neuron. 2007;55:377–91.
43. Ploner M, Lee MC, Wiech K, Bingel U, Tracey I. Flexible cerebral connectivity patterns subserve contextual modulations of pain. Cereb Cortex. 2011;21:719–26.
44. Tracey I, Ploghaus A, Gati JS, Clare S, Smith S, Menon RS, et al. Imaging attentional modulation of pain in the periaqueductal gray in humans. J Neurosci. 2002;22:2748–52.
45. Stein JL, Wiedholz LM, Bassett DS, Weinberger DR, Zink CF, Mattay VS, et al. A validated network of effective amygdala connectivity. Neuroimage. 2007;36:736–45.
46. Ohman A. The role of the amygdala in human fear: automatic detection of threat. Psychoneuroendocrinology. 2005;30:953–8.
47. Apkarian AV, Bushnell MC, Treede RD, Zubieta JK. Human brain mechanisms of pain perception and regulation in health and disease. Eur J Pain. 2005;9:463–84.
48. Kringelbach ML. The human orbitofrontal cortex: linking reward to hedonic experience. Nat Rev Neurosci. 2005;6:691–702.
49. Vandenbergh J, Dupont P, Fischler B, Bormans G, Persoons P, Janssens J, et al. Regional brain activation during proximal stomach distention in humans: a positron emission tomography study. Gastroenterology. 2005;128:564–73.
50. Bantick SJ, Wise RG, Ploghaus A, Clare S, Smith SM, Tracey I. Imaging how attention modulates pain in humans using functional MRI. Brain. 2002;125:310–9.
51. Ochsner KN, Ludlow DH, Knierim K, Hanelin J, Ramachandran T, Glover GC, et al. Neural correlates of individual differences in pain-related fear and anxiety. Pain. 2006;120:69–77.
52. Craig AD, Bushnell MC, Zhang ET, Blomqvist A. A thalamic nucleus specific for pain and temperature sensation. Nature. 1994;372:770–3.
53. Ladabaum U, Minoshima S, Hasler WL, Cross D, Chey WD, Owyang C. Gastric distention correlates with activation of multiple cortical and subcortical regions. Gastroenterology. 2001;120:369–76.
54. Vogt BA. Pain and emotion interactions in subregions of the cingulate gyrus. Nat Rev Neurosci. 2005;6:533–44.
55. Derbyshire SW, Jones AK. Cerebral responses to a continual tonic pain stimulus measured using positron emission tomography. Pain. 1998;76:127–35.
56. Derbyshire SW. Imaging visceral pain. Curr Pain Headache Rep. 2007;11:178–82.
57. Millan MJ. Descending control of pain. Prog Neurobiol. 2002;66:355–474.
58. Petrovic P, Ingvar M. Imaging cognitive modulation of pain processing. Pain. 2002;95:1–5.
59. Tracey I. Can neuroimaging studies identify pain endophenotypes in humans? Nat Rev Neurol. 2011;7:173–81.
60. Farmer AD, Coen SJ, Kano M, Paine PA, Shwahdi M, Jafari J, et al. Psychophysiological responses to pain identify reproducible human clusters. Pain. 2013;154:2266–76.
61. Farmer AD, Coen SJ, Kano M, Naqvi H, Paine PA, Scott SM, et al. Psychophysiological responses to visceral and somatic pain in functional chest pain identify clinically relevant pain clusters. Neurogastroenterol Motil. 2014;26:139–48.
62. Labus JS, Naliboff BN, Fallon J, Berman SM, Suyenobu B, Bueller JA, et al. Sex differences in brain activity during aversive visceral stimulation and its expectation in patients with chronic abdominal pain: a network analysis. Neuroimage. 2008;41:1032–43.
63. Tillisch K, Labus JS. Advances in imaging the brain-gut axis: functional gastrointestinal disorders. Gastroenterology. 2011;140(407–11):e1.
64. Lelic D, Olesen SS, Valeriani M, Drewes AM. Brain source connectivity reveals the visceral pain network. Neuroimage. 2012;60:37–46.
65. Kehoe EG, Toomey JM, Balsters JH, Bokde AL. Personality modulates the effects of emotional arousal and valence on brain activation. Soc Cogn Affect Neurosci. 2012;7:858–70.

66. Pezawas L, Meyer-Lindenberg A, Drabant EM, Verchinski BA, Munoz KE, Kolachana BS, et al. 5-HTTLPR polymorphism impacts human cingulate-amygdala interactions: a genetic susceptibility mechanism for depression. Nat Neurosci. 2005;8:828–34.
67. Ruffle JK, Farmer AD, Kano M, Giampietro V, Aziz Q, Coen SJ. The influence of extraversion on brain activity at baseline and during the experience and expectation of visceral pain. Personality Individ Differ. 2015;74:248–53.
68. Coen SJ, Kano M, Farmer AD, Kumari V, Giampietro V, Brammer M, et al. Neuroticism influences brain activity during the experience of visceral pain. Gastroenterology. 2011;141(909–17):e1.
69. Silverman DH, Munakata JA, Ennes H, Mandelkern MA, Hoh CK, Mayer EA. Regional cerebral activity in normal and pathological perception of visceral pain. Gastroenterology. 1997;112:64–72.
70. Mayer EA, Berman S, Suyenobu B, Labus J, Mandelkern MA, Naliboff BD, et al. Differences in brain responses to visceral pain between patients with irritable bowel syndrome and ulcerative colitis. Pain. Netherlands. 2005:398–409.
71. Sarkar S, Aziz Q, Woolf CJ, Hobson AR, Thompson DG. Contribution of central sensitisation to the development of non-cardiac chest pain. Lancet. 2000;356:1154–9.
72. Dimcevski G, Sami SA, Funch-Jensen P, Le Pera D, Valeriani M, Arendt-Nielsen L, et al. Pain in chronic pancreatitis: the role of reorganization in the central nervous system. Gastroenterology. 2007;132:1546–56.
73. Ritchie J. Pain from distension of the pelvic colon by inflating a balloon in the irritable colon syndrome. Gut. 1973;14:125–32.
74. Drossman DA. The functional gastrointestinal disorders and the Rome III process. Gastroenterology. 2006;130:1377–90.
75. Tillisch K, Mayer EA, Labus JS. Quantitative meta-analysis identifies brain regions activated during rectal distension in irritable bowel syndrome. Gastroenterology. 2011;140:91–100.
76. Labus JS, Mayer EA, Jarcho J, Kilpatrick LA, Kilkens TO, Evers EA, et al. Acute tryptophan depletion alters the effective connectivity of emotional arousal circuitry during visceral stimuli in healthy women. Gut. 2011;60:1196–203.
77. Craggs JG, Price DD, Verne GN, Perlstein WM, Robinson MM. Functional brain interactions that serve cognitive-affective processing during pain and placebo analgesia. Neuroimage. 2007;38:720–9.
78. Labus JS, Naliboff BD, Berman SM, Suyenobu B, Vianna EP, Tillisch K, et al. Brain networks underlying perceptual habituation to repeated aversive visceral stimuli in patients with irritable bowel syndrome. Neuroimage. 2009;47:952–60.
79. Abraham C, Cho JH. Inflammatory bowel disease. N Engl J Med. 2009;361:2066–78.
80. Tracey I, Bushnell MC. How neuroimaging studies have challenged us to rethink: is chronic pain a disease? J Pain. 2009;10:1113–20.
81. Seminowicz DA, Labus JS, Bueller JA, Tillisch K, Naliboff BD, Bushnell MC, et al. Regional gray matter density changes in brains of patients with irritable bowel syndrome. Gastroenterology. 2010;139(48–57):e2.
82. Davis KD, Pope G, Chen J, Kwan CL, Crawley AP, Diamant NE. Cortical thinning in IBS: implications for homeostatic, attention, and pain processing. Neurology. 2008;70:153–4.
83. Blankstein U, Chen J, Diamant NE, Davis KD. Altered brain structure in irritable bowel syndrome: potential contributions of pre-existing and disease-driven factors. Gastroenterology. 2010;138:1783–9.
84. Labus JS, Vianna E, Jarcho JM, Tillisch K, Bueller JA, Mayer EA. 858 reduced structural connectivity between amygdala and prefrontal cortex in patients with irritable bowel syndrome: a diffuse tensor imaging study. Gastroenterology. 2010;138:S-118.
85. Moisset X, Bouhassira D, Denis D, Dominique G, Benoit C, Sabate JM. Anatomical connections between brain areas activated during rectal distension in healthy volunteers: a visceral pain network. Eur J Pain. 2010;14:142–8.
86. Olejniczak P. Neurophysiologic basis of EEG. J Clin Neurophysiol. 2006;23:186–9.
87. Lelic D, Olesen SS, Graversen C, Brock C, Valeriani M, Drewes AM. Electrophysiology as a tool to unravel the origin of pancreatic pain. World J Gastrointestinal Pathophysiol. 2014;5:33–9.
88. Michel CM, Murray MM, Lantz G, Gonzalez S, Spinelli L, Grave de Peralta R. EEG source imaging. Clin Neurophysiol (Official Journal of the International Federation of Clinical Neurophysiology). 2004;115:2195–222.
89. Burgell RE, Lelic D, Carrington EV, Lunniss PJ, Olesen SS, Surguy S, et al. Assessment of rectal afferent neuronal function and brain activity in patients with constipation and rectal hyposensitivity. Neurogastroenterol Motil. 2013;25(260–7):e167–8.
90. Lelic D, Brock C, Softeland E, Frokjaer JB, Andresen T, Simren M, et al. Brain networks encoding rectal sensation in type 1 diabetes. Neuroscience. 2013;237:96–105.
91. Tayama J, Sagami Y, Shimada Y, Hongo M, Fukudo S. Effect of alpha-helical CRH on quantitative electroencephalogram in patients with irritable bowel syndrome. Neurogastroenterol Motil. 2007;19:471–83.
92. Bouwense SA, Olesen SS, Drewes AM, Frokjaer JB, van Goor H, Wilder-Smith OH. Is altered central pain processing related to disease stage in chronic pancreatitis patients with pain? An exploratory study. PLoS One. 2013;8:e55460.

93. Olesen SS, Hansen TM, Graversen C, Steimle K, Wilder-Smith OH, Drewes AM. Slowed EEG rhythmicity in patients with chronic pancreatitis: evidence of abnormal cerebral pain processing? Eur J Gastroenterol Hepatol. 2011;23:418–24.
94. Olesen SS, Brock C, Krarup AL, Funch-Jensen P, Arendt-Nielsen L, Wilder-Smith OH, et al. Descending inhibitory pain modulation is impaired in patients with chronic pancreatitis. Clinical Gastroenterol Hepatol (The Official Clinical Practice Journal of the American Gastroenterological Association). 2010;8:724–30.
95. Drewes AM, Gratkowski M, Sami SA, Dimcevski G, Funch-Jensen P, Arendt-Nielsen L. Is the pain in chronic pancreatitis of neuropathic origin? Support from EEG studies during experimental pain. World J Gastroenterol. 2008;14: 4020–7.
96. Drewes AM, Krarup AL, Detlefsen S, Malmstrom ML, Dimcevski G, Funch-Jensen P. Pain in chronic pancreatitis: the role of neuropathic pain mechanisms. Gut. 2008;57:1616–27.
97. Pukall CF, Strigo IA, Binik YM, Amsel R, Khalife S, Bushnell MC. Neural correlates of painful genital touch in women with vulvar vestibulitis syndrome. Pain. 2005;115:118–27.
98. Martucci KT, Shirer WR, Bagarinao E, Johnson KA, Farmer MA, Labus JS, et al. The posterior medial cortex in urologic chronic pelvic pain syndrome: detachment from default mode network. A resting-state study from the MAPP research network. Pain. 2015.
99. Farmer MA, Chanda ML, Parks EL, Baliki MN, Apkarian AV, Schaeffer AJ. Brain functional and anatomical changes in chronic prostatitis/chronic pelvic pain syndrome. J Urol. 2011;186:117–24.
100. Kutch JJ, Yani MS, Asavasopon S, Kirages DJ, Rana M, Cosand L, et al. Altered resting state neuromotor connectivity in men with chronic prostatitis/chronic pelvic pain syndrome: a MAPP: research network neuroimaging study. NeuroImage Clinical. 2015;8:493–502.
101. Luzzi GA. Chronic prostatitis and chronic pelvic pain in men: aetiology, diagnosis and management. J Eur Acad Dermatol Venereol JEADV. 2002;16:253–6.
102. Marinoff SC, Turner ML. Vulvar vestibulitis syndrome: an overview. Am J Obstet Gynecol. 1991;165:1228–33.
103. Farmer AD, Aziz Q. Mechanisms of visceral pain in health and functional gastrointestinal disorders. Scand J Pain. 2014;5:51–60.
104. Berman SM, Naliboff BD, Suyenobu B, Labus JS, Stains J, Bueller JA, et al. Sex differences in regional brain response to aversive pelvic visceral stimuli. Am J Physiol Regul Integr Comp Physiol. 2006;291:R268–76.
105. Henderson LA, Gandevia SC, Macefield VG. Gender differences in brain activity evoked by muscle and cutaneous pain: a retrospective study of single-trial fMRI data. Neuroimage. 2008;39:1867–76.
106. Linnman C, Beucke JC, Jensen KB, Gollub RL, Kong J. Sex similarities and differences in pain-related periaqueductal gray connectivity. Pain. 2012;153:444–54.
107. Labus JS, Hubbard CS, Bueller J, Ebrat B, Tillisch K, Chen M, et al. Impaired emotional learning and involvement of the corticotropin-releasing factor signaling system in patients with irritable bowel syndrome. Gastroenterology. 2013;145(1253–61):e1–3.
108. Kano M, Farmer AD, Aziz Q, Giampietro V, Brammer MJ, Williams SC, et al. Sex Differences in brain activity to anticipated and experienced visceral pain in healthy subjects. Am J Physiol Gastrointest Liver Physiol. 2013.
109. Kilpatrick LA, Mayer EA, Labus JS, Gupta A, Hamaguchi T, Mizuno T, et al. Serotonin transporter gene polymorphism modulates activity and connectivity within an emotional arousal network of healthy men during an aversive visceral stimulus. PLoS ONE. 2015;10:e0123183.
110. Caspi A, Sugden K, Moffitt TE, Taylor A, Craig IW, Harrington H, et al. Influence of life stress on depression: moderation by a polymorphism in the 5-HTT gene. Science. 2003;301:386–9.
111. Hariri AR, Mattay VS, Tessitore A, Kolachana B, Fera F, Goldman D, et al. Serotonin transporter genetic variation and the response of the human amygdala. Science. 2002;297:400–3.
112. Camilleri M, Busciglio I, Carlson P, McKinzie S, Burton D, Baxter K, et al. Candidate genes and sensory functions in health and irritable bowel syndrome. Am J Physiol Gastrointest Liver Physiol. 2008;295:G219–25.
113. Colucci R, Gambaccini D, Ghisu N, Rossi G, Costa F, Tuccori M, et al. Influence of the serotonin transporter 5HTTLPR polymorphism on symptom severity in irritable bowel syndrome. PLoS ONE. 2013;8:e54831.
114. Fukudo S, Kanazawa M, Mizuno T, Hamaguchi T, Kano M, Watanabe S, et al. Impact of serotonin transporter gene polymorphism on brain activation by colorectal distention. Neuroimage. 2009;47:946–51.
115. Fukudo S, Kanazawa M. Gene, environment, and brain-gut interactions in irritable bowel syndrome. J Gastroenterol Hepatol. 2011;26(Suppl 3):110–5.
116. Schaub N, Kano M, Farmer AD, Aziz Q, Coen SJ. Su2109 the influence of the 5-HTTLPR genotype on the cerebral processing of esophageal pain—a pilot study. Gastroenterology. 2013;144:S-560.
117. Eysenck HJ. Dimensions of personality. London: Kegan Paul; 1947.
118. Canli T. Functional brain mapping of extraversion and neuroticism: learning from individual differences in emotion processing. J Pers. 2004;72:1105–32.
119. Harkins SW, Price DD, Braith J. Effects of extraversion and neuroticism on experimental pain, clinical pain, and illness behavior. Pain. 1989;36:209–18.

120. Paine P, Kishor J, Worthen SF, Gregory LJ, Aziz Q. Exploring relationships for visceral and somatic pain with autonomic control and personality. Pain. 2009;144:236–44.
121. Paine P, Worthen SF, Gregory LJ, Thompson DG, Aziz Q. Personality differences affect brainstem autonomic responses to visceral pain. Neurogastroenterol Motil. 2009;21:1155-e98.
122. Coen SJ, Yaguez L, Aziz Q, Mitterschiffthaler MT, Brammer M, Williams SC, et al. Negative mood affects brain processing of visceral sensation. Gastroenterology 2009;137:253–61, 61 e1-2.
123. Kumari V, ffytche DH, Das M, Wilson GD, Goswami S, Sharma T. Neuroticism and brain responses to anticipatory fear. Behav Neurosci. 2007;121:643–52.
124. Coen SJ, Aziz Q, Yaguez L, Brammer M, Williams SC, Gregory LJ. Effects of attention on visceral stimulus intensity encoding in the male human brain. Gastroenterology 2008;135:2065–74, 74 e1.
125. Kong J, Jensen K, Loiotile R, Cheetham A, Wey HY, Tan Y, et al. Functional connectivity of the frontoparietal network predicts cognitive modulation of pain. Pain. 2013;154:459–67.
126. Kumari V, ffytche DH, Williams SC, Gray JA. Personality predicts brain responses to cognitive demands. J Neurosci 2004;24:10636–41.
127. Seifert F, Schuberth N, De Col R, Peltz E, Nickel FT, Maihofner C. Brain activity during sympathetic response in anticipation and experience of pain. Hum Brain Mapp. 2013;34:1768–82.
128. Ploner M, Lee MC, Wiech K, Bingel U, Tracey I. Prestimulus functional connectivity determines pain perception in humans. Proc Natl Acad Sci U S A. 2010;107:355–60.
129. Tayama J, Nakaya N, Hamaguchi T, Tomiie T, Shinozaki M, Saigo T, et al. Effects of personality traits on the manifestations of irritable bowel syndrome. Biopsychosoc Med. 2012;6:20.
130. Hazlett-Stevens H, Craske MG, Mayer EA, Chang L, Naliboff BD. Prevalence of irritable bowel syndrome among university students: the roles of worry, neuroticism, anxiety sensitivity and visceral anxiety. J Psychosom Res. 2003;55:501–5.
131. Gwee KA, Leong YL, Graham C, McKendrick MW, Collins SM, Walters SJ, et al. The role of psychological and biological factors in postinfective gut dysfunction. Gut. 1999;44:400–6.
132. Chen JY, Blankstein U, Diamant NE, Davis KD. White matter abnormalities in irritable bowel syndrome and relation to individual factors. Brain Res. 2011;1392:121–31.
133. Ramirez-Maestre C, Esteve R. Disposition and adjustment to chronic pain. Curr Pain Headache Rep. 2013;17:312.
134. Gottesman II, Gould TD. The endophenotype concept in psychiatry: etymology and strategic intentions. Am J Psychiatry. 2003;160:636–45.
135. Carlsson K, Andersson J, Petrovic P, Petersson KM, Ohman A, Ingvar M. Predictability modulates the affective and sensory-discriminative neural processing of pain. Neuroimage. 2006;32:1804–14.
136. Ploghaus A, Tracey I, Gati JS, Clare S, Menon RS, Matthews PM, et al. Dissociating pain from its anticipation in the human brain. Science. 1999;284:1979–81.
137. Bishop S, Duncan J, Brett M, Lawrence AD. Prefrontal cortical function and anxiety: controlling attention to threat-related stimuli. Nat Neurosci. 2004;7:184–8.
138. Brodersen KH, Wiech K, Lomakina EI, Lin CS, Buhmann JM, Bingel U, et al. Decoding the perception of pain from fMRI using multivariate pattern analysis. Neuroimage. 2012;63:1162–70.
139. Fairhurst M, Wiech K, Dunckley P, Tracey I. Anticipatory brainstem activity predicts neural processing of pain in humans. Pain. 2007;128:101–10.
140. Berman SM, Naliboff BD, Suyenobu B, Labus JS, Stains J, Ohning G, et al. Reduced brainstem inhibition during anticipated pelvic visceral pain correlates with enhanced brain response to the visceral stimulus in women with irritable bowel syndrome. J Neurosci. 2008;28:349–59.
141. Naliboff BD, Waters AM, Labus JS, Kilpatrick L, Craske MG, Chang L, et al. Increased acoustic startle responses in IBS patients during abdominal and nonabdominal threat. Psychosom Med. 2008;70:920–7.
142. Petrovic P, Carlsson K, Petersson KM, Hansson P, Ingvar M. Context-dependent deactivation of the amygdala during pain. J Cogn Neurosci. 2004;16:1289–301.
143. Rivkin MJ. Developmental neuroimaging of children using magnetic resonance techniques. Mental Retard Dev Disabil Res Rev. 2000;6:68–80.
144. Ulmer S, Jansen O. fMRI basics and clinical applications. Berlin: Springer; 2013.
145. Omran YA, Aziz Q. Functional brain imaging in gastroenterology: to new beginnings. Nat Rev Gastroenterol Hepatol. 2014;11:565–76.
146. Mayer EA, Berman S, Suyenobu B, Labus J, Mandelkern MA, Naliboff BD, et al. Differences in brain responses to visceral pain between patients with irritable bowel syndrome and ulcerative colitis. Pain. 2005;115:398–409.
147. Ruffle J, Coen S, Farmer A, Kano M, Giampietro V, Williams S, et al. OC-066 Influence Of Extraversion On Brain Activity At Baseline, Pain Anticipation And Visceral Pain Processing. Gut. 2014;63:A32–3.

Author Biographies

James K. Ruffle obtained his MBBS medical degree and B.Sc. (Hons) in Neuroscience at Barts and The London School of Medicine and Dentistry, UK, qualifying in 2016. He is a medical doctor at Whipps Cross University Hospital. During his time in medical school, he has been involved in neuroimaging research at the Wingate Institute of Neurogastroenterology, investigating the influence of personality on the brain processing of visceral pain.

Dr. Jens B. Frokjaer is Medical Doctor (MD) from Aarhus University, DK, in 2000. He was awarded the Ph.D. degree in 2006 from Aalborg University. Finalised the training in Diagnostic Radiology in 2009, and since 2010 Consultant and Associate Professor at Aalborg University Hospital. The main research area is within clinical and experimental gastroenterological imaging studies including neurophysiological brain imaging (EEG, MRI and fMRI). He is author of approximately 85 peer-reviewed papers.

Dr. Adam D. Farmer undertook his medical training at University College London, qualifying with the degrees of MBBS B.Sc. (Hons) in 2001. He was awarded the degree of Ph.D. at the University of London entitled 'Human Psychophysiological Responses to Pain—The Development of Human Pain Endophenotypes' in 2011. He has authored over 50 peer-reviewed academic papers and his main interests concern visceral pain and irritable bowel syndrome.

Multiple Sclerosis and Pain

14

Daniela Seixas and Daniel Teles

Abstract

In this chapter, we address a frequent and debilitating symptom—pain of one of the most common causes of neurological disability in the young adult: multiple sclerosis. We introduce multiple sclerosis and define the role of neuroimaging in the diagnosis of the disease and beyond. Pain syndromes in multiple sclerosis are described, as well as other comorbidities that may interfere or be associated with pain. We discuss the published literature in neuroimaging and pain in multiple sclerosis, and emphasize the impact of chronic pain in an already non-resilient brain.

Keywords

Plaque · Myelin · Lhermitte · Psychosocial · Default · Resting-state · Demyelination

1 Introduction to Multiple Sclerosis

Multiple sclerosis (MS) is a chronic neurological disease that causes serious morbidity and suffering, and is one of the most frequently observed neurological non-traumatic causes of progressive disability in the young adult.

D. Seixas (✉) · D. Teles
Department of Biomedicine, Faculty of Medicine, Porto University, Porto, Portugal
e-mail: dseixas@med.up.pt
URL: https://pt.linkedin.com/in/danielaseixas

D. Teles
e-mail: daniel.teles7@gmail.com

MS is triggered by environmental factors in individuals with complex genetic risk profiles, and the disease process is of autoimmune inflammatory nature, mediated mainly by T-cells that attack antigens of oligodendrocytes and myelin sheaths [1]. This results in destruction of myelin and eventually of the axons and cell bodies in the central nervous system (CNS). The characteristic histopathological lesion is the plaque, which is a zone of demyelination. Such plaques may occur anywhere in the CNS, but are most frequently found in the spinal cord, particularly in the dorsal columns, in the brainstem, and in the white matter around the ventricles in the forebrain. Apart from the white matter lesions that are easily detected by imaging techniques,

L. Saba (ed.), *Neuroimaging of Pain*, DOI 10.1007/978-3-319-48046-6_14

pathological studies showed that extensive cortical and deep gray matter areas are demyelinated in MS patients [2]. The plaque-centered view of the disease fails to explain clinical deterioration of the patients when they have reached the progressive stage of the disease. It was thus postulated during the past years that besides inflammation there is a neurodegenerative component of the disease that leads to progressive and global brain damage [3]. There is increasing evidence that the severity of the clinical manifestations of MS does not simply depend on the extent of tissue destruction, but rather represents a complex balance among tissue damage, tissue repair, and cortical reorganization.

The early course of the disease is characterized by episodes of neurological dysfunction that usually recover. However, over time the pathological changes become dominated by widespread microglial activation associated with extensive and chronic neurodegeneration, the clinical correlate of which is progressive accumulation of disability [1]. In most patients, clinical manifestations indicate the involvement of motor, sensory, visual, and autonomic systems but many other symptoms and signs can occur. MS first symptoms are frequently of the sensory type, like hypoesthesia (reduced sensitivity to cutaneous stimulation) or paresthesia (subjective cutaneous sensations experienced spontaneously) that starts in an extremity, and progress over days to involve an entire limb. Although pain is a common sensory abnormality of MS, it is rarely the presenting symptom. Symptoms usually remain stable for one or two weeks, and then resolve gradually. Other common symptoms at presentation are blurred vision, diplopia, vertigo, motor deficits, and ataxia. Few of the clinical features are disease specific, but particularly characteristic is Lhermitte's symptom (an electrical sensation running down the spine or limbs on neck flexion) and the Uhthoff phenomenon (transient worsening of symptoms and signs when core body temperature increases) [1]. The clinical evolution of MS is somewhat predictable, occurring usually in relapses in the first years of the disease, with remission of the symptoms and signs (relapsing-remitting—RR), and then becoming progressive with time (secondary progressive MS). There are also other more aggressive subtypes of the disease, like remittent-progressive MS (where the signs and symptoms of the disease do not abate completely after each relapse), and primary progressive MS, that lacks the characteristic episodic evolution, being progressive ad initium. An additional form of the disease is the denominated clinically isolated syndrome, representing the first neurological episode of the disease [4]. In all cases, the clinical course usually evolves over several decades. Death is attributable to MS in two-thirds of cases, and to the increased risk of infection and its complications in individuals with advanced neurological disability; the median time to death is around 30 years from disease onset, representing a reduction in life expectancy of 5–10 years [5].

1.1 Diagnosis of Multiple Sclerosis and the Role of Neuroimaging

There is no single clinical sign or symptom, or diagnostic test that is sufficient to diagnose MS. The diagnosis is mainly clinical, based on several criteria, in which neuroimaging has a key role.

Magnetic resonance imaging (MRI) is the neuroimaging method used in the context of MS, given its safety, availability, and high spatial resolution. Structural MRI can reveal focal or confluent lesions in the brain, both in the white and the gray matter, irreversible tissue loss (atrophy), and demonstrate inflammatory activity of the disease. Moreover, it facilitates the communication of neuroimaging results in a highly reproducible and accurate way, with reference to the brain anatomy, which is essential for the diagnosis and follow-up of the disease.

Magnetic resonance imaging reveals abnormalities in the white matter of more than 95% of patients [1]. The characteristic lesion demonstrated on MRI is the cerebral or spinal plaque. Pathologically, plaques consist of a discrete region of demyelination with a variable extent of axonal injury. Plaques suggestive of MS are typically found on MRI in the periventricular region, corpus callosum, centrum semiovale, and, to a lesser extent, deep white matter structures and basal ganglia (Fig. 1). Multiple sclerosis plaques usually have an ovoid appearance, and lesions are arranged at right angles to the corpus callosum as if radiating from this area. When viewed on sagittal images, they are referred to as Dawson fingers (Fig. 2) [6].

The most common structural MRI sequence used in the diagnosis and follow-up of MS is T2-weighted turbo spin echo (TSE), which is able to demonstrate well the white matter demyelinating plaques (as hyperintense), both in the supra and infratentorial compartment, and edema (Fig. 3), whereas T1-weighted imaging has a better correlation with clinical disability by detecting hypointense lesions ("black holes") that relate to axonal loss (Fig. 4). Fluid-attenuated inversion recovery (FLAIR) has the highest sensitivity in the detection of lesions close to the cerebrospinal fluid (CSF) in the juxtacortical and the periventricular white matter, although being less sensitive in the evaluation of the structures of the posterior fossa like the cerebellum or the brainstem (Fig. 3) [7, 8]. The double inversion recovery (DIR) pulse sequence attenuates the signal of the CSF as well as of that of the white matter, improving the ability of MRI to detect cortical and juxtacortical lesions (Fig. 5). Visualization of gray matter lesions may be further improved with the use of ultrahigh magnetic fields (7 T) [9].

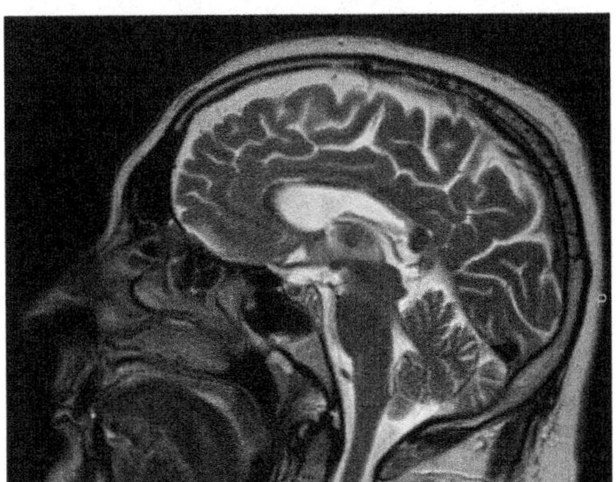

Fig. 1 Sagittal T2-weighted magnetic resonance image showing hyperintense lesions of multiple sclerosis in the corpus callosum. Note the associated atrophy of this commissure

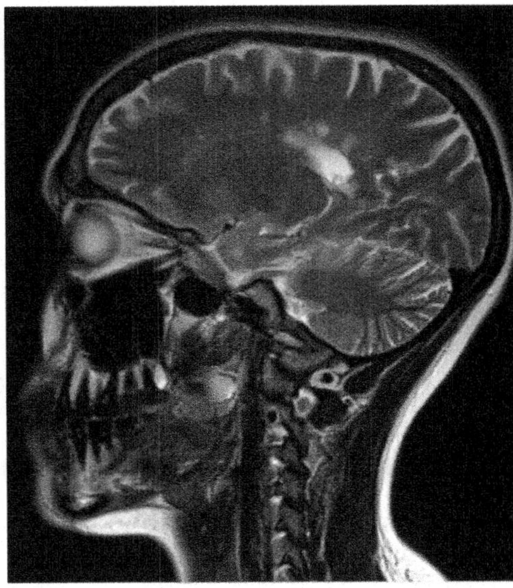

Fig. 2 Sagittal T2-weighted magnetic resonance image showing hyperintense lesions of multiple sclerosis radiating from the corpus callosum, referred to as "Dawson fingers"

Spinal cord MRI is used in studying sensory or motor symptoms in patients with spinal MS, including pain. Images of the spine in the sagittal plane correlate better with the extent of sensory impairments comparing with images in the axial plane [10], and usually include T2-weighted TSE, proton density (PD), and/or short-tau inversion recovery (STIR) sequences (Fig. 6).

Gadolinium-DTPA, a paramagnetic contrast agent that can cross only the disrupted blood–brain barrier, has been used to assess plaque activity, since the accumulation of gadolinium in plaques is associated with new or newly active plaques and with pathologically confirmed acute inflammation in MS (Fig. 7) [11]. Furthermore, gadolinium (Gd) enhancement patterns may provide clues to the diagnosis (and differential diagnosis) and underlying pathology of lesions. Concentric ring-enhancing lesions are thought to be related to accelerated disease activity and extensive tissue damage and may mark a type of inflammation characteristic of more aggressive

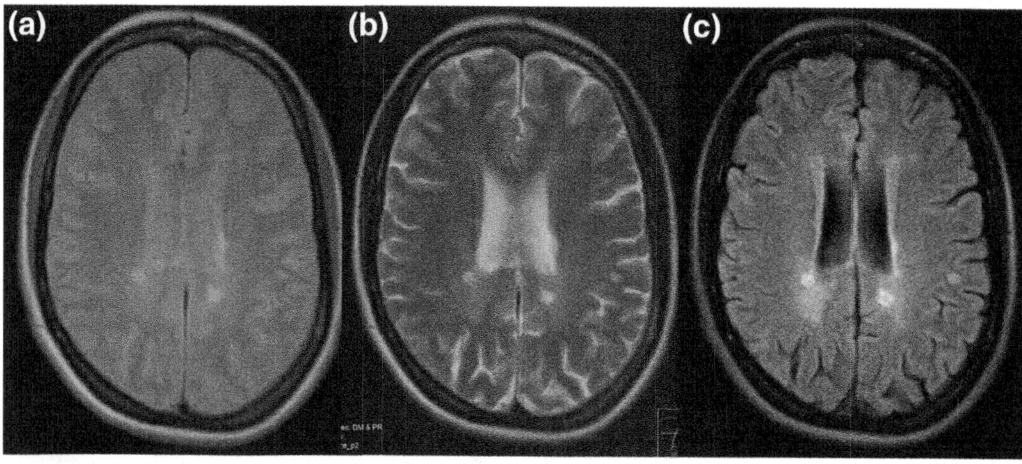

Fig. 3 Axial proton density (**a**), axial T2-weighted (**b**) and axial fluid attenuated inversion recovery (**c**) magnetic resonance images showing multiple, ovoid shaped, hyperintense foci consistent with multiple sclerosis plaques, located in the periventricular white matter

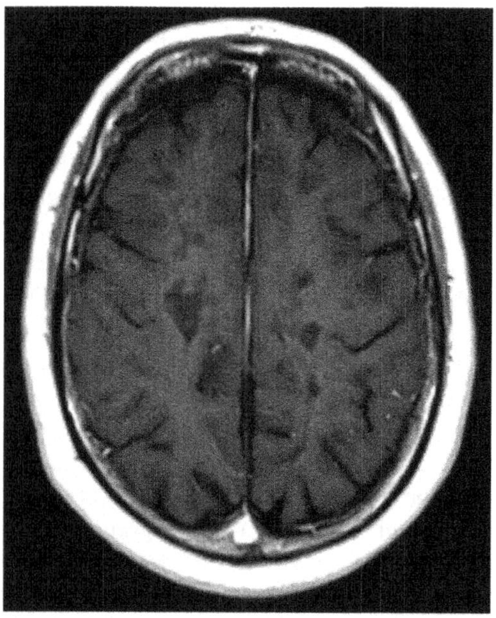

Fig. 4 Axial T1-weighted magnetic resonance image with contrast showing hypointense multiple sclerosis lesions in the centrum semiovale bilaterally, without gadolinium enhancement, the so called "black holes". These lesions are associated with axonal loss

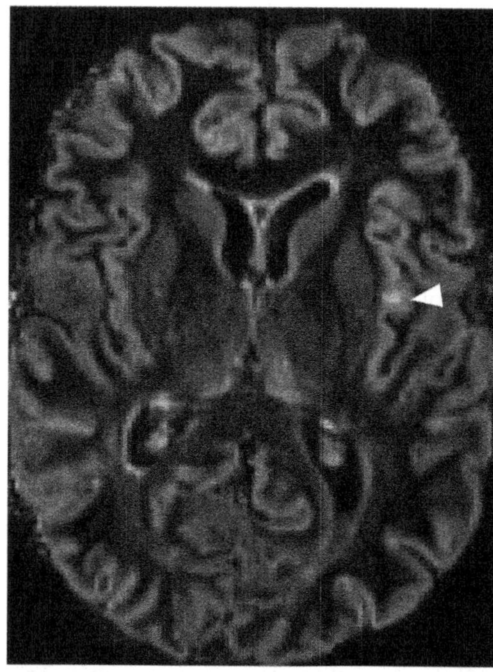

Fig. 5 Double inversion recovery magnetic resonance image revealing a multiple sclerosis plaque on the left insular cortex (*arrow*)

forms of disease [12]. Using higher doses of Gd, thinner slices or delayed imaging increases the sensitivity of Gd-enhanced MRI for the detection of active MS [13].

Not only is MRI an indicator of the anatomical dissemination of lesions, it can also show new plaques appearing over time. The core requirement for the diagnosis of MS is the

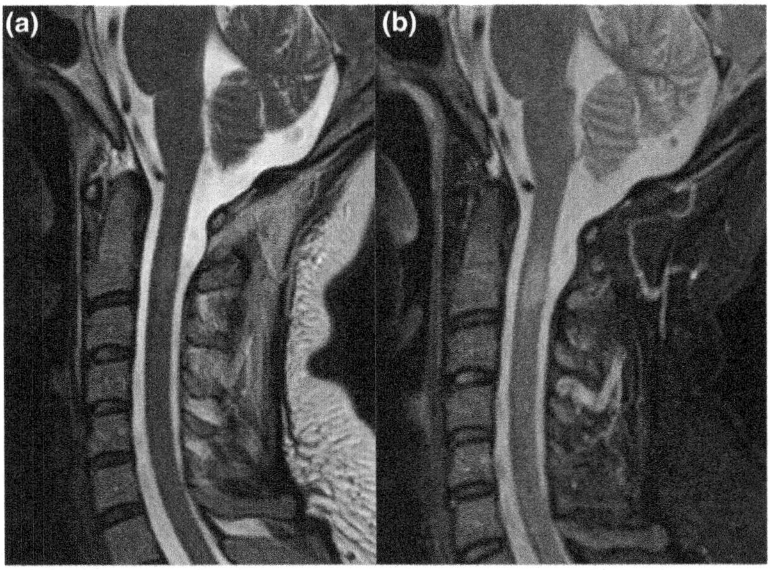

Fig. 6 T2-weighted turbo spin echo (**a**) and short-tau inversion recovery (**b**) sagittal magnetic resonance images showing a hyperintense multiple sclerosis plaque in the cervical spinal cord at the C2 level. Notice the higher sensitivity of the short-tau inversion recovery sequence compared to the T2-weighted sequence to detect lesions in the spinal cord

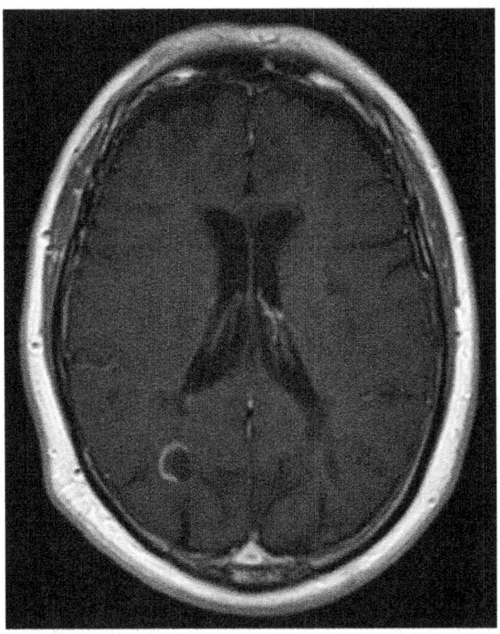

Fig. 7 Axial post-gadolinium T1-weighted magnetic resonance image showing a multiple sclerosis lesion enhancing with an open ring pattern, consistent with acute inflammation ("active" plaque)

demonstration of CNS lesion dissemination in time and space, based either in clinical findings or in a combination of clinical and MRI findings. Depending on the clinical presentation, a set of clinical, imaging, and paraclinical tests are needed to confirm the diagnosis of MS [14].

According to McDonald diagnostic criteria, dissemination in space is demonstrated with MRI by one or more T2 lesions in at least two of four MS-typical regions of the CNS (periventricular, juxtacortical, infratentorial, or spinal cord), or by other clinical attack implicating a different CNS site. For patients with brainstem or spinal cord syndromes, symptomatic MRI lesions are excluded from the criteria and do not contribute to lesion count. In its turn, dissemination in time is demonstrated with MRI by the simultaneous presence of asymptomatic Gd-enhancing and non-enhancing lesions at any time, or by a new T2 and/or Gd-enhancing lesion(s) on follow-up MRI, irrespective of its timing with reference to a baseline scan, or by the development of a second clinical attack [14]. The McDonald diagnostic criteria are presented in Table 1.

It is important to note that characteristic radiological lesions can appear in individuals without clinical signs of the disease, and many older individuals have nonspecific white matter cerebral lesions, which should not be over-interpreted. At any age, lesions detected in the spinal cord are invariably abnormal. Inevitably, diagnostic criteria do not confer absolute protection against error, because other diseases

Table 1 Diagnostic criteria for multiple sclerosis

Clinical presentation	Additional data needed for MS diagnosis
• 2 or more attacks • Objective clinical evidence of 2 or more lesions with reasonable historical evidence of a prior attack	None; clinical evidence will suffice. Additional evidence (e.g., brain MRI) desirable, but must be consistent with MS
• 2 or more attacks • Objective clinical evidence of 1 lesion	Dissemination in space demonstrated by MRI or Await further clinical attack implicating a different site
• 1 attack • Objective clinical evidence of 2 or more lesions	Dissemination in time demonstrated by MRI or second clinical attack
• 1 attack • Objective clinical evidence of 1 lesion (clinically isolated syndrome)	Dissemination in space demonstrated by MRI or await a second clinical attack implicating a different CNS site and Dissemination in time, demonstrated by MRI or second clinical attack
• Insidious neurologic progression suggestive of MS	One year of disease progression and dissemination in space, demonstrated by two of the following: • One or more T2 lesions in brain, in regions characteristic of MS • Two or more T2 focal lesions in spinal cord Positive CSF (isoelectric focusing evidence of oligoclonal bands and/or elevated IgG index)

Adapted from Polman et al. [14], diagnostic criteria for multiple sclerosis: 2010 revisions to the McDonald criteria
MS Multiple sclerosis; *CNS* Central nervous system; *MRI* Magnetic resonance imaging; *CSF* Cerebrospinal fluid; *IgG* Immunoglobulin G

can mimic MS. One of the limitations of using a conventional MRI measure in patients with MS is the discordance between the radiological extent of the lesions and the clinical presentation (clinical-radiological paradox), which other MRI techniques can help resolve [15].

1.2 Neuroimaging in Multiple Sclerosis Beyond the Diagnosis

In recent years, extensive MRI studies have had a major impact on MS, not only in diagnosis but also in the understanding of the disease [14]. By exploiting the natural history and histopathologic correlations, conventional and novel quantitative MRI techniques have demonstrated the ability to image underlying pathological processes in MS [16].

There are many MRI techniques that range from conventional MRI measures used in everyday clinical practice, to techniques more often used in investigating the mechanisms of the disease or as an outcome measure in clinical trials. Conventional MRI has contributed to the understanding of MS at the macroscopic level, but shows relatively weak relationships with clinical status [15]. Magnetic resonance imaging techniques that go beyond conventional anatomical imaging have demonstrated the ability to image underlying pathological processes in MS, and expand our knowledge on the true extent and nature of brain damage and plasticity in MS. These other measures are particularly useful in revealing diffuse damage in cerebral white and gray matter, and therefore are of help in resolving the dissociation between clinical and imaging findings. Advanced qualitative and quantitative MRI methods are thought to be more specific and sensitive for MS underlying pathology.

Quantitative MRI methods such as magnetization transfer ratio (MTR) are increasingly used to assess myelin content and axonal count in MS white matter, since MTR is significantly higher in remyelinated than demyelinated lesions [17]. Magnetization transfer contrast imaging (MTI) also increases sensitivity of Gd [18].

Diffusion-weighted imaging (DWI) is able to demonstrate differences in the magnitude and directionality of water diffusion, giving information about tissue integrity at a microscopic molecular level [19]. Diffusion tensor imaging (DTI) is the basis for white matter fiber tractography, a method to determine the pathways of anatomic white matter connectivity (Fig. 8). White matter tracts, which normally have a high degree of anisotropy due to their linear arrangement, appear with a decreased fractional anisotropy due to the injury of nerve axons or myelin sheaths. Normal-appearing white matter (NAWM) that is immediately adjacent to plaques seen on T2 imaging, may have abnormally reduced anisotropy due to either a less severe demyelization at the periphery of a centrifugally expanding plaque, or due to a continuous process of regression and repair in that area [20].

Myelin-selective MRI studies the MRI-visible water component associated with myelin. Since MS lesions show diffusely reduced NAWM when compared to healthy controls, this technique was validated as a measure of myelin density with the potential to quantitatively define the role of myelin-specific pathology in MS (Fig. 9) [21].

Magnetic resonance spectroscopy (MRS) provides insights into neurodegeneration, tissue repair, and oxidative stress in MS by detecting a range of chemical shifts that depict changes in white matter (Fig. 10) [22]. Phosphorus MRS can convey information on phospholipid metabolism, and proton MRS can generate information about other metabolic components, such as N-acetyl aspartate (NAA, a neuronal marker), creatine phosphate (Cr, an energy marker), choline (Cho, membrane components), and lactic acid (Fig. 10). Chronic MS is associated with a reduced NAA/Cr ratio within the brain, implying loss of neurons or axons. Because these findings can be correlated with disability scores, the use of MRS may prove valuable in monitoring patients after treatment and in prognosis [23].

Functional neuroimaging allows the study of the brain functions in humans in vivo. A subset of patients with MS experiences minimal clinical impairment despite significant lesions on MRI. Functional MRI (fMRI) studies detect changes in

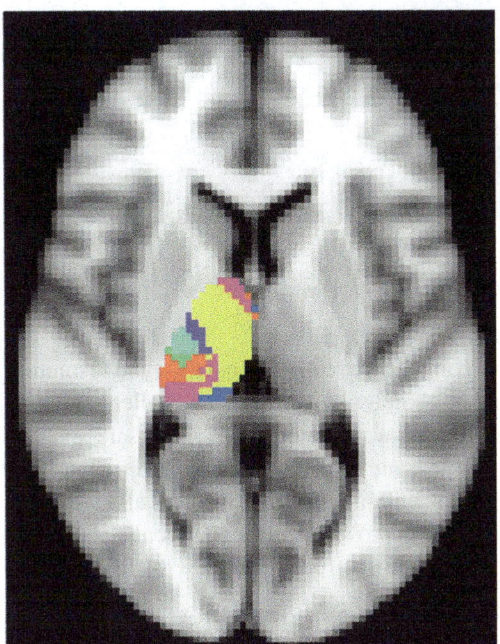

Fig. 8 Segmentation of the nuclei of right thalamus using the magnetic resonance diffusion tensor imaging technique and white matter fiber tractography

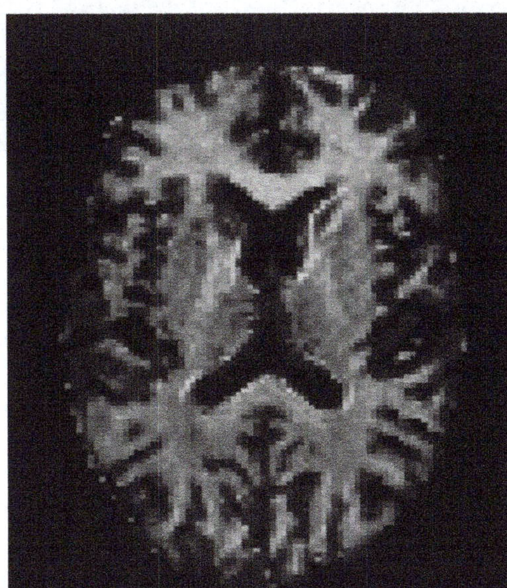

Fig. 9 Myelin water fraction map. The myelin-selective magnetic resonance imaging (MRI) techniques reveal the MRI-visible water component associated with myelin

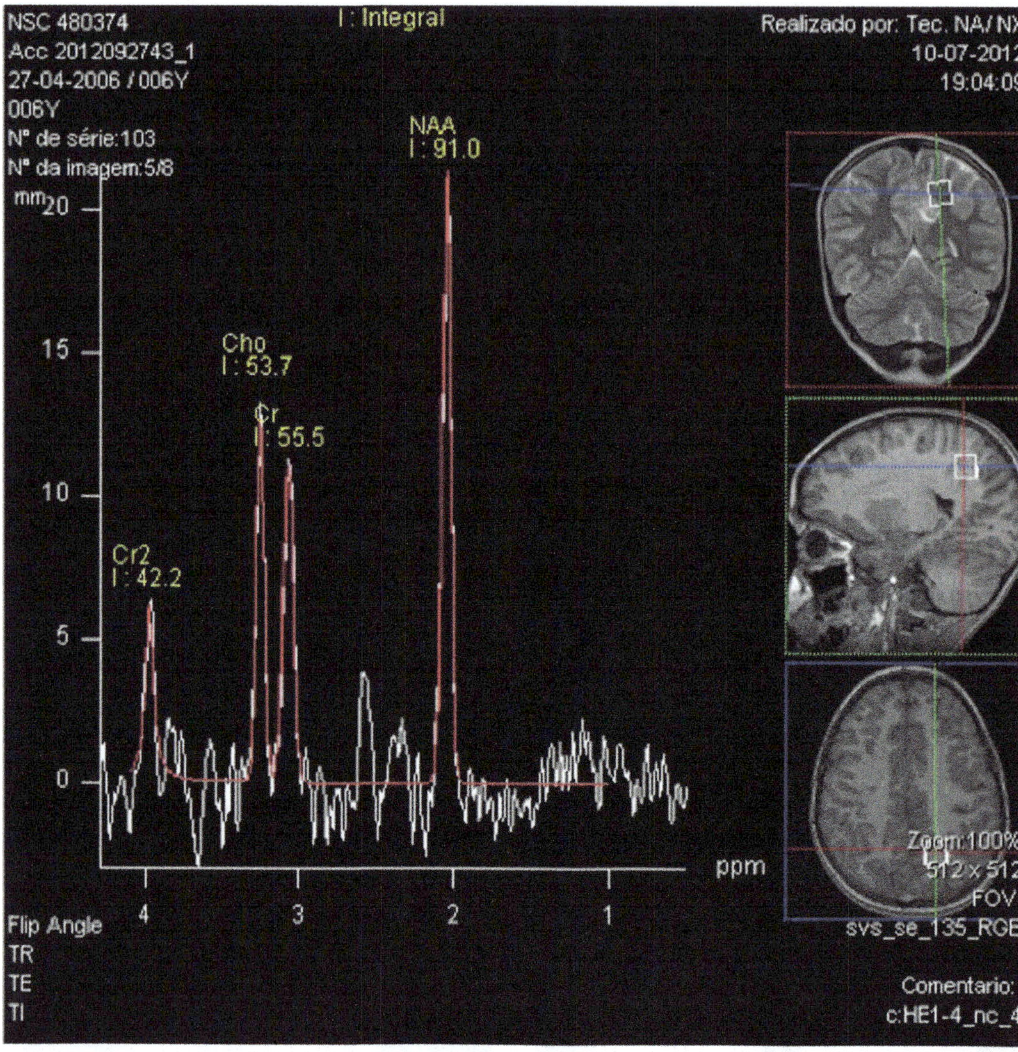

Fig. 10 Normal magnetic resonance proton spectroscopy showing the N-acetyl aspartate (NAA, a neuronal marker), creatine phosphate (Cr, an energy marker), and choline (Cho, membrane components) peaks

blood flow related to energy use by brain cells. These studies suggest that increased cognitive control recruitment in the motor system may limit the clinical manifestations of the disease in such cases [24].

Arterial spin labeling (ASL) measures cerebral perfusion using arterial water as an endogenous tracer. Brain perfusion changes have been reported in NAWM and in cortical and subcortical gray matter of MS patients [25].

MRI at ultrahigh magnetic fields (7 T) has advantages in relation to higher signal-to-noise ratio and improved image contrast and resolution, although not without technical challenges. Imaging at 7 T was demonstrated to be safe and well tolerated, and provides high-resolution anatomical images within or near the cortical layer [26]. This might prove useful for confidently classifying the location of lesions in relation to the cortical/subcortical boundary [27]. Moreover, ultrahigh field imaging has greater sensitivity to localize iron deposition [28]. New iron-based MRI contrast agents are able to track peripheral macrophages, providing complementary information on MS-related active inflammation [29]. Magnetic resonance iron-imaging has already established a link between iron deposition, gray matter damage, and clinical status [30].

FOCUS POINT: Although MRI alone cannot be used to diagnose MS, it is key in the differential diagnosis, for confirming MS and monitor disease progression.

2 Pain, Other Comorbidities, and Quality of Life in Multiple Sclerosis

2.1 Pain in Multiple Sclerosis

Pain is defined by the International Association for the Study of Pain (IASP) as "an unpleasant sensory and emotional experience associated with actual or potential tissue damage, or described in terms of such damage" [31]. The physiological purpose of pain is to protect the individual, warning of tissue damage, and most pain resolves rapidly as soon as the painful stimulus is removed. However, chronic pain may develop from poorly treated acute pain as a result of changes in the function of the CNS: the pain persists and has no protective role as it extends beyond the expected period of healing [32]. Chronic pain has traditionally been determined by an arbitrary interval of time since onset; the two most commonly used periods being 3 months and 6 months from its beginning [32]. Increasing evidence supports the idea that chronic pain could be understood not only as an altered perceptual state, but also as a consequence of maladaptive peripheral and central neuronal reorganization [33].

Pain can be classified as nociceptive when it arises from actual or threatened damage to non-neural tissue, and is due to the activation of nociceptors, i.e., a sensory receptor that is capable of transducing and encoding noxious stimuli. In turn, pain is defined as neuropathic when it is caused by a lesion or disease of the somatosensory nervous system, either in its peripheral elements (peripheral neuropathic pain) or in the CNS (central neuropathic pain) [34].

Pain is described by MS patients as one of their most important symptoms [35]. Pain is common in MS, but prevalence reports in the literature are heterogeneous. A recent systematic review and meta-analysis proposes that pain affects around 63% of adults with MS [36], comparing with the estimated 19% prevalence of chronic pain in the general population [37]. Pain in the MS population includes several pain syndromes and different mechanisms that are described in detail in the following section. Headache, followed by extremity neuropathic pain, are the most common types of pain, and trigeminal neuralgia (TN) the least frequent. Other pain syndromes include back pain, painful spasms, and the Lhermitte sign [36].

Although both neuropathic and nociceptive pain mechanisms may be in the origin of pain in MS, neuropathic pain is thought to be more prevalent than nociceptive pain [36]. In MS, causality of neuropathic pain may be difficult to establish due to the temporal and spatial complexity of the CNS lesions. The relationship between MS-related pain to disease evolution is uncertain. Headache has been described as appearing prior to MS onset [38] or related scarcely with relapses of the disease [39]. However, there is not any solid hypothesis concerning the natural history of pain during the disease course.

FOCUS POINT: Clinicians should routinely enquire MS patients about pain, and characterize existing pain syndromes.

2.2 Neuropsychiatric Abnormalities in Multiple Sclerosis

Neuropsychiatric abnormalities in MS are also frequent, and may interfere or be associated with pain; they can be broadly divided in disorders of mood, affect, and cognition [40].

Depression is the most pressing neuropsychiatric problem in MS [41], affecting nearly one in two patients during their lifetime [42], a figure three times the prevalence rate in the general population [43]. Rates of depression in MS may exceed those in other chronic medical [44] or neurological illnesses [45].

Depression and pain often co-occur in individuals with MS [46]. This coexistence can be explained by the overlap of central nociceptive and affective pathways [47], as well as the sharing of underlying neurotransmitters, with both norepinephrine and serotonin implicated in mood disorders and in the processing of pain. Moreover, there are several potential psychological and behavioral links between the two, such as the fact that pain intensity is associated with fatigue, anxiety, and sleep disturbances, which in turn are related with higher levels of depression [48]. Neuroimaging offers important clues as to the pathogenesis of depression, but psychosocial factors cannot be ignored and emerge as equally important predictors [41].

Other described concerns of mood and affect are bipolar affective disorder, euphoria, involuntary emotional expression disorder (episodes of crying or laughing that are unrelated to or out of proportion to the eliciting stimulus) and psychosis [35, 41].

2.3 Neuropsychological Abnormalities in Multiple Sclerosis

Multiple sclerosis-related cognitive dysfunction is highly prevalent, and may, as well as depression, interact with pain. In neuropsychological studies 40–65% of MS patients have shown cognitive impairment [49]. Multiple sclerosis patients do poorly in the Iowa Gambling Task (a psychological task thought to simulate real-life decision-making), probably reflecting altered decision-making capacity and emotional reactivity [50]. Their performance may relate to an increased sensitivity to immediate reward in addition to an impaired ability to evaluate the long-term consequences of decisions [51].

Pain and cognitive changes have been studied across various animal models of MS. In these models the onset of pain and cognitive dysfunction occur early, and do not coincide with the pattern of motor deficits. This is likely underpinned by a number of different mechanisms including changes in glutamate transmission, glial cell activation, and increased levels of pro-inflammatory cytokines. Changes in pain and cognition have been described as belonging to a cluster of symptoms and have been linked through centrally driven processes. In particular, the overactive immune response can induce a state of "sickness-like behaviors" that can influence both pain and cognition. Investigating the mechanism of inflammatory sickness behaviors in MS could lead to a better understanding of the links between pain and cognition [52].

FOCUS POINT: Neuropsychiatric and Neuropsychological abnormalities in MS are frequent and may interfere or be associated with pain. Depression and cognitive dysfunction are highly prevalent in MS.

2.4 Pain and Quality of Life

Pain is linked with adverse MS disease outcome —longer disease duration and higher disability [53]—and it has been associated not only with neuropsychiatric or neuropsychological factors but also with psychosocial and demographic factors, such as female sex, increased age, and lower educational level [54]. These problems often co-occur and are likely to have bidirectional effects, amplifying the impact on overall health-related quality of life (QOL) of MS patients and providing support for a biopsychosocial model of pain in MS [55].

This deterioration in QOL is manifest in daily activities, energy/vitality, mood, work, social relations, and enjoyment of life [46]. Individuals with MS who experience pain are significantly more likely to be unemployed than individuals with MS who are pain free [56], as well as a consuming more health care [54].

Psychosocial factors are more strongly associated with pain intensity than demographic and clinical variables [57]. This underlines the fact that psychosocial aspects are not additional to the experience of pain, but part of it; these factors influence how individuals react to and report pain, and result in coping strategies which may be helpful or destructive in maintaining function, particularly in chronic pain. Even though the phenotype of chronic central pain of MS does not differ psychophysically from other central neuropathic pain [58], the assessment of psychosocial factors is thus important [40].

3 Pain Syndromes in Multiple Sclerosis

Pain syndromes in MS are varied and may coexist, and may be of central neuropathic and/or nociceptive nature. Truini and co-workers recently proposed a mechanism-based classification of pain in MS, distinguishing five pain categories: nociceptive, neuropathic, psychogenic, mixed, and idiopathic [59]. Nine types of MS associated pain syndromes were identified, and their possible mechanisms are detailed in Table 2. These include headache, ongoing extremity pain, Lhermitte's phenomenon, painful tonic spasms, musculoskeletal pains, spasticity pain, pain associated with optic neuritis, TN, and treatment-induced pains [59].

Headache and ongoing extremity pain, as previously seen, are the most common types of pain in MS. Headache includes tension headache, migraine, cluster headache, or chronic daily headache [60]. Headache generally precedes the onset of MS and is not significantly modified by the disease.

Ongoing extremity pain is a kind of dysesthetic pain occurring in MS, described by patients as a "continuous burning pain" (searing, burning, tingling, piercing, electric-like), usually located in the lower extremities, mostly bilateral and that worsens with exposure to heat or weather changes [58].

Lhermitte's phenomenon is a transient short-lasting sensation related to neck movement felt in the back of the neck, lower back, or in other parts of the body usually observed in the initial stages of the disease and in patients with primary progressive MS [61].

Painful tonic spasms are seizure-like, involuntary dystonic spasms, usually brought on by movement or also by touch, hyperventilation, or emotions. They usually occur several times a day and last for less than two minutes [62].

Musculoskeletal pain, a nociceptive pain, is most often seen in the hips, legs, and arms when muscles, tendons, and ligaments remain immobile for a long time result of irregular, asymmetric movement patterns and postures, and changes in muscle strength, tone (spasticity), or length (contracture). However, it may also be a manifestation of central pain [58]. Secondary musculoskeletal pain can also be caused by treatment drugs.

Retrobulbar optic neuritis is the first symptom of MS in 20% of cases [63]. It is characterized by blurred vision or the complete loss of vision and color vision deficiency and contrast sensitivity that decrease proportionally to visual acuity loss. In most cases, it is accompanied by pain originating from behind the eye, that may even involve the whole head, and frequently preceding the disturbances of visual acuity.

Trigeminal neuralgia is a rare neuropathic pain syndrome in MS that appears in the trigeminal innervation area, spontaneously, or caused by stimuli in specific trigger areas of the face or mouth. It is characterized by paroxysms of shooting, piercing, stinging, electric-like pain, normally with a sudden onset, and often accompanied by a characteristic facial grimace [64].

Because in MS, and even in the same patient, pain may have various pathophysiological mechanisms (Table 2), it manifests with heterogeneous sensory disturbances [65]. Further refining mechanisms behind pain in MS through clinical examination, dedicated questionnaires, and procedures such as quantitative sensory testing, pain-related evoked potentials, and skin biopsy have led to the development of the so-called sensory profiles [66]. The clustering of sensory abnormalities (for example, hypo and hypersensitivity to mechanical and thermal stimuli) in a somatosensory phenotype, points to certain pathophysiological dysfunctions in afferent processing. These sensory pain-related abnormalities in patients with neuropathic pain can form different patterns, allowing sensory profiling of patients. Subgroups of patients with different somatosensory profiles may also

Table 2 Mechanism-based classification of pain in multiple sclerosis

Types of pain	Possible mechanisms
Neuropathic pains	
Ongoing extremity pain	Deafferentation pain secondary to lesions in the spino-thalamo-cortical pathways
Trigeminal neuralgia	Paroxysmal high-frequency discharges ectopically generated by intra-axial inflammatory demyelination and extra-axial mechanical demyelination of the trigeminal primary afferents
Lhermitte's phenomenon	Paroxysmal neuropathic pain due to high-frequency ectopic impulse generated by demyelination of the dorsal column primary afferents
Nociceptive pains	
Pain associated with optic neuritis	Nerve trunk pain originating from endoneural inflammation intraneural nociceptors of the nervi nervorum
Musculoskeletal pain	Nociceptive pain related to postural abnormalities secondary to motor disturbances
Back pain	Consequence of postural anomalies
Migraine	Nociceptive pain favored by predisposing factors or secondary to midbrain/periaqueductal gray matter lesions
Tension-type headache	Probably coexisting conditions
Treatment-induced pains	Interferon beta (flu-like symptoms, myalgias, and headache), glatiramer acetate (pain at the injection site), corticosteroids (osteoporosis and secondary pain)
Mixed pains	
Painful tonic spasms	High-frequency discharges ectopically generated by demyelinating lesions in the cortico-spinal pathways induce tonic spasm which, in turn, induce ischemic muscle pain
Spasticity pain	Mixed pain secondary to lesions in the central motor pathways but mediated by muscle nociceptors

Adapted from Truini et al. [59], a mechanism-based classification of pain in multiple sclerosis

respond differently to treatment [67]. Cruccu and co-authors defend that neuropathic pain should be classified according to these sensory profiles rather than etiology [65], so it could minimize the pathophysiological heterogeneity within study groups and clinical trials, thus making it easier to identify a positive treatment response and opening the way to new therapeutic approaches of pain in MS.

In this context, neurophysiologic testing becomes important in associating a specific type of sensory disturbance to specific afferent pathway damages. Evoked potentials can be useful neurophysiologic studies for evaluation of MS, including laser evoked potentials (LEP) and somatosensory evoked potentials (SEP).

Ongoing extremity pain is associated with LEP abnormalities that suggest that this type of pain is related to nociceptive pathway damage. Since MRI shows cervical or thoracic spinal cord damage, ongoing extremity pain may arise from spinothalamic tract lesions with deafferentation of thalamic nuclei [66]. Distinctively, Lhermitte's phenomenon is associated with SEP abnormalities, implying that this type of pain is related to non-nociceptive Ab-fiber pathway damage. Cervical spinal cord lesions as assessed by MRI imaging and the reported pain due to neck movement build up to the conclusion that the Lhermitte's phenomenon probably arises from a demyelinating lesion in the dorsal columns of the cervical spinal cord [66].

FOCUS POINT: Headache and ongoing extremity pain are the most common types of pain in MS. Neurophysiologic characterization of pain syndromes in MS and correlation of results with lesion location, as demonstrated by MRI, may be important for treatment selection.

4 Neuroimaging and Pain in Multiple Sclerosis

Studies investigating pain in MS with neuroimaging methods are scarce. A recent systematic review of neuroimaging studies in MS reports that most of the published articles are case reports/series aimed at describing associations between demyelinating lesions and pain syndromes, with limited impact for the knowledge of pain mechanisms in MS and for patient management [68].

More evidence on pain mechanisms in MS is warranted, considering the high relevance and impact of pain in this disease, and how little is known about its pathophysiology. In the case of central neuropathic pain, a single CNS lesion in a strategic location can be in its origin. On the other hand, it is recognized that the remainder of the lesion load and hidden pathology on conventional MRI—in the cortex and in NAWM—may contribute to MS pain and associated comorbidities.

Most neuroimaging studies in pain in MS investigated headache and facial pain [68]. Studies of migraine [69], as well as unclassified headache [70], identified abnormalities in the brainstem, a finding in line with the putative role of the brainstem in pain transmission pathways in central neuropathic pain in MS. Apart from the lesion location, the T2 lesion burden on brain MRI does not seem to account for any differences in the migraine status [71].

Studies characterizing TN and trigeminal autonomic cephalalgias (TACs) in MS focused on abnormalities associated to the trigeminal nucleus and nerve. Interestingly, there appears to be some radiological overlap between findings in these groups of headaches, which are traditionally viewed as distinct in etiology. The controversy remains, whether MS lesions in the trigeminal pathways account for TN [72–74], or if there is a simultaneous role of central and peripheral trigeminal damage [75, 76].

The role of cervical spinal imaging in investigating headache etiology (particularly, though not exclusively, when occipital, and thus hypothetically related to a cervical dermatomal distribution of pain) are also of notice [77, 78]. Cervical MRI with Gd in patients with sudden paroxysmal occipital pain might reveal a new active or new T2-weighted demyelinating C2 cervical lesion which may signal relapse of MS [78].

More frequent pain syndromes in MS, such as limb pain, have been relatively understudied, comparing, for example, with headache [68]. Considering limb and radicular pain, the limited data available suggest, as might be suspected on a neuroanatomical basis, that a spinal location of an MS plaque should be considered. Dorsal lesions in the thoracic and/or cervical cord have been associated with limb pain [79–81].

Although thalamic or cortical lesions are known to be responsible for pain syndromes, such as in post-stroke pain [82], no difference has been found regarding the presence of lesions in the thalamus, capsula interna and thalamo-cortical projections in MS patients with or without pain (Fig. 11) [83]. These studies may potentially suggest a role of spinal lesions in either directly disturbing sensory afferent pathways, or perhaps in contributing to the imbalance between spinothalamic and other sensory pathways, or dysfunction of descending inhibitory pathways [83].

FOCUS POINT: The spinal cord is a frequent origin of central pain in MS. Thalamic lesions, although common in MS, are not frequently associated with pain.

It is of note that the previously discussed describes only potential associations, rather than established causation. Moreover, pain present at multiple body sites cannot be presumed to be associated with identical radiological abnormalities as those identified in the limited studies of well-localized pain at a single site [68]. Furthermore, the current literature of neuroimaging studies of pain in MS is methodologically poor [68]. Studies tend to give emphasis to white matter pathology in MS, although histopathological and MRI research has shown that lesions are often located in the gray matter, especially in the cerebral cortex [1]. Likewise, it is important to take into account MS normal-appearing brain damage. The use of functional or molecular

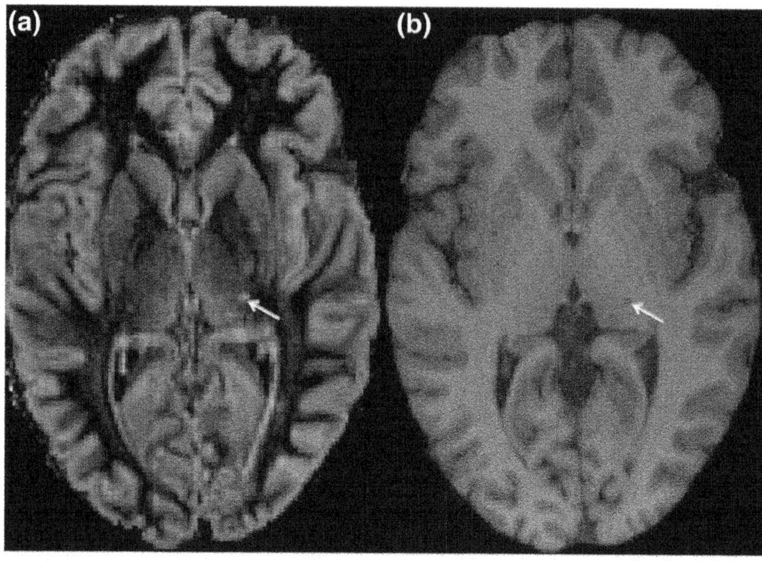

Fig. 11 Double inversion recovery (**a**) and axial T1-weighted magnetic resonance images (**b**) identifying a thalamic lesion (*arrows*) in a multiple sclerosis patient with a thalamic pain syndrome. Notice that thalamic lesions, although common in multiple sclerosis, are not frequently associated with pain in this disease

imaging techniques, serial imaging, and/or the use of intravenous contrast medium complemented by electrophysiological techniques can contribute as well to the establishment of a temporal association (and hence possible causality) between the lesion and the specific pain syndromes, bringing time to space resolution to the study of pain in MS.

Neuroimaging methods, in particular functional and advanced structural MRI techniques, are ideal to study pain noninvasively in these patients, given their already substantial contributions to both the MS and pain research fields. Functional neuroimaging is able to provide insight on critical brain regions for pain processing and to the understanding of how cognitive, emotional and contextual factors modulate the pain experience in MS. The ASL technique can measure changes in the regional cerebral blood flow (CBF) in brain areas that have been previously associated with pain perception, like the secondary somatosensory, insular and cingulate cortices [84], proving itself suitable to study pain conditions that are difficult to investigate with current fMRI, such as chronic pain. Resting-state fMRI is an MRI technique that has several potential advantages over task-activation fMRI in terms of its clinical applicability, particularly for ongoing pain states [85]. In the systematic review of Seixas and colleagues, only one study was identified investigating pain in MS using nonconventional MRI [68].

FOCUS POINT: More studies investigating pain in MS with neuroimaging methods are needed. The majority of the published articles are only case reports/series describing associations between MS plaques and pain.

4.1 Chronic Pain in Multiple Sclerosis

Neuroimaging techniques, besides allowing the study of lesion topography and its association with pain, offer as well a window to the evaluation of the consequences of chronic pain in the CNS in MS. There is evidence of brain structural and functional dysfunction in chronic pain. Studies in animal models have demonstrated that chronic pain is accompanied by molecular, neuronal, and structural changes in the brain and also in the spinal cord [86]. Chronic pain can be understood not only as an altered functional state, but also a consequence of neuronal reorganization [33, 87].

As previously discussed, in MS neuropathic pain may originate from a single lesion in the somatosensory pathways, possibly the spinal cord, and evolve into chronic pain, burdening an already

MS-damaged CNS and leading to a cycle of structural and functional brain disruption (Fig. 12). This is the context that is perhaps unique to MS, which mechanisms can be captured using a state-of-the-art imaging protocol directed at the specificities of this demyelinating disease.

The fact that MS is a demyelinating disease and changes in white matter have been identified in chronic pain conditions, suggests a link to pain chronicity in altering vulnerable or non-resilient white matter. These plastic, probably maladaptive, brain changes may be a contribution of chronic pain, and furthermore, a consequence of pain originating in the spinal cord in MS [40].

Regarding functional brain plastic changes in long-standing pain, different chronic pain conditions seem to evoke distinct brain activity patterns, which may reflect not only pain but also processes related with each disease [88]. Pain alters brain dynamics beyond pain perception by distorting brain resting-state networks (RSNs) [89–91]. These networks are brain regions that are active when the individual is not focused on the outside world and the brain is at wakeful rest. This intrinsic neuronal activity is critical for the development of synaptic connections and maintenance of synaptic homeostasis [92]. The default-mode network (DMN) (Fig. 13), one of such networks, is deactivated during demanding cognitive tasks and involved in internal modes of cognition [93]. It includes the medial temporal lobe and the medial prefrontal cortex subsystems, converging on important nodes of integration including the posterior cingulate cortex highlighting the possible adaptive role of the DMN in planning the future and in social interactions often impaired, for example, in chronic pain states [94]. A DMN dysfunction in regions subserving the reward system, the caudate nucleus and nucleus accumbens, was reported in chronic MS pain, and may be associated with altered decision-making and planning [40]. It is important to further investigate the meaning and consequences of this dysfunction in the reward system, especially because cognition and emotion disorders are also prevalent in MS.

FOCUS POINT: Chronic pain is known to induce molecular, neuronal, and structural changes in the brain and the spinal cord, which can burden an already non-resilient CNS in MS.

5 Pain Management in Multiple Sclerosis

Pharmacological treatment of pain in MS is challenging, due to the many underlying pathophysiological mechanisms [59]. It has been described the potential for several drugs in its management, including antidepressants, anticonvulsants, dextromethorphan/quinidine, opioids/opioid antagonists, and cannabinoids. Regarding invasive pain treatment, the options to relief pain include microvascular decompression for TN, CNS transcutaneous electrical nerve stimulation (motor cortex stimulation, spinal cord stimulation, and posterior nucleus of the hypothalamus stimulation). Neuroimaging methods have a role in

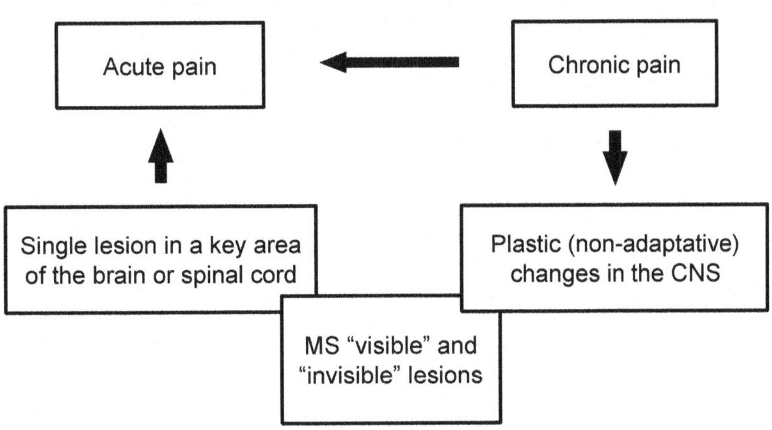

Fig. 12 The cycle of structural and functional central nervous system damage of pain associated with multiple sclerosis

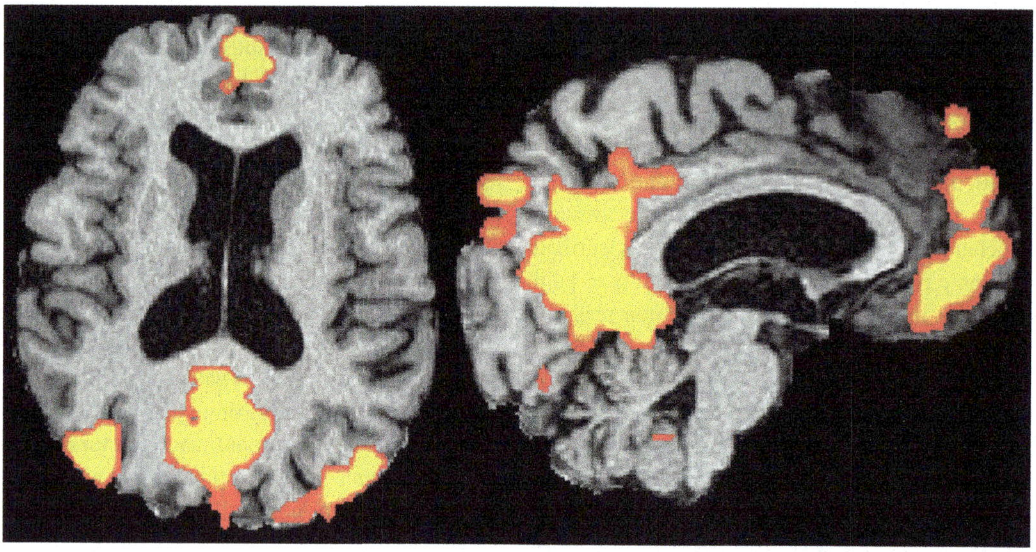

Fig. 13 The default-mode network. Regions belonging to the default-mode network include the medial prefrontal cortex, the anterior and posterior cingulate cortex, lateral and inferior parietal cortex, inferior and middle temporal gyri and mesial temporal lobe regions, the precuneus, the thalami, and cerebellar areas

invasive treatment planning, and as well as outcome measures in clinical drug trials.

6 Conclusions

Pain is a frequent and debilitating symptom of MS, which in turn is one of the most prevalent causes of neurological disability in the young adult. Pain in MS may be neuropatic or, less frequently, of nociceptive origin. Pain is still underrecognized in MS, and its mechanisms are poorly understood.

Neuroimaging techniques are key for the diagnosis and differential diagnosis in MS, and for disease follow-up. Magnetic resonance imaging, together with neurophysiological testing, has a role as well in the characterization of pain syndromes in MS, with an impact in the treatment of pain, in better targeting both drugs and interventions such as deep brain or cord stimulation.

Magnetic resonance imaging has been important in the research of pain mechanisms in humans. However, the literature is still scarce in publications investigating pain in MS using neuroimaging methods. More studies are needed, in particular addressing chronic pain and nociceptive pain of MS, and investigating the interaction of MS and comorbidities like depression and cognitive impairment. Neuroimaging methods can contribute further to the understanding of pain in MS, and to create opportunities for the recognition and effective treatment of pain in this disease.

Nonetheless, the complexity of MS, with lesions disseminating both in time and spatially in the CNS, and its invisible brain and cord damage, together with the technical complexity of the different MRI methods, warrant rigorous methodology for obtaining valid, reproducible and enlightening results in the investigation of pain syndromes in MS.

References

1. Compston A, Coles A. Multiple sclerosis. Lancet (London, England). 2008;372:1502–17. doi:10.1016/S0140-6736(08)61620-7.
2. Stadelmann C, Albert M, Wegner C, Bruck W. Cortical pathology in multiple sclerosis. Curr Opin Neurol. 2008;21:229–34. doi:10.1097/01.wco.0000318863.65635.9a.
3. Lassmann H. New concepts on progressive multiple sclerosis. Curr Neurol Neurosci Rep. 2007;7:239–44.
4. Lublin FD, Reingold SC. Defining the clinical course of multiple sclerosis: results of an International

survey. National multiple sclerosis society (USA) Advisory committee on clinical trials of new agents in multiple sclerosis. Neurology. 1996;46:907–11.

5. Bronnum-Hansen H, Koch-Henriksen N, Stenager E. Trends in survival and cause of death in Danish patients with multiple sclerosis. Brain. 2004;127:844–50. doi:10.1093/brain/awh104.
6. Traboulsee AL, Li DKB. The role of MRI in the diagnosis of multiple sclerosis. Adv Neurol. 2006;98:125–46.
7. Gawne-Cain ML, O'Riordan JI, Thompson AJ, et al. Multiple sclerosis lesion detection in the brain: a comparison of fast fluid-attenuated inversion recovery and conventional T2-weighted dual spin echo. Neurology. 1997;49:364–70.
8. Yousry TA, Filippi M, Becker C, et al. Comparison of MR pulse sequences in the detection of multiple sclerosis lesions. AJNR Am J Neuroradiol. 1997;18:959–63.
9. Filippi M, Rocca MA, Horsfield MA, et al. Imaging cortical damage and dysfunction in multiple sclerosis. JAMA Neurol. 2013;70:556–64. doi:10.1001/jamaneurol.2013.1954.
10. Fukutake T, Kuwabara S, Kaneko M, et al. Sensory impairments in spinal multiple sclerosis: a combined clinical, magnetic resonance imaging and somatosensory evoked potential study. Clin Neurol Neurosurg. 1998;100:199–204.
11. Tortorella C, Codella M, Rocca MA, et al. Disease activity in multiple sclerosis studied by weekly triple-dose magnetic resonance imaging. J Neurol. 1999;246:689–92.
12. Morgen K, Jeffries NO, Stone R, et al. Ring-enchancement in multiple sclerosis: marker of disease severity. Mult Scler. 2001;7:167–71.
13. Filippi M. Enhanced magnetic resonance imaging in multiple sclerosis. Mult Scler. 2000;6:320–6.
14. Polman CH, Reingold SC, Banwell B, et al. Diagnostic criteria for multiple sclerosis: 2010 revisions to the McDonald criteria. Ann Neurol. 2011;69:292–302. doi:10.1002/ana.22366.
15. Neema M, Stankiewicz J, Arora A, et al. T1- and T2-based MRI measures of diffuse gray matter and white matter damage in patients with multiple sclerosis. J Neuroimaging. 2007;17(Suppl 1):16S–21S. doi:10.1111/j.1552-6569.2007.00131.x.
16. Simon JH, Li D, Traboulsee A, et al. Standardized MR imaging protocol for multiple sclerosis: consortium of MS centers consensus guidelines. AJNR Am J Neuroradiol. 2006;27:455–61.
17. Schmierer K, Scaravilli F, Altmann DR, et al. Magnetization transfer ratio and myelin in post-mortem multiple sclerosis brain. Ann Neurol. 2004;56:407–15. doi:10.1002/ana.20202.
18. Silver NC, Good CD, Barker GJ, et al Sensitivity of contrast enhanced MRI in multiple sclerosis. Effects of gadolinium dose, magnetization transfer contrast and delayed imaging. Brain. 1997;120(Pt 7):1149–61.
19. Le Bihan D. Molecular diffusion, tissue microdynamics and microstructure. NMR Biomed. 1995;8:375–86.
20. Kealey SM, Kim Y, Provenzale JM. Redefinition of multiple sclerosis plaque size using diffusion tensor MRI. Am J Roentgenol. 2004;183:497–503. doi:10.2214/ajr.183.2.1830497.
21. Laule C, Leung E, Lis DKB, et al. Myelin water imaging in multiple sclerosis: quantitative correlations with histopathology. Mult Scler. 2006;12:747–53. doi:10.1177/1352458506070928.
22. Srinivasan R, Ratiney H, Hammond-Rosenbluth KE, et al. MR spectroscopic imaging of glutathione in the white and gray matter at 7 T with an application to multiple sclerosis. Magn Reson Imaging. 2010;28:163–70. doi:10.1016/j.mri.2009.06.008.
23. Arnold DL, Riess GT, Matthews PM, et al. Use of proton magnetic resonance spectroscopy for monitoring disease progression in multiple sclerosis. Ann Neurol. 1994;36:76–82. doi:10.1002/ana.410360115.
24. Colorado RA, Shukla K, Zhou Y, Wolinsky JS, Narayana P. Multi-task functional MRI in multiple sclerosis patients without clinical disability. Neuroimage. 2012;59:573–81. doi:10.1016/j.neuroimage.2011.07.065.Multi-task, https://www.ncbi.nlm.nih.gov/pubmed/21840409.
25. Adhya S, Johnson G, Herbert J, et al. Pattern of hemodynamic impairment in multiple sclerosis: dynamic susceptibility contrast perfusion MR imaging at 3.0 T. Neuroimage. 2006;33:1029–35. doi:10.1016/j.neuroimage.2006.08.008.
26. Mainero C, Benner T, Radding A, et al. In vivo imaging of cortical pathology in multiple sclerosis using ultra-high field MRI. Neurology. 2009;73:941–8. doi:10.1212/WNL.0b013e3181b64bf7.
27. Tallantyre EC, Morgan PS, Dixon JE, et al. 3 Tesla and 7 Tesla MRI of multiple sclerosis cortical lesions. J Magn Reson Imaging. 2010;32:971–7. doi:10.1002/jmri.22115.
28. Hammond KE, Metcalf M, Carvajal L, et al. Quantitative in vivo magnetic resonance imaging of multiple sclerosis at 7 Tesla with sensitivity to iron. Ann Neurol. 2008;64:707–13. doi:10.1002/ana.21582.
29. Dousset V, Brochet B, Deloire MSA, et al. MR imaging of relapsing multiple sclerosis patients using ultra-small-particle iron oxide and compared with gadolinium. AJNR Am J Neuroradiol. 2006;27:1000–5.
30. Stankiewicz J, Panter SS, Neema M, et al. Iron in chronic brain disorders: imaging and neurotherapeutic implications. Neurotherapeutics. 2007;4:371–86. doi:10.1016/j.nurt.2007.05.006.
31. Merskey H, Bogduk N. Classification of chronic pain: descriptions of chronic pain syndromes and definitions of pain terms. In: Task force on taxonomy IA for the study of pain. IASP Press; 1994.
32. Loeser JD, Bonica JJ. Bonica's management of pain. Lippincott Williams & Wilkins; 2001.

33. Baliki MN, Schnitzer TJ, Bauer WR, Apkarian AV. Brain morphological signatures for chronic pain. PLoS ONE. 2011;6:e26010. doi:10.1371/journal.pone.0026010.
34. IASP: IASP taxonomy. http://www.iasp-pain.org/Taxonomy (2012). Accessed 15 Sep 2015.
35. Heesen C, Bohm J, Reich C, et al. Patient perception of bodily functions in multiple sclerosis: gait and visual function are the most valuable. Mult Scler. 2008;14:988–91. doi:10.1177/1352458508088916.
36. Foley PL, Vesterinen HM, Laird BJ, et al. Prevalence and natural history of pain in adults with multiple sclerosis: systematic review and meta-analysis. Pain. 2013;154:632–42. doi:10.1016/j.pain.2012.12.002.
37. Breivik H, Collett B, Ventafridda V, et al. Survey of chronic pain in Europe: prevalence, impact on daily life, and treatment. Eur J Pain. 2006;10:287–333. doi:10.1016/j.ejpain.2005.06.009.
38. Vacca G, Marano E, Brescia Morra V, et al. Multiple sclerosis and headache co-morbidity. A case-control study. Neurol Sci Off J Ital Neurol Soc Ital Soc Clin Neurophysiol. 2007;28:133–5. doi:10.1007/s10072-007-0805-1.
39. Ergun U, Ozer G, Sekercan S, et al. Headaches in the different phases of relapsing-remitting multiple sclerosis: a tendency for stabbing headaches during relapses. Neurologist. 2009;15:212–6. doi:10.1097/NRL.0b013e3181906fc9.
40. Seixas D. The relationship between chronic pain and neuropsychiatric and neuropsychological disturbances in multiple sclerosis. Porto, Portugal: Faculty of Medicine of Porto University; 2012.
41. Ghaffar O, Feinstein A. The neuropsychiatry of multiple sclerosis: a review of recent developments. Curr Opin Psychiatry. 2007;20:278–85. doi:10.1097/YCO.0b013e3280eb10d7.
42. Minden SL, Schiffer RB. Affective disorders in multiple sclerosis. Review and recommendations for clinical research. Arch Neurol. 1990;47:98–104.
43. Kessler RC, Berglund P, Demler O, et al. The epidemiology of major depressive disorder: results from the National comorbidity survey replication (NCS-R). JAMA. 2003;289:3095–105. doi:10.1001/jama.289.23.3095.
44. Patten SB, Svenson LW, Metz LM. Psychotic disorders in MS: population-based evidence of an association. Neurology. 2005;65:1123–5. doi:10.1212/01.wnl.0000178998.95293.29.
45. Dalton EJ, Heinrichs RW. Depression in multiple sclerosis: a quantitative review of the evidence. Neuropsychology. 2005;19:152–8. doi:10.1037/0894-4105.19.2.152.
46. Alschuler KN, Ehde DM, Jensen MP. Co-occurring depression and pain in multiple sclerosis. Phys Med Rehabil Clin N Am. 2013;24:703–15. doi:10.1016/j.pmr.2013.06.001.
47. Bair MJ, Robinson RL, Katon W, Kroenke K. Depression and pain comorbidity: a literature review. Arch Intern Med. 2003;163:2433–45. doi:10.1001/archinte.163.20.2433.
48. Amtmann D, Askew RL, Kim J, et al. Pain affects depression through anxiety, fatigue, and sleep in multiple sclerosis. Rehabil Psychol. 2015;60:81–90.
49. Rao SM, Leo GJ, Bernardin L, Unverzagt F. Cognitive dysfunction in multiple sclerosis. I. Frequency, patterns, and prediction. Neurology. 1991;41:685–91.
50. Kleeberg J, Bruggimann L, Annoni J-M, et al. Altered decision-making in multiple sclerosis: a sign of impaired emotional reactivity? Ann Neurol. 2004;56:787–95. doi:10.1002/ana.20277.
51. Nagy H, Bencsik K, Rajda C, et al. The effects of reward and punishment contingencies on decision-making in multiple sclerosis. J Int Neuropsychol Soc. 2006;12:559–65.
52. Benson C, Kerr BJ. Pain and cognition in multiple sclerosis. Curr Top Behav Neurosci. 2014;20:201–15. doi:10.1007/7854.
53. Leonavicius R, Kalnina J. Beyond pain in multiple sclerosis. Neurol Psychiatry Brain Res. 2015;21:82–7. doi:10.1016/j.npbr.2015.01.003.
54. Hadjimichael O, Kerns RD, Rizzo MA, et al. Persistent pain and uncomfortable sensations in persons with multiple sclerosis. Pain. 2007;127:35–41. doi:10.1016/j.pain.2006.07.015.
55. Kerns RD, Kassirer M, Otis J. Pain in multiple sclerosis: a biopsychosocial perspective. J Rehabil Res Dev. 2002;39:225–32.
56. Shahrbanian S, Auais M, Duquette P, et al. Does pain in individuals with multiple sclerosis affect employment? A systematic review and meta-analysis. Pain Res Manag. 2013;18:e94–100.
57. Forbes A, While A, Mathes L, Griffiths P. Health problems and health-related quality of life in people with multiple sclerosis. Clin Rehabil. 2006;20:67–78.
58. Osterberg A, Boivie J, Thuomas K-A. Central pain in multiple sclerosis–prevalence and clinical characteristics. Eur J Pain. 2005;9:531–42. doi:10.1016/j.ejpain.2004.11.005.
59. Truini A, Barbanti P, Pozzilli C, Cruccu G. A mechanism-based classification of pain in multiple sclerosis. J Neurol. 2013;260:351–67. doi:10.1007/s00415-012-6579-2.
60. Stovner LJ, Hagen K, Jensen R, et al. The global burden of headache: a documentation of headache prevalence and disability worldwide. Cephalalgia. 2007;27:193–210. doi:10.1111/j.1468-2982.2007.01288.x.
61. Al-Araji AH, Oger J. Reappraisal of Lhermitte's sign in multiple sclerosis. Mult Scler. 2005; 11:398–402. doi:10.1191/1352458505ms1177oa.
62. Spissu A, Cannas A, Ferrigno P, et al. Anatomic correlates of painful tonic spasms in multiple sclerosis. Mov Disord. 1999;14:331–5. doi:10.1002/1531-8257(199903)14:2<331:AID-MDS1020>3.0.CO;2-H.
63. Dooley MC, Foroozan R. Optic neuritis. J Ophthalmic Vis Res. 2010;5:182–7.
64. Gronseth G, Cruccu G, Alksne J, et al. Practice parameter: the diagnostic evaluation and treatment of

trigeminal neuralgia (an evidence-based review): report of the Quality Standards Subcommittee of the American Academy of Neurology and the European Federation of Neurological Societies. Neurology. 2008;71:1183–90. doi:10.1212/01.wnl.0000326598.83183.04.
65. Cruccu G, Truini A. Sensory profiles: a new strategy for selecting patients in treatment trials for neuropathic pain. Pain. 2009;146:5–6. doi:10.1016/j.pain.2009.07.004.
66. Truini A, Galeotti F, La Cesa S, et al. Mechanisms of pain in multiple sclerosis: a combined clinical and neurophysiological study. Pain. 2012;153:2048–54. doi:10.1016/j.pain.2012.05.024.
67. Simpson DM, McArthur JC, Dworkin RH. Neuropathic pain: mechanisms, diagnosis and treatment. USA: OUP; 2012.
68. Seixas D, Foley P, Palace J, et al. Pain in multiple sclerosis: a systematic review of neuroimaging studies. NeuroImage Clin. 2014;5:322–31. doi:10.1016/j.nicl.2014.06.014.
69. Fragoso YD, Brooks JBB. Two cases of lesions in brainstem in multiple sclerosis and refractory migraine. Headache. 2007;47:852–4. doi:10.1111/j.1526-4610.2007.00823.x.
70. Haas DC, Kent PF, Friedman DI. Headache caused by a single lesion of multiple sclerosis in the periaqueductal gray area. Headache. 1993;33:452–5.
71. Kister I, Caminero AB, Monteith TS, et al. Migraine is comorbid with multiple sclerosis and associated with a more symptomatic MS course. J Headache Pain. 2010;11:417–25. doi:10.1007/s10194-010-0237-9.
72. Broggi G, Ferroli P, Franzini A, et al. Operative findings and outcomes of microvascular decompression for trigeminal neuralgia in 35 patients affected by multiple sclerosis. Neurosurgery. 2004;55:830–9.
73. Cruccu G, Biasiotta A, Di Rezze S, et al. Trigeminal neuralgia and pain related to multiple sclerosis. Pain. 2009;143:186–91. doi:10.1016/j.pain.2008.12.026.
74. Gass A, Kitchen N, MacManus DG, et al. Trigeminal neuralgia in patients with multiple sclerosis: lesion localization with magnetic resonance imaging. Neurology. 1997;49:1142–4.
75. Da Silva CJ, da Rocha AJ, Mendes MF, et al. Trigeminal involvement in multiple sclerosis: magnetic resonance imaging findings with clinical correlation in a series of patients. Mult Scler. 2005;11:282–5.
76. Eldridge PR, Sinha AK, Javadpour M, et al. Microvascular decompression for trigeminal neuralgia in patients with multiple sclerosis. Stereotact Funct Neurosurg. 2003;81:57–64. doi:75105.
77. Alstadhaug K, Breivik K, Rusic Z. Recurrent headache due to MS plaque. Headache. 2008;48:453–4. doi:10.1111/j.1526-4610.2007.01003.x.
78. De Santi L, Monti L, Menci E, et al. Clinical-radiologic heterogeneity of occipital neuralgiform pain as multiple sclerosis relapse. Headache. 2009;49:304–7. doi:10.1111/j.1526-4610.2008.01209.x.
79. Hellwig K, Lukas C, Brune N, et al. Repeat intrathecal triamcinolone acetonide application reduces acute occurring painful dysesthesia in patients with relapsing remitting multiple sclerosis. Sci World J. 2006;6:460–5. doi:10.1100/tsw.2006.86.
80. Tosi L, Righetti CA, Zanette G, Beltramello A. A single focus of multiple sclerosis in the cervical spinal cord mimicking a radiculopathy. J Neurol Neurosurg Psychiatry. 1998;64:277.
81. Burkey AR, Abla-Yao S. Successful treatment of central pain in a multiple sclerosis patient with epidural stimulation of the dorsal root entry zone. Pain Med. 2010;11:127–32. doi:10.1111/j.1526-4637.2009.00764.x.
82. Bowsher D, Leijon G, Thuomas KA. Central post-stroke pain: correlation of MRI with clinical pain characteristics and sensory abnormalities. Neurology. 1998;51:1352–8.
83. Svendsen KB, Sorensen L, Jensen TS, et al. MRI of the central nervous system in MS patients with and without pain. Eur J Pain. 2011;15:395–401. doi:10.1016/j.ejpain.2010.09.006.
84. Owen DG, Bureau Y, Thomas AW, et al. Quantification of pain-induced changes in cerebral blood flow by perfusion MRI. Pain. 2008;136:85–96. doi:10.1016/j.pain.2007.06.021.
85. Baliki MN, Chialvo DR, Geha PY, et al. Chronic pain and the emotional brain: specific brain activity associated with spontaneous fluctuations of intensity of chronic back pain. J Neurosci. 2006;26:12165–73. doi:10.1523/JNEUROSCI.3576-06.2006.
86. Julius D, Basbaum AI. Molecular mechanisms of nociception. Nature. 2001;413:203–10. doi:10.1038/35093019.
87. Rodriguez-Raecke R, Niemeier A, Ihle K, et al. Brain gray matter decrease in chronic pain is the consequence and not the cause of pain. J Neurosci. 2009;29:13746–50. doi:10.1523/JNEUROSCI.3687-09.2009.
88. Apkarian AV, Baliki MN, Geha PY. Towards a theory of chronic pain. Prog Neurobiol. 2009;87:81–97. doi:10.1016/j.pneurobio.2008.09.018.
89. Baliki MN, Geha PY, Apkarian AV, Chialvo DR. Beyond feeling: chronic pain hurts the brain, disrupting the default-mode network dynamics. J Neurosci. 2008;28:1398–403. doi:10.1523/JNEUROSCI.4123-07.2008.
90. Cauda F, D'Agata F, Sacco K, et al. Altered resting state attentional networks in diabetic neuropathic pain. J Neurol Neurosurg Psychiatry. 2010;81:806–11. doi:10.1136/jnnp.2009.188631.
91. Tagliazucchi E, Balenzuela P, Fraiman D, Chialvo DR. Brain resting state is disrupted in chronic back pain patients. Neurosci Lett. 2010;485:26–31. doi:10.1016/j.neulet.2010.08.053.
92. Pizoli CE, Shah MN, Snyder AZ, et al. Resting-state activity in development and maintenance of normal

brain function. Proc Natl Acad Sci U S A. 2011;108:11638–43. doi:10.1073/pnas.1109144108.
93. Raichle ME, MacLeod AM, Snyder AZ, et al. A default mode of brain function. Proc Natl Acad Sci U S A. 2001;98:676–82. doi:10.1073/pnas.98.2.676.
94. Buckner RL, Andrews-Hanna JR, Schacter DL. The brain's default network: anatomy, function, and relevance to disease. Ann N Y Acad Sci. 2008;1124:1–38. doi:10.1196/annals.1440.011.

Author Biographies

Daniela Seixas, MD, Ph.D., MBA is a neuroradiologist at Centro Hospitalar de Vila Nova de Gaia/Espinho and professor at the Faculty of Medicine of Porto University. She received her Ph.D. in Neuroscience collaborating with the University of Oxford, investigating pain in multiple sclerosis. Other research interests are advanced MRI techniques and neuroethics. She is an independent expert of the Directorate General for Research and Innovation of the European Commission, and a Member of the Direction of the College of Neuroradiology of the Portuguese Medical Association.

Daniel Teles is doing his master's degree in the field of pain and multiple sclerosis. Other research interests include neuroimaging and the human microbiome.

Sex and Gender Effects in Pain

15

Bettina Pfleiderer, Anika Ritzkat
and Esther Pogatzki-Zahn

Abstract

Even though it has been accepted that sex (biological factors) and gender (social roles and environmental factors) effects do exist in pain perception and response, papers assessing sex and gender effects in biomedical and clinical research, as well as clinical practice are scarce. There are even fewer imaging papers investigating sex and gender effects in neural responses to pain. This chapter reviews the existing literature and provides a comprehensive summary of the role of sex and gender in (i) pain syndromes, (ii) psychological factors in pain, (iii) the efficacy of opioid analgesics, (iv) regarding effects of the menstrual cycle and sexual hormones in pain perception, (v) pain perception and modulation under experimental conditions and (vi) imaging studies related to pain syndromes (healthy subjects, patients with fibromyalgia and irritable bowel syndrome). We hope that this overview will stimulate the inclusion of sex and gender aspects in future research designs and will lead to an increase in the number of imaging papers accordingly. Understanding the impact of sex and gender factors in pain pathophysiology and processing in more detail has the potential to lead to discoveries of new targets for treatment of pain.

Keywords

Epigenetics · Fibromyalgia · Irritable bowel syndrome · Analgesics · Opioids · Quantitative sensory testing (QST)

1 Sex and Gender

It is generally accepted that marked differences exist in the biological- (sex) and sociocultural factors (gender) of men and women which are influenced by intersecting factors like race, social class and culture. Nevertheless, both factors have rarely been accounted for in biomedical and clinical research, as well as in the daily clinical practice [1].

B. Pfleiderer (✉) · A. Ritzkat
Department of Clinical Radiology, Research Group Cognition & Gender, University Hospital Muenster, Muenster, Germany
e-mail: pfleide@uni-muenster.de

E. Pogatzki-Zahn
Department of Anesthesiology, Intensive Care and Pain Medicine, University Hospital Muenster, Muenster, Germany

L. Saba (ed.), *Neuroimaging of Pain*, DOI 10.1007/978-3-319-48046-6_15

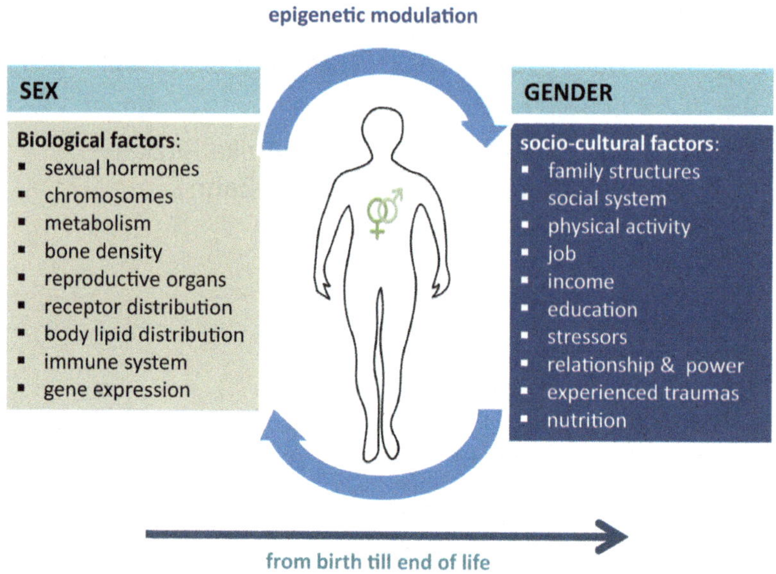

Fig. 1 Sex and gender factors and modulation by epigenetics have an impact on pain

Sex includes hereby all biological aspects of men and women such as e.g. hormones, genetics, metabolic profiles, while gender is related to factors such as sociocultural role expectations, psychological characteristics of men and women, and education. However sex and gender factors are no separate dimensions, they interact in a complex way from birth to end of life (Fig. 1) [1]. For example, hormonal levels can influence mood and sensations [2–4] and depression and anxiety can influence pain perception [5].

In addition, modulation by epigenetic processes plays an important role for sex and gender differences. The epigenetic changes, in brief, do not involve alterations in the DNA sequence itself, but it is related to modifications of the DNA structure that subsequently alters expressions of genes or differentiation of cells [6]. These changes can occur by environmental exposure e.g. as experiences of violence, nutritional factors or early childhood or perinatal stress [7–9].

1.1 Role of Sex and Gender in Pain Syndromes

It is known that sex specific differences do exist in pain and in pain related diseases [4]. Population-based studies have found a higher prevalence in women compared to men for several pain related disorders like migraines/headache, temporomandibular joint disorder, irritable bowel syndrome, rheumatoid arthritis and osteoarthritis (Fig. 2) [10–15].

In addition, women are very often more affected, for example by experiencing more severe pain or longer pain episodes. This has been shown for many pain disorders; typical examples are fibromyalgia (FMS) and irritable bowel syndrome (IBS) [16, 17].

Fibromyalgia is a syndrome in which patients present musculoskeletal pain typically in combination with other symptoms like muscle stiffness, persisting fatigue, memory problems, sleep and mood disturbances [18]. Using the criteria of the American College of Rheumatology, fibromyalgia is defined by widespread musculoskeletal pain for more than 3 months and the presence of at least 11 tender points [19]. FMS is associated with comorbid psychiatric disorders such as major depression and anxiety disorders [17].

In addition, fibromyalgia typically appears with other syndromes that have a similar pathophysiological mechanism, e.g. irritable bowel syndrome, interstitial cystitis, and tension headache [20]. Patients with fibromyalgia respond to centrally acting analgesics and nonpharmacological therapies [20]. It is assumed that a

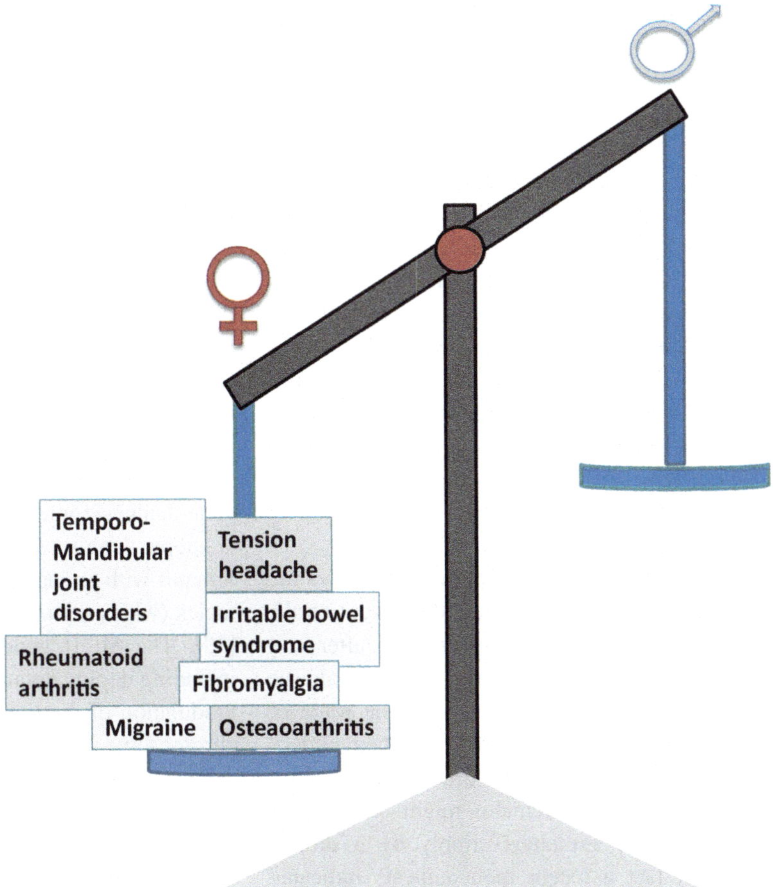

Fig. 2 Sex differences in pain: results from epidemiological studies indicate higher prevalence for several pain related disorders in female patients

sensitized or hyperactive central nervous system leads to a gain on pain and sensory processing. Many studies showed that patients with fibromyalgia have decreased pain thresholds compared to healthy controls using thermal, electrical and mechanical pain stimuli [21, 22].

Similar, women with irritable bowel syndrome present worse clinical symptoms compared to men [16]. IBS had a higher impact on women's daily lives and rate of psychiatric comorbidities [23]. Characteristics of the disease are increased visceral sensitivity together with bowel dysfunction; diagnostic criteria [24] are shown in Fig. 3.

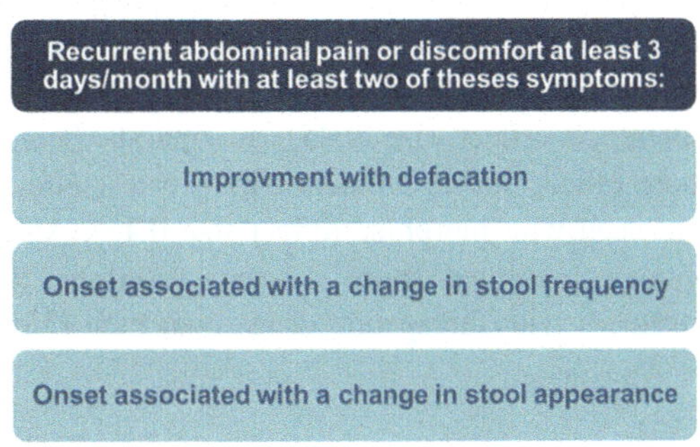

Fig. 3 Diagnostic criteria for irritable bowel syndrome according to Rome III criteria

The preference for IBS in females may be in part explained by different microbiota in the small intestine and the colon, which has been shown in male and female mice even before weaning [25, 26]. Furthermore, women and men have different gut microbiota even under the same environmental and nutritional conditions. This might also contribute to the higher prevalence of IBS in women [27, 28]. Because IBS symptoms (e.g. painful cramps) worsen within certain times of the menstrual cycle phase hormones may be relevant as well (see below).

1.2 Sex Differences in Response to Analgesics

The female sex is associated with a greater risk for experiencing severe acute pain and developing persistent pain after surgery [29–32]. Even though significant differences exist between males and females after surgery [29, 30, 33], sex differences in intensity of pain is small (for postoperative pain smaller 0.3 on a VAS from 0 to 10 between males and females [29, 30, 34] and the clinical relevance of these observed sex differences is less clear. Similar, sex differences in response to analgesics, e.g. opioids, have been studied (Fig. 4). In animals, there is a tendency towards an increased efficacy of opioids in males [35].

However, the results from animal studies are divergent and differences seem dependent on the genetic background [36]. In humans after surgery, females seem to use less doses of analgesic in many studies, (irrespectively of the route of administration); however, because they experience more pain at the same time the reasons for less requirements in females might be complex and cannot be explained simply by a greater effect. In fact a recent meta-analysis indicated that under clinical conditions, females and males do not differ with respect to opioid efficacy [37]. The same meta-analysis, however, stated that in experimental studies there seems to be a sex difference (effect size 0.35) by opioids being more efficacious in women. The greatest effect was seen with morphine (effect size 0.45). Also women presented more adverse reactions to treatment with therapeutic drugs than men [38], which might play a role under clinical conditions. For example, females may request less opioid during patient-controlled analgesia due to side effects like nausea and vomiting. This may in turn contribute to the higher pain scores in females observed after surgery [33]. Moreover, there seem to be differences in time of onset and peak of analgesia between men and women [38]. But one has to keep in mind that studies on sex differences in specific analgesic response are rare, the design of studies is heterogeneous, results are sometimes contradictory and it is still unclear whether the differences observed are clinically relevant.

Sex seems to be a relevant factor in humans in gene-analgesic interactions as well. Pentazocine, a kappa agonist, exerts an increased effect in females with an inactive MC1r gene compared to males (with an intact or an inactive MC1r gene) [39]. The MC1r gene has several other effects including depigmentation (red hair/pale skin) and modulation of the immune system, is located on chromosome 7.

1.2.1 Placebo and Nocebo Effects in Analgesia in Men and Women

A nocebo effect usually describes the appearance of reduced effect and negative side effects of a pharmacologically active substance caused by the expectations of the patient [40]. In contrast, a placebo effect is a real antinociceptive effect due to positive expectation of the patient [40]. Interestingly, it was reported that males are more susceptible to placebo analgesia than females [41–43].

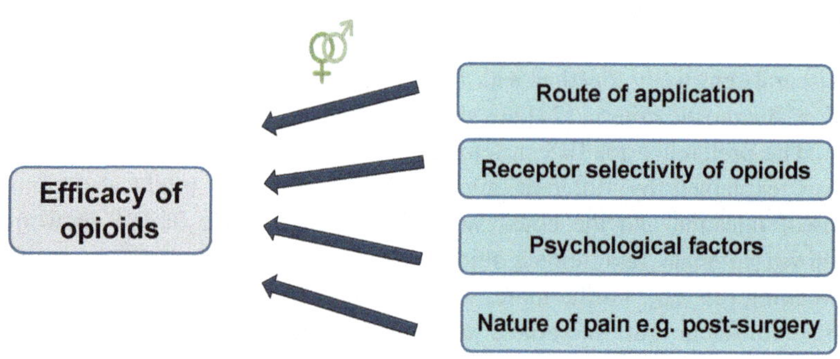

Fig. 4 Efficacy of opioids depends on many factors; here are those factors summarized where sex and gender effects have been reported

2 The Influence on Sex and Gender Factors on Pain Processing

2.1 Sex Differences in Pain Perception and Modulation Under Experimental Conditions

2.1.1 QST

During the last 15 years, assessment methods able to investigate function of the somatosensory system (including the pain pathways) have been optimized and were somehow standardized in order to enable comparison between studies and study groups. One of the most frequent and, most important, non-invasive methods to assess sensory function is termed "Quantitative Sensory Testing (QST)" [44, 45]. Most frequently, static QST is used by threshold determination for a wide variety of non-painful and painful stimuli. Dependent on the stimulus, function and dysfunction of all types of sensory fibers (large myelinated, small myelinated, and unmyelinated fibers) can be assessed, quantified and differentiated [44]. In addition to peripheral processing, cerebral pain modulation is investigated in more detail [44]. Studies using QST in healthy volunteers are performed to obtain a better understanding of the mechanisms involved in pain transduction, transmission, and perception under "normal" conditions and under pathophysiological mechanisms related to clinical pains states in human volunteers by using human models of pain [46]. These experimental models typically produce acute pain and sensitization of the peripheral and/or central nervous system related to hyperalgesia and allodynia. Under this circumstances, changes in sensory function of the pain pathway assessed for example by QST after induction of the painful state are more transferable to clinical conditions [46] and are therefore very useful.

In the past, QST have been widely used to evaluate gender differences in "normal" nociception. Whereas older studies usually investigated only one or two different pain modalities, more recent studies were designed to assess more systematically gender differences. Furthermore, many (again older but also some of the newer) studies were underpowered. To overcome this problem, a recent large (qualitative) meta-analysis including studies published from 1998 to 2008 and comparing systematically pain thresholds, pain tolerances, supra-threshold responses and unpleasantness of painful stimulation between males and females were performed. In contrast to older reviews [47], Racine et al. found only small sex differences and only for some pain modalities without a certain pattern [48]. In more detail, the data [48] suggest that pain thresholds for cold and ischemic pain are comparable in females and males; only pressure pain thresholds show clear sex differences (females showing lower thresholds than males) and the results for heat pain thresholds were equivocal. Furthermore, females were tolerating less well supra-threshold pressure, heat and cold pain, but tolerance to ischemic pain seems not to be different between men and women [48]. Finally, there seem to be no sex differences in pain intensity ratings and pain unpleasantness ratings regardless of the pain modality. These results imply that sex differences in nociception are minimal and seem to be unsystematically attributed to painful stimulation with pressure and possibly heat.

In conclusion, as summarized by Racine et al., the concept of greater generalized pain sensitivity in females [49, 50] cannot be maintained.

2.1.2 DNIC

As indicated in many experimental animal studies, pain at the spinal cord level is modulated by supra-spinal descending input which modulates pain and pain modulation. Experimentally, the descending inhibitory system can be "measured" by a paradigm termed "conditioned pain modulation (CPM)", formerly named "diffuse noxious inhibitory controls" (DNIC). The principle behind CPM is that pain intensity caused by a supra-threshold painful stimulus ("termed test stimulus", typically painful heat, but other stimuli and pain measures are used frequently as well) is reduced when applied simultaneously

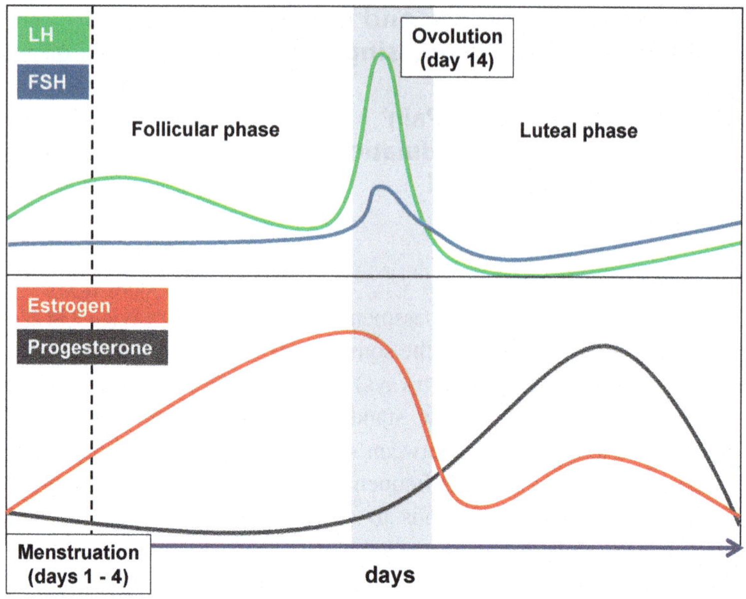

Fig. 5 Levels of female hormones during female cycle

with another painful stimulus (the conditioning stimulus, typically ice water). The second painful stimulus activates the endogenous inhibitory descending system neurons such that the response in the spinal dorsal horn (or the trigeminal nuclei) is attenuated.

In patients with chronic pain, CPM is reduced (as shown in many studies). In addition, some indications are given for a role of reduced CPM respectively endogenous inhibition being a risk factor for the development of chronic pain [51]. Interestingly, CPM seems to be lower in females compared to males in many studies, e.g. those using pain report as the CPM [52]. The largest (quantitative) meta-analysis on the role of sex for CPM reported significantly less efficient DNIC in females (mean female/male ratio = 0.54 [52]). However, sex differences were not verifiable in all CPM studies. In fact, differences between females and males related to CPM seem to vary widely due to the experimental methodology used [52].

Taken together, there seem to be reduced endogenous inhibition in females compared to males which might contribute to the greater susceptibility of females to develop chronic pain and pain of more intense severity.

2.2 Influences of Menstrual Cycle Phases and Sex Hormones on Pain Perception

The role of hormones as an influencing factor for sex differences in pain is supported by several clinical observations: for instance, females are more often affected by chronic pain syndromes especially during their reproductive years when hormones fluctuate [2, 53, 54]. Furthermore, under hormone replacement therapy in the post-menopause, some clinical pain syndromes increase in severity and some others decrease (for instance migraine). Finally, sex differences start with puberty when sex hormones start to fluctuate. The role of sex hormones seems therefore a contributor to clinical pain.

To evaluate this systematically, several studies have investigated the role of ovarian hormones and menstrual cycle phases for pain perception and modulation in females during different phases of the menstrual cycle [52, 53, 55, 56]. The menstrual cycle consists of cyclic variations in reproductive hormone production (Fig. 5).

Day 1 is normally defined as the first day of the menses; 2–5 days later the follicular phase follows for 10–14 days assuming a regular cycle of about 28 days. The plasma level of the four most important reproductive hormones, luteinizing hormone = LH, follicle stimulating hormone = FSH, estrogen and progesterone peak during menstruation. Estrogen increases around day 5 and has its first peak at the end of the follicular phase. The increase of estrogen leads to the secretion of LH which triggers ovulation (on day 14). Important is the fast decrease in estrogen concentration after ovulation. The second phase of the female menstrual cycle is termed the luteal phase. During this phase progesterone level increases and peaks around days 21–24 with a high interindividual variation in time and amount. Importantly, estrogen increases as well and peaks at the same time as progesterone; however, estrogen levels do not reach levels similar to the follicular phase. Thus, progesterone is the predominant hormone steroid hormone during the second cycle phase [57].

2.2.1 Pain and the Cycling Phases

Intensity of clinical pain as well as pain induction in pain syndromes like headache and migraine varies with respect to the menstrual cycle phase. For migraine attacks, induction occurs very frequently during low or quickly falling estrogen levels [58]. A similar association has been reported for hormones and irritable bowel syndrome (IBS) [59, 60]. The effect of the menstrual cycle phases on pain in patients with fibromyalgia is, however, inconsistent [61–63]. While some authors found that especially women with fibromyalgia describe less pain during their luteal phase [63], other studies reported that women described higher pain levels during follicular phase and menstruation [61]. However, as concluded by a recent meta-analysis, pain in patients with fibromyalgia, temporomandibular pain and other pain conditions seem to vary across the menstrual cycle in a similar way; most studies show greatest pain sensitivity in most clinical pain conditions when estrogen levels are low [2]. However, due to methodological problems (e.g. hormone levels are not assessed in most studies), reliable data on the role of hormones for most clinical pain conditions are still lacking.

Similarly, experimental studies investigating different nociceptive stimuli and modalities across the menstrual cycle in healthy volunteers were inconclusive mainly due to methodological problems. For instance, study population, timing of experiments across the menstrual cycle, nomenclature used to identify the phase of cycle, the modality of pain stimuli and outcomes evaluated vary extremely [53]. In only a few studies blood levels of female sex hormones were measured; most often volunteers were assigned to either the luteal or follicular phase by a menstrual cycle diary [64]. Many factors, including interindividual and intraindividual differences in durations of the menstrual cycle and possibly anovulatory cycles, which occur in around 20% of all cycles and prevent estrogen and progesterone increase in the luteal phase, may lead to inconsistent results [65].

2.2.2 Role of Certain Hormones on Pain

More recent studies taking many of the pitfalls into account suggest a less important role of hormones for nociception than proposed earlier; differences of pain responses between cycling phases were small and only observed for certain modalities [53, 55, 66]. Some most recent experiments suggest a relevant hormone effect. Increased responses are—if any were shown—reported in the luteal phase when progesterone and estrogen levels are both high [67]. Our own work indicates that progesterone is more relevant than estrogen for differences in Aδ-fiber mediated responses and pain and hyperalgesia after an experimental incision [68]. However, other authors found correlations of (supra-physiological levels) of estrogen with experimental pain perception and no correlations between progesterone, LH and experimental pain perception [69]. In conclusion, there seems to be no simple linear relationship between gonadal hormones and pain sensitivity, thus making it very difficult to draw a clear picture.

3 Assessment of Sex and Gender Effects in Imaging Studies Related to Pain

3.1 Sexual Dimorphism of the Brain

Brain imaging studies found differences in brain structure, function and metabolism between healthy men and women (Table 1) [70], which may have an influence in the way pain is processed and perceived [70]. However it is important to note that there exist no specific distinct female or male brain and considerable structural and functional overlap between male and female brains are present; brains are mosaic in nature [71, 72].

Nevertheless, findings suggest that one cannot automatically assume that responses to pain are sex/gender independent and it can be assumed the responses to and perception of pain are modulated by sex and gender factors. But this has not been investigated in a systematic way due to lack of studies.

3.2 Psychological Gender and Sex Differences

Beside brain structure, functionality and neurochemistry, gender and sex factors and their interaction with biological factors are of importance. Psychological factors are for instance placebo effects, anticipation of pain or attention to and distraction from pain. There are only a few studies assessing gender/sex differences of psychological factors in pain [4]. There seem to be different associations between psychological symptoms such as stress, depression, anxiety and pain in men and women [4, 5]. Stress for instance is described to increase pain sensitivity in women [81]. The interaction of pain and anxiety, however, is complex and depends on the pain-related anxiety construct (e.g. fear of pain, anxiety sensitivity) and the pain measures used in the experiments. Still, the sex of the subjects with respect to pain perception mattered across modalities and anxieties constructs used [5]. Personality traits such as neuroticism seem to be an important modulator in the perception of pain as well. It was suggested that higher neuroticism scores are associated with higher activity of brain regions known to be involved in the emotional and cognitive appraisal during anticipation of pain but reduced activity during pain [82].

It was published that, under experimental conditions, men report lower pain scores when the examining physician is female. This was explained that according to gender roles men should be pain-insensitive and strong leading to a reporting bias [4]. Taking this into account, it is feasible to assume that pain levels in men may be

Table 1 Overview on sex differences in brain structure, function and metabolism as measured by functional magnetic resonance imaging (fMRI), positron emission tomography (PET), single photon emission computed tomography (SPECT), and structural magnetic resonance imaging (MRI) [70]

Brain structure	Brain functionality/neurochemistry
Females: gray/white matter ratio higher in frontal, temporal, parietal, and occipital lobes; cingulate gyrus; and insula [73] (cognition)	Females: higher serotonin receptor sensitivity and availability (pain and depression) [74]
Females: larger Broca and Wernicke areas (language) [70]	Males: higher re-synthesis rate of serotonin [75]
Females: larger hippocampal cortex (long-term memory) [70]	Males: higher receptor affinity for glucocorticoids (twice that of women; stress) [76]
Females: larger locus coeruleus (panic and stress) [70]	Females prior to menopause: higher mu-opioid receptor binding [77, 78], resulting in a longer and stronger opioid response in women [37]
There are differences in fiber-connectivity between both sexes: men show more intrahemispheric and women more interhemispheric connections [79]	Females higher cerebral blood flow [80] (distribution of psychotropic drugs)

The brackets indicate the topics where these differences may have a modulating effect underreported, e.g. if the investigator is female. In papers publishing higher pain scores in women [83, 84], the sex of the examiner was not given.

In the same vein, effects of gender/sex were also observed in animal experiments [85]. Odor (arm sweat; androgen dependent) from male human experimenters stressed in particularly male laboratory mice/rats and they depicted reduced pain sensitivity and thus a stress induced analgesia.

3.3 Functional Imaging Studies in Pain

Central pain processing as described in the previous chapters, is influenced by multiple sensoric, affective and vegetative components which are then evaluated by cognitive processes. There are two neural systems in the pain network of the brain: the lateral pain system (sensoric-discriminative) and the medial pain system (affective). Brain areas associated with the lateral pain system are e.g. primary and secondary somatosensoric cortices and the thalamus, areas associated with the medial pain system are the anterior cingulate cortex, the amygdala and brain areas associated with memory such as the hippocampus [4]. Interaction of these areas lead to the sensation "pain", which then induces voluntary and unvoluntary motoric and psychological actions (Fig. 6) [4, 68]. The motoric dimension of pain leads to adverse-effects reflexes which can initially be processed in the spinal cord. The affective-emotional component results in unpleasant feelings. The sensoric-discriminative dimension provides information about possible nociceptive stimuli and their localization, quality and intensity. The vegetative dimension includes the autonomic nervous system. Painful stimuli normally lead to an adrenergic reaction (e.g. hyperventilation and tachycardia) or vasovagal reaction (e.g. syncope, bradycardia, hypotension) [4]. Cognitive evaluation by mainly medial-frontal brain areas gives feedback on pain intensity and plays a role in anticipation of pain. Opioid release like in placebo analgesia can be triggered by cognitive processes, too [4].

Activity of brain areas are modulated by sexual hormones (see Sect. 2.2), psychological factors (see Sect. 3.2), as well as individual pain sensitivity of subjects (Fig. 7). We could show in a post-operative human pain incision model that brain activity in areas of the medial pain system such as the amygdala was positively correlated with the individual pain sensitivity of subjects [68]. The higher the perceived pain during incision was in healthy males, the stronger was their amygdala activity.

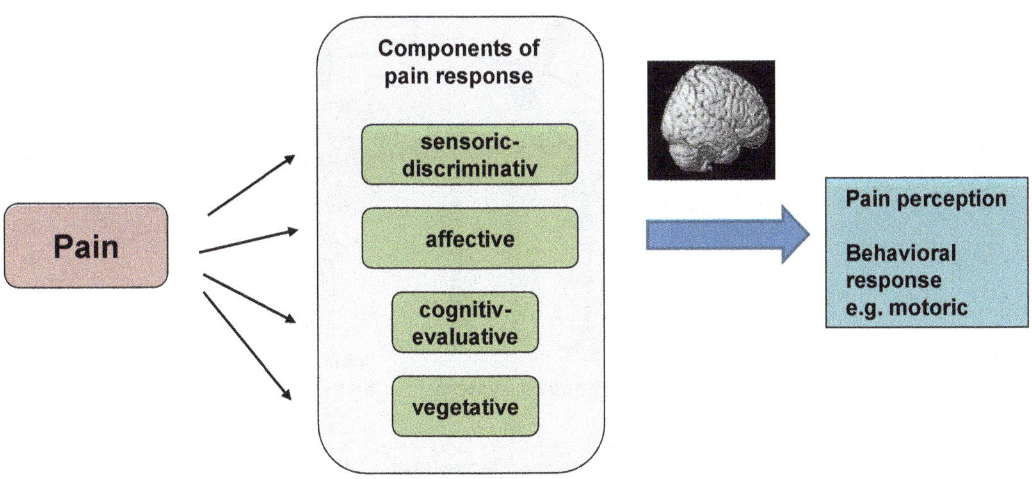

Fig. 6 There are four components of pain response to painful stimuli. This leads to the sensation of pain and related behavioral responses

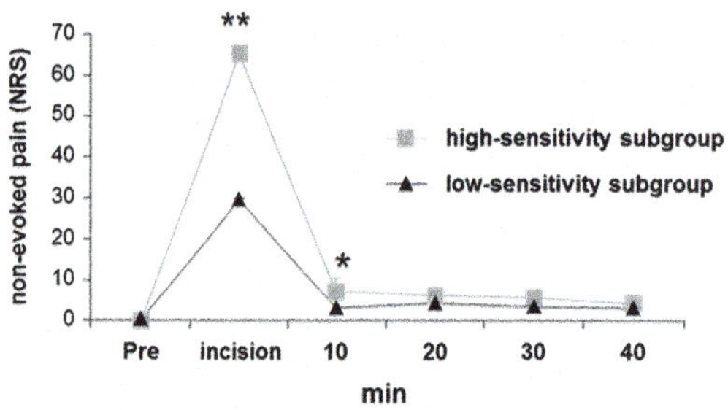

Fig. 7 Non-evoked pain (NRS 0–100) before, during and after incision in high-sensitivity and low-pain sensitivity subjects. Results are expressed as mean ± standard error of the mean, *$P < 0.05$ (Mann-Whitney). Assessment was done pre, during (0–2 min) and after incision (2–4.5, 4.5–10, 24–29, 44–49 min) at a 3T scanner using a block design. Subjective pain ratings by a numerical pain scale (NRS) were performed between the scans (unpublished data, Pfleiderer and Pogatzki-Zahn)

Pain sensitivity to surgical incision is highly variable in healthy subjects as illustrated in those male 28 subjects assessed in our human incision model [68], (Fig. 7). Subjects were divided for illustrative purposes by a mean split into two groups: one with subjects low pain sensitivity (NRS $\leq$ 47; n = 15; NRS = 29.8 + 2.8), the other group with subjects with high pain sensitivity NRS > 47; n = 13; NRS = 65.3 + 3.3). Individual pain sensitivity should be controlled for when comparing different groups of subjects in pain imaging experiments.

There are only a few functional magnetic resonance imaging (fMRI) studies investigating sex-specific differences in brain activity during experimental pain models [4, 86–88]. They found women to depict higher activities in prefrontal, somatosensory and parietal gyri as well as in insula, dorsolateral prefrontal cortex (DLPFC), cingulate cortex, para-/hippocampus, cerebellum and thalamus even when the maximal pain intensity ratings were comparable between women and men [4, 86–88]. In particular, areas related to the affective system were up-regulated in women. This is in line with the assumption that women generally seem to put more emotional weight to the experienced pain (Fig. 8).

3.4 Effect of the Menstrual Cycle

Differences in brain activation patterns in regions involved in cognitive pain modulation that depend on hormonal changes of the female

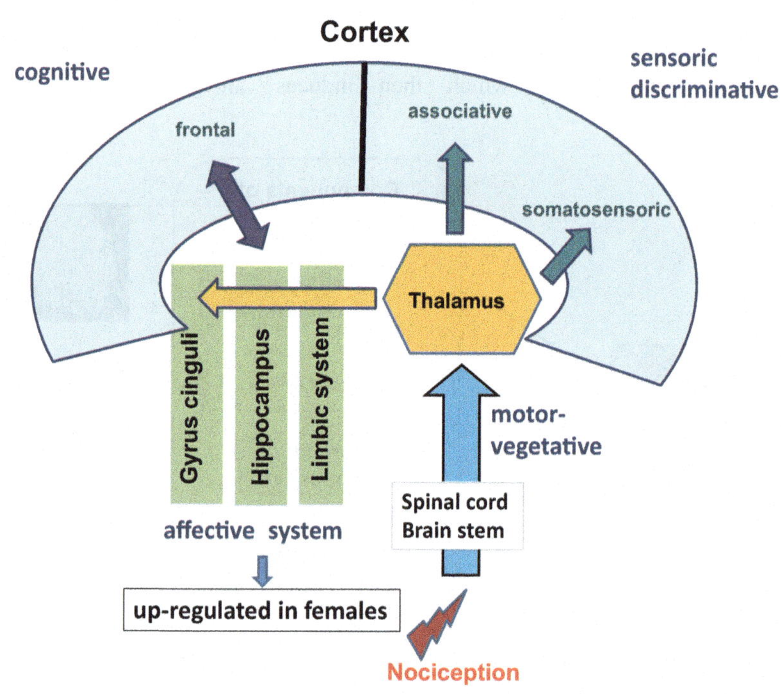

Fig. 8 Signaling pathways of pain and central pain processing. Functional imaging experiments show that in women the affective system is up-regulated was compared to men

menstrual cycle have been reported, even when behavioral pain responses remained unchanged [55]. This has been corroborated in our lab (unpublished) using a previously published human incision model [68]. Reduced activity in brain areas associated with motivational-affective and cognitive aspects of pain processing corresponded to a higher pain sensory score during incision in the luteal phase. This suggests a decreased descending inhibitory control modulated by ovarian hormones. Ovarian hormones seem thus to influence central pain processing and may play an important role in the development of acute and chronic pain states (unpublished).

4 Imaging Studies and Chronic Pain Syndromes—Impact of Sex and Gender

4.1 *Fibromyalgia*

Sex/gender effects in patients with fibromyalgia were not assessed by fMRI. However Henderson et al. investigated sex effects in a fMRI pain model in healthy subjects [86], which may provide some insight into possible sex-differences in patients. They injected hypertonic saline intramuscular and subcutaneous, whereby females reported higher sensory scores in response to the injections as compared to men, but pain intensity ratings were comparable.

Sex differences in brain activity were seen in the mid-cingulate cortex and right DLPFC, both being part of the so called medial pain system associated with the affective/motivational coding of pain [89]. Female subjects presented stronger activity during muscle and cutaneous pain in these areas, while men depicted higher activity in the cerebellar cortex. The DLPFC is well known to play a putative role of pain anticipation and is considered as “keeping pain out of mind” [90]. As the DLPFC is responsible for attention, it may be assumed that females focus more cognitively on the pain they are experiencing and respond more emotionally to it [86].

4.2 Irritable Bowel Syndrome (IBS)

Visceral pain is one of the most prominent pain conditions [16], therefore it is important to study the neural correlates in standardized experimental pain models by functional imaging. Disturbed central pain processing seems to play an important role in IBS as shown by fMRI. There are only a few imaging studies assessing sex effects in visceral pain [91–97]. Table 2 summarizes the imaging results for patients with IBS and healthy subjects.

Table 2 Sex effects in a pain induced brain activation by abdominal pain in a rectal distension model as measured by fMRI/PET

Study population	Females	Males	Comparison females and males
Healthy controls	• Negative correlation: pain thresholds and anterior cingulate cortex (ACC) and insula activity [97]	• Positive correlation: pain thresholds and insula activity [97] • Positive correlation: pain ratings and ACC activity [97]	**Higher activity females**: ACC [91], mid cingulate cortex, anterior insula, premotor cortex [92], DLPFC, middle temporal gyrus (pain anticipation) [97], cerebellum, medial frontal gyrus (pain) [97]
IBS patients	• Ventromedial prefrontal cortex [95] • Right ACC [95] • Left amygdala [95]	• Right dorsolateral prefrontal cortex [95] • Insula [93] • Dorsal pons/periaqueductal gray [95]	**Higher activity males**: DLPFC [95], insula [93, 95] **Higher activity females**: ACC [95], amygdala [95]

4.2.1 Studies in Healthy Subjects

There is one PET study in healthy subjects in a rectal distension model [91] with females depicting higher volumes of cortical activity and higher regional cerebral blood flow responses during visceral pain in the anterior cingulate cortex when compared with males [91]. In a recent fMRI study during anticipation and pain induced by esophageal distension, women presented increased brain activity in the midcingulate cortex, anterior insula, premotor cortex; areas being part of medial pain system [92]. These activities seemed to be modulated by individual pain sensitivity. Benson et al. demonstrated a relation between pain threshold and neural brain responses [97]. Women with lower pain thresholds depicted greater brain activity in the anterior cingulate and the insula.

In a very innovative psychological fMRI study investigating sex differences in a fear conditioning model with rectal pain in healthy subjects, sex differences in aversive visceral learning paradigm were seen [96]. In this study conditioned visual stimuli (CS+) were paired with painful rectal distensions. Women presented higher neural responses in brain areas related to memory during re-instatement, indicating an enhanced reactivation of fear memory when compared to men. It was speculated that women have enhanced "gut memories" [96].

4.2.2 Studies in Patients with IBS

Berman et al. showed in a PET study assessing visceral pain (rectal balloon distension) in IBS patients that the insula revealed increased regional cerebral blood flow in males [93]. The authors stated that this finding was line with the observed increased cardio-sympathetic response system in males [94]. Another more recent PET study confirmed these findings in males and also reported increased activity in the DLPFC (cognitive control), but additionally reported higher brain activity in the right anterior cingulate cortex and in the left amygdala in females [95]. The latter areas are activated by emotional stimuli and results may be explained by an automatic over-arousal and heightened affective response with a parallel suppression of cognitive control in women [98].

To summarize, the few imaging studies on sex-effects in pain syndromes indicate that females show a stronger emotional response to pain across modalities and independent of the type of pain. There are other sex differences present, but they are less consistent and depend on the type of pain stimuli and the experimental paradigm.

5 Limitations of Current Studies Assessing Sex and Gender Effects

There are currently only a few imaging studies in healthy subjects and patients' as well assessing sex and gender effects in pain research. Meta-analyses are therefore currently not feasible. The studies available are on the level of pilot studies. Limitations encompass the following:

- Nomenclature used was inconsistent. The term gender was used when sex was evaluated. Cultural background, education and psychological factors such as previous pain experiences were often not assessed and/or controlled for.
- Sex differences in psychophysics and neural responses seem to exist, but they seem to be small. Thus the number of subjects in most studies was too low to detect these differences. Many studies have been underpowered.
- There is a great experimental variety across studies, number of subjects included, experimental paradigm used, images processing and analyses applied and study population chosen.
- The sex of the experimenter in imaging studies in animals and humans are basically never stated and possible interactions have not been controlled for [85].
- In most studies the menstrual cycle has not been taken into account even though pain perception is modulated by sexual hormones. When taken into account, nomenclature of the menstrual cycle phase was inconsistent and hormones were not assessed.

To advance the field it is of importance to use pain models modeling clinical relevant pain states taking sex and gender factors in a systematic way into account to overcome the above discussed limitations. In particular the number of subjects needs to be increased in a multi-centered approach and using comparable pain models at various sites to enable meta-analyses. Moreover studies should include men and women and data should be stratified according to sex.

6 Conclusions

It has been shown that pain is a symptom of multifactorial dimension. Factors such as e.g. hormones, genetic profiles and psychological factors (e.g. anxiety, depression) as well as many social factors modify pain sensitivity in men and women in a different way. Without integrating sex and gender aspects in pain research, new classes of drugs which may only be effective in one sex or adverse effects related to one sex only may be overlooked.

Brain imaging plays an important role to investigate the neural mechanisms of sex-specific differences to painful challenges in standardized pain models and the impact of drug interventions. Differences as well as similarities between sexes are of importance and will help to design tailored pain medication and improve post-operative pain management. Pharmacological imaging in combination with a parallel assessment of genetic and other gender related risk factors in multimodal research design in humans and animals will be an innovative methodical approach in future in biomedical preclinical pain research.

References

1. Clayton JA. Studying both sexes: a guiding principle for biomedicine. FASEB J. 2015. doi:10.1096/fj.15-279554.
2. Hassan S, Muere A, Einstein G. Ovarian hormones and chronic pain: a comprehensive review. Pain. 2014;155:2448–60.
3. Schmid J, Langhorst J, Gaß F, et al. Placebo analgesia in patients with functional and organic abdominal pain: a fMRI study in IBS, UC and healthy volunteers. Gut. 2015;64:418–27.
4. Fillingim RB, King CD, Ribeiro-Dasilva MC, et al. Sex, gender, and pain: a review of recent clinical and experimental findings. J Pain. 2009;10:447–85.
5. Thibodeau MA, Welch PG, Katz J, et al. Pain-related anxiety influences pain perception differently in men and women: a quantitative sensory test across thermal pain modalities. Pain. 2013;154:419–26.
6. Danchin É, Charmantier A, Champagne FA, et al. Beyond DNA: integrating inclusive inheritance into an extended theory of evolution. Nat Rev Genet. 2011;12:475–86.
7. Baker LM, Williams LM, Korgaonkar MS, et al. Impact of early vs. late childhood early life stress on brain morphometrics. Brain Imaging Behav. 2012;7:196–203.
8. Bock J, Rether K, Groeger N, et al. Perinatal programming of emotional brain circuits: an integrative view from systems to molecules. Front Neurosci. 2014. doi:10.3389/fnins.2014.00011.
9. Ritz SA, Antle DM, Côté J, et al. First steps for integrating sex and gender considerations into basic experimental biomedical research. FASEB J. 2014;28:4–13.
10. Dworkin SF, Huggins KH, LeResche L, et al. Epidemiology of signs and symptoms in temporomandibular disorders: clinical signs in cases and controls. J Am Dent Assoc. 1990;120:273–81.
11. Verbrugge LM, Gates DM, Ike RW. Risk factors for disability among U.S. adults with arthritis. J Clin Epidemiol. 1991;44:167–82.
12. Unruh AM. Gender variations in clinical pain experience. Pain. 1996;65:123–67.
13. Lipton RB, Stewart WF, Reed M, et al. Migraine's impact today. Burden of illness, patterns of care. Postgrad Med. 2001;109(38–40):43–4.
14. Symmons DPM. Epidemiology of rheumatoid arthritis: determinants of onset, persistence and outcome. Best Pract Res Clin Rheumatol. 2002;16:707–22.
15. Mayer EA, Berman S, Chang L, et al. Sex-based differences in gastrointestinal pain. Eur J Pain. 2004;8:451–63.
16. Björkman I, Jakobsson Ung E, Ringström G, et al. More similarities than differences between men and women with irritable bowel syndrome. Neurogastroenterol Motil. 2015;27:796–804.
17. Henningsen P. Medically unexplained physical symptoms, anxiety, and depression: a meta-analytic review. Psychosom Med. 2003;65:528.
18. Goldberg DL, Bradley LA, Arnold LM, et al. Understanding fibromyalgia and its related disorders. Prim Care Companion J Clin Psychiatry. 2008;10:133–44.
19. Wolfe F, Smythe HA, Yunus MB, et al. The American college of rheumatology 1990 criteria for the classification of fibromyalgia. Arthritis Rheum. 1990;33:160–72.
20. Clauw DJ. Fibromyalgia and related conditions. Mayo Clin Proc. 2015;90:680–92.
21. Meeus M, Nijs J. Central sensitization: a biopsychosocial explanation for chronic widespread pain in patients with fibromyalgia and chronic fatigue syndrome. Clin Rheumatol. 2006;26:465–73.
22. Price DD, Staud R. Neurobiology of fibromyalgia syndrome. J Rheumatol. 2005;75:22–8.

23. Payne S. Sex, gender, and irritable bowel syndrome: making the connections. Gend Med. 2004;1:18–28.
24. Drossman DA. Rome III: new standard for functional gastrointestinal disorders. J Gastrointestin Liver Dis. 2006;15:237.
25. Lenz K, Nugent B, Haliyur R, et al. Microglia are essential to masculinization of brain and behavior. J Neurosci: Official J Soc Neurosci. 2013;33:2761–72.
26. Sugiyama MG, Hobson L, Agellon AB, et al. Visualization of sex-dimorphic changes in the intestinal transcriptome of Fabp2 gene-ablated mice. J Nutrigenet Nutrigenomics. 2012;5:45–55.
27. Markle JGM, Frank DN, Mortin-Toth S, et al. Sex differences in the gut microbiome drive hormone-dependent regulation of autoimmunity. Science. 2013;339:1084–8.
28. Markle JG, Fish EN. SeXX matters in immunity. Trends Immunol. 2014;35:97–104.
29. Gerbershagen HJ, Pogatzki-Zahn E, Aduckathil S, et al. Procedure-specific risk factor analysis for the development of severe postoperative pain. Survey Anesthesiol. 2014;58:315–6.
30. Tighe PJ, Riley JL, Fillingim RB. Sex differences in the incidence of severe pain events following surgery: a review of 333,000 pain scores. Pain Med. 2014;15:1390–404.
31. Ochroch EA, Gottschalk A, Troxel AB, et al. Women suffer more short and long-term pain than men after major thoracotomy. Clin J Pain. 2006;22:491–8.
32. Rashiq S, Dick B. Post-surgical pain syndromes: a review for the non-pain specialist. Can J Anaesth. 2014;61:123–30.
33. Schnabel A, Poepping DM, Gerss J, et al. Sex-related differences of patient-controlled epidural analgesia for postoperative pain. Pain. 2012;153:238–44.
34. Tighe PJ, Le-Wendling LT, Patel A, et al. Clinically derived early postoperative pain trajectories differ by age, sex, and type of surgery. Pain. 2015;156:609–17.
35. Rezaii T, Hirschberg AL, Carlström K, et al. The influence of menstrual phases on pain modulation in healthy women. J Pain. 2012;13:646–55.
36. Mogil JS, Bailey AL. Sex and gender differences in pain and analgesia. Prog Brain Res. 2010;186:141–57.
37. Niesters M, Dahan A, Kest B, et al. Do sex differences exist in opioid analgesia? A systematic review and meta-analysis of human experimental and clinical studies. Pain. 2010;151:61–8.
38. Miller M. Gender-based differences in the toxicity of pharmaceuticals—the food and drug administration's perspective. Int J Toxicol. 2001;20:149–52.
39. El-Serag H. Impact of irritable bowel syndrome: prevalence and effect on health-related quality of life. Rev Gastroenterol Disord Suppl. 2003;2:S3–11.
40. van Laarhoven A, Vogelaar M, Wilder-Smith O, et al. Induction of nocebo and placebo effects on itch and pain by verbal suggestions. Pain. 2011;152:1486–94.
41. Aslaksen P, Bystad M, Vambheim S, et al. Gender differences in placebo analgesia: event-related potentials and emotional modulation. Psychosom Med. 2011;73:193–9.
42. Butcher B, Carmody J. Sex differences in analgesic response to ibuprofen are influenced by expectancy: a randomized, crossover, balanced placebo-designed study. Eur J Pain. 2012;16:1005–13.
43. Flaten M, Aslaksen P, Finset A, et al. Cognitive and emotional factors in placebo analgesia. J Psychosom Res. 2006;61:81–9.
44. Arendt-Nielsen L, Yarnitsky D. Experimental and clinical applications of quantitative sensory testing applied to skin, muscles and viscera. J Pain. 2009;10:556–72.
45. Rolke R, Magerl W, Campbell KA, et al. Quantitative sensory testing: a comprehensive protocol for clinical trials. Eur J Pain. 2006;10:77–8.
46. Schmelz M. Translating nociceptive processing into human pain models. Exp Brain Res. 2009;196:173–8.
47. Riley III JL, Robinson ME, Wise EA, et al. A meta-analytic review of pain perception across the menstrual cycle. Pain. 1999;81:225–35.
48. Racine M, Tousignant-Laflamme Y, Kloda LA, et al. A systematic literature review of 10 years of research on sex/gender and experimental pain perception—part 1: are there really differences between women and men? Pain. 2012;153:602–18.
49. Burgmer M, Gaubitz M, Konrad C, et al. Decreased gray matter volumes in the cingulo-frontal cortex and the amygdala in patients with fibromyalgia. Psychosom Med. 2009;71:566–73.
50. Jensen KB, Srinivasan P, Spaeth R, et al. Overlapping structural and functional brain changes in patients with long-term exposure to fibromyalgia pain. Arthritis Rheum. 2013;65:3293–303.
51. Yarnitsky D, Crispel Y, Eisenberg E, et al. Prediction of chronic post-operative pain: pre-operative DNIC testing identifies patients at risk. Pain. 2008;138:22–8.
52. Popescu A, LeResche L, Truelove EL, et al. Gender differences in pain modulation by diffuse noxious inhibitory controls: a systematic review. Pain. 2010;150:309–18.
53. Sherman JJ, LeResche L. Does experimental pain response vary across the menstrual cycle? A methodological review. Am J Physiol Regul Integr Comp Physiol. 2006;291:R245–56.
54. Dao TT, LeResche L. Gender differences in pain. J Orofac Pain. 2000;14:169–184, discussion 184–95.
55. Iacovides S, Avidon I, Baker FC. Does pain vary across the menstrual cycle? A review. Eur J Pain. 2015;19:1389–405.
56. Pereira MP, Pogatzki-Zahn E. Gender aspects in postoperative pain. Curr Opin Anaesthesiol. 2015;28:546–58.
57. Messinis IE, Messini CI, Dafopoulos K. Novel aspects of the endocrinology of the menstrual cycle. Reprod Biomed Online. 2014;28:714–22.
58. Johannes CB, Linet MS, Stewart WF, et al. Relationship of headache to phase of the menstrual cycle among young women: a daily diary study. Neurology. 1995;45:1076–82.

59. Houghton L, Lea R, Jackson N. The menstrual cycle affects rectal sensitivity in patients with irritable bowel syndrome but not healthy volunteers. Gut. 2002;50:471–4.
60. Latman NS. Relation of menstrual cycle phase to symptoms of rheumatoid arthritis. Am J Med. 1983;74:957–60.
61. Pamuk ON. The variation in chronic widespread pain and other symptoms in fibromyalgia patients. The effects of menses and menopause. Clin Exp Rheumatol. 2005;23:778.
62. Colangelo K, Haig S, Bonner A, et al. Self-reported flaring varies during the menstrual cycle in systemic lupus erythematosus compared with rheumatoid arthritis and fibromyalgia. Rheumatology. 2011;50:703–8.
63. Alonso C, Loevinger BL, Muller D, et al. Menstrual cycle influences on pain and emotion in women with fibromyalgia. J Psychosom Res. 2004;57:451–8.
64. Allais G, Castagnoli Gabellari I, Burzio C, et al. Premenstrual syndrome and migraine. Neurol Sci. 2012;33:111–5.
65. Metcalf MG, Skidmore DS, Lowry GF, et al. Incidence of ovulation in the years after the menarche. J Endocrinol. 1983;97:213–9.
66. Klatzkin RR, Mechlin B, Girdler SS. Menstrual cycle phase does not influence gender differences in experimental pain sensitivity. Eur J Pain. 2010;14:77–82.
67. Teepker M, Peters M, Vedder H, et al. Menstrual variation in experimental pain: correlation with gonadal hormones. Neuropsychobiology. 2010;61:131–40.
68. Pogatzki-Zahn EM, Wagner C, Meinhardt-Renner A, et al. Coding of incisional pain in the brain: a functional magnetic resonance imaging study in human volunteers. Anesthesiol. 2010;112:406–17.
69. Nisenblat V, Engel-Yeger B, Ohel G, et al. The association between supra-physiological levels of estradiol and response patterns to experimental pain. Eur J Pain. 2010;14:840–6.
70. Sacher J, Neumann J, Okon-Singer H, et al. Sexual dimorphism in the human brain: evidence from neuroimaging. Magn Reson Imag. 2012.
71. Legato M. Men, women, and brains: what's hardwired, what's learned, and what's controversial. Gend Med. 2005;2:59–61.
72. Joel D, Berman Z, Tavor I, et al. Sex beyond the genitalia: the human brain mosaic. Proc Natl Acad Sci. 2015;112:15468–73.
73. Goldstein JM, Seidman LJ, Horton NJ, et al. Normal sexual dimorphism of the adult human brain assessed by in vivo magnetic resonance imaging. Cereb Cortex. 2001;11:490–7.
74. Jovanovic H, Lundberg J, Karlsson P, et al. Sex differences in the serotonin 1A receptor and serotonin transporter binding in the human brain measured by PET. NeuroImage. 2008;39:1408–19.
75. Nishizawa S, Benkelfat C, Young SN, et al. Differences between males and females in rates of serotonin synthesis in human brain. Proc Natl Acad Sci U S A. 1997;94:5308–13.
76. Turner BB, Debra AW. Sexual dimorphism of glucocorticoid binding in rat brain. Brain Res. 1985;343:16–23.
77. Zubieta J-K, Dannals RF, Frost JJ. Gender and age influences on human brain mu-opioid receptor binding measured by PET. Am J Psychiatr. 1999;156:842–8.
78. Zubieta J-K, Gorelick DA, Stauffer R, et al. Increased mu opioid receptor binding detected by PET in cocaine-dependent men is associated with cocaine craving. Nat Med. 1996;2:1225–9.
79. Ingalhalikar M, Smith A, Parker D, et al. Sex differences in the structural connectome of the human brain. Proc Natl Acad Sci. 2014;111:823–8.
80. Cosgrove KP, Mazure CM, Staley JK. Evolving knowledge of sex differences in brain structure, function, and chemistry. Biol Psychiatry. 2007;62:847–55.
81. Greenspan JD, Craft RM, LeResche L, et al. Studying sex and gender differences in pain and analgesia: a consensus report. Pain. 2007;132(Supplement 1): S26–45.
82. Coen SJ, Kano M, Farmer AD, et al. Neuroticism influences brain activity during the experience of visceral pain. Gastroenterology. 2011;141:909–17. e901.
83. Ruau D, Liu LY, Clark JD, et al. Sex differences in reported pain across 11,000 patients captured in electronic medical records. J Pain. 2012;13:228–34.
84. Aslaksen PM, Myrbakk IN, Høifødt RS, et al. The effect of experimenter gender on autonomic and subjective responses to pain stimuli. Pain. 2007;129:206–8.
85. Bateson M. Of (stressed) mice and men. Nat Meth. 2014;11:623–4.
86. Henderson LA. Gender differences in brain activity evoked by muscle and cutaneous pain: a retrospective study of single-trial fMRI data. NeuroImage. 2008;39:1867.
87. Paulson PE, Minoshima S, Morrow TJ, et al. Gender differences in pain perception and patterns of cerebral activation during noxious heat stimulation in humans. Pain. 1998;76:223–9.
88. Derbyshire SWG, Nichols TE, Firestone L, et al. Gender differences in patterns of cerebral activation during equal experience of painful laser stimulation. J Pain. 2002;3:401–11.
89. Maihöfner C, Herzner B, Otto Handwerker H. Secondary somatosensory cortex is important for the sensory-discriminative dimension of pain: a functional MRI study. Eur J Neurosci. 2006;23:1377–83.
90. Lorenz J, Minoshima S, Casey KL. Keeping pain out of mind: the role of the dorsolateral prefrontal cortex in pain modulation. Brain: J Neurol. 2003;126:1079–91.
91. Kern MK, Jaradeh S, Arndorfer RC, et al. Gender differences in cortical representation of rectal

distension in healthy humans. Am J Physiol Gastrointest Liver Physiol. 2001;281:G1512–23.
92. Kano M, Farmer AD, Aziz Q, et al. Sex differences in brain response to anticipated and experienced visceral pain in healthy subjects. Am J Physiol Gastrointest Liver Physiol. 2013;304:G687–99.
93. Berman S, Munakata J, Naliboff BD, et al. Gender differences in regional brain response to visceral pressure in IBS patients. Eur J Pain. 2000;4:157–72.
94. Dart AM, Du X-J, Kingwell BA. Gender, sex hormones and autonomic nervous control of the cardiovascular system. Cardiovasc Res. 2002;53:678–87.
95. Naliboff BD, Berman S, Chang L, et al. Sex-related differences in IBS patients: central processing of visceral stimuli. Gastroenterology. 2003;124:1738–47.
96. Benson S, Kattoor J, Kullmann JS, et al. Towards understanding sex differences in visceral pain: enhanced reactivation of classically-conditioned fear in healthy women. Neurobiol Learn Mem. 2014;109:113–21.
97. Benson S, Kotsis V, Rosenberger C, et al. Behavioural and neural correlates of visceral pain sensitivity in healthy men and women: does sex matter? Eur J Pain. 2012;16:349–58.
98. Ochsner KN, Gross JJ. The cognitive control of emotion. Trends Cogn Sci. 2005;9:242–9.

Author Biographies

Prof. Dr. Bettina Pfleiderer, Ph.D., MD is head of the research group "Cognition & Gender" at the medical faculty of Muenster (Germany). Her research focuses on sex and gender effects in brain function as assessed by various imaging methods. Her work is centered on pain disorders, pruritus, anxiety disorders, depression and memory processes. She is also developing teaching modules in gender/sex specific medicine for medical students and physicians. She have two doctor titles and she wrote two doctoral theses in physical chemistry and medicine.

Dr. Anika Ritzkat, MD studied medicine at the medical universities of Hanover (Germany) and Lausanne (Switzerland). For her medical thesis she worked in a lab at Tufts Floating Hospital for Children in Boston in 2009/2010. After clinical rotations in Switzerland, Ireland and Germany, she graduated in medicine in 2012. Since then she works at the Department of Clinical Radiology at the University Hospital Muenster (Germany).

Prof. Dr. Esther Pogatzki-Zahn, MD is an Anesthesiologist and Professor in the Department of Anesthesiology, Critical Care and Pain Medicine of the University Hospital Muenster, Germany. She is head of the Pain Service at the University Hospital of Muenster and leads the research group "Experimental and clinical pain research" which is focusing on the neurobiology of acute and chronic pain.

The Neuroimaging of Vicarious Pain 16

Esslin L. Terrighena and Tatia M.C. Lee

Abstract

The similarities between self-experienced and vicarious pain have led research to suggest that both experiences may be facilitated by shared neural representations. Indeed, neuroimaging evidence demonstrates an overlap in neural patterns during self- and other-pain. Such comparable brain activity may facilitate an empathic understanding of the current state of the individual in pain by stimulating relevant pain associations in the own sensory, affective and cognitive systems. However, research further shows the distinct contributions of neural activity during vicarious pain processing, in particular in brain regions related to perspective-taking, attention and top-down response regulation. Likewise, such activity may underpin response formation to the observed pain, such as empathic or withdrawal behaviors. This chapter reviews 31 fMRI, six EEG/MEG and four TMS studies exploring the neural correlates of vicarious pain in healthy individuals. Both shared and distinct neural contributions to stimulus and response processing during vicarious pain are discussed.

E.L. Terrighena · T.M.C. Lee
Laboratory of Neuropsychology, The University of Hong Kong, Pok Fu Lam, Hong Kong SAR

E.L. Terrighena · T.M.C. Lee
Laboratory of Social Cognitive Affective Neuroscience, The University of Hong Kong, Pok Fu Lam, Hong Kong SAR

T.M.C. Lee
The State Key Laboratory of Brain and Cognitive Sciences, The University of Hong Kong, Pok Fu Lam, Hong Kong SAR

T.M.C. Lee (✉)
Institute of Clinical Neuropsychology, The University of Hong Kong, Rm 656, The Jockey Club Tower, Pokfulam Road, Pok Fu Lam, Hong Kong SAR
e-mail: tmclee@hku.hk

L. Saba (ed.), *Neuroimaging of Pain*, DOI 10.1007/978-3-319-48046-6_16

Notably, an integrative model of vicarious pain is introduced which brings such contributions together in a comprehensive manner. Moreover, the chapter highlights inconsistencies and research gaps in current literature with the aim of stimulating further scientific investigation. This is pertinent to the detection of neurobiological markers and intervention targets for empathic deficits which characterize a wide variety of clinical health issues.

Keywords
Empathic · Perception-action model (PAM) · Motor cortex · Self-pain · Mirror neuron

Abbreviations

Brain Regions

IFG	Inferior Frontal Gyrus
IPL	Inferior Parietal Lobule
SI, SII	Primary Somatosensory Cortex, Secondary Somatosensory Cortex
PFC	Prefrontal Cortex
INS	Insula
aINS, mINS, pINS	Anterior Insula, Mid-Insula, Posterior Insula
CC, ACC, PCC	Cingulate Cortex, Anterior Cingulate Cortex, Posterior Cingulate Cortex
sgACC, rACC	subgenual ACC, rostral ACC
MCC, aMCC	Midcingulate Cortex, Anterior MCC
dlPFC, dmPFC, mPFC, rlPFC	dorsolateral PFC, dorsomedial PFC, medial PFC, rostrolateral PFC
SMA	Supplementary Motor Area

Neuroimaging Methods

fMRI	functional Magnetic Resonance Imaging
EEG	Electroencephalography
MEG	Magnetoencephalography
TMS	Transcranial Magnetic Stimulation

Models

PAM	Perception Action Model

1 Introduction

Vicarious pain is characterized by the observation of individuals who are experiencing acute pain [48, 110]. The empathic ability to relate to the affective state of these individuals has social and physical benefits as it enables observes to adjust their behavior according to the context [156]. Not only can this enhance social relationships through the display of compassion, but it also promotes adequate assessment of situational cues that require prompt withdrawal responses. Thus, stimuli that are potentially threatening can be removed before causing further harm to either the individual in pain or the observer [14, 61, 62, 75]. In line with those essential survival functions, evidence from imaging studies using electroencephalography (EEG), magnetoencephalography (MEG), transcranial magnetic stimulation (TMS), and functional magnetic resonance imaging (fMRI) suggests that the human brain is wired to facilitate empathic understanding through shared and distinct neural representations of self-experienced and observed pain [4, 67, 78, 81, 105, 110, 111, 178, 205]. Nonetheless, inconsistencies in current literature highlight that further exploration is critical to acquiring a complete understanding of the neural mechanisms underpinning vicarious pain and empathy. Research in this area is pertinent to the detection of neurobiological markers and intervention targets for empathic deficits which characterize a wide variety of clinical health issues, such as autism, schizophrenia and motor neuron disease [12, 18, 35, 172].

2 Vicarious Pain, Empathy and the Perception-Action Model

Acute self-pain is experienced in the own body directly as an "unpleasant sensory and emotional experience" of sharp quality "associated with (…) tissue damage" [94, p. 5]. In contrast, although vicarious pain is defined as the observation of pain in others, the affective, cognitive, and sensory aspects that accompany this experience are challenging to pinpoint. Particularly, empathic responding to such pain observation has not yet been clearly defined [155]. It is generally described as an understanding of affective states in others. However, more narrow definitions of empathy have specified that the platform for such understanding arises when observation or imagination of a person in a particular emotional state elicits a similar emotion in the self that remains conceptually separate from personal distress [74, 98]. Furthermore, empathy has been associated not only with emotional relatedness, but also a cognitive and somatosensory understanding of the observed pain. In line with this, research suggests that empathizing with vicarious pain can have affective, cognitive and sensory effects for an individual that are similar to self-pain [67, 156]. Reflecting this similarity, the Perception-Action Model (PAM) of empathy advances that vicarious processing is subserved by shared neural representations underlying self- and observed pain [81, 110, 111, 155, 156, 178]. It relies on the mirror neuron system which is characterized by neurons that respond both when an action is actively performed and passively observed [158]. More explicitly, during pain observation motor mirror neurons are activated that correspond to muscle groups involved in acute self-pain. This activation promotes understanding of vicarious pain through the mirroring of such pain within the own motor system [160]. In consequence, a neural network is automatically stimulated, containing learned sensory and affective information for self-pain which can be used to predict and evaluate the suffering of others. This may facilitate other-oriented empathy as well as self-oriented distress and withdrawal responses [81]. Moreover, the PAM advocates indirect sensory, affective and cognitive mirroring in pain processing regions which are not directly implicated in the mirror neuron system, such as the cingulate and prefrontal cortices. Among those, similar neural patterns during self- and vicarious pain may reflect comparable stimulus processing, resulting in shared neural pain representations. Such representations also trigger an associative pain network and facilitate a swift, concurrent holistic appraisal of the observed pain [110, 155].

Research provides evidence for both motor mirror neurons and shared neural representations in vicarious pain processing. First, studies have shown neural activity in the inferior frontal gyrus (IFG) and inferior parietal lobule (IPL) during pain observation, which are considered core regions of the human mirror neuron system [14, 65, 77, 110, 158, 171, 178, 193]. Second, brain activation in the well-established pain matrix has been reported to overlap for self- and vicarious pain forming shared neural representations of these pain experiences [48, 110]. As shown in Fig. 1, the pain matrix includes the primary

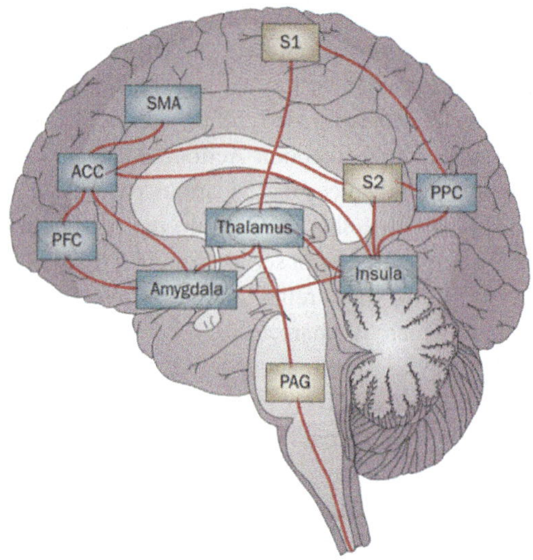

Fig. 1 Pain matrix. Abbreviations: *ACC* cingulate cortex; *PAG* periaqueductal gray; *PFC* prefrontal cortex; *SMA* supplementary motor area; *S1*, primary somatosensory cortex; *S2* secondary somatosensory cortex; *PPC* posterior parietal cortex [120, 121]

(SI) and secondary somatosensory cortices (SII), thalamus, prefrontal cortex (PFC), insula (INS), and cingulate cortex (CC).

It is suggested that SI, SII and the thalamus encode the sensory aspects of noxious stimuli, including location and intensity. In contrast, INS, CC and PFC play a role in cognitive–affective evaluation and top-down control [42, 43, 151, 186]. These regions are comparably implicated in vicarious pain. However, current methodological issues, such as coarse spatial neuroimaging resolution, prevent firm conclusions to which extent the neural patterns of self- and other-pain are identical [86, 111]. On the contrary, Zaki et al. [205] had indicated that self- and vicarious pain have differed functional connectivities across brain regions and thus, unique inter-regional communication may reflect the qualitative differences between these pain experiences. Likewise, activity in the same brain regions as self-pain may nonetheless make distinct contributions to other-pain processing [110]. Research suggests that mirroring may be neither necessary nor sufficient for empathy induction (for a critical review, see [111]. For example, individuals reported empathic distress upon learning that their partner was in pain without activation in the IFG or IPL, deeming these regions unnecessary for affective empathy [179]. Furthermore, individuals tend to exhibit appropriate empathic responses in situations where individuals are in pain, even though mirroring alone would imply that they are not [57, 112]. Accordingly, additional neural activity is required for adequate appraisal of vicarious pain [111, 149]. Thus, the involvement of aforementioned brain regions in vicarious pain is likely to go beyond mirroring [77].

Moreover, aside from appraising observed pain by means of shared and distinct neural correlates, neural patterns in observers may underpin distinct responses to the pain. The PAM predicts other-oriented empathic understanding to automatically arise from shared pain representations. Indeed, ample papers supporting the PAM report an association between empathic abilities and neural mirroring activity, proposing that the evaluation of observed pain through bottom-up mirroring induces empathy (e.g. [158, 160]). Empathy has been shown to elicit altruistic behaviors that have the aim of helping the individual in pain. These promote social relationships and their protective benefits [155, 156]. However, vicarious pain perception has also been found to induce self-oriented distress and withdrawal behaviors to the observed pain threat [65, 110, 111]. In line with this, brain regions active during pain observation have also been associated with emotional contagion, which is measured by affective distress ratings. Such contagion may elicit avoidance when observing negative affect in others and, thus, stands in opposition to the altruistic acts induced by empathy [60, 111]. While Preston and Hofelich [157] recently speculated that these concepts may also be evoked by neural mirroring, they did not extensively elaborate on the mechanisms behind this. Further contrasting the PAM, recent evidence suggests that neural activity during vicarious pain reflects top-down sensorimotor pain predictions based on higher cortical analysis rather than bottom-up mirroring. These predictions are dynamically compared with the pain observation and may evoke self-oriented motor withdrawal preparation [54, 111, 200].

Accordingly, brain responses to vicarious pain may reflect other-oriented empathy or self-oriented withdrawal. Notably, neural responses to empathy and distress have not yet been reliably teased apart. Given these issues, current findings of shared and distinct neural activation between self- and vicarious pain should be interpreted with caution [111]. At this time, the roles of each brain region involved in pain observation are not clearly established. However, this chapter will present noteworthy speculations that have been made and lend themselves to extensive future research.

The following sections will provide an overview of brain regions associated with vicarious pain processing. The chapter will first discuss evidence for shared and distinct neural substrates of self- and other-pain before exploring correlates underpinning empathic understanding and self-oriented behaviors. An integrative model will be presented that extends the well-established PAM with the aim of incorporating recent findings that offer a more rounded understanding of the complexity of vicarious pain processing. A summary of studies and reported brain activity appears in Table 1. This table presents only research directly exploring the neural correlates of vicarious pain in the healthy population and in absence of modulating factors, such as group membership [7, 91]. As can be inferred, the majority of research in this field utilizes fMRI for brain investigations (for details on fMRI methods, see [84]. To date, there are only four TMS studies [4–6, 128]; for details on TMS methods, see Rossi et al. [126, 162] and two EEG studies [27, 194]; for details on EEG methods, see [183] that investigate the neural activity during pain observation.

3 Neural Responses to Vicarious Pain

Across neuroimaging research, the most consistent brain activations during vicarious pain experience lie in the INS, CC and PFC. These regions have been implicated in the affective–cognitive processing of self- and observed pain (e.g. [49, 89, 147, 163, 179, 192]). In contrast, findings for the motor regions have been less consistent. While several studies report IFG, IPL and motor cortical activity in response to pain observation [113, 137, 193], many find none [48, 89, 179]. Likewise, activation in the somatosensory cortices remains variable [28]. Nonetheless, when contrasting brain responses to videos of limbs and objects subjected to equivalent noxious stimulation, the motor and somatosensory cortices responded exclusively to painful limbs, suggesting that sensorimotor mirroring may play an essential role in the identification of human pain [49]. Contradictory neuroimaging findings may result from the sensitivity of vicarious pain correlates to attentional focus. In an image- and coordinate-based meta-analysis, Lamm et al. [110] propose that the distinct recruitment of sensorimotor compared to affective brain areas depends on which pain components the observer highlights. Accordingly, attending to sensory factors, such as pain intensity, should be subserved by neural correlates of sensorimotor processing. On the other hand, an affective focus, such as rating pain unpleasantness, should activate affective pain substrates [110, 113]. In both cases, increased stimulus complexity should be associated with cognitive regions [88, 112, 149].

The common stimuli used to induce empathy for vicarious pain allow for such differentiation (Fig. 2). Most typically, participants are presented with images of hands or feet in painful or non-painful scenarios that are likely to evoke a sensory focus as the emphasis is on the body part (e.g., [6, 89, 98, 137]). In contrast, requiring participants to infer pain from facial expressions, abstract cues or imagination taps into affective processing [22, 98]. Furthermore, complex pain scenarios entail greater cognitive analysis of the presented context [149]. The corresponding effects of focus have been reflected in neuroimaging findings with sensory focus activating areas IFG, IPL, SI, SII and motor cortices, affective focus correlating with anterior cingulate cortex (ACC) and anterior INS (aINS) activity [4, 136, 179], and cognitive involvement eliciting anterior midcingulate cortex (aMCC) and PFC activity [88, 110]. Notably, most studies

Table 1 Studies investigating the neural correlates of vicarious pain in healthy populations

fMRI activations corresponding to vicarious pain compared to no pain in human limbs and/or faces					
Author and date	Stimuli	Task	Brain activation		
			Sensorimotor	Affective-cognitive	Functional connectivities
Images of limbs					
Corradi-Dell'Acqua et al. [48]	Images of hands in pain or no pain	Judge sensory and affective qualities of pain in observed hand	SMG	INS MCC PCC	
Gu and Han [88]	Images or cartoons of hands in pain or no pain	Attend to pain cues or count hands in image	IFG	aINS ACC	ACC and IFG (increase)
Gu et al. [89]	Images of limbs in pain or no pain	Judge if painful or not painful and laterality	SI/SII Premotor cortex	aINS	aINS and SMA aINS and IFG aINS and PFC (decrease)
Jackson et al. [98]	Images of individuals with hands or feet subjected to pain or no pain	Judge pain level in observed limbs from first-person or third-person perspective	SII IFG Thalamus Cerebellum	aINS ACC mPFC PCC TPJ	
Jackson et al. [99]	Images of individuals with hands or feet subjected to pain or no pain	Judge sensory and affective qualities of pain in observed limbs	PPC SMA Thalamus Cerebellum	aINS ACC	
Lamm et al. [113]	Images of numbed or normal hands deeply penetrated by needles	Judge sensory and affective qualities of pain in observed hand	SI/SII IPL Premotor Cortex	aINS ACC MCC	
Lamm and Decety [109]	Images of hands deeply penetrated by needles or not	Rate pain	EBA to pain and no pain		
Morrison et al. [137]	Images of hands grasping or withdrawing from painful or not painful items	Judge whether grasping or withdrawal actions are appropriate or not Rate how painful each object would be to touch Rate previous experience with object	SI/SII IPL		

(continued)

Table 1 (continued)

fMRI activations corresponding to vicarious pain compared to no pain in human limbs and/or faces					
Author and date	Stimuli	Task	Brain activation		
			Sensorimotor	Affective-cognitive	Functional connectivities
Ogino et al. [147]	Neutral, fear and pain images Pain images of limbs subjected to pain	Imagine observed pain in own body	SII PPC Cerebellum	aINS ACC	
Images of faces					
Saarela et al. [163]	Faces of chronic pain patients at resting state of chronic pain or during provoked acute pain	Focus on images	IPL IFG SMA Premotor Cortex	aINS ACC	
Simon et al. [176]	Faces of pain, anger and neutral expressions	Discriminate gender	SI SII	aINS ACC mPFC STS Amygdala	
Images of limbs and faces					
Lamm et al. [112]	Images of hands deeply penetrated by needle or touched by Q-tip and facial expressions with normal or abnormal pain responses	Judge sensory and affective qualities of pain in observed limbs	IFG SMA	aINS MCC mPFC	IFG and MCC IFG and PAG IFG and aINS (increase)
Vachon-Presseau et al. [193]	Images of individuals with hands or feet subjected to pain or no pain and facial pain or neutral expressions	Judge sensory and affective qualities of pain in observed limbs or faces	IFG IPL EBA Cerebellum	ACC aMCC PCC mPFC STS Amygdala	
Vachon-Presseau et al. [192]	Images of individuals with hands or feet subjected to pain or no pain and facial expressions of pain	Focus on images	IFG	aINS ACC	

(continued)

Table 1 (continued)

fMRI activations corresponding to vicarious pain compared to no pain in human limbs and/or faces

Author and date	Stimuli	Task	Brain activation		
			Sensorimotor	Affective-cognitive	Functional connectivities
Videos of limbs					
Akitsuki and Decety [1]	Animations of hands and feet in painful scenarios alone or with other person inflicting pain	Judge how painful stimulus is and whether pain was inflicted intentionally	SI/SII IFG SMA PAG	INS aMCC mPFC TPJ	Amygdala and aINS Amygdala and ACC Amygdala and SFG Amygdala and SMA Amygdala and OFC (increase)
Benuzzi et al. [13]	Videos on hands or feet in painful, disgusting or neutral scenarios	Rate disgust and pain of stimuli	IPL/PPC	INS ACC MCC mPFC	
Cheng et al. [36]	Animations of hands or feet in painful and non-painful scenarios	Imagine from self-, loved-one or stranger perspective	SFG	INS ACC TPJ	TPJ and INS (decrease) TPJ and SFG (increase)
Costantini et al. [49]	Video clips of hands deeply penetrated by needle or touched by Q-tip and tomatoes with equivalent stimulation	Focus on videos	SI/SII IFG Premotor Cortex	aMCC	
Morrison and Downing [134]	Short animations of a noxious or non-noxious item striking or not striking a hand	Report whether hand in animation was struck or not struck by item via button press	IFG SMA CMZ	aINS MCC PCC	
Videos of faces					
Budell et al. [26]	Video clips of facial expressions of pain	Evaluate pain experience and movement of facial muscles in observed face	IFG IPL	aINS ACC mPFC	
Lamm et al. [108]	Videos of neutral and pain faces elicited by painful sound	Judge pain intensity and unpleasantness	SMA Thalamus	INS aMCC PFC TPJ Amygdala	

(continued)

Table 1 (continued)

fMRI activations corresponding to vicarious pain compared to no pain in human limbs and/or faces

Author and date	Stimuli	Task	Brain activation		
			Sensorimotor	Affective-cognitive	Functional connectivities
Videos of limbs and faces					
Cheng et al. [37]	Videos of individuals receiving needles or Q-Tips to different body parts	Rate pain intensity and pain unpleasantness	SI/SII PAG	ACC PFC TPJ	aINS and mPFC (increase)
Osborn and Derbyshire [148]	Images or short clips of limbs or full individuals subjected to pain	Report whether any pain was felt in own body during vicarious pain	SI/SII	aINS aMCC PFC	
Direct observation					
Singer et al. [180]	Direct observation of fair and unfair confederates receiving electric pain	Rate intensity of high and low pain stimulation, liking for confederates and desire for revenge		aINS ACC	

fMRI activations corresponding to vicarious pain compared to self-pain.

Autor and date	Stimuli	Task	Brain activation		
			Sensorimotor	Affective-cognitive	Functional connectivities
Videos of limbs					
Morrison et al. [135]	Videos of hands experiencing pin prick or self-experienced pin prick	Focus on stimuli		ACC	
Morrison et al. [136]	Video of needle deeply penetrating hand or self-experienced needle penetrating hand	Rate pain unpleasantness	CMZ	ACC PCC	
Videos of faces					
Botvinick et al. [22]	Videos of faces in pain or no pain and self-experienced painful or not painful heat	Focus on stimuli		aINS ACC	
Videos of limbs and faces					

(continued)

Table 1 (continued)

fMRI activations corresponding to vicarious pain compared to self-pain.

Autor and date	Stimuli	Task	Brain activation		
			Sensorimotor	Affective-cognitive	Functional connectivities
Ochsner et al. [146]	Videos of individuals subjected to pain or self-experienced heat pain	Focus on stimuli	SPL/PPC	INS ACC MCC PFC Amygdala	
Hein et al. [91]	Videos of in-group and outgroup members receiving pain or no pain and electric self-pain	Choose to help individual in pain by taking half, not help and watch soccer video or not help and watch pain		aINS	
Zaki et al. [205]	Videos of individuals subjected to pain or self-experienced heat pain	Focus on stimuli	Precuneus	aINS ACC mPFC STS	ACC and STS ACC and PCC ACC and precuneus mPFC and ACC mPFC and aINS
Abstract cues					
Singer et al. [179]	Cue that loved one receives electric pain or self-experienced electric pain	Focus on stimuli	EBA Cerebellum	aINS ACC	

EEG/MEG/TMS responses corresponding to vicarious pain compared to no pain in human limbs and/or faces

Author and date	Method	Stimuli	Task	EEG
Images of limbs				
Cheng et al. [38]	MEG	Images of limbs in pain or no pain	Rate observed pain intensity	SI Suppression of somatosensory cortical oscillations
Fan and Han [78]	EEG	Images or cartoons of hands in pain or no pain	Attend to pain cues or count hands in image	Frontal and parietal regions Greater EEG suppression in pain than no pain showing early and late differentiation of these conditions

(continued)

Table 1 (continued)

EEG/MEG/TMS responses corresponding to vicarious pain compared to no pain in human limbs and/or faces

Author and date	Method	Stimuli	Task	EEG
Images of faces and limbs				
Perry et al. [149]	EEG	Images of hands deeply penetrated by needle or touched by Q-tip and facial expressions with normal or abnormal pain responses	Judge affective state of observed individual by imaging feeling of this individual	Frontal regions Greater EEG suppression in pain than no pain conditions During abnormal pain responses, suppression equally large in pain and no pain conditions
Videos of limbs				
Avenanti et al. [4]	TMS	Video clips of hands and feet static, being touched by a Q-tip or being penetrated deeply by a needle	Judge sensory and affective qualities of pain in observed limb	Motor cortex Inhibition of motor-evoked potential amplitudes to left and right motor cortex stimulation corresponding to muscle subjected to needle
Avenanti et al. [5]	TMS	Videos of hand penetrated by needle or static	Pay attention and focus on what stimulated individual may have felt	Motor Cortex Inhibition of motor-evoked potential amplitudes to left and right motor cortex stimulation
Avenanti et al. [6]	TMS	Video clips of hands and feet static, being touched by a Q-tip or being penetrated deeply by a needle	Judge sensory and affective qualities of pain in observed limb from third-person or first-person perspective	Motor cortex Inhibition of motor-evoked potential amplitudes to left and right motor cortex stimulation corresponding to muscle subjected to needle
Betti et al. [15]	MEG	Video clips of hand static, being touched by a Q-tip or being penetrated deeply by a needle	Imagine observed limbs are own	Somatosensory and motor cortices Increased neural synchronization between SI and primary motor cortex
Bufalari et al. [27]	SEP	Video clips of hand static, being touched by a Q-tip or being penetrated deeply by a needle	Judge sensory and affective qualities of pain in observed hand	SI Increased P45 SEP amplitude during pain condition Decreased P45 SEP amplitude during gentle touch condition

(continued)

Table 1 (continued)

EEG/MEG/TMS responses corresponding to vicarious pain compared to no pain in human limbs and/or faces

Author and date	Method	Stimuli	Task	EEG
Minio-Paluello et al. [128]	TMS	Video clips of hands without pain stimulus or being penetrated deeply by a needle.	Judge sensory and affective qualities of pain in observed hand	Motor cortex Inhibition of motor-evoked potential amplitudes to left and right motor cortex stimulation corresponding to muscle subjected to needle Inference about sensory qualities of vicarious pain inhibited during stimulation of left hemisphere motor cortex
Valeriani et al. [194]	LEP	Video clips of hand static, being touched by a Q-tip or being penetrated deeply by a needle	Judge sensory and affective qualities of pain in observed hand	Somatosensory cortex Suppression of N1/P1 LEP component

Abbreviations Brain regions: *SI/SII* primary and secondary somatosensory cortices; *SMG* supramarginal gyrus containing SI; *IFG* inferior frontal gyrus; *SFG* superior frontal gyrus; *IPL/PPC* inferior parietal lobule and posterior parietal cortex; *SPL* superior parietal lobule; *SMA* supplementary motor area; *aINS* anterior insula; *ACC* anterior cingulate cortex; *MCC* midcingulate cortex; *CMZ* caudal motor zone; *PCC* posterior cingulate cortex; *PFC/mPFC* prefrontal cortex and medial PFC; *STS* superior temporal sulcus; *TPJ* temporoparietal junction; *EBA* extrastriate body area. Methods: *EEG* electroencephalography; *MEG* magnetoencephalography; *SEP* somatosensory-evoked potentials; *LEP* laser-evoked potentials; *TMS* transcranial magnetic stimulation with sensory-focused stimuli reveal activity in both sensory and affective brain areas, suggesting that affective processing of observed pain is more readily activated than sensory processing [105, 110, 111].

4 Shared Neural Representations

As proposed by the PAM, shared neural representations of self- and vicarious pain arise from neural mirroring and facilitate an accelerated understanding of the sensory, affective and cognitive experience of the individual in pain [155, 156]. While motor and somatosensory neural activity may underpin direct pain mirroring, affective–cognitive activity reflects indirect mirroring evoked through the similar processing requirements of self- and other-pain features. Due to the current lack of evidence for cognitive mirroring, the PFC is not included in this section and its distinct contributions to vicarious pain processing will be discussed at a later point.

4.1 The Sensorimotor Regions

4.1.1 The Mirror Neuron System: Inferior Parietal Lobule and Inferior Frontal Gyrus

During vicarious pain, the IPL and IFG have been respectively implicated in sensory [131, 158] and affective mirroring [171]. Identical neurons in the IPL and IFG respond to self-performed and observed actions, thus reflecting motor mirroring of observed pain [26, 39, 45, 163, 171, 193]. Supporting this, both regions react to physical rather than abstract pain cues, indicating that they require visual perception of relevant motor information [110, 179]. Vachon-Presseau et al. [193] report greater IPL responsiveness to body parts in pain and consistent bilateral IFG activation to both facial and bodily pain cues (Fig. 3).

Furthermore, the IPL is associated with motor movement, spatial processing and increased intensity ratings of observed pain [26, 163]. As such functions require the analysis of sensory pain components, this suggests that the IPL mirrors

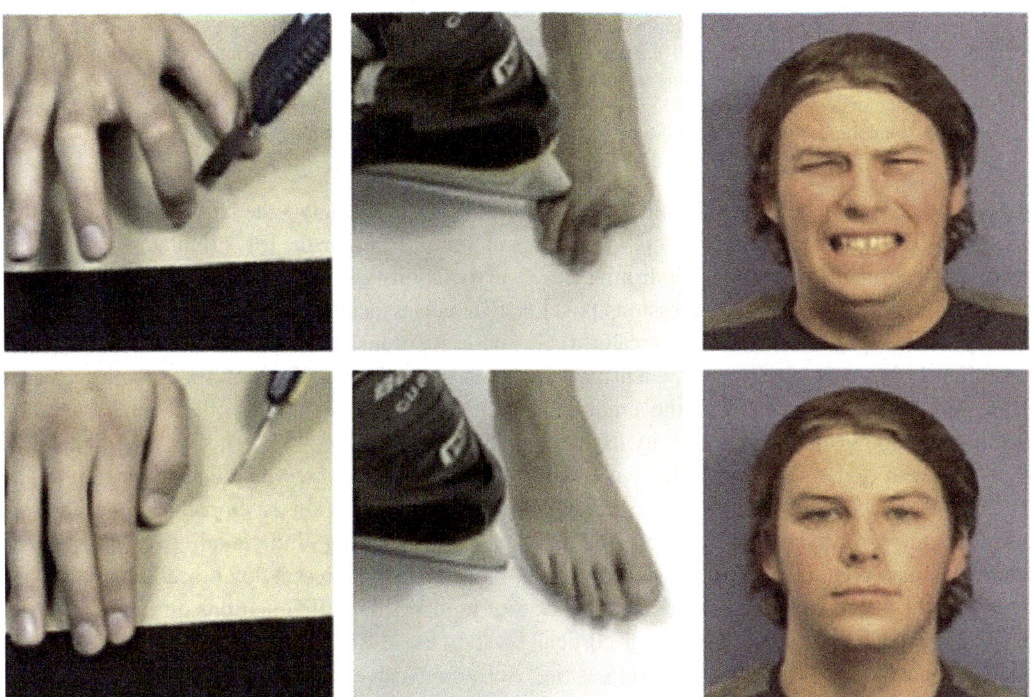

Fig. 2 Typical stimuli in vicarious pain research. facial pain and neutral expressions and hands and feet in non-painful and painful scenarios [193]

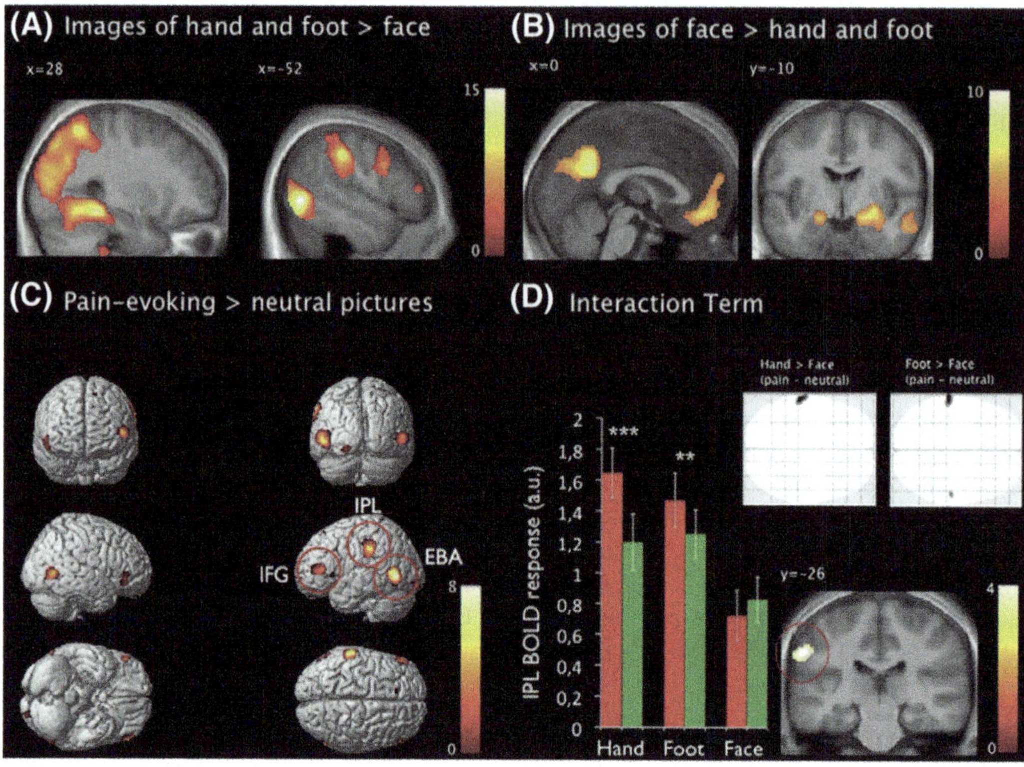

Fig. 3 **a** Observing body limbs is associated with increased activation in sensorimotor regions compared to facial expressions. **b** Observing facial expressions is associated with increased activation in medial PFC (mPFC) and Superior temporal sulcus (STS) (associated with perspective-taking) compared to limbs. **c** Vicarious pain is associated with increased activation in IFG, IPL and EBA compared to neutral images. **d** Increased activation of IPL during vicarious pain observed in limbs compared to vicarious facial pain expressions. For all images $p < 0.001$, uncorrected; error bars represent standard error of mean; $^{**}p < 0.01$; $^{***}p < 0.001$ [193]

sensorimotor aspects of vicarious pain [26, 113, 137, 146, 193]. In contrast, the IFG is involved in the extraction of affective meaning from faces, including anger and happiness expressions [141]. Correspondingly, studies indicated greater IFG activation during pain observation when participants were required to attend to the emotional meaning of pain [26, 196]. Such activity has been associated with higher self-rated affective empathy but shows no correlation with sensory pain intensity ratings, substantiating the role of the IFG in affective pain processing during vicarious pain [22, 163, 192, 193, 196]. Notably, IFG responses to pain observation can modulate higher cortical emotion centers, such as the aINS [31], suggesting that the IFG identifies motor activity in observed pain and communicates associated affective meanings to higher regions for further analysis [158]. Accordingly, the PAM advocates that motor mirroring in the IPL stimulates sensory pain associations while IFG mirroring activates affective pain associations. Thus, both provide a neural base for translating observed facial and bodily pain cues into self-correlates, creating shared representations of self- and other-pain observation [31, 147, 155, 156, 158]. This may facilitate rapid appraisal of the observed pain and corresponding emotional contagion or empathic understanding of the suffering individual [82, 171, 192]. Nonetheless, at present no intracellular recordings of human IPL or IFG neurons during pain observation exist. Reliance on vague spatial resolution of noninvasive neuroimaging techniques make it challenging to confirm that IPL and IFG activity occurs in the same neurons involved in self-pain. Thus, motor mirroring of pain in these regions is derived from previous research that is not specific to vicarious pain. Further pain-related research is needed to verify such a notion.

4.1.2 The Motor Cortices

The premotor cortex and Supplementary Motor Area (SMA) have been associated with action understanding of vicarious pain [100, 134, 163, 186]. In line with the PAM, the premotor cortex has been pinpointed as a neural correlate of motor imitation and is suggested to encode observed actions via motor mirroring (for a meta-analysis on the mirror neuron system in imitation, see [130]). Notably, somatotopical organizations within the premotor cortex facilitate the localization of perceived body parts [25]. Furthermore, the SMA has been implicated in event sequencing for the analysis and understanding of witnessed behaviors [112, 113, 136, 163]. Accordingly, these regions may analyze the motor cues of facial and bodily pain expressions, creating shared neural representations of self- and other-pain and activating relevant associations for pain evaluation [130, 155, 159]. The role of the motor cortex in motor mirroring is substantiated by TMS research. All three available TMS studies recorded an inhibition of motor-evoked potentials in participants watching videos of hands or feet being deeply penetrated by needles. This inhibition was specific to the muscle subjected to noxious stimulation. In contrast, gentle touch of humans or needle penetration of nonhuman objects had no such effect (Fig. 4) [4, 6, 128].

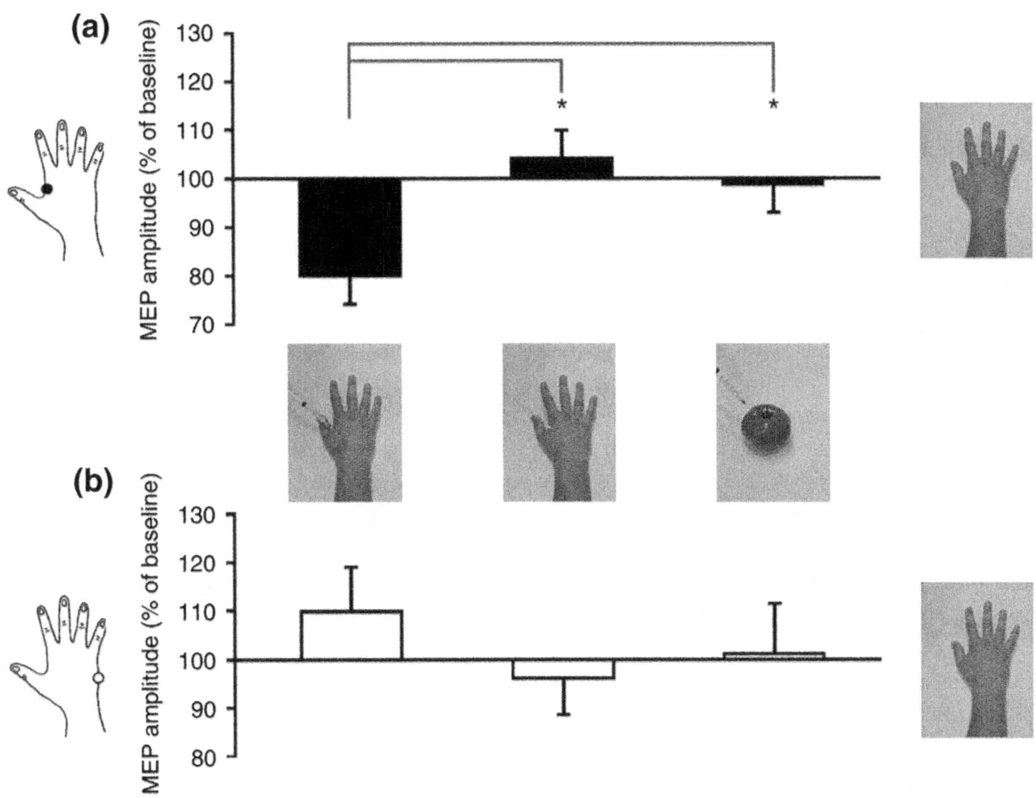

Fig. 4 Suppression of MEP amplitude in response to vicarious pain [4]. Abbreviations: *MEP* motor-evoked potentials; *FDI* first dorsal interosseous; *ADM* abductor digiti minimis. **a** MEPs recorded from FDI muscle that was penetrated by needle or touched by Q-tip. Significant amplitude decrease occurred for specific FDI when penetrated by needle compared to Q-tip ($p = 0.01$) or compared to non-corporeal object ($p = 0.01$). **b** MEPs collected from ADM muscle which was not stimulated. No significant effect was found for ADM muscle, indicating that motor suppression was specific to the muscle targeted by the observed noxious stimulation. (*) identify significant post hoc comparisons ($p < 0.02$) [4]

As equivalent motor inhibition has been noted in response to self-pain [79], these findings have been interpreted as evidence for motor mirroring that reflects a direct resonance of the witnessed pain in the motor system of the observer [5]. Furthermore, the revealed neural inhibition was associated with increased pain intensity ratings, but not with affective measures of pain or empathy measures, implicating it exclusively in sensory processing of vicarious pain [3, 4]. Such findings support the PAM, substantiating the formation of shared neural representations within the motor cortices during vicarious pain.

4.1.3 The Somatosensory Cortices

Although the involvement of the somatosensory cortices is well-established for self-pain, findings are less robust for vicarious pain. The majority of vicarious pain studies do not include somatosensory areas in regions of interest analyses, making it challenging to determine how frequently such activation takes place. Of those that do, some studies find no activation in the somatosensory areas during pain observation [22, 48, 89, 134, 135, 179], while others report neural activity in at least one of the two regions [49, 98, 137, 147, 148]. Contradicting the PAM, both Singer et al. [179] and Morrison et al. [135] reported that the somatosensory cortices were only active when participants received painful electric shocks on their own hand, but not when abstract or visual cues informed them that another individual received equivalent stimulation.

However, both a systematic review [105] and a comprehensive meta-analysis [110] highlight the need to focus on sensory aspects of vicarious pain in order to engage somatosensory processing. Evidence is provided that presenting participants with limbs in pain is more likely to activate SI and SII than using faces or abstract pain cues [110, 113]. Indeed, Bufalari et al. [27] recorded increased SI activity while witnessing body parts in pain which concurred with higher pain intensity but not unpleasantness ratings, confirming that the SI specifically encodes sensory pain components. As demonstrated in Fig. 5, such activity dissociated painful from non-painful vicarious tactile stimulation. Specifically, gentle touch decreased amplitudes in the SI, while painful touch increased them [27]. Nonetheless, SI and SII have also been implicated in the undifferentiated mirroring of sensory cues, independent from whether stimulation is noxious or not, contesting that their activity is specific to pain (for a review, see [28, 110, 165]). As fine-grained MEG analysis can differentiate painful and non-painful self-touching, such methods may aid in clarifying the pain-specificity of vicarious somatosensory responses [153].

As similar activity has been noted during self-pain, the PAM interprets SI and SII activation during vicarious pain as bottom-up sensory mirroring [158]. In line with this, the SII has been implicated in the mirror neuron system due to its connections to the IPL [194]. Furthermore, the SI and SII contain somatosensory maps that may enhance the identification of observed body parts [16, 122]. Likewise, infrequent reports of thalamus activity suggest that this region may contribute to vicarious pain processing in its role in the transmission of mirrored sensory input to the cortex [98]. These findings support the notion of shared neural pain representations which provide a reference point from which individuals interpret the pain they observe in others [16, 44]. Correspondingly, Valeriani et al. [194] found that participants rated their own heat pain higher when simultaneously witnessing other individuals receiving pain stimulation. It was suggested that pain observers map the viewed pain onto the own body, intensifying self-experienced pain. Likewise, Osborn and Derbyshire [148] found that somatosensory activity during vicarious pain was associated with subjective reports of feeling pain sensations within the own body, substantiating the concept of shared sensory pain experiences. Notably, this neural activity was absent in participants who did not report similar sensations. Activation in affective pain areas, such as aINS, was similar across groups, suggesting that sensory and affective pain mirroring are dissociable [148].

Research has yet to reveal which factors mediate the group differences in somatosensory responsiveness to observed pain. Such factors may be of significant clinical relevance as they may underpin sources of empathic deficits in clinical disorders as well as dysfunctional pain behaviors. However, at present such research is in

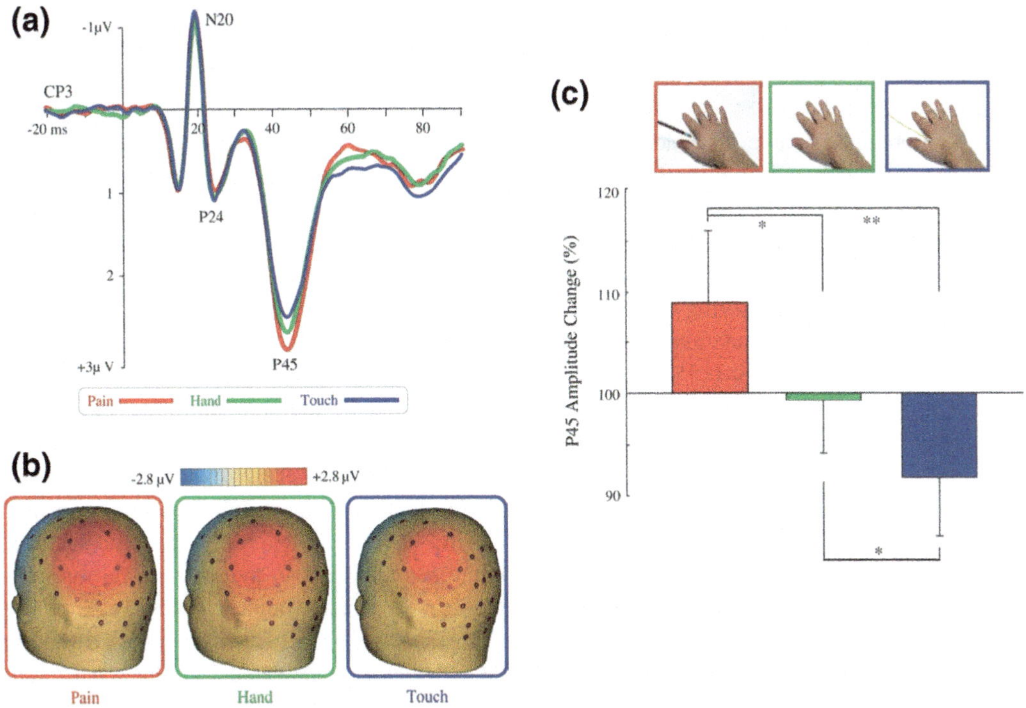

Fig. 5 Change in P45 Amplitude during vicarious pain compared to baseline according to condition. **a** P45 amplitude during vicarious pain (*red*), vicarious gentle touch (*blue*) and static hand (*green*). Significant increase of P45 amplitude during vicarious pain compared to hand or touch ($p = 0.0001$). **b** Topographic distribution of P45 during each condition. **c** Percentage change in P45 amplitude compared to baseline according to condition. $^{*}p < 0.05$; $^{**}p < 0.001$ [27]

its infancy [7, 8]. Moreover, although evidence for sensory mirroring in the somatosensory cortices, and tentatively the thalamus, during vicarious pain is provided, this neural activity is neither confirmed to unambiguously overlap with self-pain activations nor to be pain-specific. Likewise, the functions subserved by SI and SII have not been differentiated, although these regions are dissimilarly implicated in self-pain processing [184]. Thus, more research is needed to establish the manner in which shared sensory representations arise from such activation during observed pain.

4.2 The Affective Regions

4.2.1 The Insula

INS activation has been proposed to subserve interoception and affective vicarious pain processing [22, 51, 72, 98, 166, 175, 179, 192, 203]. Interoception is defined as a process by which several sources of information are integrated to form an internal representation of the current bodily state and corresponding emotional responses [50]. This is pertinent for self-pain as it allows individuals to assess their physical state. Given the comparable affective processing requirements of self- and other-pain, it is likely that the INS may play a similar role in pain observation [89, 113]. Indeed, Gu et al. [87] highlight that the INS is essential for empathy induction, having found INS lesions to inhibit empathic responding. In line with the PAM, there is evidence for an overlap in increased aINS activation during self- and vicarious pain when comparing brain activity within the same individuals [48, 146]. Likewise, between-subject paradigms have revealed consistent aINS activity during pain observation within established INS correlates of the pain matrix [88, 89, 98, 99, 112,

113, 186]. These findings provide support for indirect affective pain mirroring from which shared neural pain representations arise [22, 146, 179, 205].

Furthermore, comparable aINS activation has been demonstrated during both direct evaluative pain judgments as well as identification of limb laterality for images of limbs in pain, substantiating the notion of automaticity in such mirroring [89]. Such automatic mirroring is likely to achieve an immediate representation of the current state of the individual in pain [155, 156]. Moreover, Cheng et al. [36] instructed participants to imagine observed pain from the perspective of either a stranger or a loved one. The latter condition elicited increased aINS activation as well as greater neural overlap for self- and other pain, indicating that increased emotional attachment evokes greater affective mirroring of the observed pain within one's own system. Notably, aINS activity distinguishes between painful and non-painful vicarious stimulation [112] and responds to general negative encounters, such as self-experienced and vicarious disgust [93, 199]. Accordingly, shared neural representations in the INS may not be pain-specific, but instead encode adversity [13, 48]. aINS involvement in the anticipation of aversive stimuli and its connections to well-established emotion centers, such as the amygdala, substantiate its likely contribution to the affective processing of vicarious pain [110, 164, 166, 174]. Given the findings, it is plausible that affective mirroring gives rise to shared affective pain representations during vicarious pain, in particular when high levels of emotions are involved. However, further fine-grained analysis is required to assess to what extent neural patterns during self- and other-pain are identical.

4.2.2 The Cingulate Cortex

During self- and vicarious pain, the ACC has been has been associated with affective pain processing, [70, 76] while the midcingulate cortex (MCC) is associated with both affective and sensory pain components [43]; for review on CC see, [83, 110, 198]; for systematic review on CC in vicarious pain, see [202]. The PAM suggests that these similar roles give rise to mirroring and shared neural pain representations. However, while intracellular recordings have confirmed nociceptive neurons in the CC for self-pain, such an investigation has not yet been performed for vicarious pain [92]. In line with the role of the ACC in affective processing, neural activity in the subgenual and rostral ACC (sgACC; rACC) as well as aMCC has been found via pain observation in limbs, facial pain expressions and abstract pain cues [22, 26, 134, 163]. Supporting the PAM, the activations are reported to be overlapping and partially anterior to those commonly found during self-pain [26, 134]. Notably, Singer et al. [179] revealed ACC activation when participants received abstract information about their significant other receiving painful electric shocks, showing that emotional attachment was sufficient to induce a neural representation of pain unpleasantness. A direct comparison of neural activity confirmed that the ACC regions activated during self- and other-pain overlapped. Furthermore, similar to self-pain, greater rACC activity during other-pain was associated with increased other-pain evaluations [179]. However, to date, no correlations between affective pain measurements and ACC activity have been analyzed for facial and abstract pain cues. Thus, the extent to which the ACC subserves affective processing of vicarious pain is speculative. In contrast, studies presenting limbs in pain have investigated such correlations and confirmed that higher neural responses are indeed associated with higher unpleasantness ratings [113]. Furthermore, Jackson et al. [100] reported that ACC activity is specific to imagining the observed body part from an emotional first-person perspective and correlates with ratings of observed pain levels (Fig. 6). These findings support a role of the ACC in affective processing for vicarious pain which is comparable to affective self-pain processing and thus may contribute to shared affective pain representations [22, 48, 193]. Notably, the revealed ACC activity may not be pain-specific, but instead reflect the mirroring of a general negative affect [13].

In contrast, vicarious pain research has failed to find aMCC activity in abstract or facial cues of pain, suggesting that it may not subserve

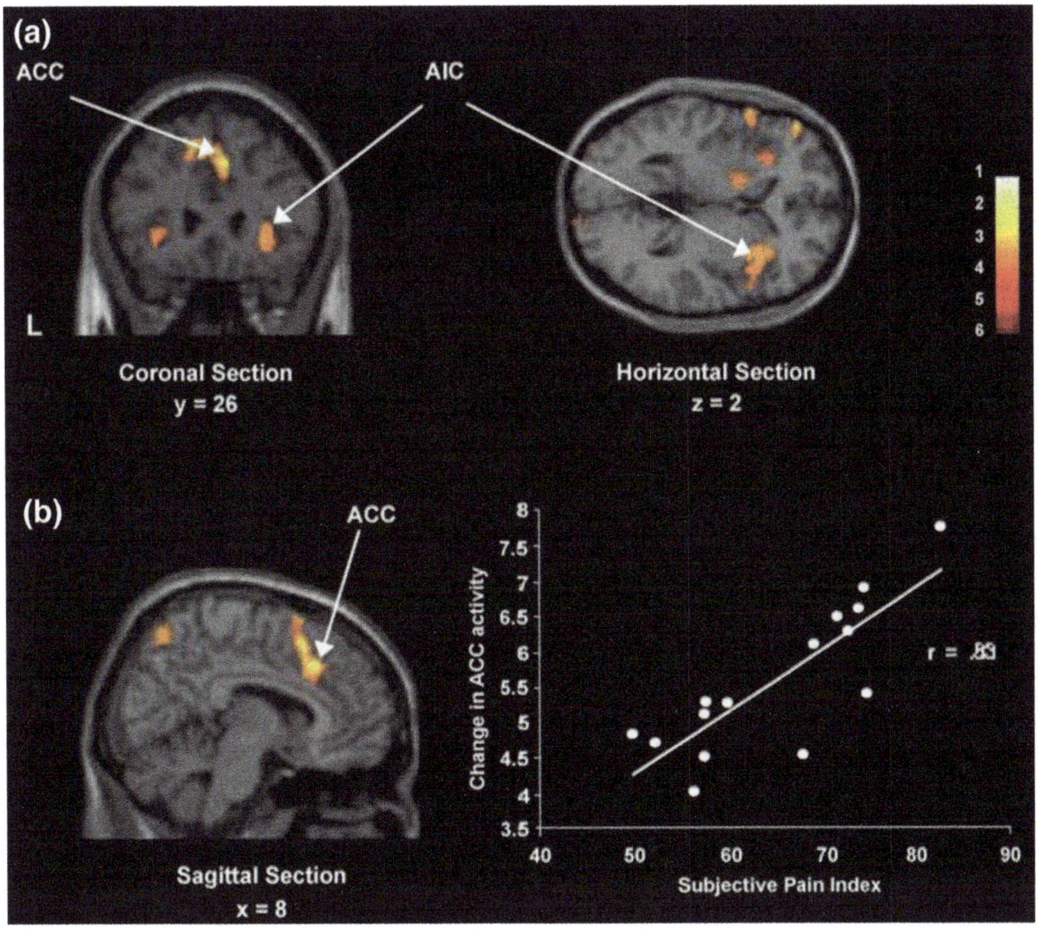

Fig. 6 ACC and aINS activation during vicarious pain. Abbreviations: *AIC* anterior insula. **a** ACC and aINS activation while observing another individual in pain. **b** Increased ACC activity was associated with increased subjective pain ratings (MNI Coordinates) [99]

affective pain representations as it does for self-pain. Instead, the aMCC has been found explicitly responsive to images of limbs in pain, regardless of whether individuals focused on the sensory or affective aspects of the presented body part [113] or imagined the pain from a first- or third-person perspective [98]. In line with its responsiveness to body parts, research proposes that the aMCC may reflect sensory rather than affective mirroring. Furthermore, it may underpin distinct roles in pain observation [147]. Indeed, lesion studies indicate that an intact aMCC is not required for affective empathy [87]. Taken together, findings for CC activation suggest that neural patterns may overlap for self- and vicarious pain, reflecting neural mirroring and shared pain representations [48, 110, 134, 170, 195]. Nonetheless, the exact functions of the CC in mirroring and distinct contributions to vicarious pain processing need further clarification.

5 Distinct Neural Contributions

Several brain regions are implicated in making distinct contributions to the cognitive processing of vicarious pain. These include self-other distinction and attentional control, which are associated with IPL, SI, SII, INS, CC and PFC, as well as contextual pain appraisal and top-down regulation, which draw on the PFC. Notably, although it is proposed that in their cognitive roles these regions reflect processing that is independent from shared neural representations, the revealed activation has not been firmly excluded from underpinning mirror processing.

5.1 Self-other Distinction

During vicarious pain, the IPL, temporoparietal junction (TPJ), somatosensory regions, aINS and PFC have been implicated in distinguishing one's own sensory and affective experiences from those of the individual in pain. Uddin et al. [188] demonstrated that TMS stimulation to the right IPL disrupted the ability to discriminate self- and other-faces, which suggests that this brain region plays a key role in maintaining a distinct sense of self. Likewise, the somatosensory cortices may contribute to such a self-other distinction. Jackson et al. [98] reported somatosensory activation exclusive to imagining noxious stimulation from a first-person but not a third-person perspective. The lack of activation in the latter condition may be a mechanism by which SI and SII separate sensory experiences observed in others from their own sensory state while a first-person perspective may contribute to shared sensory experiences [98]. Moreover, both the INS and PFC have been associated with interoceptive self-awareness during vicarious pain that establishes an understanding of the self-state as a reference point against which to compare external pain cues [169]. Brooks et al. [23] have provided evidence of a somatotopic organization in the aINS that facilitates such interoception [74, 100]. Lending tentative support, Ebisch et al. [73] showed that for vicarious touch the pINS is involved in differentiating self- and other-states; however, studies have not yet explored this specifically for vicarious pain. Furthermore, when pain observation elicits discomfort in the observer, however not in the observed individual, the PFC has been proposed to promote self-other distinction for adequate context assessment [111, 149]. Likewise, the TPJ, which has been associated with perspective-taking, in particular, is responsive in such circumstances. Cheng et al. [37] report that the TPJ subserves self-other distinction and aids the understanding of individuals when the observer does not rely on neural pain mirroring. Substantiating this, imaging the observed pain from the perspective of a stranger has been associated with decreased activation of the pain matrix and increased TPJ involvement. This indicates that greater perspective-taking is applied when individuals have less emotional attachment to the individual suffering pain (Figs. 7 and 8). Such findings are reflected in negative functional connectivity between TPJ and aINS and positive connectivity between TPJ and superior frontal gyrus during pain observation from a stranger perspective. These suggest that the aINS is less involved in the encoding of stranger pain than of loved-one pain [36, 37].

Self-other distinctions are an essential component of empathy [75, 98]. Although observing another individual in pain can have similar internal effects as self-pain for the observer, understanding that his or her pain is distinct from the self is crucial for adequate empathic responses that promote social survival. In contrast, a strong transference of other-pain to the self may trigger emotional contagion, thus evoking self-preserving withdrawal responses. Forthcoming systematic exploration may aim to further pinpoint the neural substrates of self-other distinction in the context of vicarious pain and empathy. Deficits in these correlates may be of significance for dysfunctional empathic responding [116, 197].

5.2 Salience Detection and Attention

The INS and aMCC have been associated with salience detection and attention in self- and vicarious pain processing [53, 104, 138]. Contrasting shared neural representations of the PAM, mINS and pINS activity has been reported more frequently for self- than other-pain [23, 80, 96]. Upon comparing brain activity in the same individuals, Ochsner et al. [146] found that the mINS only showed activation during application of heat pain to one's own skin, but not when the same stimulus was observed on another

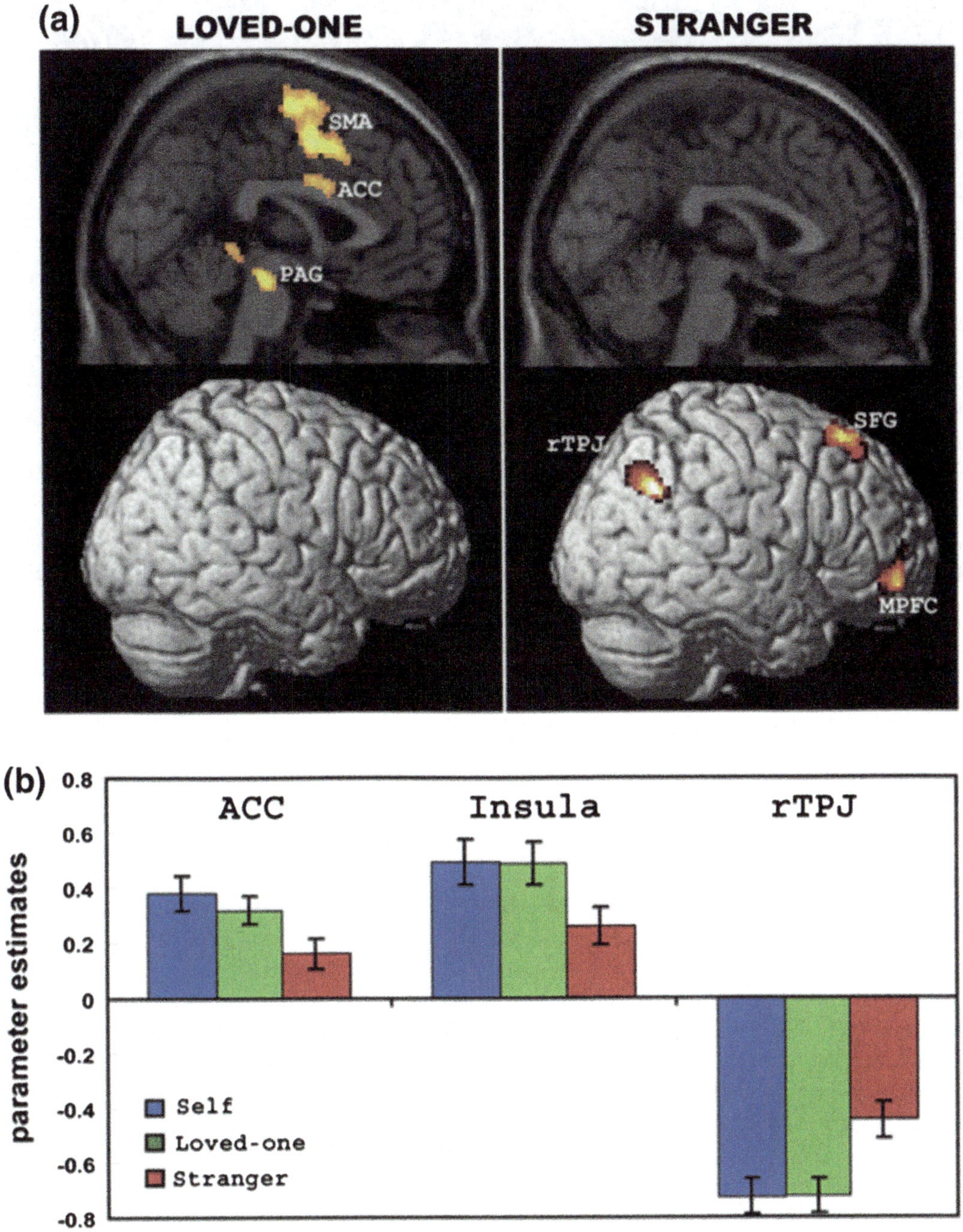

Fig. 7 Double dissociation between pain matrix and TPJ responses during vicarious pain imagined from different perspectives. Imaging pain from the perspective of a loved-on was associated with increased ACC, INS, SMA and PAG activity during pain observation, resembling self-perspective responses. Stranger perspective was associated with increased TPJ activity compared to the other two perspectives [36]

individual. It was highlighted that the aINS has connections to affective processing areas while the mINS is linked to cognitive processing regions and is implicated in attention. As self-pain may require increased attentional resources due to greater stimulus salience

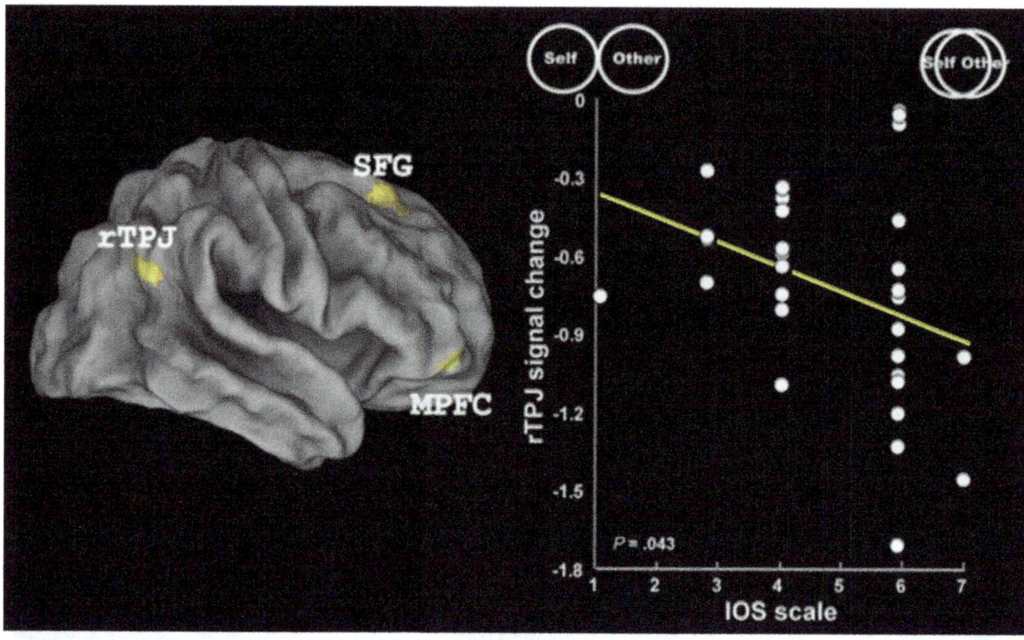

Fig. 8 Negative association between TPJ and closeness in relationships. Abbreviations: *IOS Scale* Inclusion of others in self Scale; *rTPJ* right TPJ; *SFG* superior frontal gyrus. rTPJ, SFG and mPFC showed increased activation when imagining observed pain from a stranger perspective compared to self- or loved-one perspective. Increased self-reported overlap of self and others in close relationships was associated with decreased rTPJ activity during pain observation [36]

compared to observed pain, it may specifically draw on the mINS for attention regulation [53, 104, 125, 146]. Nevertheless, fine-grained Multi-Pattern Variate Analysis (MPVA) has revealed overlapping mINS patterns for self- and vicarious pain, indicating that further research is required to clarify the role of the INS in salience detection and attention during pain observation [48]. Moreover, aMCC activation in established attention regions has been consistently reported for vicarious pain [26, 98, 99, 135–137, 147, 163, 205]. Similar to self-pain, observed pain signals threat and thus draws attention to the need for prompt responses [61, 75]. In line with the PAM, it is possible that comparable processing requirements of self- and other-pain stimuli may induce cognitive mirroring. However, it is equally likely that the aMCC directly process salient environmental hazards and regulate attention, independent from such mirroring [155, 156]. At present only three neuroimaging studies have explored attention in vicarious pain and thus no comprehensive conclusions can be drawn [78, 88, 89]. Fan and Han [78] demonstrated that participants identified and empathized with observed pain faster when pain was specifically attended to. In contrast, neural activity associated with vicarious pain processing has been found no longer significant when individuals focus away from pain cues while counting the number of presented limbs in pain. Accordingly, top-down attentional control may modulate vicarious pain processing and accordingly limit its automaticity as proposed by the PAM [88]. Moreover, when cognitive load was held constant and required similar attention levels during neutral and vicarious pain tasks, aMCC responses were equivalent for both painful and neutral images. This suggests that aMCC activation is not pain-specific but instead underpins attentional control [89]. Taken together, current research advocates that the aMCC, and tentatively the INS, facilitate attention during pain observation [9, 88, 89] and that vicarious pain processing, and thus potentially empathy, is vulnerable to resource competition from cognitive tasks [78, 88, 132]. Both premises have been established for self-pain. However, whether such processing reflects shared or distinct neural attention correlates for self- and other-pain needs to be disentangled in further research in order to gain a full understanding of the effects of salience and attention during vicarious pain [9, 59].

5.3 Context Appraisal and Top-Down Regulation

Vicarious pain studies speculate that the PFC subserves the assessment and cognitive integration of multiple inputs for adequate pain appraisal [77, 117, 185, 195]. Furthermore, it may play a role in top-down regulation of responses to the observed pain [76, 85, 117, 185, 201]; for review on PFC, see [127]. While these functions resemble those during self-pain, it has not yet been established whether PFC activity reflects shared or distinct neural representations during other-pain [26, 112, 145, 193]. Although it differentiates between painful and non-painful observations [26, 149], the majority of vicarious pain research fails to find PFC involvement. Such inconsistencies may be an effect of task differences across studies, which require varying levels of cognitive resources. For example, Lamm et al. [112] presented participants with images of individuals who had their hands penetrated by needles or touched by Q-tips and informed them that these individuals showed either normal or abnormal responses corresponding to a neurological condition. PFC activity was only found for abnormal conditions, such as when participants were told that needles elicited no pain and Q-tips elicited pain. Likewise, Perry et al. [149] reported greater EEG suppression in the frontal areas both during the observation of painful needles as well as when individuals responded with abnormal pain to non-painful Q-tip stimulations. Both studies proposed that PFC processing reflects the detection and integration of conflicts, such as pain, and thus promotes accurate appraisal of the observed context. This is particularly required when witnessed responses fail to correspond to behaviors associated with noxious stimulation [112, 149]. In contrast, passive processing of straightforward pain reactions may occur at an automatic level without significant PFC activation [112]. Corresponding to PFC involvement in more complex encoding of pain observations, the medial PFC has been specifically associated with the decoding of facial pain expressions [26, 127, 193]. As facial expressions convey less noxious information than limbs in pain, greater cognitive evaluation is required for pain assessments. Accordingly, the PFC is proposed to subserve the integration of affective pain information with stored associations about social consequences of the observed pain [26, 156, 169]. In line with this, the PFC has been implicated in the analysis of the internal states and predicted intentions of other individuals [97, 102, 169]. More specifically, the mPFC has been associated with human perspective-taking when imagining observed pain from the first- or third-person perspective, but not as an artificial object, as shown in Fig. 9 [98]. Therefore, the PFC may have the role to complement neural pain mirroring with contextual information to enable accurate response formation. Notably, the precuneus, superior temporal sulcus (STS) and cerebellum, which are similarly associated with the ability to accurately attribute internal states to others, have been infrequently shown to respond during vicarious pain (Fig. 3) [147, 161, 193]. Although this tentatively suggests that these regions also contribute to active understanding of the suffering individual during pain observation, more consistent findings are required to evaluate their role in vicarious pain. Furthermore, higher functional connectivity has been registered between dorsomedial PFC and IFG during the complex vicarious pain conditions. Given the role of the IFG in perspective-taking and the retrieval of pain-related memories [103], it is plausible that the two regions work conjunctly to infer an understanding of the current state of the individual in pain [112].

Moreover, functional connectivity analyses have revealed increased neural connectivity for the PFC with the somatosensory cortices, CC and INS, which may reflect top-down regulation of

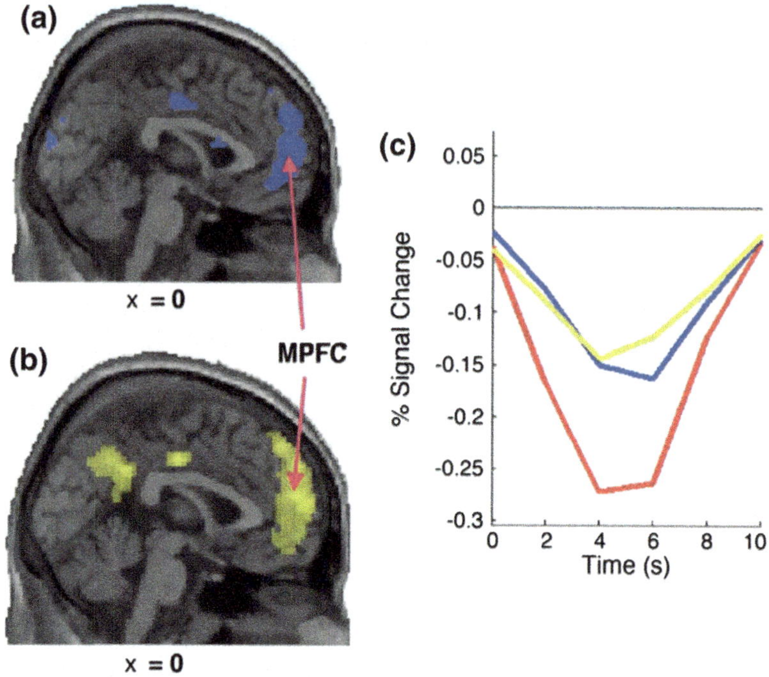

Fig. 9 mPFC activation associated with perspective-taking during vicarious pain. **a** mPFC activation when imaging presented pain from self-perspective compared to object perspective. **b** mPFC activation when imaging presented pain from other-perspective compared to object perspective. **c** mPFC responses in the three different conditions: self-pain as *blue*, other-pain as *yellow* and artificial pain as *red* [98]

responses to vicarious pain [88, 205]. More specifically, Cheng et al. [37] revealed that during pain observation, participants who are accustomed to seeing pain show greater PFC responses and decreased activity in SI, SII, INS and CC regions than controls (Fig. 10). Comparable to self-pain literature, increased activation in the PFC was associated with decreased emotional reactivity and lower pain intensity ratings [9, 24, 29, 33, 75, 154, 195]. These results indicate that the PFC exerts downregulatory control over regions involved in the processing of sensory and affective vicarious pain components.

Indeed, studies show that when cognitive analysis of the context implies that individuals in pain are to blame for the pain they are enduring [66] or are unfair to others [180], INS and CC activity is decreased and pain observers report less empathy. Such top-down regulation may occur at early pain processing stages [68]. Crucially, the revealed functional connectivity patterns between cognitive and sensory-affective regions were specific to vicarious pain and virtually absent during self-pain in the same individuals. While the role of such connectivities has not been confirmed, this substantiates that although brain regions are shared with self-pain processing, distinct neural communication may make unique contributions to vicarious pain [76, 110, 185, 205]. Furthermore, such top-down regulation may be a source of dysfunctional pain and empathy expressions. Ochsner et al. [146] advanced that the PFC encodes observed knowledge related to pain and pain-appropriate responses. In particular, increased activity in the rostrolateral PFC during pain observation has been associated with higher trait anxiety scores (Fig. 11). Such increased neural response is proposed to reflect increased encoding and learning of environmental threat cues in anxious populations, which may perpetuate dysfunctional self-pain anxiety [32, 146, 152]. This lends itself to future exploration of PFC substrates in maladjusted pain and empathy behaviors. Taken together, findings for PFC activation reinforce the role of the PFC in contextual analysis and top-down regulation during vicarious pain [2, 169]. However, as different PFC regions may contribute distinct processes, research specific to pain observation may reveal in which manner the subsections of the PFC are involved [76, 117, 201]. Likewise, it has not yet been established

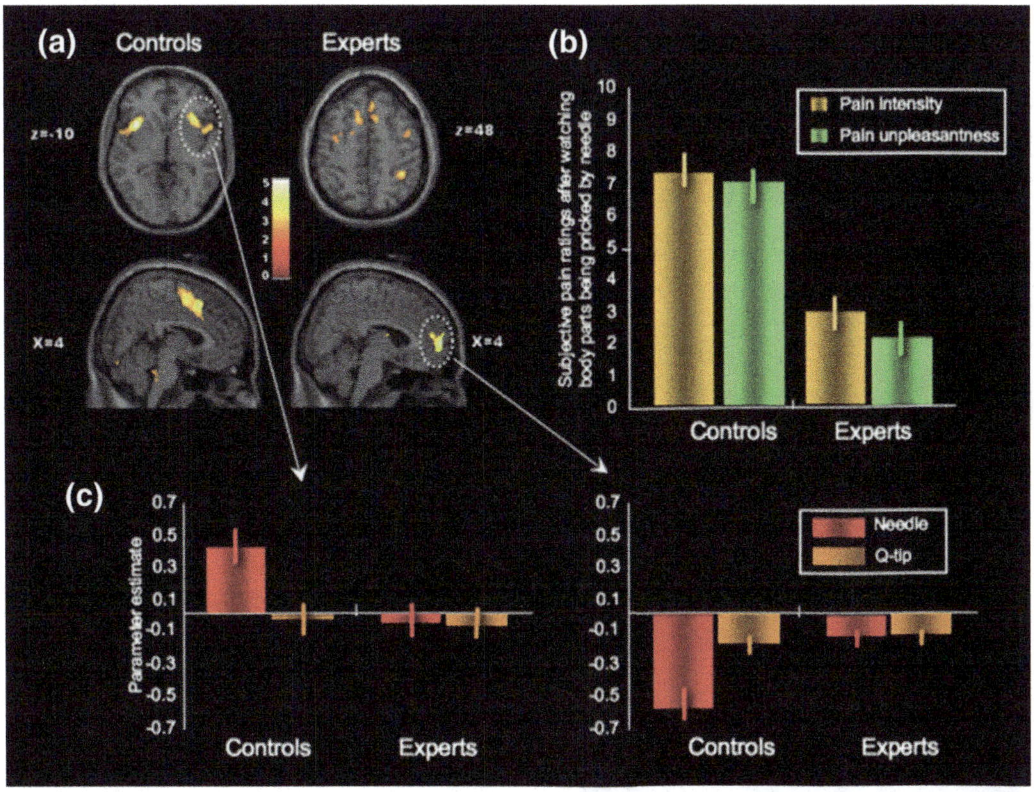

Fig. 10 Neural activity for pain experts and controls during vicarious pain through needles. **a** Non-experienced controls showed INS, PAG, ACC and SMA activation while experts showed IPL and mPFC activation. **b** Controls showed higher pain intensity and unpleasantness ratings compared to experts. **c** Compared to experts, controls showed higher aINS activation only for needles, but not for Q-Tips. Compared to controls, experts showed higher mPFC activation during needles, but not for Q-Tips [37]

which neural patterns overlap with those during self-pain, as suggested by the PAM, and which activity is independent from neural mirroring [155]; for a meta-analysis on mPFC contribution to empathy, see [169].

6 Other-Oriented Empathy and Self-oriented Withdrawal

The PAM proposes that shared neural representations of self- and other-pain evoke other-oriented empathic understanding that initiates altruistic behavior. Such behavior enhances the protective benefits of social cooperation and thus contributes to social survival. Nevertheless, the inherent threat cues of observed pain may instead induce emotional contagion and trigger self-oriented withdrawal responses. These facilitate physical survival. Nonetheless, research has not only failed to tease apart the neural patterns of these two distinct responses, but the factors mediating which response is chosen also remain unclear.

6.1 Motor Empathy and Motor Preparation

In line with the PAM, motor mirroring during vicarious pain may promote empathic response [155, 156]. However, meta-analytic evidence proposes that, similar to self-pain, the IPL, IFG, motor cortices and aMCC motor zones are actively involved in self-oriented pain predictions as well as subsequent motor preparation and

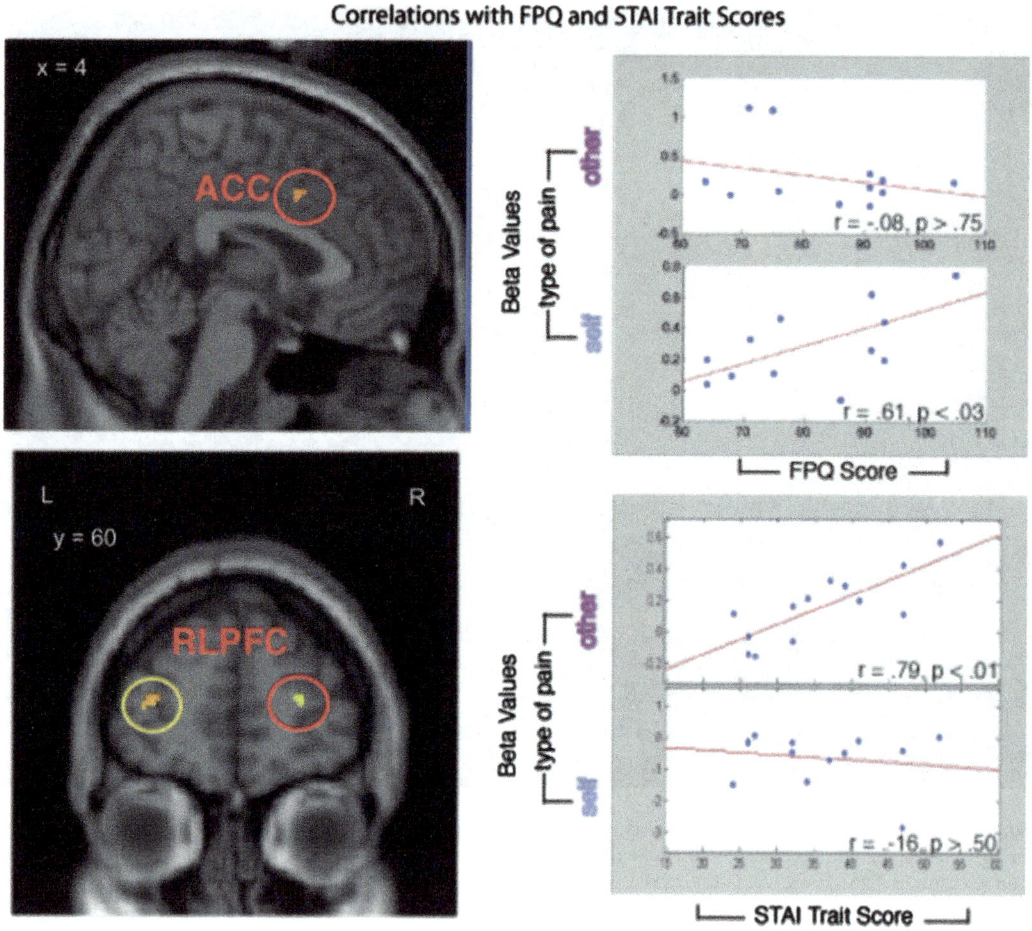

Fig. 11 Correlations between vicarious pain activity and anxiety scores. Abbreviations: *FPQ* fear of pain questionnaire; *STAI* state-trait anxiety inventory; *rlPFC* rostrolateral PFC. Higher rlPFC activity correlated with higher STAI scores during vicarious pain. No correlation with FPQ during other-pain. Higher ACC activity correlated with higher FPQ scores during self-pain. No effect for STAI during self-pain [146]

coordination [40, 46, 47, 64, 77, 139, 173, 182]. Research advances that as IPL motor mirroring typically occurs to active movement, motionless pain presentations require the IPL to derive implied movement from the observed scene [34, 46]. For example when images of hands penetrated by needles are shown, the IPL may predict the anticipated removal of the hand rather than mirroring muscle cues [101]. Likewise, in line with its involvement in serial motor prediction, the SMA may predict the motor consequences of observed pain [113, 163]. Csibra [54] proposes that such predictions result from top-down analysis and are dynamically compared against the concurrent pain context within the sensorimotor regions in order to prepare motor responses such as withdrawal. This explains findings indicating that goal-directedness is necessary to elicit premotor activity during vicarious pain as preparatory actions are goal-oriented [99]. Thus contrasting the PAM, instead of constituting the first step in stimulus analysis through bottom-up mirroring, sensorimotor activation during vicarious pain may reflect top-down predictions subsequent to higher cortical processing of the observed pain [54, 111, 155]. Accordingly, sensorimotor activity during vicarious pain may contribute to prediction and preparation of withdrawal from presented threats instead of facilitating empathy [111].

In favor of such a notion, the IPL and aMCC have been implicated in the priming of motor reactions. For example, Morrison et al. [136] displayed animations and required participants to indicate via button press whether items struck or missed hands. When noxious items struck hands, participants showed increased reaction times and increased aMCC activity compared to non-noxious items and noxious miss conditions. It was concluded that voluntary movements were facilitated when corresponding to withdrawal movements that have been triggered through vicarious pain perception (Fig. 12) [21, 136, 193]. Such interference has been reported in particular for body parts compared to faces, indicating that greater sensorimotor relevance of presented pain images elicits greater motor prediction [191]. Notably, this contrasts studies that have found the motor regions of the aMCC are only responsive to self-pain [179].

Furthermore, increased IFG activation has been reported in chronic pain patients who are more expressive about their own pain and also show increased vicarious pain responses.

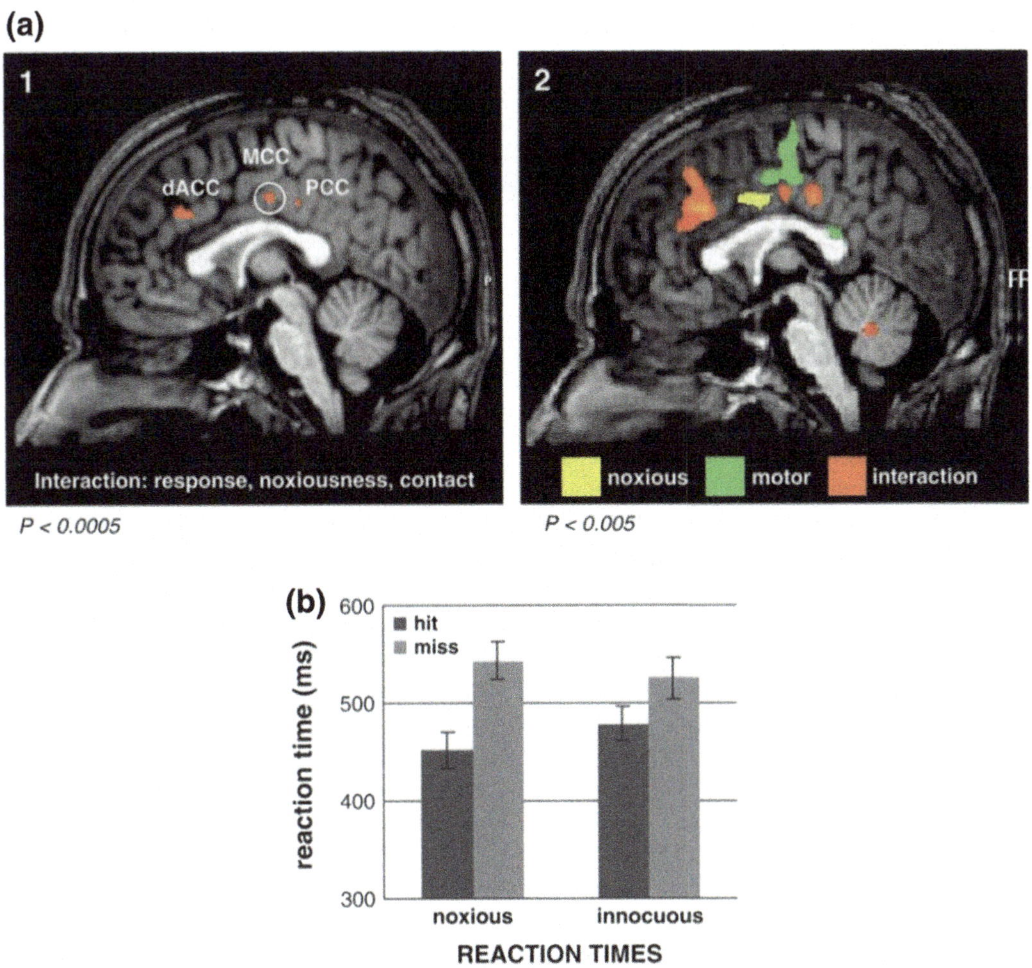

Fig. 12 Interaction between noxious items, hit or hiss conditions and motor response. **a** *1*. MCC and PCC areas of interaction associated with increased reaction times during noxious-hit conditions. *2*. Main effect of noxiousness (*yellow*), main effect of motor response (*green*), and interaction between noxiousness, motor response and hit (*red*). **b**. Decreased reaction times in milliseconds during noxious-hit conditions [136]

Although this neural activity may underpin greater receptiveness to the suffering of others mediated through self-pain sensitivity [192], alternative views propose that the overlap represents increased withdrawal urges in response to the comparable threats of both self- and other-pain. In line with this, IFG activity has been associated with self-oriented distress reactions to pain observation [111]. Such findings advocate that the sensorimotor regions do not facilitate empathy induction, but instead promote self-oriented avoidance responses [62, 64]. Substantiating this, IFG, IPL and motor activity have been seen in response to aversive stimuli, such as fear-inducing weapon images, which do not present motor cues yet trigger motor withdrawal [60]. Furthermore, imagined motor movement activates the premotor cortex and SMA in absence of motor cues, indicating that this system has functions beyond motor mirroring [118]. In line with this, SMA activity during vicarious pain has been associated with self-directed emotional distress, which may reflect greater motor preparation in response to resulting self-oriented withdrawal urges [108]. Nonetheless, Preston [155] counters that motor imagination requires an internal representation of the anticipated movement and thus utilizes similar mirroring structures as direct observation. Moreover, Avenanti et al. [5] highlight that not all motor inhibition in response to vicarious pain subserves self-oriented withdrawal responses [114, 190]. Research has shown that motor retraction reflexes to self-pain result in nonspecific suppression of all muscles in the concerned limb [79, 126]. In contrast, vicarious pain motor inhibition is specific to the exact muscle underlying the observed stimulation, corroborating its direct mirroring function. Such motor inhibition was associated with increased self-rated empathy and, in contrast, negatively correlated with personal distress. This indicates that motor inhibition reflecting bottom-up mirroring takes place during pain observation and may provide a platform for empathic understanding [5, 6]. Nonetheless, it does not exclude the possibility of concurrent top-down motor predictions within IPL, IFG and motor cortices, which facilitate self-oriented withdrawal.

Notably, sensorimotor bottom-up and top-down processing may work in conjunction. Functional connectivity analyses have revealed that sensory pain ratings were associated with increased neural synchronization between the thalamus, SI and the motor cortices during pain observation [15, 143]. Moreover, Carr et al. [31] report that the IFG receives somatosensory information from the IPL upon which it encodes the goal of the observed act. Research advances that such circuits may present reciprocal feedback loops by which sensory and motor cues mutually contribute to context-appropriate pain responses. In line with this, increased sensory signaling in self-pain has been attenuated by motor cortical stimulation resulting in a decreased pain report [19, 20, 115, 150, 187]. Likewise, similar to self-pain, increased connectivity between the aMCC motor zones and motor cortical regions during vicarious pain implicates aMCC motor zones in preparing withdrawal responses through projections to motor cortices [93, 138]. The aMCC regions for pain and motor processing have been found to lie adjacent and interact, potentially providing a direct feedback loop for motor and pain analysis [136]. Therefore, recorded sensorimotor activation during vicarious pain may reflect both initial motor mirroring and subsequent preparatory responses based on sensory and motor feedback [41, 105]. However, no confident conclusions can be drawn without further systematic investigations that pinpoint the neural pathways of these distinct functions [17, 99, 110, 111, 118, 163]. While initial motor activation may contribute to action understanding via direct mirroring of perceived motor cues within the motor regions [158, 160], feedback from higher cortical structures are likely to contribute to subsequent top-down action predictions and motor preparation [54, 119, 200].

6.2 Sensory Empathy and Sensory Preparation

Contrasting the sensory mirroring of the PAM, Morrison et al. [137] suggest that SI and SII are involved in integrating sensory information with action during pain observation in order to predict the sensory consequences of observed pain. Thus, rather than inducing empathy, SI and SII activity may be a product of higher cortical output that evokes self-oriented avoidance behaviors [54]. Such notion is supported by self-pain research that shows high SI susceptibility to top-down regulation; however, it has not been explored for vicarious pain [30, 184]. Notably, SI and SII activations have been linked to the bias of falsely reporting self-experienced tactile stimulation subsequent to observing pain in others [137]. Likewise, Valeriani et al. [194] found that participants that were concurrently subjected to self-pain while observing pain in others demonstrated increased self-pain ratings. The PAM suggests that such hypersensitivity to tactile threat is triggered when individuals evaluate noxious stimuli via their own sensory neurons. However, alternative views suggest that it results from top-down somatosensory threat predictions and facilitates both increased vigilance and withdrawal from external hazards [14, 74, 111, 137, 155]. Interestingly, somatosensory activity during pain observation also increased when limbs in pain were presented in the context of happy and pain faces compared to neutral faces. Such affective faces may increase arousal, which impacts top-down pain analysis and can lead to increased projections of threatening pain predictions to SI and SII [90]. Crucially, increased tactile hypersensitivity during vicarious pain is associated with withdrawal responses rather than empathic, altruistic behaviors [63]. Nonetheless, somatosensory activity in response to observing limbs in painful compared to non-painful scenarios has been associated with increased self-rated empathic abilities [38]. Accordingly, while SI and SII may underpin threat predictions and evoke self-oriented avoidance responses during vicarious pain, both regions also show involvement in other-oriented empathy. Future investigation may disentangle the activations corresponding to distinct bottom-up or top-down processing, instigating empathy or withdrawal.

6.3 Empathic Distress and Personal Distress

aINS, ACC and amygdala involvement in affective processing is well-established, and it is particularly challenging to tease apart which neural patterns underpin other-oriented empathy as opposed to self-oriented affective distress during vicarious pain. More specifically, aINS activity has been associated with both higher self-reported empathy and increased intensity ratings of observed pain [95, 163, 179, 193]. Research advances that the interoceptive functions of the aINS may be the mechanism by which the state of the person suffering pain is mapped onto the own state, intrinsically triggering empathic understanding [31, 89, 177]. In line with this, Hein et al. [91] found increased aINS activity during pain observation to correlate with empathic concern and, more importantly, predict helpful behavior. No association was found for aINS activity and self-oriented distress, supporting its role in other-oriented empathy. This suggests that greater responsiveness of this brain region may be a predictor of greater empathic response. Clinical findings for patients with congenital insensitivity to pain further support such a notion. This disorder is characterized by the inability to experience pain, and, thus patients cannot use sensory and affective pain mirroring to evaluate the pain experienced by another individual [57]. Alternatively, during vicarious pain the revealed aMCC and aINS activation in those individuals is more likely to reflect an empathic understanding of emotionally aversive events. However, as the stage at which deficits in congenital pain sensitivity originate is unclear, a form of pain mirroring may remain effective. Further research is required to gain a comprehensive understanding of these mechanisms [140]. Moreover, these findings do not exclude that the aINS also encodes self-oriented distress responses to observed pain [111]. The INS has been associated with increased emotional responses to self-pain, and, similarly activity during vicarious pain may reflect increased emotional distress [69, 107, 168]. Indeed, research has reported correlations between aINS activity and self-rated negative affect during pain observation, in particular when emotional attachment to the individual in pain is high [36, 179]. Likewise, Akitsuki and Decety [1] demonstrated a positive correlation between activity in the left aINS and emotional contagion. This was associated with motor preparation, implicating the aINS in a self-oriented distress response that instigates withdrawal. Furthermore, compassion training, which involves the regulation of self-experienced affect during empathic response failed to elicit aINS activation, supporting a role of this region in coding personal distress responses [106]. Thus, inconsistent findings prevent firm conclusions on whether aINS activity during vicarious pain reflects other-oriented empathy or self-oriented distress. Notably, Lamm et al. [108] suggest that the role of such activity is modulated by the perspective from which the observed pain is imagined. First-person perspectives may thus elicit personal distress while third-person perspectives elicit empathic understanding, both of which are mediated by the aINS. Furthermore, the aINS may moderate the separation of self- and other-affect during pain observation rather than underpinning one or the other [77].

Moreover, research suggests that the amygdala can differentiate other-oriented empathy from self-oriented distress during pain observation in its function in fear processing [176]. Akitsuki and Decety [1] revealed higher amygdala activation when individuals viewed images of limbs on which pain was inflicted intentionally, in particular when imagining the pain from a first-person perspective [1, 108]. As the threat value of such stimuli is high, this supports a role of the amygdala in threat detection. Substantiating this, comparisons between self- and vicarious pain have found amygdala activation specific to other-pain, independent of the perspective used to imagine the observed pain. These results indicate that observing the administration of noxious stimuli to other individuals arouses the threat detection system even when individuals do not fear the pain itself (Fig. 13) [146]. Accordingly, the danger cues of observed pain may activate a self-oriented fear response that evokes withdrawal behaviors [1, 142].

Interestingly, the amygdala shows such fear processing specifically to male but not to female pain expressions. This indicates that while male pain may be interpreted as self-directed risk, female pain may activate other-oriented empathic responses [176]. Furthermore, functional connectivity analyses indicate that the amygdala may project fear information to both the somatosensory cortices and the aINS, potentially contributing to somatosensory predictions and affective appraisal of the observed pain [176]. In turn, the aINS projects information to the ACC for further processing and response selection [89]. Notably, the ACC has been associated with increased personal distress responses during vicarious pain. In line with this, activation in this brain region decreases when pain stimulation is presented in the context of several affective facial expressions. Increasing levels of affective information should increase ACC activity that underpins empathy as this information contributes to a more comprehensive emotional appraisal of the pain observation. Thus, the decrease may instead correspond to the decrease in self-oriented affective processing of vicarious pain due to the distraction value of the presented faces [90]. Taken together, there is evidence for an involvement of the aINS, amygdala and CC in other-oriented empathy and self-oriented distress processing during observed pain. In particular, the INS may underpin empathy [31, 91, 95], while the amygdala and CC may subserve personal distress [1, 176]. Nonetheless, future research needs to pinpoint exact neural patterns for firm conclusions to be drawn. Exploring amygdala activity during vicarious pain may be especially fruitful due to its well-known connection to fear. While fear can be other-oriented, its strong evolutionary connection to self-oriented survival makes it likely to underpin personal affective distress during pain observation.

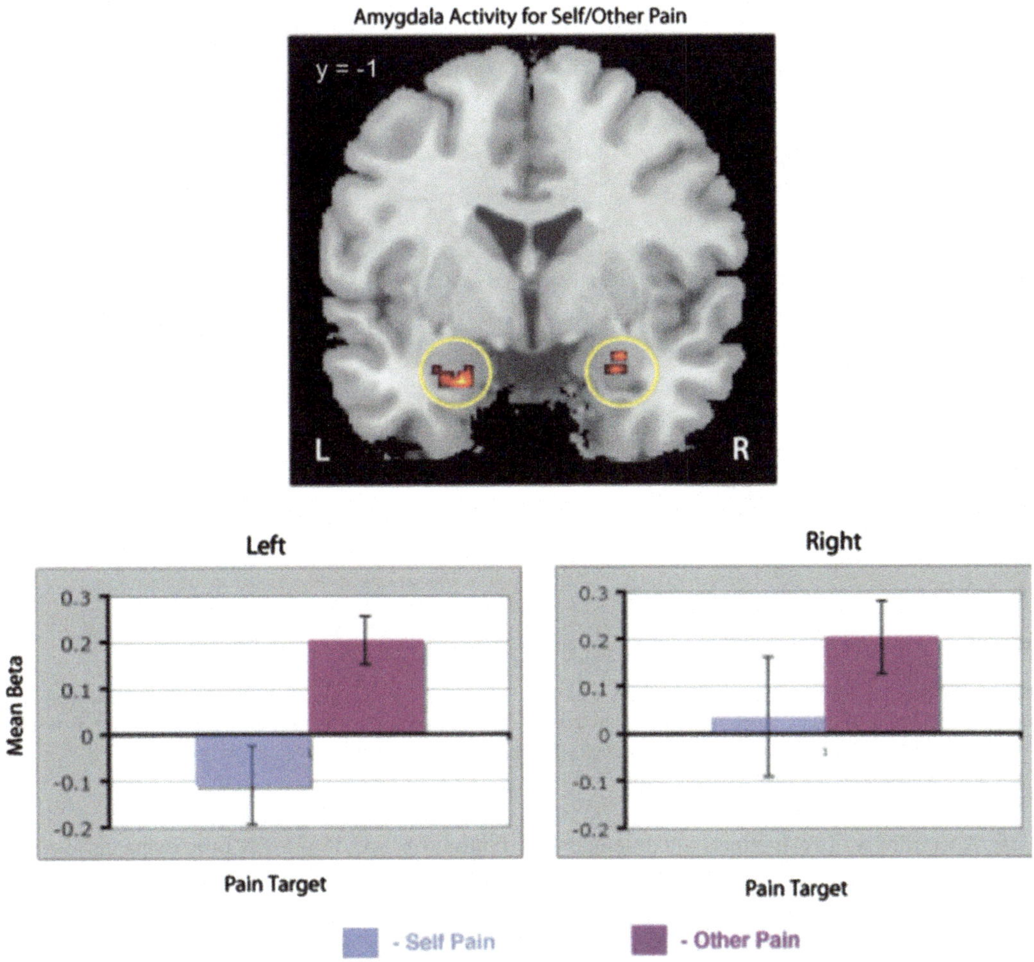

Fig. 13 Increased amygdala activity during vicarious pain compared to self-pain. Small-volume-corrected clusters of amygdala activity and their mean beta values with standard deviations as indicated by *bars* [146]

7 An Integrative Model of Vicarious Pain and Research Gaps

Available literature has demonstrated the involvement of sensorimotor, affective and cognitive processing regions in pain observation. Notably, the PAM offers essential theoretical insight into the mechanisms of vicarious pain and empathy. Nonetheless, the inclusion of distinct neural activations, empathy and withdrawal responses in such a theoretical framework may provide a more comprehensive account of vicarious pain processing. Such an integrative model of vicarious pain, as displayed in Fig. 14, can also visualize current research gaps that provide a podium for future scientific investigation.

Brain activation during vicarious pain may underpin stimulus and response processing. At the stimulus processing level, the PAM suggests that overlapping brain activation during self- and other-pain reflects sensorimotor and affective bottom-up pain mirroring that stimulates associations relevant for stimulus appraisal [155–157]. Extending this, such activation may also underpin top-down predictions about the observed pain, which are compared against the concurrent context and guide behavioral responses [54, 200]. Notably, the revealed bottom-up and top-down pain processing may form a feedback

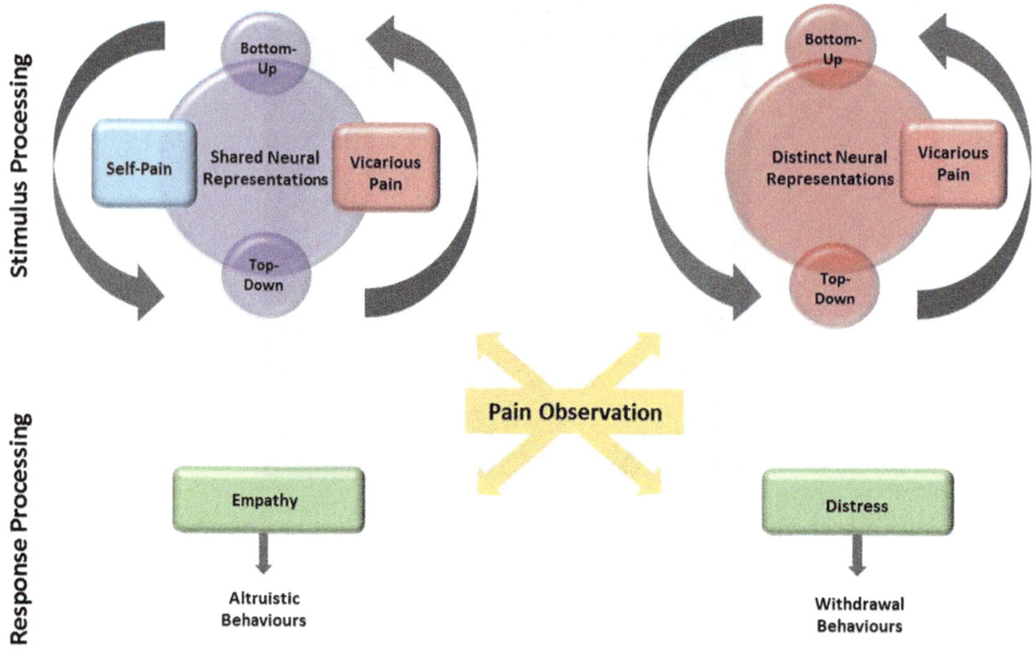

Fig. 14 Integrative model of vicarious pain. Vicarious pain is processed at stimulus level, including shared and distinct neural activations underpinning bottom-up and top-down processing of the observed pain. Two response options may result, including empathy and distress. All four concepts may interact, but further research is needed to elucidate such notion

loop. Accordingly, two distinct, yet interactive, mechanisms may be subserved by shared neural pain representations. More specifically, initial bottom-up pain processing may direct which top-down analysis is conducted. Subsequently, top-down predictions and corresponding motor preparation may be adjusted according to feedback from context comparisons [111]. Likewise, distinct brain activations during vicarious pain are indicated to contribute to both bottom-up processing as well as cognitive analysis and top-down regulation of responses to pain observation [77, 112, 188]. In particular, the qualitative differences between self-experienced and observed pain may be reflected in unique bottom-up neural pathways for each pain experience. Furthermore, complex pain contexts may be decoded in distinct neural patterns to enable individuals to adjust the representation conveyed in bottom-up neural mirroring according to supplementary cognitive information. Thus, shared and distinct neural representations of vicarious pain may have similar interactions within bottom-up and top-down processing as well as with one another. Such dynamic feedback systems are reflected in increased functional connectivities between corresponding brain regions; however, they have not yet been systematically tested [15, 143, 205]. While such interactions are a novel theoretical extension of the PAM for vicarious pain, these have been established in other fields, such as visual perception. In particular, bottom-up processing has been found to impact higher cortical activity, and top-down processing can shape the bottom-up perception of stimuli [10, 124]. Accordingly, it is likely to take effect in vicarious pain experiences and contribute to the evaluation of observed pain. Nonetheless, while there is evidence of an overlap in activation in the mirror neuron system and regions of the self-pain matrix during pain observation, the exact neural correlates have not yet been identified. Therefore, the extent to which neural correlates are shared or distinct is subject to further verification. Similarly, neural substrates of bottom-up and top-down pain processing have not yet been teased apart, thus making it challenging to investigate their interactions [77, 105, 110, 111].

At the response processing level, brain activation during vicarious pain may subserve other-oriented empathic understanding and altruistic behaviors [157] or self-oriented distress and withdrawal behaviors [36, 136, 176, 191]. However, the neural correlates of these two distinct response options have not been directly compared, and, thus no firm conclusions can be drawn. Moreover, research has shown that at early processing stages, individuals show greater neural responses and empathic responding to individuals with whom they can identify and feel positively about [66, 180]. In line with this, habituated empathic or distress responses may shape early neural processing of the observed pain, reinforcing neural pathways that may be dysfunctional [7]. Nonetheless, the factors mediating whether empathy or withdrawal is induced remain largely under-researched. Likewise, it is unclear whether these responses are evoked subsequent to the stimulus processing of pain observation or occur concurrently [111]. Concurrent stimulus and response processing of vicarious pain may contribute to remarkably complex feedback interactions. Elucidating the manner in which these responses are formed is the first step in understanding and tackling habitual dysfunctional empathy or avoidance behaviors [71]. Moreover, uncovering the interactions within and between stimulus and response processing levels is likely to contribute to a comprehensive concept of vicarious pain.

In order to shed light on the neural correlates of pain observation and their functions, behavioral and neuroimaging methods may be combined. Notably, as the roles of the different brain regions involved are not reliably confirmed for self-pain, revealing identical neural patterns for self- and other-pain may not enable research to derive their corresponding functions [93, 185, 186]. Accordingly, systematic and well-controlled paradigms may engender more robust findings. Furthermore, regions of interest analyses may be extended beyond the mirror and pain system. For example, the amygdala responds to aversive events such as facial pain expressions [31]. However, this region remains largely untouched by vicarious pain research. Pinpointing neural patterns that underlie pain observation is challenging as present neuroimaging tools are subject to coarse spatial resolution and individual brain differences can further decrease accuracy [105, 112, 113]. Likewise, these techniques may fail to detect weak levels of activation, and therefore activity in other regions cannot be confidently excluded. Ideally, single-cell recording could confirm both identical and distinct neural activation, but this is subject to immense practical and ethical restrictions [92]. Nonetheless, improvements in analytical methodology may minimize the impact of equipment issues. For example, MPVA has proven useful for detailed neural explorations and is likely to contribute to teasing apart brain responses to stimulus and response processing of vicarious pain [48]. Moreover, functional connectivity analysis as well as EEG, MEG and TMS may collectively contribute to an understanding of neural communication during pain observation. In particular, these methods can establish temporal processing sequences and elucidate the directions of such communication. Such methods may shed light on the extent and effects of shared and distinct neural activity and interaction as well as elucidate the neural correlates of empathic and withdrawal responses. Notably, although systematic paradigms are robust in identifying correlational and causal relationships, they struggle with ecological validity. Thus, findings from artificial laboratory settings may not be directly transferred to the complexity of vicarious pain in the natural environment. Nevertheless, a deeper understanding of the various aspects of vicarious pain will provide a solid basis in which more complex paradigms can be confidently rooted [204].

8 Clinical Relevance

Elucidating the neural underpinnings of vicarious pain and empathic or withdrawal responses is of significant relevance for the detection and management of clinical issues that are characterized by dysfunctional pain or empathy behaviors. Structural and functional brain abnormalities have been associated with empathic deficits in clinical disorders [35, 57, 58, 129, 133, 181]. For example, Cummins et al. [55] provided behavioral evidence for a link between motor coordination difficulties and decreased emotion recognition in others. As abnormal motor processing may be associated with defective mirror neuron systems, blunted empathy may result from inadequate motor mirroring that fails to engage evaluative associations. Indeed, motor disorders have been associated with decreased gray matter volume in the IFG [35]. Likewise, individuals with developmental and psychotic disorders, such as autism and schizophrenia, have consistently shown abnormal motor inhibition during action observation as well as decreased empathic abilities. This substantiates that dysfunctional mirroring contributes to empathic deficits [123, 129, 144]. Furthermore, abnormal aINS, aMCC and PFC functioning has been associated with decreased empathic understanding concurrent with deficits in attentional control in ADHD [189], negativity biases in depression [56, 167] and antisocial conduct [181]. Accordingly, processing deficits in affective–cognitive substrates of vicarious pain, empathy and withdrawal may underpin dysfunctional behaviors. While clinical research has focused on empathy toward facial emotional expressions, investigating potential interactions between pain and empathic understanding may provide novel intervention targets for dysfunctions in either domain. For example, based on shared neural representations between self- and other-pain, self-pain treatments may fine-tune the neural pain mirroring system and thus have spillover effects on increasing empathic understanding [52]. Notably, social and physical pain have been shown to share neural correlates during vicarious processing. Given the strong association between empathic deficits and social rejection, research may further uncover factors that perpetuate such cycles [95]. Identifying neural and behavioral sources of dysfunctional empathic behaviors may enable better management of such deficits which can increase the quality of life for many individuals suffering from clinical disorders. Moreover, such identification contributes to improved health care. As accurate empathic awareness has been found to increase diagnosis accuracy, enhanced empathic skills may facilitate both enriched treatment and patient-clinician relationships [11]. In line with this, training programs teaching self-regulation of affect during empathic understanding have been successful. Individuals reported lower self-oriented distress during pain observation, but displayed greater other-oriented altruistic behaviors than individuals without this training. As such, emotion regulation and compassion may be useful for enhancing context-appropriate responses [106]. Crucially, not only can the clinical realm benefit from establishing neuroimaging correlates of vicarious pain and empathy, but vice versa, clinical research can contribute to this by revealing dysfunctional activity associated with deficits. Such studies can confirm functional speculations derived from healthy populations.

9 Conclusion

Current vicarious pain research provides tentative evidence for shared and distinct neural representations of self- and other-pain. Nevertheless, the neural substrates of vicarious pain experiences are subject to confirmation in further systematic paradigms. These should combine neuroimaging and behavioral methodology to investigate brain responses and their corresponding functions during pain observation. Extending the PAM, an integrative model of vicarious pain is recommended as a platform for future comprehensive scientific inquiry.

Acknowledgements This work was supported by the May Endowed Professorship of The University of Hong Kong, the Research Grants Council Humanities and Social Sciences Prestigious Fellowship (Ref: HKU703-HSS-13). The funders have no role in study design, data collection and analysis, decision to publish, or preparation of the manuscript.

References

1. Akitsuki Y, Decety J. Social context and perceived agency affects empathy for pain: an event-related fMRI investigation. Neuroimage. 2009;47(2):722–34.
2. Amodio DM, Frith CD. Meeting of minds: the medial frontal cortex and social cognition. Nat Rev Neurosci. 2006;7(4):268–77.
3. Avenanti A, Aglioti SM. The sensorimotor side of empathy for pain. In: Psychoanalysis and neuroscience. Berlin: Springer; 2006. p. 235–56.
4. Avenanti A, Bueti D, Galati G, Aglioti SM. Transcranial magnetic stimulation highlights the sensorimotor side of empathy for pain. Nat Neurosci. 2005;8(7):955–60.
5. Avenanti A, Minio-Paluello I, Bufalari I, Aglioti SM. The pain of a model in the personality of an onlooker: influence of state-reactivity and personality traits on embodied empathy for pain. Neuroimage. 2009;44(1):275–83.
6. Avenanti A, Paluello IM, Bufalari I, Aglioti SM. Stimulus-driven modulation of motor-evoked potentials during observation of others' pain. Neuroimage. 2006;32(1):316–24.
7. Avenanti A, Sirigu A, Aglioti SM. Racial bias reduces empathic sensorimotor resonance with other-race pain. Curr Biol. 2010;20(11):1018–22.
8. Azevedo RT, Macaluso E, Avenanti A, Santangelo V, Cazzato V, Aglioti SM. Their pain is not our pain: brain and autonomic correlates of empathic resonance with the pain of same and different race individuals. Hum Brain Mapp. 2013;34(12):3168–81.
9. Bantick Wise RG, Ploghaus A, Clare S, Smith SM, Tracey I. Imaging how attention modulates pain in humans using functional MRI. Brain. 2002;125(2):310–9.
10. Bar M. A cortical mechanism for triggering top-down facilitation in visual object recognition. J Cogn Neurosci. 2003;15(4):600–9.
11. Bellet PS, Maloney MJ. The importance of empathy as an interviewing skill in medicine. JAMA. 1991;266(13):1831–2.
12. Benedetti F, Bernasconi A, Bosia M, Cavallaro R, Dallaspezia S, Falini A, Poletti S, Radaelli D, Riccaboni R, Scotti G. Functional and structural brain correlates of theory of mind and empathy deficits in schizophrenia. Schizophr Res. 2009;114(1):154–60.
13. Benuzzi F, Lui F, Duzzi D, Nichelli PF, Porro CA. Does it look painful or disgusting? Ask your parietal and cingulate cortex. J Neurosci. 2008;28(4):923–31.
14. Bernhardt BC, Singer T. The neural basis of empathy. Annu Rev Neurosci. 2012;35:1–23.
15. Betti V, Zappasodi F, Rossini PM, Aglioti SM, Tecchio F. Synchronous with your feelings: sensorimotor γ band and empathy for pain. J Neurosci. 2009;29(40):12384–92.
16. Bingel U, Lorenz J, Glauche V, Knab R, Gläscher J, Weiller C, Büchel C. Somatotopic organization of human somatosensory cortices for pain: a single trial fMRI study. Neuroimage. 2004;23(1):224–32.
17. Bingel U, Quante M, Knab R, Bromm B, Weiller C, Büchel C. Subcortical structures involved in pain processing: evidence from single-trial fMRI. Pain. 2002;99(1):313–21.
18. Blair RJR. Responding to the emotions of others: dissociating forms of empathy through the study of typical and psychiatric populations. Conscious Cogn. 2005;14(4):698–718.
19. Borckardt JJ, Reeves ST, Beam W, Jensen MP, Gracely RH, Katz S, Smith AR, Madan A, Patterson D, George MS. A randomized, controlled investigation of motor cortex transcranial magnetic stimulation (TMS) effects on quantitative sensory measures in healthy adults: evaluation of TMS device parameters. Clin J pain. 2011;27(6):486.
20. Borckardt JJ, Smith AR, Reeves ST, Weinstein M, Kozel FA, Nahas Z, Shelley N, Branham RK, Thomas KJ, George MS. Fifteen minutes of left prefrontal repetitive transcranial magnetic stimulation acutely increases thermal pain thresholds in healthy adults. Pain Res Manag (The Journal of the Canadian Pain Society). 2007;12(4):287.
21. Botvinick M, Cohen JD, Carter CS. Conflict monitoring and anterior cingulate cortex: an update. Trends in cognitive sciences. 2004;8(12):539–46.
22. Botvinick M, Jha AP, Bylsma LM, Fabian SA, Solomon PE, Prkachin KM. Viewing facial expressions of pain engages cortical areas involved in the direct experience of pain. Neuroimage. 2005;25(1):312–9.
23. Brooks J, Zambreanu L, Godinez A, Craig A, Tracey I. Somatotopic organisation of the human insula to painful heat studied with high resolution functional imaging. Neuroimage. 2005;27(1):201–9.
24. Brown CA, Jones AKP. A role for midcingulate cortex in the interruptive effects of pain anticipation on attention. Clin Neurophysiol. 2008;119(10):2370–9. doi:10.1016/j.clinph.2008.06.014.
25. Buccino G, Binkofski F, Fink GR, Fadiga L, Fogassi L, Gallese V, Seitz RJ, Zilles K, Rizzolatti G, Freund HJ. Action observation activates premotor and parietal areas in a somatotopic manner: an fMRI study. Eur J Neurosci. 2001;13(2):400–4.
26. Budell L, Jackson P, Rainville P. Brain responses to facial expressions of pain: emotional or motor mirroring? Neuroimage. 2010;53(1):355–63.
27. Bufalari I, Aprile T, Avenanti A, Di Russo F, Aglioti SM. Empathy for pain and touch in the human somatosensory cortex. Cereb Cortex. 2007;17(11):2553–61.
28. Bufalari I, Ionta S. The social and personality neuroscience of empathy for pain and touch. Front Human Neurosci. 2013;7.
29. Buffington AL, Hanlon CA, McKeown MJ. Acute and persistent pain modulation of attention-related

anterior cingulate fMRI activations. Pain. 2005;113(1):172–84.
30. Bushnell M, Duncan G, Hofbauer R, Ha B, Chen J-I, Carrier B. Pain perception: is there a role for primary somatosensory cortex? Proc Natl Acad Sci. 1999;96(14):7705–9.
31. Carr L, Iacoboni M, Dubeau M-C, Mazziotta JC, Lenzi GL. Neural mechanisms of empathy in humans: a relay from neural systems for imitation to limbic areas. Proc Natl Acad Sci. 2003;100(9):5497–502.
32. Carter AE, Carter G, Boschen M, Al Shwaimi E, George R. Pathways of fear and anxiety in dentistry: a review. World J Clin Cases WJCC. 2014;2(11):642.
33. Carter CS, Van Veen V. Anterior cingulate cortex and conflict detection: an update of theory and data. Cogn Affect Behav Neurosci. 2007;7(4):367–79.
34. Caspers S, Zilles K, Laird AR, Eickhoff SB. ALE meta-analysis of action observation and imitation in the human brain. Neuroimage. 2010;50(3):1148–67.
35. Cerami C, Dodich A, Canessa N, Crespi C, Iannaccone S, Corbo M, Lunetta C, Consonni M, Scola E, Falini A. Emotional empathy in amyotrophic lateral sclerosis: a behavioural and voxel-based morphometry study. Amyotrophic Lateral Scler Frontotemporal Degeneration. 2014;15(1–2):21–9.
36. Cheng Y, Chen C, Lin C-P, Chou K-H, Decety J. Love hurts: an fMRI study. Neuroimage. 2010;51(2):923–9.
37. Cheng Y, Lin C-P, Liu H-L, Hsu Y-Y, Lim K-E, Hung D, Decety J. Expertise modulates the perception of pain in others. Curr Biol. 2007;17(19):1708–13.
38. Cheng Y, Yang C-Y, Lin C-P, Lee P-L, Decety J. The perception of pain in others suppresses somatosensory oscillations: a magnetoencephalography study. Neuroimage. 2008;40(4):1833–40.
39. Chong TT-J, Cunnington R, Williams MA, Kanwisher N, Mattingley JB. fMRI adaptation reveals mirror neurons in human inferior parietal cortex. Curr Biol. 2008;18(20):1576–80.
40. Churchland MM, Byron MY, Ryu SI, Santhanam G, Shenoy KV. Neural variability in premotor cortex provides a signature of motor preparation. J Neurosci. 2006;26(14):3697–712.
41. Cioni B, Meglio M. Motor cortex stimulation for chronic non-malignant pain: current state and future prospects. In: Operative neuromodulation. Berlin: Springer; 2007. p. 45–9.
42. Coghill R. Individual differences in the subjective experience of pain: new insights into mechanisms and models. Headache: J Head Face Pain 2010;50(9):1531–5. doi:10.1111/j.1526-4610.2010.01763.x.
43. Coghill R, McHaffie JG, Yen Y-F. Neural correlates of interindividual differences in the subjective experience of pain. Proc Natl Acad Sci. 2003;100(14):8538–42.
44. Coghill R, Sang CN, Maisog JM, Iadarola MJ. Pain intensity processing within the human brain: a bilateral, distributed mechanism. J Neurophysiol. 1999;82(4):1934–43.
45. Cook R, Bird G, Catmur C, Press C, Heyes C. Mirror neurons: from origin to function. Behav Brain Sci. 2014;37(02):177–92.
46. Corbetta M, Patel G, Shulman GL. The reorienting system of the human brain: from environment to theory of mind. Neuron. 2008;58(3):306–24.
47. Corbetta M, Shulman GL. Control of goal-directed and stimulus-driven attention in the brain. Nat Rev Neurosci. 2002;3(3):201–15.
48. Corradi-Dell'Acqua C, Hofstetter C, Vuilleumier P. Felt and seen pain evoke the same local patterns of cortical activity in insular and cingulate cortex. J Neurosci. 2011;31(49):17996–8006.
49. Costantini M, Galati G, Romani GL, Aglioti SM. Empathic neural reactivity to noxious stimuli delivered to body parts and non-corporeal objects. Eur J Neurosci. 2008;28(6):1222–30.
50. Craig AD. How do you feel? Interoception: the sense of the physiological condition of the body. Nat Rev Neurosci. 2002;3(8):655–66.
51. Critchley HD, Wiens S, Rotshtein P, Öhman A, Dolan RJ. Neural systems supporting interoceptive awareness. Nat Neurosci. 2004;7(2):189–95.
52. Crockett MJ, Clark L, Hauser MD, Robbins TW. Serotonin selectively influences moral judgment and behavior through effects on harm aversion. Proc Natl Acad Sci. 2010;107(40):17433–8.
53. Crombez G, Eccleston C, Baeyens F, Eelen P. Attentional disruption is enhanced by the threat of pain. Behav Res Ther. 1998;36(2):195–204. doi:10.1016/S0005-7967(97)10008-0.
54. Csibra G. Action mirroring and action understanding: an alternative account. Sensorymotor Found Higher Cogn Attention Perform. 2008;XXII:435–59.
55. Cummins A, Piek JP, Dyck MJ. Motor coordination, empathy, and social behaviour in school-aged children. Dev Med Child Neurol. 2005;47(7):437–42.
56. Cusi AM, Nazarov A, Holshausen K, MacQueen GM, McKinnon MC. Systematic review of the neural basis of social cognition in patients with mood disorders. J Psychiatry Neurosci JPN. 2012;37(3):154.
57. Danziger N, Faillenot I, Peyron R. Can we share a pain we never felt? Neural correlates of empathy in patients with congenital insensitivity to pain. Neuron. 2009;61(2):203–12.
58. Danziger N, Prkachin KM, Willer J-C. Is pain the price of empathy? The perception of others' pain in patients with congenital insensitivity to pain. Brain. 2006;129(9):2494–507.
59. Davis KD, Taylor SJ, Crawley AP, Wood ML, Mikulis DJ. Functional MRI of pain-and attention-related activations in the human cingulate cortex. J Neurophysiol. 1997;77(6):3370–80.
60. De Gelder B, Snyder J, Greve D, Gerard G, Hadjikhani N. Fear fosters flight: a mechanism for fear contagion when perceiving emotion expressed by a whole body. Proc Natl Acad Sci U S A. 2004;101(47):16701–6.
61. De Vignemont F, Jacob P. What is it like to feel another's pain?*. Philos Sci. 2012;79(2):295–316.

62. De Vignemont F, Singer T. The empathic brain: how, when and why? Trends in cognitive sciences. 2006;10(10):435–41.
63. De Waal FB. Putting the altruism back into altruism: the evolution of empathy. Annu Rev Psychol. 2008;59:279–300.
64. Decety J. To what extent is the experience of empathy mediated by shared neural circuits? Emot Rev. 2010;2(3):204–7.
65. Decety J. Dissecting the neural mechanisms mediating empathy. Emot Rev. 2011;3(1):92–108.
66. Decety J, Echols S, Correll J. The blame game: the effect of responsibility and social stigma on empathy for pain. J Cogn Neurosci. 2010;22(5):985–97.
67. Decety J, Jackson PL. The functional architecture of human empathy. Behav Cogn Neurosci Rev. 2004;3 (2):71–100.
68. Decety J, Yang C-Y, Cheng Y. Physicians down-regulate their pain empathy response: an event-related brain potential study. Neuroimage. 2010;50(4):1676–82.
69. Derbyshire SW, Jones AK, Gyulai F, Clark S, Townsend D, Firestone LL. Pain processing during three levels of noxious stimulation produces differential patterns of central activity. Pain. 1997;73 (3):431–45.
70. Devinsky O, Morrell MJ, Vogt BA. Contributions of anterior cingulate cortex to behaviour. Brain. 1995;118(1):279–306.
71. Drwecki BB, Moore CF, Ward SE, Prkachin KM. Reducing racial disparities in pain treatment: the role of empathy and perspective-taking. Pain. 2011;152(5):1001–6.
72. Duerden EG, Arsalidou M, Lee M, Taylor MJ. Lateralization of affective processing in the insula. Neuroimage. 2013;78:159–75.
73. Ebisch SJ, Ferri F, Salone A, Perrucci MG, D'Amico L, Ferro FM, Romani GL, Gallese V. Differential involvement of somatosensory and interoceptive cortices during the observation of affective touch. J Cogn Neurosci. 2011;23(7):1808–22.
74. Eisenberg N, Fabes RA. Empathy: conceptualization, measurement, and relation to prosocial behavior. Motiv Emot. 1990;14(2):131–49.
75. Eisenberg N, Morris AS. The origins and social significance of empathy-related responding. A review of empathy and moral development: implications for caring and justice by ML Hoffman. Social Justice Res. 2001;14(1):95–120.
76. Etkin A, Egner T, Kalisch R. Emotional processing in anterior cingulate and medial prefrontal cortex. Trends Cogn Sci. 2011;15(2):85–93.
77. Fan Y, Duncan NW, de Greck M, Northoff G. Is there a core neural network in empathy? An fMRI based quantitative meta-analysis. Neurosci Biobehav Rev. 2011;35(3):903–11.
78. Fan Y, Han S. Temporal dynamic of neural mechanisms involved in empathy for pain: an event-related brain potential study. Neuropsychologia. 2008;46(1):160–73.
79. Farina S, Tinazzi M, Le Pera D, Valeriani M. Pain-related modulation of the human motor cortex. Neurol Res. 2003;25(2):130–42.
80. Farrell MJ, Laird AR, Egan GF. Brain activity associated with painfully hot stimuli applied to the upper limb: A meta-analysis. Hum Brain Mapp. 2005;25(1):129–39.
81. Gallese V. The roots of empathy: the shared manifold hypothesis and the neural basis of intersubjectivity. Psychopathology. 2003;36(4):171–80.
82. Gallese V, Keysers C, Rizzolatti G. A unifying view of the basis of social cognition. Trends cogn Sci. 2004;8(9):396–403.
83. Gasquoine PG. Localization of function in anterior cingulate cortex: from psychosurgery to functional neuroimaging. Neurosci Biobehav Rev. 2013;37 (3):340–8.
84. Gracely RH, Sundgren PC. Neuroimaging of pain. In: Functional neuroradiology. Berlin: Springer; 2012. p. 273–90.
85. Graff-Guerrero A, González-Olvera J, Fresán A, Gómez-Martín D, Méndez-Nuñez JC, Pellicer F. Repetitive transcranial magnetic stimulation of dorsolateral prefrontal cortex increases tolerance to human experimental pain. Cogn Brain Res. 2005;25 (1):153–60.
86. Grill-Spector K, Malach R. fMR-adaptation: a tool for studying the functional properties of human cortical neurons. Acta Psychol (Amst). 2001;107 (1):293–321.
87. Gu X, Gao Z, Wang X, Liu X, Knight RT, Hof PR, Fan J. Anterior insular cortex is necessary for empathetic pain perception. Brain. 2012;135(9):2726–35.
88. Gu X, Han S. Attention and reality constraints on the neural processes of empathy for pain. Neuroimage. 2007;36(1):256–67.
89. Gu X, Liu X, Guise KG, Naidich TP, Hof PR, Fan J. Functional dissociation of the frontoinsular and anterior cingulate cortices in empathy for pain. J Neurosci. 2010;30(10):3739–44.
90. Han S, Fan Y, Xu X, Qin J, Wu B, Wang X, Aglioti SM, Mao L. Empathic neural responses to others' pain are modulated by emotional contexts. Hum Brain Mapp. 2009;30(10):3227–37.
91. Hein G, Silani G, Preuschoff K, Batson CD, Singer T. Neural responses to ingroup and outgroup members' suffering predict individual differences in costly helping. Neuron. 2010;68(1):149–60.
92. Hutchison WD, Davis K, Lozano A, Tasker R, Dostrovsky J. Pain-related neurons in the human cingulate cortex. Nat Neurosci. 1999;2(5):403–5.
93. Iannetti GD, Mouraux A. From the neuromatrix to the pain matrix (and back). Exp Brain Res. 2010;205(1):1–12.
94. IASP IAftSoP. Classification of chronic pain. 2 ed. Seattle: IASP Press; 1994.
95. Immordino-Yang MH, McColl A, Damasio H, Damasio A. Neural correlates of admiration and compassion. Proc Natl Acad Sci. 2009;106 (19):8021–6.

96. Isnard J, Magnin M, Jung J, Mauguière F, Garcia-Larrea L. Does the insula tell our brain that we are in pain? Pain. 2011;152(4):946–51.
97. Isoda M, Noritake A. What makes the dorsomedial frontal cortex active during reading the mental states of others? Front Neurosci. 2013;7.
98. Jackson PL, Brunet E, Meltzoff AN, Decety J. Empathy examined through the neural mechanisms involved in imagining how I feel versus how you feel pain. Neuropsychologia. 2006;44(5):752–61.
99. Jackson PL, Meltzoff AN, Decety J. How do we perceive the pain of others? A window into the neural processes involved in empathy. Neuroimage. 2005;24(3):771–9.
100. Jackson PL, Rainville P, Decety J. To what extent do we share the pain of others? Insight from the neural bases of pain empathy. Pain. 2006;125(1):5–9.
101. Jastorff J, Begliomini C, Fabbri-Destro M, Rizzolatti G, Orban GA. Coding observed motor acts: different organizational principles in the parietal and premotor cortex of humans. J Neurophysiol. 2010;104(1):128–40.
102. Kang P, Lee J, Sul S, Kim H. Dorsomedial prefrontal cortex activity predicts the accuracy in estimating others' preferences. Front Human Neurosci. 2013;7.
103. Kelly S, Lloyd D, Nurmikko T, Roberts N. Retrieving autobiographical memories of painful events activates the anterior cingulate cortex and inferior frontal gyrus. J Pain. 2007;8(4):307–14.
104. Keogh E, Moore DJ, Duggan GB, Payne SJ, Eccleston C. The disruptive effects of pain on complex cognitive performance and executive control. PLoS ONE. 2013;8(12):e83272.
105. Keysers C, Kaas JH, Gazzola V. Somatosensation in social perception. Nat Rev Neurosci. 2010;11 (6):417–28.
106. Klimecki OM, Leiberg S, Ricard M, Singer T. Differential pattern of functional brain plasticity after compassion and empathy training. Social Cogn Affect Neurosci. 2013;nst060.
107. Kulkarni B, Bentley D, Elliott R, Youell P, Watson A, Derbyshire S, Frackowiak R, Friston K, Jones A. Attention to pain localization and unpleasantness discriminates the functions of the medial and lateral pain systems. Eur J Neurosci. 2005;21 (11):3133–42.
108. Lamm C, Batson CD, Decety J. The neural substrate of human empathy: effects of perspective-taking and cognitive appraisal. Cogn Neurosci J. 2007;19(1):42–58.
109. Lamm C, Decety J. Is the extrastriate body area (EBA) sensitive to the perception of pain in others? Cereb Cortex. 2008;18(10):2369–73.
110. Lamm C, Decety J, Singer T. Meta-analytic evidence for common and distinct neural networks associated with directly experienced pain and empathy for pain. Neuroimage. 2011;54(3):2492–502.
111. Lamm C, Majdandžić J. The role of shared neural activations, mirror neurons, and morality in empathy—a critical comment. Neurosci Res. 2015;90:15–24.
112. Lamm C, Meltzoff AN, Decety J. How do we empathize with someone who is not like us? A functional magnetic resonance imaging study. J Cogn Neurosci. 2010;22(2):362–76.
113. Lamm C, Nusbaum HC, Meltzoff AN, Decety J. What are you feeling? Using functional magnetic resonance imaging to assess the modulation of sensory and affective responses during empathy for pain. PLoS ONE. 2007;2(12):e1292.
114. Le Pera D, Graven-Nielsen T, Valeriani M, Oliviero A, Di Lazzaro V, Tonali PA, Arendt-Nielsen L. Inhibition of motor system excitability at cortical and spinal level by tonic muscle pain. Clin Neurophysiol. 2001;112(9):1633–41.
115. Lefaucheur J, Drouot X, Menard-Lefaucheur I, Keravel Y, Nguyen J. Motor cortex rTMS in chronic neuropathic pain: pain relief is associated with thermal sensory perception improvement. J Neurol Neurosurg Psychiatry. 2008;79(9):1044–9.
116. Lombardo MV, Chakrabarti B, Bullmore ET, Sadek SA, Pasco G, Wheelwright SJ, Suckling J, Baron-Cohen S, Consortium MA. Atypical neural self-representation in autism. Brain. 2010;133 (2):611–24.
117. Lorenz J, Minoshima S, Casey K. Keeping pain out of mind: the role of the dorsolateral prefrontal cortex in pain modulation. Brain. 2003;126 (5):1079–91.
118. Lotze M, Montoya P, Erb M, Hülsmann E, Flor H, Klose U, Birbaumer N, Grodd W. Activation of cortical and cerebellar motor areas during executed and imagined hand movements: an fMRI study. J Cogn Neurosci. 1999;11(5):491–501.
119. Mattiassi AD, Mele S, Ticini LF, Urgesi C. Conscious and unconscious representations of observed actions in the human motor system. J Cogn Neurosci. 2014;26(9):2028–41.
120. May A. Neuroimaging: visualising the brain in pain. Neurol Sci. 2007;28(2):S101–7.
121. May A. New insights into headache: an update on functional and structural imaging findings. Nat Rev Neurol. 2009;5(4):199–209.
122. Mazzola L, Isnard J, Mauguiere F. Somatosensory and pain responses to stimulation of the second somatosensory area (SII) in humans. A comparison with SI and insular responses. Cereb Cortex. 2006;16(7):960–8.
123. McCormick LM, Brumm MC, Beadle JN, Paradiso S, Yamada T, Andreasen N. Mirror neuron function, psychosis, and empathy in schizophrenia. Psychiatry Res: Neuroimaging. 2012;201(3):233–9.
124. Mechelli A, Price CJ, Friston KJ, Ishai A. Where bottom-up meets top-down: neuronal interactions during perception and imagery. Cereb Cortex. 2004;14(11):1256–65.
125. Menon V, Uddin LQ. Saliency, switching, attention and control: a network model of insula function. Brain Struct Funct. 2010;214(5–6):655–67.

126. Mercier C, Leonard G. Interactions between pain and the motor cortex: insights from research on phantom limb pain and complex regional pain syndrome. Physiother Can. 2011;63(3):305–14.
127. Miller EK, Cohen JD. An integrative theory of prefrontal cortex function. Annu Rev Neurosci. 2001;24(1):167–202.
128. Minio-Paluello I, Avenanti A, Aglioti SM. Left hemisphere dominance in reading the sensory qualities of others' pain? Soc Neurosci. 2006;1(3–4):320–33.
129. Minio-Paluello I, Baron-Cohen S, Avenanti A, Walsh V, Aglioti SM. Absence of embodied empathy during pain observation in Asperger syndrome. Biol Psychiatry. 2009;65(1):55–62.
130. Molenberghs P, Cunnington R, Mattingley JB. Is the mirror neuron system involved in imitation? A short review and meta-analysis. Neurosci Biobehav Rev. 2009;33(7):975–80.
131. Molnar-Szakacs I, Kaplan J, Greenfield PM, Iacoboni M. Observing complex action sequences: the role of the fronto-parietal mirror neuron system. Neuroimage. 2006;33(3):923–35.
132. Morelli SA, Lieberman MD. The role of automaticity and attention in neural processes underlying empathy for happiness, sadness, and anxiety. Front Human Neurosci. 2013;7.
133. Moriguchi Y, Decety J, Ohnishi T, Maeda M, Mori T, Nemoto K, Matsuda H, Komaki G. Empathy and judging other's pain: an fMRI study of alexithymia. Cereb Cortex. 2007;17(9):2223–34.
134. Morrison I, Downing PE. Organization of felt and seen pain responses in anterior cingulate cortex. Neuroimage. 2007;37(2):642–51.
135. Morrison I, Lloyd D, Di Pellegrino G, Roberts N. Vicarious responses to pain in anterior cingulate cortex: is empathy a multisensory issue? Cogn Affect Behav Neurosci. 2004;4(2):270–8.
136. Morrison I, Peelen MV, Downing PE. The sight of others' pain modulates motor processing in human cingulate cortex. Cereb Cortex. 2007;17(9):2214–22.
137. Morrison I, Tipper SP, Fenton-Adams WL, Bach P. "Feeling" Others' painful actions: the sensorimotor integration of pain and action information. Hum Brain Mapp. 2013;34(8):1982–98.
138. Mouraux A, Iannetti GD. Nociceptive laser-evoked brain potentials do not reflect nociceptive-specific neural activity. J Neurophysiol. 2009;101(6):3258–69.
139. Nachev P, Kennard C, Husain M. Functional role of the supplementary and pre-supplementary motor areas. Nat Rev Neurosci. 2008;9(11):856–69.
140. Nagasako EM, Oaklander AL, Dworkin RH. Congenital insensitivity to pain: an update. Pain. 2003;101(3):213–9.
141. Nakamura K, Kawashima R, Ito K, Sugiura M, Kato T, Nakamura A, Hatano K, Nagumo S, Kubota K, Fukuda H. Activation of the right inferior frontal cortex during assessment of facial emotion. J Neurophysiol. 1999;82(3):1610–4.
142. Norris CJ, Chen EE, Zhu DC, Small SL, Cacioppo JT. The interaction of social and emotional processes in the brain. J Cogn Neurosci. 2004;16(10):1818–29.
143. Nummenmaa L, Hirvonen J, Parkkola R, Hietanen JK. Is emotional contagion special? An fMRI study on neural systems for affective and cognitive empathy. Neuroimage. 2008;43(3):571–80.
144. Oberman LM, Hubbard EM, McCleery JP, Altschuler EL, Ramachandran VS, Pineda JA. EEG evidence for mirror neuron dysfunction in autism spectrum disorders. Cogn Brain Res. 2005;24(2):190–8.
145. Ochsner KN, Ludlow DH, Knierim K, Hanelin J, Ramachandran T, Glover GC, Mackey SC. Neural correlates of individual differences in pain-related fear and anxiety. Pain. 2006;120(1):69–77.
146. Ochsner KN, Zaki J, Hanelin J, Ludlow DH, Knierim K, Ramachandran T, Glover GH, Mackey SC. Your pain or mine? Common and distinct neural systems supporting the perception of pain in self and other. Social Cogn Affect Neurosci. 2008;3(2):144–60.
147. Ogino Y, Nemoto H, Inui K, Saito S, Kakigi R, Goto F. Inner experience of pain: imagination of pain while viewing images showing painful events forms subjective pain representation in human brain. Cereb Cortex. 2007;17(5):1139–46.
148. Osborn J, Derbyshire SW. Pain sensation evoked by observing injury in others. Pain. 2010;148(2):268–74.
149. Perry A, Bentin S, Bartal IB-A, Lamm C, Decety J. "Feeling" the pain of those who are different from us: modulation of EEG in the mu/alpha range. Cogn Affect Behav Neurosci. 2010;10(4):493–504.
150. Peyron R, Garcia-Larrea L, Deiber M, Cinotti L, Convers P, Sindou M, Mauguiere F, Laurent B. Electrical stimulation of precentral cortical area in the treatment of central pain: electrophysiological and PET study. Pain. 1995;62(3):275–86.
151. Peyron R, Laurent B, Garcia-Larrea L. Functional imaging of brain responses to pain. A review and meta-analysis (2000). Clin Neurophysiol. 2000;30(5):263–88.
152. Ploghaus A, Tracey I, Gati JS, Clare S, Menon RS, Matthews PM, Rawlins JNP. Dissociating pain from its anticipation in the human brain. Science. 1999;284(5422):1979–81.
153. Ploner M, Schmitz F, Freund H-J, Schnitzler A. Differential organization of touch and pain in human primary somatosensory cortex. J Neurophysiol. 2000;83(3):1770–6.
154. Porro CA, Baraldi P, Pagnoni G, Serafini M, Facchin P, Maieron M, Nichelli P. Does anticipation of pain affect cortical nociceptive systems? J Neurosci. 2002;22(8):3206–14.
155. Preston SD. A perception-action model for empathy. Empathy Mental Illness 2007;428–47.
156. Preston SD, De Waal F. Empathy: its ultimate and proximate bases. Behav Brain Sci. 2002;25(01):1–20.

157. Preston SD, Hofelich AJ. The many faces of empathy: parsing empathic phenomena through a proximate, dynamic-systems view of representing the other in the self. Emot Rev. 2012;4(1):24–33.
158. Rizzolatti G, Craighero L. The mirror-neuron system. Annu Rev Neurosci. 2004;27:169–92.
159. Rizzolatti G, Fadiga L, Gallese V, Fogassi L. Premotor cortex and the recognition of motor actions. Cogn Brain Res. 1996;3(2):131–41.
160. Rizzolatti G, Fogassi L, Gallese V. Neurophysiological mechanisms underlying the understanding and imitation of action. Nat Rev Neurosci. 2001;2 (9):661–70.
161. Roldan-Gerschcovich E, Cerquetti D, Tenca E, Leiguarda R. The impact of bilateral cerebellar damage on theory of mind, empathy and decision making. Neurocase. 2011;17(3):270–5.
162. Rossi S, Hallett M, Rossini PM, Pascual-Leone A, Group SoTC. Safety, ethical considerations, and application guidelines for the use of transcranial magnetic stimulation in clinical practice and research. Clin Neurophysiol. 2009;120(12):2008–39.
163. Saarela MV, Hlushchuk Y, Williams ACdC, Schürmann M, Kalso E, Hari R. The compassionate brain: humans detect intensity of pain from another's face. Cereb Cortex. 2007;17(1):230–7.
164. Sawamoto N, Honda M, Okada T, Hanakawa T, Kanda M, Fukuyama H, Konishi J, Shibasaki H. Expectation of pain enhances responses to nonpainful somatosensory stimulation in the anterior cingulate cortex and parietal operculum/posterior insula: an event-related functional magnetic resonance imaging study. The Journal of Neuroscience. 2000;20(19):7438–45.
165. Schaefer M, Xu B, Flor H, Cohen LG. Effects of different viewing perspectives on somatosensory activations during observation of touch. Hum Brain Mapp. 2009;30(9):2722–30.
166. Schienle A, Stark R, Walter B, Blecker C, Ott U, Kirsch P, Sammer G, Vaitl D. The insula is not specifically involved in disgust processing: an fMRI study. NeuroReport. 2002;13(16):2023–6.
167. Schneider D, Regenbogen C, Kellermann T, Finkelmeyer A, Kohn N, Derntl B, Schneider F, Habel U. Empathic behavioral and physiological responses to dynamic stimuli in depression. Psychiatry Res. 2012;200(2):294–305.
168. Schreckenberger M, Siessmeier T, Viertmann A, Landvogt C, Buchholz H-G, Rolke R, Treede R-D, Bartenstein P, Birklein F. The unpleasantness of tonic pain is encoded by the insular cortex. Neurology. 2005;64(7):1175–83.
169. Seitz RJ, Nickel J, Azari NP. Functional modularity of the medial prefrontal cortex: involvement in human empathy. Neuropsychology. 2006;20 (6):743–51. doi:10.1037/0894-4105.20.6.743.
170. Seminowicz DA, Davis KD. Pain enhances functional connectivity of a brain. J Neurophysiol. 2007;97:3651–9.
171. Shamay-Tsoory SG. The neural bases for empathy. Neuroscientist. 2011;17(1):18–24.
172. Shamay-Tsoory SG, Shur S, Harari H, Levkovitz Y. Neurocognitive basis of impaired empathy in schizophrenia. Neuropsychology. 2007;21(4):431.
173. Shima K, Tanji J. Both supplementary and presupplementary motor areas are crucial for the temporal organization of multiple movements. J Neurophysiol. 1998;80(6):3247–60.
174. Simmons A, Matthews SC, Stein MB, Paulus MP. Anticipation of emotionally aversive visual stimuli activates right insula. NeuroReport. 2004;15(14):2261–5.
175. Simmons W, Avery JA, Barcalow JC, Bodurka J, Drevets WC, Bellgowan P. Keeping the body in mind: insula functional organization and functional connectivity integrate interoceptive, exteroceptive, and emotional awareness. Hum Brain Mapp. 2013;34 (11):2944–58.
176. Simon D, Craig KD, Miltner WH, Rainville P. Brain responses to dynamic facial expressions of pain. Pain. 2006;126(1):309–18.
177. Singer T, Critchley HD, Preuschoff K. A common role of insula in feelings, empathy and uncertainty. Trends in cognitive sciences. 2009;13(8):334–40.
178. Singer T, Lamm C. The social neuroscience of empathy. Ann N Y Acad Sci. 2009;1156(1):81–96.
179. Singer T, Seymour B, O'Doherty J, Kaube H, Dolan RJ, Frith CD. Empathy for pain involves the affective but not sensory components of pain. Science. 2004;303(5661):1157–62.
180. Singer T, Seymour B, O'Doherty JP, Stephan KE, Dolan RJ, Frith CD. Empathic neural responses are modulated by the perceived fairness of others. Nature. 2006;439(7075):466–9.
181. Sterzer P, Stadler C, Poustka F, Kleinschmidt A. A structural neural deficit in adolescents with conduct disorder and its association with lack of empathy. Neuroimage. 2007;37(1):335–42.
182. Swick D, Ashley V, Turken U. Left inferior frontal gyrus is critical for response inhibition. BMC Neurosci. 2008;9(1):102.
183. Teplan M. Fundamentals of EEG measurement. Measure Sci Rev. 2002;2(2):1–11.
184. Timmermann L, Ploner M, Haucke K, Schmitz F, Baltissen R, Schnitzler A. Differential coding of pain intensity in the human primary and secondary somatosensory cortex. J Neurophysiol. 2001;86 (3):1499–503.
185. Tracey I. Imaging pain. Br J Anaesth. 2008;101 (1):32–9.
186. Tracey I, Mantyh PW. The cerebral signature for pain perception and its modulation. Neuron. 2007;55(3):377–91.
187. Tsubokawa T, Katayama Y, Yamamoto T, Hirayama T, Koyama S. Chronic motor cortex stimulation for the treatment of central pain. In: Advances in stereotactic and functional neurosurgery 9. Berlin: Springer; 1991. p. 137–9.
188. Uddin LQ, Molnar-Szakacs I, Zaidel E, Iacoboni M. rTMS to the right inferior parietal lobule disrupts self–other discrimination. Social Cogn Affect Neurosci. 2006;1(1):65–71.

189. Uekermann J, Kraemer M, Abdel-Hamid M, Schimmelmann B, Hebebrand J, Daum I, Wiltfang J, Kis B. Social cognition in attention-deficit hyperactivity disorder (ADHD). Neurosci Biobehav Rev. 2010;34(5):734–43.
190. Urban P, Solinski M, Best C, Rolke R, Hopf H, Dieterich M. Different short-term modulation of cortical motor output to distal and proximal upper-limb muscles during painful sensory nerve stimulation. Muscle Nerve. 2004;29(5):663–9.
191. Vachon-Presseau E, Martel MO, Roy M, Caron E, Jackson PL, Rainville P. The multilevel organization of vicarious pain responses: effects of pain cues and empathy traits on spinal nociception and acute pain. Pain. 2011;152(7):1525–31.
192. Vachon-Presseau E, Roy M, Martel M-O, Albouy G, Sullivan MJ, Jackson PL, Rainville P. The two sides of pain communication: effects of pain expressiveness on vicarious brain responses revealed in chronic back pain patients. J Pain. 2013;14(11):1407–15.
193. Vachon-Presseau E, Roy M, Martel M, Albouy G, Chen J, Budell L, Sullivan M, Jackson P, Rainville P. Neural processing of sensory and emotional-communicative information associated with the perception of vicarious pain. Neuroimage. 2012;63(1):54–62.
194. Valeriani M, Betti V, Le Pera D, De Armas L, Miliucci R, Restuccia D, Avenanti A, Aglioti SM. Seeing the pain of others while being in pain: a laser-evoked potentials study. Neuroimage. 2008;40(3):1419–28.
195. Valet M, Sprenger T, Boecker H, Willoch F, Rummeny E, Conrad B, Erhard P, Tolle TR. Distraction modulates connectivity of the cingulo-frontal cortex and the midbrain during pain—an fMRI analysis. Pain. 2004;109(3):399–408.
196. Van der Gaag C, Minderaa RB, Keysers C. Facial expressions: what the mirror neuron system can and cannot tell us. Soc Neurosci. 2007;2(3–4):179–222.
197. van Veluw SJ, Chance SA. Differentiating between self and others: an ALE meta-analysis of fMRI studies of self-recognition and theory of mind. Brain Imaging Behav. 2014;8(1):24–38.
198. Wager TD, Atlas LY, Lindquist MA, Roy M, Woo C-W, Kross E. An fMRI-based neurologic signature of physical pain. N Engl J Med. 2013;368(15):1388–97.
199. Wicker B, Keysers C, Plailly J, Royet J-P, Gallese V, Rizzolatti G. Both of us disgusted in < i > My </i > insula: the common neural basis of seeing and feeling disgust. Neuron. 2003;40(3):655–64.
200. Wilson M, Knoblich G. The case for motor involvement in perceiving conspecifics. Psychol Bull. 2005;131(3):460.
201. Xue G, Lu Z, Levin IP, Weller JA, Li X, Bechara A. Functional dissociations of risk and reward processing in the medial prefrontal cortex. Cereb Cortex. 2009;19(5):1019–27. doi:10.1093/cercor/bhn147.
202. Yesudas EH, Lee T. The role of cingulate cortex in vicarious pain. BioMed Res Int. 2015.
203. Zaki J, Davis JI, Ochsner KN. Overlapping activity in anterior insula during interoception and emotional experience. Neuroimage. 2012;62(1):493–9.
204. Zaki J, Ochsner KN. The neuroscience of empathy: progress, pitfalls and promise. Nat Neurosci. 2012;15(5):675–80.
205. Zaki J, Ochsner KN, Hanelin J, Wager TD, Mackey SC. Different circuits for different pain: patterns of functional connectivity reveal distinct networks for processing pain in self and others. Soc Neurosci. 2007;2(3–4):276–91.

Author Biographies

Esslin L. Terrighena is a final-year Ph.D. student in the Laboratory of Neuropsychology at the University of Hong Kong. She completed her B.Sc. Hons in Psychology at the University of Westminster (London, UK) and has received several awards and scholarships for outstanding performance. Her major research interests lie in the neural underpinnings of pain and emotion.

Professor Tatia M.C. Lee is Chair Professor of Psychology and Honorary Professor of Psychiatry and Medicine at The University of Hong Kong. During her tenure at the University, she has received numerous awards in recognition of her excellence in teaching and research. Because of her contributions to the advancement of knowledge in neuropsychology, the University has bestowed upon her an endowed professorship titled May Professorship in Neuropsychology. Professor Lee endeavors to understand how the human brain functions. Her team employs both behavioral and neuroimaging research methodologies and work in collaboration with clinicians and neuroscientists to investigate the neuropsychological mechanisms underpinning brain–behavior relationships.

Acupuncture Analgesia: A Review of Peripheral and Central Mechanisms

17

Mikiko Murakami and Albert Leung

Abstract

Acupuncture is an ancient needling modality within Traditional Chinese Medicine used for chronic pain management. Western biophysical and Chinese meridian theory views of pain differ, and so do their respective management practices. The reversible effect of naloxone on the acupuncture induced analgesia is well known. Research has also shown correlations between acupuncture and its effect on the peripheral nerve endings, connective tissue, neurotransmitters, and inflammatory mediators. Centrally, studies with functional imaging and dynamic quantitative sensory testing substantiate the modulatory role of acupuncture in the "wind-up" phenomenon of spinal wide dynamic range neurons and different brain areas related to pain perception and modulation. Despite this increased understanding in the mechanisms and the analgesic efficacy of acupuncture, controversy continues to evolve around the issues of placebo effect and its potential therapeutic role in the main stream medicine.

Keywords

Acupuncture · Pain · Acupuncture mechanism · Traditional Chinese medicine · Acupuncture analgesia · Supraspinal pain modulation · Peripheral pain modulation

1 Introduction

Chronic pain can cast a profound negative impact on quality of life in the general patient population. In the United States alone, pain affects approximately 100 million adults, costing \$560–\$635 billion per year, which is greater than the combined annual healthcare cost for heart disease, cancer, and diabetes [1]. Current pain medications can cause various negative side effects, including gastrointestinal upset, liver toxicity, cardiac toxicity, and respiratory depression. It has been estimated that in the United States alone, 44 people per day die from misusing prescription pain killers [2], and 7,000 people per day are seen for other complications related to misuse of analgesics [3]. Although various regulatory

M. Murakami · A. Leung (✉)
University of California in San Diego (UCSD), La Jolla, USA
e-mail: ayleung@ucsd.edu

M. Murakami
e-mail: miki.murakami@gmail.com

L. Saba (ed.), *Neuroimaging of Pain*, DOI 10.1007/978-3-319-48046-6_17

policies have been implemented nationwide to track aberrant behavior of patients (for example, doctor shopping), data from a recent study reveals that the United States continues to have a pain medication overdose epidemic [3]. This ongoing conundrum calls for alternate effective pain treatment modalities with fewer side effects. More and more chronic pain patients are now turning to Integrative Medicine, which combines Complementary Alternative Medicine (CAM) modalities such as acupuncture with other conventional (allopathic) modalities for pain management [4]. Among these modalities, acupuncture emerges as one of the most studied and applied methods in pain management.

Acupuncture is a treatment modality within Traditional Chinese Medicine (TCM). It is often used in combination with herbal remedies, physical manipulation, exercise, and lifestyle modification. It has been used for pain analgesia for thousands of years, with the first document dating back to about 100 BC in "The Emperor's Classic of Internal Medicine" [5]. A recent report called "Acupuncture: Review and Analysis of Reports on Controlled Clinical Trials", released by the World Health Organization (WHO) and the National Institutes of Health (NIH), states that acupuncture can be used for 43 different clinical areas including analgesia for the head and face, the locomotor system, gout, biliary and renal colic, traumatic or postoperative pain, dentistry, childbirth, surgery, and post-chemotherapy nausea and vomiting [6]. Although auricular (ear) acupuncture (Fig. 1) is less well known than conventional meridian based body acupuncture, it has gained recognition and utilization in clinical practice over the past few decades. In 1990, the WHO standardized 39 ear points, and created a standard translated nomenclature to facilitate teaching of this modality of acupuncture [7].

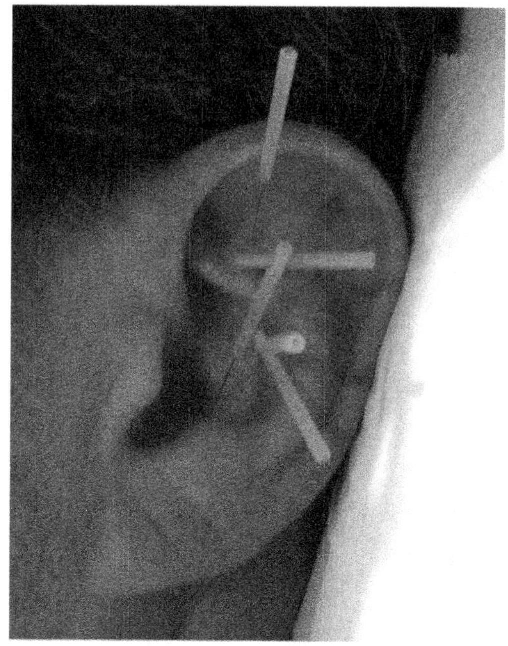

Fig. 1 Ear acupuncture

There have been multiple systematic reviews supporting the clinical efficacy for both ear [8, 9] and body acupuncture in analgesia [10] while the reported side effects are generally minor and transient [11]. Aside from efficacy assessments, there has been ongoing effort in investigating the acupuncture-related analgesia mechanisms based on the current understanding in neuroanatomy, molecular biology, and physiology. These investigations are often conducted in correlation with the traditional acupuncture practice principles.

This chapter reviews several key issues relevant to acupuncture analgesia as follows:

1. The historic background of acupuncture in the Western societies;
2. Pain from the viewpoints of both the Western Medicine and TCM;
3. The current understanding in both peripheral and central mechanisms of acupuncture analgesia;
4. The pros and cons of the methodology behind acupuncture research.

2 Background of Acupuncture

Although acupuncture has been applied for treating different illnesses or symptoms, its application in treating pain is likely more accepted by patients than any other indications. A 2010 survey of hospitals in the United States showed that the top four uses of Complementary Alternative Medicine (CAM) were pain-related [4], and that analgesia was one of the main reasons for patients to seek acupuncturists to either complement or substitute conventional care. Despite ongoing research efforts, studying acupuncture using the randomized controlled trials (RCTs) has been challenging due to a lack of consensus on placebo, and difficulty of study blinding.

2.1 Clinical Efficacy

There is an abundance of acupuncture analgesia literature that are not RCTs (see Table 1). Despite concern over placebo effect, potential conflict of interests, study biases, and conflicting data, these early investigations opened the doors for some of the better designed RCTs conducted more recently.

Table 1 List of non-RCT related to acupuncture analgesia

Positive	Negative or equivocal to comparator
Occipital neuralgia [12]	Local anesthesia for inguinal hernia repair
Cancer pain [13–15]	Labor pain [29]
Cervical radiculopathy [16]	Postpartum surgical repair (compared to lo anesthetic) [30]
Increase pain threshold during athletic training [17]	Supraspinatus tendinitis [31]
Headaches after non-penetrating blast exposure [18]	Tooth pain [32]
Acute dental pain [19]	Trigeminal neuralgia (similar to comparator carbamazepine) [33]
Rheumatoid arthritis [20, 21]	
Posterior pelvic pain and low back pain in pregnancy [22]	
Post total hip arthroplasty [23]	
Endometriosis [24]	

In a 2012 systematic review and meta-analysis of acupuncture analgesia related RCTs, Vickers et al. analyzed data from 29 clinical trials including 17,922 patients and found that acupuncture was more superior than sham acupuncture and placebo for relieving chronic back (Fig. 2) and neck pain, osteoarthritic pain, chronic headache, and shoulder pain [10]. Usichenko et al. reviewed RCTs of ear acupuncture for postoperative pain. 9 out of 23 trials fulfilled inclusion criteria, which included surgical patients with a thoracotomy, burn, hip replacement, knee replacement, oocyte aspiration, molar extraction, knee arthroscopy, or laparoscopic nephrectomy. Results were interpreted as "promising but not compelling" [9]. Another group, Asher et al., analyzed 17 trials which met inclusion criteria, and concluded that ear acupuncture may be effective to treat various types of acute and chronic pain conditions, but further well-designed studies are warranted [8].

2.2 Side Effects

The side effect profile of acupuncture is relatively low, minor, and transient compared to the possible side effects known to occur in other invasive allopathic pain treatments. A 2013 systematic literature review reported that known side effects associated with auricular acupuncture may include short-term local pain, skin irritation, minor bleeding, and dizziness [11]. In a 2015 review of adverse events for acupuncture and moxibustion, tissue, nerve and internal organ injuries (especially pneumothoraxes), though low in frequency, were reported as the main major complications. Other minor side effects may include syncope, infections, hemorrhage, allergy, burn, aphonia, hysteria, cough, thirst, fever, somnolence, and broken needles [12]. Despite theoretical concern with abortion as a possible side effect, the authors of this chapter have not found any published cases of accidental acupuncture-induced abortions in the English-language literature. To the contrary, acupuncture has been safely used for pregnant women with chemical dependence [13], fertility

Fig. 2 Acupuncture for low back pain

treatments [14] and reversal of breeched babies [15, 16].

3 Western and Traditional Chinese Medicine View of Pain

3.1 Western Medicine View of Pain

The Western medical understanding of pain is based on a biopsychosocial model containing interactions among various neuropsychological and neurosensory mechanisms [17]. The International Association for the Study of Pain defines pain as "an unpleasant sensory and emotional experience associated with actual or potential tissue damage, or described in terms of such damage" [18].

3.1.1 Peripheral and Central Pain Processing

In a broad sense, peripheral afferent neurons carry pain signals to the spinal cord where they synapse to excitatory or inhibitory secondary ascending neurons in the dorsal horn. The spinal cord then transmits pain signals up to the brain. The actual experience of pain is multifactorial and subjective. Supraspinal pain processing involves the anterior cingulate gyrus, thalamus, primary and secondary somatosensory cortices. In addition, prefrontal cortices and the amygdala are known to modulate both efferent and afferent signals. Functional magnetic resonance imaging (fMRI) studies have shown that many factors can affect pain perception. These factors include, but are not limited to: the frequency of stimuli [19], placebo analgesia [20], words [21], music [22] mood [23, 24], and even genetics [25, 26].

While peripheral signaling of pain originates distal to the spinal cord and central pain processing occurs in both spinal and supraspinal levels, these nervous systems appear to be highly interactive in pain signal processing and modulation. Central pain mechanistic alterations can often contribute to a peripheral neuropathic pain condition. For example, in painful diabetic neuropathy [27], spinal fMRI reveals that decreased central descending inhibition can contribute to the mechanical hyperalgesia and allodynia observed in the periphery [28], while the maintenance of pain requires continued peripheral noxious input [29].

3.1.2 Nociceptive Versus Neuropathic Pain

Aside from dividing pain mechanisms into central and peripheral, pain can also be broadly classified as nociceptive or neuropathic [30].

Nociceptive pain is usually caused by damage to body tissue and involves specific peripheral nociceptors that detect and transform noxious stimuli as electrical signals via nerve axons. Synaptic excitatory and inhibitory neurotransmitters including amino acids (e.g., glutamate, GABA), gasotransmittors (e.g., NO, CO), monoamines (e.g., dopamine, norepinephrine,

epinephrine, histamine, serotonin), peptides (e.g., substance P, opioid peptides), purines (e.g., ATP, adenosine), and acetylcholine can modulate the transmission of pain signals from the peripheral to the central nervous system, or vice versa.

Neuropathic pain usually occurs when there is nerve damage or recurrent pain sensitization to the peripheral and/or central nervous systems. Primary peripheral pain sensitization can occur when inflammatory mediators such as bradykinin, serotonin, cytokines, and prostaglandins are released after peripheral tissue injury. These mediators stimulate the nociceptors directly, leading to the activation threshold reduction in the afferent signal transmission. Several types of afferent sensory fibers including A-beta, A-delta, and C-fibers can be found on the skin, viscera, meninges, muscles, and joints. A-beta afferent fibers are moderately myelinated and transmit touch and pressure. A-delta-fibers are mildly myelinated and transmit pain and temperature. C-fibers are unmyelinated, and transmit mechanical, thermal, and chemical information. While acute somatic pain is mostly transmitted via A-delta and C-afferent fibers, visceral pain is largely innervated by slow conducting C-fibers. Thus, visceral pain is often poorly localized and perceived as diffuse and dull. These visceral fibers can also be closely associated with post-ganglionic autonomic changes [31], leading to symptoms such as nausea, vomiting, and changes in heart rate variability. Referred pain can occur when both somatic and visceral afferent fibers converge onto the same spinal dorsal horn resulting in the sensation of visceral pain being felt at a site distant from the visceral source [32].

3.1.3 Autonomic Nervous System Contribution to Pain

It is well known that neuropathic pain states can be mediated or augmented with increased sympathetic outflow [33]. The sympathetic system has been one of the targets for interventional pain treatments [34], and enhancing parasympathetic efferent output has been perceived as one of the ways to mitigate sympathetically mediated pain. One potential way that acupuncture may enhance the effect of the parasympathetic nervous system is by needling the outer ear due to the neuroanatomical makeup of the ear.

Various sensory nerves (Fig. 3) that innervate the ear include the lesser occipital nerve (cervical nerve roots), the greater auricular nerve (cervical nerve roots), the auricular-temporal anterior branch of the trigeminal nerve (CN-V3), the posterior auricular branch of the facial nerve

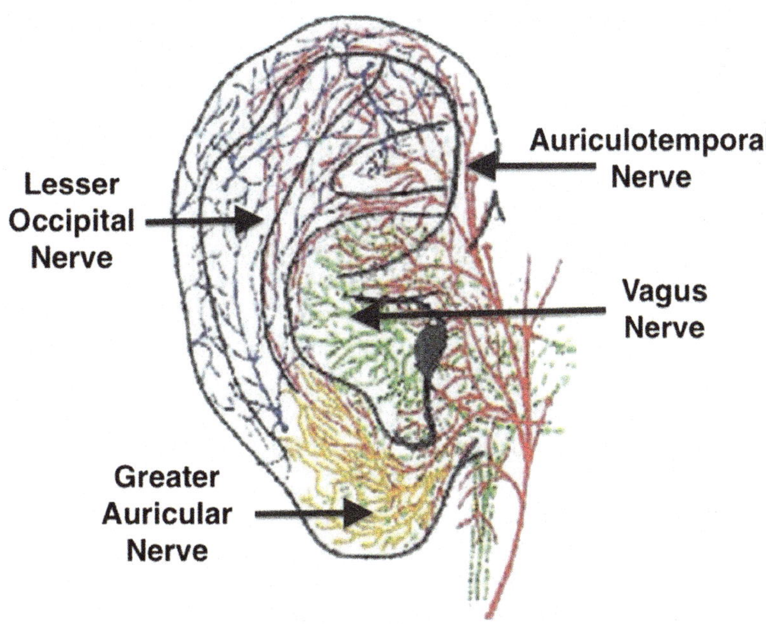

Fig. 3 Ear innervation

(CN-VII), and the auricular branch of the vagus nerve (CN-X). It is interesting to note that the ear is the only peripheral anatomical structure that directly receives vagal innervation. Therefore, directly stimulating the ear provides an effective way to enhance the vagal tone. One particular study has found that auricular acupuncture can affect cardiovascular, respiratory, and gastrointestinal systems [35]. Another study showed that stimulating areas of the external ear innervated by the auricular branch of the vagus nerve can potentially enhance the vagal effect on various organ systems [35]. As discussed earlier, there have been multiple studies showing how ear acupuncture is effective for pain analgesia [8, 36]. In addition, there have been comparative studies showing that electroacupuncture to the ear is more effective for pain control than conventional auricular acupuncture, in patients with chronic neck pain, chronic low back pain or undergoing oocyte aspiration [37].

3.1.4 "Psychogenic" Pain

Psychogenic pain is a term used to describe pain behavior or perception, predominantly caused by various psychological factors [38]. This term is being mentioned to acknowledge that pain threshold can be changed by various psychological components. Later in this chapter, acupuncture's effect on mood and simultaneous effect on pain analgesia will be revealed.

3.2 Traditional Chinese Medicine View of Pain

A TCM practitioner treats a disease state based on the entirety of the person instead of just the presenting symptoms. To illustrate this model of medical practice, 3 patients present with similar low back pain symptoms. At baseline, they have different individual characteristics. The first patient is pale, depressed, and shy. The second patient is ashen, anxious, and often sweaty. The third patient is a large football player with a deep voice and is fierce in personality. In the Western medical setting, they may all be given the same diagnosis, prescribed similar analgesic medications, and sent for physical therapy. On the contrary, a TCM provider will consider the acute complaints of the individual, along with their baseline physical attributes, and treat the imbalance of *Qi*. Understanding the key principles of Yin-yang and *Qi* is crucial in assessing the patient's baseline characteristics, and formulating diagnoses and a treatment plan in TCM.

Fig. 4 Yin-Yang

3.2.1 *Yin-Yang* Theory

Yin-yang theory encompasses the assumption that a part can only be understood in relation to its whole [39]. It describes an opposing and yet complementary duality that is not all or nothing, but rather a balance of two polar entities (Fig. 4). *Yin* and *yang* create and can transform each other [39]. Metaphysical examples of *yin-yang* include: moon–sun, inside–outside, female–male, and moist–dry. In the human body, the lower body is designated as *Yin*, and upper as *Yang* [40].

Yin-yang theory can also be used to characterize disease symptoms. For example, TCM providers will describe yin or yang imbalance as excess or deficient, as illustrated in the following examples:

- *Yin* deficit: heat sensations, possible night sweats, insomnia, dry pharynx, dry mouth, dark urine, a red tongue with scant fur, and a "fine" and rapid pulse.

- *Yang* deficit: aversion to cold, cold limbs, bright white complexion, long voidings of clear urine, diarrhea, pale and enlarged tongue, and a slightly weak, slow, and fine pulse.

In essence, TCM treatments aim to reestablish the balance of *yin-yang*. Acupuncture needling is only one of the modalities of TCM, which also includes other modalities such as herbal remedies, Tai Chi, Tui-Na, and diet and lifestyle changes.

3.2.2 Five Element Theory

Embedded in TCM is also the Five Element theory (Fig. 5). This theory presumes that the universe can be broken down into 5 elemental qualities: metal, water, fire, wood, and earth. Individuals' environmental factors or physical appearances are often used to assign their associated elemental qualities, which in turn consist of predefined interactional relationship.

3.2.3 Chinese Anatomy

Qi

Although there may not be a perfect direct English translation for *Qi*, by and large, it consists of several main connotations including energy, life force, blood, defense mechanism, and breath. In TCM, *Qi* is perceived to circulate through the meridians (to be explained below), and has branches connected to bodily organs and functions. Various attempts have been made to quantify *Qi*. Some say that it has a known frequency [41], while others say it encompasses 4 fundamental physical or energy sources: gravitational, electromagnetic, strong nuclear, and weak nuclear [42]. Although many instruments available directly to consumers claim to be able to measure energetic fields, none have been validated by research. TCM practitioners rely on their perception of *Qi* as part of the *yin-yang* system to treat patients, despite *Qi* having not been validated by quantifiable research. The term *De-Qi* refers *to* a sensation that signifies the arrival of *Qi* at a needled acupoint. The mechanistic assessments of *De-Qi* are discussed in the *De-Qi* section of this book chapter.

Blood (*Xue*) and Body Fluids (*Jinye*)

Equally as important as *Qi* in TCM are the 2 other metaphysical terms known as *Xue* and *Jinye*. *Xue* in TCM correlates with the western

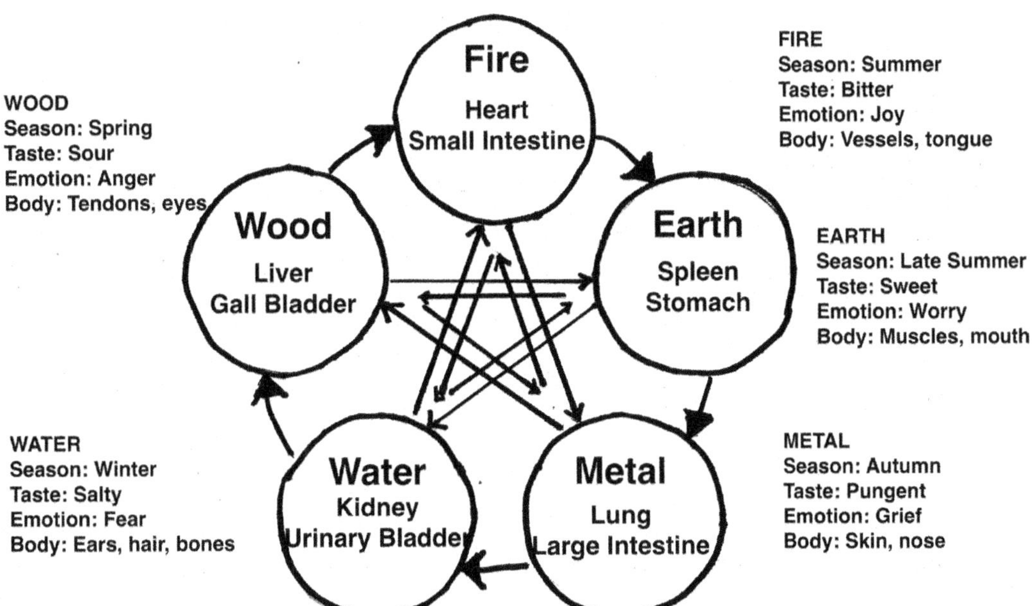

Fig. 5 Five element theory

physical form of blood, whereas, body fluid, known as *Jinye*, nurtures and moisturizes different structures of the body and also helps with secretion (e.g., tears, urine, sweat, joint fluids, gastric acid). *Jinye* is extracted from ingested food items which aids in the creation of *Xue*. Conversely, *Xue*, can also be transformed into *Jinye* [43].

Meridians

In TCM, the meridians are considered channels in which *Qi* travels (Fig. 6). The meridian network is typically divided into 2 categories, the *Jingmai* and the *Luomai* ("collaterals"). The *Jingmai* contains: 12 tendinomuscular meridians, 12 divergent meridians, 12 principal meridians, 8 extraordinary vessels as well as the *Huato* channel, and a set of bilateral low back points. The *Luomai* ("collaterals") contains 15 major arteries that connect the 12 principal meridians in various ways, in addition to interacting with their associated internal organs and other related internal structures. This collateral system also encompasses branching,and capillary-like vessels. There are 361 acupuncture points (not counting bilateral points twice) [44], most of which are situated along the major meridians each of these points is known to have specific functions in various TCM treatment algorithms, which are beyond the scope of this chapter.

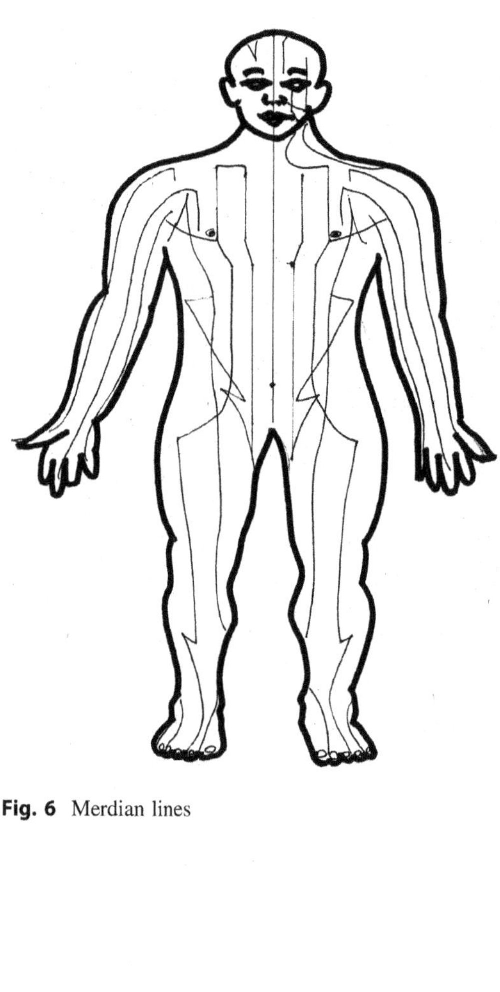

Fig. 6 Merdian lines

3.2.4 Treating Pain in TCM

Pain in TCM is generally viewed as part of the clinical presentation associated with *Qi* imbalance. In order to treat pain, a TCM practitioner may ask their patients detailed questions regarding their pain, sleep pattern, bowel movement types, emotions, and exercise tolerance. The practitioner may holistically evaluate a patient by palpating pulses at various body locations and examinating the patient's tongue and other physical characteristics and in order to formulate the diagnoses of *Qi* deficiency or stagnation, even if a patient only presents with a single complaint of pain [45].

TCM-related pain diagnoses can present as follows:

- Large Intestine Meridian Excess heat: skin lesions and potential cancers
- Kidney *Qi* deficiency: osteoporosis, kidney stones, and arthritic joints
- Spleen *Qi* Deficiency: muscle atrophy and digestive disorders
- Gallbladder *Qi* Deficiency: connective tissue and tendon conditions [46].

Although the naming of organs in the meridian system may have little to do with the actual organs themselves, *Qi*-meridian-based diagnostic approaches serve as the foundation for the acupuncture intervention to follow. Consequently, the mechanistic validation of meridian-based acupuncture treatment paradigms has been the focus of many recent investigations.

4 Mechanisms of Acupuncture Analgesia

4.1 Peripheral Mechanisms of Acupuncture Analgesia

While meridian-based acupuncture treatment has been used for thousands of years with appreciable efficacy, especially in the area of pain management, the correlated physiological and neuronal mechanisms associated with analgesia have only been recently assessed. Several important aspects of these recent discoveries are discussed in the following sections.

4.1.1 De-Qi

In acupuncture, *De-Qi* generally refers to as the "arrival of *Qi*" or the process of "obtaining *Qi*". Some acupuncturists use *De-Qi* to assess treatment efficacy as recipients report a spreading, dull, aching sensation associated with the needle manipulation. In a survey study, it was found that 7 sensations were closely associated with *De-Qi*: aching, dull, heavy, numb, radiating, spreading, and tingling. The sensations of *De-Qi* are distinguished from the 9 sensations known to be correlated with acute pain at the site of needling: burning, hot, hurting, pinching, pricking, sharp, shocking, stinging, and tenderness [47]. In another study, objective measures were used to quantify *De-Qi*. Thirty healthy volunteers (without controls) were given acupuncture treatments with *De-Qi*, and the treatments were noted to increase blood flow, displace tissue, and increase amplitude of electricity created by muscle fibers [48]. In this study, acupuncture treatment also induced fMRI signal changes in different brain regions [48].

In order to differentiate acupuncture points versus control points, a 2015 research endeavor used EEG, event-related fMRI, and resting-state functional connectivity fMRI to assess neural responses to needle stimulation of the acupuncture point ST-36 in the lower leg (Fig. 7) and 2 control point locations located in the same and different dermatomes of the acupoints. Results suggested that stimulation at acupuncture points may modulate somatosensory and saliency-processing regions (to segregate relevant information) more than non-acupuncture points. In addition, the study also suggested potential modulation of pain perception due to specific locations of acupuncture stimulation [49].

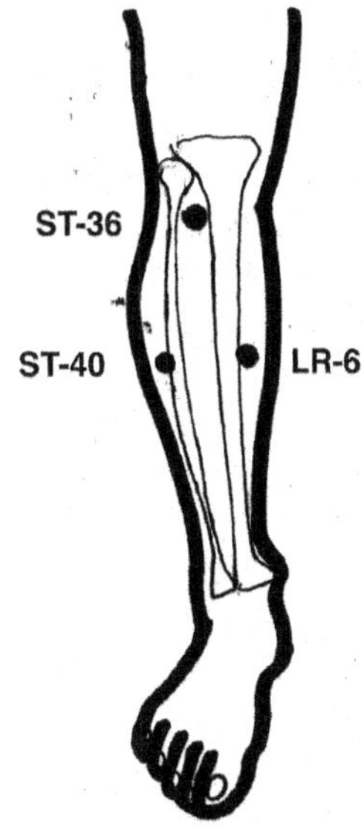

Fig. 7 ST-36

Acupuncture points have also been suspected to exhibit particular direct current, low-frequency electrical and optical properties compared to surrounding skin. A pilot study reveals that dielectric properties of acupoints differ from non-acupuncture sites [50]. Based on a review of the literature, one group concluded that the available evidence did not conclusively support the claim that acupuncture points had distinct electrical properties [51]. A counter article showed that in regards to electrophysiology, *De-Qi* can differ between manual and electroacupuncture, with an observed difference in transcutaneous conduction between true and non-acupuncture points [52].

4.1.2 Acupoints and Peripheral Nerve Endings

Given that acupuncture is perceived as a form of peripheral neuromodulation, the relationship between acupoints and peripheral nerve endings has been explored. A cadaveric dissection study in the 1970s showed that out of 324 acupoints located on meridians, 323 exhibited rich innervation mainly in the deep tissues, indicating that the relationship of the meridian systems to the peripheral nerve endings [44]. Overall, the role of C-fiber involvement in acupuncture analgesia has been controversial. Despite C-fiber activity depletion by capsaicin, it was shown in a rat model that electroacupuncture could still result in reduced analgesia compared to controls [53]. On the other hand, in a separate study done on syringomyelia patients, who suffered damage to the anterior commissure of the spinal cord and have reduced C-fiber and A-delta mediated nociception, it was shown that these patients responded to electroacupuncture poorly. This observation provides the assertion that C-fibers may be essential for acupuncture analgesia [54]. In a more recent investigation done on afferent nerve endings found within acupoints, rat hind feet were used in vivo for recordings of A-alpha, A-beta and A-delta fibers actiivities. It was found that the distribution of receptive fields was closely associated with cutaneous acupoints for both A-fibers and C-fibers. In addition, most of the muscular sensory receptors were also located in the muscular acupoints. These observations strongly suggest that acupoints are closely associated with excitable muscle/skin–nerve complexes with high density of nerve endings [55].

4.1.3 Tendinomuscular Meridians

Tendinomuscular meridians (TMM) are used for acute pain relief [56]. The treatment protocols call for stimulations of the *Ting* Points (at digits of the affected extremities) and Gathering Points (at the location of the injury in the extremity). These meridians are considered to be superficial, and not to be considered as a "root treatment", or deep organ problems. Nonetheless, the protocol is highly effective for acute pain relief [57]. In a study that aimed to characterize the role of *Ting* points in acute pain management, 13 healthy subjects were examined. It was established that a short (30 s) duration of electroacupuncture (5 Hz) treatments at SP-1 and LR-1 (Fig. 8) can result in significant warm threshold increases in the extremities suggesting that *Ting* points have an inhibitory effect on C-fiber afferents and that the analgesic results are likely A-delta mediated [58].

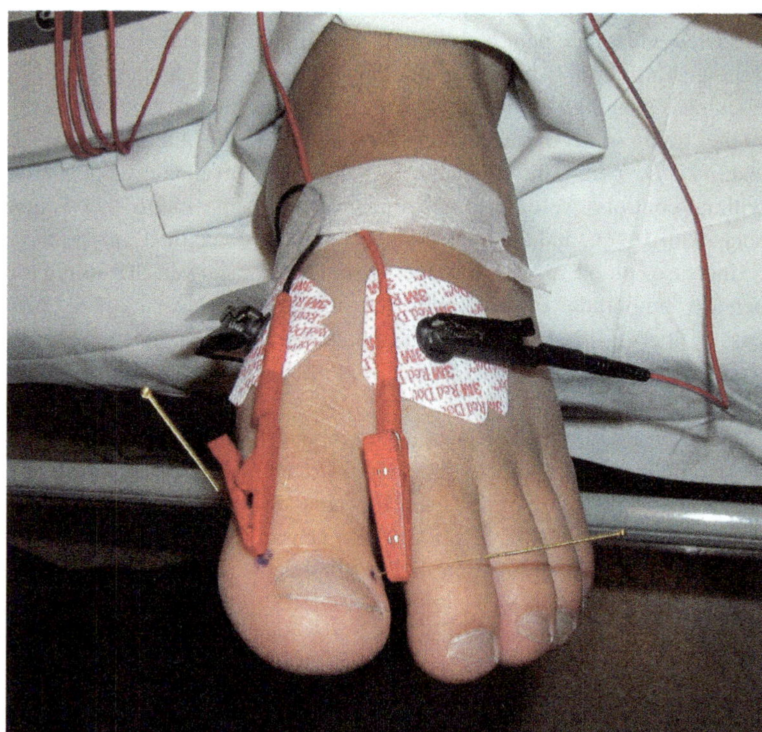

Fig. 8 Electroacupuncture at Ting Points (SP-1 and LR-1) [58]

4.1.4 Connective Tissue: Fascia

A very high degree of anatomical correlation can also be found between myofascial planes and acupuncture meridians in the truncal region. Although it has been theorized that the physical makeup of meridians may comprise of a combination of neurovascular bundles, neuromuscular attachments, sensory nerve endings, perivascular space and perineurial vessels, none of these theories have been substantially proven [59, 60]. In manual acupuncture, the main stimulation comes from a combination of the mechanical pressure of the needle being moved up and down, in conjunction with the mechanical rotation that is done by the practitioner. This motion, which leads to tissue tugging and distortion, stimulates mechanoreceptors, sends off neural signals, and results in the sensation of *De-Qi* [59, 60].

A review article in 2011 suggests that the fascial network is consistent with some of the meridian patterns, and that the efficacy of acupuncture needling relies on interactions with this network [61], which is thought to mediate active mechanical transference between the muscles and bones. The analogy has been made that a disruption in *Qi* flow constitutes disease, and a disruption in fascia physiology is associated with neurogenic inflammation and pathology [61]. In a different paper, it was noted that when acupuncture was compared to no needling physical manipulation, the positive effects on pain analgesia were apparent, whereas when acupuncture was compared to "sham acupuncture" (consisting of needling a non-acupoint), less significant differences between the 2 were noted [62]. Although this observation does not discredit analgesic effects of acupuncture needling, it does discount the specificity of acupuncture needling effects on fascia.

4.2 Neurotransmitters and Neuromodulation

Neurotransmitters and neuromodulators related to analgesia can be found within the peripheral and central nervous systems. Some of the better known acupuncture mechanistic research in this area includes inhibition of NADPH oxidase-mediated oxidative stress [63], activation of opioid receptors (reversible with naloxone) [64], activation of cholinergic muscarinic receptors, and anti-inflammatory signaling (reversible with atropine) [65]. Main investigations assessing the acupuncture needling effects on amino acids, monoamines (serotonin, dopamine, norepinephrine), peptides (opioid peptides, substance P), purines, and inflammatory markers are discussed below.

4.2.1 Amino Acids

Excitatory amino acids, such as glutamate, are abundant in the presynaptic neurons. Aberrant glutamatergic signaling disrupts normal tissue homeostasis and induces pain. Glutamate's role as a neurotransmitter at the synaptic level has been known for many decades. It has even been shown to regulate neurogenesis, neurite outgrowth, synaptogenesis, and neuron survival, playing an integral role in neuronal plasticity [66]. It has been demonstrated that electroacupuncture at the gallbladder meridian, correlated with the distribution of the median nerve (P-5 and P-6, Fig. 9), could attenuate the visceral

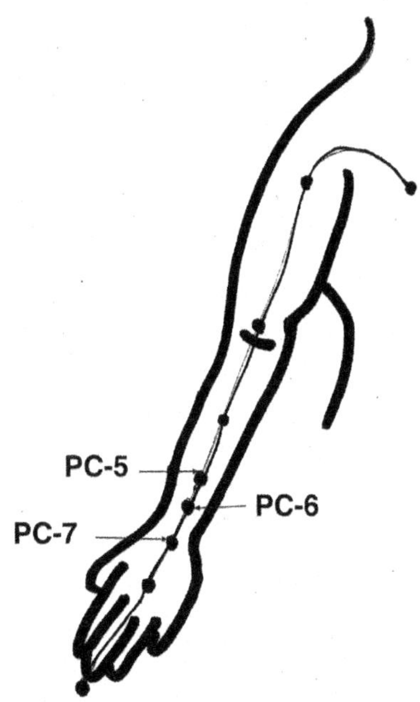

Fig. 9 PC-5 and PC-6

sympathoexcitatory reflexes with diminished bradykinin and glutamate expression/binding at the ventrolateral medulla [67]. Electroacupuncture has also been shown to upregulate Glutamate Transporter-1, and inhibit the excessive release of glutamate in the striatum after ischemic reperfusion brain injury [68]. In the rat model, it has been shown that through the action of central arginine vasopressin, glutamate induces hypothalamic paraventricular nucleus brain signaling with acupuncture analgesia [69].

Inhibitory amino acid transmitters, such as gamma-aminobutyric acid (GABA), also play an important role in the perception of pain. GABAergic neurons known to play an important role in pain perception and modulation are widely distributed throughout in the central nervous system. This neurotransmitter system has been shown to regulate control of sensory information processing in the spinal cord. In animal pain models, it has been discovered that GABA receptor agonists displayed anti-nociceptive properties [70]. In a specific rat study using a middle cerebral artery occlusion model, it was shown that acupuncture could modulate the expressions of GABA receptors in rats that have endured this occlusion [71].

There have been various research endeavors showing the effects of GABA on pain reduction. Systemic administration of a GABA-A receptor antagonist was shown to reduce acupuncture analgesia [72], whereas intrathecal diazepam (a GABA agonist) injection was shown to potentiate acupuncture analgesia [73]. Furthermore, in a research study comparing intracerebroventricular administration of GABA-B versus GABA-A, it was shown that GABA-B (but not GABA-A) receptor antagonist administration decreased acupuncture analgesia. However, the study result from a different group suggests that only GABA-B receptors in supraspinal structures contribute to acupuncture analgesia, whereas both GABA-A and GABA-B receptors in the spinal cord are associated to pain reduction via acupoints needling [74]. Although the contribution of the different GABA receptors may differ, the fact that these receptors play a role in acupuncture analgesia is well supported in the literature.

4.2.2 Monoamines

Various monoamines such as serotonin, dopamine, and norepinephrine also play a role in pain modulation, and have been researched in the context of acupuncture. Some of the investigations are discussed as follows.

Serotonin

Serotonin (5-HT) is present in central and peripheral serotonergic nerve terminals and is also released from platelets and mast cells after tissue injury. Depending on the sites of action and receptor subtypes, 5-HT can elicit pain or have analgesic effects. In regards to eliciting pain, 5-HT, acting in combination with other inflammatory mediators, may also ectopically excite and sensitize afferent nerve fibers, thus contributing to peripheral sensitization and hyperalgesia [75].

In a study that 5-HT antagonists were used to test the pain inhibitory mechanisms of electroacupuncture against the nociceptive responses in the trigeminal nucleus caudalis in rabbits, 5-HT1 (except 5-HT1A), 5-HT2 (except 5-HT2A), and 5-HT3 receptors were shown to be positively involved in electroacupuncture-induced analgesia. On the contrary, activation of 5-HT1A and 5-HT2A receptors was shown to suppress electroacupuncture-induced analgesia [76]. In rats with neuropathic pain, it was found that low-frequency (2 Hz) electroacupuncture was found to be more efficacious for treating cold allodynia than high-frequency electroacupuncture (100 Hz). Both 5-HT1A and 5-HT3, but not 5-HT2A seratonergic receptors also played important roles in mediating the pain analgesic effects of low-frequency electroacupuncture [77].

Many studies have shown that acupuncture treatments can modulate the content and the activity of central 5-HT [78]. An investigation in a rat model assessing the role of peripheral and spinal 5-HT[3] receptors in formalin-induced secondary allodynia and hyperalgesia in rats demonstrated that the stimulation of peripheral 5-HT 3 receptors induced long-term secondary allodynia and hyperalgesia [79]. In addition, electroacupuncture was also noted to activate the seratoninergic raphe-spinal neurons in the nucleus raphe magnus, one of the serotonin-releasing nuclei connecting the lower pons with the medulla in the descending pain inhibitory pathway [80].

Dopamine

The role of dopaminergic neurotransmission via serotonin or norepinephrine in pain perception and modulation is well known [81–86]. These neurotransmitters are found to interact with various pain-related supraspinal regions including the basal ganglia, insula, anterior cingulate cortex, thalamus, and periaqueductal gray in the descending pain inhibitory pathway. Abnormalities in dopaminergic neurotransmission were found in painful clinical conditions such as Parkinson's disease, burning mouth syndrome, fibromyalgia, and restless leg syndrome [85].

The effects of acupuncture on dopamine activities have been varied. In the rabbit model, D2 receptor antagonists (e.g., haloperidol, clozapine) as well as D1 receptor antagonists enhanced acupuncture analgesia [87]. In addition, chlorpromazine (dopamine antagonist) was shown to attenuate electroacupuncture analgesia [87, 88]. A separate animal study done on rats suggests that D2 receptors are involved in pain modulation, and activation of D2 receptors enhances acupuncture analgesia in the spinal cord. However, such effect is absent in D1 receptors and inactivation of the D1 receptors attenuates acupuncture analgesia [89, 90]. Additional receptor-binding studies provide further support that dopamine receptor antagonists can also potentiate electroacupuncture analgesia [91].

Norepinephrine

Norepinephrine (NE) is an important neurotransmitter known to be involved in the process of opioid dependence and pain modulation in the central nervous system. Noradrenergic neurons originate from various brain areas including the raphe nuclei, locus coeruleus, periaqueductal gray, and various nuclei of the brainstem, which then projects to the forebrain and descends along the dorsolateral tracts of the spinal cord [83, 92]. In 2015, a rat study was done to examine the role of NE on the evoked discharges of pain-excitation neurons (PENs) and pain-inhibition neurons (PINs) in the nucleus accumbens using a morphine-dependent model. Results showed that NE inhibited the evoked discharges of PENs and attenuated the inhibition of PINs. In addition, Phentolamine enhanced the evoked discharges of PENs and facilitated the inhibition of PINs. It was concluded that the inhibitory action of NE on pain modulation acted through alpha adrenoceptors in the nucleus accumbens of morphine-dependent rats [93].

The norepinephrine transporter (NET) inhibition has an additional effect on μ-opioid receptor (MOR)-mediated anti-nociception in inflammatory and neuropathic pain [94]. In a rat model, it was shown that electroacupuncture at GV-26, GV-16, PC-6, and BL-15 (Fig. 10) could upregulate the expression of both middle cervical-stellate ganglion complex NET mRNA and myocardial beta1-AR mRNA in cerebral-cardiac syndrome [95]. However, the specificity of each acupuncture point in inducing the observed effect was not assessed in the study. A different study, however, showed the opposite result with a decrease in release of NE with acupuncture. The content of NE in the nucleus reticularis paragigantocellularis lateralis (RPGL) during acupuncture treatments was studied and it was found that pain thresholds increased after 20 min of electroacupuncture, while the content of NE from the RPGL decreased, suggesting NE served as a crucial role in acupuncture induced analgesia [96].

In the spinal cord, NE may have different effects, depending on the receptor subtypes. Studies have showed that spinal alpha2-adrenergic but not alpha1-adrenergic receptors play important roles in mediating the pain relieving effects of 2 Hz electroacupuncture on cold allodynia in neuropathic rats [77].

4.2.3 Peptides

Endogenous opioid peptides have long been considered as one of the main mediators of acupuncture analgesia, with research dating back several decades. In more recent years, there has been a rise in interest regarding the role of acupuncture in regulating the pain signal mediator, substance P.

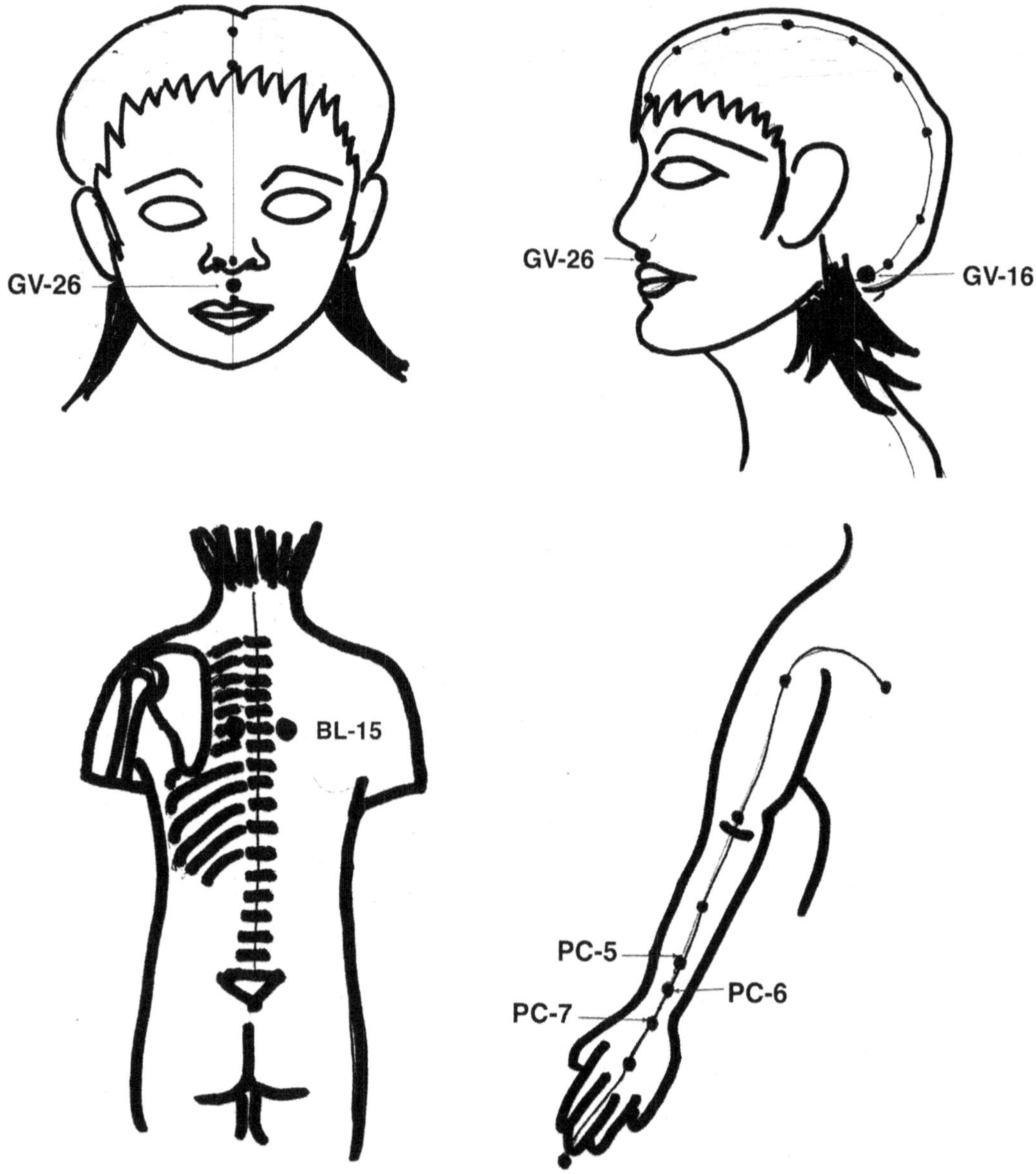

Fig. 10 GV-26, GV-16, PC-6-BL-15

Opioid Peptides

It is well known that acupuncture stimulates the release of endogenous opioids. The effect of acupuncture on both the peripheral and central opioid peptides has been investigated. In the early stages of inflammation, opioid-containing neutrophils are directed into inflamed tissue, stimulating opioid peptide release [97] and aiding in pain reduction.

In 1977, researchers were excited to find that naloxone, an opioid receptor antagonist, could partially reverse the analgesic effect of acupuncture on electrically induced pain on tooth pulp [98]. The acupuncture analgesic reversal effect of naloxone was further demonstrated in chronic pain patients in 1978 [44]. With recent advancement in molecular cloning technology, subtypes of opioid receptors including mu, delta, and kappa [99] have been identified for respective opioid peptide subtypes including beta-endorphin, enkephalins, and dynorphins.

Electroacupuncture has been found to induce long-term anti-nociception, which is blocked by anti-opioid peptide antibodies. In an animal model study, acupuncture was found to increase chemokine CXCL10 release and opioid peptide-containing macrophages expression in an inflammatory state. In control rats that did not get acupuncture, repeated injection of CXCL10 triggered opioid-mediated anti-nociception as well as increased opioid-containing macrophages. On the other hand, neutralizing CXCL10 decreased electroacupuncture-induced anti-nociception and the expression of opioid-containing macrophages [100].

Acupuncture's opioid-effect on the central nervous systems has also been investigated in rabbits. Naloxone was seen to reduce the intensity and duration of the antipyretic action of acupuncture [101]. There is even compelling evidence to support frequency-dependent acupuncture analgesia. Radioimmunoassays of the spinal perfusates from a rat were tested after electroacupuncture was applied at low (2 Hz) and high (100 Hz) frequency. It was found that low-frequency electroacupuncture facilitates the release of encephalin (mu receptor), but not dynorphin (kappa receptor). On the contrary, high-frequency electroacupuncture facilitates release of dynorphin, but not encephalin [44]. Further studies using intrathecal administration of antagonists of mu, delta, and kappa receptors have showed that low-frequency electroacupuncture analgesia is reduced by blocking mu and delta receptors, whereas high-frequency electroacupuncture analgesia is reduced by blocking kappa receptors [44, 102].

Substance P

Substance P is known to facilitate pain transmission in both the peripheral and central nervous systems. It coexists with glutamate in primary afferent fibers, and increased levels of Substance P are found in patients with various pain conditions [103], inflammation [104, 105], stress, and anxiety [106]. Various noxious stimuli can elicit Substance P release in the spinal cord. Electroacupuncture at ST-36 was found to decrease substance P and increase beta-endorphin [107]. In addition, it was noted in a rat model of irritable bowel syndrome, daily electroacupuncture at ST-25 and ST-37 decreased the number of Substance P and its receptor expression in the colon [108].

4.2.4 Inflammatory Markers

Acupuncture also modulates inflammatory processes associated with pain [109]. Peripheral tissue injury causes the release of inflammatory mediators, which in turn leads to initial peripheral pain sensitization. A subsequent chain of reactions ultimately leads to central sensitization in the spinal dorsal horn and other CNS neurons [110]. One of the widely studied inflammatory conditions is arthritis. Electroacupuncture has been shown to reduce the activities of T and B cells in the lymph nodes and enhances natural killer cells in arthritic mice [111]. Pilot data suggests that acupuncture may be a feasible and effective treatment modality for decreasing subjective pain and inflammation as measured by the expression of white blood cells, and for treating patients with acute appendicitis pain [112]. A 2012 review assessing the effect of acupuncture on knee osteoarthritic pain concluded that acupuncture provided significantly better pain relief and improvement in knee function when compared to sham acupuncture, standard care, or waiting treatment [113]. Although clinically, acupuncture continues to be used for various arthritic conditions, not all studies favor acupuncture for inflammatory pain. A 2008 review on acupuncture for rheumatoid arthritis concluded, "despite some favorable results in active-controlled trials, conflicting evidence exists in placebo-controlled trials concerning the efficacy of acupuncture for rheumatoid arthritis. Rigorous and well-controlled randomized trials are warranted" [114].

4.3 Central Mechanisms

Aside from neurotransmitters working on a central level, there are other supraspinal mechanisms that can affect the outcome of acupuncture analgesia. These include psychological and behavioral factors.

4.3.1 Placebo Effect and the Debate on Controls

Methods of control used in acupuncture studies continue to be controversial. Active acupuncture treatment is often compared to sham treatment consisting of needling non-meridian points, stimulating classic acupoints with beads, changing needling depth, using a retractable needle, or using acupressure. However, some of these sham methods such as acupressure or beads along the meridians may still have actual physical and physiological effects on the peripheral nervous/meridian systems, and therefore cannot be considered as an optimal sham. More importantly, manipulation of any kind can induce a placebo analgesic effect. Given that many analgesic investigations, not just acupuncture, are often concluded to not be better than placebo, it is important for the reader to understand the possible effects of placebo analgesia. In 1979, a study was done where intravenous placebo pain medication was given to patients after wisdom tooth extraction. The placebo response was greater if the initial pain was greater than 2.6 on the VAS. This group also reported significantly greater mean analgesia than those reported with initial lower pain values [115]. Over the past several decades, numerous studies have shown that the expectation of being treated for pain [116], physician characteristics, the color of a pill, and the medication package [117] can all affect the perception of pain.

A meta-analysis of 25 neuroimaging studies on placebo analgesia and expectancy-based pain modulation revealed that placebo effects and expectations stimulated regions of the brain which control pain pathways, and even affected the mood related supraspinal regions [116]. In a different meta-analysis involving data from multi-center RTCs for chronic hip OA pain, chronic knee OA pain and low back pain, it was found that high number of planned face-to-face visits predicted the magnitude of the placebo response [118]. Given the effectiveness of placebo analgesia, as well as the positive physiological effects of needling non-meridian points for pain reduction [116, 119], it is conceivable that if acupuncture is shown to be as effective or significantly more effective than placebo analgesia (or sham acupuncture), then the treatment modality can be considered to be clinically effective.

4.3.2 Psychological Contributions and Correlations

Since pain is a subjective sensory experience, the effect of acupuncture on emotional state can potentially affect pain perception. Although it has been shown that *De-Qi* has more of a physiological effect on acupuncture analgesia than psychological factors [120], the psychological contributors of pain perception and their effect on acupuncture analgesia cannot be ignored. It has been shown, for example, that depression independently reduces pain thresholds [121]. Clinical and experimental studies show that the onset of acupuncture effect on depression is more rapid than the effect of selective serotonin reuptake inhibitors, a class of antidepressants [122]. Studies have also shown that antidepressants suppress neuropathic pain by a peripheral beta2-adrenoceptor mediated anti-TNF-alpha mechanism [123]; these neurotransmitters are affected by acupuncture needling as discussed in the Neurotransmitter section.

Aside from depression, it is known that stress can also affect pain perception [124]. In an animal study with cold as a stressor, active ST-36 stimulation prior to cold stress significantly decreased ACTH and cortisol levels, when compared with sham acupuncture or no treatment groups. The active ST-36 treatment group was also effective at preventing stress-induced elevation of adrenal Neuropeptide Y mRNA. The authors concluded that electroacupuncture at ST-36 could block the chronic stress-induced elevations in the hypothalamic–pituitary–axis and sympathetic pathways [125]. Knowing the correlation of stress on pain perception, one can infer that acupuncture can reduce stress hormones contributing to the perception of pain analgesia.

4.3.3 Dynamic Quantitative Sensory Testing

Aside from the visual analogue scale (VAS), peripheral non-noxious and noxious sensory thresholds can be assessed via Quantitative Sensory Testing (QST) (Fig. 11) under specific established protocols [126]. QST refers to tests of sensory perception thresholds during the administration of stimuli. It has proven to be an important instrument to characterize mechanisms underlying somatic and neuropathic pain disorders, but its reliability has not been fully established in patients with visceral pain [127]. QST is also known as psychological testing and can be subdivided into Static QST and Dynamic QST. In Static QST, the states of the peripheral nervous system are measured whereas, the dynamic QST takes measurements after agitation of the pain modulation system [126].

Temporal Summation (TS) and Conditioned Pain Modulation (CPM) are dynamic QST paradigms that have been utilized in acupuncture analgesia related studies. TS and CPM represent the ascending facilitating and descending inhibitory aspects of pain perception respectively [128]. This next section will discuss these 2 outcome measures for acupuncture analgesia, which were included in a 2012 review article [126].

Temporal Summation

Temporal summation refers to increased pain perception in response to repetitive noxious stimuli over time. It correlates with the "windup" phenomenon occurring in the spinal wide dynamic range (WDR) neurons observed in the dorsal horn with repetitive C-fiber stimulation.

In a 2010 RCT that assessed the effect of acupuncture on pain temporal summation, 36 healthy volunteers were randomized into three groups: electroacupuncture (2 and 100 Hz), manual acupuncture, and sham acupuncture. These three different acupuncture treatments were delivered to ST-36 and ST-40 (Fig. 11) on the dominant leg by a blinded practitioner and pain thresholds to single and repeated electrical stimulation pulses were recorded. It was concluded that electroacupuncture induced bilateral, segmentally distributed, and prolonged analgesia for both single pain thresholds and temporal summation thresholds. On the other hand, manual acupuncture increased single pain thresholds and temporal summation thresholds, but these changes were not significantly different from the sham treatments [129]. In a separate randomized crossover pilot study, the effect of acupuncture on endogenous analgesia in chronic whiplash-associated disorders (viewed as

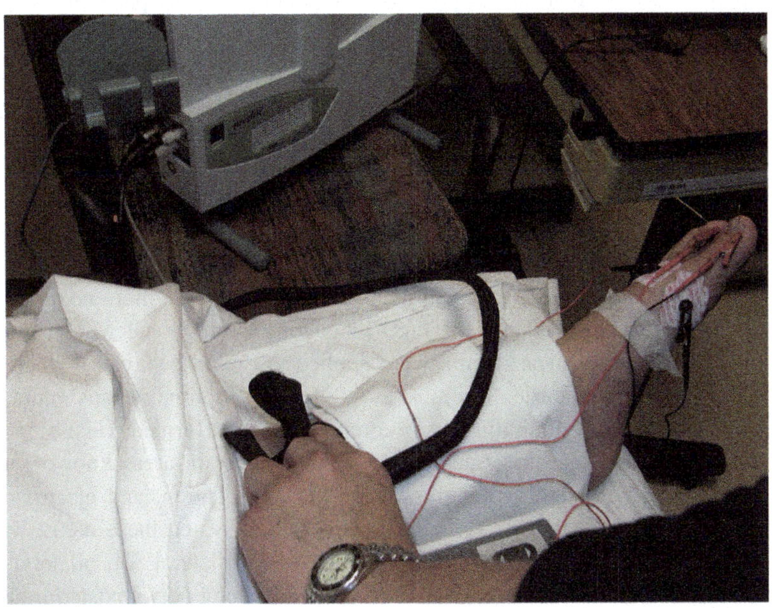

Fig. 11 QST in acupuncture analgesia research

temporal summation/chronic pain model) was investigated. Thirty-nine patients received 2 treatment sessions with an identical induration: acupuncture or relaxation therapy, and then randomly crossed over. One session of acupuncture resulted in acute improvements in pain sensitivity in the necks of patients with chronic whiplash disorder, but had no effect on conditioned pain modulation or temporal summation due to repeated pressure stimuli, suggesting the effect of acupuncture on pain temporal summation is limited [130].

Conditioned Pain Modulation

CPM is a paradigm that uses a conditioning stimulus to influence a testing stimulus. It assesses the perception of a noxious stimulus after a conditioned noxious stimulus. This "treating pain with pain" approach is often referred to as diffuse noxious inhibitory control (DNIC) [131]. The terms are used interchangably although some distinguish DNIC as a neurophysiologic process, and CPM as a behavioral correlate of this process [126].

The underlying CNS physiology of CPM is thought to be a global reduction of wide dynamic range sensitivity due to a single, heterotopic, and noxious stimulation [126]. Various conditioning stimuli have been used to research CPM. One report indicated that the approximate median magnitude of CPM represents about a 29% decrease in pain rating, regardless of the test stimulus [132]. Acupuncture analgesia studies in CPM, where needling certain points is considered the conditioning stimuli, are very limited with only 2 direct studies found comparing acupuncture analgesia to CPM or DNIC.

The first acupuncture and DNIC study was done by recording the convergent neurons in the trigeminal nucleus caudalis of rats. Innocuous and noxious mechanical stimuli were applied to one side of the muzzle. The *Zusanli* (ST-36) acupoints on the right hindlimb was compared to a sham acupuncture point. Acupuncture was also compared to noxious thermal stimulation of the left hind limb (DNIC). Acupuncture (either applied at the *Zusanli* or at the sham point) and noxious thermal stimulation induced similar strong inhibitory and long-lasting effects on the C-fiber-evoked responses of trigeminal convergent neurons. These analgesic effects were significantly reduced by systemic naloxone [64]. A separate study was done on healthy humans to investigate DNIC as a possible mechanism of acupuncture analgesia by comparing acupuncture to non-penetrating sham acupuncture (involving tapping) and cold water bath-induced DNIC. Forty-five subjects were randomized to 1 of 3 interventions and the analgesic effect was measured using pressure algometry at the second toe before and after each of the interventions. Pain pressure detection threshold was significantly increased in the DNIC test compared to acupuncture and sham. Acupuncture and sham effects did have small analgesic effects, but their effects did not significantly differ from one another. It was concluded that acupuncture does elicit acupuncture analgesia, but no different from placebo. Thus, acupuncture effects were significantly less than a DNIC-like effect [133].

Functional MRI

Although acupuncture has been shown to be clinically effective for treating pain, its site-specificity has been questioned. Functional MRI (fMRI) imaging (Fig. 12) has been utilized to show how acupuncture modulates various parts of the brain, including the limbic system [134], periaqueductal gray matter [135], cerebellum [136], motor cortex [137], amygdala [138], hypothalamus [139] basal ganglia, and the brainstem [140]. The effect of acupuncture on brain signaling will be discussed in the following sections.

In 2014, an investigation utilizing a textbook acute pain treatment paradigm [141] on the effect of thermal noxious stimuli was conducted. Functional MRI was used to correlate imaging with behavioral changes when different intensity (optimal versus minimal) electroacupuncture was performed. Heat pain had an excitatory effect on brain areas known for pain processing and perception, and electroacupuncture protocols deactivated these areas, which included the right SI, bilateral SII, bilateral frontal cortices, and bilateral dorsal posterior cingulate cortex (Fig. 13). In

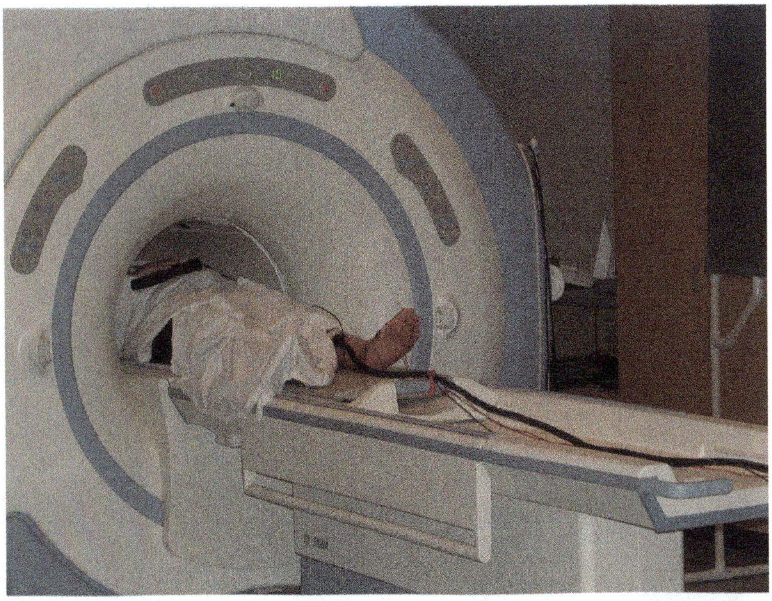

Fig. 12 Functional MRI with acupuncture

addition, optimal intensity electroacupuncture, when compared to minimal electroacupuncture, was found to elicit a more robust supraspinal effect on pain modulation and perception [142].

The same investigators also compared the effect of acupuncture needle combination on the central pain modulation. Eleven healthy subjects were divided into 2 groups that got either: (1) *Ting* points (tendinomuscular meridians used for acute pain, such as LR-1 and SP-1 or SP-2) a combination treatment of *Ting* points (LR-1 and SP-1) with Gathering points (CV-2) (Figs. 14 and 15) [143]. Thermal pain was introduced at the medial aspects of the legs. While electroacupuncture at *Ting* Points alone reduced pain, adding the Gathering Point provided a more sustainable analgesic effect. These results led the investigators to conclude that while both groups had a significant degree of deactivation in the human brain regions related to pain processing, the addition of the Gathering Point stimulation enhanced the inhibitory effect on the ascending spinoreticular pain pathway (Fig. 16) [144].

The supraspinal effect of electroacupuncture has also been compared to manual acupuncture using fMRI. One study observed the differences between subjects that received: manual acupuncture, electroacupuncture at 2 Hz (low frequency) and 100 Hz (high frequency), and tactile control stimulation was set up as sham. All 3 groups received ST-36 (Fig. 17) stimulation. In general, electroacupuncture (low more than high frequency) produced more widespread fMRI signal increases than manual acupuncture which in turn provided more signaling than the sham group. There were also specific findings between the 3 groups. Unlike sham tactile stimulation, both manual and electroacupuncture showed supraspinal activations in the anterior insula, limbic, and paralimbic structures, the cortices of the subgenual and retrosplenial cingulate, as well as the ventromedial prefrontal cortex, frontal, and temporal poles. However, only electroacupuncture produced significant signal increase in the anterior middle cingulate cortex with the 2-Hz electroacupuncture produced signal increase in the pontine raphe area [145].

In an attempt to show that needling same-meridian acupoints have similar effects on the brain, 53 healthy subjects were randomly divided in 6 groups. Two different acupoints of the liver meridians of the foot (LR-3 and LR-6), 2 stomach meridian acupoints (ST-36, ST-43), and 2 nearby sham points were tested (Fig. 18). Each subject received stimulation at one acupoint on the right side of the body. Results of fMRI data

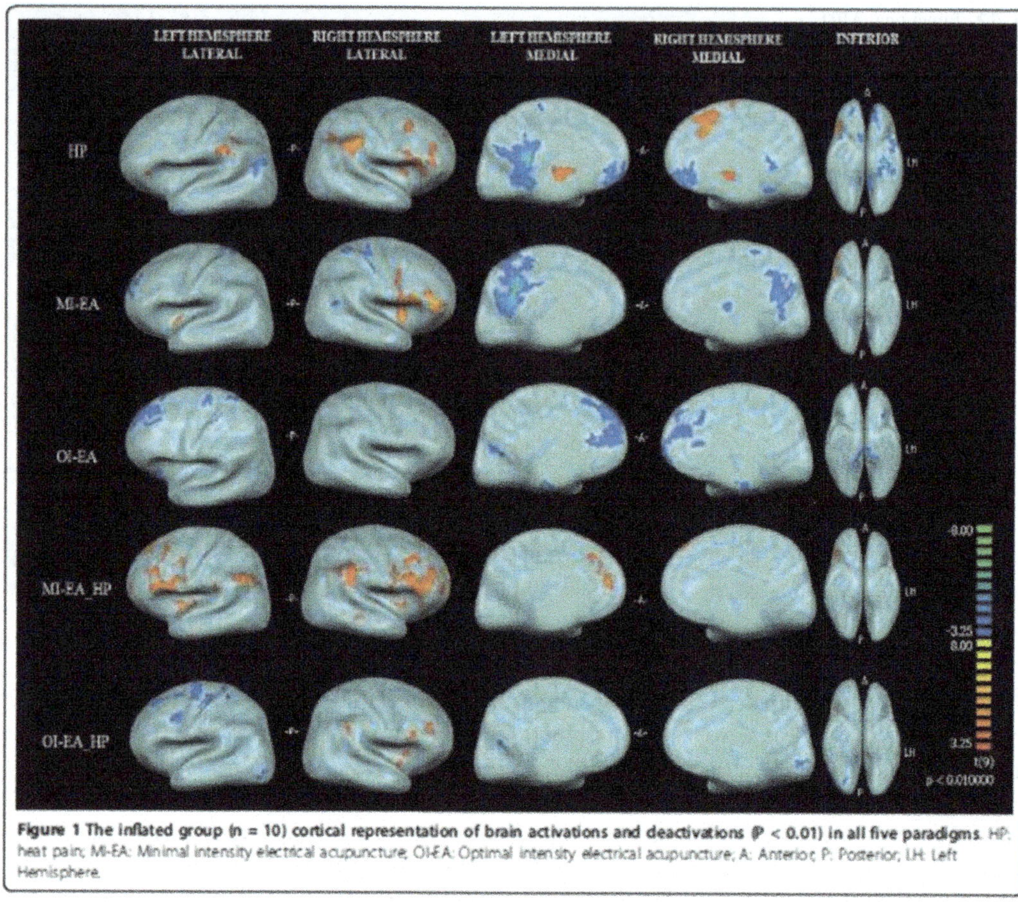

Fig. 13 Inflated cortical representation of identified brain areas of activation and deactivation and deactivation in all 5 paradigms

Fig. 14 CV-2, LR-1, SP-1

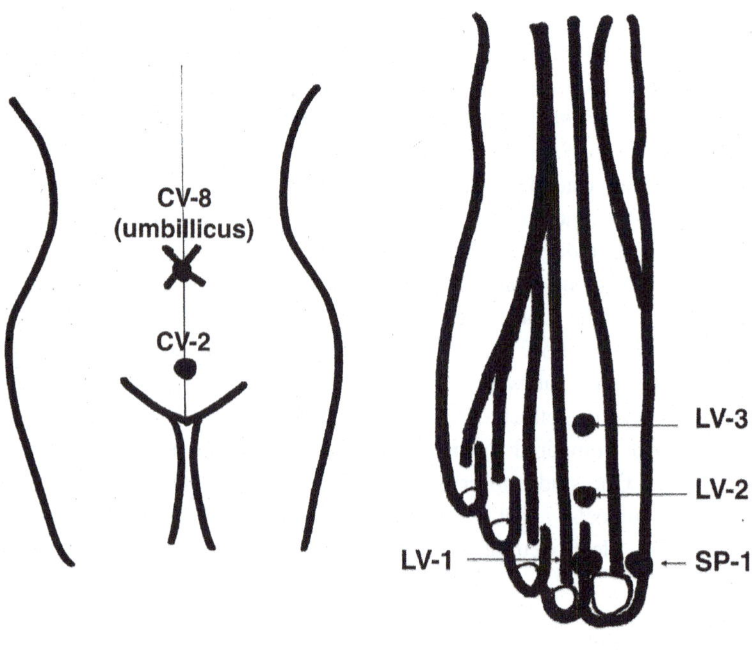

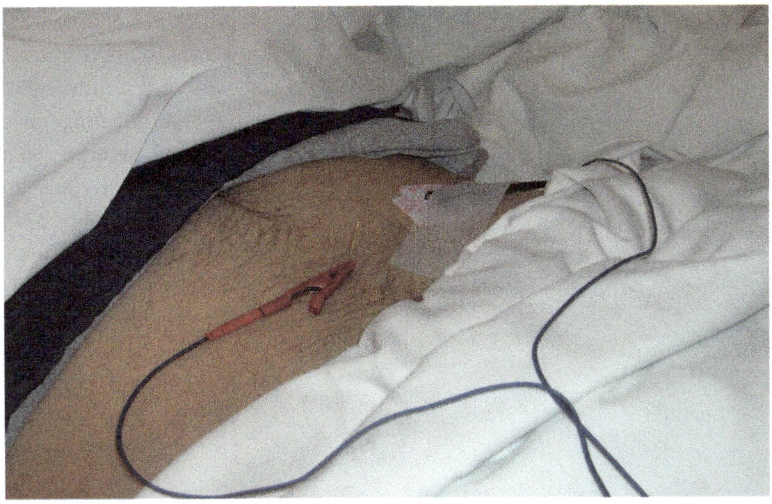

Fig. 15 Gathering point needle placement (CV-2)

analyses showed, that while stimulating both liver points evoked activation at the ipsilateral superior parietal lobe, similar stimulation given at the stomach points activated the ipsilateral middle frontal gyrus. In contrast to the activation of the brain by the sham acupoints, all 4 real acupoints

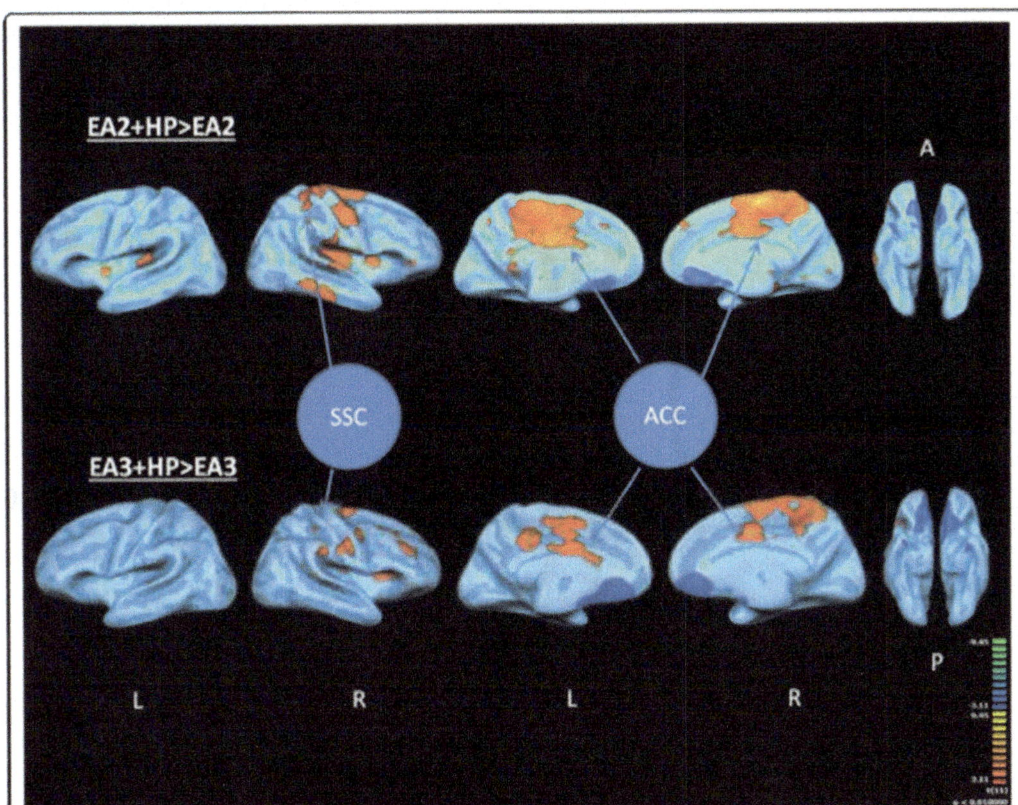

Fig. 16 Effect of Ting and gathering points with heat stimulation [144]

Fig. 17 ST-36

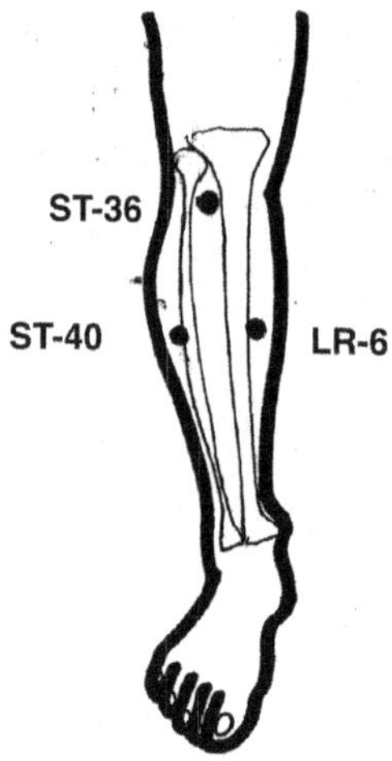

had the common effect of activating 2 specific areas of the brain, the bilateral primary somatosensory area and the ipsilateral cerebellum [146], showing that the pattern of suprapsinal activation from 2 different meridians are somewhat different, even though some brain regions are activated by stimulating either meridian.

Evidence from fMRI imaging has even shown that acupuncture modulates temporal neural responses in widely distributed brain network, which has a large overlap with pain-related areas. It was demonstrated that brain activities could vary at the different stages of acupuncture. During the needling phase, the amygdala and perigenual anterior cingulate cortex exhibited increased activities, and then signals decreased to below baseline. The periaqueductal gray and hypothalamus showed intermittent signaling during the entire fMRI session and even after the acupuncture needling was terminated [147]. The result of the study suggests that the effect of acupuncture on pain perception often outlasts the duration of the stimulation.

Other studies have also shown the site specificities of the acupoints. In one of these studies, twenty-one healthy male volunteers were enrolled into a crossover trial comparing ST-44 to a sham acupoint treatment. ST-44 stimulation affected the inferior parietal and the prefrontal cortex and the posterior insula whereas, the sham acupoint stimulation activated the anterior cingulate cortex and the anterior insula. [148].

In a 2012 systematic review and meta-analysis, the authors aimed to provide an overview of fMRI acupuncture research regarding: (1) sham versus true acupuncture, (2) effects

Fig. 18 LR-3 & LR-6

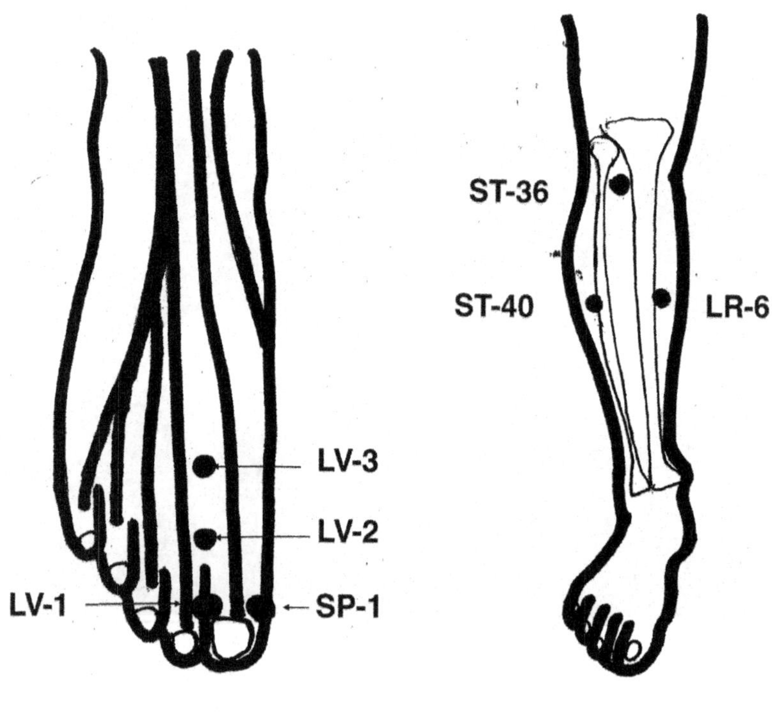

Table 2 Meta-analysis on Acupuncture with fMRI [149]

	Investigated topics	Results
1	Sham versus true acupuncture	Verum acupuncture stimuli confirmed brain activity within many areas of the brain. True versus sham acupuncture differences were noted in the middle cingulate regions. Some heterogeneity was noted, depending on how the meta-analyses was conducted
2	Effects of acupuncture needle manipulation, including electroacupuncture	Increased intensity and duration of needling was also found to increase brain response. Electroacupuncture showed more activation than manual acupuncture with low (2 Hz) versus high (100 Hz) frequencies showing different brain activity
3	Differences between healthy and non-healthy volunteers	Healthy volunteers respond differently to acupuncture compared to non-healthy volunteers
4	Brain effects from different acupuncture points	Brain maps of different acupuncture points differ. However, acupuncture points on the same meridian showed some similarities in brain signaling

of acupuncture needling manipulation, (3) differences between healthy and non-healthy volunteers and (4) the effect of different acupuncture points on the the brain. In this review, 779 papers were identified, 149 met inclusion criteria for analysis, and 34 were eligible for the meta-analysis. The main findings of the analyses are summarized in Table 2. It was concluded that the brain's response to acupuncture was consistent with somatosensory as well as affective and cognitive processing areas [149].

4.4 The Role of the Autonomic Nervous System in Acupuncture Analgesia

The discovery that cholinergic neurons in the autonomic nervous system inhibit acute inflammation has qualitatively expanded our understanding of how the nervous system modulates immune responses. It is now known that the nervous system can regulate the inflammatory response in real time, just as it controls heart rate and other vital functions. The effect of acupuncture on the ANS is known to occur both centrally and peripherally, and thus providing another line of therapeutic mechanisms related to acupuncture. [150].

Previous studies have shown that the parasympathetic nervous system can be activated by directly stimulating the vagus nerve. Stimulating the ear via acupuncture at *Shenmen* and "Point Zero" [151] in the outer ear has been shown to affect heart rate variability [152]. In addition, these acupuncture points have also been used for treating depression, epilepsy [153], and pain [37, 154–160].

Various ear acupuncture protocols including one by a military physician for pain analgesia have been developed (Fig. 19) [161–166]. Some of these protocols call for the stimulation of the auricular acupuncture "Point Zero," which serves as a conduit for activating the vagus nerve [162].

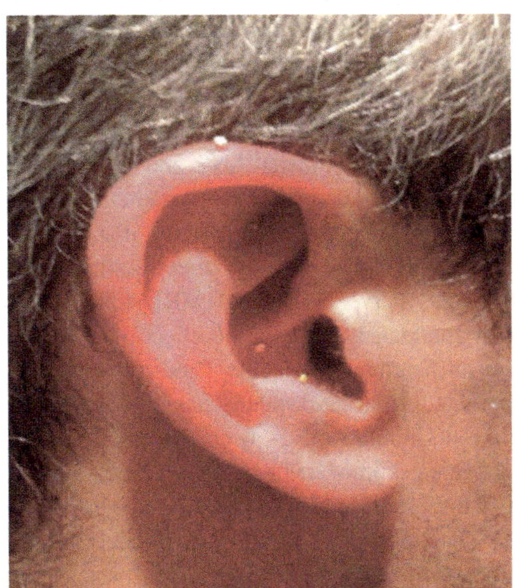

Fig. 19 Battlefield Acupuncture Protocol for Pain Analgesia using Semi-Permanent Needles

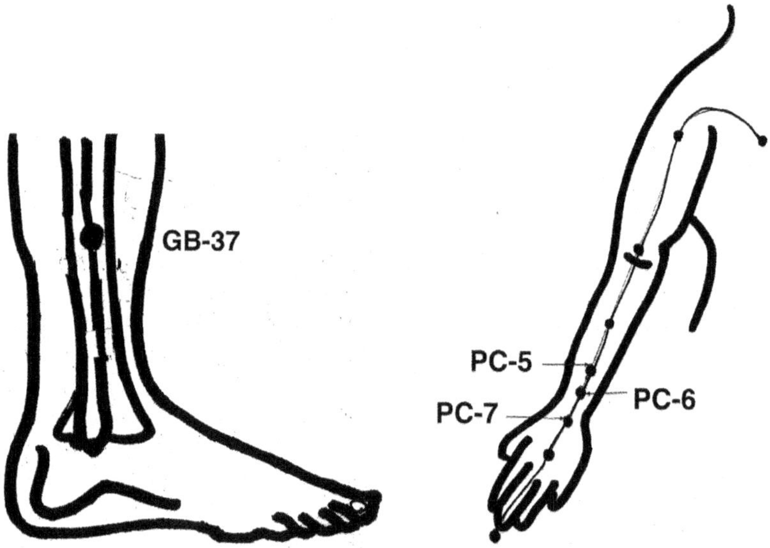

Fig. 20 PC-6,PC-7,GB-37

A different study stimulating the vagus nerve via the ear on patients with chronic pelvic pain due to endometriosis, demonstrated a significant reduction in anxiety, and reduction trend in evoked pain intensity and temporal summation of mechanical pain [167]. Vagal nerve stimulation has also been shown to increase and decrease pain threshold without affecting heart rate and blood pressure [168]. Aside from affecting pain perception, stimulating *Shen Men* in postoperative patients who receive ketamine anesthesia has been shown to reduce hallucinations at the beginning of the emergence period [169].

The vagus nerve can be directly stimulated to activate parasympathetic ANS and induce pain analgesia. However, the parasympathetic activity can also be induced by stimulating distal points in the extremities, without directly stimulating the vagus nerve. Functional MRI studies have been conducted to assess the effect of acupuncture on heart rate variability, and this has been correlated with supraspinal changes. One study showed stimulating ST-36 could induce significant changes in heart rate variability with corresponding supraspinal functional changes in the hypothalamus, the dorsal raphe nucleus, the periaqueductal gray, and the rostroventral medulla. These observations support the assertion that acupuncture needling can affect both central and peripheral autonomic nervous systems [140].

Other studies also showed that distal (i.e., not directly on the vagus nerve) acupoint activation could affect the autonomic nervous system and result in indirect pain analgesic and sedative outcomes. For example, Pericardium 6 (PC-6) is a point that is used in for various conditions, including chest tightness, palpitations, nausea, and carpal tunnel syndrome [139]. Functional MRI with stimulation at the PC-6, a point in the forearm, shows selective responses in the insula, hypothalamus, and flocculonodular lobe of the cerebellum with correlated effect on autonomic regulatory functions and pain. These effects were not observed with stimulation at the control acupoints (PC-7, GB-37) (Fig. 20) [170].

5 Discussion

Although acupuncture has been used clinically over thousands of years, its associated analgesic mechanisms have only been explored in the past few decades. Several aspects of acupuncture research including research tools, study design, choices of control, and subject and practitioner blinding, and overall limitations are worthy of discussion.

Quantitative Sensory Testing (using temporal summation and conditioned pain modulation) and fMRI has vastly improved the current understanding in the analgesic mechanisms of acupuncture. However, challenges still exist even with these advanced research tools. Issues that can affect the outcome of the QST assessment may include: (1) a lack of temporal stability ; (2) an inconsistency in testing methodology; (3) individual variabilities in TS and CPM responses; and (4) subjects' compliance or ability to follow instruction [126, 171]. In the area of fMRI related acupuncture studies, it was noted that not all acupuncture-fMRI studies met the strict methodological requirements including the choice of baseline, issues related to the interpretation of deactivations, problems with attention control and implications of different group statistics [172].

A major issue with in acupuncture study design is the focus on healthy subjects with single-session needling methods. These study treatment approaches do not necessarily reflect the TCM clinical treatment paradigms, in which patients often receive multiple sessions of acupuncture treatments consisting of multiple needles. This difference between study treatment approach and clinical practice somewhat diminishes the translational impact of the study results in the real clinical world, especially in the chronic pain setting. Although recent meta-analyses have shown positive results for acupuncture in managing chronic headache, back, neck, and shoulder pain [10], further research is required to validate the effect of acupuncture on other chronic pain conditions. In addition, treatment paradigms with clinical relevance to a specific diagnosis (e.g., lumbar radiculopathy or lumbar facet arthropathy, as opposed to lumbago) will greatly enhance the translational nature of the study result [126].

Choices of control as well as patient and practitioner blinding continue to be a topic of debate in the world of acupuncture research. TCM providers often believe that stimulating any meridian point, whether with an ear seed (small seeds held in place with adhesive tape) or acupressure (pushing on acupoints with a finger or a tool) can have a therapeutic or physiological effect and research has shown that needling anywhere in the body will elicit diffuse inhibitory control of pain [116, 119]. In addition, expectancy of receiving acupuncture alone has been shown to alter brain function associated with mood and pain perception [116]. Several approaches can potentially minimize the confounding issues of blinding in an acupuncture related RCT may include (1) sedating the research subjects during the treatment; (2) using retractable needles for sham treatments; and/or (3) adopting a blinded personal to conduct the study related assessments.

The Society of Acupuncture Research (SAR) has recently proposed some guidance for future acupuncture related research. One of the proposed ideas is the development of biomarkers that can provide meaningful correlations between animal pain analgesia and patient reported outcomes [173].

6 Conclusion

Acupuncture is an ancient modality within TCM. It continues to be used for acute and chronic pain management. This modality has recently caught the interest of consumers in the United States due to its positive efficacy, and the side effect related limitations in other currently available pain treatment modalities in Western Medicine. To fully expand the practice of this consumer-driven intervention in conventional medical practice will require credible clinical outcome data and a clear understanding in its treatment-related mechanisms. Despite the controversies surrounding acupuncture-related research, high-quality research continues to emerge to explain the effects of acupuncture on the peripheral and central nervous systems, muscle fascia, and neurotransmitter signaling, and to support its clinical efficacy in treating various pain conditions. With ongoing improved outcome measures and research tools such as fMRI and Quantitative Sensory Testing, further research on the mechanisms of acupuncture is warranted to support its clinical application in pain management.

References

1. Relieving Pain in America, A Blueprint for Transforming Prevention, Care, Education and Research. Institute of Medicine Report from the Committee on Advancing Pain Research, Care, and Education, 2011.
2. Prevention CfDCa. Injury Prevention & Control: Prescription Drug Overdose—What the Public Needs to Know about the Epidemic 2015 [updated May 5, 2015]. Available from: http://www.cdc.gov/drugoverdose/epidemic/public.html.
3. Prevention CfDCa. Injury Prevention & Control: Prescription Drug Overdose—Understanding the Epidemic 2015 [April 30, 2015]. Available from: http://www.cdc.gov/drugoverdose/epidemic/index.html.
4. Sita Ananth M. Complementary alternative medicine survey of hospitals—summary of results. Alexandria, VA: Samueli Institute; 2010. p. 2011.
5. White A, Ernst E. A brief history of acupuncture. Rheumatology. 2004;43(5):662–3.
6. Zhang X. Acupuncture: review and analysis of reports on Controlled Clinical Trials World Health Organization; 2003. Available from: http://apps.who.int/medicinedocs/pdf/s4926e/s4926e.pdf.
7. A standard international acupuncture nomenclature: memorandum from a WHO meeting. Bull World Health Organiz 1990;68(2):165–9.
8. Asher GN, Jonas DE, Coeytaux RR, Reilly AC, Loh YL, Motsinger-Reif AA, et al. Auriculotherapy for PAIN management: a systematic review and meta-analysis of randomized controlled trials. J Altern Complement Med. 2010;16(10):1097–108.
9. Usichenko TI, Lehmann C, Ernst E. Auricular acupuncture for postoperative pain control: a systematic review of randomised clinical trials. Anaesthesia. 2008;63(12):1343–8.
10. Vickers AJ, Cronin AM, Maschino AC, Lewith G, MacPherson H, Foster NE, et al. Acupuncture for chronic pain: individual patient data meta-analysis. Arch Intern Med. 2012;172(19):1444–53.
11. Tan JY, Molassiotis A, Wang T, Suen LK. Adverse events of auricular therapy: a systematic review. Evid Based Complement Altern Med eCAM. 2014;2014:506758.
12. Wu J, Hu Y, Zhu Y, Yin P, Litscher G, Xu S. Systematic review of adverse effects: a further step towards modernization of acupuncture in China. Evid Based Complement Altern Med eCAM. 2015;2015:432467.
13. Janssen PA, Demorest LC, Kelly A, Thiessen P, Abrahams R. Auricular acupuncture for chemically dependent pregnant women: a randomized controlled trial of the NADA protocol. Subst Abuse Treat Prev Policy. 2012;7:48.
14. Cochrane S, Smith CA, Possamai-Inesedy A. Development of a fertility acupuncture protocol: defining an acupuncture treatment protocol to support and treat women experiencing conception delays. J Altern Complement Med. 2011;17(4):329–37.
15. Li X, Hu J, Wang X, Zhang H, Liu J. Moxibustion and other acupuncture point stimulation methods to treat breech presentation: a systematic review of clinical trials. Chin Med. 2009;4:4.
16. Neri I, Airola G, Contu G, Allais G, Facchinetti F, Benedetto C. Acupuncture plus moxibustion to resolve breech presentation: a randomized controlled study. J Matern Fetal Neonatal Med (The Official Journal of the European Association of Perinatal Medicine, the Federation of Asia and Oceania Perinatal Societies, the International Society of Perinatal Obstet). 2004;15(4):247–52.
17. Zhang X, Cao B, Yan N, Liu J, Wang J, Tung VO, et al. Vagus nerve stimulation modulates visceral pain-related affective memory. Behav Brain Res. 2013;236(1):8–15.
18. IASP Taxonomy. Available from: http://www.iasp-pain.org/Taxonomy.
19. van den Broeke EN, Mouraux A. Enhanced brain responses to C-fiber input in the area of secondary hyperalgesia induced by high-frequency electrical stimulation of the skin. J Neurophysiol. 2014;112(9):2059–66.
20. Schmid J, Langhorst J, Gass F, Theysohn N, Benson S, Engler H, et al. Placebo analgesia in patients with functional and organic abdominal pain: a fMRI study in IBS, UC and healthy volunteers. Gut. 2015;64(3):418–27.
21. Osaka N, Osaka M, Morishita M, Kondo H, Fukuyama H. A word expressing affective pain activates the anterior cingulate cortex in the human brain: an fMRI study. Behav Brain Res. 2004;153(1):123–7.
22. Garza-Villarreal EA, Jiang Z, Vuust P, Alcauter S, Vase L, Pasaye EH, et al. Music reduces pain and increases resting state fMRI BOLD signal amplitude in the left angular gyrus in fibromyalgia patients. Front Psychol. 2015;6:1051.
23. Gosset N, Dietz N. Unlocking pain: deep brain stimulation might be the key to easing depression and chronic pain. IEEE Pulse. 2015;6(2):16–20.
24. Brain scans indicate that depression can increase pain perception. The Harvard Mental Health Letter from Harvard Medical School. 2009;25(9):7.
25. Asghar MS, Pereira MP, Werner MU, Martensson J, Larsson HB, Dahl JB. Correction: secondary hyperalgesia phenotypes exhibit differences in brain activation during noxious stimulation. PLoS ONE. 2015;10(5):e0128640.
26. Asghar MS, Pereira MP, Werner MU, Martensson J, Larsson HB, Dahl JB. Secondary hyperalgesia phenotypes exhibit differences in brain activation during noxious stimulation. PLoS ONE. 2015;10(1):e0114840.
27. Fischer TZ, Waxman SG. Neuropathic pain in diabetes–evidence for a central mechanism. Nat Rev Neurol. 2010;6(8):462–6.

28. Rempe T, Wolff S, Riedel C, Baron R, Stroman PW, Jansen O, et al. Spinal fMRI reveals decreased descending inhibition during secondary mechanical hyperalgesia. PLoS ONE. 2014;9(11):e112325.
29. Kunimoto M. [The peripheral mechanism of physiological pain]. Brain and Nerve = Shinkei kenkyu no shinpo. 2012;64(11):1205–14.
30. Chang S. The meridian system and mechanism of acupuncture—a comparative review. Part 2: mechanism of acupuncture analgesia. Taiwan J Obstet Gynecol. 2013;52(1):14–24.
31. Hanson P, Schumacker P, Debugne T, Clerin M. Evaluation of somatic and autonomic small fibers neuropathy in diabetes. Am J Phys Med Rehabil Assoc Acad Physiatr. 1992;71(1):44–7.
32. Introduction to Pain Pathways and Mechanisms.
33. Schlereth T, Birklein F. The sympathetic nervous system and pain. NeuroMol Med. 2008;10(3):141–7.
34. Abramov R. Lumbar sympathetic treatment in the management of lower limb pain. Curr Pain Headache Rep. 2014;18(4):403.
35. He W, Wang X, Shi H, Shang H, Li L, Jing X, et al. Auricular acupuncture and vagal regulation. Evid Based Complement Altern Med eCAM. 2012;2012:786839.
36. Usichenko TI, Lehmann C, Ernst E. Auricular acupuncture for postoperative pain control: a systematic review of randomised clinical trials. Anaesthesia. 2008;63(12):1343–8.
37. Sator-Katzenschlager SM, Michalek-Sauberer A. P-Stim auricular electroacupuncture stimulation device for pain relief. Expert Rev Med Devices. 2007;4(1):23–32.
38. Psychogenic Pain: Cleveland Clinic. Available from: http://my.clevelandclinic.org/services/anesthesiology/pain-management/diseases-conditions/hic-psychogenic-pain.
39. Kaptchuk TJ. The web that has no weaver. Lincolnwood, Ill.; St. Albans: Contemporary; Verulam [distributor]; 2000.
40. Wiseman N. Fundamentals of Chinese medicine. Southfield, Michigan: Readings for the Blind; 2007.
41. Hayano DS. Measuring Qi Energy Available from: http://www.equilibrium-e3.com/images/PDF/Measuring%20Qi%20Energy.pdf.
42. Hui Lin PD. Overview of the status of Chinese Chi Research International Yan Xin Qigong Association. Available from: http://www.item-bioenergy.com/infocenter/chinesechiresearch.doc.
43. Traditional Chinese Medicine: Wikipedia. Available from: https://en.wikipedia.org/wiki/Traditional_Chinese_medicine.
44. Zhao ZQ. Neural mechanism underlying acupuncture analgesia. Prog Neurobiol. 2008;85(4):355–75.
45. Shi R. Pain according to Chinese Medicine San Gabriel, Texas: Rei Shi Acupuncture Clinic. Available from: http://reishiacupuncture.com/page/pain-according-to-chinese-medicine.
46. Eades W. Ancient Chinese Medicine Meets Modern Anatomy Dissection: Acupuncture Today; 2015. January, 2015, Vol. 16, Issue 01. Available from: http://www.acupuncturetoday.com/mpacms/at/article.php?id=32980.
47. MacPherson H, Asghar A. Acupuncture needle sensations associated with De Qi: a classification based on experts' ratings. J Altern Complement Med. 2006;12(7):633–7.
48. Tian DS, Xiong J, Pan Q, Liu F, Wang L, Xu SB, et al. De qi, a threshold of the stimulus intensity, elicits the specific response of acupoints and intrinsic change of human brain to acupuncture. Evid Based Complement Altern Med eCAM. 2014;2014:914878.
49. Nierhaus T, Pach D, Huang W, Long X, Napadow V, Roll S, et al. Differential cerebral response to somatosensory stimulation of an acupuncture point vs. two non-acupuncture points measured with EEG and fMRI. Front Hum Neurosci. 2015;9:74.
50. Egot-Lemaire SJP, Ziskin MC. Dielectric properties of human skin at an acupuncture point in the 50–75 GHz frequency range A pilot study. Bioelectromagnetics. 2011;32(5):360–6.
51. Ahn AC, Wu J, Badger GJ, Hammerschlag R, Langevin HM. Electrical impedance along connective tissue planes associated with acupuncture meridians. BMC Complement Altern Med. 2005;5(1):10.
52. Leung AY, Park J, Schulteis G, Duann JR, Yaksh T. The electrophysiology of de qi sensations. J Altern Complement Med. 2006;12(8):743–50.
53. Zhu L, Li C, Yang B, Ji C, Li W. The effect of neonatal capsaicin on acupuncture analgesia–to evaluate the role of C fibers in acupuncture analgesia. Zhen ci yan jiu = Acupuncture research [Zhongguo yi xue ke xue yuan Yi xue qing bao yan jiu suo bian ji]. 1990;15(4):285–91.
54. Cao X. Scientific bases of acupuncture analgesia. Acupunct Electrother Res. 2002;27(1):1–14.
55. Li AH, Zhang JM, Xie YK. Human acupuncture points mapped in rats are associated with excitable muscle/skin-nerve complexes with enriched nerve endings. Brain Res. 2004;1012(1–2):154–9.
56. Helms J. The basic, clinical, and speculative science of acupuncture. In: Helms J, editor. Acupuncture energetics a clinical approach for physicians. Berkeley: Medical Acupuncture Publishers; 1995. p. 19–43.
57. Fuiyu Yip L. Tendinomuscular pain treatment with acupuncture. Available from: http://acupuncture-n-herbs.com/tendinomuscular-treatment-for-pain.
58. Leung A, Khadivi B, Duann JR, Cho ZH, Yaksh T. The effect of Ting point (tendinomuscular meridians) electroacupuncture on thermal pain: a model for studying the neuronal mechanism of acupuncture analgesia. J Altern Complement Med. 2005;11(4):653–61.
59. Langevin HM, Churchill DL, Cipolla MJ. Mechanical signaling through connective tissue: a mechanism for the therapeutic effect of acupuncture.

FASEB J: Official Publication of the Federation of American Societies for Experimental Biology. 2001;15(12):2275–82.
60. Langevin HM, Churchill DL, Wu J, Badger GJ, Yandow JA, Fox JR, et al. Evidence of connective tissue involvement in acupuncture. FASEB J: Official Publication of the Federation of American Societies for Experimental Biology. 2002;16 (8):872–4.
61. Yu Bai JW, Jin-peng Wu, et al. Review of evidence suggesting that the fascia network could be the anatomical basis for acupoints and meridians in the human body. Evid-Based Complement Altern Med. 2011.
62. Finando S, Finando D. Qi, acupuncture, and the fascia: a reconsideration of the fundamental principles of acupuncture. J Altern Complement Med. 2012;18(9):880–6.
63. Guo F, Song W, Jiang T, Liu L, Wang F, Zhong H, et al. Electroacupuncture pretreatment inhibits NADPH oxidase-mediated oxidative stress in diabetic mice with cerebral ischemia. Brain Res. 2014;1573:84–91.
64. Bing Z, Villanueva L, Le Bars D. Acupuncture and diffuse noxious inhibitory controls: naloxone-reversible depression of activities of trigeminal convergent neurons. Neuroscience. 1990;37(3):809–18.
65. Chung WY, Zhang HQ, Zhang SP. Peripheral muscarinic receptors mediate the anti-inflammatory effects of auricular acupuncture. Chin Med. 2011;6 (1):3.
66. Mattson MP. Glutamate and neurotrophic factors in neuronal plasticity and disease. Ann NY Acad Sci. 2008;1144:97–112.
67. Zhou W, Fu LW, Guo ZL, Longhurst JC. Role of glutamate in the rostral ventrolateral medulla in acupuncture-related modulation of visceral reflex sympathoexcitation. Am J Physiol Heart Circ Physiol. 2007;292(4):H1868–75.
68. Guo Z, Zhang L, Wu Y, Li M, Yang X, He Z, et al. The role of glutamate transporter-1 in the acquisition of brain ischaemic tolerance in rats induced by electro-acupuncture pre-treatment. Brain Inj. 2015;29(3):396–402.
69. Yang J, Liu WY, Song CY, Lin BC. Through central arginine vasopressin, not oxytocin and endogenous opiate peptides, glutamate sodium induces hypothalamic paraventricular nucleus enhancing acupuncture analgesia in the rat. Neurosci Res. 2006;54(1):49–56.
70. Enna SJ, McCarson KE. The role of GABA in the mediation and perception of pain. Adv Pharmacol. 2006;54:1–27.
71. Xu Q, Yang JW, Cao Y, Zhang LW, Zeng XH, Li F, et al. Acupuncture improves locomotor function by enhancing GABA receptor expression in transient focal cerebral ischemia rats. Neurosci Lett. 2015;588:88–94.
72. McLennan H, Gilfillan K, Heap Y. Some pharmacological observations on the analgesia induced by acupuncture in rabbits. Pain. 1977;3(3):229–38.
73. Pomeranz B, Nguyen P. Intrathecal diazepam suppresses nociceptive reflexes and potentiates electroacupuncture effects in pentobarbital-anesthetized rats. Neurosci Lett. 1987;77(3):316–20.
74. ZHU Lixia YY, MO Xiaorong, JI Changfu. The important role of activation of GABA_B receptors in acupuncture analgesia. Acupunct Res. 2002.
75. Sommer C. Serotonin in pain and analgesia: actions in the periphery. Mol Neurobiol. 2004;30(2):117–25.
76. Takagi J, Yonehara N. Serotonin receptor subtypes involved in modulation of electrical acupuncture. Jpn J Pharmacol. 1998;78(4):511–4.
77. Kim SK, Park JH, Bae SJ, Kim JH, Hwang BG, Min BI, et al. Effects of electroacupuncture on cold allodynia in a rat model of neuropathic pain: mediation by spinal adrenergic and serotonergic receptors. Exp Neurol. 2005;195(2):430–6.
78. Hu WB, Wu ZJ, Wang KM. Progress of researches on involvement of serotonin in the central nervous system in acupuncture analgesia and other effects. Zhen ci yan jiu = Acupuncture research/[Zhongguo yi xue ke xue yuan Yi xue qing bao yan jiu suo bian ji]. 2012;37(3):247–51.
79. Bravo-Hernandez M, Cervantes-Duran C, Pineda-Farias JB, Barragan-Iglesias P, Lopez-Sanchez P, Granados-Soto V. Role of peripheral and spinal 5-HT(3) receptors in development and maintenance of formalin-induced long-term secondary allodynia and hyperalgesia. Pharmacol Biochem Behav. 2012;101(2):246–57.
80. Liu X, Zhu B, Zhang SX. Relationship between electroacupuncture analgesia and descending pain inhibitory mechanism of nucleus raphe magnus. Pain. 1986;24(3):383–96.
81. Treister R, Pud D, Eisenberg E. The dopamine agonist apomorphine enhances conditioned pain modulation in healthy humans. Neurosci Lett. 2013;548:115–9.
82. Potvin S, Grignon S, Marchand S. Human evidence of a supra-spinal modulating role of dopamine on pain perception. Synapse. 2009;63(5):390–402.
83. Millan MJ. Descending control of pain. Prog Neurobiol. 2002;66(6):355–474.
84. Jarcho JM, Mayer EA, Jiang ZK, Feier NA, London ED. Pain, affective symptoms, and cognitive deficits in patients with cerebral dopamine dysfunction. Pain. 2012;153(4):744–54.
85. Wood PB. Role of central dopamine in pain and analgesia. Expert Rev Neurother. 2008;8(5):781–97.
86. Jaaskelainen SK, Lindholm P, Valmunen T, Pesonen U, Taiminen T, Virtanen A, et al. Variation in the dopamine D2 receptor gene plays a key role in human pain and its modulation by transcranial magnetic stimulation. Pain. 2014;155(10):2180–7.

87. Wang HH, Xu SF. Effect of D1 and D2 dopamine receptor antagonists on acupuncture analgesia. Sheng li xue bao: [Acta Physiologica Sinica]. 1993;45(1):61–8.
88. Dai JL, Xu SF. The attenuation effect of chlorpromazine on electro-acupuncture analgesia: involvement of dopamine system. Acupunct Electrother Res. 1991;16(3–4):101–9.
89. Liu XY, Zhou HF, Pan YL, Liang XB, Niu DB, Xue B, et al. Electro-acupuncture stimulation protects dopaminergic neurons from inflammation-mediated damage in medial forebrain bundle-transected rats. Exp Neurol. 2004;189(1):189–96.
90. Gao X, Xin BM, Zhu CB, Wu GC, Xu SF. Effect of intrathecal injection of dopamine receptor agonists/antagonists on pain and acupuncture analgesia in rats. Sheng li xue bao: [Acta Physiologica Sinica]. 1998;50(1):43–8.
91. Wang HH, Zhu YH, Xu SF. The potentiation effect of haloperidol on the binding of etorphine to brain membranes in acupuncture analgesia. Sheng li xue bao: [Acta Physiologica Sinica]. 1994;46(4):313–9.
92. Leung L. Neurophysiological basis of acupuncture-induced analgesia—an updated review. J Acupunct Meridian Stud. 2012;5(6):261–70.
93. Zhang Y, Qu H, Zhou Y, Wang Y, Zhang D, Yang X, et al. The involvement of norepinephrine in pain modulation in the nucleus accumbens of morphine-dependent rats. Neurosci Lett. 2015;585:6–11.
94. Ono H, Nakamura A, Kanbara T, Minami K, Shinohara S, Sakaguchi G, et al. Effect of the norepinephrine transporter (NET) Effect of the norepinephrine transporter (NET) Effect of the norepinephrine transporter (NET) inhibition on mu-opioid receptor (MOR)-induced anti-nociception in a bone cancer pain model. J Pharmacol Sci. 2014;125 (3):264–73.
95. Cao J, Zhou MQ, Wu SB, Wang KM, Zhou YP, Wang YL, et al. Effects of acupuncture on expression of norepinephrine transporter mRNA in the cervical sympathetic ganglion and beta1-AR mRNA in the heart in cerebral-cardiac syndrome rats. Zhen ci yan jiu = Acupuncture research/[Zhongguo yi xue ke xue yuan Yi xue qing bao yan jiu suo bian ji]. 2011;36(4):252–7.
96. Wang H, Jiang J, Can X. Changes of norepinephrine release in rat's nucleus reticularis paragigantocellularis lateralis in acupuncture analgesia. Zhen ci yan jiu = Acupuncture research/[Zhongguo yi xue ke xue yuan Yi xue qing bao yan jiu suo bian ji]. 1994;19(1):20–5.
97. Hackel D, Stolz A, Mousa SA, Brack A, Rittner HL. Recruitment of opioid peptide-containing neutrophils is independent of formyl peptide receptors. J Neuroimmunol. 2011;230(1–2):65–73.
98. Mayer DJ, Price DD, Rafii A. Antagonism of acupuncture analgesia in man by the narcotic antagonist naloxone. Brain Res. 1977;121(2):368–72.
99. Brissett DI, Whistler JL, van Rijn RM. Contribution of mu and delta opioid receptors to the pharmacological profile of kappa opioid receptor subtypes. Eur J Pain. 2012;16(3):327–37.
100. Wang Y, Gehringer R, Mousa SA, Hackel D, Brack A, Rittner HL. CXCL10 controls inflammatory pain via opioid peptide-containing macrophages in electroacupuncture. PLoS ONE. 2014;9 (4):e94696.
101. Nezhentsev M, Aleksandrov S. Effect of naloxone on the antipyretic action of acupuncture. Pharmacology. 1993;46(5):289–93.
102. Wang Y, Zhang Y, Wang W, Cao Y, Han JS. Effects of synchronous or asynchronous electroacupuncture stimulation with low versus high frequency on spinal opioid release and tail flick nociception. Exp Neurol. 2005;192(1):156–62.
103. Fearon AM, Twin J, Dahlstrom JE, Cook JL, Cormick W, Smith PN, et al. Increased substance P expression in the trochanteric bursa of patients with greater trochanteric pain syndrome. Rheumatol Int. 2014;34(10):1441–8.
104. Teodoro FC, Tronco Junior MF, Zampronio AR, Martini AC, Rae GA, Chichorro JG. Peripheral substance P and neurokinin-1 receptors have a role in inflammatory and neuropathic orofacial pain models. Neuropeptides. 2013;47(3):199–206.
105. Parenti C, Arico G, Ronsisvalle G, Scoto GM. Supraspinal injection of Substance P attenuates allodynia and hyperalgesia in a rat model of inflammatory pain. Peptides. 2012;34(2):412–8.
106. Ebner K, Singewald N. The role of substance P in stress and anxiety responses. Amino Acids. 2006;31 (3):251–72.
107. Lee HJ, Lee JH, Lee EO, Kim KH, Kim SH, Lee KS, et al. Substance P and beta-endorphin mediate electro-acupuncture induced analgesia in mouse cancer pain model. J Exp Clin Cancer Res CR. 2009;28:102.
108. Ma XP, Tan LY, Yang Y, Wu HG, Jiang B, Liu HR, et al. Effect of electro-acupuncture on substance P, its receptor and corticotropin-releasing hormone in rats with irritable bowel syndrome. World J Gastroenterol WJG. 2009;15(41):5211–7.
109. Mackey S. Mechanisms of inflammatory pain: therapeutic implications. J Clin Rheumatol Pract Reports Rheum musculoskelet Dis. 2004;10(3 Suppl):S5–11.
110. Ji RR. Peripheral and central mechanisms of inflammatory pain, with emphasis on MAP kinases. Curr Drug Targets Inflamm Allergy. 2004;3 (3):299–303.
111. Lin JG, Chen WL. Acupuncture analgesia: a review of its mechanisms of actions. Am J Chin Med. 2008;36(4):635–45.
112. Nager AL, Kobylecka M, Pham PK, Johnson L, Gold JI. Effects of acupuncture on pain and inflammation in pediatric emergency department patients with acute appendicitis: a pilot study. J Altern Complement Med. 2015;21(5):269–72.

113. Cao L, Zhang XL, Gao YS, Jiang Y. Needle acupuncture for osteoarthritis of the knee. A systematic review and updated meta-analysis. Saudi Med J. 2012;33(5):526–32.
114. Wang C, de Pablo P, Chen X, Schmid C, McAlindon T. Acupuncture for pain relief in patients with rheumatoid arthritis: a systematic review. Arthritis Rheum. 2008;59(9):1249–56.
115. Levine JD, Gordon NC, Bornstein JC, Fields HL. Role of pain in placebo analgesia. Proc Natl Acad Sci USA. 1979;76(7):3528–31.
116. Atlas LY, Wager TD. A meta-analysis of brain mechanisms of placebo analgesia: consistent findings and unanswered questions. Handb Exp Pharmacol. 2014;225:37–69.
117. Kam-Hansen S, Jakubowski M, Kelley JM, Kirsch I, Hoaglin DC, Kaptchuk TJ, et al. Altered placebo and drug labeling changes the outcome of episodic migraine attacks. Sci Transl Med. 2014;6 (218):218ra5.
118. Vase L, Vollert J, Finnerup NB, Miao X, Atkinson G, Marshall S, et al. Predictors of the placebo analgesia response in randomized controlled trials of chronic pain: a meta-analysis of the individual data from nine industrially sponsored trials. Pain. 2015;156(9):1795–802.
119. Le Bars D, Dickenson AH, Besson JM. Diffuse noxious inhibitory controls (DNIC). I. Effects on dorsal horn convergent neurones in the rat. Pain. 1979;6(3):283–304.
120. Xiong J, Liu F, Zhang MM, Wang W, Huang GY. De-qi, not psychological factors, determines the therapeutic efficacy of acupuncture treatment for primary dysmenorrhea. Chin J Integr Med. 2012;18 (1):7–15.
121. Chiu YH, Silman AJ, Macfarlane GJ, Ray D, Gupta A, Dickens C, et al. Poor sleep and depression are independently associated with a reduced pain threshold. Results of a population based study. Pain. 2005;115(3):316–21.
122. Wang XJ, Wang LL. A mechanism of endogenous opioid peptides for rapid onset of acupuncture effect in treatment of depression. Zhong xi yi jie he xue bao = J Chin Integr Med. 2010;8(11):1014–7.
123. Bohren Y, Tessier LH, Megat S, Petitjean H, Hugel S, Daniel D, et al. Antidepressants suppress neuropathic pain by a peripheral beta2-adrenoceptor mediated anti-TNFalpha mechanism. Neurobiol Dis. 2013;60:39–50.
124. Lazzeri L, Vannuccini S, Orlandini C, Luisi S, Zupi E, Nappi RE, et al. Surgical treatment affects perceived stress differently in women with endometriosis: correlation with severity of pain. Fertil Steril. 2015;103(2):433–8.
125. Eshkevari L, Permaul E, Mulroney SE. Acupuncture blocks cold stress-induced increases in the hypothalamus-pituitary-adrenal axis in the rat. J Endocrinol. 2013;217(1):95–104.
126. Kong JT, Schnyer RN, Johnson KA, Mackey S. Understanding central mechanisms of acupuncture analgesia using dynamic quantitative sensory testing: a review. Evid Based Complement Altern Med eCAM. 2013;2013:187182.
127. Olesen SS, van Goor H, Bouwense SA, Wilder-Smith OH, Drewes AM. Reliability of static and dynamic quantitative sensory testing in patients with painful chronic pancreatitis. Reg Anesth Pain Med. 2012;37(5):530–6.
128. Arendt-Nielsen L, Yarnitsky D. Experimental and clinical applications of quantitative sensory testing applied to skin, muscles and viscera. J Pain: Official Journal of the American Pain Society. 2009;10 (6):556–72.
129. Zheng Z, Feng SJ, Costa C, Li CG, Lu D, Xue CC. Acupuncture analgesia for temporal summation of experimental pain: a randomised controlled study. Eur J Pain. 2010;14(7):725–31.
130. Tobbackx Y, Meeus M, Wauters L, De Vilder P, Roose J, Verhaeghe T, et al. Does acupuncture activate endogenous analgesia in chronic whiplash-associated disorders? A randomized crossover trial. Eur J Pain. 2013;17(2):279–89.
131. Le Bars D, Villanueva L, Bouhassira D, Willer JC. Diffuse noxious inhibitory controls (DNIC) in animals and in man. Patologicheskaia fiziologiia i eksperimental'naia terapiia. 1992;4:55–65.
132. Pud D, Granovsky Y, Yarnitsky D. The methodology of experimentally induced diffuse noxious inhibitory control (DNIC)-like effect in humans. Pain. 2009;144(1–2):16–9.
133. Schliessbach J, van der Klift E, Siegenthaler A, Arendt-Nielsen L, Curatolo M, Streitberger K. Does acupuncture needling induce analgesic effects comparable to diffuse noxious inhibitory controls? Evid Based Complement Altern Med eCAM. 2012;2012:785613.
134. Hui KK, Liu J, Makris N, Gollub RL, Chen AJ, Moore CI, et al. Acupuncture modulates the limbic system and subcortical gray structures of the human brain: evidence from fMRI studies in normal subjects. Hum Brain Mapp. 2000;9(1):13–25.
135. Liu WC, Feldman SC, Cook DB, Hung DL, Xu T, Kalnin AJ, et al. fMRI study of acupuncture-induced periaqueductal gray activity in humans. NeuroReport. 2004;15(12):1937–40.
136. Yoo SS, Teh EK, Blinder RA, Jolesz FA. Modulation of cerebellar activities by acupuncture stimulation: evidence from fMRI study. NeuroImage. 2004;22(2):932–40.
137. Jeun SS, Kim JS, Kim BS, Park SD, Lim EC, Choi GS, et al. Acupuncture stimulation for motor cortex activities: a 3T fMRI study. Am J Chin Med. 2005;33(4):573–8.
138. Qin W, Tian J, Bai L, Pan X, Yang L, Chen P, et al. FMRI connectivity analysis of acupuncture effects on an amygdala-associated brain network. Mol Pain. 2008;4:55.
139. Chen Y, Wang Y, Yin Q. The role of paraventricular nucleus of hypothalamus in acupuncture analgesia in rats. Zhen ci yan jiu = Acupuncture

Research [Zhongguo yi xue ke xue yuan Yi xue qing bao yan jiu suo bian ji]. 1991;16(1):32–8.
140. Napadow V, Dhond RP, Purdon P, Kettner N, Makris N, Kwong KK, et al. Correlating acupuncture FMRI in the human brainstem with heart rate variability. Conference proceedings: annual international conference of the IEEE engineering in medicine and biology society IEEE engineering in medicine and biology society annual conference. 2005;5:4496–9.
141. Vargas CA, Helms Medical I. Medical acupuncture for physicians: a pocket clinical reference guide. Berkeley, California: Medical Acupuncture Publishers; 2004.
142. Shukla S, Torossian A, Duann JR, Leung A. The analgesic effect of electroacupuncture on acute thermal pain perception—a central neural correlate study with fMRI. Mol Pain. 2011;7:45.
143. The Eight Influential Points. Available from: http://www.acupuncture.com/education/theory/eightpts.htm.
144. Leung A, Zhao Y, Shukla S. The effect of acupuncture needle combination on central pain processing—an fMRI study. Mol Pain. 2014;10:23.
145. Napadow V, Makris N, Liu J, Kettner NW, Kwong KK, Hui KK. Effects of electroacupuncture versus manual acupuncture on the human brain as measured by fMRI. Hum Brain Mapp. 2005;24(3):193–205.
146. Li L, Liu H, Li YZ, Xu JY, Shan BC, Gong D, et al. The human brain response to acupuncture on same-meridian acupoints: evidence from an fMRI study. J Altern Complement Med. 2008;14(6):673–8.
147. Bai L, Tian J, Zhong C, Xue T, You Y, Liu Z, et al. Acupuncture modulates temporal neural responses in wide brain networks: evidence from fMRI study. Molecular Pain. 2010;6:73.
148. Usichenko TI, Wesolowski T, Lotze M. Verum and sham acupuncture exert distinct cerebral activation in pain processing areas: a crossover fMRI investigation in healthy volunteers. Brain Imaging Behav. 2015;9(2):236–44.
149. Huang W, Pach D, Napadow V, Park K, Long X, Neumann J, et al. Characterizing acupuncture stimuli using brain imaging with FMRI—a systematic review and meta-analysis of the literature. PLoS ONE. 2012;7(4):e32960.
150. Tracey KJ. The inflammatory reflex. Nature. 2002;420(6917):853–9.
151. Xudong G. Clinical study on analgesia for biliary colic with ear acupuncture at point Erzhong. Am J Acupunct. 1993;21(3):237–9.
152. Arai YC, Sakakima Y, Kawanishi J, Nishihara M, Ito A, Tawada Y, et al. Auricular acupuncture at the “shenmen” and “point zero” points induced parasympathetic activation. Evid Based Complement Altern Med eCAM. 2013;2013:945063.
153. La Marca R, Nedeljkovic M, Yuan L, Maercker A, Elhert U. Effects of auricular electrical stimulation on vagal activity in healthy men: evidence from a three-armed randomized trial. Clinical Sci (London, England: 1979). 2010;118(8):537–46.
154. Sator-Katzenschlager SM, Michalek-Sauberer A. P-Stim (TM) auricular electroacupuncture stimulation device for pain relief. Expert Rev Med Devices. 2007;4(1):23–32.
155. Najafi S, Halstead-Kenny J. Applying auricular electroacupuncture device (P-STIM) for chronic lumbar radiculopathy. Am J Phys Med Rehabil. 2014:a33-a.
156. Holzer A, Leitgeb U, Spacek A, Wenzl R, Herkner H, Kettner S. Auricular acupuncture for postoperative pain after gynecological surgery: a randomized controlled trail. Minerva Anestesiol. 2011;77(3):298–304.
157. Sator-Katzenschlager SM, Wolfler MM, Kozek-Langenecker SA, Sator K, Sator PG, Li B, et al. Auricular electro-acupuncture as an additional perioperative analgesic method during oocyte aspiration in IVF treatment. Hum Reprod. 2006;21(8):2114–20.
158. Likar R, Jabarzadeh H, Kager I, Trampitsch E, Breschan C, Szeles J. Electrical point stimulation (P-STIM) via ear acupuncture: a randomized, double-blind, controlled pilot study in patients undergoing laparoscopic nephrctomyX. Schmerz (Berlin, Germany) [Internet]. 2007; 21(2):154–9. Available from: http://onlinelibrary.wiley.com/o/cochrane/clcentral/articles/702/CN-00586702/frame.html, http://link.springer.com/article/10.1007%2Fs00482-006-0519-y.
159. Kager H, Likar R, Jabarzadeh H, Sittl R, Breschan C, Szeles J. Electrical punctual stimulation (P-STIM) with ear acupuncture following tonsillectomy, a randomised, controlled pilot study. Acute Pain. 2009;11(3–4):101–6.
160. Kong KH, Ng WW. Treatment of chronic pain with an auricular acupuncture device (P-Stim) in Singapore. Acupunct Med (Journal of the British Medical Acupuncture Society). 2009;27(4):187–8.
161. Goertz CM, Niemtzow R, Burns SM, Fritts MJ, Crawford CC, Jonas WB. Auricular acupuncture in the treatment of acute pain syndromes: a pilot study. Mil Med. 2006;171(10):1010–4.
162. Niemtzow RC. Battlefield acupuncture. J Acupunct Assoc Chart Physiother. 2010:55–9.
163. Pickett H. Battlefield acupuncture. J Chin Med. 2011;96:14–7.
164. Niemtzow RC, Litscher G, Burns SM, Helms JM. Battlefield acupuncture: update. Med Acupunct. 2009;21(1):43–6.
165. Plunkett A, Turabi A, Wilkinson I. Battlefield analgesia: a brief review of current trends and concepts in the treatment of pain in US military casualties from the conflicts in Iraq and Afghanistan. Pain Manag. 2012;2(3):231–8.
166. Soliman N. Pain: an evidence-based approach through the auricular acupuncture microsystem. Med Acupunct. 2008;20(4):263–7.

167. Napadow V, Edwards RR, Cahalan CM, Mensing G, Greenbaum S, Valovska A, et al. Evoked pain analgesia in chronic pelvic pain patients using respiratory-gated auricular vagal afferent nerve stimulation. Pain Med. 2012;13(6):777–89.
168. Laqua R, Leutzow B, Wendt M, Usichenko T. Transcutaneous vagal nerve stimulation may elicit anti- and pro-nociceptive effects under experimentally-induced pain—a crossover placebo-controlled investigation. Auton Neurosci Basic & Clinic. 2014;185:120–2.
169. Ceccherelli F, Manani G, Ambrosio F, Angel A, Valenti S, Facco E, et al. Influence of acupuncture on the postoperative complications following ketamine anesthesia. The importance of manual stimulation of point R and shen menn. Acupunct Electrother Res. 1981;6(4):255–64.
170. Bai L, Yan H, Li L, Qin W, Chen P, Liu P, et al. Neural specificity of acupuncture stimulation at pericardium 6: evidence from an FMRI study. J Magn Reson Imaging JMRI. 2010;31(1):71–7.
171. Chong PS, Cros DP. Technology literature review: quantitative sensory testing. Muscle Nerve. 2004;29 (5):734–47.
172. Beissner F, Henke C. Methodological problems in FMRI studies on acupuncture: a critical review with special emphasis on visual and auditory cortex activations. Evid Based Complement Altern Med eCAM. 2011;2011:607637.
173. Langevin HM, Wayne PM, Macpherson H, Schnyer R, Milley RM, Napadow V, et al. Paradoxes in acupuncture research: strategies for moving forward. Evid Based Complement Altern Med eCAM. 2011;2011:180805.

Author Biographies

Dr. Mikiko Murakami is Pain Medicine Fellow from the University of California in San Diego (UCSD) expected graduation in June 2017. She completed her residency training in Physical Medicine and Rehabilitation (PM&R) at the Icahn School of Medicine, Mount Sinai Hospital in New York City. She has utilized her personal experiences as a patient to create her path as a physician. The former founder of Chi Fountain Integrative Medicine, and now a Doctor of Osteopathy and acupuncturist, Dr. Murakami uses a multidisciplinary and holistic approach to optimize her patients' function.

Dr. Murakami holds a strong belief in preventing and treating the underlying cause(s) of pain. while minimizing side effects. She continues to follow her passions for creating solutions to the pain medication overdose epidemic, investigating integrative medicine modalities, and researching minimally invasive pain devices. She is a co-investigator of an ongoing clinical trial that is researching ear acupuncture for acute low back pain, and she has also presented at various national conferences. In addition, Dr. Murakami currently serves as a Board Member for the American Society of Interventional Pain Physicians.

Albert Leung is a Professor of Anesthesiology and Pain Medicine at the University of California, San Diego (UCSD). He has close to two decades of experience in evaluating, diagnosing and treating patients with complex chronic pain conditions. He is a leader in neuromodulation research currently focusing on the mechanisms and efficacy of non-invasive brain and peripheral stimulation for nerve function restoration and pain relief. He has been a Principle Investigator for the U.S. National Institute Health, Department of Defense, and Veteran Administration Merit Award funded research projects, and multiple industry sponsored clinical trails. He publishes regularly in peer-reviewed journals. Dr. Leung earned his medical degree from the University of Pittsburgh School of Medicine. After his anesthesia training, he completed a combined clinical and research pain fellowship at UCSD with additional post-graduate trainings in medical acupuncture, functional neuroimaging and transcranial magnetic stimulation from MIT and Harvard Medical School.

Neuroimaging of Paediatric Pain

18

Caroline Hartley and Rebeccah Slater

Abstract

Pain during childhood can have a major impact on a child's quality of life and development. However, given the changes in neurobiology, pharmacodynamics and pain assessment across this wide age range, from preverbal premature infants to adolescents, the question of how to effectively assess and manage pain in this population is complex. Recent research using neuroimaging techniques has advanced our understanding of paediatric pain. In this chapter, we discuss this research, including studies examining infant pain, pain in older children and the long-term effects of early life pain exposure. While there is a relative lack of neuroimaging research in paediatric pain compared with studies investigating adult pain, the early research in this field demonstrates the wealth of information that can be gained from the use of these techniques. As cortical activity is a prerequisite for pain perception, measuring pain-related brain activity may be particularly useful in children who cannot describe their pain experience. Neuroimaging studies in older children have highlighted both the vulnerability and plasticity of the developing nervous system. Understanding this plasticity may improve the treatment of chronic pain in children. Furthermore, neuroimaging studies provide an opportunity to examine how analgesics modulate neuronal activity and how this changes as the nervous system develops. In summary, neuroimaging provides a significant new direction in the complex field of paediatric pain.

C. Hartley · R. Slater (✉)
Department of Paediatrics, Level 2 Children's Hospital, John Radcliffe, University of Oxford, Oxford OX3 9DU, UK
e-mail: rebeccah.slater@paediatrics.ox.ac.uk

C. Hartley
e-mail: caroline.hartley@paediatrics.ox.ac.uk

L. Saba (ed.), *Neuroimaging of Pain*, DOI 10.1007/978-3-319-48046-6_18

Keywords
Childhood · Analgesia · Development · Noxious-evoked brain activity · Pain assessment · Chronic pain

1 Introduction

The World Health Organisation describes paediatric pain as "a public health concern of major significance in most parts of the world" [1]. Pain, whether it is acute or chronic, can be a major cause of distress to both the child and their families. Moreover, prolonged pain can lead to long-term problems such as depression and mood changes, and can cause disruption to sleep, family interactions, school attendance, and social and physical activities [1]. Effective pain management and treatment in children is therefore imperative.

The paediatric period covers a wide age range from the preterm neonate to the adolescent. Across this age range there are ongoing changes in neurobiology, pharmacodynamics and in how pain is reported [2]. All of these factors make paediatric pain management particularly challenging and may lead to under-treatment of pain in children [3]. Neuroimaging techniques provide an opportunity to investigate these factors—to examine how pain processing changes with the developing nervous system, to study how analgesics modulate activity within the developing nervous system, and to explore how pain is experienced in non-verbal populations. Research in this area could therefore ultimately aid the clinician in their treatment of pain in the paediatric population.

In this chapter we will discuss current research in the area of paediatric neuroimaging, including recent studies that have investigated nociceptive processing in the infant brain, and studies in older children that have primarily focussed on chronic pain. This chapter is divided into three broad themes: infant pain, the long-term effects of early life pain, and pain in older children. We will also discuss future directions for neuroimaging of paediatric pain.

A complete discussion of the extensive literature examining the development of nociception, including in animal models, is beyond the scope of this chapter—for comprehensive reviews see Fitzgerald et al. [4–6]. Also, more extensive reviews of paediatric pain can be found in the *Oxford Textbook of Paediatric Pain* [7], *Pain in Infants, Children and Adolescents* [3] and *Pain in Neonates and Infants* [8], and the references therein.

2 Prevalence and Types of Paediatric Pain

Estimating the prevalence of pain in children is complicated as studies differ in terms of their definitions of prevalence, the reporting period, age range investigated, the definition and measures of pain used, and methodological approaches (e.g. child self-report, parental report, clinician report, retrospective notes review, etc.) [9]. Consequently, a wide difference in prevalence is reported between studies.

2.1 Procedural Pain

Infants requiring treatment as part of their essential medical care can receive multiple painful procedures. These include routine procedures such as heel lancing, venepuncture, cannula insertion, intubation, tracheal suctioning, lumbar puncture and central line insertion [10]. Moreover, while some painful procedures performed on infants may be considered to be only mildly painful, they can be conducted multiple times throughout a hospital stay and so the frequency of these procedures is also an important factor to consider. Studies of infants in intensive care have found that they can receive an average of 1–14 painful procedures a day [11–15]. However, in many cases the infant will not receive pain relief for these procedures. For example, Carbajal et al. [11] reported an average of 12 painful procedures a day in neonates studied over the first 14 days after hospital admission. The maximum number of painful procedures received in one day by a single infant was 51, and in 79% of cases no specific analgesia (including pharmacological and non-pharmacological) was given for the procedure. More recently, Roofthooft et al. reported an average of 11 painful procedures a day, and, while all infants received non-pharmacological care designed to reduce pain and stress, pharmacological analgesia was only given to 37% of infants during the study period [14].

In a study of children up to the age of 18 years admitted as in-patients, Stevens and colleagues reported that 78% of children underwent at least one painful procedure during a 24-h study period, and on average they experienced 6.3 painful procedures. In only 28% of cases a pain management strategy was administered specifically for the painful procedure, although more positively, in 78% of cases a pain management strategy was used within the 24-h period [16]. The importance of considering the source of information was highlighted by Harrison and colleagues who found that approximately half of children they studied reported (or their caregiver reported) that a pain management strategy in the form of topical local anaesthetic or sucrose was administered prior to a painful procedure, but that this was rarely documented in medical notes [17]. More generally, the prevalence of moderate to severe pain in hospitalised children is reported to be greater than 20% [17–19], with higher prevalence in children admitted to surgical wards compared with other units [18]. Perioperative pain can delay recovery from surgery, may increase morbidity, or may lead to long-term consequences such as chronic pain, psychological problems, or altered pain sensitivity [20–22]. It is therefore important that it is adequately treated and a number of guidelines have been produced which describe best practices for pain management peri- and postoperatively [23–25].

Vaccination is the main form of medical procedural pain experienced by most children. The type and number of vaccinations varies by country; in the UK children will receive approximately 16 vaccinations by injection by the time they reach adulthood [26], and in the USA this is much higher with children currently receiving more than 30 vaccinations [27]. While injections may be thought of as a relatively minor procedure, in some children they can cause a great deal of fear and anxiety, which can lead to vaccination non-compliance and therefore have an impact on health—for the child and at a population level. In one study 63% of children reported a fear of needles, which 5% of parents said resulted in immunisation non-compliance [28]. A number of pain management techniques have been recommended [29]; however, these are frequently not used in clinical practice [30]. Better education for both parents and clinicians in pain management strategies may help alleviate problems related to needle fears in the future [30].

2.2 Chronic Pain

Chronic pain is common in children, and can have a significant impact on their quality of life. Prevalence varies largely across studies and depends on a number of factors including age and the type of pain [31]. In a comprehensive systematic review of chronic pain in children, King et al. [31] report that rates in the literature range substantially, with prevalence of headache: 8–83%; abdominal pain: 4–53%; back pain: 14–24%; musculoskeletal pain: 4–40%; multiple pains: 4–49%; and other pains: 5–88%. To date neuroimaging of paediatric chronic pain has focussed on headaches, recurrent abdominal pain and complex regional pain syndrome (see Sect. 6). Further research investigating other types of chronic pain will be beneficial.

3 Clinical Assessment of Paediatric Pain

Effective pain management in children cannot be achieved without adequate pain assessment. The type of pain assessment used varies with the child's age, and can also vary according to the type of pain, the setting, and the cognitive and language ability of the child. As pain is always subjective, self-report is often seen as the gold standard of pain assessment. From the age of about 3 years old, pain scales such as the Faces Pain Scale–Revised (which allows the child to point to the picture of the facial expression which best matches how they currently feel) can be used [32]. In older children alternative scales such as the Visual Analogue Scale (VAS) and the Numerical Rating Scale (NRS) can also be used. With these types of scales the most valuable information is gained from changes in a child's ratings, rather than a direct comparison of ratings across children, as for example, 7 out of 10 may mean something different for each child. Moreover, older children and adolescents can describe their pain experience in more detail than younger children, which aids the clinician in treatment of the child's pain.

Where a verbal pain report can be ascertained, in some cases this may differ from observer-based assessment of the child's pain. In this case, it is important to consider why the self-report may be different. For example, young children or those with developmental delay may not fully understand the instructions given to them, or some children may under rate their pain if they do not want to stay in hospital [33]. In such cases an alternative pain assessment approach may be more useful to avoid under-treatment of pain. Nevertheless, it is also important to remember that observer-based pain assessment tools are subjective and may be affected by biases of the observer [33]. If a child says they are in pain this should not be ignored.

Self-report is, however, not always possible, for example, in preverbal infants or in older children with cognitive impairments, and so the assessment of pain in these individuals is challenging. Indeed, given the subjective nature of the pain experience, it is likely we will never truly be able to answer the question of whether infants can experience pain. Such a question perhaps requires more of a philosophical discussion on the nature of consciousness [34, 35]. Nevertheless, the proper treatment of pain in non-verbal individuals is essential. Indeed, while the International Association for the Study of Pain (IASP) defines pain as "an unpleasant sensory and emotional experience associated with actual or potential tissue damage, or described in terms of such damage", therefore emphasising the subjective nature of the pain experience, they importantly note that "the inability to communicate verbally does not negate the possibility that an individual is experiencing pain and is in need of appropriate pain-relieving treatment" [36].

Historically, the inability of an infant to describe their experience, the lack of memory of events early in life, and the immaturity of an infant's nervous system led some to suggest that infants do not have the capacity to experience pain [37]. This led to the under-treatment of pain in the infant population. Indeed, due to the risks involved in giving anaesthetics to infants in the early years of their use, anaesthetics were often not given to infants for operations. Instead, a light anaesthesia was commonly achieved using a combination of muscle relaxants and nitrous oxide [37]. In landmark randomised controlled trials conducted in the 1980s, Anand and colleagues compared this standard technique with the addition of fentanyl or halothane, showing that infants who received the standard technique mounted a higher stress response, with higher hormonal and metabolite responses compared with infants receiving fentanyl or halothane. Moreover, postoperative complications were much more common in the infants that did not receive halothane or fentanyl [38, 39], and higher stress responses intra- and postoperatively were associated with increased postoperative mortality [40].

Also during the late 1980s, the case of Jeffrey Lawson came to public attention in the United States. Jeffrey Lawson was born prematurely and required cardiac surgery, for which he received a muscle relaxant but no anaesthetic. After the operation, he deteriorated and died 5 weeks later [37]. His mother, Jill Lawson, appalled by the lack of anaesthetic given to her son, began advocating for babies to be given anaesthetics for surgery [41]—a cause that became greatly publicised by the national media [42]. Following this case, and the work of Anand and colleagues, clinical practice for anaesthetic use in infants was changed. Moreover, their work also generated a dramatic increase in pain research in infants [37], with clinical staff and researchers alike realising the need to improve pain assessment and treatment in this vulnerable population.

More than 30 pain assessment tools have now been developed for infants [43]. These often incorporate physiological measures, such as change in heart rate, oxygen saturation, respiratory rate, or blood pressure; or behavioural measures, such as changes in facial expression, body movements, or length of cry [43]. The Premature Infant Pain Profile (PIPP) is an example of a tool for assessing acute procedural pain that has been well validated and used in numerous studies of infant pain [44–46]. This multidimensional pain assessment measure (and the recently revised version: the Premature Infant Pain Profile-Revised, PIPP-R [44, 46]) incorporates scores of the infant's behavioural state before the procedure, the infant's gestational age, and changes in heart rate, oxygen saturation and facial expression (based on three components—brow bulge, nasolabial furrow and eye squeeze). Other commonly used infant pain assessment tools include the Neonatal Facial Coding System (NFCS), which assesses 10 facial actions [47], and the Neonatal Infant Pain Scale (NIPS), which includes measures of facial expression, cry, breathing patterns and movement [48].

Pain assessment tools that do not require verbal report have also been developed for prolonged or postoperative pain. Examples include Échelle Douleur Inconfort Nouveau-né (EDIN), which can be used in preterm infants and incorporates scores of facial expression, movement, sleep and consolability [49]; the COMFORT scale, which can be used with children up to age 17 and includes measures such as alertness, muscle tone, blood pressure and heart rate [50]; and FLACC (Faces, Legs, Activity, Cry, Consolability), which can be used in children aged between 2 months and 7 years, and is most suitable for preverbal children [51]. For children with developmental delay or neurocognitive deficits, examples of pain scores include the Non-Communicating Children's Pain Checklist Postoperative Version (NCCPC-PV), the Paediatric Pain Profile (PPP) and a revised FLACC [52–54].

Chronic pain is more difficult to assess without self-report, and there are currently no well-established behavioural measures specifically for chronic pain. For older children, questionnaires can be used, such as the Varni/Thompson Paediatric Pain Questionnaire [55, 56] and the Bath Adolescent Pain Questionnaire [57]. Patients can also keep pain diaries, and more recently smartphone applications have been developed which allow children to easily and frequently record pain ratings [58, 59]. Neuroimaging may give us a better mechanistic understanding of nociceptive processing and chronic pain in the paediatric population, and address some of the challenges associated with pain assessment, particularly in preverbal infants.

4 Neurophysiology and Neuroimaging of Infant Pain

Reflex withdrawal of a limb—an essential protective behaviour—can occur early in development before functional connections to the cortex are formed [60]. However, while reflexes are indicative of functional processing of a stimulus at the level of the spinal cord, cortical processing is necessary for both the sensory and emotional aspects of pain perception. Indeed, nociception and reflex withdrawal can occur in the absence of pain—for example, in the presence of a spinal lesion [60]. From approximately 20 weeks' gestation, thalamic axons form connections with the transient subplate zone—a population of neurones that exists during development and largely undergoes programmed cell death by term or shortly thereafter [61–63]. In turn, subplate neurones connect to neurones in the developing cortex, so that external sensory input is processed in the developing cortex via indirect connections from the thalamus [62]. The subplate is essential for proper development, and in animals ablation of the subplate before thalamocortical connections have developed leads to weak thalamocortical connections and abnormal corticocortical connections [62, 64, 65]. Following a period of 'waiting' in the subplate, thalamic neurones form direct functional connections to the cortex [62], and from approximately 31–35 weeks' gestation (depending on the brain region) the subplate begins to disappear [63].

Functionally, electrical activity can be recorded using EEG in preterm infants, with very preterm infants exhibiting discontinuous bursts of activity, which occur with complex temporal patterns [66]. Activity becomes increasingly continuous with age, with inter-burst intervals decreasing [67, 68]. The bursts of activity can be spontaneously generated by subcortical or cortical neurones, or evoked by external sensory input [69, 70]. In very preterm infants, EEG activity has been recorded in response to visual and auditory stimuli [71, 72]. This demonstrated that external sensory inputs are processed in the preterm infant brain. However, it was only relatively recently that observations of brain activity in response to noxious stimuli in the infant were made.

In 2006 two studies were published by separate groups that investigated the cortical haemodynamic response to clinically required noxious stimulation using near-infrared spectroscopy (NIRS). Bartocci et al. [73] observed an increase in oxyhaemoglobin concentration over the somatosensory cortex in infants' aged 28–36 weeks' gestation following venepuncture. A smaller increase was also seen following tactile stimulation (cleaning the skin prior to venepuncture), and no significant changes in oxyhaemoglobin occurred over the occipital cortex, demonstrating the specificity of the response to somatosensory cortical areas. Similarly, Slater et al. demonstrated an increase in total haemoglobin concentration over the contralateral somatosensory cortex following a clinically required heel lance in infants from 25 to 45 weeks' gestation [74]. Interestingly, this response was dependent on sleep state, with higher responses in awake infants. Together, these studies demonstrated that noxious stimuli are processed by the infant cortex and marked a turning point in our understanding of infant pain. Since then, significant cortical haemodynamic changes have also been recorded following longer medical noxious and non-noxious procedures such as endotracheal tube repositioning and diaper changes [75], demonstrating the applicability of the technique to assessing different procedures.

Following on from the NIRS research, a number of other studies have been conducted examining nociceptive processing in the infant brain using EEG and, more recently, fMRI. Using these techniques has importantly allowed for both the temporal and spatial characteristics of the nociceptive response to be investigated, and developing these techniques is important for future research in order to understand different aspects of the stimulus response.

Using EEG, Slater, Fitzgerald and colleagues demonstrated that term infants exhibit a nociceptive-specific evoked pattern of brain activity in response to acute noxious stimuli. Comparing the brain activity response to a clinically required heel lance with the response to a control heel lance (where the lancet was rotated by 90° and held against the infant's foot so that when released, the spring-loaded blade did not pierce the skin but the same sound and tactile sensation was experienced by the infant) they found that while an earlier potential (occurring at ~250 ms) was evoked by both stimuli, a later potential (occurring at ~500 ms) was only evoked by the heel lance [76]. They therefore concluded that this later potential was nociceptive-specific, importantly demonstrating that a different neuronal response occurred in the infant brain to a noxious stimulus compared with a tactile stimulus.

Examining responses to clinically required heel lances across early development between 28 and 45 weeks' gestation, they also found that this nociceptive-specific brain activity is more likely to occur in infants from 35 weeks' gestation [69]. Prior to this a non-specific neuronal burst, known as a delta brush, is more likely to be observed in response to both the noxious heel lance and a non-noxious tactile stimulus. This switch from delta brushes to specific evoked potentials at approximately 35 weeks' gestation is consistent with the development of visual [72] and auditory [71] evoked responses and may be related to the development of direct thalamocortical connections and the disappearance of the subplate [62, 69].

The studies described above have examined medically required noxious stimulation. The use of experimental stimulation provides an opportunity to examine phenomenon such as intensity encoding and habituation. However, careful selection of an experimental stimulation is required as it is imperative that the infant is not distressed, and so only low-level noxious stimulation should be applied. Slater and colleagues have recently used experimental stimuli to further investigate the infant neuronal response to noxious input. The probes they used are known to activate Aδ fibres [77, 78] and adults commonly describe the sensation the stimuli evoke as sharp, pricking and mildly painful [79]. However, at low force levels the stimuli do not cause behavioural distress or clinical concern when applied to term infants [80], and in fact most infants remain asleep whilst the stimuli are applied. The stimulus can also be applied in an MRI scanner and so is a suitable tool for investigating the spatial characteristics of the brain activity response to noxious stimuli. Indeed, the feasibility of using this approach was demonstrated during fMRI scanning [81].

Applying three levels of the experimental noxious stimuli (at forces of 32, 64 and 128 mN) to term infants, Hartley et al. [80] found that they evoke nociceptive-specific brain activity, identified using EEG recordings, and that this activity is graded with the intensity of the stimulation. All three stimuli evoked activity that was smaller than that evoked by a clinically required heel lance. Furthermore, reflex withdrawal was also graded with stimulus intensity, and the magnitude of the reflex withdrawal and the nociceptive-specific brain activity were significantly correlated [80]. In a separate group of term infants, Goksan et al. investigated the blood oxygen level dependent (BOLD) responses to the same stimuli using fMRI [79]. The authors compared the responses to a group of adults who received the same stimuli (but over a larger range of forces: 32–512 mN). Adult activity was consistent with previous studies, and included activation in the pre- and postcentral gyrus, insula and thalamus, as well as other areas associated with the experience of pain. In infants, 18 of the 20 brain regions active in the adults were also active, including both sensory and affective processing regions. The two brain regions that were active in adults but not in infants were the orbitofrontal cortex and the amygdala. These regions are, respectively, known to be involved in the processing of reward and fear [82, 83], and so it may be that infants are too immature to contextualise the stimuli in this way [79]. However, activity was observed in the infant in the anterior cingulate cortex (ACC); activity in the ACC in adults parallels changes in the perceived unpleasantness of a stimulus [84], suggesting that infants may be able to experience emotional aspects of pain, as well as sensory aspects. These studies together demonstrate the spatial and temporal aspects of the infant brain response to low-level noxious stimulation. As these stimuli can be repeated multiple times in the same infant without causing distress, they may prove a useful tool for further investigation of nociceptive processing in infants.

In older infants, EEG responses to needle vaccination have been investigated [85]. Time-locked to the point of needle contact with the skin, a clear evoked response to inoculation was recorded in infants at 1–2 months old and 12 months old [85]. As vaccinations are a frequently required medical procedure performed at various ages throughout childhood, this methodology provides a valuable approach for assessing changes in nociceptive processing across childhood.

In summary, NIRS, EEG and fMRI have all been used to examine responses to acute noxious stimuli in infants. This research provides the first steps in characterising infant brain responses to noxious stimuli, and, as will be discussed in greater detail below, may provide an important methodology with which to investigate analgesic efficacy in infants.

4.1 Brain Imaging and Neonatal Pain Assessment Tools

Behavioural measures of infant pain often have low correlation with physiological measures, such as changes in heart rate and oxygen saturation [86], and there is no single pain assessment score that is considered the best to use. Behavioural measures, and in particular changes in facial expression, are considered to be more directly related to the pain experience (as they respond more selectively to painful procedures, whereas physiological measures will frequently change in response to other stimuli as well) [86, 87]. However, some infants do not mount a facial expression response to painful procedures [88], and responses have been shown to vary with factors such as infant's gestational age, prior exposure to painful procedures, and sleep state [87–90].

A lack of concordance also exists between brain activity and other measures. Indeed, while cortical haemodynamic responses are correlated with PIPP scores, with the best correlation demonstrated with the facial expression component of the score, in ~40% of infants a change in facial expression is not seen but a change in brain activity can still be observed [91]. Moreover, use of sucrose, or other sweet-tasting solution, has been shown in multiple studies to effectively reduce pain scores in response to procedures such as heel lance [92]. However, in a randomised controlled trial comparing term infants receiving sucrose or a sterile water placebo prior to a clinically required heel lance, Slater and colleagues demonstrated that while sucrose reduced PIPP scores, compared with placebo, it did not reduce the magnitude of the nociceptive-specific brain activity recorded using EEG or the magnitude of the reflex withdrawal [93].

As pain is a subjective experience, all measures of infant pain are surrogates. However, as cortical processing is necessary for the experience of pain [94] measures of brain activity may provide the best surrogate measures of infant pain. While brain activity measures currently have limited direct use in the clinical setting, where assessment tools need to be quickly and easily performed, simultaneously measuring changes in infant brain activity, physiology and behaviour provides a more detailed understanding of nociceptive processing in the infant. Identifying measures that correlate with changes in brain activity may determine the most appropriate pain assessment tools for use within the clinical setting [95]. For example, reflex withdrawal can be visually observed and its magnitude is correlated with the magnitude of the nociceptive-specific brain activity [80]. Consequently, it may be useful to incorporate reflex limb withdrawal into infant pain assessment tools. It is important to note that here we are specifically referring to responses to noxious stimulation; reflex withdrawal is not a nociceptive-specific response, or necessarily indicative of nociceptive-specific brain activity, as it can be observed in infants in response to non-noxious tactile stimulation [96]. Furthermore, this correlation has only been demonstrated in term infants and may not be true in the preterm population. Overall, understanding the complex interaction of different physiological and behavioural measures, external factors, the state of the infant and the developing nervous system will improve pain assessment and may explain the lack of behavioural response in some individuals.

5 Imaging the Long-Term Effects of Early Life Pain

There is the evidence to suggest that pain exposure early in life may have a long-term impact, both on later pain processing and also on more general neurological structure and function. In addition to genetic factors, development of the nervous system is activity-dependent [97, 98]. Whilst a lack of activity during critical developmental windows may disrupt normal development of the nervous system [97, 99–101], excessive activity during early development may also lead to maladaptation [102]. As discussed above, infants requiring neonatal care can receive multiple noxious or stressful procedures a day as part of their essential medical treatment. These procedures occur over a period of rapid neurological development during which the infant nervous system is particularly vulnerable. Furthermore, the threshold for evoking reflex responses in infants is lower than in adults, within the innocuous range, and this threshold increases with age [96, 103, 104]. The reflexes themselves are longer duration, and higher in magnitude in infants compared with adult responses [96]. Additionally, particularly in preterm infants, repeated stimulation causes sensitisation [96, 103–105] and increased responses to tactile stimulation occur following a noxious procedure [106]. This increased sensitivity to noxious stimulation in infants compared with adults may exacerbate the effect early life pain can have on the developing nervous system.

5.1 The Impact of Early Life Pain on Later Pain Responses

The literature regarding the long-term effects of early life pain in humans is multifarious, with results complicated by factors including the type and site of experimental stimulation and its relation to the site of the initial injury, the age at assessment, the type of early life pain, and the age at early life pain. In an influential study Taddio et al. [107] demonstrated increased behavioural pain responses to vaccination in infants who were circumcised early in life compared with a control group who were not circumcised. Moreover, infants who received local anaesthetic cream (EMLA) prior to circumcision had lower pain responses to the injection compared with those who received placebo [107]. The results of Peters et al. also suggested a hyperalgesic effect of early life pain—children with early life surgery had increased intraoperative and postoperative analgesic requirements and higher postoperative pain scores related to subsequent surgery performed in the same dermatome [108]. However, effects may vary depending on age. One study found that term infants who required multiple heel lances during the first day of life have increased behavioural responses to subsequent venepuncture [109]. Conversely, a study of premature infants at 32 weeks' gestation observed decreased behavioural responses to heel lance in infants who were 4 weeks postnatal age (and who consequently had multiple prior painful procedures) compared to infants who were less than 4 days old [90].

A study in rats suggests that the seemingly conflicting reports of hypoalgesia and hyperalgesia from different studies may not be mutually exclusive, and may result from differences in the time, type and location (in relation to initial injury) at which the pain response is subsequently measured [110]. Moreover, the severity of the initial pain experience appears to be important. Premature-born children have elevated heat pain thresholds, greater perceptual sensitisation to tonic heat and decreased thermal sensitivity at school age compared with term-born controls, but do not have altered responses to mechanical stimuli [20, 111]. Alterations in thermal sensitivity are more pronounced in premature-born children who have undergone early life surgery [20] and alterations in pain sensitivity in children aged 9–16 years who had a burn injury during infancy differ depending on the severity of the burn [112]. Furthermore, early injury may result in both local and global long-lasting alterations in sensory processing [112, 113]. For example, children who had cardiac surgery in infancy have altered sensory processing in the area of their thoracotomy scar and in the contralateral region [113].

Few neuroimaging studies have addressed the impact of early life experience of pain on later life pain responses. Using EEG, Slater et al. [114] demonstrated that premature-born infants studied at term-equivalent age exhibit nociceptive-specific brain activity of greater magnitude than infants born at term who are relatively pain naïve. Hohmeister et al. [115] measured the fMRI responses to tonic heat stimulation (adjusted for each child to be mildly painful) in premature-born children aged 11–16 years old compared with aged-matched term-born children who did and did not require neonatal care. The premature-born children exhibited greater levels of brain activity (with significant activity in the thalamus, anterior cingulate cortex, cerebellum, basal ganglia, and the periaquaeductal grey) in response to the painful heat stimulus compared with the term-born children who did not require neonatal care. Poorun et al. examined EEG responses to clinically required cannulation in children aged 1–12 years whilst under general anaesthesia, comparing premature-born and term-born children. They did not observe a difference in the response between the two groups; however, the stimulus was a relatively minor procedure. They hypothesise that the differences observed between term-born and premature-born children in other studies may relate to conscious pain processing [116].

5.2 The Impact of Early Life Pain on Neurological Development

Even in the absence of obvious neurological sequelae, children born very prematurely are more likely to have cognitive, behavioural and social problems later in life compared with their term-born peers [117–120]. It is also reported that they are more likely to have psychiatric disorders, including attention-deficit/hyperactivity disorder and autism spectrum disorders [121]. Many studies have examined structural neurological abnormalities in premature-born children, reporting numerous differences including decreased cerebral volumes, alterations in grey and white matter, and specific regions of vulnerability including the frontotemporal and hippocampal regions. Furthermore, these structural changes have been associated with decreased cognitive scores (reviewed by Ment and Vohr [122] and Counsell and Boardman [123]).

It is plausible that these neurological abnormalities relate to the ex-utero environment that the premature infant is exposed to. The infant receives a barrage of visual, auditory and other sensory stimuli, and neonatal units now frequently employ strategies to attempt to minimise these stimuli. Animal models provide an opportunity to directly investigate early life pain exposure without the complication of other factors such as illness severity, which is an unavoidable confounding factor in preterm human infants. In terms of cognitive effects, repetitive inflammatory pain exposure in rat pups increases cell death in cortical and subcortical areas, and diminishes cognitive abilities in the adult rat [124]. This suggests that pain alone can have a direct affect on the developing nervous system, not just in terms of pain processing, but that it may also have a wider impact.

A number of studies have now been conducted which investigate correlations between neurological outcome measures and the number of painful procedures experienced during the preterm period in humans. Smith et al. studied a group of infants born at less than 30 weeks' gestation and recorded all of the stressful procedures they received between birth and term-equivalent age [125]. Examining MRI scans conducted at term-equivalent age they found that a higher number of stressful procedures were correlated with decreased brain size in frontal and parietal regions, and altered brain diffusion and functional connectivity in the temporal lobes [125]. A series of studies by Grunau and colleagues have also investigated this question (reviewed by Ranger and Grunau [126, 127]). In preterm infants born before 33 weeks' gestation, Zwicker et al. examined corticospinal tract development using diffusion tensor imaging (DTI), with scans acquired near birth and at term-equivalent age. They found a significant interaction between the number of painful procedures during the neonatal period, and corticospinal tract development, with a slower rise in fractional anisotropy between the two scans associated with more painful procedures [128]. Brummelte et al. [129] showed that a higher number of painful procedures were significantly associated with reduced white matter maturation (reduced fractional anisotropy observed in DTI scans) and subcortical grey matter maturation (reduced *N*–acetylaspartate to choline ratio investigated using magnetic resonance spectroscopy) during the neonatal period up to term-equivalent age.

Grunau and colleagues have also demonstrated associations between the number of painful procedures in the neonatal period and neurological maturation in older children. In 7-year-old children, born at or before 32 weeks' gestation, a greater number of invasive procedures during the neonatal period are associated with reduced cortical thickness [130], reduced cerebellar volume [131] and lower white matter integrity (indicated by lower fractional anisotropy values) [132]. Furthermore, the combination of a greater number of invasive procedures and lower fractional anisotropy of the superior white matter was significantly associated with lower IQ [132]. In these studies Grunau and colleagues adjusted the number of painful procedures for clinical factors such as gestational age at birth, illness severity on the first day of life, number of days the infant received mechanical ventilation, confirmed infections and cumulative morphine exposure [127, 129, 131, 132]. Functional alterations observed in brain activity have also been linked to exposure to painful procedures in the preterm period. Doesburg et al. [133] demonstrated that altered functional brain activity (increased gamma-alpha ratio measured using EEG) is correlated with pain exposure in children born extremely prematurely, and negatively correlated with visual-perceptual ability at 7 years of age.

Interestingly, in two randomised controlled trials investigating interventions designed to reduce stress in infants receiving neonatal care, through either influencing parent or nursing interactions, these interventions appear to somewhat ameliorate structural and functional neurological alterations [134, 135]. In the first trial, infants born between 28 and 33 weeks' gestation were randomly allocated to receive either the intervention—Newborn Individualized Developmental Care and Assessment Program (NIDCAP)—or standard care. NIDCAP involves an individualised approach to assessing each infant's stress signals and behaviours, and adapting medical care in relation to these signs. At 2 weeks corrected age the intervention group had increased coherence in alpha and beta bands between frontal regions and occipital and parietal regions, and higher relative anisotropy in the left internal capsule. Furthermore, importantly behavioural function was improved at 9 months of age [134]. In the second trial, in the intervention group parents were trained to recognise signs of distress in their infant and to optimise their interactions with the infant. At term-equivalent age the intervention group had lower apparent diffusion coefficients suggesting improved white matter microstructure [135]. Future neuroimaging studies may add to our understanding of these interventions in this vulnerable population.

6 Imaging Pain in Older Children

The nervous system continues to change and mature throughout childhood and into adolescence and young adulthood. Synaptic density peaks at approximately 1–2 years of age (dependent on the brain region), and then continues to decline until around the age of 16 [136]. White matter density increases across childhood, and grey matter density peaks at about 4 years before decreasing across childhood and adolescence [137–139]. Functional changes also occur during childhood, with, for example, an increase in higher frequency activity, and a decrease in the lower frequency delta and theta bands observed in EEG recordings into adolescence [140–142]. Given these structural and functional differences, the use of neuroimaging techniques will be crucial to better understand how pain processing changes with age in the paediatric population. However, to date neuroimaging has only been used in a relatively small number of studies investigating chronic pain including migraine, recurrent abdominal pain and complex regional pain syndrome.

Clinically, migraine can be associated with background EEG abnormalities. Additionally, multiple electrophysiological studies have been conducted in adults with migraine, evaluating sensory processing, with the most noticeable finding being a lack of habituation to sensory stimuli (including visual, auditory and nociceptive stimuli) during interictal periods [143]. More recently, EEG recordings have been performed in children with headaches and have found similar results. In children, changes in visual evoked potentials [144, 145], auditory evoked potentials [146] and event-related potentials to emotional pictures [147] have been observed during interictal periods, and a lack of habituation of responses has been described [148]. Alterations in the normal pattern of age-dependent changes in evoked responses have also supported the theory of an interaction between headaches and neurological maturation [148, 149].

Zohsul et al. [150] examined self-report in response to thermal and mechanical stimuli in children with migraine aged 9–15 years, and found that whilst heat pain thresholds were not different to healthy controls, mechanical pain thresholds were reduced. In a separate study the same authors investigated the evoked neurophysiological responses to painful and non-painful mechanical stimuli in children aged 10–14 years compared with age-matched controls. Children were studied in migraine-free periods and were asked to respond to auditory stimuli that were presented in an oddball paradigm with the mechanical stimuli (i.e. children were asked to ignore the frequently repeating mechanical stimuli and respond to the rare, unpredictable, auditory tones). Children with migraine showed significantly larger P300 components (but no differences in the earlier N150 or P260 components) in response to both painful and non-painful stimuli, but no differences in their response to the auditory stimuli. As the P300 has been observed in response to stimuli that shift attention away from a task stimulus (in this case the auditory stimuli), the authors suggest that the children with migraine display an attentional bias towards painful and potentially painful stimuli [151]. Using the same paradigm, Hermann et al. [152] also demonstrated enhanced P300 responses in children aged 10–15 years with recurrent abdominal pain. Consistent with other theories of chronic pain, this attentional bias to painful stimuli may provide a mechanism through which pain syndromes become a chronic problem [151].

Rocca et al. investigated structural changes in children with migraine aged 9–17 years compared with aged-matched controls using MRI. Migraine patients demonstrated grey matter atrophy in several frontal and temporal regions (left middle temporal gyrus, right orbitofrontal gyrus, left inferior frontal gyrus and subgenual cingulum), and increased grey matter volume in the right putamen [153]. Interestingly, before puberty the prevalence of migraine in boys and girls is approximately equal. However, the prevalence in females increases across puberty and in adults migraine affects twice as many females [154]. Faria et al. examined structural and functional differences in children aged 10–16 years with migraine compared with age-matched controls. They also split the groups according to gender, and into two different age groups: 10–11 years and 14–16 years, to investigate the neural mechanisms underlying gender differences and their evolution in patients with migraine. They found a significant gender-disease interaction—female patients with migraine had significantly higher grey matter thickness in multiple cortical and subcortical regions including sensory, motor and affective regions compared with male patients with migraine and healthy controls. Females with migraine also exhibited alterations in resting state functional connectivity, with greater connectivity from the amygdala to the thalamus, anterior midcingulate cortex, and supplementary motor area, and from the precuneus to the thalamus, amygdala, caudate and putamen, compared with males with migraine and healthy controls. Moreover, the structural differences varied according to age group, demonstrating developmental as well as gender-related neurological differences [155].

In a series of studies conducted by Borsook and colleagues, brain changes related to complex regional pain syndrome (CRPS) in children have been investigated. CRPS mainly affects the lower limbs and is characterised by severe pain, hyperalgesia, allodynia, oedema and changes in skin tone, and can also involve autonomic changes such as abnormal sweating and poor circulation [156, 157]. The condition predominantly affects females (with studies reporting 85–90% of cases in girls) and often follows trauma [156, 157]. In contrast to adult patients, CRPS in children often resolves within several months to 2 years [157]. This has allowed for investigations of both the brain changes during the condition, as well as once the condition has resolved. Lebel et al. examined the functional responses to brush and cold stimuli in children and adolescents aged 9–18 years with CRPS, comparing the affected and unaffected limb during the condition and after clinical recovery. Following recovery, despite nearly complete elimination of reported evoked pain, significant differences in BOLD response still persisted between stimulation of the affected and unaffected limb [158]. Linnman et al. [159] demonstrated both transient (during the condition) and persistent (once the condition had resolved) changes in functional connectivity in children with CRPS, with increases in connectivity observed with the anterior cingulate cortex, postcentral gyrus, amygdala, caudate and putamen. Moreover, resting state functional connectivity specifically between the habenula—located within the thalamus—and multiple other brain regions is reduced in children with CRPS [160]. Importantly, persistent alterations in connectivity may allow for reoccurrence of the symptoms at a later date. Further follow-up studies may indicate how long connectivity changes persist, and whether reoccurrence might be individually predicted based on functional connectivity measures [159].

Some children with CRPS are resistant to normal treatment, and in these cases intensive treatment can often prove effective. Becerra et al. [161] examined resting state networks in children aged 10–18 years with CRPS before and after an intensive 3-week treatment program. After treatment the children's pain scores were significantly reduced. A number of resting state networks, including fronto-parietal, salience, default mode, central executive and sensorimotor networks, were significantly altered before treatment. Following treatment, although there were still some differences in networks compared with controls, these differences were reduced [161]. Simons et al. [83, 162] investigated functional connectivity of the amygdala, a brain region involved in processing fear, reward and anxiety. Before treatment, enhanced connectivity was observed between the amygdala and multiple cortical and subcortical brain regions, including the prefrontal cortex, motor cortices and thalamus. Following intensive treatment decreases in connectivity were observed with some regions, including the motor cortex and cingulate cortex [162]. Finally, Erpelding et al. examined structural brain changes, demonstrating reduced grey matter in multiple cortical and subcortical regions in CRPS patients compared with healthy controls. Patients also had increased grey matter in the mediodorsal thalamus and the posterior hippocampus. Following intensive treatment enhanced functional connectivity between the dorsolateral prefrontal cortex and periaqueductal grey was observed which may indicate that changes in the pain modulatory system are important for improvement following treatment [163]. Altogether, these studies not only demonstrate the changes that can occur in children's brains in relation to chronic pain conditions, but also the plasticity of the nervous system in relation to treatment effects, including rapid treatment effects. These studies offer the opportunity to gain a better mechanistic understanding of chronic pain and its effect on the developing nervous system, and future studies may allow for comparison between different chronic pain conditions. Moreover, brain imaging may eventually allow for an individualised and targeted approach to the treatment of chronic pain in children [164].

7 Future Directions

7.1 Analgesics

Often analgesics are not tested in children and in particular the infant population. However, it cannot be assumed that analgesics will act in the same way across all ages. Whilst ethical considerations are a priority and new treatment must be compared with the current best practice, controlled trials of analgesics are essential to provide better analgesic treatment in children.

Neuroimaging techniques provide an opportunity to gain a mechanistic insight into how analgesics modulate activity of the nervous system and how this changes as the nervous system develops. Particularly in preverbal infants these techniques provide a unique opportunity to evaluate the anti-nociceptive properties of analgesics. In adults opioid analgesics reduce the amplitude of nociceptive evoked potentials recorded using EEG [165, 166] and evaluating the magnitude of nociceptive evoked potentials in children may provide a better understanding of the anti-nociceptive properties of the intervention. As discussed previously, this approach was taken by Slater et al. [93] who compared sucrose with sterile water given before a heel lance in infants. There was no difference in the magnitude of the reflex withdrawal or the magnitude of the nociceptive-specific brain activity between the two groups suggesting that although sucrose effectively changes the facial expression responses to a heel lance, nociceptive brain processing may not be altered [93]. A similar approach will be used in a randomised controlled trial of morphine sulphate in premature infants, which is currently being conducted. The trial will investigate whether morphine reduces behavioural scores following a painful eye exam and nociceptive-specific brain activity in response to a heel lance [167].

Techniques, including machine-learning approaches, may be useful in decoupling the response to pain from the brain's responses to other stimuli [168]. Wager et al. [168] used fMRI to derive a "neurological signature" of pain in adults, which they demonstrated could discriminate between painful and non-painful heat, and physical and social pain, and was reduced by remifentanil administration. Deriving such a signature may be particularly useful in populations where self-report is not possible. Duff et al. [169] recently suggested the use of fMRI within a drug discovery pipeline, where new analgesics could be assessed against integrated data from previous studies of analgesics. A pipeline of this kind in children would be beneficial, particularly with regard to reducing the number of studies that need to be conducted in vulnerable populations and allowing rapid comparison of analgesics. In short, investigating fMRI signatures of pain and how they are modulated by analgesics in children and infants is a promising avenue for future pain research in this population.

7.2 Other Treatments

In infants, non-pharmacological interventions, including kangaroo care (skin-to-skin contact) [170], breastfeeding [171], non-nutritive sucking [172] and swaddling [172], have been shown to reduce pain scores and have the advantage of having little or no side effects. They can either be used alone, or are frequently used in combination and/or with sucrose, with this often giving further benefit then the use of a single intervention alone [173, 174]. Using brain imaging to investigate whether nociceptive processing is altered with these techniques will be beneficial.

As discussed previously, in older children with CRPS rapid treatment-induced changes have been observed. Understanding how the brain changes in chronic pain conditions, and the changes that can be induced by treatment of a particular condition, may allow for targeted therapy in other chronic pain conditions. Therapies in older children may also include psychological interventions such as cognitive behavioural therapy and mindfulness-based techniques, which could be investigated using brain imaging. Moreover, neurofeedback is an interesting area of research in adults which may be effective in the treatment of chronic pain [175]. The use of neuroimaging techniques for the treatment of chronic pain is an important area for future research in children.

7.3 Anaesthesia

The aims of anaesthesia are to cause unconsciousness, immobility, lack of memory and analgesia. To achieve these aims, multiple drugs are often used, and to avoid problems associated with anaesthesia, including under and overdosing, optimal titration of anaesthetic drugs is essential. Indeed, in recent years, particularly when considering very young infants, the possible neurotoxic effect of anaesthetic drugs has been suggested [176]. Minimising anaesthesia during surgery particularly for younger children, whilst not under dosing the patient, is therefore important.

Anaesthetists routinely use physiological measures and movement of the patient to assess anaesthetic depth. In more recent years, brain activity measures have become increasingly utilised. There are a number of commercially available software and other brain activity measures that have been used to assess depth of anaesthesia [177, 178]. However, there has been a relative lack of research in children, and use of such monitors in the paediatric setting remains particularly controversial [179]. Indeed, it should not be assumed that children will have similar measures of depth of anaesthesia to adults as the nervous system continues to change with development [180]. Furthermore, the measures may be altered in some populations, for example, children with cerebral palsy [181, 182], and in premature-born children [116].

Neurophysiological measures may also be useful for specifically assessing the anaesthetised patient's responses to stimulation (compared with the depth of anaesthesia monitors which measure ongoing brain activity). A number of studies in anaesthetised adults have identified changes in brain activity in response to noxious stimulation [183–186]. Hartley et al. [187] recently demonstrated that, at a set concentration of anaesthetic (2.5% end tidal of sevoflurane), a significant increase in delta activity can be observed in response to cannulation, experimental noxious and experimental tactile stimuli in children. A greater understanding of how these responses are generated is needed in order to properly appreciate what they mean in terms of anaesthetic dosing. However, future research in the area of neurophysiological measures in anaesthetised children may prove useful in optimising paediatric anaesthesia.

8 Conclusion

In summary, we have discussed recent research in neuroimaging of paediatric pain, including pain in infants, the long-term effects of early life pain and pain in older children. Pain is a serious clinical issue, which risks being under-treated, particularly in non-verbal populations. As cortical responses are essential for the perception of pain, neuroimaging provides an important step forward in our understanding of infant pain. Moreover, imaging studies have provided insight into the remarkable plasticity of the nervous system, demonstrating not only how the brain is changed by early life pain and chronic pain conditions, but also by treatment interventions. Neuroimaging will likely continue to play a vital role in advancing our understanding of how pain is processed at different ages and the impact of pain in this complex, developing population.

References

1. World Health Organisation. Guidelines on the pharmacological treatment of persisting pain in children with medical illnesses. 2012.
2. Walker SM. Pain in children: recent advances and ongoing challenges. Brit J Anaesth. 2008;101 (1):101–10.
3. Schechter NL, Berde CB, Yaster M, editors. Pain in infants, children, and adolescents. Lippincott Williams & Wilkins; 2003.
4. Fitzgerald M. The development of nociceptive circuits. Nat Rev Neurosci. 2005;6(7):507–20.
5. Fitzgerald M, Walker SM. Infant pain management: a developmental neurobiological approach. Nat Clin Pract Neurol. 2009;5(1):35–50.
6. Schwaller F, Fitzgerald M. The consequences of pain in early life: injury-induced plasticity in developing pain pathways. Eur J Neurosci. 2014;39(3):344–52.
7. McGrath PJ, Stevens BJ, Walker SM, Zempsky WT, editors. Oxford textbook of paediatric pain. Oxford: Oxford University Press; 2014.
8. Anand KJ, Stevens BJ, McGrath PJ, editors. Pain in neonates and infants, 3rd ed. Elsevier; 2007.
9. Stevens BJ, Zempsky WT. Prevalence and distribution of pain in children. In: McGrath PJ, Stevens BJ, Walker SM, Zempsky WT, editors. Oxford textbook of paediatric pain. Oxford: Oxford University Press; 2014. p. 12–9.
10. Anand KJ, International Evidence-Based Group for Neonatal Pain. Consensus statement for the prevention and management of pain in the newborn. Arch Pediatr Adolesc Med. 2001;155(2):173–80.
11. Carbajal R, Rousset A, Danan C, Coquery S, Nolent P, Ducrocq S, Saizou C, Lapillonne A, Granier M, Durand P, Lenclen R, Coursol A, Hubert P, de Saint Blanquat L, Boelle PY, Annequin D, Cimerman P, Anand KJ, Breart G. Epidemiology and treatment of painful procedures in neonates in intensive care units. JAMA. 2008;300 (1):60–70.
12. Johnston C, Barrington KJ, Taddio A, Carbajal R, Filion F. Pain in Canadian NICUs: have we improved over the past 12 years? Clin J Pain. 2011;27(3):225–32.
13. Kyololo OM, Stevens B, Gastaldo D, Gisore P. Procedural pain in neonatal units in Kenya. Arch Dis Child Fetal Neonatal Ed. 2014;99(6):F464–7.
14. Roofthooft DW, Simons SH, Anand KJ, Tibboel D, van Dijk M. Eight years later, are we still hurting newborn infants? Neonatology. 2014;105(3):218–26.

15. Simons SH, van Dijk M, Anand KS, Roofthooft D, van Lingen RA, Tibboel D. Do we still hurt newborn babies? A prospective study of procedural pain and analgesia in neonates. Arch Pediatr Adolesc Med. 2003;157(11):1058–64.
16. Stevens BJ, Abbott LK, Yamada J, Harrison D, Stinson J, Taddio A, Barwick M, Latimer M, Scott SD, Rashotte J, Campbell F, Finley GA, CIHR Team in Children's Pain. Epidemiology and management of painful procedures in children in Canadian hospitals. Can Med Assoc J. 2011;183(7): E403–10.
17. Harrison D, Joly C, Chretien C, Cochrane S, Ellis J, Lamontagne C, Vaillancourt R. Pain prevalence in a pediatric hospital: raising awareness during pain awareness week. Pain Res Manag. 2014;19(1):e24–30.
18. Groenewald CB, Rabbitts JA, Schroeder DR, Harrison TE. Prevalence of moderate-severe pain in hospitalized children. Paediatr Anaesth. 2012;22 (7):661–8.
19. Taylor EM, Boyer K, Campbell FA. Pain in hospitalized children: a prospective cross-sectional survey of pain prevalence, intensity, assessment and management in a Canadian pediatric teaching hospital. Pain Res Manag. 2008;13(1):25–32.
20. Walker SM, Franck LS, Fitzgerald M, Myles J, Stocks J, Marlow N. Long-term impact of neonatal intensive care and surgery on somatosensory perception in children born extremely preterm. Pain. 2009;141(1–2):79–87.
21. Argoff CE. Recent management advances in acute postoperative pain. Pain Pract. 2014;14(5):477–87.
22. Lavand'homme P. The progression from acute to chronic pain. Curr Opin Anaesthesiol. 2011;24 (5):545–50.
23. Ashburn MA, Caplan RA, Carr DB, Connis RT, Ginsberg B, Green CR, Lema MJ, Nickinovich DG, Rice LJ, Anesthesiologists Amer Soc. Practice guidelines for acute pain management in the perioperative setting an updated report by the American Society of Anesthesiologists Task Force on acute pain management. Anesthesiology. 2012;116(2):248–73.
24. Association of Paediatric Anaesthetists of Great Britain and Ireland. Good practice in postoperative and procedural pain management, 2nd ed. Paediatr Anaesth. 2012;22(Suppl 1):1–79.
25. Macintyre PE, Scott DA, Schug SA, Visser EJ, Walker SM, editors. Acute pain management: scientific evidence, 3rd ed. 2010.
26. NHS Choices. The NHS vaccination schedule. http://www.nhs.uk/conditions/vaccinations/pages/vaccination-schedule-age-checklist.aspx. Accessed 17 Aug 2015.
27. U. S. Department of Health and Human Services Recommended Immunization Schedules for Persons Aged 0 Through 18 Years, United States 2015. http://www.cdc.gov/vaccines/schedules/downloads/child/0-18yrs-child-combined-schedule.pdf. Accessed 17 Aug 2015.
28. Taddio A, Ipp M, Thivakaran S, Jamal A, Parikh C, Smart S, Sovran J, Stephens D, Katz J. Survey of the prevalence of immunization non-compliance due to needle fears in children and adults. Vaccine. 2012;30(32):4807–12.
29. Taddio A, Appleton M, Bortolussi R, Chambers C, Dubey V, Halperin S, Hanrahan A, Ipp M, Lockett D, MacDonald N, Midmer D, Mousmanis P, Palda V, Pielak K, Riddell RP, Rieder M, Scott J, Shah V. Reducing the pain of childhood vaccination: an evidence-based clinical practice guideline (summary). CMAJ. 2010;182(18):1989–95.
30. Taddio A, Chambers CT, Halperin SA, Ipp M, Lockett D, Rieder MJ, Shah V. Inadequate pain management during routine childhood immunizations: the nerve of it. Clin Ther. 2009;31(Suppl 2): S152–67.
31. King S, Chambers CT, Huguet A, MacNevin RC, McGrath PJ, Parker L, MacDonald AJ. The epidemiology of chronic pain in children and adolescents revisited: a systematic review. Pain. 2011;152 (12):2729–38.
32. Hicks CL, von Baeyer CL, Spafford PA, van Korlaar I, Goodenough B. The faces pain scale-revised: toward a common metric in pediatric pain measurement. Pain. 2001;93(2):173–83.
33. von Baeyer CL. Self-report: the primary source in assessment after infancy. In: McGrath PJ, Stevens BJ, Walker SM, Zempsky WT, editors. Oxford textbook of paediatric pain. Oxford: Oxford University Press; 2014. p. 370–8.
34. Lagercrantz H, Changeux JP. Basic consciousness of the newborn. Semin Perinatol. 2010;34(3):201–6.
35. Wilkinson DJ, Savulescu J, Slater R. Sugaring the pill: ethics and uncertainties in the use of sucrose for newborn infants. Arch Pediatr Adolesc Med. 2012;166(7):629–33.
36. Merskey H, Bogduk N IASP Task Force on Taxonomy. http://www.iasp-pain.org/Taxonomy. Accessed 21 Nov 2015.
37. Unruh AM, McGrath PJ. History of pain in children. In: McGrath PJ, Stevens BJ, Walker SM, Zempsky WT, editors. Oxford textbook of paediatric pain. Oxford: Oxford University Press; 2014. p. 3–11.
38. Anand KJ, Sippell WG, Aynsley-Green A. Randomised trial of fentanyl anaesthesia in preterm babies undergoing surgery: effects on the stress response. Lancet. 1987;1(8527):243–8.
39. Anand KJ, Sippell WG, Schofield NM, Aynsley-Green A. Does halothane anaesthesia decrease the metabolic and endocrine stress responses of newborn infants undergoing operation? Br Med J (Clin Res Ed). 1988;296(6623):668–72.

40. Anand KJ, Hansen DD, Hickey PR. Hormonal-metabolic stress responses in neonates undergoing cardiac surgery. Anesthesiology. 1990;73(4):661–70.
41. Lawson J. Letters. Birth 1986:124–5.
42. Rovner S. Surgery without anesthesia: can preemies feel pain? The Washington Post. 1986.
43. Lee GY, Stevens BJ. Neonatal and infant pain assessment. In: McGrath PJ, Stevens BJ, Walker SM, Zempsky WT, editors. Oxford textbook of paediatric pain. Oxford: Oxford University Press; 2014. p. 353–69.
44. Gibbins S, Stevens BJ, Yamada J, Dionne K, Campbell-Yeo M, Lee G, Caddell K, Johnston C, Taddio A. Validation of the premature infant pain profile-revised (PIPP-R). Early Hum Dev. 2014;90 (4):189–93.
45. Stevens B, Johnston C, Petryshen P, Taddio A. Premature infant pain profile: development and initial validation. Clin J Pain. 1996;12(1):13–22.
46. Stevens BJ, Gibbins S, Yamada J, Dionne K, Lee G, Johnston C, Taddio A. The premature infant pain profile-revised (PIPP-R) initial validation and feasibility. Clin J Pain. 2014;30(3):238–43.
47. Grunau RVE, Craig KD. Pain expression in neonates—facial action and cry. Pain. 1987;28 (3):395–410.
48. Lawrence J, Alcock D, McGrath P, Kay J, MacMurray SB, Dulberg C. The development of a tool to assess neonatal pain. Neonatal Netw. 1993;12 (6):59–66.
49. Debillon T, Zupan V, Ravault N, Magny JF, Dehan M. Development and initial validation of the EDIN scale, a new tool for assessing prolonged pain in preterm infants. Arch Dis Child. 2001;85(1): F36–40.
50. Ambuel B, Hamlett KW, Marx CM, Blumer JL. Assessing distress in pediatric intensive care environments: the COMFORT scale. J Pediatr Psychol. 1992;17(1):95–109.
51. Merkel SI, Voepel-Lewis T, Shayevitz JR, Malviya S. The FLACC: a behavioral scale for scoring postoperative pain in young children. Pediatr Nurs. 1997;23(3):293–7.
52. Breau LM, Finley GA, McGrath PJ, Camfield CS. Validation of the non-communicating children's pain checklist-postoperative version. Anesthesiology. 2002;96(3):528–35.
53. Hunt A, Goldman A, Seers K, Crichton N, Mastroyannopoulou K, Moffat V, Oulton K, Brady M. Clinical validation of the paediatric pain profile. Dev Med Child Neurol. 2004;46(1):9–18.
54. Malviya S, Voepel-Lewis T, Burke C, Merkel S, Tait AR. The revised FLACC observational pain tool: improved reliability and validity for pain assessment in children with cognitive impairment. Paediatr Anaesth. 2006;16(3):258–65.
55. Thompson KL, Varni JW, Hanson V. Comprehensive assessment of pain in juvenile rheumatoid-arthritis—an empirical-model. J Pediatr Psychol. 1987;12(2):241–55.
56. Varni JW, Thompson KL, Hanson V. The Varni Thompson pediatric pain questionnaire. 1. Chronic musculoskeletal pain in juvenile rheumatoid-arthritis. Pain. 1987;28(1):27–38.
57. Eccleston C, Jordan A, McCracken LM, Sleed M, Connell H, Clinch J. The bath adolescent pain questionnaire (BAPQ): development and preliminary psychometric evaluation of an instrument to assess the impact of chronic pain on adolescents. Pain. 2005;118(1–2):263–70.
58. Jibb LA, Stevens BJ, Nathan PC, Seto E, Cafazzo JA, Stinson JN. A smartphone-based pain management app for adolescents with cancer: establishing system requirements and a pain care algorithm based on literature review, interviews, and consensus. JMIR Res Protoc. 2014;3(1):e15.
59. Stinson JN, Jibb LA, Nguyen C, Nathan PC, Maloney AM, Dupuis LL, Gerstle JT, Alman B, Hopyan S, Strahlendorf C, Portwine C, Johnston DL, Orr M. Development and testing of a multidimensional iPhone pain assessment application for adolescents with cancer. J Med Internet Res. 2013;15(3):e51.
60. Lee SJ, Ralston HJP, Drey EA, Partridge JC, Rosen MA. Fetal pain—a systematic multidisciplinary review of the evidence. JAMA. 2005;294 (8):947–54.
61. Friauf E, McConnell SK, Shatz CJ. Functional synaptic circuits in the subplate during fetal and early postnatal development of cat visual cortex. J Neurosci. 1990;10(8):2601–13.
62. Kanold PO, Luhmann HJ. The subplate and early cortical circuits. Annu Rev Neurosci. 2010;33:23–48.
63. Kostovic I, Rakic P. Developmental history of the transient subplate zone in the visual and somatosensory cortex of the macaque monkey and human brain. J Comp Neurol. 1990;297(3):441–70.
64. Kanold PO, Kara P, Reid RC, Shatz CJ. Role of subplate neurons in functional maturation of visual cortical columns. Science. 2003;301(5632):521–5.
65. Tolner EA, Sheikh A, Yukin AY, Kaila K, Kanold PO. Subplate neurons promote spindle bursts and thalamocortical patterning in the neonatal rat somatosensory cortex. J Neurosci. 2012;32 (2):692–702.
66. Hartley C, Berthouze L, Mathieson SR, Boylan GB, Rennie JM, Marlow N, Farmer SF. Long-range temporal correlations in the EEG bursts of human preterm babies. PLoS One. 2012;7(2):e31543.
67. Andre M, Lamblin MD, d'Allest AM, Curzi-Dascalova L, Moussalli-Salefranque F, Tich SNT, Vecchierini-Blineau MF, Wallois F, Walls-Esquivel E, Plouin P. Electroencephalography in premature and full-term infants. Developmental features and glossary. Neurophysiol Clin. 2010;40 (2):59–124.

68. Niemarkt HJ, Andriessen P, Peters CH, Pasman JW, Zimmermann LJ, Bambang Oetomo S. Quantitative analysis of maturational changes in EEG background activity in very preterm infants with a normal neurodevelopment at 1 year of age. Early Hum Dev. 2010;86(4):219–24.
69. Fabrizi L, Slater R, Worley A, Meek J, Boyd S, Olhede S, Fitzgerald M. A shift in sensory processing that enables the developing human brain to discriminate touch from pain. Curr Biol. 2011;21(18):1552–8.
70. Milh M, Kaminska A, Huon C, Lapillonne A, Ben-Ari Y, Khazipov R. Rapid cortical oscillations and early motor activity in premature human neonate. Cereb Cortex. 2007;17(7):1582–94.
71. Chipaux M, Colonnese MT, Mauguen A, Fellous L, Mokhtari M, Lezcano O, Milh M, Dulac O, Chiron C, Khazipov R, Kaminska A. Auditory stimuli mimicking ambient sounds drive temporal "delta-brushes" in premature infants. PLoS One. 2013;8(11):e79028.
72. Colonnese MT, Kaminska A, Minlebaev M, Milh M, Bloem B, Lescure S, Moriette G, Chiron C, Ben-Ari Y, Khazipov R. A conserved switch in sensory processing prepares developing neocortex for vision. Neuron. 2010;67(3):480–98.
73. Bartocci M, Bergqvist LL, Lagercrantz H, Anand KJ. Pain activates cortical areas in the preterm newborn brain. Pain. 2006;122(1–2):109–17.
74. Slater R, Cantarella A, Gallella S, Worley A, Boyd S, Meek J, Fitzgerald M. Cortical pain responses in human infants. J Neurosci. 2006;26(14):3662–6.
75. Limperopoulos C, Gauvreau KK, O'Leary H, Moore M, Bassan H, Eichenwald EC, Soul JS, Ringer SA, Di Salvo DN, du Plessis AJ. Cerebral hemodynamic changes during intensive care of preterm infants. Pediatrics. 2008;122(5):E1006–13.
76. Slater R, Worley A, Fabrizi L, Roberts S, Meek J, Boyd S, Fitzgerald M. Evoked potentials generated by noxious stimulation in the human infant brain. Eur J Pain. 2010;14(3):321–6.
77. Greenspan JD, McGillis SL. Stimulus features relevant to the perception of sharpness and mechanically evoked cutaneous pain. Somatosens Mot Res. 1991;8(2):137–47.
78. Iannetti GD, Baumgartner U, Tracey I, Treede RD, Magerl W. Pinprick-evoked brain potentials: a novel tool to assess central sensitization of nociceptive pathways in humans. J Neurophysiol. 2013;110(5):1107–16.
79. Goksan S, Hartley C, Emery F, Cockrill N, Poorun R, Moultrie F, Rogers R, Campbell J, Sanders M, Adams E, Clare S, Jenkinson M, Tracey I, Slater R. fMRI reveals neural activity overlap between adult and infant pain. eLife. 2015;4. doi:10.7554/eLife.06356.
80. Hartley C, Goksan S, Poorun R, Brotherhood K, Schmidt Mellado G, Moultrie F, Rogers R, Adams E, Slater R. The relationship between nociceptive brain activity, spinal reflex withdrawal and behaviour in newborn infants. Sci Rep. 2015;5:12519.
81. Williams G, Fabrizi L, Meek J, Jackson D, Tracey I, Robertson N, Slater R, Fitzgerald M. Functional magnetic resonance imaging can be used to explore tactile and nociceptive processing in the infant brain. Acta Paediatr. 2015;104(2):158–66.
82. Kahnt T, Heinzle J, Park SQ, Haynes JD. The neural code of reward anticipation in human orbitofrontal cortex. Proc Natl Acad Sci U S A. 2010;107(13):6010–5.
83. Simons LE, Moulton EA, Linnman C, Carpino E, Becerra L, Borsook D. The human amygdala and pain: evidence from neuroimaging. Hum Brain Mapp. 2014;35(2):527–38.
84. Rainville P, Duncan GH, Price DD, Carrier B, Bushnell MC. Pain affect encoded in human anterior cingulate but not somatosensory cortex. Science. 1997;277(5328):968–71.
85. Verriotis M, Fabrizi L, Lee A, Ledwidge S, Meek J, Fitzgerald M. Cortical activity evoked by inoculation needle prick in infants up to one-year old. Pain. 2015;156(2):222–30.
86. Ranger M, Johnston CC, Anand KJS. Current controversies regarding pain assessment in neonates. Semin Perinatol. 2007;31(5):283–8.
87. Craig KD, Whitfield MF, Grunau RVE, Linton J, Hadjistavropoulos HD. Pain in the preterm neonate—behavioral and physiological indexes. Pain. 1993;52(3):287–99.
88. Johnston CC, Stevens BJ, Franck LS, Jack A, Stremler R, Platt R. Factors explaining lack of response to heel stick in preterm newborns. J Obstet Gynecol Neonatal Nurs. 1999;28(6):587–94.
89. Gibbins S, Stevens B, McGrath PJ, Yamada J, Beyene J, Breau L, Camfield C, Finley A, Franck L, Johnston C, Howlett A, McKeever P, O'Brien K, Ohlsson A. Comparison of pain responses in infants of different gestational ages. Neonatology. 2008;93(1):10–8.
90. Johnston CC, Stevens BJ. Experience in a neonatal intensive care unit affects pain response. Pediatrics. 1996;98(5):925–30.
91. Slater R, Cantarella A, Franck L, Meek J, Fitzgerald M. How well do clinical pain assessment tools reflect pain in infants? PLoS Med. 2008;5(6):e129.
92. Stevens B, Yamada J, Lee GY, Ohlsson A. Sucrose for analgesia in newborn infants undergoing painful procedures. Cochrane Database Syst Rev. 1: CD001069. doi:10.1002/14651858.CD001069.pub4.
93. Slater R, Cornelissen L, Fabrizi L, Patten D, Yoxen J, Worley A, Boyd S, Meek J, Fitzgerald M. Oral sucrose as an analgesic drug for procedural

pain in newborn infants: a randomised controlled trial. Lancet. 2010;376(9748):1225–32.
94. Tracey I, Mantyh PW. The cerebral signature for pain perception and its modulation. Neuron. 2007;55(3):377–91.
95. Hartley C, Slater R. Neurophysiological measures of nociceptive brain activity in the newborn infant—the next steps. Acta Paediatr. 2014;103(3):238–42.
96. Cornelissen L, Fabrizi L, Patten D, Worley A, Meek J, Boyd S, Slater R, Fitzgerald M. Postnatal temporal, spatial and modality tuning of nociceptive cutaneous flexion reflexes in human infants. PLoS One. 2013;8(10):e76470.
97. Katz LC, Shatz CJ. Synaptic activity and the construction of cortical circuits. Science. 1996;274 (5290):1133–8.
98. Kilb W, Kirischuk S, Luhmann HJ. Electrical activity patterns and the functional maturation of the neocortex. Eur J Neurosci. 2011;34(10):1677–86.
99. Beggs S, Torsney C, Drew LJ, Fitzgerald M. The postnatal reorganization of primary afferent input and dorsal horn cell receptive fields in the rat spinal cord is an activity-dependent process. Eur J Neurosci. 2002;16(7):1249–58.
100. Koch SC, Fitzgerald M. Activity-dependent development of tactile and nociceptive spinal cord circuits. Ann N Y Acad Sci. 2013;1279:97–102.
101. Waldenstrom A, Thelin J, Thimansson E, Levinsson A, Schouenborg J. Developmental learning in a pain-related system: evidence for a cross-modality mechanism. J Neurosci. 2003;23 (20):7719–25.
102. Granmo M, Petersson P, Schouenborg J. Action-based body maps in the spinal cord emerge from a transitory floating organization. J Neurosci. 2008;28(21):5494–503.
103. Andrews K, Fitzgerald M. The cutaneous withdrawal reflex in human neonates: sensitization, receptive fields, and the effects of contralateral stimulation. Pain. 1994;56(1):95–101.
104. Fitzgerald M, Shaw A, MacIntosh N. Postnatal development of the cutaneous flexor reflex: comparative study of preterm infants and newborn rat pups. Dev Med Child Neurol. 1988;30(4):520–6.
105. Andrews K, Fitzgerald M. Cutaneous flexion reflex in human neonates: a quantitative study of threshold and stimulus-response characteristics after single and repeated stimuli. Dev Med Child Neurol. 1999;41(10):696–703.
106. Holsti L, Grunau RE, Oberlander TF, Whitfield MF. Prior pain induces heightened motor responses during clustered care in preterm infants in the NICU. Early Hum Dev. 2005;81(3):293–302.
107. Taddio A, Katz J, Ilersich AL, Koren G. Effect of neonatal circumcision on pain response during subsequent routine vaccination. Lancet. 1997;349 (9052):599–603.
108. Peters JW, Schouw R, Anand KJ, van Dijk M, Duivenvoorden HJ, Tibboel D. Does neonatal surgery lead to increased pain sensitivity in later childhood? Pain. 2005;114(3):444–54.
109. Taddio A, Shah V, Gilbert-MacLeod C, Katz J. Conditioning and hyperalgesia in newborns exposed to repeated heel lances. JAMA. 2002;288(7):857–61.
110. Ren K, Anseloni V, Zou SP, Wade EB, Novikova SI, Ennis M, Traub RJ, Gold MS, Dubner R, Lidow MS. Characterization of basal and re-inflammation-associated long-term alteration in pain responsivity following short-lasting neonatal local inflammatory insult. Pain. 2004;110(3):588–96.
111. Hermann C, Hohmeister J, Demirakca S, Zohsel K, Flor H. Long-term alteration of pain sensitivity in school-aged children with early pain experiences. Pain. 2006;125(3):278–85.
112. Wollgarten-Hadamek I, Hohmeister J, Demirakca S, Zohsel K, Flor H, Hermann C. Do burn injuries during infancy affect pain and sensory sensitivity in later childhood? Pain. 2009;141(1–2):165–72.
113. Schmelzle-Lublecki BM, Campbell KAA, Howard RH, Franck L, Fitzgerald M. Long-term consequences of early infant injury and trauma upon somatosensory processing. Eur J Pain. 2007;11(7):799–809.
114. Slater R, Fabrizi L, Worley A, Meek J, Boyd S, Fitzgerald M. Premature infants display increased noxious-evoked neuronal activity in the brain compared to healthy age-matched term-born infants. NeuroImage. 2010;52(2):583–9.
115. Hohmeister J, Kroll A, Wollgarten-Hadamek I, Zohsel K, Demirakca S, Flor H, Hermann C. Cerebral processing of pain in school-aged children with neonatal nociceptive input: an exploratory fMRI study. Pain. 2010;150(2):257–67.
116. Poorun R, Hartley C, Goksan S, Worley A, Boyd S, Cornelissen L, Berde C, Rogers R, Ali T, Slater R. Electroencephalography during general anaesthesia differs between term-born and premature-born children. Clin Neurophysiol. 2015. doi:10.1016/j.clinph.2015.10.041.
117. de Jong M, Verhoeven M, van Baar AL. School outcome, cognitive functioning, and behaviour problems in moderate and late preterm children and adults: a review. Semin Fetal Neonat M. 2012;17(3):163–9.
118. Delobel-Ayoub M, Arnaud C, White-Koning M, Casper C, Pierrat V, Garel M, Burguet A, Roze JC, Matis J, Picaud JC, Kaminski M, Larroque B, Grp ES. Behavioral problems and cognitive performance at 5 years of age after very preterm birth: the EPIPAGE study. Pediatrics. 2009;123(6):1485–92.
119. Johnson S, Hennessy E, Smith R, Trikic R, Wolke D, Marlow N. Academic attainment and special educational needs in extremely preterm children at 11 years of age: the EPICure study. Arch Dis Child-Fetal. 2009;94(4):F283–9.
120. Marret S, Marchand-Martin L, Picaud JC, Hascoet JM, Arnaud C, Roze JC, Truffert P, Larroque B, Kaminski M, Ancei PY, Grp ES. Brain injury in

very preterm children and neurosensory and cognitive disabilities during childhood: the EPIPAGE cohort study. PLoS One. 2013;8(5):e62683.
121. Johnson S, Hollis C, Kochhar P, Hennessy E, Wolke D, Marlow N. Psychiatric disorders in extremely preterm children: longitudinal finding at age 11 years in the EPICure study. J Am Acad Child Psy. 2010;49(5):453–63.
122. Ment LR, Vohr BR. Preterm birth and the developing brain. Lancet Neurol. 2008;7(5):378–9.
123. Counsell SJ, Boardman JP. Differential brain growth in the infant born preterm: current knowledge and future developments from brain imaging. Semin Fetal Neonatal Med. 2005;10(5):403–10.
124. Anand KJ, Garg S, Rovnaghi CR, Narsinghani U, Bhutta AT, Hall RW. Ketamine reduces the cell death following inflammatory pain in newborn rat brain. Pediatr Res. 2007;62(3):283–90.
125. Smith GC, Gutovich J, Smyser C, Pineda R, Newnham C, Tjoeng TH, Vavasseur C, Wallendorf M, Neil J, Inder T. Neonatal intensive care unit stress is associated with brain development in preterm infants. Ann Neurol. 2011;70(4):541–9.
126. Grunau RE. Neonatal pain in very preterm infants: long-term effects on brain, neurodevelopment and pain reactivity. Rambam Maimonides Med J. 2013;4(4):e0025.
127. Ranger M, Grunau RE. Early repetitive pain in preterm infants in relation to the developing brain. Pain Manag. 2014;4(1):57–67.
128. Zwicker JG, Grunau RE, Adams E, Chau V, Brant R, Poskitt KJ, Synnes A, Miller SP. Score for neonatal acute physiology-II and neonatal pain predict corticospinal tract development in premature newborns. Pediatr Neurol. 2013;48(2):123–9 e121.
129. Brummelte S, Grunau RE, Chau V, Poskitt KJ, Brant R, Vinall J, Gover A, Synnes AR, Miller SP. Procedural pain and brain development in premature newborns. Ann Neurol. 2012;71 (3):385–96.
130. Ranger M, Chau CM, Garg A, Woodward TS, Beg MF, Bjornson B, Poskitt K, Fitzpatrick K, Synnes AR, Miller SP, Grunau RE. Neonatal pain-related stress predicts cortical thickness at age 7 years in children born very preterm. PLoS One. 2013;8(10):e76702.
131. Ranger M, Zwicker JG, Chau CM, Park MT, Chakravarthy MM, Poskitt K, Miller SP, Bjornson BH, Tam EW, Chau V, Synnes AR, Grunau RE. Neonatal pain and infection relate to smaller cerebellum in very preterm children at school age. J Pediatr. 2015;167(2):292–8 e291.
132. Vinall J, Miller SP, Bjornson BH, Fitzpatrick KP, Poskitt KJ, Brant R, Synnes AR, Cepeda IL, Grunau RE. Invasive procedures in preterm children: brain and cognitive development at school age. Pediatrics. 2014;133(3):412–21.
133. Doesburg SM, Chau CM, Cheung TP, Moiseev A, Ribary U, Herdman AT, Miller SP, Cepeda IL, Synnes A, Grunau RE. Neonatal pain-related stress, functional cortical activity and visual-perceptual abilities in school-age children born at extremely low gestational age. Pain. 2013;154(10):1946–52.
134. Als H, Duffy FH, McAnulty GB, Rivkin MJ, Vajapeyam S, Mulkern RV, Warfield SK, Huppi PS, Butler SC, Conneman N, Fischer C, Eichenwald EC. Early experience alters brain function and structure. Pediatrics. 2004;113 (4):846–57.
135. Milgrom J, Newnham C, Anderson PJ, Doyle LW, Gemmill AW, Lee K, Hunt RW, Bear M, Inder T. Early sensitivity training for parents of preterm infants: impact on the developing brain. Pediatr Res. 2010;67(3):330–5.
136. Huttenlocher PR, Dabholkar AS. Regional differences in synaptogenesis in human cerebral cortex. J Comp Neurol. 1997;387(2):167–78.
137. Paus T. Growth of white matter in the adolescent brain: myelin or axon? Brain Cogn. 2010;72(1):26–35.
138. Paus T, Collins DL, Evans AC, Leonard G, Pike B, Zijdenbos A. Maturation of white matter in the human brain: a review of magnetic resonance studies. Brain Res Bull. 2001;54(3):255–66.
139. Pfefferbaum A, Mathalon DH, Sullivan EV, Rawles JM, Zipursky RB, Lim KO. A quantitative magnetic resonance imaging study of changes in brain morphology from infancy to late adulthood. Arch Neurol. 1994;51(9):874–87.
140. Campbell IG, Feinberg I. Longitudinal trajectories of non-rapid eye movement delta and theta EEG as indicators of adolescent brain maturation. Proc Natl Acad Sci U S A. 2009;106(13):5177–80.
141. Clarke AR, Barry RJ, McCarthy R, Selikowitz M. Age and sex effects in the EEG: development of the normal child. Clin Neurophysiol. 2001;112(5):806–14.
142. Gasser T, Verleger R, Bacher P, Sroka L. Development of the eeg of school-age children and adolescents. I. Analysis of band power. Electroen Clin Neuro. 1988;69(2):91–9.
143. Ambrosini A, Schoenen J. The electrophysiology of migraine. Curr Opin Neurol. 2003;16(3):327–31.
144. Oelkers-Ax R, Bender S, Just U, Pfuller U, Parzer P, Resch F, Weisbrod M. Pattern-reversal visual-evoked potentials in children with migraine and other primary headache: evidence for maturation disorder? Pain. 2004;108(3):267–75.
145. Oelkers-Ax R, Parzer P, Resch F, Weisbrod M. Maturation of early visual processing investigated by a pattern-reversal habituation paradigm is altered in migraine. Cephalalgia. 2005;25(4):280–9.
146. Siniatchkin M, Kropp P, Neumann M, Gerber W, Stephani U. Intensity dependence of auditory evoked cortical potentials in migraine families. Pain. 2000;85(1–2):247–54.
147. Buodo G, Sarlo M, Battistella PA, Naccarella C, Palomba D. Event-related potentials to emotional stimuli in migrainous children. J Child Neurol. 2011;26(12):1508–15.

148. Siniatchkin M, Jonas A, Baki H, van Baalen A, Gerber WD, Stephani U. Developmental changes of the contingent negative variation in migraine and healthy children. J Headache Pain. 2010;11(2):105–13.
149. Ozkan M, Teber ST, Deda G. Electroencephalogram variations in pediatric migraines and tension-type headaches. Pediatr Neurol. 2012;46 (3):154–7.
150. Zohsel K, Hohmeister J, Oelkers-Ax R, Flor H, Hermann C. Quantitative sensory testing in children with migraine: preliminary evidence for enhanced sensitivity to painful stimuli especially in girls. Pain. 2006;123(1–2):10–8.
151. Zohsel K, Hohmeister J, Flor H, Hermann C. Altered pain processing in children with migraine: an evoked potential study. Eur J Pain. 2008;12 (8):1090–101.
152. Hermann C, Zohsel K, Hohmeister J, Flor H. Cortical correlates of an attentional bias to painful and innocuous somatic stimuli in children with recurrent abdominal pain. Pain. 2008;136(3):397–406.
153. Rocca MA, Messina R, Colombo B, Falini A, Comi G, Filippi M. Structural brain MRI abnormalities in pediatric patients with migraine. J Neurol. 2014;261(2):350–7.
154. Stewart WF, Wood C, Reed ML, Roy J, Lipton RB, Group AA. Cumulative lifetime migraine incidence in women and men. Cephalalgia. 2008;28 (11):1170–8.
155. Faria V, Erpelding N, Lebel A, Johnson A, Wolff R, Fair D, Burstein R, Becerra L, Borsook D. The migraine brain in transition: girls vs boys. Pain. 2015;156(11):2212–21.
156. Harris EJ, Schimka KE, Carlson RM. Complex regional pain syndrome of the pediatric lower extremity: a retrospective review. J Am Podiatr Med Assoc. 2012;102(2):99–104.
157. Low AK, Ward K, Wines AP. Pediatric complex regional pain syndrome. J Pediatr Orthop. 2007;27 (5):567–72.
158. Lebel A, Becerra L, Wallin D, Moulton EA, Morris S, Pendse G, Jasciewicz J, Stein M, Aiello-Lammens M, Grant E, Berde C, Borsook D. fMRI reveals distinct CNS processing during symptomatic and recovered complex regional pain syndrome in children. Brain. 2008;131(Pt 7):1854–79.
159. Linnman C, Becerra L, Lebel A, Berde C, Grant PE, Borsook D. Transient and persistent pain induced connectivity alterations in pediatric complex regional pain syndrome. PLoS One. 2013;8(3):e57205.
160. Erpelding N, Sava S, Simons LE, Lebel A, Serrano P, Becerra L, Borsook D. Habenula functional resting-state connectivity in pediatric CRPS. J Neurophysiol. 2014;111(2):239–47.
161. Becerra L, Sava S, Simons LE, Drosos AM, Sethna N, Berde C, Lebel AA, Borsook D. Intrinsic brain networks normalize with treatment in pediatric complex regional pain syndrome. Neuroimage Clin. 2014;6:347–69.
162. Simons LE, Pielech M, Erpelding N, Linnman C, Moulton E, Sava S, Lebel A, Serrano P, Sethna N, Berde C, Becerra L, Borsook D. The responsive amygdala: treatment-induced alterations in functional connectivity in pediatric complex regional pain syndrome. Pain. 2014;155(9):1727–42.
163. Erpelding N, Simons L, Lebel A, Serrano P, Pielech M, Prabhu S, Becerra L, Borsook D. Rapid treatment-induced brain changes in pediatric CRPS. Brain Struct Funct. 2014 (Epub ahead of print).
164. Tracey I. Imaging pain. Br J Anaesth. 2008;101 (1):32–9.
165. Lorenz J, Beck H, Bromm B. Cognitive performance, mood and experimental pain before and during morphine-induced analgesia in patients with chronic non-malignant pain. Pain. 1997;73(3):369–75.
166. Truini A, Panuccio G, Galeotti F, Maluccio MR, Sartucci F, Avoli M, Cruccu G. Laser-evoked potentials as a tool for assessing the efficacy of antinociceptive drugs. Eur J Pain. 2010;14(2):222–5.
167. Hartley C, Moultrie F, Juszczak E, Adams E, Slater R. Protocol 15PRT/5747:A blinded randomised placebo-controlled trial investigating the efficacy of morphine analgesia for procedural pain in infants—ISRCTN82342359. Lancet. 2016. http://www.thelancet.com/doi/story/10.1016/html.2016.01.19.2536.
168. Wager TD, Atlas LY, Lindquist MA, Roy M, Woo CW, Kross E. An fMRI-based neurologic signature of physical pain. N Engl J Med. 2013;368 (15):1388–97.
169. Duff EP, Vennart W, Wise RG, Howard MA, Harris RE, Lee M, Wartolowska K, Wanigasekera V, Wilson FJ, Whitlock M, Tracey I, Woolrich MW, Smith SM. Learning to identify CNS drug action and efficacy using multistudy fMRI data. Sci Transl Med. 2015;7(274):274ra216.
170. Johnston C, Campbell-Yeo M, Fernandes A, Inglis D, Streiner D, Zee R. Skin-to-skin care for procedural pain in neonates. Cochrane Database Syst Rev. 2014;1:CD008435. doi:10.1002/14651858.CD008435.pub2.
171. Shah PS, Herbozo C, Aliwalas LL, Shah VS. Breastfeeding or breast milk for procedural pain in neonates. Cochrane Database Syst Rev. 2012;12: CD004950. doi:10.1002/14651858.CD004950.pub3.
172. Pillai Riddell RR, Racine NM, Turcotte K, Uman LS, Horton RE, Din Osmun L, Ahola Kohut S, Hillgrove Stuart J, Stevens B, Gerwitz-Stern A. Non-pharmacological management of infant and young child procedural pain. Cochrane Database Syst Rev. 2011;(10): CD006275. doi:10.1002/14651858.CD006275.pub2.

173. Gibbins S, Stevens B, Hodnett E, Pinelli J, Ohlsson A, Darlington G. Efficacy and safety of sucrose for procedural pain relief in preterm and term neonates. Nurs Res. 2002;51(6):375–82.
174. Stevens B, Johnston C, Franck L, Petryshen P, Jack A, Foster G. The efficacy of developmentally sensitive interventions and sucrose for relieving procedural pain in very low birth weight neonates. Nurs Res. 1999;48(1):35–43.
175. Chapin H, Bagarinao E, Mackey S. Real-time fMRI applied to pain management. Neurosci Lett. 2012;520(2):174–81.
176. Sun L. Early childhood general anaesthesia exposure and neurocognitive development. Br J Anaesth. 2010;105(Suppl 1):i61–8.
177. Marchant N, Sanders R, Sleigh J, Vanhaudenhuyse A, Bruno MA, Brichant JF, Laureys S, Bonhomme V. How electroencephalography serves the anesthesiologist. Clin EEG Neurosci. 2014;45 (1):22–32.
178. Palanca BJ, Mashour GA, Avidan MS. Processed electroencephalogram in depth of anesthesia monitoring. Curr Opin Anaesthesiol. 2009;22(5):553–9.
179. Davidson AJ. Monitoring the anaesthetic depth in children—an update. Curr Opin Anaesthesiol. 2007;20(3):236–43.
180. Cornelissen L, Kim SE, Purdon PL, Brown EN, Berde CB. Age-dependent electroencephalogram (EEG) patterns during sevoflurane general anesthesia in infants. eLife. 2015;4:e06513.
181. Choudhry DK, Brenn BR. Bispectral index monitoring: a comparison between normal children and children with quadriplegic cerebral palsy. Anesth Analg. 2002;95(6):1582–5.
182. Saricaoglu F, Celebi N, Celik M, Aypar U. The evaluation of propofol dosage for anesthesia induction in children with cerebral palsy with bispectral index (BIS) monitoring. Pediatr Anesth. 2005;15 (12):1048–52.
183. Bischoff P, Kochs E, Haferkorn D, Schulte am Esch J. Intraoperative EEG changes in relation to the surgical procedure during isoflurane-nitrous oxide anesthesia: hysterectomy versus mastectomy. J Clin Anesth. 1996;8(1):36–43.
184. Kochs E, Bischoff P, Pichlmeier U, Schulte am Esch J. Surgical stimulation induces changes in brain electrical activity during isoflurane/nitrous oxide anesthesia. A topographic electroencephalographic analysis. Anesthesiology. 1994;80 (5):1026–34.
185. Morimoto Y, Matsumoto A, Koizumi Y, Gohara T, Sakabe T, Hagihira S. Changes in the bispectral index during intraabdominal irrigation in patients anesthetized with nitrous oxide and sevoflurane. Anesth Analg. 2005;100(5):1370–4 (table of contents).
186. Sleigh JW, Leslie K, Voss L. The effect of skin incision on the electroencephalogram during general anesthesia maintained with propofol or desflurane. J Clin Monit Comput. 2010;24(4):307–18.
187. Hartley C, Poorun R, Goksan S, Worley A, Boyd S, Rogers R, Ali T, Slater R. Noxious stimulation in children receiving general anaesthesia evokes an increase in delta frequency brain activity. Pain. 2014;155(11):2368–76.

Author Biographies

Dr. Caroline Hartley is a Postdoctoral Researcher at the University of Oxford. Dr. Hartley studied Mathematics at the University of Warwick, and a Ph.D. in Neuroscience at University College London. She is now a member of the Paediatric and Infant Pain and Anaesthesia (PiPA) group at the University of Oxford, and uses neuroimaging techniques to investigate sensory processing in infants and children.

Dr. Rebeccah Slater is Associate Professor of Paediatric Neuroimaging at the University of Oxford and a Wellcome Trust Career Development Fellow. Dr. Slater studied Physics at Imperial College and Neuroscience at UCL. Dr. Slater now leads the PiPA group, which is focused on understanding the development of human pain processing.

Index

L. Saba (ed.), *Neuroimaging of Pain*, DOI 10.1007/978-3-319-48046-6

Lightning Source UK Ltd.
Milton Keynes UK
UKOW07n0101140617
303257UK00004B/12/P

9 783319 480442